GENERAL RELATIVITY AND GRAVITATION
A Centennial Perspective

Explore spectacular advances in cosmology, relativistic astrophysics, gravitational wave science, mathematics, computational science, and the interface of gravitation and quantum physics with this unique celebration of the centennial of Einstein's discovery of general relativity.

Twelve comprehensive and in-depth reviews, written by a team of leading international experts, together present an up-to-date overview of key topics at the frontiers of these areas, with particular emphasis on the significant developments of the last three decades. Interconnections with other fields of research are also highlighted, making this an invaluable resource for both new and experienced researchers.

Commissioned by the International Society on General Relativity and Gravitation, and including accessible introductions to cutting-edge topics, ample references to original research papers, and informative color figures, this is a definitive reference for researchers and graduate students in cosmology, relativity, and gravitational science.

ABHAY ASHTEKAR (Editor in Chief) is the founding Director of the Institute for Gravitation and the Cosmos, and holds the Eberly Chair in Physics, at The Pennsylvania State University. He is a Fellow of the AAAS and the APS, and a former President of the International Society for General Relativity and Gravitation.

BEVERLY K. BERGER is a former Program Director for Gravitational Physics at the US National Science Foundation. She is a Fellow of the APS, and was the founder of the APS Topical Group on Gravitation.

JAMES ISENBERG is a Professor of Mathematics at the University of Oregon and former Eisenbud Professor at the Mathematical Sciences Research Institute in Berkeley. He is a Fellow of the APS.

MALCOLM MACCALLUM is Emeritus Professor of Applied Mathematics at Queen Mary University of London. He was the founding editor of *Classical and Quantum Gravity* and is a former President of the International Society for General Relativity and Gravitation.

"Expansive in scope, detailed in presentation, and rich in source referencing, *General Relativity and Gravitation* is an authoritative treatment of the tremendous impact that general relativity has had – in both theoretical and experimental arenas – one hundred years after Einstein produced his ground-breaking theory. Experts and students alike will find many things to like in this book!"

David Reitze,
California Institute of Technology

"Einstein wrote that a theory is more impressive the greater the simplicity of its premises, the more different kinds of things it relates, and the more extended its area of applicability. Nowhere could that be better said than of his own theory of relativistic gravity – general relativity – a century after its birth. General relativity has become central both for the physics of the very large – astrophysics and cosmology – and of the very small – quantum gravity. This volume reviews that achievement. The scope is broad, the editing able, the topics well chosen, and the authors are all distinguished experts. Care has been taken to make the exposition as accessible as possible. Both the diversity and the unity of a great historic achievement are on display here."

James Hartle,
University of California, Santa Barbara

GENERAL RELATIVITY AND GRAVITATION

A Centennial Perspective

Edited by

ABHAY ASHTEKAR (EDITOR IN CHIEF)
The Pennsylvania State University

BEVERLY K. BERGER
International Society on Relativity and Gravitation

JAMES ISENBERG
University of Oregon

MALCOLM MACCALLUM
Queen Mary University of London

CAMBRIDGE
UNIVERSITY PRESS

CAMBRIDGE
UNIVERSITY PRESS

University Printing House, Cambridge CB2 8BS, United Kingdom

One Liberty Plaza, 20th Floor, New York, NY 10006, USA

477 Williamstown Road, Port Melbourne, VIC 3207, Australia

314-321, 3rd Floor, Plot 3, Splendor Forum, Jasola District Centre, New Delhi - 110025, India

79 Anson Road, #06-04/06, Singapore 079906

Cambridge University Press is part of the University of Cambridge.

It furthers the University's mission by disseminating knowledge in the pursuit of
education, learning and research at the highest international levels of excellence.

www.cambridge.org
Information on this title: www.cambridge.org/9781108810241

First published 2015
First paperback edition 2020

A catalogue record for this publication is available from the British Library

ISBN 978-1-107-03731-1 Hardback
ISBN 978-1-108-81024-1 Paperback

Contents

Preface

The discovery of general relativity by Albert Einstein 100 years ago was quickly recognized as a supreme triumph of the human intellect. To paraphrase Hermann Weyl, *wider expanses and greater depths were suddenly exposed to the searching eye of knowledge, regions of which there was not even an inkling.* For 8 years, Einstein had been consumed by the tension between Newtonian gravity and the spacetime structure of special relativity. At first no one had any appreciation for his passion. Indeed, "as an older friend," Max Planck advised him against this pursuit, "for, in the first place you will not succeed, and even if you succeed, no one will believe you." Fortunately Einstein persisted and discovered a theory that represents an unprecedented combination of mathematical elegance, conceptual depth and observational success. For over 25 centuries, spacetime had been a stage on which the dynamics of matter unfolded. Suddenly the stage joined the troupe of actors. As decades passed, new aspects of this revolutionary paradigm continued to emerge. It was found that the entire universe is undergoing an expansion. Spacetime regions can get so warped that even light can be trapped in them. Ripples of spacetime curvature can carry detailed imprints of cosmic explosions in the distant reaches of the universe. A century has now passed since Einstein's discovery and yet every researcher who studies general relativity in a serious manner continues to be enchanted by its magic.

This volume was commissioned by the International Society on General Relativity and Gravitation to celebrate a century of successive triumphs of general relativity as it expanded its scientific reach. Through its 12 Chapters, divided into four Parts, the volume takes us through this voyage, highlighting the advances that have occurred during the last three decades or so, roughly since the publication of the 1979 volumes celebrating the centennial of Einstein's birth. During this period, general relativity and gravitational science have moved steadily toward the center of physics, astrophysics and cosmology and have also contributed to major advances in geometric analysis, computational science, quantum physics and several areas of technology. The next two decades should be even more exciting as new observations from gravitational wave detectors and astronomical missions open unforeseen vistas in our understanding of the cosmos.

Beginning researchers should be able to use this volume as an introduction to the entire field. More advanced researchers should be able to use individual Chapters, especially the Chapters that are outside the area of their immediate expertise, as brief overviews of the current status of those subjects. Since the scope of gravitational science has widened considerably in recent years, the volume should be useful not only to specialists in general relativity but also to researchers in related areas, especially astrophysics, cosmology, computational science, high energy physics, quantum field theory and geometric analysis. The detailed introductions to each of the four Parts should help provide a global perspective on developments in individual areas. They summarize the main advances and offer a short tour of the more detailed material that can be found in the Chapters that follow. The introductions also contain illustrative examples of outstanding important open issues, and outline important developments that could *not* be included in individual Chapters where the authors had the difficult task of covering many developments in a limited space. Each of the four editors was primarily responsible for one of the four Parts of the volume, oversaw revisions, and lightly edited the material to enhance synergy among different Chapters in that Part. In addition, the four editors jointly wrote the introductions to all four Parts.

We are fortunate that so many leading researchers not only agreed to contribute but prepared their Chapters with great care and cheerfully undertook revisions. We are grateful to them for this collaboration. As is clear from the acknowledgements in individual Chapters, many authors sought comments from experts prior to submission. In addition, the editors requested between two and six external reviews for each Chapter. These reports provided excellent, detailed feedback which led to considerable improvements and a better overall balance. The quality of the final product owes a great deal to the seriousness with which the referees undertook their job. For this, we would like to express our deepest appreciation to Ivan Agullo, Lars Andersson, Nils Andersson, Fernando Barbero, Eugenio Bianchi, Sukanta Bose, Robert Brandenberger, Bernd Bruegmann, Piotr Chruściel, Mihalis Dafermos, Ruth Durrer, Pedro Ferreira, Chris Fewster, Larry Ford, Klaus Fredenhagen, David Garfinkle, Ted Jacobson, George (Mac) Keiser, Alok Laddha, Roy Maartens, David McClelland, Sharon Morsink, Maria Alessandra Papa, Roberto Percacci, Daniel Pollack, Hans Ringström, Peter Saulson, Kelly Stelle, Iva Stavrov, and Madhavan Varadarajan. Finally, we would like to thank our CUP Editor, Simon Capelin, for his constant support and efficient help throughout this project.

Abhay Ashtekar (Editor in Chief)
Beverly K. Berger
James Isenberg
Malcolm A. H. MacCallum

Contributors

Stefanos Aretakis
Institute for Advanced Study, Princeton, NJ 08540, USA and Princeton University, Department of Mathematics, Princeton, NJ 08544, USA

Abhay Ashtekar
Institute for Gravitation & the Cosmos and Physics Department, The Pennsylvania State University, University Park, PA 16802-6300, USA

Beverly K. Berger
2131 Chateau PL, Livermore, CA 94550, USA

Alessandra Buonanno
Max Planck Institute for Gravitational Physics (Albert Einstein Institute), D-14476 Potsdam-Golm, Germany

Matthew W. Choptuik
Department of Physics and Astronomy, University of British Columbia, Vancouver, BC, Canada V6T 1Z

Karsten Danzmann
Max Planck Institute for Gravitational Physics (Albert Einstein Institute), D-30167 Hannover, Germany

George F. R. Ellis
Department of Mathematics and Applied Mathematics, University of Cape Town, Rondebosch 7701, South Africa

Henriette Elvang
Department of Physics, University of Michigan, Ann Arbor, MI 48109, USA

John Friedman
Department of Physics, University of Wisconsin–Milwaukee, Milwaukee, WI 53211, USA

Gregory J. Galloway
Department of Mathematics, University of Miami, Coral Gables, FL 33146, USA

Gabriela Gonzalez
Department of Physics and Astronomy, Louisiana State University, Baton Rouge, LA 70803, USA

Stefan Hollands
Universität Leipzig, Institut für Theoretische Physik, D-04103 Leipzig, Germany

Gary T. Horowitz
Department of Physics, UCSB, Santa Barbara, CA, 93106, USA

James Isenberg
Department of Mathematics, Fenton Hall, University of Oregon, Eugene, OR 97403-1222, USA

Eiichiro Komatsu
Max-Planck-Institut für Astrophysik, D-85741 Garching, Germany

Luis Lehner
Perimeter Institute, Ontario, Canada N2L, 2Y5

Andrea Lommen
Department of Physics and Astronomy, Franklin & Marshall College, Lancaster, PA 17604-3003, USA

Roy Maartens
Physics Department, University of the Western Cape, Cape Town 7535, South Africa and Institute of Cosmology & Gravitation, University of Portsmouth, Portsmouth PO1 3FX, UK

Malcolm A. H. MacCallum
School of Mathematical Sciences, Queen Mary University of London, Mile End Road, London E1 4NS, UK

Jeffrey E. McClintock
Harvard Smithsonian Center for Astrophysics, Cambridge, MA 02138, USA

Peter Mészáros
Department of Astronomy and Astrophysics, Department of Physics, The Pennsylvania State University, PA 16802, USA

Pengzi Miao
Department of Mathematics, University of Miami, Coral Gables, FL 33146, USA

Vincent Moncrief
Department of Physics, Yale University, New Haven, CT 06520-8120, USA

Guido Mueller
Department of Physics, University of Florida, Gainesville, FL 32611, USA

Ramesh Narayan
Harvard Smithsonian Center for Astrophysics, Cambridge, MA 02138, USA

Frans Pretorius
Department of Physics, Princeton University, Princeton, NJ 08544, USA

Martin J. Rees
Institute of Astronomy, University of Cambridge, Cambridge CB3 0HA, UK

Martin Reuter
Institute of Physics, University of Mainz, D-55099 Mainz, Germany

Igor Rodnianski
Massachusetts Institute of Technology, Department of Mathematics, Cambridge, MA 02139, United States and Princeton University, Department of Mathematics, Princeton, NJ 08544, USA

Carlo Rovelli
Centre de Physique Théorique, Université de Marseille, F-13288 Marseille, France

Albrecht Rüdiger
Albert Einstein Institute, D-30167 Hannover, Germany

Misao Sasaki
Yukawa Institute for Theoretical Physics, Kyoto University, Kyoto 606-8502, Japan

B. S. Sathyaprakash
School of Physics and Astronomy, Cardiff University, Cardiff CF24 3YB, UK

Peter Schneider
Argelander-Institut für Astronomie, D-53121 Bonn, Germany

Richard Schoen
Department of Mathematics University of California, Irvine, Irvine, CA 92697-3875, USA

Robert M. Wald
Enrico Fermi Institute and Department of Physics, University of Chicago, Chicago, IL 60637, USA

David Wands
Institute of Cosmology & Gravitation, University of Portsmouth, Portsmouth, PO1 3FX, UK

William Joseph Weber
Department of Physics, University of Trento, 38123 Povo, Trento, Italy and Trento Institute for Fundamental Physics and Application/INFN, 38123 Povo, Trento, Italy

Clifford M. Will
Department of Physics, University of Florida, Gainesville, FL 32611, USA

List of figures

List of tables

Part One

Einstein's Triumph

Introduction

Recent media attention to the centenary of the outbreak of the First World War (WWI) reminds us that it was against this backdrop that Einstein, a Swiss citizen, announced the revolutionary theory of general relativity (GR). The war affected the theory's dissemination. Eddington's report introducing GR to the English-speaking world [1] relied on information from de Sitter in neutral Holland. Inevitably, the theory's adherents were caught up in the conflict, most notably Karl Schwarzschild, who died in 1916 while serving on the Russian front.

In 1915 Einstein was already a decade on from his *annus mirabilis* of 1905, in which he had announced the theory of special relativity, explained the already well-observed photoelectric effect as due to quantization of light (a vital step towards quantum theory), and explained Brownian motion assuming the reality of atoms, an explanation experimentally confirmed by Perrin in 1908. The second of these three great papers won him the 1921 Nobel prize – and they were not all he published that year! For example, he published the famous $E = mc^2$ equation, which later gave the basis of nuclear fusion and fission (whence Einstein's intervention in the development of atom bombs). Fusion in particular explained how stars could hold themselves up against gravity as long as they do. So Einstein had already triumphed well before 1915.

However, he was aware that his work left an awkwardly unresolved question – the need for a theory of gravity compatible with special relativity that agreed with Newton's theory in an appropriate limit. Here we will not recount Einstein's intellectual development of general relativity, which resolved that problem, nor describe the interactions with friends and colleagues which helped him find the right formulation. Those are covered by some good histories of science, and biographies of Einstein, as well as his own writings.

The theory's prediction of light-bending, confirmed to good accuracy [2] by the UK's 1919 eclipse expedition led by Eddington[1] and Crommelin, brought Einstein to the attention of the general public, in particular through the famous headline in the New York Times

---

[1] How Eddington, a Quaker, while preparing for this expedition, avoided being sent to work on the land as a conscientious objector, is itself an interesting WWI story.

of November 9th. From then on, he increasingly came to be seen as the personification of scientific genius.

Why then are we calling this first part of our centennial book "Einstein's triumph"? GR had already triumphed by 1919.

The triumph since 1919 lies in GR's ever increasing relevance and importance, shown in particular by the number and range of applications to real-world observations and applications, from terrestrial use in satellite navigation systems to considerations of cosmology on the largest scales. Moreover the different applications are now interwoven, for example in the relevance of black holes in cosmology and the use of pulsars, compact relativistic stars, in strong field tests of the theory. This part of the book outlines that progress.

As Ellis describes in Chapter 1, the starting points for many later confirmations were laid in the early years of the theory: the Schwarzschild solution, leading to solar system tests and black hole theory; light-bending, which grew into gravitational lensing; and the Friedmann(–Lemaître–Robertson–Walker: FLRW) solutions, basic in cosmology. Moreover, several confirmations relate to the three "classical tests": gravitational redshift, the anomaly in the perihelion advance of Mercury as computed from Newtonian theory,[2] and light-bending: for example, the analysis of GPS (the Global Positioning System), the study of the binary and double pulsars, and the use of microlensing to detect exoplanets. The theory remains the most nonlinear of the theories of physics, prompting development in analytic and numerical technique.

Classical differential geometry as studied in introductory courses (and as briefly outlined by Ellis) is adequate to discuss the starting points of those developments. But they soon require also the proper understanding of global structure and thus of singularities and asymptotics, for example in understanding the Schwarzschild solution, black holes and the energy carried away by gravitational radiation. This increasing sophistication was reflected in the best-selling text of Hawking and Ellis [3], and further developments are described in Part Three of this book.

Much of the development of GR has come in the last half century. For its first 50 years, a time when quantum theory was making big advances, one could argue that GR remained an intellectual ornament with only some limited applications in astronomy. Even its relevance to cosmology was debatable, because Hubble's erroneous distance scale led to a conflict between the geologically known age of the Earth and the age of the universe in an FLRW model, prompting the range of alternative explanations for this discrepancy described in Bondi's book [4]. While the notion of a stagnant phase is rather belied by the many significant papers from this time which have deservedly been included in the "Golden Oldies" series of the *General Relativity and Gravitation* journal, some of them cited by Ellis, it was certainly a less dynamic period than the following 50 years of GR.

---

[2] One may note that the anomalous part is $43''$ per century in a total of around $5000''$ per century.

The changes have been partly due to the already mentioned increasing mathematical sophistication among theoretical physicists. Taub's use of symmetry groups [5] and Petrov's algebraic classification of the Weyl tensor [6] were crucial steps forward made in the 1950s. The geometric concepts of connection and curvature have become fundamental in modern gauge theories. Progress in the theory of differential equations has given a firm basis to the idea that GR is like other physical theories in that initial configuration and motion determine the future evolution. The generating techniques for stationary axisymmetric systems used to obtain exact solutions[3] relate to modern work on integrable systems. Further developments in such areas are reflected in Chapter 1 and Part Three of this book.

Another important step was introducing the theory of the matter content within FLRW models. This enabled the understanding of the formation of the chemical elements, by combining the Big Bang and stellar nucleosyntheses, the provision of evidence that there were only three types of neutrino, and the prediction of the Cosmic Microwave Background (CMB).

Progress has depended even more on advances in technology and measurement technique. The first example was the revision of Hubble's distance scale in 1952 by Baade, using the 200 inch Palomar telescope commissioned in 1950. This led to increasing belief in the FLRW models, a belief eventually cemented by the 1965 observations of the CMB, which themselves arose from developments in microwave communications technology.

The 1957 launch of the first artificial satellite, Sputnik, intensified the need for detailed calculation of orbital effects in satellite motion, in order to very accurately plan satellite projects. Such work [9] was undertaken for both the US and USSR programs and was the first practical use of GR.

Radio astronomy, by showing source counts inconsistent with the alternative Steady State theory, had provided important evidence for FLRW models. It also, combined with optical observations, led to the discovery of quasars[4] which prompted Lynden-Bell to propose that they were powered by black holes [10]: the importance black holes have subsequently assumed in our understanding of astronomy and cosmology is described by Narayan and McClintock in Chapter 3. Radio astronomy also discovered the pulsars, announced in 1968, which gave extra impetus to the already developing study of relativistic stars, discussed by Friedman in Chapter 3.

The reality of gravitational waves in the theory, which had been debated earlier, was finally clarified in the work of Bondi *et al.* in 1959 [11]. The binary and double pulsar observations, described in Chapter 2, united the understanding of compact objects and gravitational waves to provide the first strong field tests of GR.

The exquisite precision now achieved in practical and observational areas of GR has made use of the development of very high precision atomic clocks and of the burgeoning

---

[3] The construction and interpretation of exact solutions are topics not covered by this book, as they are well covered by [7] and [8] and references therein. In particular we do not consider some important techniques used in those areas, such as computer algebra and the application of local spacetime invariants.

[4] 3C48 was identified in 1960 and 3C273's redshift was found in 1963.

of electronics since the invention of the transistor in 1947. Satellite-borne telescopes in several wavebands, computers of all scales from the largest (used in numerical relativity) to mobile devices (e.g. in GPS receivers), CCD devices (based of course on the photoelectric effect), and lasers (in terrestrial gravitational wave detectors – also used, for example, in determining the exact position of the moon) have all played major roles in the observations and experiments described in the following four chapters (and in the later parts of the book).

There were fundamental aspects of gravity (e.g. the Eötvös effect) which could be and were tested on Earth, but until the 1970s the focus was on the "classical tests", complemented by the time delay measurements for satellites. Dicke initiated a more systematic analysis of the equivalence principle and its tests, as described in Chapter 2. Thorne, Will and others then developed other frameworks, notably the PPN framework, which could encompass other types of test. While the application of these ideas still relied on solar system and terrestrial tests, these became much more precise and involved much new technology (e.g. laser ranging to the moon, superconducting gravimeters on the ground, use of atomic traps and atomic clocks in terrestrial and satellite experiments), and pinned the parameters of the PPN framework down with high precision.

Tests outside the solar system consisted of the understanding of compact stars such as white dwarfs, and supernova remnants, and of cosmology (for which there was only an incomplete understanding, for reasons described below), but did not lead to new precise constraints on the theory. That changed with the discovery and observations of the (first) binary pulsar, and still further with the several now known, including the double pulsar. These give some of the most precise measurements in physics (although, perhaps surprisingly, the Newtonian constant of gravitation, G, remains the least accurately known of the fundamental constants of nature).

It is notable that the understanding of pulsars not only required GR (because of the strong fields) but also entailed the simultaneous use of quantum theory and GR (because only by taking into account quantum theory could one have adequate equations of state to model white dwarfs and neutron stars). These types of compact object, and black holes, are now the starting points for the calculation of gravitational wave sources described in Part Two.

Relativistic astrophysics then developed in a number of directions (see Chapter 3). Numerical simulations gave much more detail on relativistic stars, their properties, stability and evolution. A whole new sub-discipline of black hole astrophysics came into being, concerned with the environments of black holes, especially (for stellar size black holes) accretion from neighbouring stars and (for supermassive black holes) accretion, nearby orbits and tidal capture of stars. The improved understanding enabled us to be rather certain not only that there really are black holes in the Universe, but that they are very common.

A further direction described in Chapter 3 came about with the discovery and increasingly detailed observations of gamma ray bursts. Both their long and short varieties turned out to require models of relativistic sources, as described by Mészáros and Rees. It is interesting that there is a link with the gravitational wave detectors described in Part Two, in that the absence of gravitational waves from GRB 070201 showed that, if it had a compact binary progenitor, then that progenitor had to be behind rather than in M31 [12].

While the standard FLRW models used up to 1980 or so did very well in describing the observed isotropy and homogeneity of the universe, and explaining the evolution of the matter content which led to formation of the chemical elements and the prediction of the CMB, they failed to explain the single most obvious fact about the Universe, namely that it had a highly non-uniform density. Naturally occurring thermal fluctuations and their evolution could not give large enough variations. The inflationary paradigm altered that radically by providing reasons for a nearly flat spectrum of density fluctuations at a time sufficiently early in the universe for the subsequent linear and nonlinear phases of evolution to produce the observed structures we see. The theory is described in detail by Sasaki in Chapter 4.

The resulting standard model has been compared with a range of very high precision observations, notably those of the CMB, the baryon acoustic oscillations (BAO) and the magnitude–redshift relation for supernovae (relating distances and expansion velocities in the Universe). These, especially the CMB observations, have generated the title "precision cosmology", which, as Komatsu emphasizes in Chapter 4, required precision theory as well as precision observation. That precision in theory consists of very detailed consideration of perturbations of the FLRW models and of light propagation in perturbed models, enabling the link between the conditions produced by inflation (or some alternative to inflation providing suitable initial conditions) and the present-day observations. What is remarkable is the fine detail of those initial conditions that one can infer from observation.

To some degree, the role of GR has disappeared in the large volume of literature related to CMB, BAO and supernova, and other, observations, as almost all of it uses the FLRW models and their linearized perturbations, and may even make crucial steps using Newtonian analyses. Wands and Maartens remind us, in their introduction to Chapter 4, that GR in fact still has a crucial role to play, even in precision cosmology where its effects may be considerably larger than the very small error bars in the observations, and the correlations described in Chapter 3 imply it also has a role to play in structure formation below the scales tested by the CMB. Moreover it is essential in testing the robustness of the assumptions of the concordance model, a further topic discussed in Chapter 4.

What can we expect in the future? B-mode polarization[5] in the CMB could give evidence for primordial gravitational waves, as discussed in Chapter 4. A recent joint analysis of data [13] suggests the signal found by BICEP2 may have been due to polarized dust emission: it places only an upper limit on the gravitational wave contribution, while supporting the lensing contributions as seen by POLARBEAR. In 2015 Advanced LIGO will begin taking data (see Part Two). If such advanced gravitational wave detectors see the expected gravitational wave sources, we will have a new window for testing GR (but if no such sources are seen, that may be due only to poor astrophysical predictions). In the past, when

---

[5] There are two characteristic patterns of polarization alignments expected in the CMB. The E-mode is like that of the electric field round a charge and the B-mode like that of magnetic field round a current. Instances of these modes, with varying amplitudes and centred at random locations, will be superposed in the actual observations. For more details see Chapter 4.

new windows on the universe have opened, new and unforeseen phenomena have been found [14]; it would not be surprising if this happens again. Beyond that there are a plethora of new instruments being built or planned to study the sky in electromagnetic wavebands from low frequency radio to γ-rays: the chances of convincing funding agencies to support such work have probably been substantially enhanced by the spectacular results of recent past projects.

Gravitational lensing by galaxies seemed to surprise many when first found in 1979, even if it should not have. Now such lensing, and its stellar size counterpart, have become tools for astronomy, used for example to infer the distribution of mass within galaxies, the distribution of dark matter, the properties of distant galaxies, and the presence of new exoplanets. Recently, magnification due to microlensing was used to determine properties of a binary system containing a white dwarf and a Sun-like star [15].

We stress again that the galactic and intergalactic application is just one of the instances where different aspects of GR come together – here lensing and cosmological models.

Although the greatest challenge for GR may lie in finding and testing a good enough theory of quantum gravity, as discussed in Part Four, there are still challenges at the classical level. Cosmology provides the greatest of these, since its standard model requires three forms of matter – the inflaton, dark matter and dark energy – which have not been, and perhaps cannot be, observed in terrestrial laboratories, and whose properties are modeled only in simple and incomplete ways. It would of course be ironic if the triumph of GR in cosmology were to turn to disaster because the only way to deal with those apparently-required three forms of matter were to adopt a modified theory of gravity, but other explanations seem much more likely.

The inflaton is postulated as a way to produce the nearly flat spectrum of fluctuations required as initial data from which acoustic oscillations produce the observed CMB power spectrum. While the assumptions of inflation may be questionable, it is, as already mentioned, remarkably successful in producing the right distribution of fluctuations on present-day scales above 150 Mpc or so (a scale much larger than that of individual galaxies). Inflation theory predicts B-mode polarization due to gravitational waves, consistent with BICEP2's initial results. A definitive detection of such polarization would provide indirect evidence on quantum gravity and the quantum/classical correspondence, in that the theory assumes the quantum fluctuations of the inflationary era become classical.

The evidence for dark matter is rather securely based on observations at scales where a Newtonian approximation is good enough to show that not all the mass is visible, such as observations of galactic rotation curves and the distribution of X-ray emitting gas in clusters. It provides 25–30% of the critical energy density of the Universe, itself now known to be very close to the actual energy density (see Chapter 4). This was known before the more precise CMB and baryon acoustic oscillation (BAO) measurements [16]. Additional evidence has been provided by comparing the distribution of mass in colliding galaxies, as shown by its lensing effects, with the mass distribution of the visible gas. However, a change in the gravity theory might provide an explanation for these observations not requiring dark matter, though as yet no satisfactory such theory has been proposed.

The inference of the existence of dark energy is even more dependent on GR, in particular on the theory of perturbed FLRW models (see Chapter 4): it comes from the magnitude–redshift relation for supernovae (relating distance and expansion velocity of the Universe), CMB and BAO data. Attempted explanations within GR not requiring a new form of matter (in which we include the cosmological constant) have used both large and small scale inhomogeneities (see Chapter 4), or may arise from the astrophysics of supernovae and their environments. Or we may be able to pin down the properties of dark energy in some way independent of FLRW models, and thereby provide a further triumph for the predictions of GR.

Obtaining information about the three so far unobserved constituents of the standard model may not come from GR itself. But we would certainly like a better understanding of inhomogeneities and their back reaction and impact on light propagation. The evidence of correlations of galactic properties with central black hole masses suggests we also need to know much more about the messy nonlinear processes of galaxy and star formation and their interaction with the nonlinearities of GR.

Despite these lacunae, which may offer opportunities for future breakthroughs, when taken together the following four chapters illustrate very well the staggering extent of the triumph of Einstein's 1915 proposal of the theory of General Relativity.

References

[1] Eddington, A. 1918. *Report on the relativity theory of gravitation.* London: Fleetway Press. Report to the Physical Society of London. Reprinted 1920.
[2] Harvey, G. 1979. *Observatory*, **99**, 195–198.
[3] Hawking, S. W., and Ellis, G. F. R. 1973. *The large scale structure of space-time.* Cambridge: Cambridge University Press.
[4] Bondi, H. 1960. *Cosmology.* Cambridge: Cambridge University Press.
[5] Taub, A. 1951. *Ann. Math.*, **53**, 472. Reprinted, with editorial introduction by M. A. H. MacCallum, in *Gen. Rel. Grav.* **36**, 2689–2719 (2004).
[6] Petrov, A. 1954. *Scientific Proceedings of Kazan State University (named after V. I. Ulyanov-Lenin), Jubilee (1804–1954) Collection*, **114**, 55–69. Translation by J. Jezierski and M. A. H. MacCallum, with introduction by M. A. H. MacCallum, *Gen. Rel. Grav.* **32**, 1661–1685 (2000).
[7] Stephani, H. *et al.* 2003. *Exact solutions of Einstein's field equations, 2nd edition.* Cambridge: Cambridge University Press. Corrected paperback edition, 2009.
[8] Griffiths, J., and Podolsky, J. 2009. *Exact space-times in Einstein's general relativity.* Cambridge Monographs in Mathematical Physics. Cambridge: Cambridge University Press.
[9] Brumberg, V. 1991. *Essential relativistic celestial mechanics.* Bristol: Adam Hilger.
[10] Lynden-Bell, D. 1969. *Nature*, **22**, 690–694.
[11] Bondi, H., Pirani, F., and Robinson, I. 1959. *Proc. Roy. Soc. Lond. A*, **251**, 519.
[12] Hurley, K., *et al.* [LIGO collaboration]. 2008. *Astrophys. J.*, **681**, 1419–1428.
[13] Ade, P. A. R. *et al.* 2015. A Joint Analysis of BICEP2/Keck Array and Planck Data. arXiv:1502.00612.
[14] Harwit, M. 1984. *Cosmic discovery: the search, scope and heritage of astronomy.* Cambridge, MA: The MIT Press.
[15] Kruse, E., and Agol, E. 2014. *Science*, **344**, 275.
[16] Coles, P., and Ellis, G. R. 1997. *Is the universe open or closed? The density of matter in the universe.* Cambridge Lecture Notes in Physics, vol. 7. Cambridge: Cambridge University Press.

1

100 Years of General Relativity

GEORGE F. R. ELLIS

This chapter aims to provide a broad historical overview of the major developments in General Relativity Theory ('GR') after the theory had been developed in its final form. It will not relate the well-documented story of the discovery of the theory by Albert Einstein, but rather will consider the spectacular growth of the subject as it developed into a mainstream branch of physics, high-energy astrophysics, and cosmology. Literally hundreds of exact solutions of the full non-linear field equations are now known, despite their complexity [1]. The most important ones are the Schwarzschild and Kerr solutions, determining the geometry of the solar system and of black holes (Section 1.2), and the Friedmann–Lemaître–Robertson–Walker solutions, which are basic to cosmology (Section 1.4). Perturbations of these solutions make them the key to astrophysical applications.

Rather than tracing a historical story, this chapter is structured in terms of key themes in the study and application of GR:

1. The study of dynamic geometry (Section 1.1) through development of various technical tools, in particular the introduction of global methods, resulting in global existence and uniqueness theorems and singularity theorems.
2. The study of the vacuum Schwarzschild solution and its application to the Solar system (Section 1.2), giving very accurate tests of general relativity, and underlying the crucial role of GR in the accuracy of useful GPS systems.
3. The understanding of gravitational collapse and the nature of Black Holes (Section 1.3), with major applications in astrophysics, in particular as regards quasi-stellar objects and active galactic nuclei.
4. The development of cosmological models (Section 1.4), providing the basis for our understandings of both the origin and evolution of the universe as a whole, and of structure formation within it.
5. The study of gravitational lensing and its astronomical applications, including detection of dark matter (Section 1.5).
6. Theoretical studies of gravitational waves, in particular resulting in major developments in numerical relativity (Section 1.6), and with development of gravitational wave observatories that have the potential to become an essential tool in precision cosmology.

This Chapter will not discuss quantum gravity, covered in Part Four. It is of course impossible to refer to all of the relevant literature. I have attempted to give the reader a judicious mix of path-breaking original research articles and good review articles. Ferreira has recently discussed the historical development at greater length [2].

1.1 The study of dynamic geometry

This section deals with the study of dynamic geometry through development of various technical tools, in particular the introduction of global methods resulting in global existence and uniqueness theorems and singularity theorems.

General relativity [3–7] heralded a new form of physical effect: geometry was no longer seen as an eternal fixed entity, but as a dynamic physical variable. Thus geometry became a key player in physics, rather than being a fixed background for all that occurs. Accompanying this was the radical idea that there was no gravitational force, rather that matter curves spacetime, and the paths of freely moving particles are geodesics determined by the spacetime. This concept of geometry as dynamically determined by its matter content necessarily leads to the non-linearity of both the equations and the physics. This results in the need for new methods of study of these solutions; standard physics methods based on the assumption of linearity will not work in general.

Tensor calculus is a key tool in general relativity [8, 9]. The **spacetime geometry** is represented on some specific averaging scale and determined by the **metric** $g_{ab}(x^\mu)$. The curvature tensor R_{abcd} is given by the Ricci identities for an arbitrary vector field u^a:

$$u_{b;[cd]} = u^a R_{abcd} \qquad (1.1)$$

where square brackets denote the skew part on the relevant indices, and a semi-colon the covariant derivative. The curvature tensor plays a key role in gravitation through its contractions, the Ricci tensor $R_{ab} := R^c{}_{acb}$ and Ricci scalar $R := R^a{}_a$. The **matter** present determines the geometry, through **Einstein's relativistic gravitational field equations** ('EFE') [3] given by

$$G_{ab} \equiv R_{ab} - \tfrac{1}{2} R\, g_{ab} = \kappa T_{ab} - \Lambda\, g_{ab}. \qquad (1.2)$$

Geometry in turn determines the motion of the matter because the **twice-contracted Bianchi identities** guarantee the conservation of total energy–momentum:

$$\nabla_b G^{ab} = 0 \quad \Rightarrow \quad \nabla_b T^{ab} = 0, \qquad (1.3)$$

provided the **cosmological constant** Λ satisfies the relation $\nabla_a \Lambda = 0$, i.e., it is constant in time and space. In conjunction with suitable equations of state for the matter, represented by the stress-energy tensor T_{ab}, equations (1.2) determine the combined dynamical evolution of the model and the matter in it, with (1.3) acting as integrability conditions.

1.1.1 Technical developments

Because of the non-Euclidean geometry, coordinate freedom is a major feature of the theory, leading to the desirability of using covariant equations that are true in all coordinate systems if they are true in one. Because of the non-linear nature of the field equations, it is desirable to use exact methods and obtain exact solutions as far as possible, in order to not miss phenomena that cannot be investigated through linearized versions of the equations. A series of technical developments facilitated study of these non-linear equations.

Coordinate-free methods and general bases

The first was the use of coordinate-free methods, representing vector fields as differential operators, and generic bases rather than just coordinates bases. Thus one notes the differential geometry idea that tangent vectors are best thought of as operators acting on functions [5, 10], thus $X = X^i \frac{\partial}{\partial x^i} \Rightarrow X(f) = X^i \frac{\partial f}{\partial x^i}$, with a coordinate basis $e_i = \frac{\partial}{\partial x^i}$. Then a generic basis e_a ($a = 0, 1, 2, 3$) is given by

$$e_a = \Lambda_a^i(x^j) e_i = \Lambda_a^i(x^j) \frac{\partial}{\partial x^i}, \quad |\Lambda_a^i| \neq 0.$$

There are three important aspects of any basis. First, the commutator coefficients $\gamma^a_{bc}(x^c)$ defined by

$$\gamma^a_{bc} e_a = [e_b, e_c], \quad [X, Y] := XY - YX.$$

The basis e_a is a coordinate basis iff[1] $\gamma^a_{bc} = 0$. Second, the metric components g_{ab} are defined by

$$g_{ab} := e_a . e_b = \Lambda_a^i \Lambda_b^j g_{ij}$$

with inverses g^{ab} determined by $g^{ab} g_{bc} = \delta^a_c$. Indices are raised and lowered by g_{ab} and g^{bc}. The basis is a *tetrad* basis if the g_{ab} are constants. The two key forms of tetrad are null tetrads with two real null vectors and two complex ones, used for studying gravitational radiation, where

$$g_{ab} = \begin{pmatrix} 0 & -1 & 0 & 0 \\ -1 & 0 & 0 & 0 \\ 0 & 0 & 0 & 1 \\ 0 & 0 & 1 & 0 \end{pmatrix},$$

and orthonormal tetrads, used for studying fluid properties, where

$$g_{ab} = \begin{pmatrix} -1 & 0 & 0 & 0 \\ 0 & 1 & 0 & 0 \\ 0 & 0 & 1 & 0 \\ 0 & 0 & 0 & 1 \end{pmatrix}.$$

---

[1] iff means "if and only if".

Third, there are the rotation coefficients $\Gamma^a{}_{bc}$ characterizing the covariant derivatives of the basis vectors, defined by

$$\nabla_b e_c = \Gamma^a{}_{bc} e_a.$$

Using the standard assumptions of (1) metricity: writing $f_{,i} = \frac{\partial f}{\partial x^i}$, this is

$$\nabla_e g_{dc} = 0 \iff g_{dc,e} = \Gamma_{dec} + \Gamma_{ced}$$

and (2) vanishing torsion: writing $f_{;ab} := (f_{;a})_{;b}$, this is

$$f_{;ab} = f_{;ba} \; \forall f(x^i) \; \Leftrightarrow \; \gamma^c{}_{ab} = \Gamma^c{}_{ab} - \Gamma^c{}_{ba}$$

one obtains (3) the generalized Christoffel relations

$$\Gamma_{ced} = \tfrac{1}{2}(g_{cd,e} + g_{ec,d} - g_{de,c}) + \tfrac{1}{2}(\gamma_{edc} + \gamma_{dec} - \gamma_{ced}).$$

The first term vanishes for a tetrad, and the second for a coordinate basis. Tetrad bases were used in obtaining solutions by Levi-Civita in the 1920s. Null tetrads were the basis of the Newman–Penrose formalism used primarily to study gravitational radiation [11]; they are closely related to spinorial variables [11]. Orthonormal tetrads have been used to study fluid models and Bianchi spacetimes [12, 13]. Dual 1-form relations, using exterior derivatives, have been used by many workers (e.g. Bel, Debever, Misner, Kerr) to find exact solutions.

Tensor symmetries and the volume element

Second was a realization of the importance of tensor symmetries, for example separating a tensor T_{ab} into its symmetric and skew symmetric parts, and then separating the former into its trace and trace-free parts:

$$T_{ab} = T_{[ab]} + T_{(ab)}, \; T_{(ab)} = T_{<ab>} + \tfrac{1}{4}T g_{ab}, \; T_{<ab>} g^{ab} = 0$$

where $T_{[ab]} = \tfrac{1}{2}(T_{ab} - T_{ba})$, $T_{(ab)} = \tfrac{1}{2}(T_{ab} + T_{ba})$, $T = T_{ab} g^{ab}$. This breaks the tensor T_{ab} up into parts with different physical meanings. A tensor equation implies equality of each part with the same symmetry, for example

$$T_{ab} = W_{ab} \Leftrightarrow \left(T_{[ab]} = W_{[ab]}, \; T_{<ab>} = W_{<ab>}, \; T = W\right).$$

This plays an important role in many studies. Similar decompositions occur for more indices, for example $T_{[abc]} = \tfrac{1}{6}(T_{abc} + T_{bca} + T_{cab} - T_{acb} - T_{bac} - T_{cba})$. An example is the curvature tensor symmetries

$$R_{abcd} = R_{[ab][cd]} = R_{cdab}, \; R_{a[bcd]} = 0.$$

Arbitrarily large symmetric tensors can be split up in a similar way into trace-free parts; this is significant for gravitational radiation studies [14] and kinetic theory, in particular studies of CMB anisotropies [15, 16]. In the case of skew tensors, there is a largest possible skew tensor, namely the volume element $\eta_{abcd} = \eta_{[abcd]}$, which satisfies a key set of identities:

$$\eta_{abcd}\eta^{efgh} = -4!\delta^{[e}_{[a}\delta^f_b\delta^g_c\delta^{h]}_{d]}$$

and others that follow by contraction. These symmetries characterize invariant subspaces of a tensor product under the action of the linear group and so embody an aspect of group representation theory: how to decompose tensor representations into irreducible parts.

Symmetry groups

Third was the systematization of the use of symmetry groups in studying exact solutions. A symmetry is generated by a Killing vector field ξ, a vector field that drags the metric into itself, and so gives a zero Lie derivative for the metric tensor:

$$L_\xi g_{ab} = 0 \iff \xi_{(a;b)} = 0.$$

Such a vector field satisfies the integrability condition

$$\xi_{a;bc} = R_{abcd}\xi^d$$

showing that a solution is determined by the values $\xi_{a|P}$ and $\xi_{a;b|P}$ at a point P. The set of all Killing vectors form a Lie algebra generating the symmetry group for the spacetime. The isotropy group of a point is generated by the set of Killing vector fields vanishing at that point. Exact solutions can be characterized by the group of symmetries, together with a specification of the dimension and causal character of the surfaces of transitivity of the group [12]. The most important symmetries are when a spacetime is (a) static or stationary, (b) spherically symmetric, or (c) spatially homogeneous.

Killing vectors give integration constants for geodesics: if k^a is a geodesic tangent vector, then $E := \xi_a k^a$ is a constant along the geodesic. They also give conserved vectors from the stress tensor: $J^a = T^{ab}\xi_b$ has vanishing divergence: $J^a_{;a} = 0$.

Congruences of curves and geodesic deviation

Fourth was the study of timelike and null congruences of curves (Synge, Heckmann, Schücking, Ehlers, Kundt, Sachs, Penrose) [17–19], leading to a realisation of the importance of the geodesic deviation equation (GDE) [20] and its physical meaning in terms of expressing tidal forces and gravitational radiation [21, 22].

The kinematical properties of null and timelike vector fields are represented by their acceleration a_e, expansion θ, shear σ_{de}, and rotation ω_{de} (which are equivalent to some of the Ricci rotation coefficients for associated tetrads [12, 23]). They characterise the properties of fluid flows (the timelike case), hence are important in the dynamics of fluids, and of bundles of null geodesics (the null case with $a_e = 0$), and so are important in observations in astronomy and cosmology.

The GDE determines the second rate of change of the deviation vectors for a congruence of geodesics of arbitrary causal character, i.e., their relative acceleration. Consider the normalised tangent vector field V^a for such a congruence, parametrised by an affine parameter v. Then $V^a := \frac{dx^a(v)}{dv}$, $V_a V^a := \epsilon$, $0 = \frac{\delta V^a}{\delta v} = V^b \nabla_b V^a$, where $\epsilon = +1, 0, -1$ if the geodesics are spacelike, null, or timelike, respectively, and we define covariant derivation *along* the geodesics by $\delta T^{a\cdots}{}_{b\cdots}/\delta v := V^b \nabla_b T^{a\cdots}{}_{b\cdots}$ for any tensor $T^{a\cdots}{}_{b\cdots}$. A deviation vector $\eta^a := dx^a(w)/dw$ for the congruence, which can be thought of as linking pairs of neighbouring geodesics in the congruence, commutes with V^a, so

$$L_V \eta = 0 \Leftrightarrow \frac{\delta \eta^a}{\delta v} = \eta^b \nabla_b V^a. \tag{1.4}$$

It follows that their scalar product is constant along the geodesics: $\frac{\delta(\eta_a V^a)}{\delta v} = 0 \Leftrightarrow (\eta_a V^a) = $ constant. To simplify the relevant equations, one can choose them orthogonal: $\eta_a V^a = 0$. The general GDE takes the form

$$\frac{\delta^2 \eta^a}{\delta v^2} = -R^a{}_{bcd} V^b \eta^c V^d. \tag{1.5}$$

This shows how spacetime curvature causes focusing or defocusing of geodesics, and is the basic equation for gravitational lensing. The general solution to this second-order differential equation along any geodesic γ will have two arbitrary constant vectors (corresponding to the different congruences of geodesics that might have γ as a member). There is a *first integral* along any geodesic that relates the connecting vectors for two *different* congruences which have one central geodesic curve (with affine parameter v) in common. This is $\eta_{1a} \frac{\delta \eta_2}{\delta v}{}^a - \eta_{2a} \frac{\delta \eta_1}{\delta v}{}^a = $ constant and is completely independent of the curvature of the spacetime.

The trace of the geodesic deviation equation for timelike geodesics is the Raychaudhuri equation [24]. In the case of timelike vectors $u^a = dx^a/d\tau$ it is the fundamental equation of gravitational attraction for a fluid flow [18, 25]:

$$\frac{d\theta}{d\tau} = -\frac{1}{3}\theta^2 - 2(\omega^2 - \sigma^2) - \frac{\kappa}{2}(\rho + 3p) + \Lambda - a^b{}_{;b}. \tag{1.6}$$

In the case of null geodesics $k^a = dx^a/d\lambda$ diverging from a source (so $\varepsilon = 0$, $a_e = 0$, $\omega = 0$) it is the basic equation of gravitational focusing of bundles of light rays [5]:

$$\frac{d\theta}{d\lambda} + \frac{1}{2}\theta^2 + \sigma^2 = -R_{ab}k^a k^b. \tag{1.7}$$

These equations play a key role in singularity theorems.

The conformal curvature tensor

Fifth was a realisation, following on from this, that one could focus on the full curvature tensor R_{abcd} itself, and not just the Ricci tensor. The curvature tensor R_{abcd} is comprised of the Ricci tensor R_{ab} and the Weyl conformal curvature tensor C_{abcd}, given by

$$C_{abcd} := R_{abcd} + \tfrac{1}{2}(R_{ac}g_{bd} + R_{bd}g_{ac} - R_{ad}g_{bc} - R_{bc}g_{ad}) - \tfrac{1}{6}R(g_{ac}g_{bd} - g_{ad}g_{bc}). \tag{1.8}$$

This has the same symmetries as the curvature tensor but in addition is trace-free: $C^c{}_{acd} = 0$. The Ricci tensor is determined pointwise by the matter present through the field equations, but the Weyl tensor is not so determined: rather it is fixed by matter elsewhere plus boundary conditions. Its value at any point is determined by the Bianchi identities

$$R_{ab[cd;e]} = 0 \tag{1.9}$$

which are integrability conditions for the curvature tensor that must always be satisfied. In four dimensions this gives both the divergence identities $\nabla^d(R_{cd} - \frac{1}{2}Rg_{cd}) = 0$ for the Ricci tensor, which imply matter conservation (see (1.3)), and divergence relations for C_{abcd}:

$$\nabla^d C_{abcd} = \nabla_{[a}(-R_{b]c} + \tfrac{1}{6}Rg_{b]c}). \tag{1.10}$$

Substituting from the field equations (1.2), matter tensor derivatives are a source for the divergence of the Weyl tensor. Thus one can think of the Weyl tensor as the free gravitational field (it is not determined by the matter at a point), being generated by matter inhomogeneities and then propagating to convey information on distant gravitating matter to local systems. It will then affect local matter behaviour through the geodesic deviation equation. Thus these are Maxwell-like equations governing tidal forces and gravitational radiation effects [10, 26].

This geometrical and physical significance of the Weyl tensor has led to studies of its algebraic structure (the Petrov Classification) inter alia by Petrov [27], Pirani, Ehlers, Kundt, and by Penrose using a spinor formalism [11, 28], and use of the Weyl tensor components as auxiliary variables in studies of exact solutions, gravitational radiation, and cosmology. One can search for vacuum solutions of a particular Petrov type [1], and relate asymptotic power series at large distances to outgoing radiation conditions [11, 19, 23].

Junction conditions

Sixth, many solutions have different domains with different properties, for example vacuum and fluid-filled. An important question then is how to join two different such domains together without problems arising at the join. Lichnerowicz (using a coordinate choice) and Darmois (using coordinate-free methods) showed how to join domains smoothly together, and Israel [29] showed how to assign properties to shock waves, boundary surfaces, and thin shells that could lead to such junctions with a surface layer occurring. The Darmois–Lichnerowicz case is included as the special situation where there is no surface layer.

Generation techniques

Finally, generation techniques that generated new exact solutions from old ones, for example fluid solutions from vacuum ones, were discovered. These are discussed extensively in [1].

1.1.2 Exact theorems and global structure

The results mentioned so far are local in nature, but there has been a major development of global results also (building on the local methods mentioned above). This is covered in detail in Chapter 9.

Global properties and causality

The first requirement is a careful use of coordinate charts as parts of atlases that cover the whole spacetime considered and so avoid coordinate singularities. This enables study of global topology, which may often not be what was first expected, and is closely related to the causal structure of the manifold. It is clear that closed timelike lines can result if a spacetime is closed in the timelike direction, but Gödel showed that closed timelike lines could occur for exact solutions of the field equations for pressure-free matter that are simply connected [30]. This occurs basically because, due to global rotation of the matter, light cones tip over as one moves further from the origin. This paper led to an intensive study of causation in curved spacetimes by Penrose, Carter, Geroch, Hawking, and others. The field equations of classical general relativity do not automatically prevent causality violation: so various causality conditions have been proposed as extra conditions to be imposed in addition to the Einstein equations, the most physically relevant being stable causality (no closed timelike lines exist even if the spacetime is perturbed) [5]. The global structures of examples such as the Gödel universe and Taub–NUT space were crucial in seeing the kinds of pathologies that can occur.

Conformal diagrams and horizons

Studying the conformal structure of a spacetime is greatly facilitated by using conformal diagrams. Penrose pioneered this method [31] and showed that one can rescale the conformal coordinates so that the boundary of spacetime at infinite distance is represented at a finite coordinate value, hence one can represent the entire spacetime and its boundary in this way [31]. For example Minkowski space has null infinities I_- and I_+ for incoming and outgoing null geodesics, an infinity i_0 for spacelike geodesics, and past and future infinities i_-, i_+ for timelike geodesics, and, perhaps surprisingly, the points i_0, i_- and i_+ have to be identified. Penrose diagrams are now a standard tool in general relativity studies, particularly in cosmology, where they make the structure of particle horizons and visual horizons very clear, and in studying black holes [32].

In the case of cosmology, there was much confusion about the nature of horizons until Rindler published a seminal paper that clarified the matter [33]. Penrose then showed that the nature of particle horizons could be well understood by characterising the initial singularity in cosmology as spacelike [31]. Nowadays they are widely used in studying inflationary cosmology (but papers on inflation often call the Hubble radius the horizon, when it is not; the particle horizon is non-locally defined). The key feature of black holes is event horizons, which are null surfaces bounding the regions that can send information to infinity from those which cannot; they occur when the future singularity is spacelike. In the case of non-rotating black holes, their nature and relation to the various spacetime domains (inside and outside the event horizon), the existence of two spacelike singularities, and the occurrence of two separate asymptotically flat exterior regions, is fully clarified by the associated Penrose diagrams (Section 1.3). Carter developed the analogous but much more complex diagrams for the Reissner–Nordström charged solution and the Kerr rotating black hole solution [5]. Without these diagrams, it would be very hard indeed to understand their global structure.

Initial data, existence and uniqueness theorems

Provided there are no closed timelike lines, suitable initial data on a spacelike surface S satisfying the initial-value equations [34], together with suitable equations of state for whatever matter may be present, determines a unique solution for the Einstein equations within the future domain of dependence of S: that is, the region of spacetime such that all past timelike and null curves intersect S. Applied to the gravitational equations, the spacetime developing from the initial data is called the future Cauchy development of the data on S [5, 35]. An important technical point is that one can prove the existence of timelike and null geodesics from S to every point in this domain of dependence. The spacetime is called globally hyperbolic if the future and past domains of dependence cover the entire spacetime; then data on S determines the complete spacetime structure, and S is called a Cauchy surface for the spacetime.

Arnowitt, Deser, and Misner [36] developed the ADM (Hamiltonian-based) formalism showing precisely what initial data was needed on S, and what constraint equations it had to satisfy, in order that the spacetime development from that data would be well defined. The initial-value problem is to determine what initial data satisfies these constraints. They also formulated the evolution equations needed to determine the time development of this data. Lichnerowicz, Choquet-Bruhat, and Geroch showed that the existence and uniqueness of a maximal Cauchy development can be proved using functional analysis techniques based on Sobolev spaces. These existence and uniqueness theorems are discussed in Chapter 8.

The study of asymptotically flat spacetimes leads to positive-mass theorems. Associated with these theorems are non-linear stability results for the lowest-energy solutions of Einstein's equations (the Minkowski and de Sitter spacetimes) that are discussed in Chapter 9.

1.1.3 Singularity theorems

Singularities – an edge to spacetime – occur in the Schwarzschild solution and in the standard FLRW models of cosmology. A key issue is whether these are a result of the high symmetry of these spacetimes, in which case they might disappear in more realistic models of these situations. Many attempts to prove theorems in this regard by direct analysis of the field equations and examination of exact solutions failed. The situation was totally transformed by a highly innovative paper by Roger Penrose in 1965 [37] that used global methods and causal analysis to prove that singularities will occur in gravitational collapse situations where closed trapped surfaces occur, a causality condition is satisfied, and suitable energy conditions are satisfied by the matter and fields present. A closed trapped surface occurs when the gravitational field is so strong that outgoing null rays from a 2-sphere converge – which occurs inside $r = 2m$ in the Schwarzschild solution. There are various positive energy conditions that play a key role in these theorems, for example the null energy condition $\rho + p \geq 0$, where ρ is the energy density of the matter and p its pressure [5]. Instead of characterising a singularity by divergence of a scalar field such as the energy density, it was characterised in this theorem by geodesic incompleteness: that is, some timelike or null geodesics could not be extended to infinite affine parameter values, showing that there is an edge to spacetime that can be reached in a finite time by freely moving articles or photons. Thus their possible future or past is finite. Penrose's paper proved that the occurrence of black hole singularities is not due to special symmetries, but is generic.

This paper opened up entirely new methods of analysis and showed their utility in key questions. Its methods were extended by Hawking, Geroch, Misner, Tipler, and others; in particular Hawking proved similar theorems for cosmology, in effect using the fact that time-reversed closed trapped surfaces occur in realistic cosmological models; indeed their existence can be shown to be a consequence of the existence of the cosmic microwave blackbody radiation [5]. A unifying singularity theorem was proved by Hawking and Penrose [38]. The nature of energy conditions and the kinds of singularities that might exist were explored, leading to characterization of various classes of singularities (scalar, non-scalar, and locally regular), and alternative proposals for defining singularities were given particularly by Schmidt.

A 'major crisis for physics'

John Wheeler emphasized that the existence of spacetime singularities – an edge to spacetime, where not just space, time, and matter cease to exist, but even the laws of physics themselves no longer apply – is a major crisis for physics:

The existence of spacetime singularities represents an end to the principle of sufficient causation and so to the predictability gained by science. How could physics lead to a violation of itself – to no physics? [39]

This is of course a prediction of the classical theory. It is still not known if quantum gravity solves this issue or not.

1.1.4 Conclusion

Because GR moves spacetime from being a fixed geometrical background arena within which physics takes place to being a spacetime that is a dynamical participant in physics, and replaces Euclid's Parallel Postulate by the geodesic deviation equation, it radically changes our understanding of the nature of spacetime geometry. Because it conceives of gravity and inertia as being locally indistinguishable from each other, it radically changes our view of the nature of the gravitational force. The resulting theory has been tested to exquisite precision [40, 41]; see also Chapter 2. Because the curvature of spacetime allows quite different global properties than in flat spacetime, it is possible for closed timelike lines to occur. Because it allows for a beginning and end to spacetime, where not just matter but even spacetime and the laws of physics cease to exist, it radically alters our views on the nature of existence. What it does not do is give any account of how spacetime might have come into existence: that is beyond its scope.

1.2 The Schwarzschild solution and the solar system

Karl Schwarzschild developed his vacuum solution of the Einstein equations in late 1915, even before the General Theory of Relativity was fully developed [42]. This is one of the most important solutions in general relativity because of its application to the Solar system (this section), giving very accurate tests of general relativity, because of the remarkable properties of its maximal analytic extension, and because of its application to black hole theory (next sections).

1.2.1 The Schwarzschild exterior solution

Consider a vacuum, spherically symmetric solution of the Einstein Field Equations. To model the exterior field of the Sun in the Solar system, or of any static star, we look for a solution that is

(1) spherically symmetric (here we ignore the rotation of the sun and its consequent oblateness, leading to a slightly non-spherical exterior field, as well as the small perturbations due to the gravitational fields of the planets),
(2) vacuum outside some radius r_S representing the surface of a central star or other massive object.

Thus we consider the *exterior solution*: $R_{bf} = 0$, ignoring the gravitational field of dust particles, the solar wind, planets, comets, etc. It can be attached at r_S to a corresponding interior solution for the star that generates the field, for example the Schwarzschild interior solution [4] which has constant density (but does not have a realistic equation of state).

We can choose coordinates for which the metric form is manifestly spherically symmetric:

$$ds^2 = -A(r,t)dt^2 + B(r,t)dr^2 + r^2(d\theta^2 + \sin^2\theta \, d\phi^2) \qquad (1.11)$$

where $x^0 = t$, $x^1 = r$, $x^2 = \theta$, $x^3 = \phi$ (the symmetry group $SO(3)$ acts on the 2-spheres $\{r = const, t = const\}$). A major result is the *Jebsen–Birkhoff theorem*: the solution necessarily has an extra symmetry (there is a further Killing vector) and so is necessarily static in the exterior region: $A = A(r), B = B(r)$, independent of the time coordinate t [4, 5]. This is a local result: provided the solution is spherically symmetric, it does not depend on boundary conditions at infinity. It implies that spherical objects cannot radiate away their mass (that is, there is no dilaton in general relativity theory). Define $m \equiv MG/c^2 > 0$ (mass in geometrical units, giving the one essential constant of the solution); then, setting $c = 1$,

$$ds^2 = -\left(1 - \frac{2m}{r}\right)dt^2 + \left(1 - \frac{2m}{r}\right)^{-1}dr^2 + r^2(d\theta^2 + \sin^2\theta \, d\phi^2) \qquad (1.12)$$

This is the Schwarzschild (exterior) solution [43]. It is an *exact* solution of the EFE (no linearization was involved) for the exterior field of a central massive object. It is valid for $r > r_S$, where r_S is the coordinate radius of the surface of the object; we require that

$r_S > r_G \equiv 2m \equiv 2MG/c^2$, where r_G is the gravitational radius or *Schwarzschild radius* of the object. If $r_S < r_G$, we would have a black hole and the interior/exterior matching would be impossible.

It is *asymptotically flat*: as $r \to \infty$, (1.12) becomes the metric of Minkowski spacetime in spherically symmetric coordinates. Note that we did not have to put this in as an extra condition: it automatically arose as a property of the exact solution.

1.2.2 Effects on particles

The spacetime geometry determines how particles move in the spacetime.

Particle orbits and light rays

The importance of this solution is that it determines the paths of particles and light rays in the vicinity of the Sun. To determine them one needs to solve the geodesic equations for timelike and lightlike curves. The Lagrangian for geodesics is $L = g_{ab}(x^c)\dot{x}^a\dot{x}^b$, where $\dot{x}^a = dx^a/dv$, with v an affine parameter along the geodesics. Applying this to timelike paths, one determines orbits for planets around the Sun, obtaining standard bound and unbound planetary orbits but with perihelion precession, as confirmed by observations [4]; applying it to photon orbits, one gets equations both for gravitational redshift and for gravitational lensing, again as confirmed by observations [4].

Putting them together, one gets the predictions for laser ranging experiments that have confirmed the predictions of general relativity theory with exquisite precision [40]. One can also determine the proper time measured along any world line and combine it with the gravitational redshift and Doppler shift predictions for signals from a satellite to Earth, thereby providing the basis for precision GPS systems [44]. Thus GR plays a crucial role in the accuracy of useful GPS systems.

Spin precession

Parallel transport along a curve corresponds to Fermi–Walker transport along the curve, leading to the prediction of spin precession, or frame dragging. This can be measured by gyroscope experiments, as tested recently by gravity probe B [45]. The effect is also observable in binary pulsars [46].

1.2.3 Solar system tests of general relativity

This set of results enables precision tests of General Relativity in the Solar system. Clifford Will and others have set up a Parameterized Post-Newtonian (PPN) formalism [40], whereby general relativity can be compared to other gravitational theories through this set of experiments, that is 1. Planetary orbits; 2. Light bending; 3. Radar ranging; 4. Frame-dragging experiment; 5. Validation of GPS systems. These Solar system tests of GR establish its validity on Solar system scales to great accuracy. They are discussed in detail in Chapter 2.

Together these tests confirm that Einstein's gravitational theory – a radical revolution in terms of viewing how gravitation works, developed by pure thought from careful analysis of the implications of the equivalence principle – is verified to extremely high accuracy. There is no experimental evidence that it is wrong.

1.2.4 Reissner–Nordström, Kottler, and Kerr solutions

There are three important generalizations of the Schwarzschild solution.

Reissner–Nordström solution

Firstly, there is the Reissner–Nordström solution, which is the charged version of the Schwarzschild solution [47, 48], where in (1.11) the factor $A(r,t) = 1 - 2m/r + Q^2/r^2 = 1/B(r,t)$. It is spherically symmetric and static, but has a non-zero electric field due to a charge Q (in geometrized units) at the centre. It is of considerable theoretical interest, but is not useful in astrophysics, as stars are not charged (if they were, electromagnetism rather than gravitation would dominate astronomy).

Kottler solution

Secondly, the generalization of this spacetime to include a cosmological constant Λ is the Kottler solution [49], where in (1.11) the factor $A(r,t) = 1 - 2m/r - \Lambda r^2/3 = 1/B(r,t)$, which is a Schwarzschild solution with cosmological constant. It gives the field of a spherical body imbedded in a de Sitter universe. This is relevant to the present-day universe, because astronomical observations have detected the effect of an effective cosmological constant in cosmology at recent times.

Kerr solution

Finally there is the Kerr solution [50, 51], which is the rotating version of the Schwarzschild solution. It is a vacuum solution that is stationary rather than static and axially symmetric rather than spherically symmetric [4, 5]. It is of considerable importance because most astrophysical objects are rotating. There is one important difference from the Schwarzschild solution: while we can construct exact interior solutions to match the Schwarszchild exterior solution, that is not the case for the Kerr solution. It has a complex and fascinating structure that is still giving new insights [52].

The geodesic structure of these spaces can be examined as in the case of the Schwarzschild solution. Carter showed that in the case of the Kerr solution, there were hidden integrals for geodesics associated with the existence of Killing tensors [53, 54].

1.2.5 Maximal extension of the Schwarzschild solution

The previous section considered the solution exterior to the surface of a star, and this was all known and understood early on, certainly by the mid 1930s. The behaviour at the coordinate singularity at $r = 2m$ is another matter: despite useful contributions by Eddington, Finkelstein, Szekeres, and Synge, it was only fully understood in the late 1960s through papers by Kruskal and Fronsdal (see [55, 56] for the early historical development).

Consider now the Schwarzschild solution as a vacuum solution with no central star. It is then apparently singular at $r = 2m$, so it has to be restricted to $r > 2m$; then both ingoing and outgoing null geodesics are incomplete, because they cannot be extended beyond $r = 2m$. However, the scalar invariant $R_{abcd}R^{abcd} = 48m^2/r^6$ is finite there, and it turned out that this is just a coordinate singularity: one can find regular coordinates that extend the solution across this surface, which is in fact a null surface where the solution changes from being static but spatially inhomogeneous (for $r > 2m$) to being spatially homogeneous but time-varying (for $0 \leq r < 2m$).

Eddington, Szekeres, Finkelstein, and Novikov showed how one could find regular coordinates that cross the surface $r = 2m$. To do so, one can note that radial null geodesics are given by $dt/dr = \pm 1/(1 - 2m/r) \Leftrightarrow t = \pm r^* + const$, where $dr^* = dr/(1 - 2m/r) \Leftrightarrow r^* = r + 2m \ln[r/(2m) - 1]$. Defining $v = t + r^*$, $w = t - r^*$, one can obtain three null forms for the Schwarzschild metric. Changing to coordinates (v, r, θ, ϕ), the metric is

$$ds^2 = -\left(1 - \frac{2m}{r}\right) dv^2 + 2\, dv\, dr + r^2\, d\Omega^2 \qquad (1.13)$$

which is the Eddington–Finkelstein form of the metric [57]. The coordinate transformation has succeeded in getting rid of the singularity at $r = 2m$, and shows how infalling null geodesics can cross from $r > 2m$ to $r < 2m$, but the converse is not possible, as can be seen in the Eddington–Finkelstein diagram (Figure 1.1), which shows how the light cones tip over at different radii. This is thus a *black hole* because light cannot escape from the interior region $r < 2m$ to the exterior region $r > 2m$. The coordinate surfaces $\{r = const\}$ are timelike for $r > 2m$, null for $r = 2m$, and spacelike for $r < 2m$; the coordinate surfaces $\{t = const\}$ are spacelike for $r > 2m$, undefined for $r = 2m$, and timelike for $r < 2m$. This warns us not to expect that a coordinate t will necessarily be a good time coordinate everywhere just because we call it "time".

The coordinate transformation (which is singular at $r = 2m$) extends the original space-time region **I**, defined by $2m < r < \infty$, to a new spatially homogeneous but time dependent region **II**, defined by $0 < r < 2m$. Ingoing null geodesics are then complete; but outgoing ones are not complete in the past. Use of coordinates (w, r, θ, ϕ) by contrast adds a new region **II**′ that makes the outgoing geodesics complete but the ingoing ones incomplete. Using coordinates (v, w, θ, ϕ) and then conformally rescaling these coordinates makes both sets of null geodesics complete, but to make the geodesics in regions **II** and **II**′ complete one must add a further region **I**′ that completes the past-directed null geodesics of **II** and the future-directed null geodesics of **II**′.

Thus this shows that the maximally extended solution [58] consists of these four domains separated by null surfaces $r = 2m$ that form the event horizons of the black hole, which has a non-trivial topology: two asymptotically flat regions back to back,

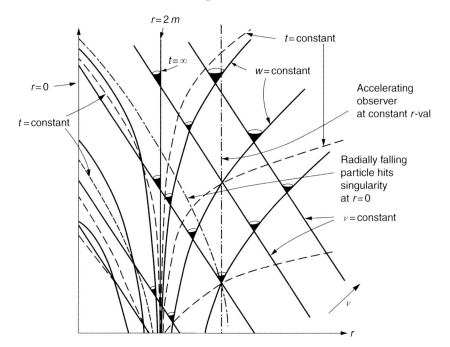

Figure 1.1 The Eddington–Finkelstein extension of the Schwarschild solution. Reproduced, with permission, from [5].

connected by "wormholes" [4]; it is maximally extended because all geodesics either extend to infinity, or end on a past or future singularity (Figure 1.2). Taking a cross section $t' = const$, for large negative values of t' one has two separated asymptotically flat regions **I, I'**; for $-2m < t' < 2m$, the two asymptotically flat regions are connected by a throat; for $t' > 2m$ they are again separated from each other. This opening and closing bridge between them is a wormhole joining the two asymptotically flat regions.

The singularities are spacelike boundaries to spacetime (one in the future and one in the past), not timelike world lines as one would expect. The coordinate singularities in the original Schwarzschild form of the metric are because the time coordinate t diverges there. Use of conformal transformations of the null coordinates yields the Penrose diagram of the maximally extended spacetime, including its boundaries at infinity [4, 5].

Symmetries

The solution is invariant under

(a) A left–right symmetry, where regions **I** and **I'** are identical to each other, while regions **II** and **II'** are individually symmetric under the interchange $t' \to -t'$;
(b) A time symmetry, where regions **II** and **II'** are identical to each other, while regions **I** and **I'** are individually symmetric under the interchange $x' \to -x'$;

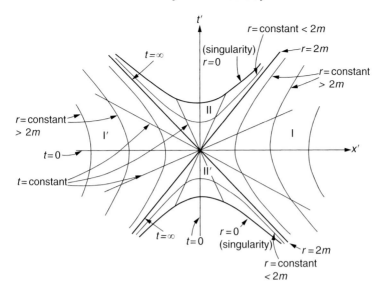

Figure 1.2 The Kruskal maximal extension of the Schwarzschild vacuum solution. Reproduced, with permission, from [5].

(c) The boost symmetry that is shown to exist by Birkhoff's theorem, with timelike orbits in regions **I** and **I′**, spacelike orbits in regions **II** and **II′**, and null orbits in the four null horizons (Killing vector orbits) that bifurcate at the central 2-sphere, which is a set of fixed points of the group. This saddle-point behaviour is a generic property of bifurcate Killing horizons, where the affine parameter and Killing vector parameter are exponentially related to each other [59].

Reissner–Nordström, Kottler, and Kerr solutions

As shown by Carter [60] and Boyer and Lindquist [61], one can obtain similar maximal extensions of the Reissner–Nordström and Kerr solutions [5, 60]. They are far more complex than those for the Schwarzschild solution, having many horizons and asymptotically flat regions. Lake and Roeder have given the maximal extension for the Kottler spacetime [62].

1.2.6 Conclusion

The Schwarzschild solution enables us to model the geometry of the Solar system with exquisite precision, giving a more accurate description of Solar system dynamics than Newtonian theory does. This enables us to test the static aspects of general relativity to very high accuracy, and so confirm its correctness as one of the fundamental theories of physics.

The maximally extended Schwarzschild solution is an extraordinary discovery. The very simple looking metric (1.12) implies the existence of two asymptotically flat spacetime regions connected by a wormhole; event horizons (the null surfaces $r = 2m$) separating the interior ($r < 2m$) and exterior ($r > 2m$) regions; and two singularities that are spatially homogeneous in the limit $r \to 0$. It is impossible for a maximally extended solution to be static. There is no central world line, as a point-particle picture suggests. Thus just as quantum physics implied a radical revision of the idea of a particle, so does general-relativity: there is no general relativity version of the Newtonian idea of a point particle.

None of this is obvious. The global topology is not optional; it follows from the way the Einstein equations for this vacuum curve spacetime. And the nature of this solution emphasizes why one should always try to determine exact properties of solutions in general relativity: the global properties of the linearised form of the Schwarzschild solution (which does not exactly satisfy the field equations) will be radically different.

1.3 Gravitational collapse and black holes

Astrophysical black holes exist because they arise by the collapse of matter due to its gravitational self-attraction.

1.3.1 Spherically symmetric matter models

Spherically symmetric fluid-filled models enable one to investigate how gravitational attraction makes matter aggregate from small perturbations into major inhomogeneities that enter the non-linear regime and form black holes. Pressure-free models one can use to investigate this are the Lemaître–Tolman–Bondi models [63–65] which evolve inhomogeneously with spherical shells each obeying a version of the Friedmann equation of cosmology. One can use these as interior solutions, with the Schwarzschild solution as the exterior (vacuum) solution, joined across a timelike surface where the standard junction conditions (Section 1.1.1) are fulfilled. More realistic models with pressure obey the Oppenheimer–Volkov equation of evolution [4]. Astrophysical studies show that if the collapsing object is massive enough (its mass is greater than the Chandrasekhar limit), there is no physical pressure that will halt the collapse: the final state will be a black hole [4, 66]. Using pressure-free models to investigate gravitational collapse, Oppenheimer and Snyder showed this would eventually occur: "*we see that for a fixed value of R as t tends toward infinity, τ tends to a finite limit, which increases with R. After this time τ₀ an observer comoving with the matter would not be able to send a light signal from the star; the cone within which a signal can escape has closed entirely.*" [67].

1.3.2 Causal diagrams

Basically the situation is like Figure 1.1, but the solution's central part of the diagram is no longer vacuum; it is cut off by the infalling fluid. Initially the surface of the fluid is at

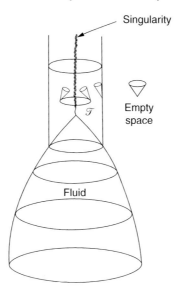

Figure 1.3 Collapse to a black hole. Reproduced, with permission, from [5].

$r_S > 2m$ so there are no closed trapped surfaces. Then the fluid surface crosses the value $r_S = 2m$ and thereafter lies inside the event horizon. Light emitted from the surface of the star at later times is trapped behind the event horizon and cannot emerge to the outside world.

The spacetime diagram showing how this occurs is given in Figure 1.3: the event horizon at $r = 2m$ is a null surface, so light emitted at that radius never moves inwards or outwards; the gravitational attraction of the mass at the centre holds it at a constant distance from the centre. From the outside, the collapse never seems to end; there is always light arriving at infinity from just outside the event horizon, albeit with ever increasing redshift and hence ever decreasing intensity.

However, this diagram is misleading in some ways: it suggests that the central singularity is a timelike world line, which is not the case; it is spacelike because it exists in the part of spacetime corresponding to region II in Figure 1.2. The Penrose diagram for what happens is shown in Figure 1.4; it shows that the outer regions are the same as region I in Figure 1.3, but the existence of the infalling star leads to a regular centre rather than another asymptotically flat region I'. A key feature, pointed out by Penrose, is that closed trapped surfaces exist for region II given by $r < 2m$: the area spanned by outgoing null geodesics from each 2-sphere decreases as one goes to the future. This is the reason that the occurrence of a singularity in the future is inevitable [37, 39].

The question then is, how generic is this situation where the final singularity is hidden behind the horizon and so is invisible to the external world: does this only occur in spherically symmetric spacetimes? A very innovative uniqueness theorem by Israel [68] stated that black holes must necessarily be spherically symmetric if a regular event horizon forms.

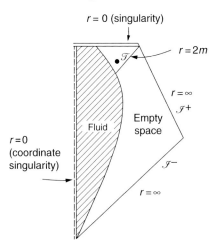

Figure 1.4 Penrose diagram of collapse to a black hole. Reproduced, with permission, from [5].

But then the question is, how general is formation of an event horizon? Penrose formulated the cosmic censorship hypothesis, that such horizons would indeed form in the generic case [69]. This conjecture is still unresolved; it is discussed in Chapter 9.

Much effort has been expended in showing that the Kerr solution is the likely final state of gravitational collapse of a rotating object. Work by Hawking, Carter, Robinson and others shows that this indeed seems to be the case [70]. Black hole uniqueness theorems are discussed in Chapter 9.

An important feature in such collapse is the Penrose inequality relating mass and black hole horizon areas, discussed in Chapter 8, and Hawking's area theorem, which states that the area of cross sections of a black hole horizon is non-decreasing towards the future. This leads to the proposed laws of black hole thermodynamics, associating a temperature and entropy to black holes in a way that is parallel to the usual laws of thermodynamics [71], in particular the temperature is $T_H = \kappa/(2\pi)$, where κ is the surface gravity of the hole, and the black hole entropy is proportional to the area of its event horizon divided by the Planck area: $S_{BH} = kA/(4\pi l_P^2)$, where l_P is the Planck length. This led on to Hawking's famous discovery of the emission of black body radiation by black holes through quantum field theory processes; that development, however, lies outside the scope of this chapter.

1.3.3 Accretion disks and domains

Black holes will generally be surrounded by accretion disks that will emit X-rays due to viscous heating as the matter falls in towards the event horizon (see Chapter 3). Rotating black holes have three important domains affecting this process [72]:

1. *The Event Horizon*: That radius inside of which escape from the black hole is not possible.

2. *The Ergosphere*: That radius inside of which negative-energy states are possible (giving rise to the potentiality of tapping the energy of the black hole).
3. *Innermost Stable Circular Orbit* (ISCO): That radius inside of which free circular orbital motion is not possible.

These are the geometric features on which the theory of accretion disks is based.

1.3.4 Powerhouses in astrophysics

Black holes are intriguing aspects of general relativity, and theoretical studies showed they were indeed likely to form in astrophysical situations. But are they relevant to the universe out there?

It seems indeed so. They were discovered theoretically as an unexpected consequence of the maximal extensions of the Schwarzschild solution. They became central to high energy astrophysics, as shown by Lynden Bell and Rees, being the key to understanding of QSOs [66, 73]. Furthermore supermassive black holes occur in many galaxies, surrounded by accretion disks [74], while stellar-mass black holes occur as the endpoint of the lives of massive stars [75]. There is much astronomical evidence for their existence [76], including one at the centre of our own galaxy [77, 78]. The accretion disks that emit radiation whereby we can detect them can be modelled in a general-relativity way [72]. This is all discussed in Chapter 3 on relativistic astrophysics, considering the observational status of black holes, and their relation to gamma-ray bursts. Chapter 7 discusses how numerical simulations have provided insights into the critical phenomenon at the threshold of black hole formation, gravitational collapse, mergers of black holes, and mergers of black holes and neutron stars.

1.3.5 Conclusion

Black holes were theoretically predicted to occur because they are solutions of the Einstein Field Equations, but have turned out to play a key role in high-energy astrophysics [66]. They occur as the endpoint of evolution of massive stars, and occur at the centre of galaxies, where they play a key role in galaxy dynamics and provide the powerhouse for high-energy astrophysical phenomena that are otherwise inexplicable. Furthermore, black hole interactions are expected to provide the source of gravitational waves that will enable us to probe the extremely early universe with great precision (Section 1.6.5). They have no analogue in Newtonian gravitational theory because there is no limit to the speed of propagation of signals in that theory (where the speed of light plays no special role).

1.4 Cosmology and structure formation

The development of cosmological models based in General Relativity theory provides the basis for our understanding both of the evolution of the universe as a whole and of structure formation within it. The start was Einstein's static universe model of 1917 [79], followed by the static de Sitter universe in 1917, Friedmann's expanding models of 1922 [80] and 1924 [81], and Lemaître's expanding model of 1927 [82, 83]. These expanding models were ignored until Eddington's proof of the instability of the Einstein static universe in 1930 [84]. The idea of the expanding universe was then generally accepted, and canonised in Robertson's fine review article in 1933 [85] (which gives annotated references to all the earlier papers). It is noteworthy that the first Newtonian cosmological models were developed only in 1935 by Milne and McCrea – some 18 years after the first general relativistic models.

Today the 'standard model of cosmology', based in the way general relativity shows how matter curves spacetime, is highly successful in describing precision observations of the large-scale properties of the universe, although some significant problems concerning the nature of the matter and energy fields controlling the dynamics of the universe remain unresolved [10, 86, 87].

1.4.1 FLRW models and the Hot Big Bang

The expanding universes of Friedmann and Lemaître ('FL') are a family of exact solutions of the Einstein field equations that are spatially homogeneous and isotropic. The geometry of these standard models was clarified by Robertson and Walker ('RW') in 1935. The metric is characterised by a scale factor $a(t)$ representing the time change of the relative size of the universe:

$$ds^2 = -dt^2 + a^2(t)d\sigma^2, \quad d\sigma^2 = dr^2 + f^2(r)d\Omega^2$$

with $f(r) = \{\sin r, r, \sinh r\}$ if $k = \{+1, 0, -1\}$ and $d\Omega^2 = d\theta^2 + \sin^2\theta\, d\phi^2$; thus the 3-space metric $d\sigma^2$ represents a 3-space of constant curvature k. Spatial homogeneity together with isotropy implies that a multiply transitive group of isometries G_6 acts on the surfaces $\{t = const\}$, and consequently all physical and geometric quantities depend only on t. The 4-velocity $u^a = dx^a/dt$ of preferred fundamental observers is

$$u^a = \delta_0^a \Rightarrow u^a u_a = -1$$

so the time parameter t represents proper time measured along fundamental world lines $x^a(t)$ with u^a as tangent vector; the distances between these world lines scale with $a(t)$, so volumes scale as $a^3(t)$. The relative expansion rate of matter is represented by the Hubble parameter $H(t) = (1/a)(da/dt)$, and the rate of slowing down by the deceleration parameter $q := -(1/a)(d^2a/dt^2)$. These spacetimes are conformally flat: $C_{abcd} = 0$ (there is no free gravitational field, so no tidal forces or gravitational waves occur in these models).

Dynamics

The behaviour of matter and expansion of the universe are governed by three related equations. First, the energy conservation equation (1.3) becomes

$$d\rho/dt + 3H(\rho + p) = 0 \tag{1.14}$$

relating the rate of change of the energy density $\rho(t)$ to the pressure $p(t)$. It becomes determinate when we are given an equation of state $p = p(\rho, \phi)$ determining p in terms of ρ and possibly some internal variable ϕ (which, if present, will have to have its own dynamical equations). The normalized density parameter is $\Omega := \kappa\rho/(3H^2)$. The Raychaudhuri equation (1.6) becomes

$$\frac{3}{a}\frac{d^2a}{dt^2} = -\frac{\kappa}{2}(\rho + 3p) + \Lambda \tag{1.15}$$

which directly gives the deceleration due to matter, and shows (i) the active gravitational mass density is $\rho_{grav} := \rho + 3p$, which is positive for all ordinary matter, and (ii) a cosmological constant Λ causes acceleration iff $\Lambda > 0$. Its first integral is the Friedmann equation

$$3H^2 = \kappa\rho + \Lambda - \frac{3k}{a^2}, \tag{1.16}$$

where $3k/a^2$ is the curvature of the 3-spaces; this is just the Gauss–Codazzi equation relating the curvature of imbedded 3-spaces to the curvature of the imbedding 4-dimensional spacetime [18, 25]. The evolution of $a(t)$ is determined by any two of (1.14), (1.15), and (1.16).

The basic behaviour has been known since the 1930s [85]. Normal matter can be represented by the equation of state $p = (\gamma - 1)\rho$, where $\gamma = 1$ for pressure-free matter ('dust' or 'baryonic matter'); $\gamma = 4/3$ for radiation; and $\gamma = -1$ is equivalent to a cosmological constant. There is always a singular start to the universe at time $t_0 < 1/H_0$ ago if $\rho + 3p > 0$, which will be true for ordinary matter plus radiation, and $\Lambda \leq 0$; however, one can get a bounce if both $\Lambda > 0$ and $k = +1$ (for matter with $\rho > 0$, bounces can occur only if $k = +1$). When $\Lambda = 0$, if $k = +1$, $\Omega > 1$ the universe recollapses; if $k = -1$, $\Omega < 1$ it expands forever; and $k = 0$ ($\Omega = 1$) is the critical case separating these behaviours, that just succeeds in expanding forever. When $\Lambda > 0$ and $k = +1$, a static solution is possible (the Einstein static universe [88]), and the universe can bounce or 'hover' close to a constant radius (these are the Eddington–Lemaître models). All these behaviours can be illuminatingly represented by appropriate dynamical systems phase planes [89, 90].

Observational relations

Observational relations can be worked out for these models, based on the fact that photons move on null geodesics $x^a(v)$ with tangent vector $k^a(v) = dx^a/dv$: $k^a k_a = 0$, $k^a_{;b}k^b = 0$ [91]. This shows that the radial coordinate value of radial null geodesics through the origin (which are generic null geodesics, because of the spacetime symmetry) is given by

$$\left\{ds^2 = 0,\ d\theta = 0,\ d\phi = 0\right\}$$
$$\Rightarrow u(t_0, t_1) := \int_0^{r_{emit}} dr = \int_{t_{emit}}^{t_{obs}} \frac{dt}{a(t)} = \int_{a_{emit}}^{a_{obs}} \frac{1}{H(t)}\frac{da}{a^2(t)}. \tag{1.17}$$

Substitution from the Friedmann equation (1.16) shows how the cosmological dynamics affects $u(t_0, t_1)$. The key variables resulting from this are the observed redshift z, given by

$$1 + z = \frac{(k^a u_a)_{emit}}{(k^b u_b)_{obs}} = \frac{a(t_{obs})}{a(t_{emit})},$$

and the area distance r_0, which up to a redshift factor is the same as the luminosity distance D_L:

$$D_L = r_0(1 + z);$$

this is the reciprocity theorem [25, 92]. One can work out observational relations for galaxy number counts versus magnitude (n, m) and the magnitude–redshift relation (m, z), which determines the present deceleration parameter q_0 when applied to "standard candles" [93]. A power-series derivation of observational relations in generic cosmological models is given by Kristian and Sachs [91].

Major observational programmes have examined these relations and determined H_0 and q_0, showing that, assuming the geometry is indeed that of a Robertson–Walker spacetime, the universe is accelerating at recent times $(q_0 < 0)$. This means some kind of dark energy is present such that at recent times, $\rho + 3p < 0$ [10, 86, 87]. The simplest interpretation is that this is due to a cosmological constant $\Lambda > 0$ that dominates the recent dynamics of the universe.

A key finding is the existence of limits both to causation, represented by particle horizons, and to observations, represented by visual horizons. The issue is whether $u(t_0, t_1)$ converges or diverges as $t_0 \to 0$; and, for ordinary matter and radiation, it converges, representing a limit to how far causal effects can propagate since the start of the universe. Much confusion about their nature was cleared up by Rindler in a classic paper [33], with further clarity coming from use of Penrose causal diagrams for these models [31]. This showed that particle horizons would occur if and only if the initial singularity was spacelike. There are many statements in the literature that such horizons represent motion of galaxies away from us at the speed of light, but that is not the case; they occur due to the integrated behaviour of light from the start of the universe to the present day [94], with the visual horizon, determined by the most distant matter we can detect by electromagnetic radiation, lying inside the particle horizon. This is why the visual horizon size can be 42 billion light years in an Einstein–de Sitter model with a Hubble scale of 14 billion years. Event horizons relate to the ultimate limits of causation in the future universe, that is whether $u(t_0, t_1)$ converges or diverges as $t_1 \to \infty$; while they play a key role as regards the nature of black holes (Section 1.2.5), they are irrelevant to observational cosmology.

Cosmological physics

Cosmological models started off as purely geometrical, but then a major realisation was that standard physics could be applied to the properties of matter in the early universe.

First was the application of atomic physics to the expanding universe, resulting in prediction of a Hot Big Bang ('HBB') early phase of the universe with ionised matter and radiation in equilibrium with each other because of tight coupling between electrons and radiation. This phase ended when the temperature dropped below the ionisation temperature of the matter, resulting in cosmic blackbody radiation being emitted at the Last Scattering Surface at about $T_{emit} = 4000$ K. A major theoretical result, consequent on the reciprocity theorem, is that the blackbody spectrum will be preserved but with temperature $T_{obs} = T_{emit}/(1 + z)$ [25]; hence this radiation is observed today as cosmic microwave blackbody radiation (CMB) with a temperature of 2.73 K. There then follows a complex interaction of matter and radiation in the expanding universe [95, 96]. A key feature is the evolution of the speed of sound with time, as well as diffusion effects leading to damping of fine-scale structure [97].

Second was the application of nuclear physics to the epoch of nucleosynthesis at a temperature of about $T = 10^8$ K, leading to predictions of light-element abundances resulting from nuclear interactions in the early universe [95, 98, 99]. The key point here is that the Friedmann equation for early times, when radiation dominates the dynamics and curvature is negligible, gives the temperature–time relation [100]

$$T = \frac{T_0}{t^{1/2}}, \quad T_0 := 0.74 \left(\frac{10.75}{g_*} \right)^{1/2} \tag{1.18}$$

with no free parameters. Here t is time in seconds, T is the temperature in MeV, and g_* is the effective number of particle species, which is 10.75 for the standard model of particle physics, where there are contributions of 2 from photons, 7/2 from electron–positron pairs and 7/4 from each neutrino flavor. It is this relation that determines the course of the nuclear reactions, leading to formation of light elements (up to lithium) in the early universe. These element abundances agree with those determined by astronomical observations, up to some unresolved worries about lithium. It was this theory that established cosmology as a solid branch of physics [100].

In summary

These models are the opposite of the Schwarzschild vacuum solution. Those models represent the dynamics of pure vacuum (there is no matter tensor); these models represent the dynamics of spacetime governed purely by matter (there is no free gravitational field). Hence the outcome depends irrevocably on the type of matter present, as shown for example in the dependence (1.18) of $T(t)$ on g_*. They are remarkably simple, and remarkably good models of the real universe we observe by means of astronomical observations. They also represent a key conundrum: they predict a start to the universe, and indeed all of physics, at a spacetime singularity. One key issue is whether that conclusion can somehow be avoided.

1.4.2 More general dynamics: inflation

The previous section was based in well-established physics. At earlier times more speculative physics will necessarily be involved, involving interactions that are not yet testable by particle accelerators.

More general matter dynamics will lead to more general behaviour of the cosmological model at early times. In particular a scalar field $\phi(t)$ with potential $V(\phi)$ may be present, obeying the Klein–Gordon Equation

$$d^2\phi/dt^2 + 3H\phi + dV/d\phi = 0. \tag{1.19}$$

It will have an energy density $\rho = \frac{1}{2}(d\phi/dt)^2 + V(\phi)$ and pressure $p = \frac{1}{2}(d\phi/dt)^2 - V(\phi)$, so the inertial and gravitational energy densities are

$$\rho + p = (d\phi/dt)^2 > 0, \quad \varrho + 3p = (d\phi/dt)^2 - V(\phi). \tag{1.20}$$

Hence the active gravitational mass can be negative in the 'slow roll' case: $(d\phi/dt)^2 < V(\phi)$. So scalar fields can cause an exponential acceleration when ϕ stays at a constant value ϕ_0 because of the friction term $3H\phi$ in (1.19) resulting in $\varrho + 3p \approx -V(\phi_0) = const$. This is the physical basis of the *inflationary universe* idea of an extremely rapid exponential expansion at very early times that smooths out and flattens the universe, also causing any matter or radiation content to die away towards zero. The inflaton field itself dies away at the end of inflation when slow rolling comes to an end and the inflaton gets converted to matter and radiation by a process of reheating.

It is now broadly agreed that there was indeed such a period of inflation in the very early universe, but the details are not clear: there are over 100 different variants [101], including single-field inflation, multiple-field inflation, and models where matter is not described by a scalar field, such as, for example, vector inflation. As the potential is not tied in to any specific physical field, one can run the field equations backwards to determine the effective inflaton potential from the desired dynamic behaviour [102, 103].

1.4.3 Perturbed FLRW models and the growth of structure

The real universe is only approximately a Robertson–Walker spacetime. Structure formation in an expanding universe can be studied by using linearly perturbed FLRW models at early times, plus numerical simulations at later times when the inhomogeneities have gone non-linear [104, 105]. The development of the theory of structure formation, a major theoretical achievement, is described illuminatingly by Longair ([97], Chapter 15), which gives full references: it is only possible to highlight a few of these papers in this brief section (and see also Chapter 4 in this volume).

Lemaître initiated the use of inhomogeneous models to study structure formation [106–108]. The study of general linear perturbations of an expanding universe was initiated by Lifshitz [109], and the path-breaking paper by Sachs and Wolfe [110] first studied their effect on microwave background radiation anisotropies. That paper was very careful about gauge freedom. However, many others were not, being plagued by the gauge problem: because of the absence of a fixed background spacetime, one can encounter unphysical gauge modes that are essentially just coordinate fluctuations. A path-breaking paper by Bardeen [111] developed a gauge-covariant approach to linear perturbations that enabled distinguishing of physical from coordinate modes, and this approach [112, 113] is widely used. An alternative 1+3 covariant approach was initiated by Hawking [114] and then developed by Ellis, Bruni, and collaborators [115, 116]. Both methods have been used to study both structure formation and associated CMB anisotropies [10, 86, 87, 117]. The perturbation equations can, for example, be used to determine the dynamical effects of the cosmological constant [118].

Key aspects of CMB anisotropies are the baryon acoustic oscillations before decoupling [104, 119, 120], which lead to the observed CMB power spectrum peaks, and the Sunyaev–Zel'dovich effect (scattering of the CMB by ionised hot gas in clusters of galaxies) after decoupling [96, 121], which acts as a sensitive test of conditions at the time of scattering. The perfection of the calculations by Silk and Wilson [122] and Bond and Efstathiou [123], and the subsequent explosion of precision calculations (in addition to measurements) in the 1990s, based on the growth of fluctuations seeded by quantum fluctuations in the inflationary era [112, 113], is one of the great success stories in cosmology ([97]: Chapter 13). A key realisation was that introduction of a cosmological constant would allow a cold dark matter scenario to match observations of an almost flat universe [124].

1.4.4 The Standard Model of cosmology

Use of the perturbed FLRW models in conjunction with extraordinary observational advances has led to an era of precision cosmology, combining physics and astrophysics with general relativity to predict the background model evolution together with the growth of structure in it. Key features are: the dynamical evolution of the background model, with an inflationary era followed by a hot big bang era, a matter-dominated era, and a dark-energy-dominated late era; physical modelling of the hot big bang era, including nucleosynthesis and matter–radiation decoupling leading to a CMB with a precise black body spectrum; models of galaxy and other source numbers with respect to distance as well as luminosity versus distance relations; modelling of structure formation, related to matter power spectra and Baryon Acoustic Oscillations; and prediction of CMB angular power spectra. Together they comprise the concordance model of cosmology [10, 86, 87, 117].

The puzzles are that we do not know the nature of the inflaton, for which there are over 100 models [101]; we do not know the nature of the dark matter that is indicated to exist by dynamical studies, and is far more abundant than baryonic matter; and we do not know the nature of the dark energy causing an acceleration of the expansion of the universe at late times. It is possible that some of these issues may be indicating we need a different theory of gravity than general relativity, for example MOND or a scalar–tensor theory.

1.4.5 More general geometries: LTB and Bianchi models

There are two other classes of exact solutions that have been used quite widely in cosmological studies.

LTB spherically symmetric models

Firstly, the growth of inhomogeneities may be studied by using exact spherically symmetric solutions, enabling study of non-linear dynamics. The zero-pressure such models are the Lemaître–Tolman–Bondi exact solutions [64, 65] where the time evolution of spherically symmetric shells of matter is governed by a radially dependent Friedmann equation. The solutions generically have a matter density that is radially dependent, as well as a spatially varying bang time. These models can be used to study (i) the way a spherical mass with low enough kinetic energy breaks free from the overall cosmic expansion and recollapses, hence putting limits on the rate of growth of inhomogeneities [125], and (ii) the way that any observed (m, z) and (N, z) relations can be obtained in a suitable LTB model where one runs the EFE backwards to determine the free functions in the metric from observations, for any value whatever of the cosmological constant Λ [10, 126]. This opens up the possibility of doing away with the need for dark energy if we live in an inhomogeneous universe, where the data usually taken to indicate a change of expansion rate in time due to dark energy are in fact due to a variation of expansion rate in space. However, although the supernova observations can be explained in this way, detailed observational studies based in the kinematic Sunyaev–Zel'dovich effect show this is unlikely [10].

The Bianchi models and phase planes

Secondly, following the work of Gödel, Taub, and Heckmann and Schücking, there is a large literature examining the properties of spatially homogeneous anisotropically expanding models. These models are generically invariant under 3-dimensional continuous Lie groups of symmetries, whose Lie algebra was first investigated by Luigi Bianchi using projective geometry; much simpler methods are now available [13]. Special cases allow higher symmetries (Locally Rotationally Symmetric models [12]). These allow a rich variety of non-linear behaviour, including anisotropic expansion at early and late times even if the present behaviour is nearly isotropic; different expansion rates at the time of nucleosynthesis than in FLRW models, leading to different primordial element abundances; complex anisotropy patterns in the CMB sky; and much more complex singularity behaviour than in FLRW models, including cigar singularities, pancake singularities (where particle horizons may be broken in specific directions), chaotic ('mixmaster') type behaviour [127] characterised by 'billiard ball' dynamics, and non-scalar singularities if the models are tilted. Dynamical systems methods can be used to show the dynamical behaviour of solutions and the relations of families of such models to each other [90]. If a cosmological model is generic, it should include Bianchi anisotropic modes as well as inhomogeneous modes, and may well show mixmaster behaviour [128].

1.4.6 Cosmological success and puzzles

Standard cosmology is a major application of GR showing how matter curves spacetime and spacetime determines the motion of matter and radiation. It is both a major success, showing how the dynamical nature of spacetime underlies the evolution of the universe itself, with this theory tested by a plethora of observations [10], and a puzzle, with three major elements of the standard model unknown (Dark matter, Dark energy, the Inflaton). While much of cosmological theory (the epoch since decoupling) follows from Newtonian gravitational theory (NGT), this is not true of the dynamics of the early universe, where pressure plays a key role in gravitational attraction: thus for example NGT cannot give the correct results for nucleosynthesis. The theory provides a coherent view of structure formation (with a few puzzles), and hence of how galaxies come into existence. The theory raises the issue that the universe not only evolves but (at least classically) had a beginning, whose dynamics lies outside the scope of standard physics because it lies outside of space and time. This is all discussed in depth in Chapter 4.

1.5 Gravitational lensing and dark matter

The Schwarzschild solution predicts bending of light by massive objects: that is, gravitational lensing will occur. Indeed any mass concentration will differentially attract light rays; this is nowadays an important effect in astronomy and cosmology.

The bending of light by a gravitational field was predicted by Einstein in 1911 from the equivalence of a uniform gravitational field with an accelerated reference frame. In 1912 he derived an equation for gravitational lensing [129], well before he deduced the gravitational field equations. In 1915, he applied the gravitational field equations and found the deflection angle is twice the result obtained from the equivalence principle, the factor of two arising because of the curvature of space. He predicted bending of light by the Sun in 1916, as famously verified by Eddington's Expedition in 1919, but he only published his pioneering paper on the bending of light by stars and galaxies in 1936 [130]. He stated there that the bending would not be observable in these cases because of the small angular scales involved, but recognised that lensing would cause fluctuations in the brightness of distant objects that might be detectable.

Gravitational lensing is now a key part of astronomy and cosmology [131, 132].

1.5.1 Calculating lensing

One can calculate deflection of light, and associated image displacement, distortion, and brightening, either through a weak-field approximation, which will be valid for any mass distribution, or by determining geodesics in an exact solution, for example a spherically symmetric solution. They of course have to agree in the weak-field approximation. Both show that for weak lensing, the angle of deflection of light caused by a spherical mass M is:

$$\hat{\alpha}(\xi) = \frac{4GM}{c^2} \frac{\xi}{|\xi|^2}$$

where ξ is the impact parameter, that is, the closest distance the light beam would reach from the centre of the lensing mass if there were no bending. Because the Schwarzschild radius of the Sun is 2.95 km and the Solar radius is 6.96×10^5 km, a light ray grazing the limb of the Sun is deflected by an angle $(5.9/7.0) \times 10^{-5}$ radians $= 1.''7$.

For more complex mass distributions, we represent the mass projected onto a mass sheet called the lens plane. Following Blandford and Narayan [133], a light ray from a source S at redshift z is incident on a deflector or lens L at redshift z_d with impact parameter ξ relative to some fiducial lens center. Assuming the lens is thin compared to the total path length, its influence can be described by a deflection angle $\hat{\alpha}(\xi)$ (a two-vector) for the ray when it crosses the lens plane. The mass projected onto this plane is characterized by its surface mass density

$$\Sigma(\xi) = \int \rho(\xi, z) dz$$

where ξ is a two-dimensional vector in the lens plane. The deflection angle at position ξ is the sum of the deflections due to all the mass elements in the plane:

$$\hat{\alpha}(\xi) = \frac{4\pi G}{c^2} \int \frac{(\xi - \xi')\Sigma(\xi')}{|\xi - \xi'|^2} d^2\xi'.$$

In the cosmological context, the intrinsic lensing angle $\hat{\alpha}(\theta)$ must be corrected to give the observed lensing angle α by using the angular diameter distances D_d, D_s, and D_{ds} between the source, deflector, and observer, which are affected by cosmological parameters. When the deflected ray reaches the observer, she sees the image of the source apparently at position θ on the sky. The true direction of the source, i.e. its position on the sky in the absence of the lens, is given by β. Now $\theta D_s = \beta D_s - \alpha D_{ds}$, where β is the angle from the source centre to the lensing element. Therefore, the positions of the source and the image are related through the lensing equation

$$\beta = \theta - \alpha(\theta)$$

where $\alpha(\theta) = (D_{ds}/D_s)\hat{\alpha}(D_d\theta)$. This light deflection leads to image distortion and amplification, characterised by the image magnification and shear. The shear is caused by the Weyl tensor the light encounters; the magnification is caused directly by the matter it encounters, and indirectly by the cumulative effect of the shear [5, 23, 91] (this follows from (1.5) with (1.8), and leads to (1.7)). The more mass (and the closer to the center of mass), the more the light is bent, and the more the image of a distant object is displaced, distorted, and perhaps magnified. These are the basic equations from which the cosmological applications follow.

1.5.2 Strong lensing

Lensing by galaxies can bend the light of a point background source by a large enough angle that it can be observed as several separate images. If the source, lensing object, and observer lie in a straight line, the source can appear as a ring around the lensing object (this is the case of Einstein rings). Misalignment will result in an arc segment instead. Lensing masses such as galaxy groups and clusters can result in the source being seen as many arcs around the lens. The observer may then see multiple distorted images of the same source, with time delays detectable between them if the source varies. The time delay between images is proportional to the difference in the absolute lengths of the light paths, which in turn is proportional to H_0. The number and shape of the arcs depends on the relative positions of the source, the lens, and the observer, as well as the gravitational well of the lensing object. The number and position of images is due to caustics in the light sheet that can be characterised as cusps, folds, etc. on using catastrophe theory [134]. Details are given in Chapter 3.

This effect enables detecting very distant galaxies that are otherwise unobservable. For example combining observations from the Hubble Space Telescope, Spitzer Space Telescope, and gravitational lensing by cluster Abell 2218, allowed discovery of the galaxy MACS0647-JD, that is roughly 13 billion light-years away. The galaxy appears 20 times larger and over three times brighter than typically lensed galaxies.

1.5.3 Weak lensing

Weak lensing results in image distortion, but the shear may be too small to be seen directly. Also, apparent shear may be due to the galaxy's distinct shape, or an angle of view that makes it appear elongated, or the telescope, or the detector, or the atmosphere, so one cannot deduce weak lensing from images of a single object. However, the faint distortions due to lensing of images of a set of galaxies can be worked out statistically, and the average shear due to lensing by some massive objects in front can be computed. Thus in the weak lensing case, measuring statistical distortion is the key to measuring the mass of the lensing object. Weak lensing surveys use this method to determine the intervening mass distribution. This is an important method in detecting dark matter [131]. Indeed the most important result from weak lensing concerns the proof of existence of collisionless dark matter in the 'bullet cluster' (see Chapter 3). An exciting recent application is detection of weak lensing in observations of CMB anisotropy patterns by the Planck satellite.

1.5.4 Microlensing

For point objects, it may happen that no distortion in shape can be seen but the amount of light received from a distant object changes in time, with a characteristic shape of the resulting light curve. The lensing objects may be stars in the Milky Way, with the source being stars in a remote galaxy, or a distant quasar. The method has been used to discover extrasolar planets, which is one of its most important applications.

1.5.5 Conclusion

Gravitational lensing was predicted very early on by Einstein and provided the first new observational evidence that vindicated general relativity. It is a confirmation of the curving of spacetime by matter and allows us to detect matter which has no other observational signature. It therefore plays a key role in the mapping of dark matter in the universe. It also acts a lens allowing us to see far further than otherwise possible. It is thus a key tool in cosmology, discussed in Chapter 3.

1.6 Gravitational waves

Einstein also predicted the existence of gravitational waves very early on [88]. However, because of the coordinate freedom of general relativity, there was confusion as to whether these were real waves, or just coordinate waves. This was sorted out by the understanding, particularly due to Pirani, that the gravitational waves could be seen as Weyl tensor waves [22], which could carry away energy and momentum. They could therefore be indirectly observed by their effect on binary pulsar orbits, which can be measured to extremely high precision. Direct detection is much more difficult, but spectacular development in detector design promises that they may be detected in the next decades, with gravitational wave observatories having the potential to become an essential tool in precision cosmology. Crucial to this project are major developments in numerical relativity allowing us to predict the nature of waves expected to be emitted from binary black hole mergers and other strong-field sources.

1.6.1 Weak-field formulation

Gravitational waves exist because GR is a relativistic theory of gravitation; unlike Newtonian theory, influences cannot be exerted instantaneously. Using a weak-field approximation $g_{\mu\nu} = \eta_{\mu\nu} + h_{\mu\nu}$, $h_{\mu\nu} \ll 1$, and harmonic coordinates, Einstein predicted the existence of waves in the metric tensor. But this may just be a wave in the coordinates! One can avoid this problem by using geometric variables, for example the 1+3 covariant representation of the Weyl tensor, whereby one obtains Maxwell-like equations for the electric and magnetic parts of the Weyl tensor [25, 26, 114].

Nevertheless, usual calculations use the weak-field method of Einstein [88]. The perturbation h_{ab} is tracefree and transverse, so gravitational waves are transverse and, given their direction of propagation k, have two degrees of freedom (two polarisations) as represented by tracefree 2-tensors orthogonal to k. That is, they are spin-2 fields. Defining $\bar{h}_{ab} \equiv h_{ab} - \frac{1}{2}\eta_{ab}h$ and choosing coordinates so that $\bar{h}^{ab}{}_{,b} = 0$, then, on defining $z' = z - ct$, the two polarisation modes are $h_+(z')$ and $h_-(z')$, with weak field metric form

$$ds^2 = -dt^2 + dx^2 + dy^2 + dz^2 + h_{\alpha\beta}(z')dx^\alpha dx^\beta \ (h \ll 1),$$

$$h_{ab} = \begin{pmatrix} 0 & 0 & 0 \\ 0 & h_+ & h_- \\ 0 & h_- & -h_+ \end{pmatrix}.$$

One can get plane-wave solutions of this form, characterised by their amplitude h, frequency f, wavelength λ, and speed c, related by the usual equation $c = \lambda f$. They can be linearly polarized, with quadrupole polarization components h_+ and $h_\times$ in canonical coordinates, rotated by 45 degrees relative to each other. One can also have circularly polarized waves.

In the slow-motion approximation for a weak metric perturbation for a source at distance r, the wave radiated by a local source is given by

$$h_{\mu\nu} = \frac{2G}{c^4 r} \frac{d^2 I_{\mu\nu}}{dt^2}$$

where $I_{\mu\nu}$ is the reduced quadrupole moment defined as

$$I_{\mu\nu} = \int \rho(\mathbf{r}) \left(x_\mu x_\nu - \frac{1}{3}\delta_{\mu\nu} r^2 \right) dV.$$

Hence gravitational waves are radiated by objects whose motion involves acceleration, provided that the motion is not perfectly spherically symmetric (the case of an expanding or contracting sphere). There are no spherical gravitational waves! An example is two gravitating objects that form a binary system. Seen from the plane of their orbits, the quadrupole term $I_{\mu\nu}$ will have non-zero second derivative.

1.6.2 Exact gravitational waves

One can find exact gravitational wave solutions for cases with high symmetries.

Cylindrical gravitational waves

Einstein and Rosen [135] derived the exact solution for cylindrical gravitational waves. They state, "The rigorous solution for cylindrical gravitational waves is given. After encountering relationships which cast doubt on the existence of rigorous solutions for undulatory gravitational fields, we investigate rigorously the case of cylindrical gravitational waves. It turns out that rigorous solutions exist and that the problem reduces to the usual cylindrical waves in euclidean space."

Plane gravitational waves

Plane gravitational waves are Petrov type N vacuum solutions that allow arbitrary information to be carried at the speed of light. They are exact solutions of the empty space field equations that can have high symmetries as studied by Ehlers and Kundt, and by Bondi, Pirani, and Robinson [136]. They have intriguing global properties due to the way they focus null geodesics [137]. One can find exact solutions for two colliding plane gravitational waves [138].

1.6.3 Asymptotic flatness and gravitational radiation

Gravitational waves can carry energy, momentum, and information, and it is useful to investigate this in general cases without symmetry. This can be done by asymptotic expansions in flat spacetimes, using multipole expansions clarified nicely by Thorne [14]. Following Trautman, outgoing radiation conditions have been formulated by Bondi, Sachs, Newman and Penrose, and others, leading to peeling-off theorems showing how successively different Petrov types occur at larger distances, and a 'news function' related to the derivative of shear is related to mass loss [19, 23, 139]. Penrose showed how to express all this using his conformal representation of infinity [11]. Positive mass theorems aim to show that one cannot radiate so much mass away that the mass becomes negative [140, 141].

1.6.4 Emission

The linear perturbation formulae given above suffice for gravitational wave emission in weak-field situations like the binary pulsar, and there is a large literature on analytic perturbations of black hole solutions [142] and the black hole perturbation approach to gravitational radiation [143, 144] as well as studies of gravitational waves from gravitational collapse [145] and from post-Newtonian sources and inspiralling compact binaries [146]. However, the real progress has come from major developments in numerical methods for examining processes such as coalescence of two black holes, and associated numerical codes [147–150]. Progress in this area has particularly been due to breakthroughs by Frans Pretorius. Chapter 7 discusses probing strong-field gravity through numerical simulations.

1.6.5 Gravitational wave detection

As shown by Pirani, the effect of gravitational waves on local systems is via the (generalised) geodesic deviation equation [21, 22], which shows how the wave will tend to move particles transverse to the direction of motion in a quadrupole mode. Joseph Weber pioneered the effort to build gravitational detectors based on this effect, but it is enormously difficult technically because the strain is so small: typically of amplitude $h \simeq 10^{-20}$. Indirect detection via the effects of energy loss is easier.

Indirect detection: binary pulsars

As noted above, emission requires anisotropic motion of a mass. This occurs in astronomical binary systems. If two masses m_1 and m_2 are in orbit around each other, separated by a distance r, the power radiated by this system is:

$$P = \frac{dE}{dt} = -\frac{32}{5}\frac{G^4}{c^5}\frac{(m_1 m_2)^2(m_1 + m_2)}{r^5} \tag{1.21}$$

This energy loss will be reflected in a change in the orbital period, which is measurable in the case of binary pulsars because their precise pulse timing allows very accurate orbital tracking. Observations of binary pulsar decay rates by Hulse and Taylor confirmed this effect, and so confirmed the existence of gravitational radiation carrying energy away from the system [151, 152].

Indirect detection: inflation

Gravitational waves are expected to be emitted in the inflationary era in the very early universe and so CMB observations can provide evidence for their existence [153]. The relation between scalar contributions S to the quadrupole and tensor contributions T determines the amplitude (T/S). This is related to the tilt n_T by the key relation

$$n_T = -\frac{1}{7}\frac{T}{S} \tag{1.22}$$

which not only provides a consistency check of inflation, but also allows direct detection of gravity waves, as it relates the overall amplitude to the tilt [154]. The ratio T/S must be greater than 0.2 for a statistically significant detection of tensor perturbations. Gravitational waves will imprint as B-modes in the CMB anisotropy patterns, and so should be detectable by CMB polarisation measurements [86, 123]. The value of the tensor-to-scalar ratio determined by observations such as those by BICEP2 strongly constrains models of inflation [155].

Direct detection by bars or interferometers

Ground-based direct detection is possible in principle via bar detectors or interferometers. Because of the very small size of the strain, this requires immense skill in technical development, discussed in Chapter 5; in particular it requires quantum non-demolition methods pioneered by Braginski. Gravitational-wave experiments with interferometers and with resonant masses can search for stochastic backgrounds of gravitational waves of cosmological origin. The sensitivity of these detectors as a function of frequency has been carefully explored in relation to the expected astronomical sources, and this method has the prospect of opening up a new window on the universe because it will reach back to much earlier times than any other method [156, 157]. Direct detection of the inflationary gravitational wave background constrains inflationary parameters and complements CMB polarization measurements [158].

Pulsar timing arrays

Gravitational waves will affect the time a pulse takes to travel from a pulsar to the Earth. A pulsar timing array uses millisecond pulsars to search for effects of gravitational waves on measurements of pulse arrival times, giving delays of less than 10^{-6} seconds. Three pulsar timing array projects are searching for patterns of correlation and anti-correlation between signals from an array of pulsars that will signal the effects of gravitational waves.

1.6.6 Conclusion

The theory of gravitational waves extends the idea of transverse wave propagation from the spin-1 field of electromagnetism (Maxwell's theory) to the spin-2 field of gravitation. They are an essentially relativistic phenomenon: they cannot occur in Newtonian gravitational theory. Gravitational waves can carry energy and arbitrary information, and indeed convey information to us from the earliest history of the universe that we can access. They provide the ultimate limit of our possible access to knowledge about the early universe. Their direct detection is a formidable technological problem: the detectors being constructed are a triumph of theory realised in practice. Chapter 6 discusses "Sources of gravitational waves: theory and observations" while Chapter 5 discusses current and future ground and space based laser interferometric gravitational wave observatories, pulsar timing arrays and the Einstein Telescope. It summarizes how these efforts will provide a brand new window on the universe.

1.7 Generalisations

Classical generalisations of general relativity include scalar–tensor theories, higher-derivative theories, theories with torsion, bimetric theories, unimodular theories, and higher-dimensional theories. However, if one demands only second-order equations in four dimensions and with one spacetime metric, general relativity is the unique gravitational theory based in Riemannian geometry, as shown by Lovelock [159]. The theory was derived not because of experiment, but as the result of pure thought; but it has survived all experimental tests [40]: see Chapter 2. There is no observational or experimental reason to modify or abandon the field equations.

However, there is a significant problem in terms of the relation of general relativity to quantum field theory calculations that predict existence of a vacuum energy density vastly greater than the observed value of the cosmological constant [160]. Possible solutions include either the existence of a multiverse combined with observational selection effects, or some form of unimodular gravity leading to the trace-free form of the Einstein equations. In the latter case the exactly constant vacuum energy density does not gravitate, and this major problem is fully solved, with no change to all the results mentioned above in this chapter.

General relativity theory, and all its applications (most of which he pioneered), are yet another testament to Albert Einstein's extraordinary creativity and physical insight.

References

[1] Stephani, H. *et al.* 2003. *Exact solutions of Einstein's field equations, Second edition.* Cambridge: Cambridge University Press. Corrected paperback reprint, 2009.
[2] Ferreira, P. G. 2014. *The perfect theory: a century of geniuses and the battle over general relativity.* London: Little, Brown.
[3] Einstein, A. 1915. *Sitzungsber. Preuss. Akad. Wiss. Berlin (Math. Phys.),* 844–847.
[4] Misner, C. W., Thorne, K. S., and Wheeler, J. A. 1973. *Gravitation.* San Francisco, CA: W. H. Freeman and Co.
[5] Hawking, S. W., and Ellis, G. F. R. 1973. *The large scale structure of space-time.* Cambridge: Cambridge University Press.
[6] Wald, R. M. 1984. *General relativity.* Chicago, IL: University of Chicago Press.
[7] Stephani, H. 1990. *General relativity: an introduction to the theory of the gravitational field (2nd edition).* Cambridge: Cambridge University Press.

[8] Synge, J. L., and Schild, A. 1949. *Tensor calculus.* Toronto: University of Toronto Press. Reprinted 1961.

[9] Schouten, E. 1954. *Ricci calculus: an introduction to tensor analysis and its geometrical applications (2nd edition).* Die Grundlehren der Mathematischen Wissenschaften, vol. X. Berlin: Springer.

[10] Ellis, G. F. R., Maartens, R., and MacCallum, M. A. H. 2012. *Relativistic cosmology.* Cambridge: Cambridge University Press.

[11] Penrose, R., and Rindler, W. 1984. *Spinors and space-time I: two-spinor calculus and relativistic fields.* Cambridge: Cambridge University Press.

[12] Ellis, G. F. R. 1967. *J. Math. Phys.,* **8**, 1171–1194.

[13] Ellis, G. F. R., and MacCallum, M. A. H. 1969. *Commun. Math. Phys.,* **12**, 108–141.

[14] Thorne, K. S. 1980. *Rev. Mod. Phys.,* **52**, 299–340.

[15] Challinor, A., and Lasenby, A. 1999. *Astrophys. J.,* **513**, 1–22.

[16] Lewis, A., Challinor, A., and Lasenby, A. 2000. *Astrophys. J.,* **538**, 473–476.

[17] Synge, J. L. 1937. *Proc. Lond. Math. Soc.,* **43**, 376. Reprinted as *Gen. Rel. Grav.* **41**, 2177 (2009).

[18] Ehlers, J. 1961. *Akad. Wiss. Lit. Mainz, Abh. Math.-Nat. Kl.,* 11. English translation by G. F. R. Ellis and P. K. S. Dunsby, in *Gen. Rel. Grav.* **25**, 1225–1266 (1993).

[19] Sachs, R. K. 1962. *Proc. Roy. Soc. Lond. A,* **270**, 103–126.

[20] Synge, J. L. 1934. *Ann. Math.,* **35**, 705–713. Reprinted as *Gen. Rel. Grav.* **41**, 1206 (2009).

[21] Pirani, F. A. E. 1956. *Acta. Phys. Polon.,* **15**, 389. Reprinted as *Gen. Rel. Grav.* **41** 1216 (2009).

[22] Pirani, F. A. E. 1957. *Phys. Rev.,* **105**, 1089.

[23] Newman, E. T., and Penrose, R. 1962. *J. Math. Phys.,* **3**, 566.

[24] Raychaudhuri, A. K. 1955. *Phys. Rev.,* **98**, 1123–1126. Reprinted as *Gen. Rel. Grav.* **32**, 749 (2000).

[25] Ellis, G. F. R. 1971. Relativistic cosmology. Pages 104–182 of: Sachs, R. K. (ed), *General relativity and cosmology.* Proceedings of the International School of Physics "Enrico Fermi", vol. XLVII. New York and London: Academic Press. Reprinted as *Gen. Rel. Grav.* **41**, 581–660 (2009).

[26] Maartens, R., and Bassett, B. A. C. C. 1998. *Class. Quant. Grav.,* **15**, 705–717.

[27] Petrov, A. Z. 1954. *Scientific Proceedings of Kazan State University (named after V. I. Ulyanov-Lenin), Jubilee (1804–1954) Collection,* **114**, 55–69. Translation by J. Jezierski and M. A. H. MacCallum, with introduction by M. A. H. MacCallum, *Gen. Rel. Grav.* **32**, 1661–1685 (2000).

[28] Penrose, R. 1960. *Annals of Physics,* **10**, 171–201.

[29] Israel, W. 1966. *Nuovo Cim. B,* **44**, 1–14. Erratum: *Nuovo Cim. B* **48**, 463 (1967).

[30] Gödel, K. 1949. *Rev. Mod. Phys.,* **21**, 447–450. Reprinted as *Gen. Rel. Grav.* **32**, 1409 (2000).

[31] Penrose, R. 1964. Conformal treatment of infinity. Pages 565–584 of: DeWitt, B., and Dewitt, C. (eds), *Relativity, groups and topology.* New York: Gordon and Breach. Reprinted as *Gen. Rel. Grav.* **43**, 901–922 (2011).

[32] Tipler, F. J., Clarke, C. J. S., and Ellis, G. F. R. 1980. Singularities and horizons: a review article. Page 97 of: Held, A. (ed), *General relativity and gravitation: one hundred years after the birth of Albert Einstein,* vol. 2. New York: Plenum.

[33] Rindler, W. 1956. *Mon. Not. Roy. Astr. Soc.,* **116**, 662. Reprinted as *Gen. Rel. Grav.* **34**, 133 (2002).

[34] Stellmacher, K. 1938. *Math. Ann.,* **115**, 136–52. Reprinted as *Gen. Rel. Grav.* **42**, 1769–1789 (2010).

[35] Rendall, A. D. 2005. *Living Rev. Rel.,* **8**, 6.

[36] Arnowitt, R., Deser, S., and Misner, C. W. 1962. The dynamics of general relativity. Pages 227–265 of: Witten, L. (ed), *Gravitation: an introduction to current research.* New York and London: Wiley. Reprinted in *Gen. Rel. Grav.* **40**, 1997 (2008).

[37] Penrose, R. 1965. *Phys. Rev. Lett.,* **14**, 579.

[38] Hawking, S. W., and Penrose, R. 1970. *Proc. R. Soc. Lond. A,* **314**, 529–548.

[39] Curiel, E., and Bokulich, P. 2012. Singularities and black holes. In: Zalta, E. N. (ed), *The Stanford encyclopedia of philosophy (Fall 2012 edition)*. Stanford, CA: Stanford.

[40] Will, C. M. 2006. *Living Rev. Rel.*, **9**.

[41] Will, C. M. 1979. The confrontation between gravitational theory and experiment. In: Hawking, S., and Israel, W. (eds), *General relativity: an Einstein centenary survey*. Cambridge: Cambridge University Press.

[42] Fromholz, P., Poisson, E., and Clifford, C. M. 2014. *Amer. J. Phys.*, **82**, 295.

[43] Schwarzschild, K. 1916. *Sitzungsber. Preuss. Akad. Wiss.*, 189–196. Reprinted as *Gen. Rel. Grav.* **35**, 951 (2003).

[44] Ashby, N. 2003. *Living Rev. Rel.*, **6**, 1.

[45] Will, C. M. 2011. *Physics*, **4**, 43.

[46] Breton, R. P. *et al.* 2008. *Science*, **321**, 104–107.

[47] Reissner, H. 1916. *Annalen der Physik*, **50**, 106–120.

[48] Nordström, G. 1918. *Proc. Kon. Ned. Akad. Wet.*, **20**, 1238.

[49] Kottler, F. 1918. *Annalen der Physik*, **56**, 410–461.

[50] Kerr, R. P. 1963. *J. Math. Mech.*, **12**, 33.

[51] Kerr, R. P., and Schild, A. 1965. A new class of vacuum solutions of the Einstein field equations. Pages 1–12 of: *Atti del convegno sulla relatività generale: problemi dell'energia e onde gravitazionali*. Florence: Edizioni G. Barbèra. Reprinted as *Gen. Rel. Grav.* **41**, 2485–2499 (2009).

[52] Abdelqader, M., and Lake, K. 2013. *Phys. Rev. D*, **88**, 064042.

[53] Carter, B. 1968. *Commun. Math. Phys.*, **10**, 280–310.

[54] Carter, B. 1973. Black hole equilibrium states, part I analytic and geometric properties of the Kerr solutions. Pages 61–124 of: deWitt, C., and deWitt, B. (eds), *Black holes – les astres occlus*. New York: Gordon and Breach. Reprinted as *Gen. Rel. Grav.* **41**, 2874–2938 (2009).

[55] Eisenstaedt, J. 1982. *Arch. Hist. Exact Sci.*, **27**, 157–198.

[56] Eisenstaedt, J. 1987. *Arch. Hist. Exact Sci.*, **37**, 275–357.

[57] Eddington, A. S. 1924. *Nature*, **113**, 192.

[58] Kruskal, M. D. 1960. *Phys. Rev.*, **119**, 1743–1745.

[59] Boyer, R. H. 1969. *Proc. R. Soc. Lond. A*, **311**, 245–252.

[60] Carter, B. 1968. *Phys. Rev.*, **174**, 1559–1571.

[61] Boyer, R. H., and Lindquist, R. W. 1967. *J. Math. Phys.*, **8**, 265.

[62] Lake, K., and Roeder, R. C. 1977. *Phys. Rev. D*, **15**, 3513–3519.

[63] Tolman, R. 1934. *Proc. Nat. Acad. Sci.*, **20**, 169. Reprinted in *Gen. Rel. Grav.*, **29**, 935–943 (1997).

[64] Bondi, H. 1947. *Mon. Not. Roy. Astr. Soc.*, **107**, 410. Reprinted as *Gen. Rel. Grav.* **31**, 1777–1781 (1999).

[65] Krasiński, A. 2006. *Inhomogeneous cosmological models (2nd edition)*. Cambridge: Cambridge University Press.

[66] Begelman, M., and Rees, M. J. 2010. *Gravity's fatal attraction: black holes in the Universe*. Cambridge: Cambridge University Press.

[67] Oppenheimer, J. R., and Snyder, H. 1939. *Phys. Rev.*, **56**, 455–459.

[68] Israel, W. 1968. *Commun. Math. Phys.*, **8**, 245–260.

[69] Penrose, R. 1999. *J. Astrophys. Astr.*, **17**, 213–231.

[70] Robinson, D. C. 1975. *Phys. Rev. Lett.*, **34**, 905–906.

[71] Bardeen, J. M., Carter, B., and Hawking, S. W. 1973. *Commun. Math. Phys.*, **31**, 161–170.

[72] Abramowicz, M. A., and Fragile, P. C. 2013. *Living Rev. Rel.*, **16**, 1.

[73] Rees, M. J. 1984. *Ann. Rev. of Astron. Astrophys.*, **22**, 471–506.

[74] Ferrarese, L., and Merritt, D. 2002. *Phys. World*, **15(6)**, 41–46.

[75] Penrose, R. 1996. *J. Astrophys. Astr.*, **17**, 213–231.

[76] Celotti, A., Miller, J. C., and Sciama, D. W. 1999. *Class. Quant. Grav.*, **16**, A3–A21.

[77] Schödel, R. *et al.* 2002. *Nature*, **419**, 694–696.

[78] Melia, F. 2007. *The Galactic supermassive black hole*. Princeton, NJ: Princeton University Press.

[79] Einstein, A. 1917. *Sitzungsber. Preuss. Akad. Wiss.*, 142–152. English translation in *The principle of relativity* by H. A. Lorentz, A. Einstein, H. Minkowski and H. Weyl (Dover: New York) 1923.

[80] Friedmann, A. 1922. *Z. Phys.*, **10**, 377–386. Reprinted as *Gen. Rel. Grav.* **31**, 1991 (1999).

[81] Friedmann, A. 1924. *Z. Phys.*, **21**, 326–332. Reprinted as *Gen. Rel. Grav.* **31**, 12 (1999).

[82] Lemaître, G. 1931. *Nature*, **127**, 706.

[83] Lemaître, G. 1933. *C. R. Acad. Sci. Paris*, **196**, 1085.

[84] Ellis, G. F. R. 1990. Innovation, resistance and change: the transition to the expanding universe. Pages 97–114 of: Bertotti, B., Balbinot, R., Bergia, S., and Messina, A. (eds), *Modern cosmology in retrospect*. Cambridge: Cambridge University Press.

[85] Robertson, H. P. 1933. *Rev. Mod. Phys.*, **5**, 62–90. Reprinted as *Gen. Rel. Grav.* **44**, 2115–2144 (2012).

[86] Dodelson, S. 2003. *Modern cosmology: anisotropies and inhomogeneities in the Universe.* Amsterdam: Academic Press.

[87] Peter, P., and Uzan, J.-P. 2013. *Primordial cosmology*. Oxford Graduate Texts. Oxford: Oxford University Press.

[88] Einstein, A. 1918. *Sitzungsber. Preuss. Akad. Wiss.*, 154–167.

[89] Ehlers, J., and Rindler, W. 1989. *Mon. Not. Roy. Astr. Soc.*, **238**, 503–521.

[90] Wainwright, J., and Ellis, G. F. R. 1997. *Dynamical systems in cosmology*. Cambridge: Cambridge University Press.

[91] Kristian, J., and Sachs, R. K. 1966. *Astrophys. J.*, **143**, 379–399.

[92] Etherington, I. M. H. 1933. *Phil. Mag.*, **15**, 761. Reprinted as *Gen. Rel. Grav.* **39**, 1055 (2007).

[93] Sandage, A. 1961. *Astrophys. J.*, **133**, 355–392.

[94] Ellis, G. F. R., and Rothman, T. 1993. *Amer. J. Phys.*, **61**, 883–893.

[95] Wagoner, R. V., Fowler, W. A., and Hoyle, F. 1967. *Astrophys. J.*, **148**, 3–50.

[96] Sunyaev, R. A., and Zel'dovich, Ya. B. 1970. *Astrophys. Space Sci.*, **7**, 21–30.

[97] Longair, M. 2013. *The cosmic century*. Cambridge: Cambridge University Press.

[98] Doreshkevich, A., Zel'dovich, Ya. B., and Novikov, I. D. 1967. *Sov. Astron. AJ*, **11**, 233–239. Translated from *Astronomicheskii Zhurnal* **44**, 295–303.

[99] Peebles, P. J. E. 1966. *Astrophys. J.*, **146**, 542–552.

[100] Beringer, J. *et al.* [Particle Data Group]. 2012. *Phys. Rev. D*, **86**, 010001.

[101] Martin, J., Ringeval, C., and Vennin, V. 2013. *Encyclopaedia inflationaris*. arXiv:1303.3787.

[102] Ellis, G. F. R., and Madsen, M. 1991. *Class. Quant. Grav.*, **8**, 667–676.

[103] Lidsey, J. E. *et al.* 1997. *Rev. Mod. Phys.*, **69**, 373–410.

[104] Peebles, P. J. E., and Yu, J. T. 1970. *Astrophys. J.*, **162**, 815.

[105] Peebles, P. J. E. 1980. *The large-scale structure of the Universe*. Princeton, NJ: Princeton University Press.

[106] Lemaître, G. 1933. *Ann. Soc. Sci. Bruxelles A*, **53**, 51. Translation by M. A. H. MacCallum in *Gen. Rel. Grav.*, **29**, 641–680 (1997).

[107] Lemaître, G. 1949. *Rev. Mod. Phys.*, **21**, 357.

[108] Lemaître, G. 1958. *Pontifical Acad. Sci., Scripta Varia*, **16**, 475–488.

[109] Lifshitz, E. M. 1946. *Acad. Sci. USSR J. Phys.*, **10**, 116.

[110] Sachs, R. K., and Wolfe, A. M. 1967. *Astrophys. J.*, **147**, 73. Reprinted as *Gen. Rel. Grav.* **39**, 1944 (2007).

[111] Bardeen, J. M. 1980. *Phys. Rev. D*, **22**, 1882–1905.

[112] Mukhanov, V. F., Feldman, H. A., and Brandenberger, R. H. 1992. *Phys. Reports*, **215**, 203–333.

[113] Kodama, H., and Sasaki, M. 1984. *Prog. Theor. Phys. Suppl.*, **78**, 1–166.

[114] Hawking, S. W. 1966. *Astrophys. J.*, **145**, 544–554.

[115] Ellis, G. F. R., and Bruni, M. 1989. *Phys. Rev. D*, **40**, 1804–1818.

[116] Bruni, M., Dunsby, P. K. S., and Ellis, G. F. R. 1992. *Astrophys. J.*, **395**, 34–53.

[117] Mukhanov, V. F. 2005. *Physical foundations of cosmology*. Cambridge: Cambridge University Press.

[118] Lahav, O., Lilje, P. B., Primack, J. R., and Rees, M. J. 1991. *Mon. Not. Roy. Astr. Soc.*, **251**, 128–136.
[119] Sakharov, A. D. 1965. *Zh. Eksp. Teor. Fiz.*, **49**, 345–358. Translated as *Sov. Phys. JETP* **22**, 241–249 (1966).
[120] Sunyaev, R. A., and Zel'dovich, Ya. B. 1970. *Astrophys. Space Sci.*, **7**, 3.
[121] Sunyaev, R. A., and Zel'dovich, Ya. B. 1980. *Ann. Rev. Astron. Astrophys.*, **18**, 537–560.
[122] Silk, J., and Wilson, M. L. 1979. *Astrophys. J.*, **228**, 641–646.
[123] Bond, J. R., and Efstathiou, G. 1984. *Astrophys. J. Lett.*, **285**, L45–L48.
[124] Efstathiou, G., Sutherland, W. J., and Maddox, S. J. 1990. *Nature*, **348**, 705–707.
[125] Bonnor, W. B. 1956. *Z. Astrophys.*, **39**, 143. Reprinted as *Gen. Rel. Grav.* **30** 1113–1132 (1998).
[126] Mustapha, N., Hellaby, C. W., and Ellis, G. F. R. 1999. *Mon. Not. Roy. Astr. Soc.*, **292**, 817–830.
[127] Misner, C. W. 1969. *Phys. Rev. Lett.*, **22**, 1071–1074.
[128] Lifshitz, E. M., and Khalatnikov, I. M. 1964. *Sov. Phys. Usp.*, **6**, 495–522.
[129] Renn, J., Sauer, T., and Stachel, J. 1997. *Science*, **275**, 184–186.
[130] Einstein, A. 1936. *Science*, **84**, 506–7.
[131] Schneider, P., Ehlers, J., and Falco, E. E. 1992. *Gravitational lenses*. New York: Springer-Verlag.
[132] Wambsganss, J. 2001. *Living Rev. Rel.*, **1**, 12.
[133] Blandford, R. D., and Narayan, R. 1992. *Ann. Rev. Astron. Astrophys.*, **30**, 311–358.
[134] Perlick, V. 2004. *Living Rev. Rel.*, **7**, 9.
[135] Einstein, A., and Rosen, N. J. 1937. *J. Franklin Inst.*, **223**, 43.
[136] Bondi, H., Pirani, F. A. E., and Robinson, I. 1959. *Proc. Roy. Soc. Lond. A*, **251**, 519–533.
[137] Penrose, R. 1965. *Rev. Mod. Phys.*, **37**, 215.
[138] Szekeres, P. 1972. *J. Math. Phys.*, **13**, 286–294.
[139] Bondi, H., van der Burg, M. G. J., and Metzner, A. W. K. 1962. *Proc. Roy. Soc. Lond. A*, **269**, 21.
[140] Schoen, R., and Yau, S.-T. 1979. *Commun. Math. Phys.*, **65**, 45–76.
[141] Witten, E. 1981. *Commun. Math. Phys.*, **80**, 381–402.
[142] Teukolsky, S. A. 1973. *Astrophys. J.*, **185**, 635–648.
[143] Sasaki, M., and Tagoshi, H. 2003. *Living Rev. Rel.*, **6**, 6.
[144] Kokkotas, K. D., and Schmidt, B. 1999. *Living Rev. Rel.*, **2**, 2.
[145] Fryer, C. L., and New, K. C. B. 2011. *Living Rev. Rel.*, **14**, 1.
[146] Blanchet, L. 2006. *Living Rev. Rel.*, **9**, 4.
[147] Owen, R. *et al.* 2011. *Phys. Rev. Lett.*, **106**, 151101.
[148] Nichols, D. A. *et al.* 2011. *Phys. Rev. D*, **84**, 124014.
[149] Zhang, F. *et al.* 2012. *Phys. Rev. D*, **86**, 084049.
[150] Nichols, D. A. *et al.* 2012. *Phys. Rev. D*, **86**, 104028.
[151] Taylor, J. H., Fowler, L. A., and McCulloch, P. M. 1979. *Nature*, **277**, 437–440.
[152] Taylor, J. H., and Weisberg, J. M. 1989. *Astrophys. J.*, **345**, 434–450.
[153] Turner, M. S., White, M. J., and Lidsey, J. E. 1993. *Phys. Rev. D*, **48**, 4613–4622.
[154] Turner, M. S. 1996. *Phys. Rev. D*, **55**, 435–439.
[155] Martin, J., Ringeval, C., Trotta, R., and Vennin, V., 2014. *Phys. Rev. D*, **90**, 063501.
[156] Krauss, L. M., Dodelson, S., and Meyer, S. 2010. *Science*, **328**, 989–992.
[157] Sathyaprakash, B. S., and Schutz, B. F. 2009. *Living Rev. Rel.*, **12**, 2.
[158] Kuroyanagi, S., Gordon, C., Silk, J., and Sugiyama, N. 2010. *Phys. Rev. D*, **81**, 083524.
[159] Lovelock, D. 1971. *J. Math. Phys.*, **12**, 498.
[160] Weinberg, S. 1989. *Rev. Mod. Phys.*, **61**, 1–23.

2

Was Einstein Right? A Centenary Assessment

CLIFFORD M. WILL

2.1 Introduction

When general relativity was born 100 years ago, experimental confirmation was almost a side issue. Admittedly, Einstein did calculate observable effects of general relativity, such as the perihelion advance of Mercury, which he knew to be an unsolved problem, and the deflection of light, which was subsequently verified. But compared to the inner consistency and elegance of the theory, he regarded such empirical questions as almost secondary. He famously stated that if the measurements of light deflection disagreed with the theory he would "feel sorry for the dear Lord, for the theory *is* correct!".

By contrast, today at the centenary of Einstein's towering theoretical achievement, experimental gravitation is a major component of the field, characterized by continuing efforts to test the theory's predictions, both in the solar system and in the astronomical world, to detect gravitational waves from astronomical sources, and to search for possible gravitational imprints of phenomena originating in the quantum, high-energy or cosmological realms.

The modern history of experimental relativity can be divided roughly into four periods: Genesis, Hibernation, a Golden Era, and the Quest for Strong Gravity. The Genesis (1887–1919) comprises the period of the two great experiments which were the foundation of relativistic physics – the Michelson–Morley experiment and the Eötvös experiment – and the two immediate confirmations of general relativity – the deflection of light and the perihelion advance of Mercury. Following this was a period of Hibernation (1920–1960) during which theoretical work temporarily outstripped technology and experimental possibilities, and, as a consequence, the field stagnated and was relegated to the backwaters of physics and astronomy.

But beginning around 1960, astronomical discoveries (quasars, pulsars, cosmic background radiation) and new experiments pushed general relativity to the forefront. Experimental gravitation experienced a Golden Era (1960–1980) during which a systematic, world-wide effort took place to understand the observable predictions of general relativity, to compare and contrast them with the predictions of alternative theories of gravity, and to perform new experiments to test them. New technologies – atomic clocks, radar and laser ranging, space probes, cryogenic capabilities, to mention only a few – played a central role in this golden era. The period began with an experiment to confirm the gravitational frequency shift of light (1960) and ended with the reported decrease in the orbital period of the Hulse–Taylor binary pulsar at a rate consistent with the general relativistic prediction of gravitational-wave energy loss (1979). The results all supported general relativity, and most alternative theories of gravity fell by the wayside (for a popular review, see [1]).

Since that time, the field has entered what might be termed a Quest for Strong Gravity. Much like modern art, the term "strong" means different things to different people. To one steeped in general relativity, the principal figure of merit that distinguishes strong from weak gravity is the quantity $\epsilon \sim GM/(Rc^2)$, where G is the Newtonian gravitational constant, M is the characteristic mass scale of the phenomenon, R is the characteristic distance scale, and c is the speed of light. Near the event horizon of a non-rotating black hole, or for the expanding observable universe, $\epsilon \sim 1$; for neutron stars, $\epsilon \sim 0.2$. These are the regimes of strong gravity. For the solar system, $\epsilon < 10^{-5}$; this is the regime of weak gravity.

An alternative view of "strong" gravity comes from the world of particle physics. Here the figure of merit is $GM/(R^3c^2) \sim \ell^{-2}$, where the Riemann curvature of spacetime associated with the phenomenon, represented by the left-hand side, is comparable to the inverse square of a favorite length scale ℓ. If ℓ is the Planck length, this would correspond to the regime where one expects conventional quantum gravity effects to come into play. Another possible scale for ℓ is the TeV scale associated with many models for unification of the forces, or models with extra spacetime dimensions. From this viewpoint, strong gravity is where the curvature is comparable to the inverse length squared. Weak gravity is where the curvature is much smaller than this. The universe at the Planck time is strong gravity. Just outside the event horizon of an astrophysical black hole is weak gravity.

Considerations of the possibilities for new physics from either point of view have led to a wide range of questions that will motivate new tests of general relativity as we move into its second century:

- Are the black holes that are in evidence throughout the universe truly the black holes of general relativity?
- Do gravitational waves propagate with the speed of light and do they contain more than the two basic polarization states of general relativity?
- Does general relativity hold on cosmological distance scales?
- Is Lorentz invariance strictly valid, or could it be violated at some detectable level?
- Does the principle of equivalence break down at some level?
- Are there testable effects arising from the quantization of gravity?

In this centenary assessment of the experimental basis of general relativity, we will summarize the current status of experiments, and attempt to chart the future of the subject. We will not provide complete references to early work done in this field but instead will refer the reader to selected recent papers and to the appropriate review articles and monographs. For derivations of many of the effects discussed in this article we will refer to *Theory and Experiment in Gravitational Physics* [2], hereafter referred to as TEGP; references to TEGP will be by chapter or section, e.g., "TEGP 8.9". For a more comprehensive review, we refer readers to the author's "Living Review" [3] (or its recent update [4]).[1] The "Resource Letter" by the author [5], contains 100 key references for experimental gravity.

2.2 The foundations of gravitation theory

2.2.1 The Einstein equivalence principle

The principle of equivalence has historically played an important role in the development of gravitation theory. Newton regarded this principle as such a cornerstone of mechanics that he devoted the opening paragraph of the *Principia* to it. In 1907, Einstein used the principle as a basic element in his development of general relativity. We now regard the principle of equivalence as the foundation, not of Newtonian gravity or of general relativity, but of the broader idea that spacetime is curved. Much of this viewpoint can be traced back to Robert Dicke, who contributed crucial ideas about the foundations of gravitation theory between 1960 and 1965. These ideas were summarized in his influential Les Houches lectures of 1964 [6], and resulted in what has come to be called the Einstein equivalence principle (EEP).

One elementary equivalence principle is the kind Newton had in mind when he stated that the property of a body called "mass" is proportional to the "weight", and is known as the weak equivalence principle (WEP). An alternative statement of WEP is that the trajectory of a freely falling "test" body (one not acted upon by such forces as electromagnetism, too small to be affected by tidal gravitational forces, and spinless) is independent of its internal structure and composition. In the simplest case of dropping two different bodies in a gravitational field, WEP states that the bodies fall with the same acceleration. This is often termed the Universality of Free Fall, or UFF.

The Einstein equivalence principle (EEP) is a more powerful and far-reaching concept; it has three components:

1. *Weak Equivalence Principle*. The trajectory of a freely falling "test" body is independent of its internal structure and composition.
2. *Local Lorentz Invariance*. The outcome of any local non-gravitational experiment is independent of the velocity of the freely falling reference frame in which it is performed.
3. *Local Position Invariance*. The outcome of any local non-gravitational experiment is independent of where and when in the universe it is performed.

The Einstein equivalence principle is the heart and soul of gravitational theory, for it is possible to argue convincingly that if EEP is valid, then gravitation must be a "curved spacetime" phenomenon, in other words, gravity must be governed by a "metric theory of gravity", whose postulates are:

---

[1] This chapter was prepared roughly in parallel with the author's Living Review update, and may be considered a streamlined version of the latter. This will account for essentially identical prose between the two texts in numerous places.

1. Spacetime is endowed with a symmetric metric.
2. The trajectories of freely falling test bodies are geodesics of that metric.
3. In local freely falling reference frames, the non-gravitational laws of physics are those written in the language of special relativity.

General relativity is a metric theory of gravity, but then so are many others, including the Brans–Dicke theory and its generalizations. It is not uncommon for modern variants of general relativity, especially those motivated by quantum gravity, unification, or extra dimensions, to introduce weak, *effective* violations of the metric postulates. Accordingly, it is important to test the various aspects of the Einstein equivalence principle thoroughly. We first survey the experimental tests, and describe some of the theoretical formalisms that have been developed to interpret them.

Tests of the weak equivalence principle

A direct test of WEP is the comparison of the acceleration of two laboratory-sized bodies of different composition in an external gravitational field. If the principle were violated, then the accelerations of different bodies would differ. A measurement or limit on the fractional difference in acceleration between two bodies then yields a quantity called the "Eötvös ratio" given by $\eta \equiv 2|a_1 - a_2|/|a_1 + a_2|$.

Many high-precision Eötvös-type experiments have been performed, from the pendulum experiments of Newton, Bessel, and Potter to the classic torsion-balance measurements of Eötvös, Dicke, Braginsky, and their collaborators. In modern torsion-balance experiments, two objects of different composition are connected by a rod or placed on a tray and suspended in a horizontal orientation by a fine wire. If the gravitational acceleration of the bodies differs, and this difference has a component perpendicular to the suspension wire, there will be a torque induced on the wire, related to the angle between the wire and the direction of the gravitational acceleration. Beginning in the late 1980s, numerous experiments were carried out primarily to search for a "fifth force", but their null results also constituted tests of WEP. The "Eöt-Wash" experiments carried out at the University of Washington used a sophisticated torsion-balance tray to compare the accelerations of various materials toward local topography on Earth, movable laboratory masses, the Sun and the galaxy, and have reached levels of 2×10^{-13} [7].

The recent development of atom interferometry has yielded tests of WEP, albeit to modest accuracy, comparable to that of the original Eötvös experiment. In these experiments, one measures the local acceleration of the two separated wavefunctions of an atom such as cesium by studying the interference pattern when the wavefunctions are combined, and compares that with the acceleration of a nearby macroscopic object of different composition [8, 9]. A claim [9] that these experiments test the gravitational redshift was subsequently shown to be incorrect [10]. Further improvements are anticipated [11].

The resulting upper limits on η are summarized in Fig. 2.1 (for an extensive bibliography of experiments up to 1991, see [12]).

A number of projects are in the development or planning stage to push the bounds on η even lower. The project MICROSCOPE is designed to test WEP to 10^{-15}. It is being

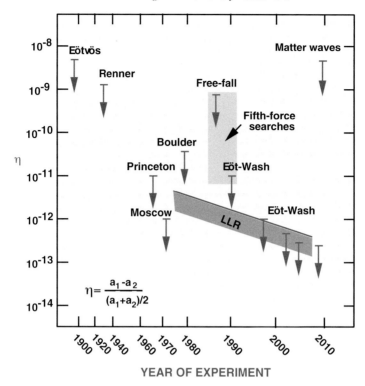

Figure 2.1 Selected tests of the weak equivalence principle, showing bounds on η. The free-fall and Eöt-Wash experiments were originally performed to search for a fifth force (green region, representing many experiments). The blue band shows evolving bounds on η for gravitating bodies from lunar laser ranging (LLR).

developed by the French space agency CNES for launch in late 2015, for a one-year mission. The drag-compensated satellite will be in a Sun-synchronous polar orbit at 700 km altitude, with a payload consisting of two differential accelerometers, one with elements made of the same material (platinum), and another with elements made of different materials (platinum and titanium). Other concepts for future improvements include advanced space experiments (Galileo-Galilei, STEP), lunar laser ranging (see Section 2.5.3), binary pulsar observations, and experiments with anti-hydrogen. For a review of past and future tests of WEP, see Vol. 29, Issue 18 of *Classical and Quantum Gravity* [13] and the review by Adelberger *et al.* [14].

Tests of local Lorentz invariance

Although special relativity itself never benefited from the kind of "crucial" experiments, such as the perihelion advance of Mercury and the deflection of light, that contributed so much to the initial acceptance of general relativity and to the fame of Einstein, the steady accumulation of experimental support, together with the successful merger of special relativity with quantum mechanics, led to its being accepted by mainstream physicists by the late 1920s, ultimately to become part of the standard toolkit of every working physicist. This accumulation included

- the classic Michelson–Morley experiment and its descendents;
- the Ives–Stillwell, Rossi–Hall, and other tests of time-dilation;
- tests of the independence of the speed of light of the velocity of the source, using both binary X-ray stellar sources and high-energy pions;
- tests of the isotropy of the speed of light.

In addition to these direct experiments, there was the Dirac equation of quantum mechanics and its prediction of anti-particles and spin; later would come the stunningly successful relativistic theory of quantum electrodynamics. For a pedagogical review on the occasion of the 2005 centenary of special relativity, see [15].

In 2015, on the 110th anniversary of special relativity, one might ask "what is there left to test?" Special relativity has been so thoroughly integrated into the fabric of modern physics that its validity is rarely challenged, except by cranks and crackpots. It is ironic then, that during the past 15 years, a vigorous theoretical and experimental effort has been launched to find violations of special relativity. The motivation for this effort is not a desire to repudiate Einstein, but to look for evidence of new physics "beyond" Einstein, such as apparent, or "effective", violations of Lorentz invariance that might result from certain models of quantum gravity. Quantum gravity asserts that there is a fundamental length scale given by the Planck length, $\ell_{Pl} = (\hbar G/c^3)^{1/2} = 1.6 \times 10^{-33}$ cm, but since length is not an invariant quantity (Lorentz–FitzGerald contraction), there could be a violation of Lorentz invariance at some level in quantum gravity. In brane world scenarios, while physics may be locally Lorentz invariant in the higher-dimensional world, the confinement of the interactions of normal physics to our four-dimensional "brane" could induce apparent Lorentz-invariance-violating effects. And in models such as string theory, the presence of additional vector and tensor long-range fields that couple to matter of the standard model could induce effective violations of Lorentz symmetry. These and other ideas have motivated a serious reconsideration of how to test Lorentz invariance with better precision and in new ways.

A simple and useful way of interpreting some of these modern experiments, called the c^2-formalism, is to suppose that the electromagnetic interactions suffer a slight violation of Lorentz invariance, through a change in the speed of electromagnetic radiation c relative to the limiting speed of material test particles (c_0, made to take the value unity via a choice of units), in other words, $c \neq 1$. Such a violation necessarily selects a preferred universal rest frame, presumably that of the cosmic background radiation, through which we are moving at about 370 km s^{-1}. Such a Lorentz-non-invariant electromagnetic interaction would cause shifts in the energy levels of atoms and nuclei that depend on the orientation of the quantization axis of the state relative to our universal velocity vector, and on the quantum numbers of the state. The presence or absence of such energy shifts can be examined by measuring the energy of one such state relative to another state that is either unaffected or is affected differently by the supposed violation. The magnitude of these

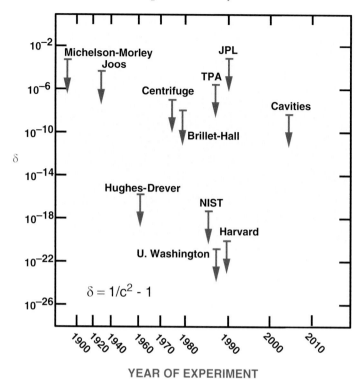

Figure 2.2 Selected tests of local Lorentz invariance showing the bounds on the parameter δ, which measures the degree of violation of Lorentz invariance in electromagnetism. The Michelson–Morley, Joos, Brillet–Hall and cavity experiments test the isotropy of the round-trip speed of light. The centrifuge, two-photon absorption (TPA) and JPL experiments test the isotropy of the light speed using one-way propagation. The most precise experiments test the isotropy of atomic energy levels.

"clock anisotropies" turns out to be proportional to $\delta \equiv |c^{-2} - 1|$ (see TEGP 2.6(a) for details).

The earliest clock anisotropy experiments were the Hughes–Drever experiments, performed in the period 1959–60, yielding limits on the parameter δ shown in Fig. 2.2. Dramatic improvements were made in the 1980s using laser-cooled trapped atoms and ions, which made it possible to reduce the broadening of resonance lines caused by collisions. Also included for comparison in Fig. 2.2 is the corresponding limit obtained from Michelson–Morley-type experiments and their modern variants such as the Brillet–Hall experiment, and comparisons between cavity oscillators and atomic clocks.

The c^2 framework focuses exclusively on classical electrodynamics. It has been extended to the entire standard model of particle physics by Kostelecký and collaborators. Called the Standard Model Extension (SME), it takes the standard $SU(3) \times SU(2) \times U(1)$ field theory of particle physics, and modifies the terms in the action by inserting a variety of tensorial quantities in the quark, lepton, Higgs, and gauge boson sectors that could explicitly violate local Lorentz invariance (LLI). The modified terms split naturally into those that are odd under CPT (i.e. that violate CPT) and terms that are even under CPT. The result is a rich and complex framework, with many parameters to be analyzed and tested by experiment. Experimentalists have risen to the challenge, and have placed interesting bounds on many of the SME parameters. A variety of clock anisotropy experiments have been carried out to bound the electromagnetic parameters of the SME framework. Other testable effects of Lorentz-invariance violation include arrival-time variations in TeV gamma-ray bursts from blazars, threshold effects in particle reactions, birefringence in photon propagation, gravitational Čerenkov radiation, and neutrino oscillations. The details of the SME and other approaches to testing LLI are beyond the scope of this article; the reader is referred to the reviews by Mattingly [16], Liberati [17] and Kostelecký and Russell [18]. The last article gives "data tables" showing experimental bounds on all the various parameters of the model. The SME has also been extended to a parametrization of local Lorentz violations in gravity (see, for example, [19]).

Tests of local position invariance

The principle of local position invariance (LPI), the third part of EEP, can be tested by the gravitational redshift experiment, the first experimental test of gravitation proposed by Einstein, eight years before the full theory of general relativity. Despite the fact that Einstein regarded this as a crucial test of general relativity, we now realize that it does not distinguish between general relativity and any other metric theory of gravity, but is only a test of EEP. A typical gravitational redshift experiment measures the frequency or wavelength shift $Z \equiv \Delta \nu / \nu = -\Delta \lambda / \lambda = \Delta U / c^2$ between two identical frequency standards (clocks) placed at rest at different heights in a static gravitational potential U. If LPI is not valid, then it turns out that the shift can be written

$$Z = (1 + \alpha) \frac{\Delta U}{c^2},   \tag{2.1}$$

where the parameter α may depend upon the nature of the clock whose shift is being measured (see TEGP 2.4(c) for details).

The first successful, high-precision redshift measurement was the series of Pound–Rebka–Snider experiments of 1960–1965 that measured the frequency shift of gamma-ray photons from ^{57}Fe as they ascended or descended the Jefferson Physical Laboratory tower at Harvard University. The high accuracy achieved – one percent – was obtained by making use of the Mössbauer effect to produce a narrow resonance line whose shift could be accurately determined. Other experiments since 1960 measured the shift of spectral lines in the Sun's gravitational field and the change in rate of atomic clocks transported aloft on aircraft, rockets and satellites. Figure 2.3 summarizes the important test of LPI that have been performed since 1960.

After almost 50 years of inconclusive or contradictory measurements, the gravitational redshift of solar spectral lines was finally measured reliably. During the early years of general relativity, the failure to measure this effect in solar lines was seized upon by some

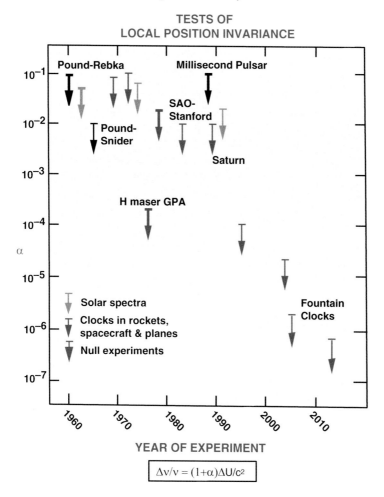

Figure 2.3 Selected tests of local position invariance via gravitational redshift experiments, showing bounds on α, which measures the degree of deviation of redshift from the formula $\Delta \nu / \nu = \Delta U / c^2$. In null redshift experiments, the bound is on the difference in α between different kinds of clocks.

as reason to doubt the theory. Unfortunately, the measurement is not simple. Solar spectral lines are subject to the "limb effect", a variation of line wavelengths between the center of the solar disk and its edge or "limb"; this effect is actually a Doppler shift caused by complex convective and turbulent motions in the photosphere and lower chromosphere, and is expected to be minimized by observing at the solar limb, where the motions are predominantly transverse. The secret is to use strong, symmetrical lines, leading to unambiguous wavelength measurements. Successful measurements were finally made in 1962 and 1972. In 1991, LoPresto *et al.* measured the solar shift in agreement with LPI to about 2 percent by observing the oxygen triplet lines both in absorption in the limb and in emission just off the limb [20].

The most precise standard redshift test to date was the Vessot–Levine rocket experiment known as Gravity Probe-A (GPA) that took place in June 1976. A hydrogen-maser clock was flown on a rocket to an altitude of about 10,000 km and its frequency compared to a similar clock on the ground. Analysis of the data yielded a limit $|\alpha| < 2 \times 10^{-4}$.

A "null" redshift experiment performed in 1978 tested whether the *relative* rates of two different clocks – hydrogen masers and cavity stabilized oscillators – depended upon the diurnal and annual variations of the solar gravitational potential at the location of the laboratory. This bound was subsequently improved using more stable frequency standards, such as atomic fountain clocks. The best current bounds, from comparing various pairings of rubidium, cesium, hydrogen and isotopes of dysprosium [21–23] hover around the one part per million mark.

The Atomic Clock Ensemble in Space (ACES) project will place both a cold trapped atom clock based on Cesium called PHARAO (Projet d'Horloge Atomique par Refroidissement d'Atomes en Orbit), and an advanced hydrogen maser clock on the International Space Station to measure the gravitational redshift to parts in 10^6, as well as to carry out a number of fundamental physics experiments and to enable improvements in global timekeeping. Launch is currently scheduled for May 2016.

Modern navigation using Earth-orbiting atomic clocks and accurate time-transfer must routinely take gravitational redshift and time-dilation effects into account. For example, the Global Positioning System (GPS) provides absolute positional accuracies of around 15 m (even better in its military mode), and 50 nanoseconds in time-transfer accuracy, anywhere on Earth. Yet the difference in rate between satellite and ground clocks as a result of relativistic effects is a whopping 39 *microseconds* per day (46 μs from the gravitational redshift, and −7 μs from time dilation). If these effects were not accurately accounted for, GPS would fail to function at its stated accuracy. This represents a welcome practical application of general relativity! For the role of general relativity in GPS, see [24]; for a popular essay, see [25].

Local position invariance also refers to position in time. If LPI is satisfied, the fundamental constants of non-gravitational physics should be constants in time. Table 2.1 shows current bounds on cosmological variations in selected dimensionless constants. For discussion and references to early work, see TEGP 2.4(c) or [26]. For a comprehensive recent review both of experiments and of theoretical ideas that underlie proposals for varying constants, see [27].

Experimental bounds on varying constants come in two types: bounds on the present rate of variation, and bounds on the difference between today's value and a value in the distant past. The main example of the former type is the clock comparison test, in which highly stable atomic clocks of different fundamental type are intercompared over periods ranging from months to years (variants of the null redshift experiment). The second type of bound involves measuring the relics of or signal from a process that occurred in the distant past and comparing the inferred value of the constant with the value measured in the laboratory today. One sub-type uses astronomical measurements of spectral lines at

Table 2.1 *Bounds on cosmological variation of fundamental constants of non-gravitational physics. For an in-depth review, see [27].*

Constant k	Limit on $\dot{k}/k$ (yr^{-1})	Redshift	Method
Fine structure	1.3×10^{-16}	0	Clock comparisons
constant	0.5×10^{-16}	0.15	Oklo natural reactor
($\alpha_{EM} = e^2/(\hbar c)$)	3.4×10^{-16}	0.45	^{187}Re decay in meteorites
	1.2×10^{-16}	0.4–2.3	Spectra in distant quasars
Weak interaction	1×10^{-11}	0.15	Oklo natural reactor
constant	5×10^{-12}	10^9	Big Bang nucleosynthesis
($\alpha_W = G_f m_p^2 c/\hbar^3$)			
Electron–proton mass ratio	3.3×10^{-15}	0	Clock comparisons
	3×10^{-15}	2.6–3.0	Spectra in distant quasars

large redshift, while the other uses fossils of nuclear processes on Earth to infer values of constants early in geological history, notably from the Oklo natural nuclear reactor in Gabon. Despite continuing reports by one group [28] of a variation of the fine structure constant over cosmic time as seen in spectral lines, most groups find no significant variation (e.g. [29, 30]).

EEP, particle physics, and the search for new interactions

There is mounting theoretical evidence to suggest that EEP is likely to be violated at some level, whether by quantum gravity effects, by effects arising from string theory, or by hitherto undetected interactions. As a result, EEP and related tests are now viewed as ways to discover or place constraints on new physical interactions, or as a branch of "non-accelerator particle physics", searching for the possible imprints of high-energy particle effects in the low-energy realm of gravity. One example of this is the "fifth-force" episode of the middle 1980s, in which Eötvös-type experiments and tests of the gravitational inverse-square law were used to set limits on the existence of intermediate-range forces that could arise from particle interactions beyond the standard model. Tests of the inverse square law at sub-mm distance scales continue in an effort to search for evidence of extra dimensions or light scalar particles [14]. Anomalies in the orbit of the Pioneer 10 and 11 spacecraft at 20 to 70 astronomical units from the Sun were touted for a time as evidence for new physics, but were ultimately shown to be the result of anisotropic emission of heat from the spacecraft [31, 32].

2.2.2 Metric theories of gravity and the strong equivalence principle

The Einstein equivalence principle is strongly supported by empirical evidence, as we have seen. This tells us that the only theories of gravity that have a hope of being viable are metric theories, or possibly theories that at worst admit very weak or short-range non-metric couplings. Therefore for the remainder of this chapter, we will turn our attention exclusively to metric theories of gravity.

The property that all non-gravitational fields should couple in the same manner to a single gravitational field is sometimes called "universal coupling". Because of it, one can discuss the metric as a property of spacetime itself rather than as a field over spacetime. This is because its properties may be measured and studied using a variety of different experimental devices, composed of different non-gravitational fields and particles, and, because of universal coupling, the results will be independent of the device. Thus, for instance, the proper time between two events is a characteristic of spacetime and of the location of the events, not of the clocks used to measure it.

Mathematically, if EEP is valid, the non-gravitational laws of physics may be formulated by taking their special relativistic forms in terms of the Minkowski metric η and simply "going over" to new forms in terms of the curved spacetime metric g, using the mathematics of differential geometry. The details of this "going over" can be found in standard textbooks (see [33], TEGP 3.2).

In any metric theory of gravity, matter and non-gravitational fields respond only to the spacetime metric g. In principle, however, there could exist other gravitational fields besides the metric, such as scalar fields, vector fields, and so on. The role of such fields is to mediate the manner in which matter and non-gravitational fields generate gravitational fields and produce the metric; once determined, however, the metric alone acts back on the matter in the manner prescribed by EEP.

What distinguishes one metric theory from another, therefore, is the number and kind of gravitational fields it contains in addition to the metric, and the equations that determine the structure and evolution of these fields. The fields could be dynamical, in that they obey their own set of dynamical equations, or they could be "fixed" or non-dynamical – an example of the latter is a background Minkowski metric.

General relativity is a purely dynamical theory since it contains only one gravitational field, the metric itself, and its structure and evolution are governed by partial differential equations (Einstein's equations). Brans–Dicke theory and its generalizations are purely dynamical theories; the field equation for the metric involves the scalar field (as well as the matter as source), and that for the scalar field involves the metric.

Consider now a local gravitating system, as described by a chosen metric theory. The system could be a star, a black hole, the solar system, or a Cavendish experiment. Because the metric is coupled directly or indirectly to the other fields of the theory, the gravitational environment in which the local gravitating system resides can influence the metric generated by the local system via the boundary values of the auxiliary fields. Those boundary values typically depend on the distribution of matter external to the system. Consequently, the results of local gravitational experiments may depend on the location and velocity of the frame relative to the external environment. Of course, local *non*-gravitational experiments are unaffected since the gravitational fields they generate are assumed to be negligible, and since those experiments couple only to the metric, whose form can always be made locally Minkowskian at a given spacetime event.

A theory which contains only the metric g (such as general relativity) yields local gravitational physics which is independent of the location and velocity of the local system. This follows from the fact that the only field coupling the local system to the environment is g, and it is always possible to find a coordinate system in which g takes the Minkowski form at the boundary between the local system and the external environment (neglecting inhomogeneities in the external gravitational field). Thus, apart from standard tidal effects, the external environment cannot act back on the local gravitating system.

A theory which contains the metric g and dynamical scalar fields φ_A yields local gravitational physics which may depend on the location of the frame but which is independent of the velocity of the frame. This follows from the asymptotic Lorentz invariance of the Minkowski metric and of the scalar fields, but now the asymptotic values of the scalar fields may depend on the location of the frame. For example, in scalar–tensor theories, the effective gravitational coupling strength depends on the scalar field, which may vary in space and time, depending on the external environment.

A theory which contains the metric g and additional dynamical vector or tensor fields or prior-geometric fields yields local gravitational physics which may have both location- and velocity-dependent effects. These ideas can be summarized in the strong equivalence principle (SEP), which states that:

1. WEP is valid for self-gravitating bodies as well as for test bodies.
2. The outcome of any local test experiment is independent of the velocity of the (freely falling) apparatus.
3. The outcome of any local test experiment is independent of where and when in the universe it is performed.

The distinction between SEP and EEP is the inclusion of bodies with self-gravitational interactions (planets, stars) and of experiments involving gravitational forces (Cavendish experiments, gravimeter measurements). Note that SEP contains EEP as the special case in which local gravitational forces are ignored.

General relativity seems to be the only viable metric theory that embodies SEP completely. In Section 2.5.3, we will discuss experimental evidence for the validity of SEP.

2.3 The parametrized post-Newtonian formalism

Despite the possible existence of long-range gravitational fields in addition to the metric, matter and non-gravitational fields are completely oblivious to them in metric theories of gravity. The only gravitational field that enters the equations of motion is the metric g. Thus the metric and the equations of motion for matter become the primary entities for calculating observable effects, and all that distinguishes one metric theory from another is the particular way in which matter and possibly other gravitational fields generate the metric.

The comparison of metric theories of gravity with each other and with experiment becomes particularly simple when one takes the slow-motion, weak-field limit. This approximation, known as the post-Newtonian limit, is sufficiently accurate to encompass most solar-system tests that can be performed in the foreseeable future. In this limit, the spacetime metric g predicted by a broad class of metric theories of gravity has the same structure. It can be written as an expansion about the Minkowski metric ($\eta_{\mu\nu} = \mathrm{diag}(-1, 1, 1, 1)$) in terms of dimensionless gravitational potentials of varying degrees of smallness. These potentials are constructed from the matter variables, such as mass density ρ, energy density $\rho\Pi$, pressure p, four-velocity u^α, ordinary velocity $v^j \equiv u^j/u^0$, etc., in imitation of the Newtonian gravitational potential

$$U(\mathbf{x}, t) \equiv G \int \rho(\mathbf{x}', t)|\mathbf{x} - \mathbf{x}'|^{-1} d^3x'. \qquad (2.2)$$

The "order of smallness" is determined according to the rules $U/c^2 \sim (v/c)^2 \sim \Pi/c^2 \sim p/\rho c^2 \sim \epsilon$, $v^i/c \sim |d/d(ct)|/|d/dx| \sim \epsilon^{1/2}$, and so on.

A consistent post-Newtonian limit requires determination of g_{00} correct through $\mathcal{O}(\epsilon^2)$, g_{0i} through $\mathcal{O}(\epsilon^{3/2})$, and g_{ij} through $\mathcal{O}(\epsilon)$ (for details see TEGP 4.1 or [34]). The only way that one metric theory differs from another is in the numerical values of the coefficients that appear in front of the metric potentials. The parametrized post-Newtonian (PPN) formalism inserts parameters in place of these coefficients, parameters whose values depend on the theory under study. In the current version of the PPN formalism, ten parameters are used, chosen in such a manner that they measure or indicate general properties of metric theories of gravity (see Table 2.2). Under reasonable assumptions about the kinds of potentials that can be present at post-Newtonian order (basically only Poisson-like potentials), one finds that ten PPN parameters exhaust the possibilities.

The parameters γ and β are the usual Eddington–Robertson–Schiff parameters used to describe the "classical" tests of general relativity, and are in some sense the most important; they are the only non-zero parameters in general relativity and scalar–tensor gravity. The parameter ξ is non-zero in a class of theories of gravity that predict preferred-location effects such as a galaxy-induced anisotropy in the local gravitational constant G_L (also called "Whitehead" effects); $\alpha_1, \alpha_2, \alpha_3$ measure whether or not the theory predicts post-Newtonian preferred-frame effects; $\alpha_3, \zeta_1, \zeta_2, \zeta_3, \zeta_4$ measure whether or not the theory predicts violations of global conservation laws for total momentum. In Table 2.2 we show the values these parameters take in general relativity; in any theory of gravity that possesses conservation laws for total momentum, called "semi-conservative" (any theory that is based on an invariant action principle is semi-conservative); and in any theory that in addition possesses six global conservation laws for angular momentum, called "fully conservative" (such theories automatically predict no post-Newtonian preferred-frame effects). Semi-conservative theories have five free PPN parameters ($\gamma, \beta, \xi, \alpha_1, \alpha_2$) while fully conservative theories have three (γ, β, ξ).

The PPN formalism was pioneered by Kenneth Nordtvedt [35], who studied the post-Newtonian metric of a system of gravitating point masses, extending earlier work by

Table 2.2 *The PPN parameters and their significance (note that α_3 has been shown twice to indicate that it is a measure of two effects).*

Parameter and what it measures relative to GR		Value in GR	Value in semiconservative theories	Value in fully conservative theories
γ	How much space-curvature produced by unit rest mass?	1	γ	γ
β	How much "nonlinearity" in the superposition law for gravity?	1	β	β
ξ	Preferred-location effects?	0	ξ	ξ
α_1	Preferred-frame effects?	0	α_1	0
α_2		0	α_2	0
α_3		0	0	0
α_3	Violation of conservation	0	0	0
ζ_1	of total momentum?	0	0	0
ζ_2		0	0	0
ζ_3		0	0	0
ζ_4		0	0	0

Eddington, Robertson and Schiff (TEGP 4.2). Will [36] generalized the framework to perfect fluids. A general and unified version of the PPN formalism was developed by Will and Nordtvedt [37]. The canonical version, with conventions altered to be more in accord with standard textbooks such as [38], is discussed in detail in TEGP, Chapter 4.

2.4 Competing theories of gravity

One of the important applications of the PPN formalism is the comparison and classification of alternative metric theories of gravity. The population of viable theories has fluctuated over the years as new effects and tests have been discovered, largely through the use of the PPN framework, which eliminated many theories thought previously to be viable. The theory population has also fluctuated as new, potentially viable theories have been invented.

In this review, we will focus on general relativity, the general class of scalar–tensor modifications of it, of which the Jordan–Fierz–Brans–Dicke theory (Brans–Dicke, for short) is the classic example, and vector–tensor or scalar–vector–tensor theories. The reasons are several-fold:

- A full compendium of alternative theories circa 1981 is given in TEGP, Chapter 5.
- Many alternative metric theories developed during the 1970s and 1980s could be viewed as "straw-man" theories, invented to prove that such theories exist or to illustrate

particular properties. Few of these could be regarded as well-motivated theories from the
point of view, say, of field theory or particle physics.

- A number of theories fall into the class of "prior-geometric" theories, with absolute
  elements such as a flat background metric in addition to the physical metric. Most of
  these theories predict "preferred-frame" effects, that have been tightly constrained by
  observations (see Section 2.5.3). An example is Rosen's bimetric theory.
- A large number of alternative theories of gravity predict gravitational wave emission
  substantially different from that of general relativity, in strong disagreement with observations of the binary pulsar (see Section 2.8).
- Scalar–tensor modifications of general relativity have become very popular in unification
  schemes such as string theory, and in cosmological model building. Because the scalar
  fields could be massive, the potentials in the post-Newtonian limit could be modified by
  Yukawa-like terms.
- Theories that also incorporate vector fields have attracted recent attention, in the spirit
  of the SME (see Section 2.2.1), as models for violations of Lorentz invariance in the
  gravitational sector, and as potential candidates to account for phenomena such as galaxy
  rotation curves without resorting to dark matter.

2.4.1 General relativity

The metric g is the sole dynamical field, and the theory contains no arbitrary functions
or parameters, apart from the value of the Newtonian coupling constant G, which is measurable in laboratory experiments. Throughout this chapter, we ignore the cosmological
constant Λ_C. We do this despite recent evidence, from supernova data, of an accelerating
universe, which would indicate either a non-zero cosmological constant or a dynamical
"dark energy" contributing about 70 percent of the critical density. Although Λ_C has significance for quantum field theory, quantum gravity, and cosmology, on the scale of the
solar system or of stellar systems its effects are negligible, for the values of Λ_C inferred
from supernova observations. On the other hand, the conundrum of accelerated expansion
has motivated the development of alternative theories of gravity, notably of the so-called
$f(R)$ type.

The field equations of general relativity are derivable from an invariant action principle
$\delta I = 0$, where

$$I = \frac{c^3}{16\pi G} \int R(-g)^{1/2} \, d^4x + I_{\rm m}(\psi_{\rm m}, g_{\mu\nu}), \tag{2.3}$$

where R is the Ricci scalar, and $I_{\rm m}$ is the matter action, which depends on matter fields $\psi_{\rm m}$
universally coupled to the metric g. By varying the action with respect to $g_{\mu\nu}$, we obtain
the field equations

$$G_{\mu\nu} \equiv R_{\mu\nu} - \frac{1}{2}g_{\mu\nu}R = \frac{8\pi G}{c^4}T_{\mu\nu}, \tag{2.4}$$

Table 2.3 *Metric theories and their PPN parameter values ($\alpha_3 = \zeta_i = 0$ for all cases). The parameters γ', β', α'_1, and α'_2 denote complicated functions of the arbitrary constants and matching parameters.*

Theory	Arbitrary functions or constants	Cosmic matching parameters	PPN parameters				
			γ	β	ξ	α_1	α_2
General relativity	none	none	1	1	0	0	0
Scalar–tensor							
Brans–Dicke	ω_{BD}	ϕ_0	$\dfrac{1 + \omega_{BD}}{2 + \omega_{BD}}$	1	0	0	0
General, $f(R)$	$A(\varphi), V(\varphi)$	φ_0	$\dfrac{1 + \omega}{2 + \omega}$	$1 + \dfrac{\lambda}{4 + 2\omega}$	0	0	0
Vector–tensor							
Unconstrained	$\omega, c_1, c_2, c_3, c_4$	u	γ'	β'	0	α'_1	α'_2
Einstein-Æther	c_1, c_2, c_3, c_4	none	1	1	0	α'_1	α'_2
Khronometric	$\alpha_K, \beta_K, \lambda_K$	none	1	1	0	α'_1	α'_2
Tensor–vector–scalar	k, c_1, c_2, c_3, c_4	ϕ_0	1	1	0	α'_1	α'_2

where $T_{\mu\nu}$ is the matter energy–momentum tensor. General covariance of the matter action implies the equations of motion $T^{\mu\nu}{}_{;\nu} = 0$; varying I_m with respect to ψ_m yields the matter field equations of the Standard Model. By virtue of the *absence* of prior-geometric elements, the equations of motion are also a consequence of the field equations via the Bianchi identities $G^{\mu\nu}{}_{;\nu} = 0$.

The general procedure for deriving the post-Newtonian limit of metric theories is spelled out in TEGP 5.1, and is described in detail for general relativity in TEGP 5.2, or Chapter 8 of [34]. The PPN parameter values are listed in Table 2.3.

2.4.2 Scalar–tensor theories

These theories contain the metric g, a scalar field φ, a potential function $V(\varphi)$, and a coupling function $A(\varphi)$ (generalizations to more than one scalar field have also been carried out [39]). For some purposes, the action is conveniently written in a non-metric representation, sometimes denoted the "Einstein frame", in which the gravitational action looks exactly like that of general relativity:

$$\tilde{I} = \frac{c^3}{16\pi G} \int \left[\tilde{R} - 2\tilde{g}^{\mu\nu} \partial_\mu \varphi \, \partial_\nu \varphi - V(\varphi) \right] (-\tilde{g})^{1/2} \, d^4 x + I_m \left(\psi_m, A^2(\varphi) \tilde{g}_{\mu\nu} \right), \quad (2.5)$$

where $\tilde{R} \equiv \tilde{g}^{\mu\nu} \tilde{R}_{\mu\nu}$ is the Ricci scalar of the "Einstein" metric $\tilde{g}_{\mu\nu}$. This representation is a "non-metric" one because the matter fields ψ_m couple to a combination of φ and $\tilde{g}_{\mu\nu}$.

Despite appearances, however, it is a metric theory, because it can be put into a metric representation by identifying the "physical metric"

$$g_{\mu\nu} \equiv A^2(\varphi)\tilde{g}_{\mu\nu}. \tag{2.6}$$

The action can then be rewritten in the metric form

$$I = \frac{c^3}{16\pi G} \int \left[\phi R - \phi^{-1}\omega(\phi)g^{\mu\nu}\,\partial_\mu\phi\,\partial_\nu\phi - \phi^2 V \right] (-g)^{1/2}\,d^4x + I_{\mathrm{m}}(\psi_{\mathrm{m}}, g_{\mu\nu}), \tag{2.7}$$

where

$$\phi \equiv A(\varphi)^{-2},$$
$$3 + 2\omega(\phi) \equiv \alpha(\varphi)^{-2}, \tag{2.8}$$
$$\alpha(\varphi) \equiv \frac{d(\ln A(\varphi))}{d\varphi}.$$

The Einstein frame is useful for discussing general characteristics of such theories, and for some cosmological applications, while the metric representation is most useful for calculating observable effects. The field equations, post-Newtonian limit and PPN parameters are discussed in TEGP 5.3, and the values of the PPN parameters are listed in Table 2.3.

The parameters that enter the post-Newtonian limit are

$$\omega \equiv \omega(\phi_0), \qquad \lambda \equiv \left[\frac{\phi\,d\omega/d\phi}{(3 + 2\omega)(4 + 2\omega)} \right]_{\phi_0}, \tag{2.9}$$

where ϕ_0 is the value of ϕ today far from the system being studied, as determined by appropriate cosmological boundary conditions. In Brans–Dicke theory ($\omega(\phi) \equiv \omega_{\mathrm{BD}} = $ const.), the larger the value of ω_{BD}, the smaller the effects of the scalar field, and in the limit $\omega_{\mathrm{BD}} \to \infty$ ($\alpha_0 \to 0$), the theory becomes indistinguishable from general relativity in all its predictions. In more general theories, the function $\omega(\phi)$ could have the property that, at the present epoch, and in weak-field situations, the value of the scalar field ϕ_0 is such that ω is very large and λ is very small (theory almost identical to general relativity today), but that for past or future values of ϕ, or in strong-field regions such as the interiors of neutron stars, ω and λ could take on values that would lead to significant differences from general relativity.

Damour and Esposito-Farèse [39] have adopted an alternative parametrization of scalar–tensor theories, in which one expands $\ln A(\varphi)$ about a cosmological background field value φ_0:

$$\ln A(\varphi) = \alpha_0(\varphi - \varphi_0) + \frac{1}{2}\beta_0(\varphi - \varphi_0)^2 + \dots \tag{2.10}$$

A precisely linear coupling function produces Brans–Dicke theory, with $\alpha_0^2 = 1/(2\omega_{\mathrm{BD}} + 3)$, or $1/(2 + \omega_{\mathrm{BD}}) = 2\alpha_0^2/(1 + \alpha_0^2)$. The function $\ln A(\varphi)$ acts as a potential for the scalar field φ within matter, and, if $\beta_0 > 0$, then during cosmological evolution, the scalar field naturally evolves toward the minimum of the potential, i.e. toward $\alpha_0 \approx 0$, $\omega \to \infty$, or toward a theory close to, though not precisely, general relativity [40, 41]. Estimates of the expected relic deviations from general relativity today in such theories depend on the cosmological model, but range from 10^{-5} to a few times 10^{-7} for $|\gamma - 1|$.

Negative values of β_0 correspond to a "locally unstable" scalar potential (the overall theory is still stable in the sense of having no tachyons or ghosts). In this case, objects such as neutron stars can experience a "spontaneous scalarization", whereby the interior values of φ can take on values very different from the exterior values, through nonlinear interactions between strong gravity and the scalar field, dramatically affecting the stars' internal structure and leading to strong violations of SEP. On the other hand, in the case $\beta_0 < 0$, one must confront that fact that, with an unstable φ potential, cosmological evolution would presumably drive the system away from the peak where $\alpha_0 \approx 0$, toward parameter values that could be excluded by solar system experiments.

Scalar fields coupled to gravity or matter are also ubiquitous in particle-physics-inspired models of unification, such as string theory. In some models, the coupling to matter may lead to violations of EEP, which could be tested or bounded by the experiments described in Section 2.2.1. In many models the scalar field could be massive; if the Compton wavelength is of macroscopic scale, its effects are those of a "fifth force". Only if the theory can be cast as a metric theory with a scalar field of infinite range or of range long compared to the scale of the system in question (solar system) can the PPN framework be strictly applied. If the mass of the scalar field is sufficiently large that its range is microscopic, then, on solar-system scales, the scalar field is suppressed, and the theory is essentially equivalent to general relativity. For a detailed review of scalar–tensor theories see [42].

2.4.3 $f(R)$ *theories*

These are theories whose action has the form

$$I = \frac{c^3}{16\pi G} \int f(R)(-g)^{1/2} d^4x + I_{\rm m}(\psi_{\rm m}, g_{\mu\nu}), \qquad (2.11)$$

where f is a function chosen so that at cosmological scales, the universe will experience accelerated expansion without resorting to either a cosmological constant or dark energy. However, it turns out that such theories are equivalent to scalar–tensor theories: replace $f(R)$ by $f(\chi) - f_{,\chi}(\chi)(R - \chi)$, where χ is a dynamical field. Varying the action with respect to χ yields $f_{,\chi\chi}(R - \chi) = 0$, which implies that $\chi = R$ as long as $f_{,\chi\chi} \neq 0$. Then defining a scalar field $\phi \equiv f_{,\chi}(\chi)$ once puts the action into the form of a scalar–tensor theory given by Eq. (2.7), with $\omega(\phi) = 0$ and $\phi^2 V = \phi\chi(\phi) - f(\chi(\phi))$. As we will see, this value of ω would ordinarily strongly violate solar-system experiments, but it turns out that in many models, the potential $V(\phi)$ has the effect of giving the scalar field a large effective mass in the presence of matter (the so-called "chameleon mechanism"), so that the scalar field is suppressed at distances that extend outside bodies like the Sun and Earth. In this way, with only modest fine tuning, $f(R)$ theories can claim to obey standard tests, while providing interesting, non general-relativistic behavior on cosmic scales. For detailed reviews of this class of theories, see [43, 44].

2.4.4 Vector–tensor theories

These theories contain the metric g and a dynamical, typically timelike, four-vector field u^μ. In some models, the four-vector is unconstrained, while in others, called Einstein-Æther theories, it is constrained to be timelike with unit norm. The most general action for such theories that is quadratic in derivatives of the vector is given by

$$I = \frac{c^3}{16\pi G} \int \left[(1 + \omega u_\mu u^\mu) R - K^{\mu\nu}_{\alpha\beta} \nabla_\mu u^\alpha \nabla_\nu u^\beta + \lambda (u_\mu u^\mu + 1) \right]$$
$$\times (-g)^{1/2} d^4 x + I_m (\psi_m, g_{\mu\nu}), \tag{2.12}$$

where

$$K^{\mu\nu}_{\alpha\beta} = c_1 g^{\mu\nu} g_{\alpha\beta} + c_2 \delta^\mu_\alpha \delta^\nu_\beta + c_3 \delta^\mu_\beta \delta^\nu_\alpha - c_4 u^\mu u^\nu g_{\alpha\beta}. \tag{2.13}$$

The coefficients c_i are arbitrary. In the unconstrained theories, $\lambda \equiv 0$ and ω is arbitrary. In the constrained theories, λ is a Lagrange multiplier, and by virtue of the constraint $u_\mu u^\mu = -1$, the factor $\omega u_\mu u^\mu$ in front of the Ricci scalar can be absorbed into a rescaling of G; equivalently, in the constrained theories, we can set $\omega = 0$. Note that the possible term $u^\mu u^\nu R_{\mu\nu}$ can be shown under integration by parts to be equivalent to a linear combination of the terms involving c_2 and c_3.

Unconstrained theories were studied during the 1970s as "straw-man" alternatives to general relativity. In addition to having up to four arbitrary parameters, they also left the magnitude of the vector field arbitrary, since it satisfies a linear homogeneous vacuum field equation of the form $\mathcal{L}u^\mu = 0$ ($c_4 = 0$ in all such cases studied). Indeed, this latter fact was one of most serious defects of these theories.

General vector–tensor theory; $\omega, \tau, \epsilon, \eta$ (TEGP 5.4)

The gravitational Lagrangian for this class of theories had the form $R + \omega u_\mu u^\mu R + \eta u^\mu u^\nu R_{\mu\nu} - \epsilon F_{\mu\nu} F^{\mu\nu} + \tau \nabla_\mu u_\nu \nabla^\mu u^\nu$, where $F_{\mu\nu} = \nabla_\mu u_\nu - \nabla_\nu u_\mu$, corresponding to the values $c_1 = 2\epsilon - \tau$, $c_2 = -\eta$, $c_1 + c_2 + c_3 = -\tau$, $c_4 = 0$ and $\lambda = 0$. In these theories γ, β, α_1, and α_2 are complicated functions of the parameters and of $u^2 = -u^\mu u_\mu$, while the rest vanish.

Will–Nordtvedt theory [37]

This is the special case $c_1 = -1$, $c_2 = c_3 = c_4 = \lambda = 0$. In this theory, the PPN parameters are given by $\gamma = \beta = 1$, $\alpha_2 = u^2/(1 + u^2/2)$, and zero for the rest.

Hellings–Nordtvedt theory; ω [45]

This is the special case $c_1 = 2$, $c_2 = 2\omega$, $c_1 + c_2 + c_3 = 0 = c_4$, $\lambda = 0$. Here γ, β, α_1 and α_2 are complicated functions of u^2, while the rest vanish.

Einstein-Æther theories; c_1, c_2, c_3, c_4 [46–50]

These theories were motivated in part by a desire to explore possibilities for violations of Lorentz invariance in gravity, in parallel with similar studies in matter interactions, such as the SME. Here $\gamma = \beta = 1$, α_1 and α_2 are complicated functions of the c_k

parameters, and the rest vanish. By requiring that gravitational wave modes have real (as opposed to imaginary) frequencies, one can impose the bounds $c_1/(c_1 + c_4) \geq 0$ and $(c_1 + c_2 + c_3)/(c_1 + c_4) \geq 0$. Considerations of positivity of energy impose the constraints $c_1 > 0$, $c_1 + c_4 > 0$ and $c_1 + c_2 + c_3 > 0$.

Khronometric theory; α_K, β_K, λ_K [51–54]

This is the low-energy limit of "Hořava gravity", a proposal for a gravity theory that is power-counting renormalizable. The vector field is required to be hypersurface orthogonal, so that higher spatial derivative terms could be introduced to effectuate renormalizability. A "healthy" version of the theory can be shown to correspond to the values $c_1 = -\epsilon$, $c_2 = \lambda_K$, $c_3 = \beta_K + \epsilon$ and $c_4 = \alpha_K + \epsilon$, where the limit $\epsilon \to \infty$ is to be taken.

2.4.5 Tensor–vector–scalar (TeVeS) theories

This class of theories was invented to provide a fully relativistic theory of gravity that could mimic the phenomenological behavior of so-called Modified Newtonian Dynamics (MOND), whereby in a weak-field regime, Newton's laws hold, namely $a = Gm/r^2$ where m is the mass of a central object, as long as a is large compared to some fundamental scale a_0, but in a regime where $a < a_0$, the equations of motion would take the form $a^2/a_0 = Gm/r^2$. With such a behavior, the rotational velocity of a particle far from a central mass would have the form $v \sim \sqrt{ar} \sim (Gma_0)^{1/4}$, thus reproducing the flat rotation curves observed for spiral galaxies, without invoking a distribution of dark matter.

Devising such a theory turned out to be no simple matter, and the final result, TeVeS, was rather complicated [55]. Furthermore, it was shown to have unexpected singular behavior that was most simply cured by incorporating features of the Einstein-Æther theory [56]. The extended theory is based on an "Einstein" metric $\tilde{g}_{\mu\nu}$, related to the physical metric $g_{\mu\nu}$ by

$$g_{\mu\nu} \equiv e^{-2\phi}\tilde{g}_{\mu\nu} - 2u_\mu u_\nu \sinh(2\phi), \tag{2.14}$$

where u^μ is a vector field, and ϕ is a scalar field. The action for gravity is the standard general relativity action of Eq. (2.3), but defined using the Einstein metric $\tilde{g}_{\mu\nu}$, while the matter action is that of a standard metric theory, using $g_{\mu\nu}$. These are supplemented by the vector action, given by that of Einstein-Æther theory, Eq. (2.12), and a scalar action, given by

$$I_S = -\frac{c^3}{2k^2\ell^2 G}\int \mathcal{F}(k\ell^2 h^{\mu\nu}\phi_{,\mu}\phi_{,\nu})(-g)^{1/2}\,d^4x, \tag{2.15}$$

where k is a constant, ℓ is a distance, and $h^{\mu\nu} \equiv \tilde{g}^{\mu\nu} - u^\mu u^\nu$, indices being raised and lowered using the Einstein metric. The function $\mathcal{F}(y)$ is chosen so that $\mu(y) \equiv d\mathcal{F}/dy$ is unity in the high-acceleration, or normal Newtonian, regime, and nearly zero in the MOND regime.

The PPN parameters of the theory [57] have the values $\gamma = \beta = 1$ and $\xi = \alpha_3 = \zeta_i = 0$, while the parameters α_1 and α_2 are complicated functions of k, c_k and the asymptotic value ϕ_0 of the scalar field.

For reviews of TeVeS, MOND and their confrontation with the dark-matter paradigm, see [58, 59].

2.4.6 Other theories

Numerous alternative theories of gravity have been developed and studied extensively in recent years, most motivated from the direction of particle physics, quantum gravity or unification.

Massive gravity theories attempt to give the putative "graviton" a mass. The simplest attempts fall afoul of the so-called van Dam–Veltman–Zakharov discontinuity, which leads to a violation of experiment. Attempts to avoid this problem involve treating nonlinear aspects of the theory at the fundamental level; many models incorporate a second tensor field in addition to the metric. For a recent review, see [60]. Quadratic gravity is a recent incarnation of an old idea of adding to the action of general relativity terms quadratic in the Riemann and Ricci tensors or the Ricci scalar, as an effective model for quantum gravity corrections. Chern–Simons gravity adds a parity-violating term proportional to $^*R^{\alpha\beta\gamma\delta}R_{\alpha\beta\gamma\delta}$ to the action of general relativity, where $^*R^{\alpha\beta\gamma\delta}$ is the dual Riemann tensor.

2.5 Tests of post-Newtonian gravity

2.5.1 Tests of the parameter γ

With the PPN formalism in hand, we are now ready to confront gravitation theories described in Section 2.4 with the results of solar-system experiments. In this section we focus on tests of the parameter γ, consisting of the deflection of light and the time delay of light.

The deflection of light

A light ray (or photon) which passes the Sun at a distance d is deflected by an angle

$$\delta\theta = \frac{1}{2}(1 + \gamma)\frac{4GM_\odot}{dc^2}\frac{1 + \cos\Phi}{2} \tag{2.16}$$

(TEGP 7.1), where $M_\odot$ is the mass of the Sun and Φ is the angle between the Earth–Sun line and the incoming direction of the photon. For a grazing ray, $d \approx d_\odot$, $\Phi \approx 0$, and

$$\delta\theta \approx \frac{1}{2}(1 + \gamma)1.''7505, \tag{2.17}$$

independent of the frequency of light. In practice, one measures how the relative angular separation between an observed source of light and a nearby reference source evolves as the Sun moves across the sky, as seen from Earth.

It is interesting to note that the classic derivations of the deflection that use only the corpuscular theory of light, by Cavendish in 1784 and von Soldner in 1803 [61], or that use only the principle of equivalence, by Einstein in 1911, yield only the "1/2" part of the coefficient in front of the expression in Eq. (2.16). But the result of these calculations is the deflection of light relative to local straight lines, as established for example by rigid rods; however, because of space curvature around the Sun, determined by the PPN parameter γ, local straight lines are bent relative to asymptotic straight lines far from the Sun by just enough to yield the remaining factor "$\gamma/2$". The first factor "1/2" holds in any metric theory, the second "$\gamma/2$" varies from theory to theory. Thus, calculations that purport to derive the full deflection using the equivalence principle alone are incorrect.

The prediction of the full bending of light by the Sun was one of the great successes of Einstein's general relativity. Eddington's confirmation of the bending of optical starlight observed during a solar eclipse in the first days following World War I helped make Einstein famous. However, the experiments of Eddington and his co-workers had only 30 percent accuracy, and succeeding experiments were not much better: The results were scattered between one half and twice the Einstein value (see Fig. 2.4), and the accuracies were low.

However, the development of radio interferometery, and later of very-long-baseline radio interferometry (VLBI), produced greatly improved determinations of the deflection of light. These techniques now have the capability of measuring angular separations and changes in angles to accuracies better than 100 microarcseconds. Early measurements took advantage of the fact that certain groups of strong quasistellar radio sources annually pass very close to the Sun (as seen from the Earth), and by 1975 were reaching accuracies at the level of a percent.

In recent years, transcontinental and intercontinental VLBI observations of quasars and radio galaxies have been made primarily to monitor the Earth's rotation ("VLBI" in Fig. 2.4). These measurements are sensitive to the deflection of light over almost the entire celestial sphere (at 90° from the Sun, the deflection is still 4 milliarcseconds). A 2004 analysis of almost 2 million VLBI observations of 541 radio sources, made by 87 VLBI sites between 1979 and 1999, yielded $(1 + \gamma)/2 = 0.99992 \pm 0.00023$, or equivalently, $\gamma - 1 = (-1.7 \pm 4.5) \times 10^{-4}$ [62]. Analyses that incorporated data through 2010 yielded $\gamma - 1 = (-0.8 \pm 1.2) \times 10^{-4}$ [63].

Analysis of observations made by the Hipparcos optical astrometry satellite yielded a test at the level of 0.3 percent, while the GAIA mission, launched in 2013, is expected to reach the parts per million level.

Finally, a remarkable measurement of γ on galactic scales was reported in 2006 [64]. It used data on gravitational lensing by 15 elliptical galaxies, collected by the Sloan Digital Sky Survey. The Newtonian potential U of each lensing galaxy (including the contribution from dark matter) was derived from the observed velocity dispersion of stars in the galaxy. Comparing the observed lensing with the lensing predicted by the models provided a 10 percent bound on γ, in agreement with general relativity. Unlike the much tighter bounds described previously, which were obtained on the scale of the solar system, this bound was obtained on a galactic scale.

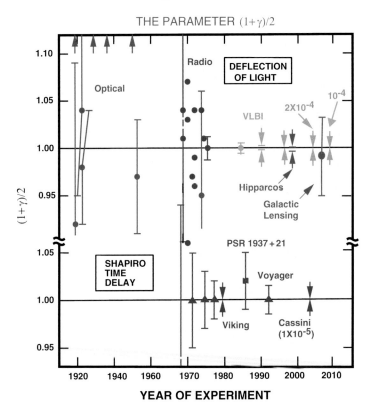

Figure 2.4 Measurements of the coefficient $(1 + \gamma)/2$ from light-deflection and time-delay measurements. Its general relativity value is unity. The arrows at the top denote anomalously large values from early eclipse expeditions. The Shapiro time-delay measurements using the Cassini spacecraft yielded an agreement with general relativity to 10^{-3} percent, and VLBI light-deflection measurements have reached 0.02 percent. Hipparcos denotes the optical astrometry satellite, which reached 0.1 percent.

The results of light-deflection measurements are summarized in Fig. 2.4.

The time delay of light

A radar signal sent across the solar system past the Sun to a planet or satellite and returned to the Earth suffers an additional non-Newtonian delay in its round-trip travel time, given by

$$\delta t = 2(1 + \gamma)\frac{GM_{\odot}}{c^3} \ln \left(\frac{(r_{\oplus} + \mathbf{x}_{\oplus} \cdot \mathbf{n})(r_{\mathrm{e}} - \mathbf{x}_{\mathrm{e}} \cdot \mathbf{n})}{d^2} \right), \tag{2.18}$$

where $\mathbf{x}_{\mathrm{e}}$ ($\mathbf{x}_{\oplus}$) are the vectors, and r_{e} ($r_{\oplus}$) are the distances from the Sun to the source (Earth), respectively (TEGP 7.2). For a ray which passes close to the Sun,

$$\delta t \approx \frac{1}{2}(1 + \gamma) \left(240 - 20 \ln \left(\frac{d^2}{r} \right) \right) \text{ μs,} \qquad (2.19)$$

where d is the distance of closest approach of the ray in solar radii, and r is the distance of the planet or satellite from the Sun, in astronomical units.

In the two decades following Irwin Shapiro's 1964 discovery of this effect as a theoretical consequence of general relativity, several high-precision measurements were made using radar ranging to targets passing through superior conjunction. Since one does not have access to a "Newtonian" signal against which to compare the round-trip travel time of the observed signal, it is necessary to do a differential measurement of the variations in round-trip travel times as the target passes through superior conjunction, and to look for the logarithmic behavior of Eq. (2.19). In order to do this accurately, however, one must take into account the variations in round-trip travel time due to the orbital motion of the target relative to the Earth. This is done by using radar-ranging (and possibly other) data on the target taken when it is far from superior conjunction (i.e. when the time-delay term is negligible) to determine an accurate ephemeris for the target, using the ephemeris to predict the PPN coordinate trajectory $\mathbf{x}_e(t)$ near superior conjunction, then combining that trajectory with the trajectory of the Earth $\mathbf{x}_\oplus(t)$ to determine the Newtonian round-trip time and the logarithmic term in Eq. (2.19). The resulting predicted round-trip travel times in terms of the unknown coefficient $\frac{1}{2}(1 + \gamma)$ are then fit to the measured travel times using the method of least-squares, and an estimate obtained for $\frac{1}{2}(1 + \gamma)$.

The targets employed included planets, such as Mercury or Venus, used as passive reflectors of the radar signals, and artificial satellites, such as Mariners 6 and 7, Voyager 2, the Viking Mars landers and orbiters, and the Cassini spacecraft to Saturn, used as active retransmitters of the radar signals.

The results for the coefficient $\frac{1}{2}(1 + \gamma)$ of all radar time-delay measurements performed to date (including a measurement of the one-way time delay induced by the Sun on signals from the millisecond pulsar PSR 1937+21) are shown in Fig. 2.4.

The best bound comes from Doppler tracking of the Cassini spacecraft while it was on its way to Saturn [65], with a result $\gamma - 1 = (2.1 \pm 2.3) \times 10^{-5}$. This was made possible by the ability to do Doppler measurements using both X-band (7175 MHz) and Ka-band (34316 MHz) radar, thereby significantly reducing the dispersive effects of the solar corona. In addition, the 2002 superior conjunction of Cassini was particularly favorable: with the spacecraft at 8.43 astronomical units from the Sun, the distance of closest approach of the radar signals to the Sun was only $1.6 R_\odot$.

The Shapiro time delay plays a key role in determining the parameters of binary pulsar orbits, notably in the "double pulsar" system (see Section 2.6.2).

2.5.2 The perihelion shift of Mercury

The explanation of the anomalous perihelion shift of Mercury's orbit was another of the triumphs of general relativity. This had been an unsolved problem in celestial mechanics for over half a century, since the announcement by Le Verrier in 1859 that, after the perturbing effects of the planets on Mercury's orbit had been accounted for, there remained an unexplained advance in the perihelion of Mercury. The modern value for this discrepancy is 42.98 arcseconds per century. A number of *ad hoc* proposals were made in an attempt to account for this excess, including, the existence of a new planet Vulcan near the Sun, a ring of planetoids, a solar quadrupole moment and a deviation from the inverse-square law of gravitation, but none was successful. General relativity accounted for the anomalous shift in a natural way without disturbing the agreement with other planetary observations.

The predicted advance per orbit $\Delta \tilde{\omega}$, including both relativistic PPN contributions and the Newtonian contribution resulting from a possible solar quadrupole moment, is given by

$$\Delta \tilde{\omega} = 6\pi \frac{Gm}{pc^2} \left(\frac{2 + 2\gamma - \beta}{3} \right) + 3\pi J_2 \left(\frac{R}{p} \right)^2 , \qquad (2.20)$$

where m is the total mass of the Sun–Mercury system; $p \equiv a(1 - e^2)$ is the semi-latus rectum of the orbit, with the semi-major axis a and the eccentricity e; R is the mean radius of the Sun; and J_2 is a dimensionless measure of its quadrupole moment (for details of the derivation see TEGP 7.3). We have here restricted attention to fully-conservative theories of gravity.

Through observations of the normal modes of solar oscillations (helioseismology) it is now known that $J_2 = (2.2 \pm 0.1) \times 10^{-7}$, comparable to what would be estimated from a uniformly rotating solar model. Substituting standard orbital elements and physical constants for Mercury and the Sun we obtain the rate of perihelion shift $\dot{\omega}$, in seconds of arc per century,

$$\dot{\omega} = 42.''98 \left(\frac{2 + 2\gamma - \beta}{3} + 3 \times 10^{-4} \frac{J_2}{10^{-7}} \right). \qquad (2.21)$$

The most recent fits to planetary data include data from the Messenger spacecraft that orbited Mercury, thereby significantly improving knowledge of its orbit. Adopting the Cassini bound on γ *a priori*, these analyses yield a bound on β given by $|\beta - 1| = (-4.1 \pm 7.8) \times 10^{-5}$ [66, 67]. Further analysis could push this bound even lower, although knowledge of J_2 would have to improve simultaneously.

Laser tracking of the Earth-orbiting satellite LAGEOS II led to a measurement of its relativistic perigee precession (3.4 arcseconds per year) in agreement with general relativity to two percent [68].

2.5.3 Tests of the strong equivalence principle

The next class of solar-system experiments that test relativistic gravitational effects may be called tests of the strong equivalence principle (SEP). In Section 2.2.2 we pointed out that many metric theories of gravity (perhaps all except general relativity) can be expected to violate one or more aspects of SEP. Among the testable violations of SEP are a violation of the weak equivalence principle for gravitating bodies that leads to perturbations in the

Earth–Moon orbit, preferred-location and preferred-frame effects in orbital dynamics or in the structure of bodies such as the Earth, and possible variations in the gravitational constant over cosmological timescales.

The Nordtvedt effect

In a pioneering calculation using his early form of the PPN formalism, Kenneth Nordtvedt [69] showed that many metric theories of gravity predict that massive bodies violate the weak equivalence principle – that is, fall with different accelerations depending on their gravitational self-energy. Dicke [70] argued that such an effect would occur in theories with a spatially varying gravitational constant, such as scalar–tensor gravity. In the PPN framework, the acceleration of a massive body in an external gravitational potential U has the form

$$\mathbf{a} = \left(1 - \eta_{\mathrm{N}} \frac{E_{\mathrm{g}}}{mc^2}\right) \nabla U,$$
$$\eta_{\mathrm{N}} = 4\beta - \gamma - 3 - \frac{10}{3}\xi - \alpha_1 + \frac{2}{3}\alpha_2 - \frac{2}{3}\zeta_1 - \frac{1}{3}\zeta_2,$$

(2.22)

where E_{g} is the negative of the gravitational self-energy of the body ($E_{\mathrm{g}} > 0$). This violation of the massive-body equivalence principle is known as the "Nordtvedt effect". The effect is absent in general relativity ($\eta_{\mathrm{N}} = 0$) but present in scalar–tensor theory ($\eta_{\mathrm{N}} = (1 + 2\lambda)/(2 + \omega)$). The existence of the Nordtvedt effect does not violate the results of laboratory Eötvös experiments, since for laboratory-sized objects $E_{\mathrm{g}}/(mc^2) \leq 10^{-27}$, far below the sensitivity of current or future experiments. However, for astronomical bodies, $E_{\mathrm{g}}/(mc^2)$ may be significant (3.6×10^{-6} for the Sun, 4.6×10^{-10} for the Earth, 0.2×10^{-10} for the Moon, ~ 0.2 for neutron stars). If the Nordtvedt effect is present ($\eta_{\mathrm{N}} \neq 0$) then the Earth should fall toward the Sun with a slightly different acceleration than the Moon. This perturbation in the Earth–Moon orbit leads to a polarization of the orbit that is directed toward the Sun as it moves around the Earth–Moon system, as seen from Earth. This polarization represents a perturbation in the Earth–Moon distance of the form $\delta r = 13.1\eta_{\mathrm{N}} \cos[(\omega_0 - \omega_{\mathrm{s}})t]$ meters, where ω_0 and ω_{s} are the angular frequencies of the orbits of the Moon and Sun around the Earth (see TEGP 8.1 for detailed derivations).

Since August 1969, when the first successful acquisition was made of a laser signal reflected from the Apollo 11 retroreflector on the Moon, the LLR experiment has made regular measurements of the round-trip travel times of laser pulses between a network of observatories and the lunar retroreflectors, with accuracies that are approaching the 5 ps (1 mm) level. These measurements are fit using the method of least-squares to a theoretical model for the lunar motion that takes into account perturbations due to the Sun and the other planets, tidal interactions, and post-Newtonian gravitational effects. The predicted round-trip travel times between retroreflector and telescope also take into account the librations of the Moon; the orientation of the Earth, the location of the observatories, and atmospheric effects on the signal propagation. The "Nordtvedt" parameter η_{N} along with several other important parameters of the model are then estimated in the least-squares method. For a review of lunar laser ranging, see [71].

Numerous ongoing analyses of the data find no evidence, within experimental uncertainty, for the Nordtvedt effect [72]. These results represent a limit on a possible violation of WEP for massive bodies of about 1.4 parts in 10^{13} (see Fig. 2.1).

However, at this level of precision, one cannot regard the results of LLR as a "clean" test of SEP until one eliminates the possibility of a compensating violation of WEP for the two bodies, because the chemical compositions of the Earth and Moon differ in the relative fractions of iron and silicates. To this end, the Eöt-Wash group carried out an improved test of WEP for laboratory bodies whose chemical compositions mimic that of the Earth and Moon. The resulting bound of 1.4 parts in 10^{13} [73] from composition effects reduces the ambiguity in the LLR bound, and establishes the firm SEP test at the level of about 2 parts in 10^{13}. These results can be summarized by the Nordtvedt parameter bound $|\eta_N| = (4.4 \pm 4.5) \times 10^{-4}$.

APOLLO, the Apache Point Observatory for Lunar Laser ranging Operation, a joint effort by researchers from the Universities of Washington, Seattle, and California, San Diego, has achieved mm ranging precision using enhanced laser and telescope technology, together with a good, high-altitude site in New Mexico. However, models of the lunar orbit must be improved in parallel in order to achieve an order-of-magnitude improvement in the test of the Nordtvedt effect [74]. This effort will be aided by the fortuitous 2010 discovery by the Lunar Reconnaissance Orbiter of the precise landing site of the Soviet Lunokhod I rover, which deployed a retroreflector in 1970. Its uncertain location made it effectively "lost" to lunar laser ranging for almost 40 years. Its location on the lunar surface will make it useful in improving models of the lunar libration.

Tests of the Nordtvedt effect for neutron stars have also been carried out using a class of systems known as wide-orbit binary millisecond pulsars (WBMSPs), which are pulsar–white-dwarf binary systems with very small orbital eccentricities. In the gravitational field of the galaxy, a non-zero Nordtvedt effect can induce an apparent anomalous eccentricity pointed toward the galactic center, which can be bounded using statistical methods, given enough WBMSPs. Using data from 21 WBMSPs, including recently discovered highly circular systems, Stairs *et al.* [75] obtained the bound $\Delta < 5.6 \times 10^{-3}$, where $\Delta = \eta_N(E_g/M)_{NS}$. Because $(E_g/M)_{NS} \sim 0.1$ for typical neutron stars, this bound does not compete with the bound on η_N from LLR; on the other hand, it does test SEP in the strong-field regime because of the presence of the neutron star in each system. The 2013 discovery of a millisecond pulsar in orbit with *two* white dwarfs in very circular, coplanar orbits [76] may lead to a test of the Nordtvedt effect in the strong-field regime that surpasses the precision of lunar laser ranging by a substantial factor.

Preferred-frame and preferred-location effects

Some theories of gravity violate SEP by predicting that the outcomes of local gravitational experiments may depend on the velocity of the laboratory relative to the mean rest frame of the universe (preferred-frame effects) or on the location of the laboratory relative to a

Table 2.4 *Current limits on the PPN parameters.*

Parameter	Effect	Limit	Remarks
$\gamma - 1$	time delay	2.3×10^{-5}	Cassini tracking
	light deflection	2×10^{-4}	VLBI
$\beta - 1$	perihelion shift	8×10^{-5}	$J_{2\odot} = (2.2 \pm 0.1) \times 10^{-7}$
	Nordtvedt effect	2.3×10^{-4}	$\eta_N = 4\beta - \gamma - 3$ assumed
ξ	spin precession	4×10^{-9}	millisecond pulsars
α_1	orbital polarization	10^{-4}	lunar laser ranging
		4×10^{-5}	PSR J1738+0333
α_2	spin precession	2×10^{-9}	millisecond pulsars
α_3	pulsar acceleration	4×10^{-20}	pulsar $\dot{P}$ statistics
ζ_1	—	2×10^{-2}	combined PPN bounds
ζ_2	binary acceleration	4×10^{-5}	$\ddot{P}_p$ for PSR 1913+16
ζ_3	Newton's 3rd law	10^{-8}	lunar acceleration
ζ_4	—	—	not independent

nearby gravitating body (preferred-location effects). In the post-Newtonian limit, preferred-frame effects are governed by the values of the PPN parameters α_1, α_2, and α_3, and some preferred-location effects are governed by ξ (see Table 2.2).

The most important such effects are variations and anisotropies in the locally-measured value of the gravitational constant which lead to anomalous Earth tides and variations in the Earth's rotation rate, anomalous contributions to the orbital dynamics of planets, the Moon and binary pulsars; self-accelerations of isolated pulsars; and anomalous torques on spinning bodies such as the Sun or pulsars (see TEGP 8.2, 8.3, 9.3, and 14.3(c)). A tight bound on α_3 was obtained from the period derivatives of 21 millisecond pulsars [75]. The best bound on α_1, comes from the orbit of the pulsar–white-dwarf system J1738+0333; while the best bounds on α_2 and ξ come from bounding torques on the solitary millisecond pulsars B1937+21 and J1744–1134 [77–79]. Because these bounds involved systems with strong internal gravity of the neutron stars, they should strictly speaking be regarded as bounds on "strong field" analogues of the PPN parameters. Here we will treat them as bounds on the standard PPN parameters, as shown in Table 2.4.

Constancy of the Newtonian gravitational constant

Most theories of gravity that violate SEP predict that the locally measured Newtonian gravitational constant may vary with time as the universe evolves. For the scalar–tensor theories listed in Table 2.3, for example, the predictions for $\dot{G}/G$ can be written in terms of time derivatives of the asymptotic scalar field. Where G does change with cosmic evolution, its rate of variation should be related to the expansion rate of the universe, i.e. $\dot{G}/G \sim H_0$, where H_0 is the Hubble expansion parameter and is given by $H_0 = 73 \pm 3$ km s^{-1} Mpc^{-1} $\approx 0.75 \times 10^{-10}$ yr^{-1}.

Table 2.5 *Constancy of the gravitational constant.*

Method	$\dot{G}/G$ $(10^{-13} \text{ yr}^{-1})$	Reference
Mars ephemeris	0.1 ± 1.6	[80]
Lunar laser ranging	4 ± 9	[81]
Binary pulsars	-7 ± 33	[82, 83]
Helioseismology	0 ± 16	[84]
Big Bang nucleosynthesis	0 ± 4	[85, 86]

Several observational constraints can be placed on $\dot{G}/G$, one kind coming from bounding the present rate of variation, another from bounding a difference between the present value and a past value. The first type of bound typically comes from LLR measurements, planetary radar-ranging measurements, and pulsar timing data. The best limits come from improvements in the ephemeris of Mars using range and Doppler data from the Mars Global Surveyor (1998–2006), Mars Odyssey (2002–2008), and Mars Reconnaissance Orbiter (2006–2008), together with improved data and modeling of the effects of the asteroid belt [80]. Since the bound is actually on variations of $GM_\odot$, any future improvements in $\dot{G}/G$ beyond a part in 10^{13} will have to take into account models of the actual mass loss from the Sun, due to radiation of light and neutrinos ($\sim 0.7 \times 10^{-13} \text{ yr}^{-1}$) and due to the solar wind ($\sim 0.2 \times 10^{-13} \text{ yr}^{-1}$). The second type of bound comes from studies of the evolution of the Sun, stars and the Earth, Big Bang nucleosynthesis, and analyses of ancient eclipse data. Selected results are shown in Table 2.5; a thorough review is given in [27].

2.5.4 The search for gravitomagnetism

According to general relativity, moving or rotating matter should produce a contribution to the gravitational field that is the analogue of the magnetic field of a moving charge or a magnetic dipole. From the geometrical point of view, rotating matter produces a "dragging of inertial frames" around the body, also called the Lense–Thirring effect. One consequence of this phenomenon is a precession of a gyroscope's spin $\mathbf{S}$ according to

$$\frac{d\mathbf{S}}{dt} = \mathbf{\Omega}_{\text{LT}} \times \mathbf{S}, \qquad \mathbf{\Omega}_{\text{LT}} = -\frac{1}{2}\left(1 + \gamma + \frac{1}{4}\alpha_1\right)\frac{G\left(\mathbf{J} - 3\mathbf{n}(\mathbf{n} \cdot \mathbf{J})\right)}{c^2 r^3}, \qquad (2.23)$$

where $\mathbf{n}$ is a unit radial vector, and r is the distance from the center of the body with angular momentum $\mathbf{J}$ (TEGP 9.1).

Another effect on the spin of a gyroscope arises from curved spacetime around the central body, called the "geodetic precession", given by

$$\frac{d\mathbf{S}}{d\tau} = \mathbf{\Omega}_{\text{G}} \times \mathbf{S}, \qquad \mathbf{\Omega}_{\text{G}} = \left(\gamma + \frac{1}{2}\right)\frac{\mathbf{v} \times \nabla U}{c^2}, \qquad (2.24)$$

where **v** is the velocity of the gyroscope, and U is the Newtonian gravitational potential of the source (TEGP 9.1).

The Relativity Gyroscope Experiment (Gravity Probe B or GPB), carried out by Stanford University, NASA and Lockheed–Martin Corporation, was a space mission designed to detect both precessional effects. Gravity Probe B will very likely go down in the history of science as one of the most ambitious, difficult, expensive, and controversial relativity experiments ever performed.[2] It was almost 50 years from inception to completion, although only about half of that time was spent as a full-fledged, approved space program.

The GPB spacecraft was launched on April 20, 2004 into an almost perfectly circular polar orbit at an altitude of 642 km, with the orbital plane parallel to the direction of a guide star known as *IM Pegasi* (HR 8703). The spacecraft contained four spheres made of fuzed quartz, all spinning about the same axis (two were spun in the opposite direction), which was oriented to be in the orbital plane, pointing toward the guide star. An onboard telescope pointed continuously at the guide star, and the direction of each spin was compared with the direction to the star, which was at a declination of 16° relative to the Earth's equatorial plane. With these conditions, the predicted precessions were 6630 milliarcsecond per year for the geodetic effect, and 38 milliarcsecond per year for frame dragging, the former in the orbital plane (in the north–south direction) and the latter perpendicular to it (in the east–west direction).

In order to reduce the non-relativistic torques on the rotors to an acceptable level, the rotors were fabricated to be both spherical and homogeneous to better than a few parts in 10 million. Each rotor was coated with a thin film of niobium, and the experiment was conducted at cryogenic temperatures inside a dewar containing 2200 litres of superfluid liquid helium. As the niobium film becomes a superconductor, each rotor develops a magnetic moment parallel to its spin axis. Variations in the direction of the magnetic moment relative to the spacecraft were then measured using superconducting current loops surrounding each rotor. As the spacecraft orbits the Earth, the aberration of light from the guide star causes an artificial but predictable change in direction between the rotors and the on-board telescope; this was an essential tool for calibrating the conversion between the voltages read by the current loops and the actual angle between the rotors and the guide star.

The mission ended in September 2005, as scheduled, when the last of the liquid helium boiled off. Although all subsystems of the spacecraft and the apparatus performed extremely well, they were not perfect. Calibration measurements carried out during the mission, both before and after the science phase, revealed unexpectedly large torques on the rotors. Numerous diagnostic tests worthy of a detective novel showed that these were caused by electrostatic interactions between surface imperfections ("patch effect") on the niobium films and the spherical housings surrounding each rotor. These effects and other anomalies greatly contaminated the data and complicated its analysis, but finally, in October 2010, the Gravity Probe B team announced that the experiment had successfully measured

---

[2] Full disclosure: The author served as Chair of an external NASA Science Advisory Committee for Gravity Probe B from 1998 to 2010.

both the geodetic and frame-dragging precessions. The outcome was in agreement with general relativity, with a precision of 0.3 percent for the geodetic precession, and 20 percent for the frame-dragging effect [87].

Another way to look for frame-dragging is to measure the precession of orbital planes of bodies circling a rotating body. One implementation of this idea is to measure the relative precession, at about 31 milliarcseconds per year, of the line of nodes of a pair of laser-ranged geodynamics satellites (LAGEOS), ideally with supplementary inclination angles; the inclinations must be supplementary in order to cancel the huge (126 degrees per year) nodal precession caused by the Earth's Newtonian gravitational multipole moments. Unfortunately, the two existing LAGEOS satellites are not in appropriately inclined orbits. Nevertheless, Ciufolini *et al.* [88, 89] combined nodal precession data from LAGEOS I and II with improved models for the Earth's multipole moments provided by two recent orbiting geodesy satellites, Europe's CHAMP (Challenging Minisatellite Payload) and NASA's GRACE (Gravity Recovery and Climate Experiment), and reported a 5–10 percent confirmation of general relativity.

On February 13, 2012, a third laser-ranged satellite, known as LARES (Laser Relativity Satellite) was launched by the Italian Space Agency. Its inclination was very close to the required supplementary angle relative to LAGEOS I, and its eccentricity was very nearly zero. However, because its semi-major axis is only 2/3 that of either LAGEOS I or II, and because the Newtonian precession rate is proportional to $a^{-3/2}$, LARES does not provide a cancellation of the Newtonian precession. Combining data from all three satellites with continually improving Earth data from GRACE, the LARES team hopes to achieve a test of frame-dragging at the one percent level.

2.6 Binary-pulsar tests of gravitational theory

The 1974 discovery of the binary pulsar B1913+16 by Joseph Taylor and Russell Hulse during a routine search for new pulsars was a milestone in the history of general relativity. It led to the first confirmation of the existence of gravitational radiation and to a Nobel Prize for Taylor and Hulse, and it opened a new field for testing gravitational theories in the strong-field and gravitational-wave regimes.

Following that discovery, only two new binary pulsars were discovered with interesting relativistic properties during the next 15 years (out of a total of 14 systems), but thanks to improved radio-telescope sensitivity, better techniques for removing the effects of interstellar dispersion, better timing capabilities (exploiting GPS time transfer, for example) and improved detection algorithms, the pace of discovery of binary pulsars picked up steadily. Today, over 220 binary pulsars are known, with 23 discovered in 2012 alone. To be sure, the vast majority of these are of little relativistic interest, because they are so widely separated that relativistic effects are negligible, because they involve significant mass-transfer from the companion (X-ray binary pulsars), or because the timing characteristics of the pulsar do not meet the stability demanded to measure relativistic effects.

Nevertheless, close to a dozen binary pulsars are relativistically interesting for one reason or another, and this zoo of systems contains some fascinating beasts, including the famous "double pulsar", systems with white-dwarf companions, a system with the most massive neutron star known, and systems with extraordinarily circular orbits.

2.6.1 The binary pulsar and general relativity

We will begin by reviewing the iconic Hulse–Taylor binary pulsar, by far the best studied of all the systems. Pulse arrival-time measurements have been made regularly (apart from a four-year break during upgrading of the Arecibo radio telescope) for over 30 years. Table 2.6 lists the current values of the key orbital and relativistic parameters, from analysis of data through 2006 [90].

The observational parameters that are obtained from a least-squares solution of the arrival-time data fall into three groups:

1. non-orbital parameters, such as the pulsar period and its rate of change (defined at a given epoch), and the position of the pulsar on the sky;
2. five "Keplerian" parameters, most closely related to those appropriate for standard Newtonian binary systems, such as the eccentricity e, the orbital period P_b, and the semi-major axis of the pulsar projected along the line of sight, $a_p \sin i$; and
3. five "post-Keplerian" parameters.

Table 2.6 *Parameters of the binary pulsar B1913+16. The numbers in parentheses denote errors in the last digit.*

Parameter	Symbol (units)	Value
(i) Astrometric and spin parameters:		
Right Ascension	α	$19^h 15^m 27.^s 99928(9)$
Declination	δ	$16°06'27.''3871(13)$
Pulsar period	P_p (ms)	$59.0299983444181(5)$
Derivative of period	$\dot{P}_p$	$8.62713(8) \times 10^{-18}$
(ii) "Keplerian" parameters:		
Projected semimajor axis	$a_p \sin i$ (s)	$2.341782(3)$
Eccentricity	e	$0.6171334(5)$
Orbital period	P_b (day)	$0.322997448911(4)$
Longitude of periastron	ω_0 (°)	$292.54472(6)$
Julian date of periastron	T_0 (MJD)	$52144.90097841(4)$
(iii) "Post-Keplerian" parameters:		
Mean rate of periastron advance	$\langle \dot{\omega} \rangle$ (° yr^{-1})	$4.226598(5)$
Redshift/time dilation	γ' (ms)	$4.2992(8)$
Orbital period derivative	$\dot{P}_b$ (10^{-12})	$-2.423(1)$

The five post-Keplerian parameters are: $\langle \dot{\omega} \rangle$, the average rate of periastron advance; γ', the amplitude of delays in arrival of pulses caused by the varying effects of the gravitational redshift and time dilation as the pulsar moves in its elliptical orbit at varying distances from the companion and with varying speeds (not to be confused with the PPN parameter γ); $\dot{P}_b$, the rate of change of orbital period, caused predominantly by gravitational radiation damping; and r and $s = \sin i$, respectively the "range" and "shape" of the Shapiro time delay of the pulsar signal as it propagates through the curved spacetime region near the companion, where i is the angle of inclination of the orbit relative to the plane of the sky. An additional 14 relativistic parameters are measurable in principle [91].

In general relativity, the five post-Keplerian parameters can be related to the masses of the two bodies and to measured Keplerian parameters by the equations (TEGP 12.1, 14.6(a))

$$\langle \dot{\omega} \rangle = \frac{6\pi}{P_b} \left(\frac{2\pi G m \omega_b}{c^3} \right)^{2/3} \frac{1}{1-e^2}, \tag{2.25}$$

$$\gamma' = e \frac{P_b}{2\pi} \left(\frac{2\pi G m \omega_b}{c^3} \right)^{2/3} \frac{m_2}{m} \left(1 + \frac{m_2}{m} \right), \tag{2.26}$$

$$\dot{P}_b = -\frac{192\pi}{5} \left(\frac{2\pi G \mathcal{M} \omega_b}{c^3} \right)^{5/3} F(e), \tag{2.27}$$

$$s = \sin i, \tag{2.28}$$

$$r = m_2, \tag{2.29}$$

where $\omega_b \equiv 2\pi/P_b$ is the orbital angular frequency; m_1 and m_2 are the pulsar and companion masses, respectively; m is the total mass; $\mathcal{M}$ is the so-called "chirp" mass, given by $\mathcal{M} \equiv (m_1 m_2/m^2)^{3/5} m$; and $F(e) = (1-e^2)^{-7/2}(1 + 73e^2/24 + 37e^4/96)$.

Notice that, by virtue of Kepler's third law, $(2\pi G m \omega_b/c^3)^{2/3} = Gm/(ac^2) \sim \epsilon$, thus the first two post-Keplerian parameters can be seen as $\mathcal{O}(\epsilon)$, or 1PN corrections to the underlying variable, while the third is an $\mathcal{O}(\epsilon^{5/2})$, or 2.5PN correction. The parameters r and s are not separately measurable with interesting accuracy for B1913+16 because the orbit's 47° inclination does not lead to a substantial Shapiro delay.

Because P_b and e are separately measured parameters, the measurements of the three post-Keplerian parameters provide three constraints on the two unknown masses. The periastron shift measures the total mass of the system, $\dot{P}_b$ measures the chirp mass, and γ' measures a complicated function of the masses. General relativity passes the test if it provides a consistent solution to these constraints, within the measurement errors.

From the intersection of the $\langle \dot{\omega} \rangle$ and γ' constraints we obtain the values $m_1 = 1.4398 \pm 0.0002\,M_\odot$ and $m_2 = 1.3886 \pm 0.0002\,M_\odot$. The third of Eqs. (2.29) then predicts the value $\dot{P}_b = -2.402531 \pm 0.000014 \times 10^{-12}$. In order to compare the predicted value for $\dot{P}_b$ with the observed value of Table 2.6, it is necessary to take into account the small kinematic effect of a relative acceleration between the binary pulsar system and the solar system caused by the differential rotation of the galaxy. Using data on the location and proper motion of the pulsar, combined with the best information available on galactic rotation; the current value of this effect is $\dot{P}_b^{gal} \simeq -(0.027 \pm 0.005) \times 10^{-12}$. Subtracting this from the observed $\dot{P}_b$ (see Table 2.6) gives the corrected $\dot{P}_b^{corr} = -(2.396 \pm 0.005) \times 10^{-12}$, which agrees with the prediction within the errors. In other words,

$$\frac{\dot{P}_b^{corr}}{\dot{P}_b^{GR}} = 0.997 \pm 0.002. \tag{2.30}$$

The consistency among the measurements is displayed in Fig. 2.5, in which the regions allowed by the three most precise constraints have a single common overlap. Uncertainties in the parameters that go into the galactic correction are now the limiting factor in the accuracy of the test of gravitational damping.

A third way to display the agreement with general relativity is by plotting the cumulative shift of the time of periastron passage caused by the changing orbital period. Figure 2.6 shows the results: the dots are the data points, while the curve is the predicted difference using the measured masses and the quadrupole formula for gravitational radiation damping [90].

The consistency among the constraints provides a test of the assumption that the two bodies behave as "point" masses, without complicated tidal effects, obeying the general

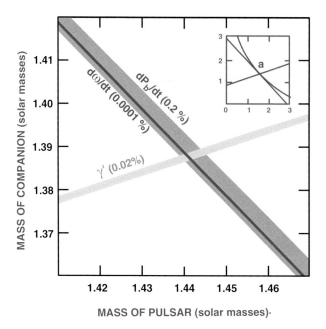

Figure 2.5 Constraints on masses of the pulsar and its companion from data on B1913+16, assuming general relativity to be valid. The width of each strip in the plane reflects observational accuracy, shown as a percentage. An inset shows the three constraints on the full mass plane; the intersection region (a) has been magnified 400 times for the full figure.

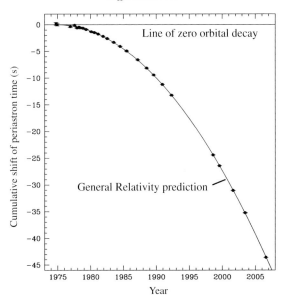

Figure 2.6 Plot of the cumulative shift of the periastron time from 1975–2005. The points are data, the curve is the general relativity prediction. The gap during the middle 1990s was caused by a closure of Arecibo for upgrading. From [90]. © AAS. Reproduced with permission.

relativistic equations of motion including gravitational radiation. It is also a test of strong gravity, in that the highly relativistic internal structure of the neutron stars does not influence their orbital motion, as predicted by the SEP of general relativity.

Observations of variations in the pulse profile suggest that the pulsar is undergoing geodetic precession on a 300-year timescale as it moves through the curved spacetime generated by its companion (see Section 2.5.4). The amount is consistent with general relativity, assuming that the pulsar's spin is suitably misaligned with the orbital angular momentum [92, 93]. Unfortunately, the evidence suggests that the pulsar beam may precess out of our line of sight by 2025.

2.6.2 A zoo of binary pulsars

Over 70 binary neutron star systems with orbital periods less than a day are now known. While some are less interesting for testing relativity, some have yielded interesting tests, and others, notably the recently discovered "double pulsar" are likely to continue to produce significant results well into the future. Here we describe some of the more interesting or best studied cases.

The "double" pulsar: J0737–3039A, B. This binary pulsar system, discovered in 2003 [94], was already remarkable for its extraordinarily short orbital period (0.1 days) and large periastron advance ($16.8995° \text{ yr}^{-1}$), but then the companion was also discovered to

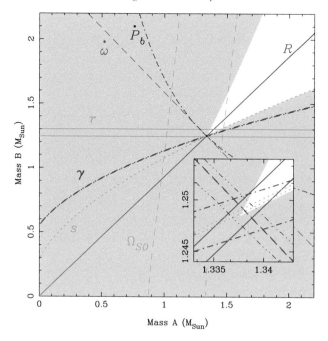

Figure 2.7 Constraints on masses of the pulsar and its companion from data on J0737–3039A,B, assuming general relativity to be valid. The inset shows the intersection region magnified by a factor of 80. (Figure by permission of Michael Kramer.)

be a pulsar [95]. Because two projected semi-major axes could be measured, the mass ratio was obtained directly from the ratio of the two values of $a_p \sin i$, and thereby the two masses could be obtained by combining that ratio with the periastron advance, assuming general relativity. The results are $m_A = 1.3381 \pm 0.0007 \, M_\odot$ and $m_B = 1.2489 \pm 0.0007 \, M_\odot$, where A denotes the primary (first) pulsar. From these values, one finds that the orbit is nearly edge-on, with $\sin i = 0.9997$, a value which is completely consistent with that inferred from the Shapiro delay parameter. In fact, the five measured post-Keplerian parameters plus the ratio of the projected semi-major axes give six constraints on the masses (assuming general relativity): as seen in Fig. 2.7, all six overlap within their measurement errors [96]. Because of the location of the system, galactic proper motion effects play a significantly smaller role in the interpretation of $\dot{P}_b$ measurements than they did in B1913+16; this and the reduced effect of interstellar dispersion means that the accuracy of measuring the gravitational-wave damping may soon beat that from the Hulse–Taylor system. The geodetic precession of pulsar B's spin axis has also been measured by monitoring changes in the patterns of eclipses of the signal from pulsar A, with a result in agreement with general relativity to about 13 percent [97]; the constraint on the masses from that effect (assuming general relativity to be correct) is also shown in Fig. 2.7.

J1738+0333: A white-dwarf companion. This is a low-eccentricity, 8.5-hour period system in which the white-dwarf companion is bright enough to permit detailed spectroscopy, allowing the companion mass to be determined directly to be $0.181\,M_\odot$. The mass ratio is determined from Doppler shifts of the spectral lines of the companion and of the pulsar period, giving the pulsar mass $1.46\,M_\odot$. Ten years of observation of the system yielded both a measurement of the apparent orbital period decay, and enough information about its parallax and proper motion to account for the substantial galactic kinematic effect to give a value of the intrinsic period decay of $\dot{P}_b = (-25.9 \pm 3.2) \times 10^{-15}\,\mathrm{s\,s^{-1}}$ in agreement with the predicted effect [98]. But because of the asymmetry of the system, the result also places a significant bound on the existence of dipole radiation, predicted by many alternative theories of gravity (see Section 2.6.3 below for discussion). Data from this system were also used to place the tight bound on the PPN parameter α_1 shown in Table 2.4.

J1141–6545: A white-dwarf companion. This system is similar in some ways to the Hulse–Taylor binary: short orbital period (0.20 days), significant orbital eccentricity (0.172), rapid periastron advance (5.3 degrees per year) and massive components ($M_{pulsar} = 1.4\,M_\odot$, $M_{companion} = 1.0\,M_\odot$). The key difference is that the companion is again a white dwarf. The intrinsic orbit period decay has been measured to be in agreement with general relativity to about six percent, again placing limits on dipole gravitational radiation [99].

J0348+0432: The most massive neutron star. Discovered in 2011, this is another neutron-star white-dwarf system, in a very-short-period (0.1-day), low-eccentricity (2×10^{-6}) orbit. Timing of the neutron star and spectroscopy of the white dwarf have led to mass values of $0.172\,M_\odot$ for the white dwarf and $2.01 \pm 0.04\,M_\odot$ for the pulsar, making it the most massive accurately measured neutron star yet. Along with an earlier measurement of a $2\,M_\odot$ pulsar, this ruled out a number of heretofore viable soft equations of state for nuclear matter. The orbit period decay agrees with the general relativity prediction within 20 percent and is expected to improve steadily with time.

J0337+1715: Two white-dwarf companions. This remarkable system was reported in 2014 [76]. It consists of a 2.73 millisecond pulsar ($M = 1.44\,M_\odot$) with extremely good timing precision, accompanied by *two* white dwarfs in coplanar circular orbits. The inner white dwarf ($M = 0.1975\,M_\odot$) has an orbital period of 1.629 days, with $e = 6.918 \times 10^{-4}$, and the outer white dwarf ($M = 0.41\,M_\odot$) has a period of 327.26 days, with $e = 3.536 \times 10^{-2}$. This is an ideal system for testing the Nordtvedt effect in the strong-field regime. Here the inner system is the analogue of the Earth–Moon system, and the outer white dwarf plays the role of the Sun. Because the outer semi-major axis is about 1/3 of an astronomical unit, the basic driving perturbation is comparable to that provided by the Sun. However, the self-gravitational binding energy per unit mass of the neutron star is almost a billion times larger than that of the Earth, greatly amplifying the size of the Nordtvedt effect. Depending on the details, this system could exceed lunar laser ranging in testing the Nordtvedt effect by several orders of magnitude.

Other binary pulsars. Two of the earliest binary pulsars, B1534+12 and B2127+11C, discovered in 1990, failed to live up to their early promise despite being similar to the

Hulse–Taylor system in most respects (both were believed to be double neutron-star systems). The main reason was the significant uncertainty in the kinematic effect on $\dot{P}_b$ of local accelerations, galactic in the case of B1534+12, and those arising from the globular cluster that was home to B2127+11C.

2.6.3 Binary pulsars and alternative theories

Soon after the discovery of the binary pulsar it was widely hailed as a new testing ground for relativistic gravitational effects. As we have seen in the case of general relativity, in most respects, the system has lived up to, indeed exceeded, the early expectations.

In another respect, however, the system has only partially lived up to its promise, namely as a direct testing ground for alternative theories of gravity. The origin of this promise was the discovery that alternative theories of gravity generically predict the emission of dipole gravitational radiation from binary star systems. In general relativity, there is no dipole radiation because the "dipole moment" (center of mass) of isolated systems is uniform in time (conservation of momentum), and because the "inertial mass" that determines the dipole moment is the same as the mass that generates gravitational waves (SEP). In other theories, while the inertial dipole moment may remain uniform, the "gravity wave" dipole moment need not, because the mass that generates gravitational waves depends differently on the internal gravitational binding energy of each body than does the inertial mass (violation of SEP).

In theories that violate SEP, the difference between gravitational wave mass and inertial mass is a function of the internal gravitational binding energy of the bodies. This additional form of gravitational radiation damping could, at least in principle, be significantly stronger than the usual quadrupole damping, because it depends on fewer powers of the orbital velocity v, and it depends on the gravitational binding energy per unit mass of the bodies, which, for neutron stars, could be as large as 20 percent (see TEGP 10 for further details). As one fulfillment of this promise, Will and Eardley worked out in detail the effects of dipole gravitational radiation in the bimetric theory of Rosen, and, when the first observation of the decrease of the orbital period was announced in 1979, the Rosen theory suffered a terminal blow. A wide class of alternative theories also failed the binary pulsar test because of dipole gravitational radiation (TEGP 12.3).

On the other hand, the early observations of PSR 1913+16 already indicated that, in general relativity, the masses of the two bodies were nearly equal, so that, in theories of gravity that are in some sense "close" to general relativity, dipole gravitational radiation would not be a strong effect, because of the apparent symmetry of the system. Thus, despite the presence of dipole gravitational radiation, the Hulse–Taylor binary pulsar provides at present only a weak test of Brans–Dicke theory, not competitive with solar-system tests.

However, in some generalized scalar–tensor theories, the strong internal gravity of the neutron star can induce a phenomenon called "spontaneous scalarization" whereby the behavior can be significantly different from either general relativity or from a scalar–tensor theory with a nominally large value of ω_0. This is true of theories with negative values of the parameter β_0 in Eq. (2.10) [100]. Furthermore, the recently discovered mixed binary pulsar systems J1141–6545 and J1738+0333 have been exploited using precise timing of the pulsar, spectroscopic observations of the white-dwarf companion, and the strong dipole gravitational radiation effect to yield stringent bounds. Indeed, the latter system surpasses the Cassini bound for $\beta_0 > 1$ and $\beta_0 < -2$, and is close to that bound for the pure Brans–Dicke case $\beta_0 = 0$ [98].

2.7 Testing general relativity in its second century

2.7.1 Gravitational-wave tests

Shortly after the centenary year of general relativity, a new opportunity for testing relativistic gravity will be realized, when a worldwide network of advanced kilometer-scale, laser interferometric gravitational wave observatories in the U.S. (LIGO project), and Europe (VIRGO and GEO600 projects), begins regular detection and analysis of gravitational wave signals from astrophysical sources. These will be followed by advanced detectors in Japan and India. They will have the capability of detecting and measuring the gravitational waveforms from astronomical sources in a frequency band between about 10 Hz and 5000 Hz, with a maximum sensitivity to strain at around a hectahertz (100 Hz). The most promising source for detection and study of the gravitational wave signal is the "inspiralling compact binary" – a binary system of neutron stars or black holes (or one of each) in the final minutes of a death spiral leading to a violent merger. Such is the fate, for example, of the Hulse–Taylor binary pulsar B1913+16 in about 300 Myr, or the "double pulsar" J0737–3039 in about 85 Myr. Given the expected sensitivity of "advanced LIGO" (around 2016), which could see such sources out to many hundreds of megaparsecs, it has been estimated that from 10 to several hundred annual inspiral events could be detectable. Other sources, such as binary black-hole mergers, supernova core collapse events, instabilities in rapidly rotating newborn neutron stars, signals from non-axisymmetric pulsars, and a stochastic background of waves, may be detectable. For a review of gravitational-wave theory and detection, see [101].

In addition, plans are being developed for an orbiting laser interferometer space antenna. Such a system, consisting of three spacecraft orbiting the Sun in a triangular formation separated from each other by a million kilometers, would be sensitive primarily in the very low frequency band between 10^{-4} and 10^{-1} Hz [102]. Such a mission, dubbed eLISA, is a leading candidate to address the "gravitational universe" science theme that has been selected by the European Space Agency for a mission to be launched around 2034.

A third approach that focuses on the ultra-low-frequency band (nanohertz) is that of Pulsar Timing Arrays (PTAs), whereby a network of highly stable millisecond pulsars is monitored in a coherent way using radio telescopes, in the hope of detecting the fluctuations in arrival times induced by passing gravitational waves.

In addition to opening a new astronomical window, the detailed observation of gravitational waves by such observatories may provide the means to test general relativistic predictions for the polarization and speed of the waves, for gravitational radiation damping and for strong-field gravity.

Polarization of gravitational waves

Like electromagnetic waves, gravitational waves can be characterized by polarizations. In general metric theories of gravity there are six independent polarization modes. Three are transverse to the direction of propagation, with two representing quadrupolar deformations and one representing a monopolar "breathing" deformation. Three modes are longitudinal, with one an axially symmetric stretching mode in the propagation direction, and one quadrupolar mode in each of the two orthogonal planes containing the propagation direction. Figure 2.8 shows the displacements induced on a ring of freely falling test particles by each of these modes. General relativity predicts only the first two transverse quadrupolar modes (a) and (b) independently of the source. Massless scalar–tensor gravitational waves can in addition contain the transverse breathing mode (c). More general metric theories predict additional longitudinal modes, up to the full complement of six (TEGP 10.2).

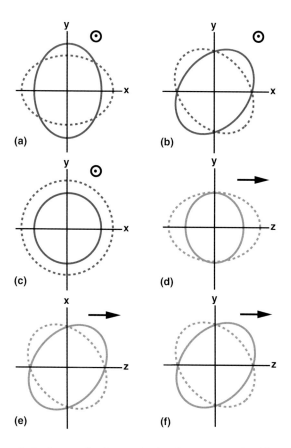

Figure 2.8 The six polarization modes for gravitational waves permitted in any metric theory of gravity. Shown is the displacement that each mode induces on a ring of test particles. The wave propagates in the $+z$ direction. There is no displacement out of the plane of the picture. In (a), (b), and (c), the wave propagates out of the plane; in (d), (e), and (f), the wave propagates in the plane.

A suitable array of gravitational antennas could delineate or limit the number of modes present in a given wave; a space antenna could do likewise by virtue of its changing orientation during the passage of a sufficiently long wave train. If distinct evidence were found of any mode other than the two transverse quadrupolar modes of general relativity, the result would be disastrous for general relativity.

Gravitational radiation back-reaction

In the binary pulsar, a test of general relativity was made possible by measuring at least three relativistic effects that depended upon only two unknown masses. The evolution of the orbital phase under the damping effect of gravitational radiation played a crucial role. Another situation in which measurement of orbital phase can lead to tests of general relativity is that of the inspiralling compact binary system. The key differences are that here gravitational radiation itself is the detected signal, rather than radio pulses, and the phase evolution alone carries all the information. In the binary pulsar, the first derivative of the binary frequency $\dot{\omega}_b$ was measured; here the full nonlinear variation of ω_b as a function of time is measured.

Broad-band laser interferometers are especially sensitive to the phase evolution of the gravitational waves, which carry the information about the orbital phase evolution. The analysis of gravitational wave data from such sources will involve some form of matched filtering of the noisy detector output against an ensemble of theoretical "template" waveforms which depend on the intrinsic parameters of the inspiralling binary, such as the component masses, spins, and so on, and on its inspiral evolution.

But for laser interferometric observations of gravitational waves, the bottom line is that, in order to measure the astrophysical parameters of the source and to test the properties of the gravitational waves, it is necessary to derive the gravitational waveform and the resulting radiation back-reaction on the orbit phasing to many post-Newtonian orders beyond the quadrupole approximation. Two decades of intensive work by many groups have led to the development of waveforms in general relativity that are accurate to 3.5PN order, and for some specific effects, such as those related to spin, to 4.5PN order (see [103] for a thorough review).

On the other hand, alternative theories of gravity are likely to predict rather different gravitational waveforms, notably via the addition of dipole gravitational radiation. Thus tests of alternative theories might be possible using the gravitational-wave signals. Several attempts have been made to parametrize the gravitational-wave signal from binary inspiral, in order to encompass alternative theories of gravity, in the spirit of the PPN framework [104, 105].

Speed of gravitational waves

According to general relativity, in the limit in which the wavelength of gravitational waves is small compared to the radius of curvature of the background spacetime, the waves propagate along null geodesics of the background spacetime, i.e. they have the same speed c as light (in this section, we do not set $c = 1$). In other theories, the speed could differ from c because of coupling of gravitation to other gravitational fields, or if gravitation were propagated by a massive field (a massive graviton). In the latter case v_g would be given by, in a local inertial frame,

$$\frac{v_g^2}{c^2} = 1 - \frac{m_g^2 c^4}{E^2}, \tag{2.31}$$

where m_g and E are the graviton rest mass and energy, respectively.

The most obvious way to test this is to compare the arrival times of a gravitational wave and an electromagnetic wave from the same event, e.g., a supernova, or a binary inspiral with an associated electromagnetic signal. For a source at a distance D, the resulting value of the difference $1 - v_g/c$ is

$$1 - \frac{v_g}{c} = 5 \times 10^{-17} \left(\frac{200 \text{ Mpc}}{D}\right)\left(\frac{\Delta t}{1 \text{ s}}\right), \tag{2.32}$$

where $\Delta t \equiv \Delta t_a - (1+Z)\Delta t_e$ is the "time difference", where Δt_a and Δt_e are the differences in arrival time and emission time of the two signals, respectively, and Z is the redshift of the source. In many cases, Δt_e is unknown, so that the best one can do is employ an upper bound on Δt_e based on observation or modelling. The result will then be a bound on $1 - v_g/c$.

The foregoing discussion assumes that the source emits *both* gravitational and electromagnetic radiation in detectable amounts, and that the relative time of emission can be established to sufficient accuracy, or can be shown to be sufficiently small.

However, there is a situation in which a bound on the graviton mass can be set using gravitational radiation alone [106]. That is the case of the inspiralling compact binary. Because the frequency of the gravitational radiation sweeps from low frequency at the initial moment of observation to higher frequency at the final moment, the speed of the gravitons emitted will vary, from lower speeds initially to higher speeds (closer to c) at the end. This will cause a distortion of the observed phasing of the waves that can be detected or bounded through matched filtering.

These and other tests of gravitational theory using gravitational waves are thoroughly reviewed by [107] and by [108].

2.7.2 Strong-field tests

One of the central difficulties of testing general relativity in the strong-field regime is the possibility of contamination by uncertain or complex physics. In the solar system, weak-field gravitational effects can in most cases be measured cleanly and separately from non-gravitational effects. The remarkable cleanliness of binary pulsars permits precise measurements of gravitational phenomena in a partially strong-field context.

Unfortunately, nature is rarely so kind. Still, under suitable conditions, qualitative and even quantitative strong-field tests of general relativity could be carried out.

One example is the exploration of the spacetime near black holes and neutron stars. Studies of certain kinds of accretion known as advection-dominated accretion flow (ADAF) in low-luminosity binary X-ray sources may yield the signature of the black hole event horizon [109]. The spectrum of frequencies of quasi-periodic oscillations (QPO) from galactic black hole binaries may permit measurement of the spins of the black holes [110]. Aspects of strong-field gravity and frame-dragging may be revealed in spectral shapes of iron fluorescence lines from the inner regions of accretion disks [111, 112]. Using submm VLBI, a collaboration dubbed the Event Horizon Telescope could image our galactic center black hole SgrA* and the black hole in M87 with horizon-scale angular resolution; observation of accretion phenomena at these angular resolutions could provide tests of the spacetime geometry very close to the black hole [113]. Tracking of hypothetical stars whose orbits are within a fraction of a milliparsec of SgrA* could test the black hole "no-hair" theorem, via a direct measurement of both the angular momentum J and the quadrupole moment Q of the black hole, and a test of the requirement that $Q = -J^2/M$ [114].

Because of uncertainties in the detailed models, the results to date of studies like these are suggestive at best, but the combination of future higher-resolution observations and better modelling could lead to striking tests of strong-field predictions of general relativity.

For a detailed review of strong-field tests of general relativity using electromagnetic observations, see [115].

2.8 Conclusions

General relativity has held up under extensive experimental scrutiny. The question then arises, why bother to continue to test it? One reason is that gravity is a fundamental interaction of nature, and as such requires the most solid empirical underpinning we can provide. Another is that all attempts to quantize gravity and to unify it with the other forces suggest that the standard general relativity of Einstein may not be the last word. Furthermore, the predictions of general relativity are fixed; the pure theory contains no adjustable constants so nothing can be changed. Thus every test of the theory is either a potentially deadly test or a possible probe for new physics. Although it is remarkable that this theory, born 100 years ago out of almost pure thought, has managed to survive every test, the possibility of finding a discrepancy will continue to drive experiments for years to come. These experiments will search for new physics beyond Einstein at many different scales: the large distance scales of the astrophysical, galactic, and cosmological realms; scales of very short distances or high energy; and scales related to strong or dynamical gravity.

Having spent almost half of the century of general relativity's existence being astonished by its continuing agreement with observation, I might be permitted a personal reflection at this point on the future of the subject: *It would not at all surprise me if general relativity turned out to be perfectly valid at all scales, from the cosmological to the astrophysical to the microscopic, failing only at the Planck scale where one naturally expects quantum gravity to take over.*

Of course, having made that statement, I am reminded of the story about Yogi Berra, the famous New York Yankees baseball player and manager (one of my childhood heroes), and one blessed with the most sublimely illogical mind. At some point after his retirement from baseball, his wife Carmen asked him: "Yogi, you were born in St. Louis, you played ball in New York and we now live in New Jersey. If you should die first, where would you like me to bury you?" Yogi's answer: "Surprise me!"

Acknowledgments

This work was supported in part by the National Science Foundation, Grant Numbers PHY 12-60995 and 13-06069. I am grateful for the hospitality of the Institut d'Astrophysique de Paris, where this article was prepared.

References

[1] Will, C. M. 1993. *Was Einstein Right?: Putting General Relativity to the Test*. 2nd edn. New York: Basic Books.
[2] Will, C. M. 1993. *Theory and Experiment in Gravitational Physics*. 2nd edn. Cambridge; New York: Cambridge University Press.
[3] Will, C. M. 2006. *Living Rev. Relativity*, **9**. http://www.livingreviews.org/lrr-2006-3.
[4] Will, C. M. 2014. *Living Rev. Relativity*, **17**. http://www.livingreviews.org/lrr-2014-4.
[5] Will, C. M. 2010. *Am. J. Phys.*, **78**, 1240–1247.
[6] Dicke, R. H. 1964. Experimental relativity. Pages 165–313 of: DeWitt, C., and DeWitt, B. (eds), *Relativity, Groups and Topology*. New York; London: Gordon and Breach.
[7] Schlamminger, S. *et al.* 2008. *Phys. Rev. Lett.*, **100**, 041101.
[8] Merlet, S. *et al.* 2010. *Metrologia*, **47**, L9–L11.
[9] Müller, H., Peters, A., and Chu, S. 2010. *Nature*, **463**, 926–929.
[10] Wolf, P. *et al.* 2011. *Class. Quantum Grav.*, **28**, 145017.
[11] Hogan, J. M., Johnson, D. M. S., and Kasevich, M. A. 2009. Light-pulse atom interferometry. Page 411 of: Arimondo, E., Ertmer, W., Schleich, W. P., and Rasel, E. M. (eds), *Atom Optics and Space Physics: Proceedings of the International School of Physics "Enrico Fermi", Course 168*. IOS Press Amsterdam.
[12] Fischbach, E. *et al.* 1992. *Metrologia*, **29**, 213–260.
[13] Speake, C. C., and Will, C. M. 2012. *Class. Quantum Grav.*, **29**, 180301.
[14] Adelberger, E. G. *et al.* 2009. *Prog. Part. Nucl. Phys.*, **62**, 102–134.
[15] Will, C. M. 2006. Special relativity: A centenary perspective. Page 33 of: Damour, T., Darrigol, O., Duplantier, B., and Rivasseau, V. (eds), *Einstein, 1905–2005: Poincaré Seminar 2005*. Basel: Birkäuser Verlag.
[16] Mattingly, D. 2005. *Living Rev. Relativity*, **8**. http://www.livingreviews.org/lrr-2005-5.
[17] Liberati, S. 2013. *Class. Quantum Grav.*, **30**, 133001.
[18] Kostelecký, V. A., and Russell, N. 2011. *Rev. Mod. Phys.*, **83**, 11–31.
[19] Kostelecký, V. A., and Tasson, J. D. 2011. *Phys. Rev. D*, **83**, 016013.
[20] LoPresto, J. C., Schrader, C., and Pierce, A. K. 1991. *Astrophys. J.*, **376**, 757–760.
[21] Guéna, J. *et al.* 2012. *Phys. Rev. Lett.*, **109**, 080801.
[22] Peil, S. *et al.* 2013. *Phys. Rev. A*, **87**, 010102.
[23] Leefer, N. *et al.* 2013. *Phys. Rev. Lett.*, **111**, 060801.
[24] Ashby, N. 2003. *Living Rev. Relativity*, **6**. http://www.livingreviews.org/lrr-2003-1.
[25] Will, C. M., 2000. *Einstein's relativity and everyday life*. http://www.physicscentral.com/writers/writers-00-2.html.

[26] Dyson, F. J. 1972. The fundamental constants and their time variation. Pages 213–236 of: Salam, A., and Wigner, E. (eds), *Aspects of Quantum Theory*. Cambridge; New York: Cambridge University Press.

[27] Uzan, J.-P. 2011. *Living Rev. Relativity*, **14**. http://www.livingreviews.org/lrr-2011-2.

[28] King, J. A. *et al.* 2012. *Mon. Not. Roy. Astr. Soc.*, **422**, 3370–3414.

[29] Kanekar, N. *et al.* 2012. *Astrophys. J. Lett.*, **746**, L16.

[30] Lentati, L. *et al.* 2013. *Mon. Not. Roy. Astr. Soc.*, **430**, 2454–2463.

[31] Turyshev, S. G., and Toth, V. T. 2010. *Living Rev. Relativity*, **13**. http://www.livingreviews.org/lrr-2010-4.

[32] Turyshev, S. G. *et al.* 2012. *Phys. Rev. Lett.*, **108**, 241101.

[33] Schutz, B. F. 2009. *A First Course in General Relativity*. Cambridge: Cambridge University Press.

[34] Poisson, E., and Will, C. M. 2014. *Gravity: Newtonian, Post-Newtonian, Relativistic*. Cambridge: Cambridge University Press.

[35] Nordtvedt Jr., K. 1968. *Phys. Rev.*, **169**, 1017–1025.

[36] Will, C. M. 1971. *Astrophys. J.*, **163**, 611–628.

[37] Will, C. M., and Nordtvedt Jr., K. L. 1972. *Astrophys. J.*, **177**, 757–774.

[38] Misner, C. W., Thorne, K. S., and Wheeler, J. A. 1973. *Gravitation*. San Francisco: W. H. Freeman.

[39] Damour, T., and Esposito-Farèse, G. 1992. *Class. Quantum Grav.*, **9**, 2093–2176.

[40] Damour, T., and Nordtvedt Jr., K. 1993. *Phys. Rev. Lett.*, **70**, 2217–2219.

[41] Damour, T., and Nordtvedt Jr., K. 1993. *Phys. Rev. D*, **48**, 3436–3450.

[42] Fujii, Y., and Maeda, K.-I. 2007. *The Scalar–Tensor Theory of Gravitation*. Cambridge: Cambridge University Press.

[43] Sotiriou, T. P., and Faraoni, V. 2010. *Rev. Mod. Phys.*, **82**, 451–497.

[44] De Felice, A., and Tsujikawa, S. 2010. *Living Rev. Relativity*, **13**. http://www.livingreviews.org/lrr-2010-3.

[45] Hellings, R. W., and Nordtvedt Jr., K. 1973. *Phys. Rev. D*, **7**, 3593–3602.

[46] Jacobson, T., and Mattingly, D. 2001. *Phys. Rev. D*, **64**, 024028.

[47] Mattingly, D., and Jacobson, T. A. 2002. Relativistic gravity with a dynamical preferred frame. Pages 331–335 of: Kostelecký, V. (ed), *CPT and Lorentz Symmetry II*. Singapore; River Edge: World Scientific.

[48] Jacobson, T., and Mattingly, D. 2004. *Phys. Rev. D*, **70**, 024003.

[49] Eling, C., and Jacobson, T. 2004. *Phys. Rev. D*, **69**, 064005.

[50] Foster, B. Z., and Jacobson, T. 2006. *Phys. Rev. D*, **73**, 064015.

[51] Hořava, P. 2009. *Phys. Rev. D*, **79**, 084008.

[52] Blas, D., Pujolàs, O., and Sibiryakov, S. 2010. *Phys. Rev. Lett.*, **104**, 181302.

[53] Blas, D., Pujolàs, O., and Sibiryakov, S. 2011. *J. High Energy Phys.*, **4**, 18.

[54] Jacobson, T. 2014. *Phys. Rev. D*, **89**, 081501.

[55] Bekenstein, J. D. 2004. *Phys. Rev. D*, **70**, 083509.

[56] Skordis, C. 2008. *Phys. Rev. D*, **77**, 123502.

[57] Sagi, E. 2009. *Phys. Rev. D*, **80**, 044032.

[58] Skordis, C. 2009. *Class. Quantum Grav.*, **26**, 143001.

[59] Famaey, B., and McGaugh, S. S. 2012. *Living Rev. Relativity*, **15**. http://www.livingreviews.org/lrr-2012-10.

[60] Hinterbichler, K. 2012. *Rev. Mod. Phys.*, **84**, 671–710.

[61] Will, C. M. 1988. *Am. J. Phys.*, **56**, 413–415.

[62] Shapiro, S. S., Davis, J. L., Lebach, D. E., and Gregory, J. S. 2004. *Phys. Rev. Lett.*, **92**, 121101.

[63] Lambert, S. B., and Le Poncin-Lafitte, C. 2011. *Astron. Astrophys.*, **529**, A70.

[64] Bolton, A. S., Rappaport, S., and Burles, S. 2006. *Phys. Rev. D*, **74**, 061501.

[65] Bertotti, B., Iess, L., and Tortora, P. 2003. *Nature*, **425**, 374–376.

[66] Fienga, A. *et al.* 2011. *Cel. Mech. Dyn. Astron.*, **111**, 363–385.

[67] Verma, A. K. *et al.* 2014. *Astron. Astrophys.*, **561**, A115.

[68] Lucchesi, D. M., and Peron, R. 2014. *Phys. Rev. D*, **89**, 082002.
[69] Nordtvedt Jr., K. 1968. *Phys. Rev.*, **169**, 1014–1016.
[70] Dicke, R. H. 1970. *Gravitation and the Universe.* Memoirs of the American Philosophical Society. Jayne Lecture for 1969, vol. 78. Philadelphia: American Philosophical Society.
[71] Merkowitz, S. M. 2010. *Living Rev. Relativity*, **13**. http://www.livingreviews.org/lrr-2010-7.
[72] Williams, J. G., Turyshev, S. G., and Murphy Jr., T. W. 2004. *Int. J. Mod. Phys. D*, **13**, 567–582.
[73] Baeßler, S. *et al.* 1999. *Phys. Rev. Lett.*, **83**, 3585–3588.
[74] Murphy, Jr., T. W. *et al.* 2012. *Class. Quantum Grav.*, **29**, 184005.
[75] Stairs, I. H. *et al.* 2005. *Astrophys. J.*, **632**, 1060–1068.
[76] Ransom, S. M. *et al.* 2014. *Nature*, **505**, 520–524.
[77] Shao, L., and Wex, N. 2012. *Class. Quantum Grav.*, **29**, 215018.
[78] Shao, L. *et al.* 2013. *Class. Quantum Grav.*, **30**, 165019.
[79] Shao, L., and Wex, N. 2013. *Class. Quantum Grav.*, **30**, 165020.
[80] Konopliv, A. S. *et al.* 2011. *Icarus*, **211**, 401–428.
[81] Williams, J. G., Turyshev, S. G., and Boggs, D. H. 2004. *Phys. Rev. Lett.*, **93**, 261101.
[82] Deller, A. T., Verbiest, J. P. W., Tingay, S. J., and Bailes, M. 2008. *Astrophys. J. Lett.*, **685**, L67–L70.
[83] Lazaridis, K. *et al.* 2009. *Mon. Not. Roy. Astr. Soc.*, **400**, 805–814.
[84] Guenther, D. B., Krauss, L. M., and Demarque, P. 1998. *Astrophys. J.*, **498**, 871–876.
[85] Copi, C. J., Davis, A. N., and Krauss, L. M. 2004. *Phys. Rev. Lett.*, **92**, 171301.
[86] Bambi, C., Giannotti, M., and Villante, F. L. 2005. *Phys. Rev. D*, **71**, 123524.
[87] Everitt, C. W. F. *et al.* 2011. *Phys. Rev. Lett.*, **106**, 221101.
[88] Ciufolini, I., and Pavlis, E. C. 2004. *Nature*, **431**, 958–960.
[89] Ciufolini, I., Pavlis, E. C., and Peron, R. 2006. *New Ast.*, **11**, 527–550.
[90] Weisberg, J. M., Nice, D. J., and Taylor, J. H. 2010. *Astrophys. J.*, **722**, 1030–1034.
[91] Damour, T., and Taylor, J. H. 1992. *Phys. Rev. D*, **45**, 1840–1868.
[92] Kramer, M. 1998. *Astrophys. J.*, **509**, 856–860.
[93] Weisberg, J. M., and Taylor, J. H. 2002. *Astrophys. J.*, **576**, 942–949.
[94] Burgay, M. *et al.* 2003. *Nature*, **426**, 531–533.
[95] Lyne, A. G. *et al.* 2004. *Science*, **303**, 1153–1157.
[96] Kramer, M. *et al.* 2006. *Science*, **314**, 97–102.
[97] Breton, R. P. *et al.* 2008. *Science*, **321**, 104.
[98] Freire, P. C. C. *et al.* 2012. *Mon. Not. Roy. Astr. Soc.*, **423**, 3328–3343.
[99] Bhat, N. D. R., Bailes, M., and Verbiest, J. P. W. 2008. *Phys. Rev. D*, **77**, 124017.
[100] Damour, T., and Esposito-Farèse, G. 1998. *Phys. Rev. D*, **58**, 042001.
[101] Creighton, J. D. E., and Anderson, W. G. 2011. *Gravitational-Wave Physics and Astronomy: An Introduction to Theory, Experiment and Data Analysis.* Cambridge; New York: Wiley.
[102] Amaro-Seoane, P. *et al.* 2012. *Class. Quantum Grav.*, **29**, 124016.
[103] Blanchet, L. 2014. *Living Rev. Relativity*, **17**. http://www.livingreviews.org/lrr-2014-2.
[104] Yunes, N., and Pretorius, F. 2009. *Phys. Rev. D*, **80**, 122003.
[105] Mishra, C. K., Arun, K. G., Iyer, B. R., and Sathyaprakash, B. S. 2010. *Phys. Rev. D*, **82**, 064010.
[106] Will, C. M. 1998. *Phys. Rev. D*, **57**, 2061.
[107] Gair, J. R., Vallisneri, M., Larson, S. L., and Baker, J. G. 2013. *Living Rev. Relativity*, **16**. http://www.livingreviews.org/lrr-2013-7.
[108] Yunes, N., and Siemens, X. 2013. *Living Rev. Relativity*, **16**. http://www.livingreviews.org/lrr-2013-9.
[109] Narayan, R., and McClintock, J. E. 2008. *New Astron. Rev.*, **51**, 733–751.
[110] Psaltis, D. 2004. Measurements of black hole spins and tests of strong-field general relativity. Pages 29–35 of: Kaaret, P., Lamb, F., and Swank, J. (eds), *X-Ray Timing 2003: Rossi and Beyond.* AIP Conference Proceedings, vol. 714. Melville: American Institute of' Physics.

[111] Reynolds, C. S. 2013. *Space Sci. Rev. On Line*, 1–18.
[112] Reynolds, C. S. 2013. *Class. Quantum Grav.*, **30**, 244004.
[113] Doeleman, S. *et al.* 2009. Imaging an event horizon: Submm-VLBI of a supermassive black hole. Page 68 of: *Astro2010: The Astronomy and Astrophysics Decadal Survey.*
[114] Will, C. M. 2008. *Astrophys. J. Lett.*, **674**, L25–L28.
[115] Psaltis, D. 2008. *Living Rev. Relativity*, **11**. http://www.livingreviews.org/lrr-2008-9.

3

Relativistic Astrophysics

3.1 Introduction
John Friedman

The star has to go on radiating and radiating and contracting and contracting until, I suppose, it gets down to a few km. radius, when gravity becomes strong enough to hold in the radiation, and the star can at last find peace. Dr. Chandrasekhar had got this result before, but he has rubbed it in in his latest paper; and, when discussing it with him, I felt driven to the conclusion that this was almost a reductio ad absurdum *of the relativistic degeneracy formula.*

(A. S. Eddington [1])

The emphasis of this chapter is on four parts of relativistic astrophysics in which general relativity plays a fundamental role. After briefly reviewing the early history of the subject, we discuss

The structure and stability of relativistic stars
Observational evidence for black holes
Gamma-ray bursts
Gravitational lensing

General relativistic astrophysics encompasses a broader arena, and separate chapters or parts of chapters in this volume are devoted to cosmology, gravitational waves, the inspiral and merger of compact binaries, and black-hole stability.

Relativistic astrophysics began in 1916 on the Russian front, where Karl Schwarzschild wrote two papers, one reporting the solution to the Einstein equation for an incompressible spherical star, the other presenting the celebrated vacuum Schwarzschild spacetime. Schwarzschild was dead within the year, and for the next 47 years his solutions had a twilight existence. In no known stars did general relativity play a significant role, and only a handful of papers in astronomy or astrophysics mentioned the work.

Although sparsely distributed, the exceptions to this neglect were remarkable. In 1931, shortly before Chadwick's discovery of the neutron and shortly after the first paper by Chandrasekhar [2] (following approximate computations by Anderson [3] and Stoner [4]) on an upper mass limit of white dwarfs, Landau [5] submitted a paper that independently argued that there was an upper limit on the mass of a collection of degenerate fermions and speculated on the existence of stars with cores of nuclear density.[1] After Chadwick, Landau was thinking about stars with neutron cores, now called Thorne–Żytkow objects [8], but it was Baade and Zwicky [9] who proposed "that supernovae represent the transition from ordinary stars into neutron stars, which in their final stages consist of closely packed neutrons." The connection between the limiting mass of a degenerate core (or a white dwarf) and the collapse to a neutron star did not appear in print until 1939 articles by Gamow [10] and Chandrasekhar [11]. In the same year, Tolman [12] and Oppenheimer and Volkoff [13] obtained the hydrostatic equilibrium (TOV) equation. The latter paper found the family of spherical stars with the equation of state (EOS) of an ideal neutron gas and showed that the maximum mass of a neutron star for this EOS was $0.7 M_\odot$. Remarkably, they also considered the possibility of repulsive interactions at high density, but they restricted a stiffer EOS to $\rho > 10^{15}$ g/cm^3 and thereby enforced a maximum mass of order $M_\odot$, below the Chandrasekhar limit. This may have undermined the proposal that neutron stars are formed in supernovae, and another two decades passed before Cameron [14], following Harrison *et al.* [15] but using a stiffer EOS based on a Skyrme potential, found a maximum mass of about 2 $M_\odot$ and reopened the argument: "As a result of an examination of the physics of supernova explosions and of the formation of the elements, the writer has concluded that neutron stars are probable products of the supernova process."

It is clear from the opening quote of this chapter that, by 1935, Chandrasekhar and Eddington had inferred from the upper limit on the mass of degenerate matter that a degenerate star above the mass limit would contract to within its Schwarzschild radius. Calling this a *reductio ad absurdum*, Eddington famously renounced the relativistic degeneracy formula, and the astronomy community followed Eddington. The physicists did not, and Oppenheimer and Snyder [16], introducing their study of collapsing dust, wrote, "When all thermonuclear sources of energy are exhausted, a sufficiently heavy star will collapse. Unless fission due to rotation, the radiation of mass, or the blowing off of mass by radiation, reduce the star's mass to the order of that of the sun, this contraction will continue indefinitely." This last phrase referred to an external observer, and the authors noted that a comoving observer would collapse in finite time beyond the radius for which "the cone within which a signal can escape has closed entirely . . . the star thus tends to close itself off from any communication with a distant observer; only its gravitational field persists." Despite this insight, only after Finkelstein [17] had rediscovered coordinates that are smooth on the future Schwarzschild horizon[2] did the nature of the black hole itself finally become clear:

---

[1] Léon Rosenfeld [6] gives a widely repeated description that appears to be inaccurate: "when the news on the neutron's discovery reached Copenhagen, we had a lively discussion on the same evening about the prospects opened by this discovery. In the course of it Landau improvised the conception of neutron stars – "unheimliche Sterne," weird stars, which would be invisible and unknown to us unless by colliding with visible stars they would originate explosions, which might be supernovae." A careful historical study by Yakovlev *et al.* [7], however, finds that Landau was in Copenhagen in 1931 *before* Chadwick's experiments – and just after Landau had submitted his paper on nuclear-density cores.

[2] Eddington [18] had found coordinates that are smooth on the past horizon but does not mention the fact that the singular behavior of metric components at $r = 2m$ is gone or that the $r = 2m$ surface has the character of a horizon. (In Eq. (2), defining a new time coordinate, $\log(r - m)$ should be replaced by $\log(r - 2m)$.) Lemaître [19] and Synge [20] also obtained coordinates that are smooth at one or both horizons, apparently without knowledge of Eddington's coordinates and without elucidating the role of the horizon. The Kruskal–Szekeres coordinates were yet to be published, although Finkelstein includes a note in print referring to Kruskal's work.

The Schwarzschild surface $r = 2m$ is not a singularity but acts as a perfect unidirectional membrane: causal influences can cross it but only in one direction.

Finkelstein's work, Kerr's discovery four years later of the solution governing a rotating black hole, and Wheeler's insistence on the importance of collapse fortuitously presaged a decade of spectacular observational discovery. Schmidt's (1962) observation of quasars, followed quickly by the recognition of their cosmological redshift, spurred Robinson, Schild, and Schucking to organize a conference on a field whose name they invented – relativistic astrophysics. At this first Texas Symposium, held in December, 1963, Harlan Smith announced that the enormous luminosity of the quasar 3C273 had sharply changed in the course of one week [21, 22]: With the implication that 10^{47} erg/s was emitted by a region less than one light-week across, collapsed objects became candidates for the engine of quasars.

One year later, x-ray satellites found a set of bright sources concentrated in the direction of the Galaxy. Shklovsky [23] suggested that Sco X-1 might be a neutron star and that synchrotron radiation from relativistic electrons might power the Crab nebula [24]. In the next two years, Pacini [25] proposed the magnetic field of a rotating neutron star as the energy source of the Crab and other supernova remnants, and Hewish *et al.* [26] announced the discovery of pulsars. With the discovery of the Crab pulsar [27, 28] and its 33 Hz period [29], modeling pulsars as oscillating white dwarfs was no longer possible [30], and the subsequent measurement of its spin-down [31] was consistent with the luminosity of the Crab nebula, if the energy was the rotational energy of a 33 Hz neutron star.

Narayan and McClintock describe the much longer time taken to make the connection between black holes and Reber's first observation of radio galaxies. The primary reason was the time before observations revealed the distance to radio sources and the astonishingly small size of their engine. But there was also a theoretical bias against collapse, partly related to a lack of understanding of the nature of the event horizon that persisted for several years after Finkelstein's clarification. At that first Texas Meeting, Hoyle and Fowler [32] were the proponents of the gravitational energy of collapse as the engine of quasars, but even they still thought the $r = 2m$ surface was a worrisome singularity at which "world lines can be broken."

Zel'dovich and Novikov [33] suggested that accretion onto a neutron star powered the recently observed x-ray sources, but they had in mind accretion from the interstellar medium. Burbidge [34] describes the way the "binary hypothesis" emerged from a discussion at the Noordwijk Symposium in August 1966 that included G. and M. Burbidge, Ginzburg, Shklovsky, Savedoff, Woltjer, Prendergast, and Herbig. After Rossi had presented observations suggesting that Sco X-1 was a binary, "several of us then realized that ... the very powerful x-ray source might naturally arise through mass exchange between the secondary and primary ..." Articles with accretion from a companion to a neutron star [35] and white dwarf [36, 37] quickly followed.

Finally, in 1971, Cygnus X-1 was found to be a binary system whose compact star was above the upper mass limit on neutron stars. Salpeter [38] had suggested accretion onto supermassive black holes in connection with quasars, and with accretion onto a neutron star by now the leading model for a class of x-ray binaries, detailed models of accretion onto black holes were soon explored. Blandford and Thorne [39] discuss models prior to 1978, while Abramowicz and Fragile [40] and Camenzind [41] review more recent work. Other models were initially proposed – accretion onto differentially rotating white dwarfs, for example [42, 43], but by the mid 1970s, evidence for both galactic and stellar black holes had largely converted the astronomy community.

The mathematicians and physicists who studied the vacuum Schwarzschild solution and first modeled neutron stars were too far from astronomy to predict the ways they would ultimately be observed. In contrast, a decade before the Vela satellites' detection of gamma-ray bursts was announced in 1971, Colgate had warned that a gamma-ray burst from a supernova's shock wave might be detected and misinterpreted as a violation of the Nuclear Test Ban Treaty, which was then being negotiated [44]; but after 1971 a supernova was only one of many guesses that included *relativistic BBs* – grains of iron interacting with sunlight as they enter the solar system (named for the pellets shot by a BB gun). After the isotropy of the burst distribution had become apparent, a cosmological distance for many of the bursts was plausible, and Paczyński [45], citing the recently observed inspiral of the Hulse–Taylor pulsar, suggested that a class of bursts came from the coalescence of double neutron-star systems. More serious studies of supernovae and coalescence of compact binaries as burst sources followed the first observation of an optical counterpart, finding a cosmological distance; and these are reviewed by Mészáros and Rees.

Gamma-ray bursts are the most luminous events observed in the universe, but tens to billions of times more energy is emitted by still-invisible coalescing binary black holes. Most are likely to be observable only by detection of gravitational waves, either by ground-based observatories or, for coalescence involving supermassive black holes, by a future space-based interferometric antenna.

In contrast to the unexpected way black holes and neutron stars were reified by the dawn of radio and x-ray astronomy, double images, image magnification (Einstein, 1912 cited in [46]), Einstein rings [47, 48], and the imaging of galaxies [49] were anticipated long before the discovery of the first double quasar. What had not been anticipated and what Schneider reviews in Section 3.2 is the way gravitational lensing has become "an indispensible tool in astrophysics."

3.1.1 Relativistic stars

For years Fritz [Zwicky] had been pushing his ideas about neutron stars to anyone who would listen and had been universally ignored. I believe that part of the problem was his personality, which implied strongly that people were idiots if they did not believe in neutron stars.

(A. G. W. Cameron (1959), quoted in [7])

The relativistic stars of nature have a complex composition, spanning fifteen orders of magnitude in density.[3] Thought to consist primarily of a gas of neutrons with a gradually varying density of free protons, electrons, and muons, they are surrounded by a crust of

---

[3] Parts of Section 3.1.1 are taken with permission from Stergioulas and Friedman [50].

ordinary matter. Their cores may hold hyperons, pion or kaon condensates, or possibly free quarks, although recent observations mentioned below have made these alternatives somewhat less attractive.

In the conventional neutron-star model, a 1 km crust surrounds an interior in which neutrons and protons form a two-component superfluid. High magnetic fields, whose strength in some cases is likely to exceed 10^{14} G, are observed and thought to extend in quantized flux tubes through the superfluid interior. The angular velocities of observed millisecond pulsars range up to 716 Hz and the vorticity of their velocity fields is similarly thought to be confined, in the neutron stars' interiors, to quantized tubes.

Soon after formation, neutrino emission cools the star to 10^{10} K $\simeq$ 1 MeV, well below the $\sim$ 60 MeV Fermi energy of nucleons at nuclear density. Because isolated and accreting neutron stars are both cold in this sense, they are governed by a zero-temperature equation of state, $p = p(\rho)$, $\epsilon = \epsilon(\rho)$, where p is the pressure, ρ the rest-mass density, ϵ the energy density, all measured by a comoving observer.

Departures from local isotropy are associated with the crust, the vortex and magnetic flux tubes, and with heat flow and viscosity. Nevertheless, a neutron star *in equilibrium* is accurately approximated by a stationary self-gravitating *perfect fluid*, its structure determined by a balance between its intense gravity, the pressure of its degenerate particles, and its rotation. In particular, departures from perfect fluid equilibrium due to a solid crust are expected to be smaller than one part in $\sim 10^{-3}$, corresponding to the maximum strain that an electromagnetic lattice can support [51]; and this estimate is supported by observations of pulsar glitches, which are consistent with departures from a perfect fluid equilibrium of order 10^{-5}.

Similarly, on scales of meters or larger, a single rotational velocity field u^α describes the averaged superfluid motion [52–54]. The error in computing the gravitational field is much smaller than errors in the fluid model, because the characteristic length over which a potential varies is much larger than the distance between vortices. Although the assumption of a perfect fluid is adequate for describing equilibrium configurations, studies of neutron-star dynamics – of formation, oscillations, and stability, and of the interaction of binaries during and just prior to merger – require a more detailed knowledge of the stars' microphysics.

A rotating star is accurately described by a stationary, axisymmetric solution to the Einstein–Euler system – to the Einstein field equation

$$G^{\alpha\beta} = 8\pi T^{\alpha\beta}, \tag{3.1}$$

with a perfect-fluid stress-energy tensor,

$$T^{\alpha\beta} = \epsilon u^\alpha u^\beta + p q^{\alpha\beta}. \tag{3.2}$$

Here $q^{\alpha\beta} := g^{\alpha\beta} + u^\alpha u^\beta$ is the projection operator orthogonal to the fluid trajectories. The metric,

$$ds^2 = -e^{2\nu} dt^2 + e^{2\psi} (d\phi - \omega \, dt)^2 + e^{2\mu} (dr^2 + r^2 \, d\theta^2), \tag{3.3}$$

has four independent potentials, each a function of r and θ.

A star rotating with uniform angular velocity Ω has 4-velocity

$$u^\alpha = U(t^\alpha + \Omega\phi^\alpha), \tag{3.4}$$

where t^α and ϕ^α are the symmetry vectors generating time translations and rotations. The relativistic Euler equation,

$$(\epsilon + p)u^\beta \nabla_\beta u^\alpha = q^{\alpha\beta} \nabla_\beta p, \tag{3.5}$$

has the first integral (the relativistic Poincaré–Wavre equation)

$$\frac{h}{U} = \frac{\epsilon + p}{\rho\, U} = \mathcal{E}, \tag{3.6}$$

where the constant $\mathcal{E}$ is the *injection energy*, the energy per unit rest mass needed to inject matter with zero entropy into the star; U^{-1} is a redshift factor, reducing the energy of a fluid element at infinity by the amount lost in lowering it to a point in the star; and the quantity h is a relativistic specific enthalpy, the enthalpy per baryon mass.

In constructing numerical models of rotating stars, one solves Eq. (3.6) together with four components of the field equation for the four independent metric potentials. Numerical solutions [55–65] use variants of standard iterative methods for solving nonlinear elliptic equations; these are summarized in [50]. Two public-domain codes, RNS [66] and LORENE [67], compute accurate stellar models from an arbitrary EOS.

The uniformly rotating perfect-fluid equilibria comprise a two-dimensional family, parametrized by central density ϵ_c and angular velocity Ω. It is bounded at large angular velocity by the sequence of models rotating at the Kepler (or *mass-shedding*) limit with $\Omega = \Omega_K$, the angular velocity of a particle in circular orbit at the equator. The sequence of configurations of maximum mass at constant angular momentum J is a ridge of turning points; stars on the high-density side of this ridge are unstable to collapse, and the ridge of turning points nearly coincides with the boundary of the set of stars stable against collapse. A line of minimum-mass neutron stars (with masses of about $0.1\ M_\odot$) is similarly a lower bound on the stable configurations: Stars on the low-density side are unstable.

For a given EOS, the minimum period found from numerical models has the approximate value [68–70]

$$P_{\min} \simeq 0.82 \left(\frac{M_\odot}{M_{\mathrm{sph}}^{\max}}\right)^{1/2} \left(\frac{R_{\mathrm{sph}}^{\max}}{10\,\mathrm{km}}\right)^{3/2} \ \mathrm{ms}, \tag{3.7}$$

in terms of the mass and radius of the maximum mass *nonrotating* model. (See [71] for a more accurate empirical formula with an additional compactness parameter.)

Limits set by causality on mass, spin, radius and redshift

A theoretical maximum mass of neutron stars is obtained by using a maximally stiff EOS consistent with causality above a matching density ϵ_m, together with a presumed known EOS for $\epsilon < \epsilon_m$. Maximally stiff means that $c_{\mathrm{sound}} = \sqrt{dp/d\epsilon} = c$. (Here c_{sound} is the speed of sound and the speed associated with the characteristics of the relativistic Euler equation.) For *nonrotating stars*, the maximum mass has the form

$$M_{\text{max}} = 4.8 \left(\frac{2 \times 10^{14} \, \text{g/cm}^3}{\epsilon_m} \right)^{1/2} M_\odot, \tag{3.8}$$

given by Hartle and Sabbadini [72], following earlier work by Rhoades and Ruffini [73]. If one optimistically assumes that the equation of state is known up to about twice nuclear density, this would give an upper limit of about 3.5 $M_\odot$. Rapid uniform rotation can raise the upper limit set by causality by about 20% [74, 75]:

$$M_{\text{max}}^{\text{rot}} = 6.1 \left(\frac{2 \times 10^{14} \, \text{g/cm}^3}{\epsilon_m} \right)^{1/2} M_\odot. \tag{3.9}$$

In Section 3.3, Narayan and McClintock use an upper limit of 2.9 $M_\odot$, given by Kalogera and Baym [76], who assume a spherical star and use a particular nuclear EOS below twice nuclear density.

A corresponding minimum period, associated with a *maximally compact* EOS consistent with causality and with the additional constraint that the EOS allow a spherical star of mass $M \geq M_{\text{max,obs}}$, is given by Koranda *et al.* [75] (following Glendenning [77]):

$$P_{\text{min}} = 0.39 \, \text{ms} \, \frac{M_{\text{max,obs}}}{2 \, M_\odot}, \tag{3.10}$$

where $M_{\text{max,obs}}$ is the largest observed neutron-star mass.

The equation of state that minimizes the spin is maximally *soft* ($p = 0$, $c_{\text{sound}} = 0$) below some core density ϵ_C, and maximally stiff ($c_{\text{sound}} = 1$) above ϵ_C. A soft EOS yields the most compact stars and thus allows the most rapid rotation; but a soft EOS has a small maximum mass, and a stiff core is needed to allow a mass as large as $M_{\text{max,obs}}$.

Lattimer [78] uses this maximally compact EOS to obtain a lower limit on neutron-star radius and a corresponding upper limit on redshift, with value similar to that obtained earlier by Lindblom [79], who used causality and a match to a low-density nuclear EOS:

$$R > 8.25 \, \text{km} \, \frac{M_{\text{max,obs}}}{2 \, M_\odot}, \qquad z = (1 - 2M/r)^{-1/2} - 1 < 0.85. \tag{3.11}$$

The causality-enforced upper mass limit can be violated by proto-neutron stars and by stars formed in the merger of a double neutron-star system, because, in these cases, the stars are hot and differentially rotating. Differential rotation can increase the mass of cold neutron stars by more than a factor of two [80], and there is, in principle, no upper limit on the mass of a hot star. Stellar models above the upper mass limit for uniform rotation with a given cold EOS are called *hypermassive*, and merger simulations find remnants whose largest mass is in this hypermassive range but that can still avoid prompt collapse (see, for example, [81–83] and references therein). Depending on the neutron-star EOS, remnants with mass between 30% and 70% above the upper limit for cold, spherical stars (for the given EOS) can survive until they cool by neutrino emission; by that time magnetic braking is expected to enforce uniform rotation, and the remnant will collapse to a black hole.

Maximum observed mass and the neutron-star EOS

The strongest observational constraint on the neutron-star EOS comes from recent observations of two 2.0 $M_\odot$ neutron stars. Demorest *et al.* [84], using accurate pulsar timing to find orbital parameters and Shapiro time delay from the binary millisecond pulsar PSR J1614–2230, obtain a mass 1.97 ± 0.04 for the pulsar; Antoniadis *et al.* [85] find, with similar claimed precision, a mass $2.01 \pm 0.04\, M_\odot$ for a pulsar–white dwarf system.

One can systematize the observational constraints on the neutron-star EOS by introducing a parametrized EOS above nuclear density with a set of parameters large enough to encompass the wide range of candidate EOSs and small enough that the number of parameters is smaller than the number of relevant observations. Read *et al.* [86] found that one can match a representative set of EOSs above nuclear density to within about 3% rms accuracy with a four-parameter EOS that uses three linear segments to match $\log p$ as a function of $\log \rho$.

EOSs for which $\log p$ is a linear function of $\log \rho$ are called polytropic, and the parametrized EOS thus uses a piecewise-polytropic form to approximate the universe of candidate EOSs. On each of the three density intervals $\rho_0 \leq \rho \leq \rho_1, \rho_1 \leq \rho \leq \rho_2$, and $\rho_2 > \rho$, the parametrized EOS is then $p(\rho) = K_i \rho^{\Gamma_i}$, where Γ_i is constant on $[\rho_{i-1}, \rho_i]$, and the polytropic constant K_i is fixed by continuity. The first law of thermodynamics, in the form $d(\epsilon/\rho) = -p\, d(1/\rho)$, determines the energy density. The dividing densities are $\rho_1 = 10^{14.7}$ g/cm^3 $= 1.85\rho_{\mathrm{nuc}}$, $\rho_2 = 10^{15.0}$ g/cm^3. Following the observation by Lattimer and Prakash [87] that neutron-star radii are closely tied to the pressure somewhat above nuclear density, the value $p_1 = p(\rho_1)$ of the pressure is taken as one parameter, and the remaining parameters can be chosen as the slopes $\Gamma_1, \Gamma_2, \Gamma_3$ or as pressures p_i (with an additional pressure needed above ρ_2).

The region in the EOS parameter space allowed by causality and by the largest observed neutron-star mass is shown in Fig. 3.1, due to B. Lackey, where the darker shaded region is forbidden by causality, the lighter shaded region by the requirement that the maximum mass exceed $2.0\, M_\odot$. Observing a star with a mass of 2.3 or 2.6 $M_\odot$ would restrict the allowed region to lie above surfaces outlined by the higher labeled lines. The piecewise-polytropic form of the EOS is designed to match candidate EOSs only up to the highest central density found in a neutron star – up to the central density of the maximum-mass neutron star for a given EOS. The parametrized EOS is therefore required to be causal only up to that density. As a result, the causality constraint allows large values of p_1; that is, when p_1 is large (when the EOS is stiff near nuclear density), the maximum-mass star occurs before the density ϵ is high enough for the neutrons to be relativistic. Formally, p/ϵ is too small for $c_{\mathrm{sound}}^2 \equiv \Gamma p/(\epsilon + p) > 1$. If one were to demand that the parametrized EOS be causal at all densities, causality would restrict the parameter space to the much smaller region lying in front of the solid black outline that starts as a nearly vertical line on the left at $\Gamma_3 \approx 2$.

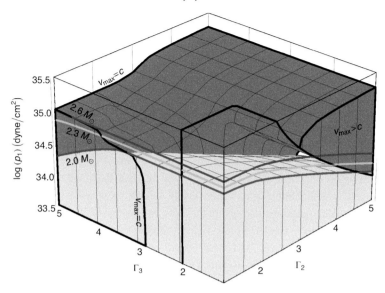

Figure 3.1 Constraints imposed by causality and a maximum mass above $2.0\,M_\odot$ restrict the parameters p_1, Γ_2 and Γ_3 of a parametrized space of equations of state to the part of parameter space lying outside the shaded regions. Labeled solid lines show the more stringent boundaries corresponding to larger maximum masses. The EOS is required to be causal only up to the maximum density of a stable neutron star. A black outline that starts as a nearly vertical line on the left at $\Gamma_3 \approx 2$ marks off a much smaller region in front of it that would be allowed if the parametrized EOS were required to be causal at all densities. The text explains the difference between the two regions.

The extra degrees of freedom associated with strange quarks soften the EOS above the critical density at which they appear (in hyperons or mesons or as unconfined quarks), leading to smaller maximum masses. When $2\,M_\odot$ neutron stars were first observed, most candidate EOSs with hyperons and free quarks were ruled out. There is, however, more than enough uncertainty in the quark-matter EOS to accomodate the stiffness needed for a $2\,M_\odot$ star (see, for example, [88] and [89]), vector mesons may be able to supply a repulsive force large enough to support a $2\,M_\odot$ star with hyperons in its core (see [90] and references therein), and three-nucleon interactions may also be able to stiffen the EOS enough to allow hyperons or quarks [91] (but see [92], for example, for an opposing view).

On the low-mass end, an analysis by Özel *et al.* [93] gives masses of $1.037 \pm 0.085\,M_\odot$ for SMC X-1 and $1.073 \pm 0.358\,M_\odot$ for Her X-1, respectively. Their review of masses is consistent with a range of birth masses peaked near $1.3\,M_\odot$ and a dispersion of about $\frac{1}{4}\,M_\odot$; and it is consistent with the highest-mass neutron stars having gained their extra mass by accretion.

Measurements of neutron-star radii have been made using a variety of observations, and Lattimer [78] provides a detailed summary. The greatest attention has been given to thermally emitting neutron stars (isolated or with negligible accretion), observed as x-ray sources; and to x-ray bursts whose source is the sudden fusion of accreted helium (see, for example, [94] and references therein). In the first case, one deduces a combination of mass and radius from the redshifted flux from a blackbody, corrected for composition-dependent atmospheric effects. In the second, one assumes that, after the burst, the photosphere settles down to the radius of the star. Özel *et al.* [95] use mass and radius measurements to constrain a parametrized EOS similar to that described above; a subsequent analysis by Steiner *et al.* [96] finds a radius between 10.4 and 12.9 km for a 1.4 $M_\odot$ neutron star.

Future gravitational-wave measurements of neutron-star–neutron-star and neutron-star–black-hole inspiral and coalescence by the generation of detectors now under construction are likely to attain similar accuracy. The accuracy of electromagnetic measurement will have improved by then, but the gravitational-wave measurements are model-independent.

If one were able to measure mass and radius for a wide range of neutron-star masses, one could recover the neutron-star EOS [97]. This could be done in the context of a piecewise-polytrope with several segments, but Lindblom and Indik [98, 99] show one can obtain better accuracy with a spectral representation of $\log \Gamma$, by writing $\log \Gamma = \sum_k \gamma_k (\log H/H_0)^k$, where $H = \int_0^p dp/(\epsilon + p) = \log h$. Assuming high-precision measurements of mass and radius for three stars that span a range of neutron-star masses, the method recovers candidate EOSs to within a few percent.

Stability of relativistic stars

The limits on mass and rotation of uniformly rotating stars that we have just discussed look only what equilibrium configurations exist, not at whether they are stable. Instability to collapse sets an upper limit on neutron-star mass that, for rotating stars, is slightly smaller than the most massive equilibrium configuration; and a nonaxisymmetric (CFS) instability driven by gravitational waves sets an upper limit on their rotation that may be related to the fastest observed neutron-star spins. We first review these instabilities and possible physical implications and then turn to a more formal discussion of stability criteria and recent advances.

Axisymmetric instability The best-known instability result in general relativity is the statement that instability to collapse sets in at a point of maximum mass along a sequence of spherical barotropic models (models with a one-parameter EOS that governs the equilibrium and its perturbations). This *turning-point* instability and its generalization to rotating stars can be understood from the first law of thermodynamics for relativistic stars [100]: The difference in mass between two nearby uniformly rotating equilibria is related to the change in entropy S, angular momentum J, and baryon mass M_0 by

$$\delta M = \frac{T}{U} \delta S + \Omega \, \delta J + \mathcal{E} \, \delta M_0. \tag{3.12}$$

For spherical zero-temperature stars, the first law becomes

$$\delta M = \mathcal{E} \, \delta M_0. \tag{3.13}$$

This relation implies that the maximum-mass configuration is also an extremum of baryon mass M_0 and hence that there are models on opposite sides of the turning point (the maximum-mass configuration) with the same baryon mass. At the maximum mass, $\mathcal{E}$ is a decreasing function of central density, implying that models on the *high-density* side of the maximum-mass instability point have larger M for the same M_0 and are therefore unstable. At the minimum mass, it is the *low-density* side that is unstable.

A similar secular instability criterion holds for uniformly rotating stars [101]: Instability to collapse is implied by a point of maximum mass and maximum baryon mass, along a sequence of uniformly rotating barotropic models with fixed angular momentum. As in the spherical case, stars with higher central density than that of the maximum-mass configuration are unstable. For rotating stars, however, the turning point is a sufficient but not a necessary condition for instability: The onset of instability is at a configuration with slightly lower central density (for fixed angular momentum) than that of the maximum-mass star [102]. This is due to the fact that collapse leads to differential rotation, and the turning point identifies only nearby uniformly rotating configurations with lower energy. Rapidly rotating stars are therefore likely to be unstable to collapse at densities slightly lower than the turning-point density.

Although differential rotation in neutron stars is quickly removed by magnetic braking, the stability of differentially rotating stars against collapse is important in understanding the formation of neutron stars and the evolution of stars formed by the coalescence of a double neutron-star system after inspiral. For differentially rotating stars, while there is no simple turning-point criterion to diagnose stability, one can use a criterion based on the sign of the perturbation's canonical energy, and this will be discussed below. The method that has been used in practice, however, is simply to evolve initial data with an accurate code and observe whether or not the star collapses. Because a merged remnant, above the upper mass limit for a uniformly rotating cold neutron star, can be temporarily stabilized by large differential rotation and heat, it should be possible to observe a train of gravitational waves from its post-merger oscillations; see [103] for a first relativistic study, [104], [105] and references therein for more recent work. Temporary stability may also explain the puzzle of supernova 1987A, in which the dramatic observation of a 10-s burst of neutrinos from the birth of a neutron star was followed by a lack of any evidence in the luminosity of the supernova remnant of energy from a rotating, magnetized neutron star [106].

Nonaxisymmetric instability A hard upper limit on the spin of uniformly rotating neutron stars is the Kepler (or mass-shedding) limit Ω_K at which the angular velocity of the star is the angular velocity of a satellite in orbit at its equator. A nonaxisymmetric instability driven by gravitational radiation, the CFS instability [107–109], may impose a more stringent upper limit. If one ignores viscosity, all uniformly rotating perfect-fluid stars are unstable to nonaxisymmetric perturbations driven by gravitational waves. In contrast to axisymmetric instability, however, the growth time for the CFS instability is set by the power radiated in gravitational waves. For slowly rotating stars, the growth time is longer than the viscous damping time, and the instability of all modes is damped out. On the other hand, for rapidly rotating neutron stars, the growth time may be short enough to limit the rotation of old neutron stars spun up by accretion or of nascent neutron stars.

The nature of the instability can be understood from the angular momentum of a mode with angular and time dependence of the form $\cos(m\phi + \omega t)e^{-t/\tau}$. The frequency measured by a corotating observer is

$$\omega_r = \omega + m\Omega, \tag{3.14}$$

and the mode's angular momentum J is negative if the pattern speed $-m/\omega_r$ is negative. If the star rotates with angular velocity greater than $|\omega_r/m|$, this backward-going mode is dragged *forward* relative to an observer at infinity, and ω_r and ω have opposite signs: $\omega_r\omega < 0$. Because the pattern speed seen from infinity is now positive, the mode radiates positive angular momentum. As it does so, J becomes increasingly negative, implying that the amplitude of the mode grows in time: *Gravitational radiation now drives the mode instead of damping it.*

The instability is strikingly different for modes of polar and axial parity.[4] Axial perturbations of a spherical star are time-independent velocity fields – zero-frequency modes. Modes of a rotating star whose spherical limit is axial are called *r*-modes (Rossby modes). In a rotating star, they acquire a nonzero frequency proportional to the star's angular velocity Ω, a frequency whose Newtonian limit has, for a mode with $\ell = m$, the simple form $\omega = -[(\ell - 1)(\ell + 2)/(\ell + 1)]\Omega$. The equation implies that the *r*-mode associated with every nonaxisymmetric multipole obeys the instability condition for every value of Ω: It is forward moving in an inertial frame and backwards moving relative to a rotating observer: $\omega_r = [2/(\ell + 1)]\Omega$, with sign opposite to that of ω. Because the rate at which energy is radiated is greatest for the $\ell = m = 2$ *r*-mode, that is the mode whose instability grows most quickly and which determines whether an axial-parity instability can outpace viscous damping.

The *r*-mode instability was pointed out by Andersson [111], with a mode-independent proof for relativistic stars given by Friedman and Morsink [112]. The first computations of the growth and evolution were reported by Lindblom *et al.* [113] and by Andersson *et al.* [114], with effects of a crust first discussed by Lindblom *et al.* [115]; related references to the substantial body of subsequent work can be found in Bondarescu *et al.* [116].

The instability of low-multipole *r*-modes for arbitrarily slow rotation differs sharply from the behavior of the low-multipole *f*- and *p*-modes, which are unstable only for large values of Ω: Because the frequencies of *f*- and *p*-modes are high for spherical stars, a high angular velocity is needed before a mode that moves backward relative to the star is dragged forward relative to an inertial observer at infinity. Instability points of the polar modes with fastest growth rates are shown in Fig. 3.2.

Evidence for an upper limit on neutron-star spin smaller than the Keplerian frequency Ω_K comes from observed angular velocities populating a range of frequencies below 716 Hz. Chakrabarty [118] argues that the class of sources whose pulses are seen in nuclear

---

[4] Perturbations associated with an ℓ, m angular harmonic have *polar* (*axial*) parity if they have the same (opposite) parity as the function $Y_{\ell m}$, namely $(-1)^\ell$.

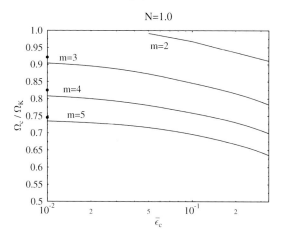

Figure 3.2 Critical angular velocity Ω/Ω_K vs. the dimensionless central energy density $\bar{\epsilon}_c$ for the $m = 2, 3, 4$ and 5 neutral modes of $N = 1.0$ polytropes. The filled circles on the vertical axis are the Newtonian values of the neutral points for each mode. (From [117]. © AAS. Reproduced with permission.)

bursts (nuclear-powered accreting millisecond x-ray pulsars) constitute a sample without significant bias. Although there have in the past been selection biases against detection of the fastest *radio* pulsars, recent searches are sensitive to sub-millisecond periods. They have dramatically increased the numbers of known millisecond radio pulsars, and the absence of radio pulsars above 800 Hz has also become increasingly significant.

A magnetic field of order 10^8 G can limit the spin of an accreting millisecond pulsar. Because matter within the magnetosphere corotates, only matter that accretes from outside the magnetosphere can spin up the star, leading to an equilibrium period given approximately by Ghosh and Lamb [119]:[5] $P_{eq} \sim 2 \times 10^{-3} \, \text{s} \, (B/10^8 \, \text{G})^{6/7} \times (\dot{M}/10^{-10} \, M_\odot \, \text{yr}^{-1})^{-3/7}$. Because this period depends on the magnetic field, a sharp cutoff in the frequency of accreting stars is not an obvious prediction of magnetically limited spins. For a magnetically set maximum rotation rate of order 700–800 Hz the range of magnetic fields would need to have a corresponding minimum cutoff value of about 10^8 G; and the highest observed spin rates should be correlated with the lowest magnetic fields. The required cutoff and a fairly narrow range of observed frequencies have made gravitational-wave-limited spin a competitive possibility for accreting neutron stars. Arguments for and against this based on available observations are given by White [121] and by Patruno [122], respectively.

Under what circumstances the CFS instability could limit the spin of recycled pulsars has now been studied in a large number of papers. References to this work can be found

---

[5] Shapiro and Teukolsky [120] give a clear, simplified version, and this equation is their Eq. (15.2.22), with $M = 1.4 M_\odot$, $R = 10$ km, and a ratio ω_s of the angular velocity to Ω_K at the inner edge of the disk set to 1.

in the treatment in [50] on which the present review is based and in comprehensive earlier discussions by Stergioulas [123], by Andersson and Kokkotas [124], and by Kokkotas and Ruoff [125], while briefer reviews of more recent work are given by Andersson *et al.* [126] and Owen [127].

Whether the instability survives the complex physics of a real neutron star has been the focus of most recent work, but it remains an open question. Studies have focused on dissipation from bulk and shear viscosity and superfluid mutual friction; magnetic field wind-up; nonlinear evolution and the saturation amplitude; and a continuous spectrum possibly replacing *r*-modes in relativistic stars.

When viscosity is included, the growth-time or damping time of an oscillation with time dependence $\cos(\omega t)e^{-t/\tau}$ has the form $\tau^{-1} = \tau_{GR}^{-1} + \tau_b^{-1} + \tau_s^{-1}$, with $-\tau_{GR}$ the growth time due to gravitational radiation and τ_b and τ_s the damping times due to bulk and shear viscosity. Bulk viscosity arises from nuclear reactions driven by the changing density of an oscillating fluid element and is large at high temperatures. Shear viscosity, in contrast, increases as the temperature drops. As shown in Fig. 3.3, this leaves a window of opportunity in which a star with large enough angular velocity can be unstable. The highest solid curves on left and right mark the critical angular velocity Ω_c above which the $\ell = m = 2$ *r*-mode is unstable. The curves on the left show the effect of shear viscosity at low temperature, allowing instability when $\Omega < \Omega_K$ only for $T > 10^6$ K; the curve

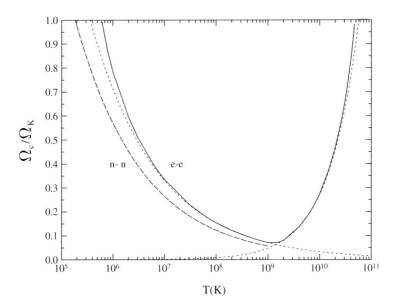

Figure 3.3 Critical angular velocity for the onset of the *r*-mode instability as a function of temperature (for a 1.5 $M_\odot$ neutron-star model). The solid line corresponds to the $O(\Omega^2)$ result using electron–electron shear viscosity and modified URCA bulk viscosity. The dashed line corresponds to the case of neutron–neutron shear viscosity. Dotted lines are $O(\Omega)$ approximations. (From [134]. Reproduced with permission © ESO.)

on the right shows the corresponding effect of bulk viscosity, cutting off the instability at temperatures above about 4×10^{10} K (see also [128–130]).[6]

A quark or hyperon core would add an additional set of nuclear reactions that dissipate energy and increase the bulk viscosity [135–144]. Shear viscosity may be greatly enhanced after formation of the crust in a boundary layer (Ekman layer) between crust and core [115, 145–148]. The enhancement depends on the unknown extent to which the core participates in the oscillation.

Both r-modes and f-modes may be unstable in nascent neutron stars that are rapidly rotating at birth, and recent work on f-modes in relativistic models [149–151] finds growth times substantially shorter than previously computed Newtonian values. In a typical scenario, a star with rotation near the Kepler limit becomes unstable within a minute of formation, when the temperature has dropped below 10^{11} K. As the temperature drops further, the instability grows to saturation amplitude in days or weeks. Loss of angular momentum to gravitational waves spins down the star until the critical angular velocity is reached, below which the star is stable, at or before the time at which the core becomes a superfluid. The evolution of r-modes in nascent stars is similar, but with longer damping times. For reviews and references, see [50, 124, 152].

The r-mode instability of neutron stars spun up by accretion from a companion has been more intensively studied in connection with the observed spins of LMXBs. Wagoner [153], following Papaloizou and Pringle [154], suggested accretion could spin up a star until it becomes unstable, reaching a steady state, with the angular momentum gained by accretion equal to the angular momentum lost to gravitational waves. Although mutual friction appears to rule out the steady-state picture for f-modes, it remains a possibility for r-modes [146, 155–157]. Levin [158] and (independently) Spruit [159], however, pointed out that viscous heating of the neutron star by its unstable oscillations will lower the shear viscosity and so increase the mode's growth rate, leading to a runaway instability. The resulting scenario is a cycle in which a cold, stable neutron star is spun up over a few million years until it becomes unstable; the star then heats up, the instability grows, and the star spins down until it is again stable, all within a few months; the star then cools, and the cycle repeats.

This scenario would rule out r-modes in LMXBs as a source of detectable gravitational waves because the stars would radiate for only a small fraction of the cycle. A small saturation amplitude, however, lengthens the time spent in the cycle, possibly allowing observability [160]. In fact, a small nonlinear saturation amplitude for the r-mode (of order 10^{-4}) appears likely, based on second-order perturbative calculations that take into account three-mode couplings [161–164].

The steady state itself remains a possible alternative in stars whose core contains hyperons or free quarks [136, 138, 139, 141, 157, 165]. Bondarescu *et al.* [166] study nonlinear

---

[6] At temperatures above roughly 10^{10} K, another complication appears: neutrino absorption increases with increasing temperature [131, 132], and the modified URCA bulk viscosity no longer rises, but is reduced by an order of magnitude between 10^{10} K and 10^{11} K, apparently allowing the instability to operate in very hot proto-neutron stars [133].

evolutions, restricted to three coupled modes, that include neutrino cooling, shear viscosity, hyperon bulk viscosity and crust–core coupling. Uncertainty in the magnitudes of these dissipative mechanisms allows alternative scenarios – steady state, cycle, and fast and slow runaways. In all cases, the *r*-mode amplitude remains very small, but, because of the long duration of the instability, such systems are candidates for gravitational wave detection by aLIGO-class interferometers [127, 166, 167].

The time over which a CFS instability is active depends on the saturation amplitude, the cooling rate, and, for young stars, the superfluid transition temperature, and all of these have large uncertainties. The time at which a superfluid transition occurs could be shorter than a year, but recent analyses of the cooling of a neutron star in Cassiopeia A [168, 169] suggest a superfluid transition time for that star of order 100 years. On the other hand, if rapid initial spin leads to a magnetar with a rapidly generated poloidal magnetic field of order 10^{15} G, magnetic spin-down could spin the star down to critical frequency in hours or less.

Instability criteria and formal advances Before discussing the formal criteria for stability, it is helpful to distinguish an instability of a perfect-fluid star from an instability that is present only when there is dissipation other than gravitational radiation (typically shear or bulk viscosity, mutual friction, or heat flow). We adopt (with slightly different terminology) a definition in recent work of Green *et al.* [170].

A stationary solution to the Einstein–Euler equations is *weakly dynamically unstable* if some smooth, asymptotically flat solution to the linearized Einstein-fluid equations grows without bound (in some suitable gauge).

Secular instabilities are instabilities with longer-than-dynamical timescales. As we are using the term, weak dynamical instabilities include a class of *secular* instabilities driven by gravitational waves (CFS instabilities), with longer-than-dynamical timescales.

Weak dynamical instability is governed by the sign of the energy of a perturbation that preserves the total baryon number and entropy of the star and the angular momentum of each ring of fluid. One obtains an action for stellar perturbations by introducing a Lagrangian displacement ξ^α joining each unperturbed fluid trajectory (the unperturbed worldline of a fluid element) to the corresponding trajectory of the perturbed fluid. Using conservation of baryons and assuming that the perturbation is adiabatic, one can express the perturbation in all fluid variables, p, ϵ, and u^α, in terms of ξ^α and the metric perturbation $h_{\alpha\beta}$. An action for the Einstein–Euler system has the form

$$\int \left(\frac{1}{16\pi}{}^4R - \epsilon\right)\sqrt{|g|}d^4x. \qquad (3.15)$$

From Eq. (3.15), one obtains an action for the perturbed system,

$$I^{(2)} = \int d^4x \, \mathcal{L}^{(2)}. \qquad (3.16)$$

The conserved canonical energy E_c associated with the timelike Killing vector of the background spacetime is the Hamiltonian of the perturbation, expressed in terms of configuration space variables,

$$E_c = \int_S d^3x \, (\Pi^\alpha \, \mathcal{L}_t \xi_\alpha + \pi^{\alpha\beta} \, \mathcal{L}_t h_{\alpha\beta} - \mathcal{L}^{(2)}), \qquad (3.17)$$

where Π^α and $\pi^{\alpha\beta}$ are the momenta conjugate to ξ^α and $h_{\alpha\beta}$. The symbol $\mathcal{L}_t$ is the Lie derivative along the symmetry vector t^α; with components written in t, r, θ, ϕ coordinates, it is simply ∂_t.

Along a family Σ_t of asymptotically null hypersurfaces, E_c decreases monotonically in time due to the radiation of energy to future null infinity. This leads to a criterion for stability, but one must first address a subtlety associated with a gauge freedom in the choice of the Lagrangian displacement ξ^α. There is a set of trivial displacements that leave all fluid variables unchanged, but which change the value of E_c; one eliminates the gauge freedom by demanding that ξ^α be *canonical*, symplectically orthogonal to all trivial displacements. The stability criterion can then be stated as follows.

1. If $E_c < 0$ for some canonical data on Σ, then the configuration is unstable or marginally stable: There exist perturbations that do not die away in time along a family of asymptotically null hypersurfaces that foliate the spacetime.
2. If $E_c \geq 0$ for all canonical data on Σ, the magnitude of E_c is bounded in time and only finite energy can be radiated.

For radial perturbations of a spherical star the criterion was in effect used by Chandrasekhar [171] in his early paper on dynamical stability of spherical relativistic stars. In that case, because there is no gravitational radiation, the growth time of an unstable radial mode is dynamical, not dissipative; and for rotating stars the contribution of radiation to the growth time remains negligible for the corresponding fundamental quasi-radial mode.

Note that this *dynamical* instability point does not exactly coincide with the instability located by the turning-point criterion: A sequence of stars with increasing central density is *secularly* unstable before the point of dynamical instability is reached. This is because the turning-point method identifies nearby equilibria that have lower energy but are not accessible to an evolution that obeys the dynamical equations with an adiabatic perturbation (conserving the entropy of each fluid element). Violating conservation of entropy allows nuclear reactions to occur. Similarly, for rotating stars, one must violate conservation of circulation to make a transition to lower-energy uniformly rotating equilibria. Because the timescales for nuclear reactions and heat flow are vastly shorter than the age of observed neutron stars, it is the less stringent secular stability line that bounds the region of neutron stars stable against collapse.

The precise statement of the turning-point criterion for rotating stars is the following result, a corollary to a theorem of Sorkin [172]:

Theorem 3.1.1 *[101]. Consider a two-parameter family of uniformly rotating stellar models having an equation of state of the form $p = p(\epsilon)$. Suppose that along a continuous* *sequence of models labeled by a parameter* λ, *there is a point* λ_0 *at which both* $\dot{M}_0$ *and* $\dot{J}$ *vanish and where* $(\dot{\Omega}\dot{J}+\dot{\mathcal{E}}\dot{M}_0)^{\cdot}\neq 0$. *Then the part of the sequence for which* $\dot{\Omega}\dot{J}+\dot{\mathcal{E}}\dot{M}_0 > 0$ *is unstable for* λ *near* λ_0.

In two recent papers, Green *et al.* [170] and Schiffrin and Wald [173] re-examine axisymmetric stability and the turning-point theorem, extending the canonical energy criterion to perfect-fluid spacetimes with black holes, and clarifying the relation between thermodynamic and dynamic stability. They define thermodynamic stability as follows: A star in thermodynamic equilibrium is *thermodynamically stable* if S is a maximum; it is thermodynamically unstable if there is a perturbation with $\delta M = \delta J = \delta M_0 = 0$ that raises the entropy at quadratic order: $\delta^2 S > 0$. They prove that a necessary condition for thermodynamic stability with respect to axisymmetric perturbations is positivity of E_c on all axisymmetric linearized solutions within the Lagrangian framework that have $\delta J = 0$.

The criterion for the nonaxisymmetric (CFS) instability is again $E_c < 0$ for perfect-fluid models. This is independent of the existence of discrete quasinormal modes, but for a normal mode of a uniformly rotating star it coincides with the criterion discussed above, that the real part of a mode's frequency in an inertial frame has sign opposite to its value in a rotating frame. More precisely [50]:

Theorem 3.1.2 *Consider an outgoing mode that varies smoothly along a family of uniformly rotating perfect-fluid equilibria, labeled by* λ. *Assume that it has t and* ϕ *dependence of the form* $e^{i(m\phi+\sigma t)}$, *that* $\omega = \mathrm{Re}\,\sigma$ *satisfies* $\omega_r \equiv \omega/m + \Omega > 0$ *for all* λ, *and that the sign of* ω/m *is positive for* $\lambda < \lambda_0$ *and negative for* $\lambda > \lambda_0$. *Then, in a neighborhood of* λ_0, $\mathrm{Im}\,\sigma \leq 0$; *and, if the mode has at least one nonzero asymptotic multipole moment with* $l \geq 2$ *at future null infinity, the mode is unstable* ($\mathrm{Im}\,\sigma < 0$) *for* $\lambda > \lambda_0$.

References

[1] Eddington, A. S. 1935. *The Observatory*, **58**, 37–39.
[2] Chandrasekhar, S. 1931. *Astrophys. J.*, **74**, 81.
[3] Anderson, W. 1929. *Z. Phys.*, 851–856.
[4] Stoner, E. C. 1930. *Phil. Mag.*, **9**, 944.
[5] Landau, L. D. 1932. *Phys. Z. Sowjetunion*, **1**, 285.
[6] Rosenfeld, L. 1974. Page 174 of: *Astrophysics and gravitation, Proc. Solvay Conference on Physics, 16th*. Brussels: Univ. Bruxelles.
[7] Yakovlev, D. G., Haensel, P., Baym, G., and Pethick, C. 2013. *Sov. Phys. Uspekhi*, **56**, 289–295.
[8] Thorne, K. S., and Żytkow, A. N. 1975. *Astrophys. J. Lett.*, **199**, L19–L24.
[9] Baade, W., and Zwicky, F. 1934. *Phys. Rev.*, **4**, 138.
[10] Gamow, G. 1939. *Phys. Rev.*, **55**, 718–725.
[11] Chandrasekhar, S. 1939. Pages 17–23 of: *Conférences du Collège de France, Colloque International d'Astrophysique*. Paris: Hermann.
[12] Tolman, R. C. 1939. *Phys. Rev.*, **55**, 364–373.
[13] Oppenheimer, J. R., and Volkoff, G. M. 1939. *Phys. Rev.*, **55**, 374–381.
[14] Cameron, A. G. 1959. *Astrophys. J.*, **130**, 884.
[15] Harrison, B. K., Wakano, M., and Wheeler, J. A. 1958. In: *Onzième Conseil de Physique Solvay, La Stucture et l'evolution de l'universe*. Brussels: Stoop.

[16] Oppenheimer, J. R., and Snyder, H. 1939. *Phys. Rev.*, **56**, 455–459.
[17] Finkelstein, D. 1958. *Phys. Rev.*, **110**, 965–967.
[18] Eddington, A. S. 1924. *Nature*, **113**, 192.
[19] Lemaître, G. 1933. *Annales de la Société Scientifique de Bruxelles A*, **53**, 51–85.
[20] Synge, J. L. 1950. *Proceedings of the Royal Irish Academy. Section A: Mathematical and Physical Sciences*, **53**, 83–114.
[21] Smith, H. J., and Hoffleit, D. 1963. *Nature*, **198**, 650–651.
[22] Sharov, A. S., and Efremov, Y. N. 1963. *Information Bulletin on Variable Stars*, **23**, 1.
[23] Shklovsky, I. S. 1965. *Sov. Astron.*, **9**, 224.
[24] Shklovskii, I. S. 1966. *Sov. Astron.*, **10**, 6.
[25] Pacini, F. 1967. *Nature*, **216**, 567–568.
[26] Hewish, A. *et al.* 1968. *Nature*, **217**, 709–713.
[27] Staelin, D. H., and Reifenstein, III, E. C. 1968. *Science*, **162**, 1481–1483.
[28] Reifenstein, E. C., Brundage, W. D., and Staelin, D. H. 1969. *Phys. Rev. Lett.*, **22**, 311–311.
[29] Lovelace, R., Sutton, J., and Craft, H. 1968. *IAU Astronomical Telegram Circular*, **2113**.
[30] Gold, T. 1968. *Nature*, **218**, 731–732.
[31] Richards, D. W., and Comella, J. M. 1969. *Nature*, **222**, 551.
[32] Hoyle, F., and Fowler, W. A. 1965. Report on the properties of massive objects. Page 17 of: Robinson, I., Schild, A., and Schucking, E. L. (eds), *Quasi-stellar sources and gravitational collapse*.
[33] Zel'dovich, Ya. B., and Novikov, I. D. 1966. *Sov. Phys. Uspekhi*, **8**, 522–577.
[34] Burbidge, G. 1972. *Comments Astrophys. Space Phys.*, **4**, 105.
[35] Shklovsky, I. S. 1967. *Astrophys. J.*, **148**, L1.
[36] Cameron, A. B. W., and Mock, M. 1967. *Nature*, **215**, 464–466.
[37] Prendergast, K. H., and Burbidge, G. R. 1968. *Astrophys. J. Lett.*, **151**, L83.
[38] Salpeter, E. E. 1964. *Astrophys. J.*, **140**, 796–800.
[39] Blandford, R. D., and Thorne, K. S. 1979. Pages 454–503 of: Hawking, S. W., and Israel, W. (eds), *General relativity: an Einstein centenary survey*. Cambridge: Cambridge University Press.
[40] Abramowicz, M. A., and Fragile, P. C. 2013. *Living Rev. Rel.*, **16**.
[41] Camenzind, M. 2007. *Compact objects in astrophysics: white dwarfs, neutron stars and black holes*. Berlin: Springer-Verlag.
[42] Brecher, K., and Morrison, P. 1973. *Astrophys. J. Lett.*, **180**, L107.
[43] Lamb, D. Q., and van Horn, H. M. 1973. *Astrophys. J.*, **183**, 959–966.
[44] Wheeler, J. C. 2000. *Cosmic catastrophes*. Cambridge: Cambridge University Press.
[45] Paczyński, B. 1986. *Astrophys. J. Lett.*, **308**, L43–L46.
[46] Renn, J., Sauer, T., and Stachel, J. 1997. *Science*, **275**, 184–186.
[47] Chwolson, O. 1924. *Astron. Nachr.*, **221**, 329.
[48] Einstein, A. 1936. *Science*, **84**, 506–507.
[49] Zwicky, F. 1937. *Phys. Rev. Lett.*, **51**, 290.
[50] Friedman, J. L., and Stergioulas, N. 2013. *Rotating relativistic stars*. Cambridge: Cambridge University Press.
[51] Chugunov, A. I., and Horowitz, C. J. 2010. *Mon. Not. R. Astron. Soc.*, **407**, L54–L58.
[52] Baym, G., and Chandler, E. 1983. *J. Low Temp. Phys.*, **50**, 57–87.
[53] Sonin, E. B. 1987. *Rev. Mod. Phys.*, **59**, 87–155.
[54] Lindblom, L., and Mendell, G. 1994. *Astrophys. J.*, **421**, 689–704.
[55] Wilson, J. R. 1972. *Astrophys. J.*, **176**, 195–204.
[56] Bonazzola, S., and Schneider, S. 1974. *Astrophys. J.*, **191**, 195–290.
[57] Butterworth, E. M., and Ipser, J. R. 1976. *Astrophys. J.*, **204**, 200–223.
[58] Komatsu, H., Eriguchi, Y., and Hachisu, I. 1989. *Mon. Not. R. Astron. Soc.*, **237**, 355–379.
[59] Komatsu, H., Eriguchi, Y., and Hachisu, I. 1989. *Mon. Not. R. Astron. Soc.*, **239**, 153–171.
[60] Cook, G. B., Shapiro, S. L., and Teukolsky, S. A. 1992. *Astrophys. J.*, **398**, 203–223.
[61] Cook, G. B., Shapiro, S. L., and Teukolsky, S. A. 1994. *Astrophys. J.*, **422**, 227–242.
[62] Cook, G. B., Shapiro, S. L., and Teukolsky, S. A. 1994. *Astrophys. J.*, **424**, 823–845.

[63] Bonazzola, S., Gourgoulhon, E., Salgado, M., and Marck, J.-A. 1993. *Astron. Astrophys.*, **278**, 421–443.
[64] Bonazzola, S., Gourgoulhon, E., and Marck, J.-A. 1998. *Phys. Rev. D*, **58**, 104020.
[65] Ansorg, M., Kleinwächter, A., and Meinel, R. 2002. *Astron. Astrophys.*, **381**, L49–L52.
[66] Stergioulas, N. 1997. *RNS,* available at www.gravity.phys.uwm.edu/rns.
[67] Gourgoulhon, E. *et al.* 1997. *LORENE,* available at http://www.lorene.obspm.fr.
[68] Haensel, P., and Zdunik, J. L. 1989. *Nature*, **340**, 617–619.
[69] Friedman, J. L., Ipser, J. R., and Parker, L. 1989. *Phys. Rev. Lett.*, **62**, 3015–3019.
[70] Haensel, P., Salgado, M., and Bonazzola, S. 1995. *Astron. Astrophys.*, **296**, 746–751.
[71] Lasota, J., Haensel, P., and Abramowicz, M. A. 1996. *Astrophys. J.*, **456**, 300–304.
[72] Hartle, J. B., and Sabbadini, A. G. 1977. *Astrophys. J.*, **213**, 831–835.
[73] Rhoades, C. E., and Ruffini, R. 1974. *Phys. Rev. Lett.*, **32**, 324–327.
[74] Friedman, J. L., and Ipser, J. R. 1987. *Astrophys. J.*, **314**, 594–597.
[75] Koranda, S., Stergioulas, N., and Friedman, J. L. 1997. *Astrophys. J.*, **488**, 799–806.
[76] Kalogera, V., and Baym, G. 1996. *Astrophys. J. Lett.*, **470**, L61.
[77] Glendenning, N. K. 1992. *Phys. Rev. D*, **46**, 4161–4168.
[78] Lattimer, J. M. 2012. *Ann. Rev. Nuclear Particle Sci.*, **62**, 485–515.
[79] Lindblom, L. 1984. *Astrophys. J.*, **278**, 364–368.
[80] Baumgarte, T. W., Shapiro, S. L., and Shibata, M. 2000. *Astrophys. J. Lett.*, **528**, L29–L32.
[81] Bauswein, A., Baumgarte, T. W., and Janka, H.-T. 2013. *Phys. Rev. Lett.*, **111**, 131101.
[82] Hotokezaka, K. *et al.* 2011. *Phys. Rev. D*, **83**, 124008.
[83] Hotokezaka, K. *et al.* 2013. *Phys. Rev. D*, **88**, 044026.
[84] Demorest, P. B. *et al.* 2010. *Nature*, **467**, 1081–1083.
[85] Antoniadis, J. *et al.* 2013. *Science*, **340**, 448.
[86] Read, J. S., Lackey, B. D., Owen, B. J., and Friedman, J. L. 2009. *Phys. Rev. D*, **79**, 124032.
[87] Lattimer, J. M., and Prakash, M. 2001. *Astrophys. J.*, **550**, 426.
[88] Alford, M. *et al.* 2007. *Nature*, **445**.
[89] Kurkela, A., Romatschke, P., Vuorinen, A., and Wu, B., 2010. arXiv:1006.4062.
[90] Weissenborn, S., Chatterjee, D., and Schaffner-Bielich, J. 2012. *Phys. Rev. C*, **85**, 065802.
[91] Gandolfi, S., Carlson, J., and Reddy, S. 2012. *Phys. Rev. C*, **85**, 032801.
[92] Schulze, H.-J., and Rijken, T. 2011. *Phys. Rev. C*, **84**, 035801.
[93] Özel, F., Psaltis, D., Narayan, R., and Santos Villarreal, A. 2012. *Astrophys. J.*, **757**, 55.
[94] Güver, T., Psaltis, D., and Özel, F. 2012. *Astrophys. J.*, **747**, 76.
[95] Özel, F., Baym, G., and Güver, T. 2010. *Phys. Rev. D*, **82**, 101301.
[96] Steiner, A. W., Lattimer, J. M., and Brown, E. F. 2013. *Astrophys. J. Lett.*, **765**, L5.
[97] Lindblom, L. 1992. *Astrophys. J. Lett.*, **398**, 569–573.
[98] Lindblom, L. 2010. *Phys. Rev. D*, **82**, 103011.
[99] Lindblom, L., and Indik, N. M. 2014. *Phys. Rev. D*, **89**, 064003.
[100] Bardeen, J. M. 1970. *Astrophys. J.*, **162**, 71–95.
[101] Friedman, J. L., Ipser, J. R., and Sorkin, R. D. 1988. *Astrophys. J.*, **325**, 722–724.
[102] Takami, K., Rezzolla, L., and Yoshida, S. 2011. *Mon. Not. R. Astron. Soc.*, **416**, L1–L5.
[103] Shibata, M., and Uryū, K. 2000. *Phys. Rev. D*, **61**, 064001.
[104] Hotokezaka, K. *et al.* 2013. *Phys. Rev. D*, **88**, 044026.
[105] Bauswein, A., Janka, H.-T., Hebeler, K., and Schwenk, A. 2012. *Phys. Rev. D*, **86**, 063001.
[106] Prakash, M., Lattimer, J. M., Sawyer, R. F., and Volkas, R. R. 2001. *Ann. Rev. Nuclear Particle Sci.*, **51**, 295–344.
[107] Chandrasekhar, S. 1970. *Phys. Rev. Lett.*, **24**, 611–615.
[108] Friedman, J. L., and Schutz, B. F. 1978. *Astrophys. J.*, **222**, 281–296.
[109] Friedman, J. L. 1978. *Commun. Math. Phys.*, **62**, 247–278.
[110] Vallis, G. K. 2006. *Atmospheric and oceanic fluid dynamics: fundamentals and large-scale circulation.* Cambridge: Cambridge University Press.
[111] Andersson, N. 1998. *Astrophys. J.*, **502**, 708–713.
[112] Friedman, J. L., and Morsink, S. M. 1998. *Astrophys. J.*, **502**, 714–720.
[113] Lindblom, L., Owen, B. J., and Morsink, S. M. 1998. *Phys. Rev. Lett.*, **80**, 4843–4846.

[114] Andersson, N., Kokkotas, K. D., and Schutz, B. F. 1999. *Astrophys. J.*, **510**, 846–853.
[115] Lindblom, L., Owen, B. J., and Ushomirsky, G. 2000. *Phys. Rev. D*, **62**, 084030.
[116] Bondarescu, R., Teukolsky, S. A., and Wasserman, I. 2009. *Phys. Rev. D*, **79**, 104003.
[117] Stergioulas, N., and Friedman, J. L. 1998. *Astrophys. J.*, **492**, 301–322.
[118] Chakrabarty, D. 2008. The spin distribution of millisecond X-ray pulsars. Pages 67–74 of: Wijnands, R., *et al.* (eds), *A decade of accreting millisecond X-ray pulsars.* AIP Conference Proceedings, vol. 1068.
[119] Ghosh, P., and Lamb, F. K. 1979. *Astrophys. J.*, **232**, 259–276.
[120] Shapiro, S. L., and Teukolsky, S. A. 1983. *Black holes, white dwarfs, and neutron stars.* New York: Wiley.
[121] White, N. E., and Zhang, W. 1997. *Astrophys. J.*, **490**, L87–L90.
[122] Patruno, A., Haskell, B., and D'Angelo, C. 2012. *Astrophys. J.*, **746**, 9.
[123] Stergioulas, N. 2003. *Living Rev. Rel.*, **6**, 3.
[124] Andersson, N., and Kokkotas, K. D. 2001. *Int. J. Mod. Phys. D*, **10**, 381–441.
[125] Kokkotas, K. D., and Ruoff, J. 2002. Instabilities of relativistic stars. In: *2001: A relativistic spacetime Odyssey.* 25th Johns Hopkins Workshop. Florence 2001.
[126] Andersson, N. *et al.* 2011. *Gen. Rel. Grav.*, **43**, 409–436.
[127] Owen, B. J. 2010. *Phys. Rev. D*, **82**, 104002.
[128] Ipser, J. R., and Lindblom, L. 1991. *Astrophys. J.*, **379**, 285–289.
[129] Ipser, J. R., and Lindblom, L. 1991. *Astrophys. J.*, **373**, 213–221.
[130] Yoshida, S., and Eriguchi, Y. 1995. *Astrophys. J.*, **438**, 830–840.
[131] Lai, D., and Shapiro, S. L. 1995. *Astrophys. J.*, **442**, 259–272.
[132] Bonazzola, S., Frieben, J., and Gourgoulhon, E. 1996. *Astrophys. J.*, **460**, 379–389.
[133] Lai, D. 2001. Secular bar-mode evolution and gravitational waves from neutron stars. Pages 246–257 of: *Astrophysical sources for ground-based gravitational wave detectors.* AIP Conference Proceedings, vol. 575.
[134] Kokkotas, K. D., and Stergioulas, N. 1999. *Astron. Astrophys.*, **341**, 110–116.
[135] Jones, P. B. 2001. *Phys. Rev. Lett.*, **86**, 1384–1384.
[136] Lindblom, L., and Owen, B. J. 2002. *Phys. Rev. D*, **65**, 063006.
[137] Haensel, P., Levenfish, K. P., and Yakovlev, D. G. 2002. *Astron. Astrophys.*, **381**, 1080–1089.
[138] Nayyar, M., and Owen, B. J. 2006. *Phys. Rev. D*, **73**, 084001.
[139] Haskell, B., and Andersson, N. 2010. *Mon. Not. R. Astron. Soc.*, **408**, 1897–1915.
[140] Madsen, J. 1998. *Phys. Rev. Lett.*, **81**, 3311–3314.
[141] Andersson, N., Jones, D. I., and Kokkotas, K. D. 2002. *Mon. Not. R. Astron. Soc.*, **337**, 1224–1232.
[142] Jaikumar, P., Rupak, G., and Steiner, A. W. 2008. *Phys. Rev. D*, **78**, 123007.
[143] Rupak, G., and Jaikumar, P. 2010. *Phys. Rev. C*, **82**, 055806.
[144] Alford, M., Mahmoodifar, S., and Schwenzer, K. 2012. *Phys. Rev. D*, **85**, 024007.
[145] Bildsten, L., and Ushomirsky, G. 2000. *Astrophys. J.*, **529**, L33–L36.
[146] Andersson, N., Jones, D. I., Kokkotas, K. D., and Stergioulas, N. 2000. *Astrophys. J.*, **534**, L75–L78.
[147] Glampedakis, K., and Andersson, N. 2006. *Phys. Rev. D*, **74**, 044040.
[148] Glampedakis, K., and Andersson, N. 2006. *Mon. Not. R. Astron. Soc.*, **371**, 1311–1321.
[149] Passamonti, A., Gaertig, E., Kokkotas, K. D., and Doneva, D. 2013. *Phys. Rev. D*, **87**, 084010.
[150] Gaertig, E., and Kokkotas, K. D. 2011. *Phys. Rev. D*, **83**, 064031.
[151] Gaertig, E., Glampedakis, K., Kokkotas, K. D., and Zink, B. 2011. *Phys. Rev. Lett.*, **107**, 101102.
[152] Alford, M. G., and Schwenzer, K. 2014. *Astrophys. J.*, **781**, 26.
[153] Wagoner, R. V. 1984. *Astrophys. J.*, **278**, 345–348.
[154] Papaloizou, J., and Pringle, J. E. 1978. *Mon. Not. R. Astron. Soc.*, **184**, 501–508.
[155] Bildsten, L. 1998. *Astrophys. J.*, **501**, L89–L93.
[156] Andersson, N., Kokkotas, K. D., and Stergioulas, N. 1999. *Astrophys. J.*, **516**, 307–314.
[157] Wagoner, R. V. 2002. *Astrophys. J.*, **578**, L63–L66.
[158] Levin, Y. 1999. *Astrophys. J.*, **517**, 328–333.

118   *Peter Schneider*

[159] Spruit, H. C. 1999. *Astron. Astrophys.*, **341**, L1–L4.
[160] Heyl, J. 2002. *Astrophys. J.*, **574**, L57–L60.
[161] Morsink, S. M. 2002. *Astrophys. J.*, **571**, 435–446.
[162] Arras, P. *et al.* 2003. *Astrophys. J.*, **591**, 1129–1151.
[163] Schenk, A. K. *et al.* 2002. *Phys. Rev. D*, **65**, 024001.
[164] Brink, J., Teukolsky, S. A., and Wasserman, I. 2004. *Phys. Rev. D*, **70**, 121501.
[165] Reisenegger, A., and Bonacić, A. 2003. *Phys. Rev. Lett.*, **91**, 201103.
[166] Bondarescu, R., Teukolsky, S. A., and Wasserman, I. 2007. *Phys. Rev. D*, **76**, 064019.
[167] Watts, A. L., and Krishnan, B. 2009. *Adv. Space Res.*, **43**, 1049–1054.
[168] Page, D., Prakash, M., Lattimer, J. M., and Steiner, A. W. 2011. *Phys. Rev. Lett.*, **106**, 081101.
[169] Shternin, P. S. *et al.* 2011. *Mon. Not. R. Astron. Soc.*, **412**, L108–L112.
[170] Green, S. R., Schiffrin, J. S., and Wald, R. M. 2014. *Class. Quant. Grav.*, **31**, 035023.
[171] Chandrasekhar, S. 1964. *Astrophys. J.*, **140**, 417.
[172] Sorkin, R. D. 1981. *Astrophys. J.*, **249**, 254–257.
[173] Schiffrin, J. S., and Wald, R. M. 2014. *Class. Quant. Grav.*, **31**, 035024.

3.2 Gravitational lensing
Peter Schneider

3.2.1 Introduction

Gravitational lensing, the effects caused by gravitational light deflection in the weak-field regime,[7] began to flourish as a field after the discovery of the 'double QSO' in 1979 (Walsh, Carswell & Weymans 1979), the first multiply imaged quasi-stellar object (QSO) lensed by a massive foreground elliptical galaxy, soon followed by several more serendipitously found multiply imaged objects. Giant luminous arcs in galaxy clusters were found in the mid 1980s [1, 2], followed by spectroscopic verification as images of background sources [3], demonstrating their lensing nature (see Fig. 3.4 for an example). Several ring-shaped images, first seen in radio observations [4], later in infrared observations of multiply imaged active galaxies, showed the existence of Einstein rings. Galactic microlensing surveys commenced in the early 1990s and showed that at most a small fraction of the dark matter in our Milky Way can be in the form of compact objects. The statistical methods of weak gravitational lensing were first applied to the galaxy population and clusters in the mid 1990s, and in 2000, weak lensing by the large-scale structure was first discovered. Galactic microlensing surveys detected extrasolar planets starting from 2005, and QSO microlensing resolved the innermost emission regions of active galaxies. By now, gravitational lensing is an indispensable tool in astrophysics, which allows one to study properties of the lens (e.g., its mass), the source, and the properties of spacetime in between.

This brief contribution cannot do justice to the richness of the field and its broad range of applications. The fundamentals of gravitational lensing are described in the monograph [5], as well as in [6] and the review [7]. The Saas Fee lectures [8] provide a broad overview over

---

[7] Light deflections near neutron stars and black holes are not included in the field of gravitational lensing.

Figure 3.4 The cluster of galaxies A 2218 at redshift $z_d = 0.175$ contains one of the most spectacular arc systems, i.e., the highly elongated images seen here. The majority of the galaxies visible in the image are associated with the cluster, and the redshifts of many of the strongly distorted arcs have now been measured. These sources are located behind the cluster, some of them at a rather high redshift, and the observed shapes are strongly distorted by light deflection in the tidal gravitational field of the cluster (Credit: W. Couch/University of New South Wales, R. S. Ellis/Cambridge University and NASA).

much of the field. References to reviews on specific topics will be given in due course; we also refer to the resource letter [9] for a very useful compilation of literature on gravitational lensing.

3.2.2 Basic theory

In gravitational lensing, the gravitational fields occurring are assumed to be weak, so that the superposition principle can be applied. Hence, the deflection angle $\hat{\alpha}$ a light ray is subject to if it passes a mass concentration is a convolution of the density distribution with the Einstein deflection angle for a point mass. If, in addition, the linear extent of the deflector is much smaller than its distance D_d from the observer, and the distance D_{ds} from the lens to the source, the deflection angle of a ray with impact vector ξ,

$$\hat{\alpha}(\xi) = \frac{4G}{c^2} \int d^2\xi' \, \Sigma(\xi') \frac{\xi - \xi'}{|\xi - \xi'|^2} , \qquad (3.18)$$

depends only on the surface mass density $\Sigma(\xi')$ of the deflector. This 'geometrically thin lens' approximation is valid for (almost) all known lens systems except for lensing by the large-scale structure (see below); however, the large-scale inhomogeneous matter distribution between source and lens, and lens and observer, adds some small contributions to the effective deflection, as will be discussed later.

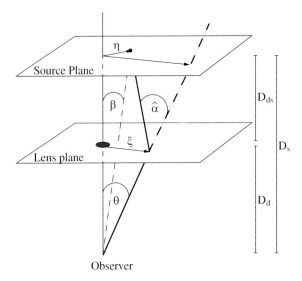

Figure 3.5 Geometry of a gravitational lens system. The source is located at a distance D_s from us, behind a mass concentration at distance D_d. An optical axis is defined as a line through the observer and the 'center' of the mass concentration and used to define the origin in the source and lens planes, planes perpendicular to the optical axis. A source at location η in the source plane is seen at location ξ in the lens plane, with corresponding angular position $\beta = \eta/D_s$, $\theta = \xi/D_d$. The relation (3.19) between these two angles follows from simple geometry (adapted from: P. Schneider, J. Ehlers & E. E. Falco 1992, *Gravitational Lenses*, Springer-Verlag).

The deflection defines a mapping from the direction θ at which a light ray from a source reaches us, to the direction β where the source would be seen in the absence of light deflection, reading (see Fig. 3.5)

$$\beta = \theta - \frac{D_{ds}}{D_s}\,\hat{\alpha}(D_d\theta) \equiv \theta - \alpha(\theta)\,, \qquad (3.19)$$

where D_s is the distance to the source (all distances D are angular diameter distances, according to their definition), and α is the scaled deflection angle, related to the deflection potential $\psi(\theta) = \pi^{-1} \int d^2\theta'\,\kappa(\theta')\,\ln|\theta - \theta'|$ by $\alpha = \nabla\psi$. Here,

$$\kappa(\theta) := \frac{\Sigma(D_d\theta)}{\Sigma_{cr}} \quad \text{with} \quad \Sigma_{cr} = \frac{c^2}{4\pi G}\frac{D_s}{D_d\,D_{ds}}\,; \qquad (3.20)$$

is the dimensionless surface mass density, or *convergence*, and we defined the critical surface mass density Σ_{cr} which depends only on the distances involved. Lenses with $\kappa \sim 1$ at some points are called *strong lenses*, those with $\kappa \ll 1$ everywhere *weak lenses*. Multiple solutions θ_i of the lens equation (3.19) for a given source position β correspond to multiple images of the source. The deflection potential satisfies the two-dimensional Poisson equation $\nabla^2\psi = 2\kappa$.

Since surface brightness is unaffected by light deflection, the fluxes of the images of an infinitesimally small source are changed, relative to the flux of the unlensed source, by the magnification $\mu = \det^{-1}(\mathcal{A})$, where

$$\mathcal{A}(\boldsymbol{\theta}) \equiv \frac{\partial \boldsymbol{\beta}}{\partial \boldsymbol{\theta}} = \begin{pmatrix} 1 - \kappa - \gamma_1 & -\gamma_2 \\ -\gamma_2 & 1 - \kappa + \gamma_1 \end{pmatrix} = (1 - \kappa) \begin{pmatrix} 1 - g_1 & -g_2 \\ -g_2 & 1 + g_1 \end{pmatrix} \quad (3.21)$$

is the Jacobian of the lens mapping, $\gamma = \gamma_1 + i\gamma_2 = (\psi_{,11} - \psi_{,22})/2 + i\psi_{,12}$ is the shear, describing the tidal component of the deflection field, and $g = \gamma/(1 - \kappa)$ is the reduced shear. The magnification of extended sources is the surface-brightness-weighted average of the point-source magnification. The sign of $\det(\mathcal{A})$ yields the parity of the image. To first order, a circular source is mapped onto an ellipse, with its axes determined by the eigenvalues λ_i of $\mathcal{A}$, where $\lambda_{1,2} = 1 - \kappa \pm |\gamma|$. The orientation of the image ellipse is determined by the phase of γ.

Curves on which $\det \mathcal{A}(\boldsymbol{\theta}) = 0$ are called critical curves; their images in the source plane under the lens mapping are called caustics. The number of images changes by ± 2 if the source position changes across a caustic. Whereas $\det \mathcal{A} = 0$ formally implies infinite magnification, the actual magnifications are of course finite, due to the finite source size.

3.2.3 Strong lensing by galaxies

Currently, several hundred strong lens systems where a galaxy lenses a background source, either an active galactic nucleus (AGN) or a normal galaxy (see Fig. 3.6 for examples), are known; see [10] and [11] for extended reviews on the subject. In most cases, a compact source is mapped onto two or four observable images,[8] whereas extended sources can lead to more complex image geometries, including large arcs or even complete rings. The characteristic image separation in these systems is about 1".

Properties of the mass distribution

Multiple images of background sources allow constraints on the mass distribution of the deflector. Specifically, if N images of an unresolved source are observed, the mass distribution κ must be such as to satisfy the $(N - 1)$ independent two-dimensional equations $\boldsymbol{\theta}_i - \boldsymbol{\alpha}(\boldsymbol{\theta}_i) = \boldsymbol{\theta}_j - \boldsymbol{\alpha}(\boldsymbol{\theta}_j)$. For extended sources, the constraints on κ are such that points $\boldsymbol{\theta}_i$ on the sky that are associated via the lens equation have the same surface brightness, after effects of the point-spread function and noise are accounted for. The number of constraints is rather limited in general, and therefore the resulting mass distribution is not unique. In particular, if κ is constant on confocal ellipses, the deflection at some point $\boldsymbol{\theta}$ depends solely on the mass distribution inside the ellipse on which $\boldsymbol{\theta}$ lies, and thus the density distribution is constrained only in the inner region of the deflector where the multiple images are located.

---

[8] A general theorem states that a transparent, smooth, non-singular lens produces an odd number of images; the absence of the third or fifth image in most observed systems is attributed to the small magnification μ expected for this odd, central image, making it unobservable.

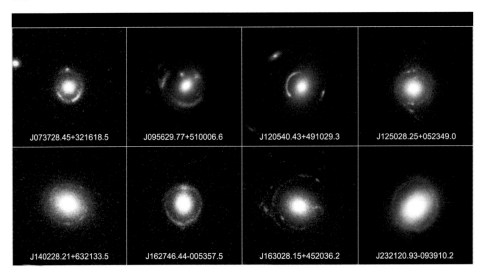

Figure 3.6 Eight strong lens systems, as observed by the Hubble Space Telescope (HST). In all cases, the lens is an elliptical galaxy, and background sources are multiply imaged, where the images are highly elongated or even mapped onto a full Einstein ring. Such extended images can probe the gravitational potential of the lens at far more locations than a few point-like images, and thus potentially provide more information about the mass distribution (Credit: NASA, ESA, and the SLACS Survey team: A. Bolton (Harvard/Smithsonian), S. Burles (MIT), L. Koopmans (Kapteyn), T. Treu (UCSB), and L. Moustakas (JPL/Caltech))

Typically, the mass distribution of the lens is parametrized, using plausible mass models compatible with observations. The most robust quantity obtained from such modeling is the mass inside the Einstein ring, a circle centered on the lens galaxy which traces the multiple images.[9] Depending on the image configuration – four images yield more robust constraints than two-image systems – the Einstein radius θ_E can be determined very precisely, to within a few percent or less. The mass inside the Einstein radius is given by $M(< \theta_E) = \pi D_d^2 \theta_E^2 \Sigma_{cr}$, and can thus be readily inferred from the Einstein radius and the redshifts of source and lens. This mass depends on the cosmological parameters, but the current small uncertainties in H_0 and the density parameters render this the most precise mass determination in extragalactic astronomy. We point out that this is the mass projected inside a cylinder, not the mass inside a sphere with angular radius θ_E, as is true for all mass estimates from lensing. Estimating the stellar mass from the luminosity and spectral properties of lens galaxies, one finds that about half (somewhat dependent on the assumed stellar initial mass function) of the mass inside the Einstein radius is dark, with an increased dark matter fraction for more massive lens galaxies.

---

[9] A source located directly behind a sufficiently compact axi-symmetric lens forms a ring-like image; its radius θ_E is called the Einstein radius.

Combining this lensing mass estimate with stellar dynamics, which typically estimate the mass at the effective radius[10] of an early-type lens galaxy, the mean slope of the density profile between the Einstein and effective radius can be estimated. It turns out to be nearly isothermal, i.e., $\rho \propto r^{-\gamma}$, with $\gamma \approx 2$, and some indication that the mass profile is flatter for higher-redshift galaxies; see [12] and references therein.

Time delay and Hubble constant

The light travel time from the source to the observer in general is different for the various multiple images. The corresponding difference – the time delay – between images i and j can be written in terms of the Fermat potential $\tau(\boldsymbol{\theta}) := (\boldsymbol{\theta} - \boldsymbol{\beta})^2/2 - \psi(\boldsymbol{\theta})$ as

$$\Delta t = \frac{D_\mathrm{d} D_\mathrm{s}}{c\, D_\mathrm{ds}}(1 + z_\mathrm{d})\left[\tau(\boldsymbol{\theta}_i) - \tau(\boldsymbol{\theta}_j)\right], \tag{3.22}$$

and hence is proportional to the Hubble radius c/H_0, since all cosmological distances determined from redshift are proportional to c/H_0. Thus, from the measurement of the time delay, which is possible for sources which vary intrinsically, as almost all QSOs do, one can estimate the Hubble constant [13], provided the density profile of the deflector is sufficiently well constrained. Indeed, time delays are currently known for more than a dozen lens systems, and for some of them, very detailed lens models have been constructed, using extended multiple image components mapped with high angular resolution; see [14] and references therein. However, the lens models are not unique; indeed, it is easy to show that the mass model $\kappa(\boldsymbol{\theta})$ and the family of mass models $\kappa_\lambda(\boldsymbol{\theta}) := \lambda\kappa(\boldsymbol{\theta}) + (1 - \lambda)$ predict exactly the same multiple image positions, magnification (and thus flux) ratios, and image shapes, for all λ. However, the predicted time delay is $\propto \lambda$. This so-called mass-sheet degeneracy (MSD) leaves the Einstein radius (and thus the mass it encloses) invariant, as well as the critical curves and the shape of the isodensity contours; therefore the angular structure (such as ellipticity) of lenses can be well constrained in systems with extended image structure. The MSD cannot be broken from lensing observations alone (unless the lens system contains two or more sources at vastly different redshifts, or if the magnification can be determined, e.g., from independent knowledge about the source luminosity), but requires additional mass estimates, such as from stellar dynamics. Unfortunately, those are currently not accurate enough to yield precision estimates for H_0, unless additional assumptions are employed [15]. On the other hand, combining accurate measurements of H_0 from other methods with time delays allows the most precise estimates of the slope of the total mass profile in the inner regions of galaxies, at about the transition radius between baryon and dark matter dominance.

Substructure in lens systems

Whereas simple parametrized mass models can reproduce the observed image positions very accurately (even down to $\sim 10^{-3}\theta_\mathrm{E}$ precision) in most systems, they typically fail

---

[10] The radius inside which half of the total light of the galaxy is emitted.

to explain the observed flux ratios. For sources located close to a caustic, the predicted flux ratio of the corresponding pair or triplet of highly magnified images follows from the universal mapping properties near critical points and is independent of the chosen mass model, provided it is 'smooth'. Observed violations of this universal property thus indicate the presence of mass substructure whose minimum mass scale depends on the source size [16]. Since the deflection angle depends on the first derivatives of ψ, whereas the magnification contains second derivatives, a small-scale density fluctuation in the lens can affect the magnification without appreciably changing the image positions. In particular, the magnification $\mu = 1/(\lambda_1\lambda_2)$ of images near a critical curve, for which one of the eigenvalues λ_i of $\mathcal{A}$ is close to zero, can be strongly affected by a small additional contribution to κ or γ.

Whereas for the optical and X-ray emission in QSOs, the graininess of the mass provided by stars in the lens galaxies can significantly affect the magnification (see below), stars cannot account for flux ratio anomalies in radio sources, due to their larger source size. Hence, the characteristic mass scale of the substructure must exceed that of stars. There is a long and as yet unsettled debate in the literature about whether the flux ratio anomalies are due to the dark matter subhalos which are predicted from cold dark matter (CDM) models [17] to be more abundant than the observed satellite galaxies (which has frequently been considered a problem for the CDM model) – see, e.g., [18–20], and references therein.

QSO microlensing

As mentioned before, the graininess of the mass distribution caused by stars in lensing galaxies can affect the magnification of images, provided their extent is not much larger than the Einstein radius of stars, corresponding to a few micro-arcseconds. The peculiar velocity of source, lens and observer, and the random motion of stars in the lens galaxy, cause the magnification to change in time, and these changes are uncorrelated between the different images – in contrast to intrinsic variability, which is seen in all images with a corresponding time delay. The properties of these microlensing light curves depend on the relative peculiar velocities, stellar masses and stellar surface mass densities, but the shortest time-scale is defined by the time it takes the source to cross a caustic. Within the statistical uncertainties of the estimated transverse velocities, the size of the emitting region in QSOs can thus be estimated, so that microlensing yields a microscopic view of the innermost region of the central engine.

The original idea of 'star disturbances' on image magnifications in lens systems was presented in [21]; detailed discussions of the astrophysical applications of microlensing, and ray-tracing simulations for obtaining synthetic light curves are found in [22–24]. A quantitative method for comparing these model light curves with observed ones is described in [25]. Results from such microlensing studies, which require long-term monitoring of the lensed QSO images, include (see [26] and references therein): The size of the optical–ultraviolet continuum emitting region is consistent with the standard optically thick accretion disk model (for luminous QSOs, this corresponds to a few light-days), whereas the X-ray emission comes from a more compact source, not much larger than corresponding to the last stable orbit around the supermassive black hole. Broad emission lines are only weakly affected by microlensing, indicating a larger size (as expected), with the high-ionization lines being most affected – i.e., being smallest.

3.2.4 Galactic microlensing

If the dark matter in our Milky Way was composed of compact objects, like stars or black holes, they could lead to observable lensing effects. B. Paczyński suggested an experimental method to search for astrophysical dark matter in the Galaxy [27]: If a large sample of distant compact sources is photometrically monitored, some of these will experience a lensing event by a compact object which moves through its line-of-sight due to relative transverse velocities. Whereas the image splitting is too small to be observable (of order a milliarcsecond), time-dependent magnification leads to a characteristic variability. The corresponding lightcurve is characterized by a mere four parameters – unlensed flux, time of maximum, maximum magnification, and duration t_E of the event – where only the latter is of astrophysical interest, depending on a combination of lens mass M, relative transverse velocity v, and distance D of the lens (with the source distance being known), $t_E \propto \sqrt{MD}/v$. For characteristic values of these quantities, t_E is about one month – hence, such microlensing events can be followed in one or a few observing seasons. The probability for any distant source to undergo a microlensing event is small, $\tau \sim 10^{-7}$, even if the dark matter in the Milky Way is entirely composed of Massive Astrophysical Compact Halo Objects (MACHOs). Hence, one needs to monitor a large set of sources, for which the stars in the Magellanic Clouds are the best targets. To distinguish microlensing events from stellar variability, the characteristic shape of the light curve and its achromaticity can be employed.

Starting in the 1990s, such microlensing surveys were conducted, with first events reported in 1993 by three different teams [28–30]. Since then, the field has flourished, although with a shift of emphasis. Some of the most significant results are summarized below; the reader is referred to the reviews [31–33].

Whereas about 20 microlensing events have been found towards the Magellanic Clouds, the inferred mass density of MACHOs is far smaller than the dark matter density needed to explain the Galactic rotation curve. Since only the combination t_E of relevant quantities can be measured for any single event, there is ambiguity regarding the location of the lens. At least in some of the cases, the lens is either close to us, i.e., a member of the Galactic stellar disk, or located in the Magellanic Clouds themselves, causing so-called self-lensing. One can rule out any appreciable MACHO contribution to the Galactic dark matter. Some ten thousand microlensing events have been found towards the Galactic bulge, where the lens is either a star in the disk or in the bulge itself. The distribution of these microlensing events on the sky led to the independent discovery that our bulge contains a bar, oriented some 30° away from our line-of-sight, in agreement with results from infrared photometry. Statistics of event durations t_E can be used to measure the stellar mass function.

Some 10% of the events deviate from the simple light curve prediction, due to a number of reasons: Many stars are members of a binary system, and if the binary separation is comparable to the Einstein radius of the lens, the light curves become significantly more complex as the source moves through the caustic pattern. During caustic crossing, the finite size of the source star becomes important, and can be resolved; indeed, limb-darkening effects were observed, providing an observational test of stellar atmosphere theory. Model fitting to such binary events yields the mass ratio of the binary components and their separation (in units of the Einstein radius). When the time-scale t_E of the event is much longer than a month, the motion of the Earth around the Sun affects the shape of the light curve,[11] and introduces another length scale into the problem, which allows breaking of the degeneracy between M, v and D. Such events were used to infer the self-lensing nature of several MACHO candidates. Similarly, if the source star is a member of a binary, its orbital motion can as well affect the light curve.

Much of the recent emphasis in microlensing has shifted to the search for extrasolar planets. In this context, a star with a planet is a binary with very large mass ratio. Its presence is indicated by a disturbance of the lightcurve which occurs on small time-scales. In order to detect these disturbances, the lightcurve of microlensing events needs to be monitored with high time resolution, which is achieved by employing several telescopes distributed in geographical longitude (so that 24 hour coverage is possible).

The selection function of planets via microlensing differs from that of other methods; in particular, microlensing is more sensitive to low-mass planets than the radial velocity method. Some 20 extrasolar planets have been found so far, including at least two with masses below 10 Earth masses. Furthermore, whereas other methods search planets around stars, microlensing offers the opportunity to detect free-floating planets, i.e., those without a host star; several candidates have been discovered [34].

3.2.5 Weak gravitational lensing

The Jacobian $\mathcal{A}$ of the mapping (3.19) distorts image shapes and sizes. The latter effect is described by the magnification; the shape distortion is caused by the shear γ. A source with ellipticity ϵ^s is mapped onto an image with ellipticity

$$\epsilon = \frac{\epsilon^s + g}{1 + g^* \epsilon^s} , \quad \text{where} \quad g = \frac{\gamma}{1 - \kappa} \tag{3.23}$$

is the reduced shear, and the ellipticity of an image with elliptical isophotes of constant axis ratio $r \leq 1$ reads $\epsilon = [(1 - r)/(1 + r)] e^{2i\varphi}$, where φ is the polar angle of the major axis and the asterisk denotes complex conjugation. Since no direction is preferred, the expectation

---

[11] In this case, the relative motion of source, lens and observer is no longer linear, as assumed for deriving the 'standard' microlensing lightcurves.

value $\langle \epsilon^s \rangle = 0$, due to phase averaging, yielding $\langle \epsilon \rangle = g$: Each measured image ellipticity yields an unbiased estimate of the reduced shear along its line-of-sight. The estimate though is noisy, due to the intrinsic ellipticity distribution of sources. Therefore, in order to beat the noise, weak lensing necessarily requires a statistical approach, employing large samples of images. Hence, weak lensing studies require very wide fields and/or deep observations to increase the number (density) of sources. Faint distant sources are, however, small, necessitating observations at the best astronomical sites on Earth or from space.

A major challenge of weak lensing is a proper measurement of image ellipticities on noisy, pixelized images affected by the point-spread function, in combination with the fact that the isophotes of distant galaxies are not elliptical in general. There is an ongoing world-wide effort to improve on current methods; see [35] and references therein. With increasing statistical power of weak lensing surveys, the systematic effects need to be reduced or controlled accordingly.

Besides the shear effect, weak lensing can also be studied using the magnification. For $|\gamma| \ll 1, |\kappa| \ll 1$, the magnification differs little from unity, $\mu = 1 + \delta\mu$. The magnification affects the source counts $n(> S)$ of distant objects, relative to the unlensed counts $n_0(> S)$, by $n(> S) = n_0(> S/\mu)/\mu$, where the first term describes the change of the unlensed flux threshold S under the impact of magnification, whereas the second factor accounts for the stretching of the observed sky. For small $\delta\mu$, a first-order Taylor expansion yields

$$\frac{n(> S)}{n_0(> S)} \approx 1 + (\alpha - 1)\,\delta\mu \approx 1 + 2(\alpha - 1)\kappa , \qquad (3.24)$$

where in the second step we used $\mu \approx 1 + 2\kappa$, and $\alpha = -d \ln n_0/d \ln S$ is the local slope of the unlensed counts. Hence, the source counts can either be enhanced or depleted by magnification, depending on α. For steep counts and $\kappa > 0$, the number density of observed sources is increased. Source counts are easier to obtain than shape measurements; furthermore, magnification effects break the MSD. The drawback of this method is the necessity to exclude all possible sources which might be physically related (i.e., correlated) with the lens (population). Magnification effects have been observed, using high-redshift Lyman-break galaxies as sources, since for those contamination by lower-redshift galaxies is minimal; see, e.g., [36] and references therein.

The main applications of weak lensing will be summarized in this section. For extensive reviews of the field, see [37–40].

Galaxy clusters

Since the shear around massive clusters of galaxies is rather strong, they can be studied individually with weak lensing. By combining results from strong and weak lensing, the mass distribution of clusters can be obtained over a wide range of radii; see, e.g., [41–43] and references therein.

The estimate of the reduced shear from image ellipticities allows a parameter-free recon-struction of the surface mass density κ of clusters. Since both κ and γ are second-order derivatives of ψ, they are linearly related, and the relation is most easily expressed in Fourier space, $\hat{\gamma}(\boldsymbol{\ell}) = \hat{\kappa}(\boldsymbol{\ell})\,e^{2i\beta}$, where β is the polar angle of $\boldsymbol{\ell}$ [44]. From an estimate of the shear field γ, κ can be reconstructed, up to an additive constant (which is due to the fact that the foregoing relation is undefined for $\boldsymbol{\ell} = \mathbf{0}$). Likewise, from an estimate of the reduced shear g, one can reconstruct $\ln(1-\kappa)$ up to an additive constant, i.e., κ up to the previously mentioned MSD. Since lensing measures the projected mass density, foreground and background density fluctuations due to the large-scale structure of the matter in the Universe contribute to the signal, acting as a source of noise. This effect provides a limit to the accuracy with which the mass properties of individual clusters can be determined.

Arguably, the most important weak lensing result from clusters concerns the direct proof of the existence of collisionless dark matter in so-called 'bullet clusters' [45]. Due to a recent collision of two clusters, the X-ray emitting intracluster gas is displaced from the two galaxy concentrations, as gas is affected by friction. Since the intracluster gas contains ~ 5 times more mass than the stars in galaxies, in the absence of dark matter the mass would be concentrated on the gas. Weak lensing mass maps, however, show that the matter concentrations are coincident with those of the galaxies, showing that dark matter behaves similarly to the galaxies, i.e., the behavior is collisionless. Alternative gravity models, which have been suggested to avoid the need for the presence of dark matter to explain the rotation curves of spiral galaxies, cannot account for the lensing results of these bullet clusters.

In contrast to galaxies, the mass distribution in clusters is expected to be less subject to baryonic effects, and hence their density profile should follow approximately the universal density profile of cold dark matter halos. Combining strong lensing in the inner part of clusters, using multiple images and highly elongated arcs (for which the high-resolution and deep HST images yielded spectacular results), with weak lensing measurements, one finds that the mean projected density profile of clusters is compatible with the universal density profile. Whereas the concentration is on average higher than the CDM predictions, this is to a large extent due to a combination of halo triaxiality, projection and selection effects; see, e.g., [46].

The abundance of giant arcs depends on the number density and the mass distribution of clusters, which in turn are cosmology-dependent. Hence, by comparing the observed number of arcs with theoretical expectations, cosmological constraints can be derived. Whereas the first results from arc statistics seemed to be in conflict with models [47], accounting for the complex mass structure of clusters relieved that tension. The current state of arc statistics is summarized in [48].

Cosmic shear: lensing by the large-scale structure (LSS)

The inhomogeneous mass distribution in the Universe leads to small statistical distortions of light bundles. These distortions are correlated, due to the correlation of the LSS; hence, the correlation of the distortions yields information about the statistical properties of the LSS, as pointed out by [49] and [50].

Any second-order statistics of the shear (such as the shear two-point correlation function) can be expressed as a weighted integral over the power spectrum $P_\delta(k, z)$ of the density fluctuations, which in turn depends on the cosmological model. Furthermore, the weight function depends on the distance–redshift relation, and thus on the expansion history of the Universe. If one has redshift information for the source galaxies, these second-order shear statistics can be obtained for different redshift bins, in which case one obtains several differently-weighted integrals over the power spectrum. These observables together therefore probe the primordial power spectrum, the growth factor of density fluctuations, and the expansion history. The latter two depend also on the equation-of-state of dark energy, parametrized in the form $p = w\rho c^2$, with $w = -1$ corresponding to a cosmological constant; indeed, cosmic shear is viewed as one of the most promising methods to constrain the properties of dark energy [51, 52]. Based on these findings, the European Space Agency has selected Euclid as the next cosmology mission [53]. In addition, higher-order shear statistics carry additional independent cosmological information. The best studied ones of those are the three-point shear correlation and the shear peak abundance. Shear peaks are circular shear patterns similar to those caused by a mass concentration. Therefore, the sensitivity of the abundance of shear peaks to cosmology is similar to that of the cluster abundance.

The first detections of cosmic shear were reported in 2000, from four independent groups [54–57], based on little more than 1 deg^2 of imaging data. Larger surveys, with more refined analysis methods, soon followed. The current state-of-the-art in this field is defined by the COSMOS survey from space [58], and the CFHTLenS survey from the ground (see [59] and references therein), the latter providing ~ 150 deg^2 of five-waveband high-quality imaging data. The main quantitative result from current surveys concerns the normalization σ_8 of the power spectrum, or, more precisely, a combination of the form $\sigma_8 \Omega_m^\alpha$, with $\alpha \approx 0.5$. Combined with cosmic microwave background (CMB) anisotropy data, the degeneracy between these two parameters can be broken, and constraints on w can be obtained. The ~ 100 times larger and deeper Euclid survey, together with its higher angular resolution and near-infrared photometry (important for estimating galaxy redshifts), will substantially narrow down the allowed range of w to within $\sim 1\%$.

Beside issues of shear estimates, several other challenges need to be mastered. These include the possible intrinsic alignment of galaxies with the tidal gravitational field (which can give rise to spurious ellipticity correlations), a galaxy selection bias [60], and in particular the comparison with model predictions, due to the impact of less well understood baryonic effects. Furthermore, the calculation of realistic covariances will provide a major challenge for cosmological simulations, in particular when higher-order shear information is included in the analysis; therefore, efficient data compression is required.

Whereas nearly all weak lensing studies employ faint optical galaxies as sources, most recently the weak lensing effect was detected in the cosmic microwave background. Lensing induces non-Gaussian features on its temperature distribution, which can be analyzed and used to measure the lensing power spectrum [61]. The lensing nature of the signal was verified by correlating the signal with source populations of different redshift distributions.

The measurement of the lensing effect contributed greatly to the cosmological results from the Planck satellite, as it allows one to break degeneracies which are present when considering only the CMB temperature power spectrum.

The future Square Kilometer Array (SKA) will allow the observation of a large number density of radio sources on the sky; together with the well-controlled point-spread function of the radio interferometer, the SKA may become an extremely powerful observatory for weak gravitational lensing studies.

The relation between mass and light: biasing

Whereas galaxy clusters are massive enough for us to study them individually with weak lensing, this is not the case for galaxies and groups. However, the mean weak lensing signal of a population of galaxies or groups can be measured by superposition, measuring the mean shear around this population. This method is known as galaxy–galaxy lensing (GGL). First detected in 1996 [62], it is considered the most direct measurement of the (mean) mass profiles of galaxies and groups, and can be followed out to large radii.

One can roughly distinguish between three different regimes: On small scales, GGL probes the mean density profile of galaxies, i.e., their concentrated baryonic component and the dark matter halo in which they are embedded. The signal clearly shows that the dark matter halo of galaxies is far more extended than their stellar distribution, with a halo size compatible with CDM models. On intermediate scales, the fact that most galaxies are members of groups or clusters becomes visible, and the lensing signal can be dominated by this host component. On even larger scales, one notes that dark matter halos are correlated, and the shear signal becomes dominated by these correlated structures. At these large scales, the GGL measures the correlation between galaxies and the LSS. The correlation properties of galaxies can be different from that of the overall mass distribution, an effect called galaxy biasing. Combining the GGL signal with measured galaxy correlation functions, this bias can be probed directly, as a function of galaxy luminosity, type, and redshift.

Due to its large sky coverage and the availability of spectroscopic redshifts for $\sim 10^6$ galaxies, the Sloan Digital Sky Survey has been most important for GGL; see [63] and references therein. Furthermore, GGL in the HST COSMOS survey allowed GGL measurements to high redshifts [64]. Results include that the bias factor increases with galaxy luminosity, or stellar mass, and that the stellar-to-total mass ratio exhibits a pronounced maximum at a mass corresponding to the cut-off scale of the stellar mass function. The bias is expected to approach a constant for large spatial scales, in which case the combination of GGL and galaxy clustering can be used to determine the clustering properties of the total mass distribution and thus to derive cosmological constraints, complementing cosmology from cosmic shear. In addition, GGL provides a valuable tool for mitigating intrinsic alignment effects in cosmic shear.

From measuring the shear signal around the position of groups and clusters, the mean mass of these systems is obtained, from which the mass–luminosity relation is derived; see, e.g., [65]. This method can thus be used to calibrate the scaling relations between observables of clusters (like X-ray luminosity) and their mass, an important ingredient to cluster cosmology.

3.2.6 Natural telescopes

Whereas the magnification of any real source remains finite even if it is located on a caustic, due to the finite source size, the magnification can be very large in some cases. The best evidence for this is the occurrence of giant luminous arcs whose length-to-width (L/W) ratio can exceed 50. For typical mass profiles, the radial magnification is close to unity, so that the total magnification is of the order of L/W. For some microlensing events, magnifications well in excess of 100 or more were observed. This magnification is of specific astrophysical interest, as it allows one to study in detail sources which, without magnification, would be too faint – a magnification μ reduces the required exposure time for a spectrum at fixed signal-to-noise by a factor of μ^2.

From the universal behavior of the lens mapping near critical curves it can be shown that the probability density for magnification μ behaves like $p(\mu) \propto \mu^{-3}$, up to a cut-off given by the properties of the lens and the source size. The observed source counts $n(> S)$ are obtained by convolving $p(\mu)$ with the unlensed counts $n_0(> S)$, $n(> S) = \int d\mu\, \mu^{-1} p(\mu)\, n_0(> S/\mu)$. The luminosity functions of many astrophysical objects display a power-law behavior with an exponential cut-off (a Schechter-like function). For fluxes corresponding to luminosities many times larger than this cut-off, the observed source counts are then dominated by highly magnified objects. Indeed, all bright Lyman-break galaxies at $z \sim 3$ are highly magnified by a lens, as is true for all of the brightest sub-millimeter galaxies, and for many of the apparently most luminous QSOs.

One can systematically employ this effect, by searching for faint distant objects in regions of the sky known to exhibit large magnifications – the central regions of massive clusters. Some of the deepest surveys (in different wavebands) were conducted in these regions of the sky to detect the most distant sources. Indeed, many of the highest-redshift candidate galaxies known have been detected behind lensing clusters; e.g., [66] and references therein.

3.2.7 Concluding remarks

Gravitational light deflection has evolved from a crucial test of General Relativity in 1919 to an indispensable tool for astrophysics and cosmology. Some of its applications have long been foreseen, others came as welcome surprises. The field continues to flourish, and its various facets have evolved into individual research fields with their own communities, methods, and sometimes jargon. Furthermore, the use of the lensing effect has spread out to many other communities, from extrasolar planets, stellar astrophysics, galaxies, active galactic nuclei, galaxy clusters, and cosmology. Upcoming surveys with existing telescopes and new observatories will guarantee the continued fruitful exploitation of this basic, but nevertheless powerful, consequence of General Relativity.

References

[1] Lynds, R., and Petrosian, V. 1989. *Astrophys. J.*, **336**, 1–8.
[2] Soucail, G., Fort, B., Mellier, Y., and Picat, J. P. 1987. *Astron. Astrophys.*, **172**, L14–L16.
[3] Soucail, G. *et al.* 1988. *Astron. Astrophys.*, **191**, L19–L21.
[4] Hewitt, J. N. *et al.* 1988. *Nature*, **333**, 537–540.
[5] Schneider, P., Ehlers, J., and Falco, E. E. 1992. *Gravitational lenses*. Astronomy and Astrophysics Library. Berlin, Heidelberg and New York: Springer-Verlag.
[6] Seitz, S., Schneider, P., and Ehlers, J. 1994. *Class. Quant. Grav.*, **11**, 2345–2373.
[7] Bartelmann, M. 2010. *Class. Quant. Grav.*, **27**, 233001.
[8] Meylan, G. *et al.* (eds). 2006. *Gravitational lensing: strong, weak and micro*. Berlin and Heidelberg: Springer-Verlag.
[9] Treu, T., Marshall, P. J., and Clowe, D. 2012. *Amer. J. Phys.*, **80**, 753–763.
[10] Kochanek, C. S. 2006. Part 2: Strong gravitational lensing. Pages 91–268 of: Meylan, G. *et al.* (eds), *Saas-Fee Advanced Course 33: gravitational lensing: strong, weak and micro*.
[11] Treu, T. 2010. *Ann. Rev. Astron. Astrophys.*, **48**, 87–125.
[12] Bolton, A. S. *et al.* 2012. *Astrophys. J.*, **757**, 82.
[13] Refsdal, S. 1964. *Mon. Not. Roy. Astr. Soc.*, **128**, 307.
[14] Suyu, S. H. *et al.* 2013. *Astrophys. J.*, **766**, 70.
[15] Schneider, P., and Sluse, D. 2013. *Astron. Astrophys.*, **559**, A37.
[16] Mao, S., and Schneider, P. 1998. *Mon. Not. Roy. Astr. Soc.*, **295**, 587.
[17] Moore, B. *et al.* 1999. *Astrophys. J. Lett.*, **524**, L19–L22.
[18] Dalal, N., and Kochanek, C. S. 2002. *Astrophys. J.*, **572**, 25–33.
[19] Metcalf, R. B., and Amara, A. 2012. *Mon. Not. Roy. Astr. Soc.*, **419**, 3414–3425.
[20] Xu, D. D. *et al.*, 2015. *Mon. Not. Roy. Astr. Soc.*, **447**, 3189–3206.
[21] Chang, K., and Refsdal, S. 1979. *Nature*, **282**, 561–564.
[22] Paczyński, B. 1986. *Astrophys. J.*, **301**, 503–516.
[23] Kayser, R., Refsdal, S., and Stabell, R. 1986. *Astron. Astrophys.*, **166**, 36–52.
[24] Schneider, P., and Weiss, A. 1987. *Astron. Astrophys.*, **171**, 49–65.
[25] Kochanek, C. S. 2004. *Astrophys. J.*, **605**, 58–77.
[26] Mosquera, A. M. *et al.* 2013. *Astrophys. J.*, **769**, 53.
[27] Paczyński, B. 1986. *Astrophys. J.*, **304**, 1–5.
[28] Alcock, C. *et al.* 1993. *Nature*, **365**, 621–623.
[29] Aubourg, E. *et al.* 1993. *Nature*, **365**, 623–625.
[30] Udalski, A. *et al.* 1993. *Acta Astron.*, **43**, 289–294.
[31] Mao, S. 2012. *Res. Astron. Astrophys.*, **12**, 947–972.
[32] Wambsganss, J. 2006. Part 4: Gravitational microlensing. Pages 453–540 of: Meylan, G. *et al.* (eds), *Saas-Fee Advanced Course 33: gravitational lensing: strong, weak and micro*.
[33] Gaudi, B. S. 2012. *Ann. Rev. Astron. Astrophys.*, **50**, 411–453.
[34] Sumi, T. *et al.* 2011. *Nature*, **473**, 349–352.
[35] Mandelbaum, R. *et al.* 2014. *Astrophys. J. Suppl.*, **212**, 5.
[36] Hildebrandt, H. *et al.* 2013. *Mon. Not. Roy. Astr. Soc.*, **429**, 3230–3237.
[37] Bartelmann, M., and Schneider, P. 2001. *Phys. Reports*, **340**, 291–472.
[38] Schneider, P. 2006. Part 3: Weak gravitational lensing. Pages 269–451 of: Meylan, G. *et al.* (eds), *Saas-Fee Advanced Course 33: gravitational lensing: strong, weak and micro*.
[39] Hoekstra, H., and Jain, B. 2008. *Ann. Rev. Nucl. Particle Sci.*, **58**, 99–123.

[40] Munshi, D., Valageas, P., van Waerbeke, L., and Heavens, A. 2008. *Phys. Reports*, **462**, 67–121.

[41] Kneib, J.-P., and Natarajan, P. 2011. *Astron. Astrophys. Rev.*, **19**, 47.

[42] Massey, R., Kitching, T., and Richard, J. 2010. *Rep. Prog. Phys.*, **73**, 086901.

[43] Hoekstra, H. *et al.* 2013. *Space Sci. Rev.*, **177**, 75–118.

[44] Kaiser, N., and Squires, G. 1993. *Astrophys. J.*, **404**, 441–450.

[45] Clowe, D. *et al.* 2006. *Astrophys. J. Lett.*, **648**, L109–L113.

[46] Oguri, M. *et al.* 2012. *Mon. Not. Roy. Astr. Soc.*, **420**, 3213–3239.

[47] Bartelmann, M. *et al.* 1998. *Astron. Astrophys.*, **330**, 1–9.

[48] Meneghetti, M., Bartelmann, M., Dahle, H., and Limousin, M. 2013. *Space Sci. Rev.*, **177**, 31–74.

[49] Blandford, R. D., Saust, A. B., Brainerd, T. G., and Villumsen, J. V. 1991. *Mon. Not. Roy. Astr. Soc.*, **251**, 600–627.

[50] Kaiser, N. 1992. *Astrophys. J.*, **388**, 272–286.

[51] Albrecht, A. *et al.*, 2006. arXiv:astro-ph/0609591.

[52] Peacock, J. A. *et al.* 2006. arXiv:astro-ph/0610906.

[53] Laureijs, R. *et al.*, 2011. arXiv:1110.3193.

[54] Van Waerbeke, L. *et al.* 2000. *Astron. Astrophys.*, **358**, 30–44.

[55] Wittman, D. M. *et al.* 2000. *Nature*, **405**, 143–148.

[56] Bacon, D. J., Refregier, A. R., and Ellis, R. S. 2000. *Mon. Not. Roy. Astr. Soc.*, **318**, 625–640.

[57] Kaiser, N., Wilson, G., and Luppino, G. A., 2000. arXiv:astro-ph/0003338.

[58] Schrabback, T. *et al.* 2010. *Astron. Astrophys.*, **516**, A63.

[59] Heymans, C. *et al.* 2012. *Mon. Not. Roy. Astr. Soc.*, **427**, 146–166.

[60] Hartlap, J., Hilbert, S., Schneider, P., and Hildebrandt, H. 2011. *Astron. Astrophys.*, **528**, A51.

[61] Ade, P. A. R. *et al.* [Planck collaboration]. 2015. *Astron. Astrophys.*, **571**, A17.

[62] Brainerd, T. G., Blandford, R. D., and Smail, I. 1996. *Astrophys. J.*, **466**, 623.

[63] Mandelbaum, R. *et al.* 2013. *Mon. Not. Roy. Astr. Soc.*, **432**, 1544–1575.

[64] Leauthaud, A. *et al.* 2012. *Astrophys. J.*, **744**, 159.

[65] Johnston, D. E. *et al.*, 2007. arXiv:0709.1159.

[66] Coe, D. *et al.* 2013. *Astrophys. J.*, **762**, 32.

3.3 Observational evidence for black holes

Ramesh Narayan and Jeffrey E. McClintock

3.3.1 Historical introduction

Astronomers have discovered two populations of black holes: (i) stellar-mass black holes with masses in the range 5 to 30 solar masses, millions of which are present in each galaxy in the universe, and (ii) supermassive black holes with masses in the range 10^6 to 10^{10} solar masses, one each in the nucleus of every galaxy. There is strong circumstantial evidence that all these objects are true black holes with event horizons. The measured masses of supermassive black holes are strongly correlated with properties of their host galaxies, suggesting that these black holes, although extremely small in size, have a strong influence on the formation and evolution of entire galaxies. Spin parameters have recently been measured for a number of black holes. Based on the data, there is an indication that the kinetic power of at least one class of relativistic jet ejected from accreting black holes may

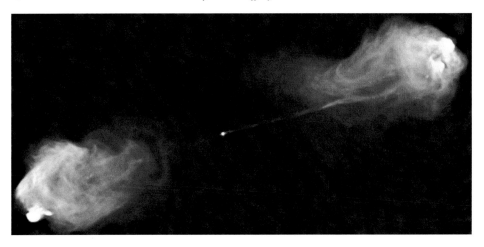

Figure 3.7 A modern radio image of Cyg A. The compact bright dot at the center of the image is the presumed supermassive black hole located in the nucleus of a giant elliptical galaxy. The two broken lines extending out on either side are relativistic jets carrying enormous amounts of energy in twin collimated beams. The jets are stopped by the intergalactic medium and then spread out into two giant lobes. All of the observed radio emission is due to synchrotron radiation from relativistic electrons spiraling in magnetic fields. (Figure courtesy of C. Carilli and R. Perley, NRAO.)

be correlated with black hole spin. If verified, it would suggest that these jets are powered by a generalized Penrose process mediated by magnetic fields.

The first astrophysical black hole to be discovered was Cygnus A, which stood out already as a bright localized radio source in the pioneering radio sky map of Grote Reber [1, 2]. The recognition that Cyg A is a black hole, however, had to wait a few decades. It required identifying the source with a distant galaxy [3, 4]; resolving its radio image into a pair of radio lobes [5] with a compact source at the center [6] (Fig. 3.7); the discovery of quasars [7]; and the growing realization that all of these objects require an extremely powerful but extraordinarily compact engine. The only plausible explanation is that Cyg A, like quasars and other active galactic nuclei (AGN), is powered by a supermassive black hole.

The first stellar-mass black hole to be discovered was Cyg X–1 (also, coincidentally, in the constellation of Cygnus), which was catalogued in the early days of X-ray astronomy as a bright X-ray point source [8]. Evidence of its black hole nature came relatively soon. The optical counterpart was confirmed to be a 5.6-day binary star system in our Galaxy [9], and dynamical observations of the stellar component showed that Cyg X–1 has a mass of at least several solar masses [10, 11], making it too massive to be a neutron star. It was therefore recognized as a black hole.

Cyg A and Cyg X–1 are members of two large but distinct populations of black holes in the universe. We briefly review our current knowledge of the two populations, and summarize the reasons for identifying their members as black holes.

3.3.2 Supermassive black holes

Mass and size estimates

The radiation we see from supermassive black holes is produced by accretion, the process by which gas spirals into the black hole from a large radius. As the gas falls into the potential well, it converts a part of the released potential energy to thermal energy, and ultimately into radiation. A characteristic luminosity of any gravitating object is the Eddington limit at which outward radiative acceleration is balanced by the inward pull of gravity:

$$L_E = \frac{4\pi G M m_p c}{\sigma_T} = 1.25 \times 10^{38} \frac{M}{M_\odot} \text{ erg s}^{-1}, \quad (3.25)$$

where M is the mass of the object, $M_\odot = 1.99 \times 10^{33}$ g is the mass of the Sun, m_p is the mass of the proton, and $\sigma_T = 6.65 \times 10^{-25}$ cm^2 is the Thomson cross-section for electron scattering. A spherical object in equilibrium cannot have a luminosity $L > L_E$. Since bright quasars have typical luminosities $L \sim 10^{46}$ erg s^{-1}, they must thus be very massive: $M > 10^8 \, M_\odot$.

A large mass by itself does not indicate that an object is a black hole. The second piece of information needed is its size. In the case of quasars (and other AGN), a number of observations indicate that their sizes are not very much larger than the gravitational radius of a black hole of mass M:

$$R_g = \frac{GM}{c^2} = 1.48 \times 10^5 \frac{M}{M_\odot} \text{ cm.} \quad (3.26)$$

The earliest indication for a small size came from the fact that quasars show noticeable variability on a time scale of days. Since an object cannot have large-amplitude variations on a time scale shorter than its light-crossing time, it was deduced that quasars are no more than a light-day across, i.e., their sizes must be $< 10^2 R_g$. Modern limits are tighter. For instance, gravitational microlensing observations of the quasar RXJ 1131–1231 indicate that the X-ray emission comes from a region of size $\sim 10 R_g$ [12]. Tighter and more direct limits ($<$ few R_g) are obtained from observations of the Kα line of iron, which show that gas orbits the central object at a good fraction of the speed of light [13, 14]. The only astrophysically plausible object satisfying these mass and size limits is a supermassive black hole.

The mass of Sagittarius A*

Quasars are too distant for direct measurements of their mass. The situation is more favorable for supermassive black holes in the nuclei of nearby galaxies. The majority of these black holes have very low accretion luminosities and are thus very dim. This is an advantage. Without the glare of a bright central source, it is possible to carry out high-resolution imaging and spectroscopic observations relatively close to the black hole and thereby estimate the black hole mass via dynamical methods. The most spectacular results have been obtained in our own Milky Way galaxy.

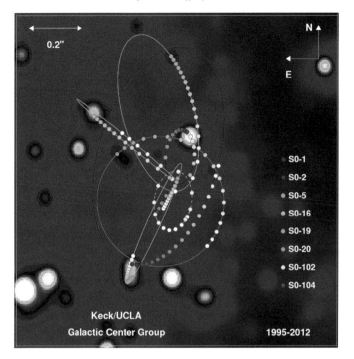

Figure 3.8 Orbital tracks on the plane of the sky over the period 1995–2012 of 8 bright stars (S0–1, S0–2, ..., S0–104) at the center of our Galaxy. Keplerian fits to these orbits give the position and mass of the supermassive black hole in the nucleus of the Galaxy. (Figure courtesy of A. Ghez and her research team at UCLA, based on data obtained with the W. M. Keck Telescopes.)

Over the past twenty years, two different groups have successfully used the largest telescopes on Earth to obtain diffraction-limited infrared images of our Galactic Center, and have thereby mapped the trajectories of stars orbiting the Galactic nucleus. Remarkably, all the stars move on Keplerian orbits around a common focus [15, 16] (see Fig. 3.8) containing a dark mass of $4.4 \pm 0.4 \times 10^6 \, M_\odot$ [17]. Since the dark mass must be interior to the pericenter of the most compact stellar orbit, its radius must be $< 10^3 R_g$.

In fact, a much tighter limit can be placed on the size. At the center of our Galaxy is a radio source called Sagittarius A* (Sgr A*). This source has been shown to be essentially at rest with respect to the Galaxy, moving with a speed less than about $1 \, \mathrm{km\,s^{-1}}$ [18]. Given the huge velocities of stars orbiting in its vicinity (the fastest stars mentioned in the previous paragraph move with speeds up to $10^4 \, \mathrm{km\,s^{-1}}$), equipartition arguments imply that Sgr A* must be more massive than $10^5 \, M_\odot$. The only plausible interpretation is that Sgr A* is identical to the $4.4 \times 10^6 \, M_\odot$ object inferred from stellar orbits.

Meanwhile, Sgr A* is a bright radio source and has been imaged with exquisite precision using interferometric techniques. The most recent observations indicate that emission at 1.3 mm wavelength comes from within a radius of a few R_g [19, 20]. This robust size constraint makes it virtually certain that Sgr A* must be a supermassive black hole.

Other nearby supermassive black holes

Occasionally, the orbiting gas in the accretion disk around a supermassive black hole produces radio maser emission from transitions of the water molecule. If the galaxy is sufficiently nearby, the maser emitting "blobs" can be spatially resolved by interferometric methods and their line-of-sight velocities can be measured accurately by the Doppler technique. The most spectacular example is the nearly edge-on disk in the nucleus of the galaxy NGC 4258 [21, 22], where the measured velocities follow a perfect Keplerian profile. The required black hole mass is $4.00 \pm 0.09 \times 10^7 \, M_\odot$ [23].

Maser disks are relatively rare. A more widely used method employs high-spatial-resolution observations in the optical band with the Hubble Space Telescope (see [24] for a comprehensive review). By simultaneously fitting the spatial brightness distribution and two-dimensional line-of-sight velocity distribution of stars in the vicinity of a galactic nucleus, and using advanced three-integral dynamical models for stellar orbits, it is possible to estimate the mass of a compact central object, if one is present. Several tens of black hole masses have been measured by this method, with uncertainties typically at about a factor of two. In the majority of cases, models without a compact central mass are ruled out with high statistical significance. From these studies it has become clear that essentially every galaxy in the universe hosts a supermassive black hole in its nucleus.

Other less direct methods are available for measuring masses of more distant black holes. Two methods in particular, one based on reverberation mapping [25] and the other on an empirical linewidth–luminosity relation [26], deserve mention.

A remarkable correlation

Unquestionably, the most dramatic discovery to come out of the work described in the previous subsection is the fact that supermassive black hole masses are correlated strongly with the properties of their host galaxies. Figure 3.9 (from [24]) shows two such correlations: (a) between the black hole mass M and the luminosity (in this case infrared luminosity) of the bulge of the galaxy [27], and (b) between M and the stellar velocity dispersion σ_e of the bulge [28, 29].

At first sight, these correlations are baffling. The mass of the black hole is typically 10^3 times smaller than that of the bulge, and its size (R_g) is 10^8 times smaller. How could such an insignificant object show such a strong correlation with its parent galaxy? The answer can be understood by considering a more relevant quantity than either mass or radius alone: the binding energy GM^2/R. In terms of binding energy, the black hole is actually "stronger" (by quite a bit) than the entire galaxy. Indeed, the current paradigm, and a major area of research, is that supermassive black holes exert considerable "feedback" on their host galaxies during the formation and growth of both entities. As a natural consequence, their parameters become strongly correlated in the manner shown in Fig. 3.9 (e.g. [30–32]).

As an important practical application, the above correlations can be used to estimate the masses of high-redshift black holes using the luminosities or stellar velocity distributions of their host galaxies.

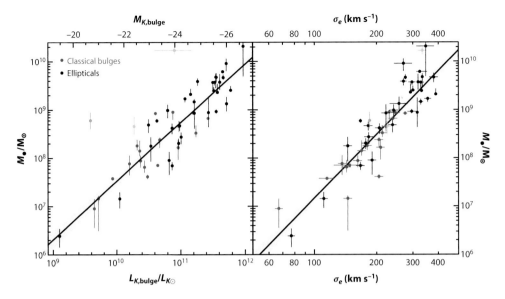

Figure 3.9 Observed correlations between supermassive black hole mass $M_\bullet$ and (left) the infrared luminosity of the bulge of the host galaxy in units of solar luminosities (represented in the top axis by the absolute magnitude $M_{K,bulge}$), and (right) the velocity dispersion σ_e of the stars in the bulge. (Reproduced, with permission, from The Annual Review of Astronomy and Astrophysics, Volume 51 (©) 2013, pgs 511–653 [24] by Annual Review; www.annualreviews.org.)

3.3.3 Stellar-mass black holes

Many millions of stellar-mass black holes are inferred to be present in our Milky Way galaxy, and in every other galaxy in the universe, but the existence of only 24 of them has been confirmed via dynamical observations. These 24, whose masses are in the range $M = 5$–$30\ M_\odot$, are located in X-ray binary systems, 21 of which are sketched to scale in Fig. 3.10. The X-rays are produced by gas that flows from the companion star on to the black hole via an accretion disk. Close to the black hole (radii $\sim 10R_g$), the accreting gas reaches a typical temperature of $\sim 10^7$ K and emits X-rays.

The 24 black hole binaries divide naturally into two classes: (i) 5 of these black holes are *persistent* X-ray sources, which are fed by winds from their massive companion stars. (ii) The remaining 19 binaries are *transient* sources, whose X-ray luminosities vary widely, ranging from roughly the Eddington luminosity L_E down to as low as $\sim 10^{-8}L_E$. A typical transient black hole is active for about a year and then remains quiescent for decades.

Mass measurements

The masses of stellar-mass black holes are measured by employing the same methodologies that have been used for over a century to measure the masses of ordinary stars in binary systems. Most important is the radial velocity curve of the companion star, which is derived from a collection of spectroscopic observations that span an orbital cycle. These velocity

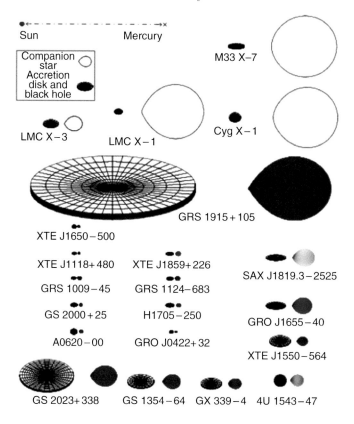

Figure 3.10 Sketches of 21 black hole binaries (see scale and legend in the upper-left corner). The tidally-distorted shapes of the companion stars are accurately rendered in Roche geometry. The black holes are located at the centers of the disks. A disk's tilt indicates the inclination angle i of the binary, where $i = 0$ corresponds to a system that is viewed face-on; e.g., $i = 21°$ for 4U 1543–47 (bottom right) and $i = 75°$ for M33 X–7 (top right). The size of a system is largely set by the orbital period, which ranges from 33.9 days for the giant system GRS 1915+105 to 0.2 days for tiny XTE J1118+480. Three systems hosting persistent X-ray sources – M33 X–7, LMC X–1 and Cyg X–1 – are located at the top. The other 18 systems are transient sources. (Figure courtesy of J. Orosz.)

data deliver two key parameters: the orbital period P and the semi-amplitude of the velocity curve K, which in turn determine the value of the mass function:

$$f(M) \equiv \frac{PK^3}{2\pi G} = \frac{M \sin^3 i}{(1+q)^2}, \tag{3.27}$$

where i is the orbital inclination angle of the binary (Fig. 3.10) and q is the ratio of the companion star mass to that of the black hole.

An inspection of the above equation shows that the value of the observable $PK^3/(2\pi G)$, which can be accurately measured, is the absolute minimum mass of the black hole. For ten out of the total sample of 24 stellar-mass black holes, this minimum mass ranges from 3 to 8 $M_\odot$ (see Table 2 in [33] and [34]). In comparison, the maximum stable mass of a

Table 3.1 *Masses and spins, determined via the continuum-fitting method, for a selected sample of seven black holes. By the No-Hair Theorem (Section 3.3.4), the data constitute complete descriptions of these black holes[a].*

System	$M/M_\odot$	a_*	References
Persistent			
Cyg X–1	14.8 ± 1.0	> 0.95	[38]; [39]
LMC X–1	10.9 ± 1.4	$0.92^{+0.05}_{-0.07}$	[40]; [41]
M33 X–7	15.65 ± 1.45	0.84 ± 0.05	[42]; [43]
Transient			
GRS 1915+105	10.1 ± 0.6	$> 0.95^{b}$	[34]; [44]
GRO J1655–40	6.3 ± 0.5	0.70 ± 0.10^{b}	[45]; [56]
XTE J1550–564	9.1 ± 0.6	$0.34^{+0.20}_{-0.28}$	[47]; [48]
A0620–00	6.6 ± 0.25	0.12 ± 0.19	[49]; [50]

Notes:
[a] Errors are quoted at the 68% level of confidence, except for the two spin limits, which are estimated to be at the 99.7% level of confidence.
[b] Uncertainties are greater than those in papers cited because early error estimates were crude.

neutron star is widely agreed to be less than about 3 $M_\odot$ [35, 36]. Therefore, on the basis of a single robust observable, one can conclude that these ten compact X-ray sources must be black holes.

In order to obtain the actual masses of these and other stellar-mass black holes, one must additionally determine q and i. The mass ratio q is usually estimated by measuring the rotational velocity of the companion star. The inclination angle i can be constrained in several ways; commonly, one models the light curve of the tidally-distorted companion star. Selected mass measurements for seven black holes are given in Table 3.1. For further details on measuring the masses of black holes, see the references cited in the table and [37].

Spin estimates

Spin is difficult to measure because its effects manifest only near the black hole ($R < 10R_g$). Not only must one make discerning observations in this tiny region of space-time, but also one needs a reliable model of the accretion flow in strong gravity. Two fortunate circumstances come to the rescue: (i) We do have at least one simple black hole accretion model, viz., the thin accretion disk model [51, 52]. (ii) Among the several distinct and long-lived accretion states observed in individual stellar-mass black holes [53], one particularly simple state, called the *thermal* state, is dominated by emission from an optically thick accretion disk, and is well-described by the thin-disk model.

According to the thin-disk model, the inner edge of the accretion disk is located at the radius of the innermost stable circular orbit R_{ISCO}. Moreover, R_{ISCO}/R_g is a monotonic function of the dimensionless black hole spin parameter $a_* = cJ/(GM^2)$ [54], where J is the angular momentum of the black hole (note, $|a_*| < 1$). In the continuum-fitting method of measuring spin [44, 55, 56], one observes radiation from the accreting black hole when it is in the thermal state. One then estimates R_{ISCO}, and hence a_*, by fitting the thermal continuum spectrum to the thin-disk model. The method is simple: It is strictly analogous to using the theory of blackbody radiation to measure the radius of a star whose flux, temperature and distance are known. By this analogy, it is clear that to measure R_{ISCO} one must measure the flux and temperature of the radiation from the accretion disk, which one obtains from X-ray observations. One must also measure the source distance D and the disk inclination i (an extra parameter that is not needed for a spherical star). Additionally, one must know M in order to scale R_{ISCO} by R_g to determine a_*. The uncertainties in all these ancillary measurements contribute to the overall error budget.

The spins of ten black holes have been measured by the continuum-fitting method, seven of which are presented in Table 3.1. The robustness of the method is demonstrated by the dozens or hundreds of independent and consistent measurements of spin that have been obtained for several black holes, and through careful consideration of many sources of systematic error. For a review of the continuum-fitting method and a summary of results, see [57].

An alternative method of measuring black hole spin, in which one determines R_{ISCO}/R_g by modeling the profile of the broad and skewed fluorescence Fe Kα line, has been widely practiced since its inception [58]. However, obtaining reliable results for stellar-mass black holes is challenging because one must use data in states other than the thermal state, where the disk emission is strongly Comptonized and harder to model. Furthermore, the basic geometry of the disk is poorly constrained, and it is even doubtful that the inner edge of the disk is located at R_{ISCO}.

To date, the Fe-line method has been used to estimate the spins of more than a dozen stellar-mass black holes. A few of these black holes have been studied using both the continuum-fitting and Fe-line methods, and there is reasonable agreement between the two independent spin estimates. The Fe-line method is especially important in the case of supermassive black holes [59], where it is difficult to apply the continuum-fitting method.

Intermediate-mass black holes

Are there black holes of intermediate mass, i.e., black holes that are too massive ($M > 100\,M_\odot$) to have formed from ordinary stars but, at the same time, are not in the nucleus of a galaxy? Such objects, referred to as intermediate-mass black holes (IMBHs), would represent a new and distinct class of black hole. The leading IMBH candidates are the brightest "ultraluminous" X-ray sources in external galaxies, whose observed luminosities can be up to ~ 100–1000 times the Eddington luminosity of a $10\,M_\odot$ black hole. Although there are some promising candidates (e.g. [60, 61]), none has been confirmed because of the difficulties of obtaining a firm dynamical measurement of mass.

3.3.4 Physics of astrophysical black holes

Are they really black holes?

The astrophysical black holes discussed so far are technically only black hole candidates. True, they are sufficiently massive and compact that we cannot match the observations with any object in stable equilibrium other than a black hole. However, this by itself does not prove that the objects are true black holes, defined as objects with event horizons. Black hole candidates could, in principle, be exotic objects made of some kind of unusual matter that enables them to have a surface (no horizon), despite their extreme compactness.

Astronomers have devised a number of tests to check whether black hole candidates have a "surface". In brief, all the evidence to date shows that black hole candidates do not have normal surfaces that are visible to distant observers (see [62, 63] and references therein). The arguments are sufficiently strong that – barring scenarios that are more bizarre than a black hole – they essentially "prove" that the astrophysical black hole candidates discussed here possess event horizons. However, the proof is still indirect [64].

Spinning black holes and the Penrose process

It is a remarkable consequence of black hole theory that a spinning black hole has free energy available to be tapped. Penrose [65] showed via a simple toy model that particles falling into a spinning hole on negative energy orbits can extract some of the black hole's spin energy. Energy extraction is allowed by the Area Theorem, which states that the horizon area A of a black hole cannot decrease with time: $dA/dt \geq 0$ (e.g., [66]). For a black hole of mass M and dimensionless spin a_*, the area is given by $A = 8\pi R_g^2 [1 + (1 - a_*^2)^{1/2}]$. Therefore, a black hole can lose energy and mass, and thus reduce the magnitude of R_g, provided the spin parameter a_* also decreases by a sufficient amount to satisfy the Area Theorem. In effect, the hole spins down and gives up some of its spin energy to infinity. Penrose's negative-energy particles are a conceptually transparent way of demonstrating this effect.

An astrophysically more promising scenario for the extraction of spin energy makes use of magnetized accretion flows, as outlined in early papers [67, 68]. In this mechanism, the dragging of space-time by a spinning hole causes magnetic field lines to be twisted, resulting in an outflow of energy and angular momentum along field lines. Does this actually happen anywhere in the universe? Astronomers have long hypothesized that relativistic jets such as those in Cygnus A (Fig. 3.7) might be explained by some such process.

In recent years, general relativistic magnetohydrodynamic (MHD) numerical simulations of accreting spinning black holes have been carried out that show relativistic jets forming naturally from fairly generic initial conditions. More importantly, the simulations show unambiguously that, in at least some cases, the jet receives its power from the spin energy of the black hole and not from the accretion disk [69–71]. Specifically, energy and angular momentum flow directly from the black hole through the jet to the external universe, and the mass and spin of the black hole consequently decrease. The jet power varies approximately as a_*^2, and is thus largest for the most rapidly spinning holes. In brief,

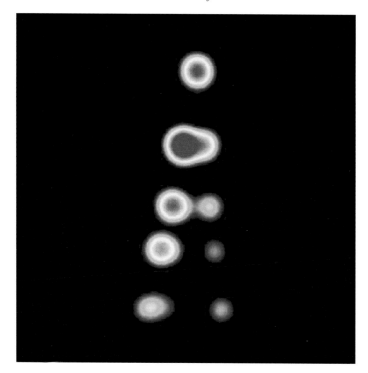

Figure 3.11 A sequence of radio images [72] of the transient black hole X-ray binary GRS 1915+105 during the period 18 March 1994 (uppermost image consisting of a single blob) to 16 April 1994 (lowermost image with two widely separated blobs). Two radio-emitting blobs were ejected from the source around the time of the first observation, and they subsequently moved ballistically outward from the source. The blob on the left has an apparent speed on the sky greater than the speed of light (superluminal motion), which is a relativistic effect. The Lorentz factor of each blob is estimated to be $\gamma \approx 2.6$. (Figure courtesy of F. Mirabel. © Nature Publishing Group.)

a generalized MHD version of the Penrose process operates naturally and efficiently in idealized simulations on a computer.

The observational situation is less clear. Radio-emitting relativistic jets have been known in AGN for many decades (e.g., Fig. 3.7), and more recently, jets have been discovered also in stellar-mass black holes (Fig. 3.11 shows a famous example). A very interesting relation has been found between the radio luminosity L_R, which measures jet power, the X-ray luminosity L_X, which measures accretion power, and the black hole mass M [73–75]. This relation, called the "fundamental plane of black hole activity" (Fig. 3.12), extends over many decades of the parameters, connecting the most massive and luminous AGN with stellar-mass black holes. The relatively tight correlation, which is further emphasized in recent work [76], implies that the jet power depends primarily on the black hole mass and accretion rate, leaving little room for an additional dependence on black hole spin. However, most of the black holes plotted on the fundamental plane do not have spin measurements, so the argument for a lack of spin-dependence is somewhat indirect.

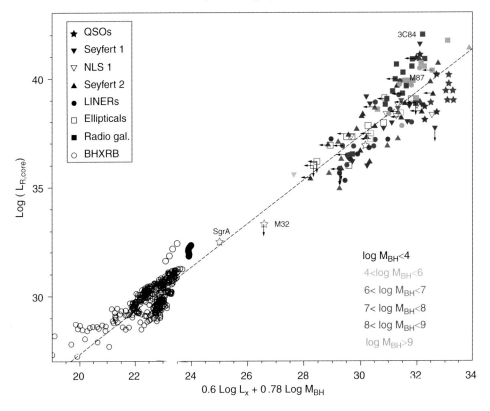

Figure 3.12 The "fundamental plane of black hole activity" [73]. The correlation extends over many decades of black hole mass and accretion luminosity, and includes many different source types. (Figure courtesy of A. Merloni.)

Meanwhile, tentative but direct observational evidence for a correlation between black hole spin and jet power has been found in a sample of stellar-mass black holes for which spins had previously been measured [77, 78] (Fig. 3.13). The evidence is still controversial [57, 79], in large part because of the small size of the sample. In addition, the correlation is restricted to stellar-mass black holes that accrete at close to the Eddington limit and produce so-called "episodic" or "ballistic" jets (e.g., Fig. 3.11), which are different from the jets considered for the fundamental plane (for a discussion see [80]).

In summary, theory and numerical simulations suggest strongly that relativistic jets are powered by black hole spin, i.e., by a generalized Penrose process. However, observational evidence is limited to that shown in Fig. 3.13 and is in comparison weak.

Testing the No-Hair Theorem

The No-Hair Theorem states that stationary black holes, such as those discussed herein, are completely described by the Kerr metric, which has only two parameters: black hole

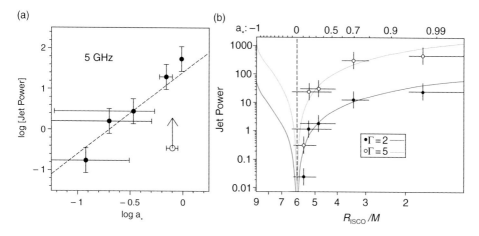

Figure 3.13 (a) Plot of jet power, estimated from 5 GHz radio flux at light-curve maximum, versus black hole spin, measured via the continuum-fitting method, for five transient stellar-mass black holes [77, 78]. The dashed line has slope fixed to 2 (as predicted by theoretical models [68]) and is not a fit. (b) Plot of jet power versus R_{ISCO}/R_g. Here jet power has been corrected for beaming assuming jet Lorentz factor $\Gamma = 2$ (filled circles) or $\Gamma = 5$ (open circles). The two solid lines correspond to fits of a relation of the form "Jet Power" $\propto \Omega_H^2$, where Ω_H is the angular frequency of the black hole horizon. Note that the jet power varies by a factor of $\sim 10^3$ among the five objects shown.

mass M and spin parameter a_*.[12] Testing this theorem requires measuring M and a_* of a black hole with great accuracy and demonstrating that no additional parameter is needed to explain any observable. At the present time, mass and spin measurements of stellar-mass and supermassive black holes are not accurate enough, nor are there a sufficient number of independent observables, to permit such a test.

The most promising system for testing the No-Hair Theorem is Sgr A*, the supermassive black hole in our Galaxy (Section 3.3.2). Within the next decade, ultra-high resolution interferometric observations are planned at millimeter wavelengths with the "Event Horizon Telescope" [81], which will produce direct images of the accreting gas in Sgr A* on length scales comparable to the horizon. These measurements could potentially be used to test the No-Hair Theorem (e.g. [82]).

3.3.5 *Conclusion*

The dawning that black holes are real occurred at the midpoint of this century of General Relativity, at the First Texas Symposium on Relativistic Astrophysics in 1963. There, Roy Kerr announced his solution, Jesse Greenstein described Maarten Schmidt's discovery of quasars, and Harlan Smith reported on the rapid variability of these objects [83]. Today,

---

[12] In principle, a black hole can have a third parameter, electric charge, but the black holes studied in astrophysics are unlikely to have sufficient charge for it to be dynamically important.

black hole astrophysics is advancing at a breathtaking rate. Tomorrow, spurred on by the commissioning of the Event Horizon Telescope and the advent of gravitational wave astronomy, it is reasonable to expect the discovery of many new unimaginable wonders.

References

[1] Reber, G. 1944. *Astrophys. J.*, **100**, 279–287.
[2] Hey, J. S., Parsons, S. J. and Phillips, J. W. 1946. *Nature*, **158**, 234.
[3] Smith, F. G. 1951. *Nature*, **168**, 555.
[4] Baade, W. and Minkowski, R. 1954. *Astrophys. J.*, **119**, 206–214.
[5] Jennison, R. C. and Das Gupta, M. K. 1953. *Nature*, **172**, 996–997.
[6] Hargrave, P. J. and Ryle, M. 1974. *Mon. Not. Roy. Astr. Soc.*, **166**, 305–327.
[7] Schmidt, M. 1963. *Nature*, **197**, 1040.
[8] Bowyer, S., Byram, E. T., Chubb, T. A. and Friedman, H. 1965. *Science*, **147**, 394–398.
[9] Tananbaum, H., Gursky, H., Kellogg, E., Giacconi, R. and Jones C. 1972. *Astrophys. J.*, **177**, L5–L10.
[10] Bolton, C. T. 1972. *Nature*, **235**, 271–273.
[11] Webster, B. L. and Murdin, P. 1972. *Nature*, **235**, 37–38.
[12] Dai, S., Kochanek, C. S., Chartas, G., *et al.* 2010. *Astrophys. J.*, **709**, 278–285.
[13] Tanaka, Y., Nandra, K., Fabian, A.C., *et al.* 1995. *Nature*, **375**, 659–661.
[14] Fabian, A. C., Iwasawa, K., Reynolds, C. S. and Young, A. J. 2000. *Publ. Astron. Soc. Pacific*, **112**, 1145–1161.
[15] Schödel, R., Ott, T., Genzel, R., *et al.* 2002. *Nature*, **419**, 694–696.
[16] Ghez, A. M., Salim, S , Horrnstein, S. D., *et al.* 2005. *Astrophys. J.*, **620**, 744–757.
[17] Meyer, L., Ghez, A. M., Schödel, R., *et al.* 2012. *Science*, **338**, 84–87.
[18] Reid, M. J. and Brunthaler, A. 2004. *Astrophys. J.*, **616**, 872–884.
[19] Doeleman, S. S., Weintroub, J., Rogers, A. E. E., *et al.* 2008. *Nature*, **455**, 78–80.
[20] Fish, V. L., Doeleman, S. S., Beaudoin, C., *et al.* 2011. *Astrophys. J. Lett.*, **727**, L36–L41.
[21] Miyoshi, M., Moran, J., Herrnstein, J., *et al.* 1995. *Nature*, **373**, 127–129.
[22] Greenhill, L. J., Jiang, D. R., Moran, J. M., *et al.* 1995. *Astrophys. J.*, **440**, 619–627.
[23] Humphreys, E. M. L., Reid, M. J., Moran, J. M., Greenhill, L. J. and Argon, A. L. 2013. *Astrophys. J.*, **775**, 13.
[24] Kormendy, J. and Ho, L. C. 2013. *Ann. Rev. Astron. Astrophys.*, **51**, 511–653.
[25] Peterson, B. M., Ferrarese, L., Gilbert, K. M., *et al.* 2004. *Astrophys. J.*, **613**, 682–699.
[26] Vestergaard, M. and Peterson, B. M. 2006. *Astrophys. J.*, **641**, 689–709.
[27] Magorrian, J., Tremaine, S., Richstone, D., *et al.* 1998. *Astron. J.*, **115**, 2285–2305.
[28] Ferrarese, L. and Merritt, D. 2000. *Astrophys. J.*, **539**, L9–L12.
[29] Gebhardt, K., Bender, R., Bower, G., *et al.* 2000. *Astrophys. J.*, **539**, L13–L16.
[30] Di Matteo, T., Springel, V. and Hernquist, L. 2005. *Nature*, **433**, 604–607.
[31] King, A. 2003. *Astrophys. J.*, **596**, L27–L29.
[32] King, A. 2005. *Astrophys. J.*, **635**, L121–L123.
[33] Özel, F., Psaltis, D., Narayan, R., *et al.* 2010. *Astrophys. J.*, **725**, 1918–1927.
[34] Steeghs, D., McClintock, J. E., Parsons, S. G., *et al.* 2013. *Astrophys. J.*, **768**, 75.
[35] Rhoades, C. E. and Ruffini, R. 1974. *Phys. Rev. Lett.*, **32**, 324–327.
[36] Kalogera, V. and Baym, G. 1996. *Astrophys. J.*, **470**, L61–L64.
[37] Charles, P. A. and Coe, M. J. 2006. Optical, ultraviolet and infrared observations of X-ray binaries. in *Compact stellar X-ray sources*, eds. W. H. G. Lewin and M. van der Klis, 215–265.
[38] Orosz, J. A., McClintock, J. E., Aufdenberg, J. P., *et al.* 2011. *Astrophys. J.*, **742**, 84.
[39] Gou, L., McClintock, J. E., Reid, M. J., *et al.* 2011. *Astrophys. J.*, **742**, 85.
[40] Orosz, J. A., Steeghs, D., McClintock, J. E., *et al.* 2009. *Astrophys. J.*, **697**, 573–579.
[41] Gou, L., McClintock, J. E., Liu, J., *et al.* 2009. *Astrophys. J.*, **701**, 1076–1090.

[42] Orosz, J. A., McClintock, J. E., Narayan, R., *et al.* 2007. *Nature*, **449**, 872–875.
[43] Liu, J.,McClintock, J. E., Narayan, R., *et al.* 2008. *Astrophys. J.*, **679**, L37–L40 (Erratum: **719**, L109).
[44] McClintock, J. E., Shafee, R. and Narayan, R. 2006. *Astrophys. J.*, **652**, 518–539.
[45] Greene, J., Bailyn, C. D. and Orosz, J. A. 2001. *Astrophys. J.*, **554**, 1290–1297.
[47] Orosz, J. A., Steiner, J. F., McClintock, J. E., *et al.* 2011. *Astrophys. J.*, **730**, 75.
[48] Steiner, J. F., Reis, R. C., McClintock, J. E., *et al.* 2011. *Mon. Not. Roy. Astr. Soc.*, **416**, 941–958.
[49] Cantrell, A. G., Bailyn, C. D. and Orosz, J. A. 2010. *Astrophys. J.*, **710**, 1127–1141.
[50] Gou, L., McClintock, J. E., Steiner, J. F., *et al.* 2010. *Astrophys. J.*, **718**, L122–L126.
[51] Shakura, N. I. and Sunyaev, R. A. 1973. *Astron. Astrophys.*, **24**, 337–355.
[52] Novikov, I. D. and Thorne, K. S. 1973. Pages 343–450 of: C. DeWitt and B. S. DeWitt (eds.), *Black holes (Les astres occlus)*, New York, London and Paris: Gordon and Breach.
[53] Remillard, R. A. and McClintock, J. E. 2006. *Ann. Rev. Astron. Astrophys.*, **44**, 49–92.
[54] Bardeen, J. M., Press, W. H. and Teukolsky, S. A. 1972. *Astrophys. J.*, **178**, 347–370.
[55] Zhang, S. N., Cui, W. and Chen, W. 1997. *Astrophys. J. Lett.*, **482**, L155–L158.
[56] Shafee, R., McClintock, J. E. and Narayan, R. 2006. *Astrophys. J.*, **636**, L113–L116.
[57] McClintock, J. E., Narayan, R. and Steiner, J. F. 2014. *Space Sci. Rev.*, **183**, 295–332.
[58] Fabian, A. C., Rees, M. J., Stella, L., *et al.* 1989. *Mon. Not. Roy. Astr. Soc.*, **238**, 729–736.
[59] Reynolds, C. S. 2014. *Space Sci. Rev.*, **183**, 277–294.
[60] Feng, H. and Kaaret, P. 2010. *Astrophys. J. Lett.*, **712**, L169–L173.
[61] Davis, S. W., Narayan, R., Zhu, Y., *et al.* 2011. *Astrophys. J.*, **734**, 111.
[62] Narayan, R. and McClintock, J. E. 2008. *New Astron. Rev.*, **51**, 733–751.
[63] Broderick, A. E., Loeb, A. and Narayan, R. 2009. *Astrophys. J.*, **701**, 1357–1366.
[64] Abramowicz, M. A., Kluźniak, W. and Lasota, J.-P. 2002. *Astron. Astrophys.*, **396**, L31–L34.
[65] Penrose, R. 1969. *Riv. Nuovo Cim.*, **I**, 252.
[66] Bardeen, J. M., Carter, B. and Hawking, S. W. 1973. *Commun. Math. Phys.*, **31**, 161–170.
[67] Ruffini, R. and Wilson, J. R. 1975. *Phys. Rev. D*, **12**, 2959–2962.
[68] Blandford, R. D. and Znajek, R. L. 1977. *Mon. Not. Roy. Astr. Soc.*, **179**, 433–456.
[69] Tchekhovskoy, A., Narayan, R. and McKinney, J. C. 2011. *Mon. Not. Roy. Astr. Soc.*, **418**, L79–L83.
[70] Penna, R. F., Narayan, R. and Sadowski, A. 2013. *Mon. Not. Roy. Astr. Soc.*, **436**, 3741–3758.
[71] Lasota, J.-P., Gourgoulhon, E., Abramowicz, M., Tchekhovskoy, A. and Narayan, R. 2014. *Phys. Rev. D*, **89**, 024041.
[72] Mirabel, I. F. and Rodriguez, L. F. 1994. *Nature*, **371**, 46–48.
[73] Merloni, A., Heinz, S. and di Matteo, T. 2003. *Mon. Not. Roy. Astr. Soc.*, **345**, 1057–1076.
[74] Heinz, S. and Sunyaev, R. A. 2003. *Mon. Not. Roy. Astr. Soc.*, **343**, L59–L64.
[75] Falcke, H., Körding, E. and Markoff, S. 2004. *Astron. Astrophys.*, **414**, 895–903.
[76] van Velzen, S. and Falcke, H. 2013. *Astron. Astrophys.*, **557**, L7.
[77] Narayan, R. and McClintock, J. E. 2012. *Mon. Not. Roy. Astr. Soc.*, **419**, L69–L73.
[78] Steiner, J. F., McClintock, J. E. and Narayan, R. 2013. *Astrophys. J.*, **762**, 104–113.
[79] Russell, D. M., Gallo, E. and Fender, R. P. 2013. *Mon. Not. Roy. Astr. Soc.*, **431**, 405–414.
[80] Fender, R. P. and Belloni, T. M. 2004. *Ann. Rev. Astron. Astrophys.*, **42**, 317–364.
[81] Doeleman, S., Agol, E., Backer, D., *et al.* 2008. Imaging an event horizon: submm-VLBI of a supermassive black hole. Astro2010: The astronomy and astrophysics decadal survey, science white papers, No. 68. arXiv:0906.3899.
[82] Johannsen, T. and Psaltis D. 2010. *Astrophys. J.*, **718**, 446–454.
[83] Schucking, E. L. 1989. *Phys. Today*, **42**, 46–52.

3.4 Gamma-ray bursts

Peter Mészáros and Martin J. Rees

3.4.1 Historical overview and basic concepts

Gamma-Ray Bursts (GRBs) were serendipitously discovered in the late 1960s by the military Vela satellites which were monitoring the Nuclear Test Ban Treaty between the US and the Soviet Union. The announcement was postponed for several years, until the possibility of their having a human origin had been excluded and it had been ascertained that they were outside the immediate solar system [1]. In a matter of a few years more than a hundred models had been proposed to explain their astrophysical origin [2], ranging from comet infalls, through stellar cataclysmic events, to events associated with supermassive black holes at the center of galaxies. The problem in making the first steps towards a theoretical understanding was that the gamma-ray instruments of the time had poor positional accuracy, transmitted to Earth only many hours after the trigger, so that only wide-field, insensitive telescopes could attempt to follow up the bursts to look for counterparts at other wavelengths.

In the 1990s the Compton Gamma Ray Observatory (CGRO) was launched, one of whose main objectives was the detection of GRBs. The Burst and Transient Source Experiment (BATSE) onboard CGRO obtained, over a decade, the positions of ~ 3000 GRBs to within several degrees. This showed that they were uniformly distributed over the sky [3], indicating either an extragalactic or a 'galactic-halo' origin. The gamma-ray light curves detected with BATSE exhibited a diversity of variability timescales, with total durations ranging from less than a second to tens of minutes (Fig. 3.14). BATSE also found that GRBs can be classified into two duration classes, short and long GRBs, with a dividing line at ~ 2 s [4].

The search for GRB counterparts at other wavelengths remained unsuccessful for almost 25 years, until in 1997 the Beppo-SAX satellite localized with greater accuracy the first long-lasting X-ray afterglows [5], which in turn enabled the first optical host galaxy identification and redshift measurement [6]. The long bursts were found to be associated with galaxies where active star formation was taking place, typically at redshifts $z \sim 1$–2, and in some cases a supernova of type Ic was detected associated with the bursts, confirming the stellar origin of this class. The power-law time decay of the light curve was also observed, in a number of cases, to exhibit a steepening after ~ 0.5–1 day, suggesting (for reasons explained below) that the emission was collimated into a jet, of typical opening half-angle $\sim 5°$, which eased the energy requirements. Even so, at cosmological distances this implied a total time-integrated energy output of $\sim 10^{50}$–10^{51} erg. This is roughly 10^{-3} of a solar rest mass, emitted over tens of seconds. This is more than our Sun emits over its ten-billion-year lifetime, and about as much as the entire Milky Way emits over a hundred years – and that is mainly concentrated into gamma rays.

Well before the CGRO and Beppo-SAX observations, early theoretical ideas about the origin of GRBs had converged towards an energy source provided by the gravitational potential of a compact stellar source, the latter being suggested by the short duration (tens

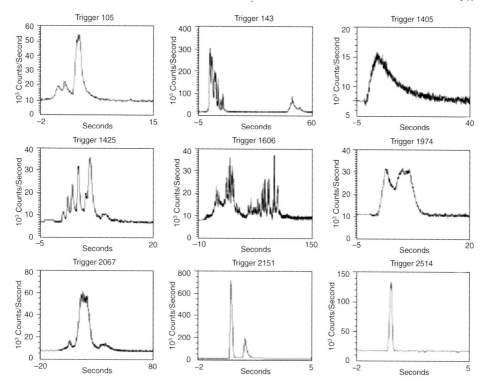

Figure 3.14 A diversity of gamma-ray light curves from the BATSE instrument on the Compton Gamma Ray Observatory. Variability timescales from milliseconds to more than a second are found, the light-curve morphology ranging from spiky to relatively smooth.

of seconds) and fast variability $\gtrsim 10^{-3}$ s of the γ-ray emission, using a simple causality argument $R \lesssim c\,\Delta t \lesssim 10$–100 km. The large energies liberated in a small volume and in a short time, as well as the observed hard spectrum ($\gtrsim$ MeV) would then produce abundant electron–positron pairs via photon–photon interactions, creating a hot fireball which would expand, eventually reaching relativistic bulk velocities [7].

Among the first stellar sources discussed which could be responsible for GRBs were binary double neutron star (DNS) mergers, or black hole–neutron star (BH–NS) mergers, whose occurrence rate as well as the expected energy liberated $\sim GM^2/R$ appeared sufficient for powering even extragalactic GRBs [8–11]. These are nowadays the leading candidates for the short gamma-ray bursts, as shown by Swift and other observations, e.g. [12]. Another candidate stellar source was the core collapse of massive stars and the accretion into the resulting black hole [13, 14]. Initially it was thought that this would result in a GRB and a failed supernova, but later observations, e.g. [15] and others, showed an unusually luminous core collapse supernova of type Ic associated with some GRBs; these supernovae have since been referred to as hypernovae. The core collapse model, referred to as a collapsar, is currently well established as the source of most long GRBs.

The predicted rate of occurrence of binary mergers and of hypernovae is sufficient to account for the number of bursts observed, even if the gamma-rays are beamed to the extent that only one event in 100–1000 is observed. (We expect less than one observable burst per million years from a typical galaxy, but the detection rate can nonetheless be of order one per day because they are so powerful that they can be detected out to the Hubble radius).

3.4.2 CGRO results and basic models

The dynamics of the expected relativistic fireball expansion were investigated by [16, 17]. The fact that photons of over 100 MeV are detected provides compelling evidence for ultra-relativistic expansion. To avoid degradation of the spectrum via photon–photon inter-actions to energies below the electron–positron formation threshold $m_e c^2 = 0.511$ MeV the outward flow must have a bulk Lorentz factor Γ high enough that the relative angle at which the photons collide is less than Γ^{-1}, thus diminishing the pair-production threshold [18, 19].

Since each baryon in the outflow must be given an energy exceeding 100 times its rest mass, a key requirement of the central engine is that it must concentrate a lot of its energy into a very small fraction of its total mass. This favours models where magnetic fields and Poynting flux are important.

The observed spectrum extends to high energies, generally in a broken power-law shape, i.e., highly nonthermal. Two initial problems [9, 20] with the first expanding fireball models were that (a) they are initially optically thick and the photon spectrum escaping from the Thomson-scattering photosphere would be expected to be an approximate blackbody, and (b) most of the initial fireball energy would be converted into kinetic energy of expansion, with a concomitantly reduced energy in the observed photons, i.e. a very low radiative efficiency.

A simple way to achieve a high efficiency and a nonthermal spectrum, which is currently the most widely invoked explanation, is by reconverting the kinetic energy of the flow into random energy via shocks, after the flow has become optically thin [25], see Fig. 3.15. Two different types of shocks may be expected. There will be an external shock, when the expanding fireball runs into the external interstellar medium or a pre-ejected stellar wind, and a reverse shock propagating back into the ejecta. As in supernova remnants, Fermi acceleration of electrons into a relativistic power distribution in the turbulent magnetic fields boosted in the shock leads to synchrotron emission [25, 26], resulting in a broken power-law spectrum, where the high-energy photon spectral slope fits easily the observations, and the single-electron low-energy photon slope $-2/3$ can, with a distribution of minimum-energy electrons γ_{min}, reproduce the observed average low-energy photon slope values of -1 (see also [27, 28]). The reverse shock would lead to optical photons, while inverse Compton emission in the forward blast wave would produce photons in the GeV–TeV range [29].

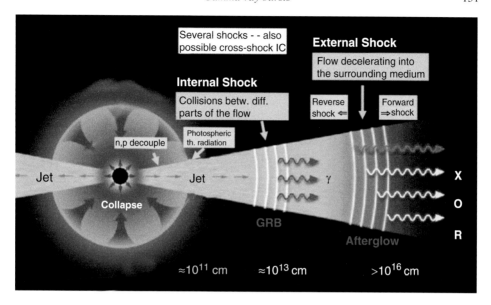

Figure 3.15 Schematic GRB jet from a collapsing star. As the jet expands, it becomes optically thin at a scattering photosphere, whose thermal spectrum can have non-thermal additions due to proton–neutron collisions or shocks [21, 22]. At intermediate radii internal shocks can produce non-thermal gamma-rays [23], and further out, as the jet is decelerated by the swept-up external matter in an external shock, a longer-wavelength afterglow is produced [24], which decays with time. Figure from P. Meszaros, *Science*, **291**, 79 (2001). © AAAS.

There could, additionally, be dissipation and acceleration within the outflowing jet itself. If the jet is unsteady, internal shocks [23, 30] can form as faster portions of the flow catch up with slower portions. And if magnetic stresses are important within the jets (i.e. they are Poynting-dominated outflows [31], instead of the usual baryonic-inertia-dominated outflows) then magnetic reconnection can provide efficient mechanical conversion of bulk into random energy [32] (see also [33, 34]). Any of these models provides a generic scenario for explaining the radiation spectrum, largely independent of the specific nature of the progenitor.

Internal shocks continue to be the model most widely used by observers to interpret the prompt MeV emission, with continuing theoretical discussions about their spectrum and efficiency, e.g. [35]. The external shock model is the favored interpretation for the long-term afterglows starting at high energies and phasing into gradually longer wavelengths over periods of days to months. Coincidentally, the detection of the afterglows was preceded, a few weeks earlier, by the publication of quantitative predictions of the power-law spectral and time dependence of X-ray, optical and radio afterglows [24], in general agreement with observations. Prompt optical afterglows were first detected in 1999 [36], while multi-GeV emission was reported by CGRO-EGRET [37], and more recently and in greater detail by Fermi (see Section 3.4.5).

3.4.3 Beppo-SAX and HETE-2 results and issues

The evidence for a jet outflow is based on the observed steepening of the light curve after
$\sim$ a day [38], which is attributed to the transition between the afterglow early relativistic
expansion, when the relativistic beaming angle is narrower than the jet opening half-angle
θ_j and the late expansion, when the relativistic beaming angle has become wider than the
jet, $\Gamma^{-1} \geq \theta_j$, leading to a drop in the effective flux [39–41]. A jet opening half-angle
$\theta_j \sim$ 3–5 degrees is inferred, which reduces the total energy requirement to about
10^{51}–10^{52} ergs. This, even allowing for substantial inefficiencies, is compatible with
currently favored scenarios based on a stellar collapse or a compact merger, e.g. [12] and
Section 3.4.1.

Observations with the Beppo-SAX and HETE-2 satellites indicated the existence of a
sub-class of GRBs called X-ray flashes (XRFs), whose spectrum peaks at energies 30–80
keV instead of the 300 keV–1 MeV of classical GRBs, and with wider jet opening angles,
e.g. [42]. The relative frequencies of XRFs versus GRBs led to considerations about a
possible continuum distribution of angles, as well as about the jet angular shape, including
departures from a simple top-hat (abrupt cut-off), including an inverse power law or a
Gaussian dependence on the angle [43–46].

A problem with simple internal shock synchrotron models of the prompt MeV emission
is that the low-energy photon number spectral slope, which is expected to be $-2/3$, is found
to be flatter in a fraction of BATSE bursts [47]. In addition, the synchrotron cooling time
can be typically shorter than the dynamical time, which would lead to slopes $-3/2$ [48].
Both in internal shock Fermi acceleration and in magnetic reconnection schemes, a number
of effects can modify the simple synchrotron spectrum to satisfy these constraints. Another
solution involves a photospheric component, discussed below.

A natural question is whether the fact that the spectral peak energies are observed
typically in the 0.1–0.5 MeV range is intrinsic or due to observational selection effects
[49, 50]. A preferred peak energy may be attributed to a blackbody spectrum at the
comoving pair recombination temperature in the fireball photosphere [51]. A photospheric
component can address also the above low-energy spectral slope issue with its steep
Rayleigh–Jeans part of the spectrum, at the expense of the high-energy power law.
This was generalized [52] to a photospheric blackbody spectrum at low energies with a
Comptonized photospheric component and possibly an internal shock or other dissipation
region outside it producing Fermi accelerated electrons and synchrotron photons at high
energies. Photospheric models with moderate scattering depth can in fact lead to a Compton
equilibrium which gives spectral peaks in the right energy range [53] and positive low-
energy slopes as well as high-energy power-law slopes (the positive low-energy slopes can
always be flattened through a distribution of peak energies). A high radiative efficiency
can be a problem if the photosphere occurs beyond the saturation radius $r_{sat} \sim r_0\eta$, where
r_0 is the base of the outflow and $\eta = L/(\dot{M}c^2)$ is the asymptotic bulk Lorentz factor [52].
However, a high radiation efficiency with adequate low- and high-energy slopes can be
obtained in all cases if significant dissipation (whether by magnetic reconnection, shocks
or nuclear collisions) is present in the photosphere [21, 54, 55]. This can also address the

phenomenological Amati [56] and Ghirlanda [57] relations between spectral peak energy and burst fluence [54, 58].

3.4.4 Bursts in the Swift era

The launch of the Swift satellite in 2004 ushered in a new era of extensive data collection and analysis on GRBs, at wavelengths ranging from optical to MeV energies. This resulted in a number of interesting new discoveries, which have motivated various refinements and reappraisals as well as new work on theoretical models, as discussed at greater length in the next sections.

Swift is equipped with three instruments: the Burst Alert Telescope (BAT), the X-Ray Telescope (XRT) and the UV Optical Telescope (UVOT). The BAT detects bursts and locates them to about 2 arcminutes accuracy. This position is then used to automatically slew the spacecraft, typically within less than a minute, re-pointing the high-angular-resolution XRT and UVOT instruments towards the event. The positions are also rapidly sent to Earth so that ground telescopes can follow the afterglows.

A surprising new result achieved by Swift was that in a large fraction of the bursts the X-ray afterglow shows an initial very steep time decay, starting after the end of the prompt γ-ray emission. This then is generally followed by a much shallower time decay, often punctuated by abrupt, large-amplitude X-ray flares, lasting sometimes for up to $\sim$ 1000 s, which then steepens into a power-law time decay with the more usual (pre-Swift) slope of index roughly -1.2 to -1.7 [59, 60]. A final further steepening is sometimes detected, which is ascribed to the beaming angle Γ^{-1} exceeding the finite jet opening angle (Fig. 3.16). The initial steep decay may be ascribed to the evanescent radiation from high latitudes $\theta > \Gamma^{-1}$ relative to the line of sight [61, 62], while the ensuing shallow decay phase may be due to continued outflow of material after the prompt emission has ended [63], which may undergo occasional internal shocks resulting in X-ray flares, e.g. [12, 59, 64]. The subsequent steepening can be ascribed to the previously known forward-shock gradual deceleration and the beaming-induced jet break. These structures in the X-ray afterglow light curves are present both in long and short bursts.

Long GRBs (LGRBs) are found in galaxies where massive stars are forming, being present over a large redshift range from $z = 0.0085$ to $z > 9$. Most LGRBs that occur near enough for supernova detection have an accompanying Type Ib or Ic supernova, supporting the growing evidence that LGRBs are caused by "collapsars" where the central core of a massive star collapses to a compact object such as a black hole or possibly a magnetar.

The number of GRB redshifts obtained underwent a rapid expansion after the launch of Swift (currently in excess of 300), thanks to the rapid localization allowing large ground-based telescopes to acquire high-quality spectra while the afterglow was still bright. The most distant ones are intrinsically the brightest, typically $E_{iso} \gtrsim 10^{55}$ erg, the current record holder being GRB090423 at a spectroscopically confirmed redshift $z = 8.2$ [66], and GRB090429B, at a photometric redshift $z \sim 9.4$ [67].

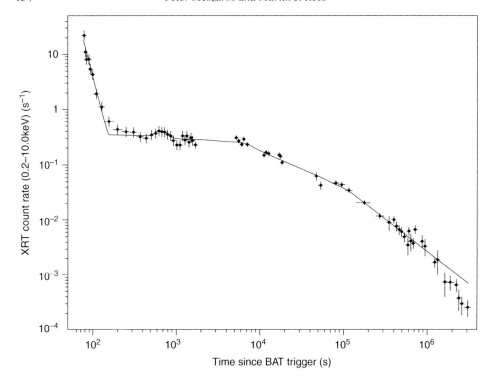

Figure 3.16 X-ray light curve of GRB060428A observed by the Swift XRT [65], showing the initial steep decay, the subsequent shallow decay punctuated by occasional flares, followed by a power-law decay with a final steepening expected after the relativistic causal angle $\theta \sim \Gamma^{-1}$ exceeds the jet opening angle. (Adapted from [65], by permission of the Royal Society.)

With the increasing statistics, LGRBs are contributing to a better understanding of the high-redshift universe. They provide spectroscopic information about the chemical composition of the intervening intergalactic medium at epochs when the Universe was as little as 1/20th its present age. Also, since LGRBs are the endpoints of the lives of massive stars, their rate is approximately proportional to the star formation rate. This gives information at high redshift where the rate is highly uncertain. There can be evolutionary biases, such as a dependence of LGRBs on the metalicity of host galaxies, which must be taken into account [68, 69].

Swift succeeded in finally localizing the host galaxies of a number of short GRBs (SGRBs), e.g. [70, 71]. Unlike long GRBs, the SGRBs typically originate in host galaxies with a wide range of star formation properties, including low formation rates. The host properties are substantially different than those of LGRBs [72–74], indicating a different origin. Furthermore, nearby SGRBs show no evidence for simultaneous supernovae [75], as do many long bursts. These results reinforce the interpretation that SGRBs arise from an old population of stars, probably due to mergers of compact binaries such as double neutron star or neutron star–black hole binaries [8, 9, 75].

Short GRBs are found to have generally a lower isotropic-equivalent luminosity and total energy output E_{iso} than LGRBs, typically $E_{iso} \sim 10^{50}$ ergs, with a weak afterglow, and in the few cases where a jet break has been measured, the jet opening angle appears to be wider than in LGRBs, $\theta_j \sim 5°$–$25°$ [76, 77]. Another new result was the discovery, in about 25% of SGRBs, of a longer (~ 100 s) light-curve tail with a spectrum softer than the initial episode [78, 79]. This is puzzling in the context of double neutron star or neutron star–black hole mergers, since numerical simulations suggest that the disk of disrupted matter is accreted in at most a few seconds, e.g. [75]. A longer accretion timescale, however, may occur if the disk is highly magnetized [80], or if the compact merger results in a temporary magnetar whose magnetic field holds back the accretion disk until the central object collapses to a black hole [81].

Other types of bursts

The demarcation into two classes of bursts is, however, too simplistic to be the whole story. Some bursts fit neither category. For example, some bursts detected by Swift are extreme magnetar flares – caused by the sudden readjustment (and release of stored energy) in the magnetosphere of a highly magnetized ($\gtrsim 10^{14}$ G) neutron star. These are of interest for phenomenologists but are a confusing complication for those seeking correlations between the observable parameters of classical bursts.

But the Swift spacecraft has revealed another type of object that is of great interest, and which was a surprise: bursts characterized by unusually persistent and prolonged emission, and located at the centre of the host galaxy. These are interesting both to astrophysicists and to relativists, as they may be triggered by a long-predicted effect that has not before been conclusively detected: the tidal disruption of a star by a massive black hole.

Tidal capture and disruption of stars attracted interest back in the 1970s, when theorists started to address the dynamics of stars concentrated in a high-density 'cusp' surrounding the kind of black hole expected to exist in the centres of galaxies (and perhaps in some globular star clusters as well). It was recognized that stars could be captured and swallowed by the central black hole if they were in a 'loss cone' of near-radial orbits.

If the central black hole is sufficiently massive, tidal forces at the horizon may be too gentle to disrupt the star while it is still in view, in which case it is captured without any conspicuous display. For a solar-type star, this requires $\sim 10^8 \, M_\odot$; for white dwarfs the corresponding mass is $\sim 10^4 \, M_\odot$. (And neutron stars are swallowed whole by black holes with masses above about $10 \, M_\odot$ – this is important for the gravitational wave signal in coalescing binary stars, as discussed elsewhere in this volume). For a spinning black hole, the cross-section for capture, and the tidal radius for disruption, depend on the relative orientation of the orbital and spin angular momenta. (Stars on orbits counter-rotating with respect to the black hole are preferentially captured: this is a process that would reduce the spin of a black hole in a galactic nucleus.)

When stars are swallowed before disruption, they can be treated as point-mass particles moving in the gravitational field of the black hole; their interactions among themselves can be treated the same way, except insofar as star–star collisions are important. But the physics is much messier in the cases when the tidal radius is outside the black hole and the star is disrupted rather than swallowed whole. This phenomenon has been studied since the 1970s, first via analytic models (e.g. [82, 83]) and subsequently by progressively more powerful numerical simulations (e.g. [84, 85], etc.). In the Newtonian approximation the tidal radius is $R_t \sim R_*(M_{BH}/M_*)^{1/3}$. There are several key parameters: the type of star; the pericentre of the star's orbit relative to the tidal radius, and the orientation of the orbit relative to the black hole's spin axis. In most astrophysical contexts, the captured stars would be on highly eccentric orbits (i.e. the orbital binding energy would be small compared to that of a circular orbit at the tidal radius). If the pericentre is of order R_t, the star will be disrupted, and the debris will continue on eccentric orbits, but with a spread of energies of order the binding energy of the original star. Indeed nearly half the debris will escape from the black hole's gravitational field completely; the rest will be on more tightly bound (but still eccentric) orbits, and would be fated to dissipate further, forming a disk, much of which would then be accreted into the black hole. A pericentre passage at (say) 2 or 3 times R_t would not disrupt a star completely, but would remove its envelope, and induce internal oscillations, thereby extracting orbital energy and leaving the star vulnerable on further passages. On the other hand, as first discussed by Carter and Luminet [82], a star that penetrates far inside the tidal radius (but not so close to the black hole that it spirals in) will be drastically distorted and compressed by the tidal forces, perhaps to the extent that a nuclear explosion occurs, leading to a greater spread in the energy of the debris than would result from straight gas dynamics.

There have in recent years been detailed computations of these processes, and also of the complicated and dissipative gas dynamics that leads to the accretion of the debris, and the decline of the associated luminosity as the dregs eventually drain away. There are two generic predictions: the debris enveloping the black hole should initially have a thermal emission with a power comparable to the Eddington luminosity of the black hole; and at late times, when the emission comes from the infall of debris from orbits with large apocentre, the luminosity falls approximately as $L \propto t^{-5/3}$.

There has been much debate about the role of tidal capture in the growth of supermassive black holes, and the fueling of AGN emission, and many calculations of the expected rate, taking account of what has been learnt about the masses of black holes, and the properties of the stellar populations surrounding them. Some flares in otherwise quiescent galactic nuclei, where the X-ray luminosity surges by a factor $\gtrsim 100$, have been attributed to tidal disruptions.

But tidal disruption is included in this chapter mainly because of a remarkable burst detected by Swift, Sw J1644+57 [86], which was located at the centre of its host galaxy, and which was exceptionally prolonged in its emission. This is perhaps the best candidate so far for an event triggered by tidal capture of a star. The high-energy radiation, were this model correct, would come from a jet generated near the black hole. Modeling is still tentative, and is difficult because there is no reason to expect alignment between the angular momentum vectors of the black hole and of the infalling material. But the inner disk (and therefore the inner jet) would be expected to align with the black hole, though it is possible that the jet is deflected further out by material with different alignment (c.f. [87]).

Be that as it may, this exceptional burst offers model-builders an instructive 'missing link' between the typical long ('Type 1') burst, involving a massive star, and the jets in AGNs) which are generated by processes around supermassive black holes.

3.4.5 *Bursts at energies above GeV: Fermi and beyond*

The Fermi satellite, launched in 2008, has two instruments: the Gamma-ray Burst Monitor (GBM, [88]) and the Large Area Telescope (LAT, [89]). The GBM measures the spectra of GRB in the energy range from 8 keV to 40 MeV, determining their position to $\sim 5°$ accuracy. The LAT measures the spectra in the energy range from 20 MeV to 300 GeV, locating the source positions to an accuracy of $< 1°$. The GBM detects GRBs at a rate of ~ 250 per year, of which on average 20% are short bursts, while the LAT detects bursts at a rate of ~ 8 per year. The great strength of this combination is that it provides the large field of view and high detection rate of the GBM, extending to energies as low as the BAT in Swift, with the very-high-energy window of the LAT, which opens up a whole new vista into the previously almost unexplored GeV to sub-TeV range of GRBs.

Two unexpected features of the GeV emission of bursts were soon discovered by the Fermi-LAT. One is that the onset of the GeV emission is invariably delayed relative to the onset of the MeV emission (by a few seconds in LGRBs, and fractions of a second in SGRBs), e.g. [90–93]. The other is that the GeV emission generally lasts for much longer then the MeV emission, decaying as a power law in time and lasting up to 1000 s in some cases, i.e. well into the afterglow phase, including both LGRBs and SGRBs. The fact that GeV emission has been detected from a number of SGRBs is, in itself, also new. Remarkably, the GeV behavior of LGRBs and SGRBs is quite similar. This is not unexpected, since most of the GeV emission is produced in the afterglow phase, which is essentially a self-similar process. What is more unexpected is that the ratio of the total energy in the GeV range to that in the MeV range is ~ 0.1–0.5 for LGRBs, while it is $\gtrsim 1$ for SGRBs.

While in some GRBs observed by the LAT the spectrum is well fitted by a simple broken power-law spectrum (the classical "Band function") with a break near MeV energies and the high-energy slope extending into the tens of GeVs, e.g. GRB080916C [90], in several other bursts, e.g. GRB090902B [93], a second, harder spectral component was evident, at 5σ significance. Another burst with a high-energy second component was GRB 090926A [94], which also showed a well-defined turnover of this second high-energy power law (see Fig. 3.17).

Bursts detected with the LAT have spanned a range of redshifts extending up to $z = 4.3$, with photon energies (in the burst rest frame) up to 10–130 GeV, the highest value so far being that found for GRB 130427A [95], at a redshift $z \sim 0.33$. This is encouraging for the planned large Cherenkov Telescope Array (CTA) [96, 97], whose energy threshold may be as low as 50 GeV and whose detection rate of GRBs is estimated to be in the range

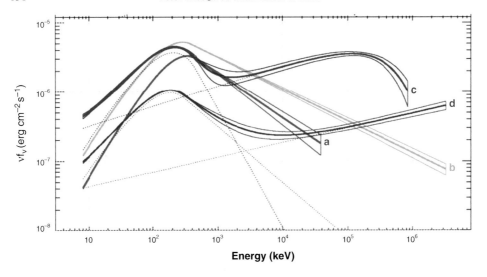

Figure 3.17   Spectra of GRB090926A from the Fermi GBM at low energies and the Fermi LAT at high energies, at four different time intervals, a, [0.0–3.3 s], b, [3.3–9.7 s], c, [9.7–10.5 s], d, [10.5–21.6 s] [94]. The initial time bins show a simple broken power-law spectrum up to $\sim$ GeV, but the latter time bins show a second, rising power-law component extending to tens of GeV, which could be due to inverse Compton scattering or hadronic interactions. © AAS. Reproduced with permission.

0.7–1.6 per year, based on the rate of Swift triggers. A roughly similar rate of detection is also expected for the High Altitude Water Cherenkov (HAWC) detector [98, 99], whose threshold is expected to be 10–20 GeV.

3.4.6 *GRBs in non-photonic channels?*

Two types of non-photonic signals that may be expected from GRBs are gravitational waves (GWs) and high-energy neutrinos (HENUs). The most likely GW emitters are short GRBs [100], if these indeed arise from merging compact objects [12]. The Swift and Fermi localization of a short GRB would help to narrow the search window for gravitational waves from that object [101]. The detection of gravitational waves from a well-localized GRB would lead to a great scientific payoff for understanding the merger physics, the progenitor types, and the neutron star equation of state. The rates of compact merger GW events in the advanced LIGO and VIRGO detectors may be at least several per year [102]. However, even if these events all give rise to gamma-ray bursts, only a small fraction would be beamed towards us. Long GRBs, more speculatively, might be detectable in GWs if they go through a magnetar phase [103], or if their core collapse breaks up into substantial blobs [104]; more detailed numerical calculations of collapsar (long) GRBs lead to GW prospects which range from pessimistic [105] to modest [106].

High-energy neutrinos may also be expected from baryon-loaded GRBs, if sufficient protons are co-accelerated in the shocks. The most widely considered paradigm involves proton acceleration and $p\gamma$ interactions in internal shocks, resulting in prompt ~ 100 TeV HENUs [107, 108]. Other interaction regions considered are external shocks, with $p\gamma$ interactions on reverse-shock UV photons leading to EeV HENUs [109]; and pre-emerging or choked jets in collapsars resulting in HENU precursors [110]. An EeV neutrino flux is also expected from external shocks in very massive Pop. III magnetically dominated GRBs [111]. Current IceCube observations [112] are putting significant constraints on the simplest internal shock neutrino emission model. More careful modeling of internal shocks [113] reveals that several years of observations will be needed for reliably testing such models, while other types of models, such as photospheric models [114] or modified internal shock models [35], are yet to be tested. However, the excitement in this field is palpable, especially since the announcement of the detection by IceCube of PeV neutrinos [115] whose origin is almost certainly astrophysical.

References

[1] Klebesadel, R. W., Strong, I. B., and Olson, R. A. 1973. *Astrophys. J. Lett.*, **182**, L85.
[2] Ruderman, M. 1975. Pages 164–180 of: Bergman, P. G., Fenyves, E. J., and Motz, L. (eds), *Seventh Texas Symposium on Relativistic Astrophysics*. Annals of the New York Academy of Sciences, vol. 262. New York Academy of Sciences.
[3] Meegan, C. A. *et al.* 1992. *Nature*, **355**, 143.
[4] Kouveliotou, C. *et al.* 1993. *Astrophys. J. Lett.*, **413**, L101.
[5] Costa, E. *et al.* 1997. *Nature*, **387**, 783.
[6] van Paradijs, J. *et al.* 1997. *Nature*, **386**, 686.
[7] Cavallo, G., and Rees, M. J. 1978. *Mon. Not. Roy. Astr. Soc.*, **183**, 359.
[8] Eichler, D., Livio, M., Piran, T., and Schramm, D. N. 1989. *Nature*, **340**, 126.
[9] Paczyński, B. 1990. *Astrophys. J.*, **363**, 218.
[10] Mészáros, P., and Rees, M. J. 1992. *Astrophys. J.*, **397**, 570.
[11] Narayan, R., Paczyński, B., and Piran, T. 1992. *Astrophys. J. Lett.*, **395**, L83.
[12] Gehrels, N., Ramirez-Ruiz, E., and Fox, D. B. 2009. *Ann. Rev. Astron. Astrophys.*, **47**, 567.
[13] Woosley, S. E. 1993. *Astrophys. J.*, **405**, 273.
[14] Paczyński, B. 1998. Pages 783–787 of: Meegan, C. A., Preece, R. D., and Koshut, T. M. (eds), *Gamma-ray bursts, 4th Huntsville Symposium*. American Institute of Physics Conference Series, vol. 428. American Institute of Physics.
[15] Galama, T. J. *et al.* 1998. *Nature*, **395**, 670.
[16] Paczyński, B. 1986. *Astrophys. J. Lett.*, **308**, L43.
[17] Goodman, J. 1986. *Astrophys. J. Lett.*, **308**, L47.
[18] Fenimore, E. E., Epstein, R. I., and Ho, C. 1993. *Astron. Astrophys. Suppl.*, **97**, 59.
[19] Harding, A. K., and M. G. Baring, 1994. Page 520 of: Fishman, G. J. (ed), *Gamma-ray bursts*. American Institute of Physics Conference Series, vol. 307. American Institute of Physics.
[20] Shemi, A., and Piran, T. 1990. *Astrophys. J. Lett.*, **365**, L55.
[21] Beloborodov, A. M. 2010. *Mon. Not. Roy. Astr. Soc.*, **407**, 1033.
[22] Mészáros, P., and Rees, M. J. 2011. *Astrophys. J. Lett.*, **733**, L40.
[23] Rees, M. J., and Mészáros, P. 1994. *Astrophys. J. Lett.*, **430**, L93.
[24] Mészáros, P., and Rees, M. J. 1997. *Astrophys. J.*, **476**, 232.
[25] Rees, M. J., and Mészáros, P. 1992. *Mon. Not. Roy. Astr. Soc.*, **258**, 41P.
[26] Mészáros, P., and Rees, M. J. 1993. *Astrophys. J.*, **405**, 278.
[27] Katz, J. I. 1994. *Astrophys. J. Lett.*, **432**, L107.
[28] Sari, R., and Piran, T. 1995. *Astrophys. J. Lett.*, **455**, L143.
[29] Mészáros, P., and Rees, M. J. 1993. *Astrophys. J. Lett.*, **418**, L59.

[30] Paczyński, B., and Xu, G. 1994. *Astrophys. J.*, **427**, 708.
[31] Mészáros, P., and Rees, M. J. 1997. *Astrophys. J. Lett.*, **482**, L29.
[32] Lyutikov, M., and Blandford, R., 2003. arXiv:astro-ph/0312347.
[33] Usov, V. V. 1994. *Mon. Not Roy. Astron. Soc.*, **267**, 1035.
[34] Thompson, C. 1994. *Mon. Not Roy. Astron. Soc.*, **270**, 480.
[35] Murase, K., Asano, K., Terasawa, T., and Mészáros, P. 2012. *Astrophys. J.*, **746**, 164.
[36] Akerlof, C. *et al.* 1999. *Nature*, **398**, 400.
[37] Hurley, K. *et al.* 1994. *Nature*, **372**, 652.
[38] Frail, D. A. *et al.* 2001. *Astrophys. J. Lett.*, **562**, L55.
[39] Rhoads, J. E. 1997. *Astrophys. J. Lett.*, **487**, L1.
[40] Kulkarni, S. R. *et al.* 1999. *Nature*, **398**, 389.
[41] Mészáros, P., and Rees, M. J. 1999. *Mon. Not. Roy. Astr. Soc.*, **306**, L39.
[42] D'Alessio, V., Piro, L., and Rossi, E. M. 2006. *Astron. Astrophys.*, **460**, 653.
[43] Rossi, E., Lazzati, D., and Rees, M. J. 2002. *Mon. Not Roy. Astron. Soc.*, **332**, 945.
[44] Kumar, P., and Granot, J. 2003. *Astrophys. J.*, **591**, 1075.
[45] Zhang, B., Dai, X., Lloyd-Ronning, N. M., and Mészáros, P. 2004. *Astrophys. J. Lett.*, **601**, L119.
[46] Lamb, D. Q., Donaghy, T. Q., and Graziani, C. 2005. *Astrophys. J.*, **620**, 355.
[47] Preece, R. D. *et al.* 1998. *Astrophys. J. Lett.*, **506**, L23.
[48] Ghisellini, G., and Celotti, A. 1999. *Astron. Astrophys. Suppl.*, **138**, 527.
[49] Preece, R. D. *et al.* 1998. *Astrophys. J.*, **496**, 849.
[50] Dermer, C. D., Böttcher, M., and Chiang, J. 1999. *Astrophys. J. Lett.*, **515**, L49.
[51] Eichler, D., and Levinson, A. 2000. *Astrophys. J.*, **529**, 146.
[52] Mészáros, P., and Rees, M. J. 2000. *Astrophys. J.*, **530**, 292.
[53] Pe'er, A., and Waxman, E. 2004. *Astrophys. J.*, **613**, 448.
[54] Rees, M. J., and Mészáros, P. 2005. *Astrophys. J.*, **628**, 847.
[55] Pe'er, A., Mészáros, P., and Rees, M. J. 2006. *Astrophys. J.*, **642**, 995.
[56] Amati, L. *et al.* 2002. *Astron. Astrophys.*, **390**, 81.
[57] Ghirlanda, G., Ghisellini, G., and Lazzati, D. 2004. *Astrophys. J.*, **616**, 331.
[58] Thompson, C., Mészáros, P., and Rees, M. J. 2007. *Astrophys. J.*, **666**, 1012.
[59] Zhang, B. *et al.* 2006. *Astrophys. J.*, **642**, 354.
[60] Nousek, J. A. *et al.* 2006. *Astrophys. J.*, **642**, 389.
[61] Kumar, P., and Panaitescu, A. 2000. *Astrophys. J. Lett.*, **541**, L51.
[62] Zhang, B.-B., Zhang, B., Liang, E.-W., and Wang, X.-Y. 2009. *Astrophys. J. Lett.*, **690**, L10.
[63] Liang, E.-W., Zhang, B.-B., and Zhang, B. 2007. *Astrophys. J.*, **670**, 565.
[64] Falcone, A. D. *et al.* 2007. *Astrophys. J.*, **671**, 1921.
[65] Burrows, D. N. *et al.* 2007. *Phil. Trans. Roy. Soc. Lond. A*, **365**, 1213–1226.
[66] Tanvir, N. R. *et al.* 2009. *Nature*, **461**, 1254.
[67] Cucchiara, A. *et al.* 2011. *Astrophys. J.*, **736**, 7.
[68] Kistler, M. D. *et al.* 2009. *Astrophys. J. Lett.*, **705**, L104.
[69] Robertson, B. E., and Ellis, R. S. 2012. *Astrophys. J.*, **744**, 95.
[70] Berger, E. *et al.* 2005. *Nature*, **438**, 988.
[71] Fox, D. B. *et al.* 2005. *Nature*, **437**, 845.
[72] Fong, W. *et al.* 2013. *Astrophys. J.*, **769**, 56.
[73] Fong, W., Berger, E., and Fox, D. B. 2010. *Astrophys. J.*, **708**, 9.
[74] Leibler, C. N., and Berger, E. 2010. *Astrophys. J.*, **725**, 1202.
[75] Nakar, E. 2007. *Phys. Rep.*, **442**, 166.
[76] Burrows, D. N. *et al.* 2006. *Astrophys. J.*, **653**, 468.
[77] Fong, W. *et al.* 2012. *Astrophys. J.*, **756**, 189.
[78] Norris, J. P., and Bonnell, J. T. 2006. *Astrophys. J.*, **643**, 266.
[79] Gehrels, N. *et al.* 2006. *Nature*, **444**, 1044.
[80] Proga, D., and Zhang, B. 2006. *Mon. Not. Roy. Astr. Soc.*, **370**, L61.
[81] Rezzolla, L. *et al.* 2011. *Astrophys. J. Lett.*, **732**, L6.
[82] Carter, B., and Luminet, J.-P. 1983. *Astron. Astrophys.*, **121**, 97.

[83] Rees, M. J. 1988. *Nature*, **333**, 523.
[84] Kobayashi, S., Laguna, P., Phinney, E. S., and Mészáros, P. 2004. *Astrophys. J.*, **615**, 855.
[85] De Colle, F., Guillochon, J., Naiman, J., and Ramirez-Ruiz, E. 2012. *Astrophys. J.*, **760**, 103.
[86] Burrows, D. N. *et al.* 2011. *Nature*, **476**, 421.
[87] McKinney, J. C., Tchekhovskoy, A., and Blandford, R. D. 2013. *Science*, **339**, 49.
[88] Meegan, C. *et al.* 2009. *Astrophys. J.*, **702**, 791.
[89] Atwood, W. B. *et al.* 2009. *Astrophys. J.*, **697**, 1071.
[90] Abdo, A., *et al.* [Fermi collaboration]. 2009. *Science*, **323**, 1688.
[91] Abdo, A. A., *et al.* [Fermi collaboration]. 2009. *Nature*, **462**, 331.
[92] Ackermann, M., *et al.* [Fermi collaboration]. 2010. *Astrophys. J.*, **716**, 1178.
[93] Abdo, A. A., *et al.* [Fermi collaboration]. 2009. *Astrophys. J. Lett.*, **706**, L138.
[94] Ackermann, M., *et al.* [Fermi collaboration]. 2011. *Astrophys. J.*, **729**, 114.
[95] Zhu, S. *et al.* 2013. *Gamma-ray Burst* Coordinates Network Circular 14471.
[96] Bouvier, A. *et al.*, 2011. arXiv:1109.5680.
[97] Funk, S., and Hinton, J., 2012. arXiv:1205.0832.
[98] Abeysekara, A. U., *et al.* [HAWC collaboration]. 2012. *Astroparticle Phys.*, **35**, 641.
[99] Taboada, I., and Gilmore, R. C., 2013. arXiv:1306.1127.
[100] Centrella, J., Nissanke, S., and Williams, R., 2011. arXiv:1111.1701.
[101] Finn, L. S., Mohanty, S. D., and Romano, J. D. 1999. *Phys. Rev. D*, **60**, 121101.
[102] Leonor, I. *et al.* 2009. *Class. Quant. Grav.*, **26**, 204017.
[103] Corsi, A., and Mészáros, P. 2009. *Astrophys. J.*, **702**, 1171.
[104] Kobayashi, S., and Mészáros, P. 2003. *Astrophys. J.*, **589**, 861.
[105] Ott, C. D. *et al.* 2011. *Phys. Rev. Lett.*, **106**, 161103.
[106] Kiuchi, K., Shibata, M., Montero, P. J., and Font, J. A., 2011. *Phys. Rev. Lett.*, **106**, 251102.
[107] Waxman, E., and Bahcall, J. 1997. *Phys. Rev. Lett.*, **78**, 2292.
[108] Murase, K., and Nagataki, S. 2006. *Phys. Rev. D*, **73**, 063002.
[109] Waxman, E., and Bahcall, J. N. 2000. *Astrophys. J.*, **541**, 707.
[110] Mészáros, P., and Waxman, E. 2001. *Phys. Rev. Lett.*, **87**, 171102.
[111] Gao, S., Toma, K., and Mészáros, P. 2011. *Phys. Rev. D*, **83**, 103004.
[112] Abbasi, R. *et al.* 2012. *Nature*, **484**, 351.
[113] Hümmer, S., Baerwald, P., and Winter, W. 2012. *Phys. Rev. Lett.*, **108**, 231101.
[114] Gao, S., Asano, K., and Mészáros, P. 2012. *J. Cosm. Astroparticle Phys.*, **11**, 58.
[115] Aartsen, M. G. *et al.* 2013. *Phys. Rev. Lett.*, **111**, 021103.

4

Cosmology

4.1 Introduction

David Wands and Roy Maartens

Our Universe provides the grandest arena in which to test General Relativity as a theory of space, time and gravity. It becomes essential to consider both the causal propagation of matter and radiation through space-time and the dynamical evolution of space-time if we are to construct a consistent theoretical framework in which to interpret astronomical observations on the largest observable scales. Einstein himself originally tried to construct a static cosmology but it was soon appreciated that the field equations of General Relativity naturally accommodate dynamical and evolving space-times. Einstein's own static model of a 3-sphere, balancing the gravitational pull of matter against a positive spatial curvature and a cosmological constant was shown to be poised between expansion and collapse and hence unstable to infinitesimal disturbance.

Friedmann and Lemaître showed that Einstein's field equations admit expanding-universe solutions, which have become the basis for modern cosmology, despite Einstein's initial dismissal of the solutions. However persuasive the theoretical models, empirical observations are, of course, necessary to determine the actual dynamics of our observable Universe; the work of Slipher and Hubble in the 1920s [1] persuaded scientists that in fact our Universe is expanding. The logical consequence of this expansion is that either our Universe was hotter and denser in the past (coming ultimately from a Hot Big Bang) or, perhaps, that energy had to be continually created as the universe expanded (the Steady State model). The discovery of the Cosmic Microwave Background (CMB) radiation by Penzias and Wilson in 1965 [2] convinced most astronomers that the Universe did in fact begin at a Hot Big Bang, a finite time in the past. This hot dense plasma, in thermal equilibrium at early times, also provides a setting for the freezing out of the light atomic nuclei as the universe cools below several million degrees Kelvin [3], although heavier elements must be formed later in stars.

The remarkably uniform and black-body CMB radiation, and the inferred primordial abundance of light elements from Big Bang Nucleosynthesis (BBN), remain two cornerstones of the Hot Big Bang model as reviewed in the 1979 Einstein Centenary Survey [4]. At that time, various different models were proposed for the origin and growth of cosmological structures, possibly from topological defects in the vacuum manifold [5] and/or primordial explosions [6]. Indeed the uniformity of the CMB requires an extremely efficient growth of structure, not only due to baryonic matter (protons, neutrons and nuclei), but also implying the existence of non-baryonic dark matter which decoupled from radiation much earlier, and began to coalesce even before the last scattering of the CMB photons.

A new era in cosmology began with the measurement of small primordial anisotropies in the CMB temperature by NASA's COBE satellite in 1992 [7]. Although initially a weak, but statistically significant, signal on the largest angular scales, the pattern of temperature anisotropies at higher angular scales was soon refined by a series of ground- and balloon-based experiments mapping smaller patches of the CMB sky. The angular power spectrum they helped reveal shows striking peaks and dips [8], characteristic of the different phases on different scales of acoustic oscillations at the time of last scattering [9, 10], supporting the existence of a primordial distribution of 'passive' density perturbations (in the adiabatic growing mode) extending to scales beyond the causal horizon size at last-scattering, in contrast to the active perturbations generated by topological defects or other causal mechanisms.

This new CMB cosmology culminated in results from NASA's WMAP satellite [11] and ESA's Planck satellite [12], whose multi-frequency instruments have enabled astronomers to construct all-sky maps of the CMB in exquisite detail, probing both the temperature and the polarisation of the primordial radiation, revealing structure on arc-minute scales. These results are complemented by high-resolution images from ground-based experiments such as the South Pole Telescope [13] and the Atacama Cosmology Telescope [14]. Precise measurements of the acoustic peak structure have enabled percent-level measurements of the parameters of the basic cosmological model, as reviewed by Komatsu in this volume, while opening a new window onto the primordial density perturbation and its possible origin in the very early universe, discussed by Sasaki.

4.1.1 The Hot Big Bang

Although cosmological observations have improved enormously in recent years in terms of both quality and quantity of data, the basic cosmological space-times hark back to the early pioneers of General Relativity: Friedmann, Lemaître, Einstein and de Sitter.

The standard cosmological framework is based on the models of Friedmann and Lemaître, written in the form given by Robertson and Walker. This requires that at a given cosmic time, space is homogeneous and isotropic, but has an overall scale-factor, $a(t)$, that evolves with time. The interval between neighbouring events in the most general spatially homogeneous and isotropic spacetime is given by the FLRW metric

$$ds^2 = -c^2 \, dt^2 + a^2(t)\gamma_{ij} \, dx^i \, dx^j \,, \tag{4.1}$$

where γ_{ij} is the metric on the maximally symmetric 3-space with constant curvature K. At a fixed cosmic time t the three-dimensional spatial hypersurfaces thus have a uniform intrinsic (Ricci) curvature $6K/a^2$. Although 3-spheres with positive curvature necessarily have a finite volume, flat or hyperbolic spaces, $K \leq 0$, can have open or compact spatial sections. There is currently no evidence for any upper bound on the spatial size of our Universe [15], and the latest CMB data imply that the physical curvature scale, $a_0/\sqrt{6|K|}$, is much larger than our observable universe [12].

The physical distance, d, between two observers "at rest" with respect to the cosmic expansion can be given as $d(t) = a(t)r$, where r is the comoving distance, which remains constant for particles comoving with the cosmological expansion. Nonetheless light emitted from distant galaxies travels through the expanding space and its physical wavelength is redshifted by an amount

$$1 + z_e \equiv \frac{\lambda_0}{\lambda_e} = \frac{a_0}{a_e}. \tag{4.2}$$

For small distances $z_e \simeq H_0 d_e/c$, where the observed Hubble "constant", H_0, is just the present value of the Hubble expansion rate, $H = \dot{a}/a$. The present value of the scale factor is usually taken to be $a_0 = 1$ and comoving distances then correspond to the physical distance at the present cosmic time.

In the era of precision cosmology, one of the biggest uncertainties remains the value of H_0 and hence the actual physical scale of the universe; H_0 is often expressed in terms of the dimensionless parameter h, where $H_0 = 100h \, \mathrm{km \, s^{-1} \, Mpc^{-1}}$. One of the key science missions of the Hubble Space Telescope was to measure the Hubble constant to 10-percent accuracy, returning a value of $H_0 = 72 \pm 8 \, \mathrm{km \, s^{-1} \, Mpc^{-1}}$ [16]. More recent data reduces the error to $73.8 \pm 2.4 \, \mathrm{km \, s^{-1} \, Mpc^{-1}}$ [17]. However, recent CMB results from ESA's Planck satellite appear to require either a lower value of $67.3 \pm 1.2 \, \mathrm{km \, s^{-1} \, Mpc^{-1}}$ [12] in the minimal cosmological model, or additional parameters in the model. It is a measure of the progress in observational cosmology that a 10-percent agreement is no longer good enough.

The Hubble expansion rate defines the extrinsic curvature of the three-dimensional spatial hypersurfaces at fixed cosmic time t embedded in the four-dimensional space-time (4.1). Einstein's equations relate this expansion rate to the total energy density, ρc^2, and pressure, p, which therefore must also be homogeneous and isotropic. In particular the Friedmann constraint equation requires

$$H^2 = \frac{8\pi G}{3}\rho + \frac{\Lambda c^2}{3} - \frac{Kc^2}{a^2}, \tag{4.3}$$

while the acceleration is given by

$$\frac{\ddot{a}}{a} = -\frac{4\pi G}{3}\left(\rho + 3\frac{p}{c^2}\right) + \frac{\Lambda c^2}{3}. \tag{4.4}$$

G is Newton's constant and a cosmological constant, Λ, is allowed for. In the absence of spatial curvature or a cosmological constant, a given Hubble expansion, H, would require a corresponding critical density

$$\rho_{\mathrm{crit}} \equiv \frac{3H^2}{8\pi G}, \quad \rho_{\mathrm{crit}0} = 1.88 \times 10^{-29} \, h^2 \, \mathrm{g \, cm^{-3}}. \tag{4.5}$$

The contributions of different terms to the cosmic expansion are often given relative to this critical density, defining the dimensionless density parameters $\Omega_I = \rho_I/\rho_{\text{crit}}$, such that (4.3) is written as

$$\sum_I \Omega_I + \Omega_K + \Omega_\Lambda = 1 , \tag{4.6}$$

where $\Omega_K = -Kc^2/(a^2 H^2)$ and $\Omega_\Lambda = \Lambda c^2/(3H^2)$.

The two most important terms driving the present expansion are found to be that of a cosmological constant and non-relativistic (cold) matter In the minimal "concordance" cosmological model ($\Omega_m + \Omega_\Lambda = 1$) this gives [12]

$$\Omega_{\Lambda 0} = 0.685 \pm 0.017 . \tag{4.7}$$

If the cosmological parameter space is extended to allow for spatial curvature the resulting contribution to the Friedmann equation is constrained (at 95% confidence) to be [12]

$$\Omega_{K0} = -0.037^{+0.043}_{-0.049} . \tag{4.8}$$

The density of non-relativistic matter is diluted by the cosmic expansion such that the matter density decreases in an expanding universe as $\rho_m \propto a^{-3}$ while the additional red-shifting of photons (and other relativistic species) leads the density of radiation to decrease as $\rho_r \propto a^{-4}$. Thus, although radiation contributes only a small fraction of the energy density in the universe today, $\Omega_{r0} \sim 10^{-4}$, radiation is expected to dominate the early expansion, $\Omega_r \simeq 1$. The black-body spectrum of the CMB photons suggests that we have a hot plasma in thermal equilibrium state at early times. The early universe thus appears to be remarkably simple and we can in principle calculate the abundance of different particle species in the late universe by following their equilibrium abundance as interaction rates drop below the Hubble expansion rate as the early universe expands and cools [18]. We do know that baryons have a non-equilibrium abundance when they freeze out, requiring a primordial baryon asymmetry laid down at some very early time. Traditionally the baryon–photon ratio ($n_B/n_\gamma \sim 10^{-9}$) is the only free parameter in the standard cosmological model required to calculate the primordial abundances of the light elements in BBN. However, in recent years CMB experiments such as WMAP and Planck have provided independent determinations of this ratio, and BBN has become a parameter-free theory whose predictions for the primordial helium abundance, Y_P, can in turn be compared against CMB data which is sensitive to both Y_P and the number of relativistic species [3].

The existence of dark matter has for a long time been inferred from dynamical studies of galaxies and clusters of galaxies, but CMB experiments have also confirmed the existence of non-baryonic dark matter which decoupled from the CMB photons significantly before the time of last scattering. In the context of particle physics beyond the standard electroweak model it is not difficult to imagine that there should be additional particle species and that at least one of these might be stable (or very long-lived), perhaps due to the existence of a new conserved quantum number. In particular, supersymmetric theories lead to the concept of R-parity (distinguishing standard-model particles from their super-partners) and the lightest supersymmetric particle (LSP), a super-partner of our standard model particles, would necessarily be stable if R-parity is conserved [19]. A weakly coupled particle whose mass (~ 100 GeV) is close to the supersymmetry-breaking scale would provide roughly the critical density required to dominate cosmic expansion after radiation domination. Whether or not supersymmetry proves correct, it would not be surprising if there exists a new dark matter particle with an abundance close to $\Omega_{m0} \sim 0.3$. Nonetheless, until there is a direct detection of such a particle (or confirmation of supersymmetry, perhaps at the LHC) there will remain interest in theories that might explain galactic dynamics through modification of gravity, such as modified Newtonian dynamics or MOND [20]. Such theories, however, need a broader relativistic framework (such as tensor–vector–scalar gravity, TeVeS [21]) if they are also to mimic the role of dark matter in the CMB anisotropies and the growth of structure in the cosmos.

4.1.2 Dark energy

Super-luminous exploding stars (supernovae) have turned out to be a valuable tool in mapping the cosmological expansion out to redshifts of order one. While galaxies vary in size and luminosity by many orders of magnitude, exploding stars in a specific class, type Ia supernovae, are found to have a characteristic maximum luminosity, with a relatively small variation in luminosity which can be correlated with decay-time [22]. Automated surveys have now mapped hundreds of such supernovae, and it was this method that led the Supernova Cosmology Project and the High-z Supernova Team to announce in 1998 [23, 24] that our Universe is not only expanding, but expanding at an accelerating rate.

The discovery confounded many cosmologists' theoretical prejudice that (cold dark) matter should dominate the present cosmological expansion at late times in an expanding universe, after radiation is redshifted away, $\Omega_r/\Omega_m \propto a^{-1} \to 0$ as $a \to \infty$. On the other hand, in the presence of a cosmological constant we have $\Omega_\Lambda/\Omega_m \propto a^3$ and hence any nonzero cosmological constant would be expected to dominate over matter eventually. The current minimal cosmological model is an FLRW model whose expansion is dominated by a cosmological constant and non-relativistic (cold) dark matter, hence it is known as ΛCDM.

The apparent late-time acceleration has led some authors to re-consider the validity of a homogeneous background cosmology at late times and this possibility is addressed later in this Chapter by MacCallum. However, in an FLRW geometry an accelerating expansion (4.4) requires either a cosmological constant or a dominant energy density with negative pressure, $p < -\rho c^2/3$, or some correction to Einstein's General Relativity. A cosmological constant can be seen either as an otherwise undetermined parameter in Einstein's theory, or as a universal vacuum energy density $\rho_V c^2$ which remains even when the particle number density (matter or radiation) is zero. The two are equivalent in General Relativity: $\Lambda c^2 = 8\pi G \rho_V$. Either is perfectly acceptable from a classical point of view. The absolute value of a universal vacuum energy would have no physical effect apart from its gravitational effect.

And in quantum field theory one would expect the zero-point fluctuations of the vacuum to have an energy density. The cosmological constant problem [25] is the remarkably small observational value, $\rho_V \sim (10^{-3} \text{ eV})^4 c^{-5} \hbar^{-3}$, far smaller than the characteristic scale of supersymmetry, for example, which could regularise the otherwise divergent contribution to the vacuum energy from high-energy modes.

This has led cosmologists to consider dynamical models of dark energy [26] where the present vacuum energy is obtained only at late times. Indeed there are simple scalar field models where the scalar field potential energy scales with other matter energy densities through the (decelerating) radiation and matter-dominated eras. The problem is then how and why the vacuum energy comes to dominate and accelerate the universe at late times, close to the present cosmic epoch – the coincidence problem. Such classical evolutions do not solve the problem of why the quantum vacuum state has such a small energy density, but they do provide a range of predictive models to test against observational data and thus explore the properties of dark energy, such as the equation of state, sound speed or growth rate of linear perturbations.

An alternative approach is to consider modifications of General Relativity [21] which might either drive acceleration, assuming the vacuum energy is exactly zero, or could disguise the true vacuum energy and lead to a much smaller cosmic acceleration rate than predicted in GR. This has led to a resurgence of interest in modified gravity theories such as scalar–tensor gravity or higher-order gravity (e.g., $f(R)$ Lagrangians which generalise the linear Einstein–Hilbert action). There have also been new developments in the study of massive gravity [27], which might be expected to modify gravity only on very long wavelengths, below the graviton mass scale. Many of these developments have taken advantage of higher-dimensional realisations of our four-dimensional cosmology. In particular self-gravitating branes, singular surfaces in a higher-dimensional bulk, offer a novel perspective on cosmological expansion as motion in a bulk space-time [28]; four-dimensional curvature induced on a brane-world may be non-trivially related to the energy density in the full space-time.

4.1.3 Inhomogeneous geometry and inflation

The FLRW metric provides an idealised description of our universe as perfectly homogeneous and isotropic, but the observed universe about us evidently does have inhomogeneous variations in density and temperature. In many cases these can usefully be described as perturbations about a spatially homogeneous background, as first considered by Lifshitz. The density can, for example, be written as $\tilde{\rho}(t, x^i) = \rho(t) + \delta\rho(t, x^i)$, while the space-time metric may be written as [29–31]

$$ds^2 = -c^2(1 + 2A)dt^2 + 2a\left(B_{|i} - S_i\right) c\, dt\, dx^i$$
$$+ a^2(t) \left[(1 + 2C)\, \gamma_{ij} + 2E_{|ij} + 2F_{(i|j)} + h_{ij}\right] dx^i\, dx^j, \tag{4.9}$$

where $X_{i|j}$ denotes a covariant derivative with respect to the maximally-symmetric 3-metric γ_{ij}. Here we have split the metric perturbations into scalars (A, B, C, E), transverse vectors

(S_i and F_i) and transverse and tracefree tensor parts (h_{ij}) [29]. These scalar, vector and tensor parts can then be further decomposed into Fourier modes (or, more generally, eigenvalues of the spatial Laplacian) with comoving wavenumber k. These then evolve independently under the linearised perturbation equations which describe the freely propagating modes.

Breaking the symmetry of the background space-time breaks some important simplifications of the FLRW cosmology, reintroducing some of the complexity of the full relativistic theory. In particular, perturbations in quantities such as density or pressure become gauge-dependent. The FLRW background has a preferred choice of time coordinate, corresponding to homogeneous hypersurfaces at fixed cosmic time, but in the presence of inhomogeneities there is no unique choice of time coordinate and we are free to redefine our time coordinate, $t \rightarrow t + \delta t(t, x^i)$, leading to a change at first order in the local density perturbation and in metric perturbations such as C:

$$\delta\rho \rightarrow \delta\rho - \dot{\rho}\,\delta t, \quad C \rightarrow C - H\,\delta t. \tag{4.10}$$

Gauge-dependence (including spatial gauge transformations) allows us to eliminate two scalar variables and one vector perturbation in a particular gauge. Common choices of gauge include the Poisson or conformal Newtonian gauge (in which $S_i = 0$ and $B = E = 0$) or the comoving-orthogonal gauge in which $B = 0$ and the matter 3-momentum is set to zero.

Alternatively one can construct gauge-invariant combinations [29] such as the conformal Newtonian metric potentials (following the convention used in the section by Komatsu)

$$\Phi = C + aH\left(B - a\dot{E}\right), \quad \Psi = A + \frac{d}{dt}\left[a\left(B - a\dot{E}\right)\right], \tag{4.11}$$

or the gauge-invariant combination of density and curvature perturbations [32]

$$\zeta = C - \frac{H\,\delta\rho}{\dot{\rho}}. \tag{4.12}$$

ζ corresponds to the scalar metric perturbation C in a uniform-density gauge in which $\delta\rho = 0$, or a dimensionless density perturbation, $-H\,\delta\rho/\dot{\rho}$, in a gauge in which $C = 0$. In slow-roll inflation, ζ is the same as the comoving metric perturbation, $\mathcal{R}$, used by Sasaki later in this Chapter. In the concordance model, or its generalisations with isotropic dark energy, $\Psi = -\Phi$ for linear perturbations in the era of structure formation, whereas modified gravity typically has $\Psi \neq -\Phi$.

ζ (and its nonlinear generalisation [33]) has the useful property that it remains constant for adiabatic perturbations on scales much larger than the Hubble scale. (This follows directly from the perturbed continuity equation, independently of the gravity theory [34].) It is therefore commonly used to describe the primordial density perturbation at early times in the Hot Big Bang model, where all astronomical scales were larger than the Hubble scale at that time. It can be used to set the amplitude of the adiabatic growing mode for the Newtonian potentials, $\Phi = -\Psi = (2/3)\zeta$ deep in the radiation-dominated era or $\Phi = -\Psi = (3/5)\zeta$ on large scales in the matter-dominated era [30].

CMB experiments reveal primordial density perturbations described by a power spectrum $\mathcal{P}_\zeta \simeq 2 \times 10^{-9}$ on comoving scales from 10 to 1000 Mpc [12], justifying the use of perturbation theory in the early universe. However, the origin of the primordial density perturbation is completely undetermined in the standard Big Bang model, where it must be specified at early times on scales larger than the causal horizon.

As discussed by Sasaki later in this chapter, models of cosmological inflation in the very early universe offer an explanation for the origin of the primordial density perturbation, arising from initial quantum fluctuations on microscopic scales that are stretched by the accelerated expansion up to arbitrarily large scales. This is the cosmological analogue of Hawking radiation emitted by black holes – a consequence of quantum field theory in curved (in this case cosmological) space-time.[1] Light scalar fields (with an effective mass less than the Hubble scale) acquire an almost scale-invariant spectrum of fluctuations, which can be described by a Gaussian distribution for a weakly coupled field, i.e., neglecting nonlinear interactions. The precise scale-dependence of the power spectrum depends on the effective mass of the field and the time-dependence of the Hubble rate during inflation. The precision of CMB experiments has revealed a weak, but significant, scale-dependence, as predicted in most inflation models. The value of this spectral tilt [12],

$$n_s \equiv 1 + \frac{d \ln \mathcal{P}_\zeta}{d \ln k} = 0.9603 \pm 0.0073 \,, \tag{4.13}$$

is now an important constraint on the dynamics of inflation models, in addition to the overall amplitude of the primordial density perturbations.

Fields other than those driving inflation could play an important role in the cosmological dynamics, including reheating the universe, after inflation [35], and non-adiabatic perturbations in these isocurvature fields during inflation could leave distinctive imprints on the primordial density perturbations, including non-adiabatic and non-Gaussian features which are all now tightly constrained by the latest observations. Many of the proposed alternatives to inflation as the origin of large-scale structure share the same basic picture of vacuum fluctuations [36–41] (or thermal fluctuations [42]) being transferred to super-Hubble scales. Attention has recently turned to the statistics of the primordial density perturbations [43]. Nonlinear interactions of the fields driving inflation, or nonlinear evolution of the perturbations during or after inflation, would provide valuable clues to the physical processes at work during inflation and the origin of structure.

In addition to scalar field perturbations, inflation excites the free (transverse, tracefree) tensor metric perturbations, h_{ij} in (4.9). The amplitude of these primordial gravitational waves generated during inflation is directly proportional to the energy scale of inflation. While primordial density perturbations were required to explain existing observations of large-scale structure, an almost scale-invariant spectrum of primordial gravitational waves was an original prediction of the inflationary paradigm. It is no surprise then that there is currently great excitement following the claim by the BICEP2 experiment to have detected

---

[1] A theory discussed in Chapter 10.

primordial gravitational waves through the distinctive B-mode pattern of polarisation on degree scales [44]. If confirmed, this would fix the energy scale during inflation (around 10^{16} GeV), and present a challenge to alternative models for the origin of structure.[2]

4.1.4 Mapping the matter distribution on the past light cone

In parallel with the development of CMB experiments over the last 30 years, surveys of distant galaxies have evolved from the pioneering CfA survey of hundreds of galaxies [45], to large automated projects with dedicated telescopes mapping hundreds of thousands of galaxies with the two-degree field (2dF) galaxy redshift survey [46] and now millions of galaxies with the Sloan Digital Sky Survey (SDSS I–III) [47, 48], or with the photometric-redshift Dark Energy Survey (DES: see http://www.darkenergysurvey.org).

As discussed by Komatsu later in this Chapter, the CMB offers an all-sky snapshot of the primordial universe with exquisite precision, where nonlinear effects can be accurately treated by second-order perturbations. However, its limitation is that it views the universe primarily at one cosmic time – the epoch of last scattering. By contrast, galaxy surveys have the ability to explore the universe over a range of redshifts, thus probing a far greater volume than the CMB. Earlier galaxy surveys covered a relatively small fraction of the sky at effectively one low redshift. The trend now is towards larger sky area and higher redshifts, so that surveys will provide a map of the galaxy distribution as it evolves in time. Future surveys by instruments such as the Large Synoptic Survey Telescope (LSST) [49], ESA's Euclid satellite [50] and the Square Kilometre Array (SKA) [51], will map unprecedented cosmological volumes.

Counting galaxies and recording their angular and redshift coordinates leads to a map of galaxies on the past light cone in observed (redshift) space. This evolving galaxy distribution encodes a rich tapestry of information on the expansion dynamics of the background universe, the primordial perturbations that seeded the growth of large-scale structure, and the properties of the dark matter and dark energy that influence this growth. Acoustic oscillations in the primordial plasma, observed in the CMB at decoupling when they freeze, leave a fossil imprint in the galaxy distribution in the form of the baryon acoustic oscillation (BAO) scale $r_a \sim 100h^{-1}/(1 + z)$ Mpc. This appears as a local maximum in the two-point correlation function, first measured by the SDSS and 2dF [52].

The BAO feature along (Δz) and transverse ($\Delta \theta$) to the line of sight provides in principle a standard ruler at each redshift, giving independent measures of the Hubble rate $H(z) = \Delta z/[r_a(1 + z)]$ and angular diameter distance $d_A(z) = r_a/\Delta \theta$. BAO is thus a powerful constraint on dark energy and the geometry, and has emerged as one of the most important probes of cosmology from galaxy surveys. Peculiar velocities of galaxies induce an anisotropic distortion of the galaxy correlations in redshift space, turning the BAO sphere into an ellipsoid. However, this distortion itself contains information about the

---

[2] A subsequent joint analysis by BICEP and Planck teams suggests that galactic dust contributes a significant part of the B-mode signal seen by BICEP2, yielding only an upper limit on the amplitude of primordial gravitational waves (see ref [13] in the introduction to this part).

velocities and allows for a measure of the growth rate of structure $f = -d\ln D/d\ln(1+z)$, where $D(z)$ is the growing mode of the linear matter overdensity in comoving-orthogonal gauge, $\delta_c = \delta\rho_c/\rho$.

Cross-correlations of large-scale structure with the CMB anisotropies have provided another crucial test through the detection of the integrated Sachs–Wolfe effect in the CMB [53], arising from the decay of the gravitational perturbation $\Phi - \Psi$ as dark energy (or modified gravity) comes to dominate.

The galaxy power spectrum P_g provides a biased tracer of the matter power spectrum P_m on linear scales: $\delta_g = b\delta_c$ and $P_g = b^2 P_m$. The nonlinear process of halo formation due to small-scale density fluctuations reflects the imprint of large-scale linear modes of the matter overdensity, $\delta_c \propto k^2\Phi$, which provide an effective collapse threshold that varies in space. Galaxy number counts may carry a signature of deviations from a Gaussian primordial density perturbation, not only in the higher-order statistics, but also in the linear power spectrum P_g. An additional coupling between large- and small-scale modes in the primordial potential is characteristic of local-type non-Gaussianity (discussed later by Sasaki)

$$\Phi_p = \phi_p + f_{NL}\left(\phi_p^2 - \langle\phi_p^2\rangle\right), \tag{4.14}$$

where ϕ_p is Gaussian. This leads to a modulation of the small-scale matter power by the long-wavelength potential and hence a scale-dependent as well as redshift-dependent galaxy bias [54, 55]

$$b(z) \to b(z) + \Delta b(z,k), \quad \Delta b \propto f_{NL}k^{-2}, \tag{4.15}$$

which modifies the linear power spectrum $P_g = b^2 P_m$ on very large scales. Current constraints on f_{NL} from galaxy surveys are still significantly weaker than those from the Planck CMB experiment. However, future surveys of much greater volume are expected to outperform the CMB as a probe of non-Gaussianity.

In addition to number counts, we can measure the apparent shapes and sizes of galaxies, to uncover the tiny effects of weak gravitational lensing shear and convergence by the intervening large-scale structure. Weak lensing has the great advantage of being independent of galaxy bias, providing a direct measure of the total matter. It directly probes the combination $\Phi - \Psi$, and is therefore a powerful tool for testing dark energy and modified gravity. The first detections of cosmic shear [56, 57] opened up a new window on the universe, and weak lensing is now a major cosmological probe – one of the more obvious examples of General Relativity at play in cosmology.

On large scales, $>10\,\text{Mpc}$ today, a perturbative description of the growth of structure is still possible, but on smaller scales the density field becomes non-perturbative. The nonlinear matter evolution is commonly treated via a combination of Newtonian N-body numerical simulations and Newtonian analytical methods, calibrated by simulations. A long-standing question that then comes to the forefront is whether (or how) it is consistent to employ Newtonian simulations to describe the growth of structure in General Relativity. This is often assumed to be valid for non-relativistic motion on smaller scales where gravity is well described by a (weak) Newtonian potential and effects such as gauge dependence can be neglected. However, simulations are increasingly being used, and observations are being made, on scales comparable with the Hubble scale (especially at high redshifts where the Hubble scale is smaller). A relativistic interpretation then becomes essential. For linear perturbations the question is relatively straightforward. The Einstein constraint equations can be written in the familiar Newtonian form

$$\nabla^2 \Phi = -4\pi G \rho \delta_c, \tag{4.16}$$

where Φ is in the Newtonian gauge (4.11) and δ_c is in the comoving-orthogonal gauge. Thus one can construct a "dictionary" to relate Newtonian variables in simulations to the correct relativistic quantities [58, 59].

At higher order (or non-perturbatively), there are of course differences between causal, relativistic dynamics and Newtonian action-at-a-distance. Nonetheless, in some cases, in particular in a ΛCDM universe where matter is non-relativistic, it may be possible to use Newtonian evolution equations [60]. Thus one may consistently extract relativistic quantities from Newtonian simulations [61, 62]. In general, however, one needs to impose on the initial data the correct general-relativistic constraints which, unlike the Poisson equation (4.16), are nonlinear [63]. In particular, this relativistic correction leads to an effective local primordial non-Gaussianity $f_{\rm NL}^{\rm rel} = -5/3$ in large-scale structure [64], reflecting the fact that the intrinsic nonlinearity of General Relativity generates non-Gaussianity on the largest scales, even if the primordial perturbations are exactly Gaussian. Due to this correction, the local-type non-Gaussianity (4.14) that can be observed in large-scale structure is related to the primordial value measured by the CMB as

$$f_{\rm NL}^{\rm LSS\ obs} = f_{\rm NL} + f_{\rm NL}^{\rm rel} = f_{\rm NL} - \frac{5}{3}. \tag{4.17}$$

Planck has shown that $f_{\rm NL} = 2.7 \pm 5.8\ (1\sigma)$. One implication is to rule out early-universe models that generate large non-Gaussianity. Another is that the relativistic correction (4.17) cannot be ignored [65].

In addition to these second-order effects, there are also first-order relativistic effects in the observed number overdensity in galaxy surveys, $\delta_g^{\rm obs} = \delta n_g^{\rm obs}/n_g$. Early relativistic approaches to galactic observations formulated the problem on the past light cone of a general universe [66, 67]. Recent work has computed the effects in detail for a perturbed flat FLRW universe [68–71]. These effects arise since we observe galaxies on the past light cone (and not on a constant-time surface), and since we have to compute the physical, observed, overdensity to avoid gauge-dependence on very large scales. In terms of variables in the Newtonian gauge this gives

$$\delta_g^{\rm obs} = (b + \Delta b)\delta_c - \frac{(1+z)}{H}(n^i\,\partial_i)^2 V - 2(1-Q)\kappa$$
$$+ (3 - b_e)\frac{H}{(1+z)}V + \left[b_e - 2Q - 1 - \frac{\dot{H}}{H^2} - 2\frac{(1-Q)(1+z)}{\chi H}\right]n^i\,\partial_i V$$
$$+ \left[2Q - b_e + 1 + \frac{\dot{H}}{H^2} + 2\frac{(1-Q)(1+z)}{\chi H}\right]\Psi + 2(1-Q)\Phi - \frac{1}{H}\dot{\Phi}$$

$$+ 2 \frac{(1-Q)}{\chi} \int_0^\chi d\tilde{\chi} \, (\Psi - \Phi)$$

$$- \left[b_e - 2Q - 1 - \frac{\dot{H}}{H^2} - 2\frac{(1-Q)(1+z)}{\chi H} \right] \int_0^\chi d\tilde{\chi} \frac{(\dot{\Psi} - \dot{\Phi})}{(1+z)}, \tag{4.18}$$

where $\chi = \int dz/H(z)$ is the comoving radial distance. The first line of (4.18) contains the matter overdensity (in comoving-orthogonal gauge in order to give the correct definition of bias), the standard Kaiser redshift distortions (where $\partial_i V$ is the matter peculiar velocity and n^i defines the line-of-sight direction) and the lensing contribution (where κ is the convergence and Q is the magnification bias). The Kaiser and lensing terms are of course relativistic terms, which can be significant on sub-Hubble scales. The following lines give the relativistic effects that can become significant on scales near and beyond $H^{-1}(z)$. The second line has relativistic Doppler terms, where $b_e = -d \ln N_g/d \ln(1+z)$ is the galaxy evolution bias and N_g is the comoving number density. The third line contains Sachs–Wolfe type contributions and the last two lines show the integrated terms of time-delay and ISW type.

4.1.5 Outlook: general relativity and cosmology

The successes of cosmology as it develops into a data-driven science have of necessity shifted the focus to observational, computational and phenomenological issues, sometimes obscuring the role of General Relativity or creating the impression that a Newtonian approach is sufficient. The reality is that General Relativity is at the very foundation of cosmology. When a Newtonian approach is effective, it is only as an approximation to General Relativity, and it is clear that the approximation can at most operate at the level of dynamics, and then only when relativistic species can be neglected. No Newtonian approximation can describe the CMB, or more generally the propagation of lightrays in the inhomogeneous universe, which is crucial for the interpretation of cosmological observations.

The growing precision of CMB observations has required a commensurate advance in second-order general relativistic analysis – including the computation of the scalar contribution to CMB lensing (detected by Planck), and the careful identification of secondary contamination to the primordial non-Gaussianity in the bi- and trispectrum. CMB B-mode polarization is a direct probe of inflationary gravitational waves. More accurate data will be vital to help tie down the tensor–scalar ratio – and a corresponding theoretical effort will be needed to consider possible contaminants that could bias the signal.

The careful relativistic treatment of primordial non-Gaussianity is crucial for constraining models of inflation. Future ground-based CMB data will not be able to improve significantly on Planck constraints on non-Gaussianity, and the new frontier will be massive galaxy surveys. The relativistic corrections (4.17) and (4.18) for these surveys will be needed to accurately measure primordial non-Gaussianity. In particular, to rule out the simplest single-field inflation models, it will be necessary to show that the effective f_{NL} deviates from the general-relativistic prediction $-5/3$.

Future galaxy surveys covering huge volumes will begin to deliver the statistical power necessary for stringent tests of the concordance model, of alternative dynamical models of dark energy, and of modified gravity. Cosmology has become an arena for testing General Relativity itself on the largest scales – which will require careful accounting of all relativistic effects involved in galaxy surveys.

In addition to counting radio-luminous galaxies, the SKA will perform intensity mapping of the total neutral hydrogen (HI) 21 cm emission in patches of the sky, thus probing the matter distribution on large scales, with automatic and very accurate redshift information. These intensity map surveys can probe both HI galaxies out to redshifts $z \sim 3$–5, and the evolution of uncollapsed HI during the epoch of reionization $z \gtrsim 6$. Ultimately, with generation-after-next sensitivity, it will be possible to probe the HI during the 'dark ages', before the first stars are formed.

References

[1] Raifeartaigh, C. O., 2012. arXiv:1212.5499.
[2] Penzias, A. A., and Wilson, R. W. 1965. *Astrophys. J.*, **142**, 419.
[3] Steigman, G. 2012. *Adv. High Energy Phys.*, 268321.
[4] Hawking, S. W., and Israel, W. (eds). 1979. *General relativity: an Einstein centenary survey.* Cambridge: Cambridge University Press.
[5] Kibble, T. W. B. 1980. *Phys. Reports*, **67**, 183.
[6] Ostriker, J. P., and Cowie, L. L. 1981. *Astrophys. J.*, **243**, L127–L131.
[7] Smoot, G. F. *et al.* 1992. *Astrophys. J.*, **396**, L1–L5.
[8] Jaffe, A. H., et al. [Boomerang Collaboration]. 2001. *Phys. Rev. Lett.*, **86**, 3475–3479.
[9] Peebles, P. J. E., and Yu, J. T. 1970. *Astrophys. J.*, **162**, 815.
[10] Sunyaev, R. A., and Zel'dovich, Ya. 1970. *Astrophys. Space Sci.*, **7**, 3–19.
[11] Hinshaw, G., et al. [WMAP]. 2013. *Astrophys. J. Suppl.*, **208**, 19.
[12] Ade, P. A. R., et al. [Planck Collaboration]. 2014. *Astron. Astrophys.*, **571**, A16.
[13] Keisler, R. et al. 2011. *Astrophys. J.*, **743**, 28.
[14] Sievers, J. L., et al. [Atacama Cosmology Telescope]. 2013. *JCAP*, **1310**, 060.
[15] Ade, P. A. R., et al. [Planck Collaboration]. 2014. *Astron. Astrophys.*, **571**, A26.
[16] Freedman, W. L., et al. [HST Collaboration]. 2001. *Astrophys. J.*, **553**, 47–72.
[17] Riess, A. G. et al. 2011. *Astrophys. J.*, **730**, 119.
[18] Kolb, E. W., and Turner, M. S. 1990. *Front. Phys.*, **69**, 1–547.
[19] Olive, K. A., 1999. arXiv:hep-ph/9911307.
[20] Milgrom, M. 1983. *Astrophys. J.*, **270**, 365–370.
[21] Clifton, T., Ferreira, P. G., Padilla, A., and Skordis, C. 2012. *Phys. Reports*, **513**, 1–189.
[22] Phillips, M. M. 1993. *Astrophys. J.*, **413**, L105–L108.
[23] Riess, A. G., et al. [Supernova Search Team]. 1998. *Astron. J.*, **116**, 1009–1038.
[24] Perlmutter, S., et al. [Supernova Cosmology Project]. 1999. *Astrophys. J.*, **517**, 565–586.
[25] Weinberg, S. 1989. *Rev. Mod. Phys.*, **61**, 1–23.
[26] Copeland, E. J., Sami, M., and Tsujikawa, S. 2006. *Int. J. Mod. Phys.*, **D15**, 1753–1936.
[27] Hinterbichler, K. 2012. *Rev. Mod. Phys.*, **84**, 671–710.
[28] Maartens, R. 2004. *Living Rev. Rel.*, **7**, 7.
[29] Bardeen, J. M. 1980. *Phys. Rev.*, **D22**, 1882–1905.
[30] Mukhanov, V. F., Feldman, H. A., and Brandenberger, R. 1992. *Phys. Reports*, **215**, 203–333.
[31] Malik, K. A., and Wands, D. 2009. *Phys. Reports*, **475**, 1–51.
[32] Bardeen, J. M., Steinhardt, P. J., and Turner, M. 1983. *Phys. Rev.*, **D28**, 679.
[33] Lyth, D. H., Malik, K. A., and Sasaki, M. 2005. *JCAP*, **0505**, 004.

[34] Wands, D., Malik, K. A., Lyth, D. H., and Liddle, A. 2000. *Phys. Rev.*, **D62**, 043527.
[35] Bassett, B. A., Tsujikawa, S., and Wands, D. 2006. *Rev. Mod. Phys.*, **78**, 537–589.
[36] Gasperini, M., and Veneziano, G. 1993. *Astropart. Phys.*, **1**, 317–339.
[37] Wands, D. 1999. *Phys. Rev.*, **D60**, 023507.
[38] Finelli, F., and Brandenberger, R. 2002. *Phys. Rev.*, **D65**, 103522.
[39] Khoury, J., Ovrut, B. A., Steinhardt, P. J., and Turok, N. 2001. *Phys. Rev.*, **D64**, 123522.
[40] Rubakov, V. A. 2009. *JCAP*, **0909**, 030.
[41] Hinterbichler, K., and Khoury, J. 2012. *JCAP*, **1204**, 023.
[42] Brandenberger, R. H., and Vafa, C. 1989. *Nucl. Phys.*, **B316**, 391.
[43] Bartolo, N., Komatsu, E., Matarrese, S., and Riotto, A. 2004. *Phys. Reports*, **402**, 103–266.
[44] Ade, P. A. R., *et al.* [BICEP2 Collaboration]. 2014. *Phys. Rev. Lett.*, **112**, 241101.
[45] V. de Lapparent, M. J. G., and Huchra, J. P. 1986. *Astrophys. J.*, **302**, L1.
[46] Cole, S., *et al.* [2dFGRS Collaboration]. 2005. *Mon. Not. Roy. Astron. Soc.*, **362**, 505–534.
[47] Abazajian, K. N., *et al.* [SDSS Collaboration]. 2009. *Astrophys. J. Suppl.*, **182**, 543–558.
[48] Dawson, K. S., *et al.* [BOSS Collaboration]. 2013. *Astron. J.*, **145**, 10.
[49] Abell, P. A., *et al.* [LSST Science Collaborations, LSST Project]. 2009. arXiv:0912.0201.
[50] Amendola, L., *et al.* [Euclid Theory Working Group]. 2013. *Living Rev. Rel.*, **16**, 6.
[51] Blake, C. A., Abdalla, F. B., Bridle, S. L., and Rawlings, S. 2004. *New Astron. Rev.*, **48**, 1063–1077.
[52] Percival, W. J. *et al.* 2007. *Mon. Not. Roy. Astron. Soc.*, **381**, 1053–1066.
[53] Boughn, S., and Crittenden, R. 2004. *Nature*, **427**, 45–47.
[54] Dalal, N., Dore, O., Huterer, D., and Shirokov, A. 2008. *Phys. Rev.*, **D77**, 123514.
[55] Matarrese, S., and Verde, L. 2008. *Astrophys. J.*, **677**, L77–L80.
[56] Wittman, D. M. *et al.* 2000. *Nature*, **405**, 143–149.
[57] Bacon, D. J., Refregier, A. R., and Ellis, R. S. 2000. *Mon. Not. Roy. Astron. Soc.*, **318**, 625.
[58] Chisari, N. E., and Zaldarriaga, M. 2011. *Phys. Rev.*, **D83**, 123505.
[59] Green, S. R., and Wald, R. M. 2012. *Phys. Rev.*, **D85**, 063512.
[60] Hwang, J.-C., Noh, H., and Gong, J.-O. 2012. *Astrophys. J.*, **752**, 50.
[61] Bruni, M., Thomas, D. B., and Wands, D. 2014. *Phys. Rev.*, **D89**, 044010.
[62] Adamek, J., Daverio, D., Durrer, R., and Kunz, M. 2013. *Phys. Rev.*, **D88**, 103527.
[63] Bruni, M., Hidalgo, J. C., Meures, N., and Wands, D. 2014. *Astrophys. J.*, **785**, 2.
[64] Verde, L., and Matarrese, S. 2009. *Astrophys. J.*, **706**, L91–L95.
[65] Camera, S., Santos, M. G., and Maartens, R. 2015. *Mon. Not. Roy. Astr. Soc.*, **448**, 1035.
[66] Kantowski, R., and Sachs, R. K. 1966. *J. Math. Phys.*, **7**, 443–446.
[67] Ellis, G. F. R. *et al.* 1985. *Phys. Reports*, **124**, 315–417.
[68] Yoo, J., Fitzpatrick, A. L., and Zaldarriaga, M. 2009. *Phys. Rev.*, **D80**, 083514.
[69] Yoo, J. 2010. *Phys. Rev.*, **D82**, 083508.
[70] Bonvin, C., and Durrer, R. 2011. *Phys. Rev.*, **D84**, 063505.
[71] Challinor, A., and Lewis, A. 2011. *Phys. Rev.*, **D84**, 043516.

4.2 Inflationary cosmology

Misao Sasaki

4.2.1 Introduction

One of the most successful applications of the theory of general relativity is cosmology. The big-bang theory of the universe, that the universe was born in an extremely hot and dense state, expanded explosively and cooled down to the present state, was observationally tested from various aspects and by the end of the last century it was firmly established.

According to the big-bang theory, our universe is about 14 Gigayears old, and the universe was initially radiation-dominated. It became matter-dominated when the universe was about 100,000 years old, which happens to be about the same time as when the photons decoupled from baryons, and started to travel freely. Today they are observed as the cosmic microwave background (CMB) radiation. The epoch when the CMB photons were scattered last before they reach observers at present forms a three-dimensional hypersurface, and it is called the last scattering surface (LSS). These CMB photons turned out to carry extremely valuable information about the very early universe. The CMB frequency spectrum, which was found to be an almost exact Planck distribution by COBE FIRAS [1] is one such piece of information. It gave the final confirmation of the big-bang theory.

Despite its tremendous success, however, there were still a few very basic problems that the big-bang theory could not explain. One of them is the horizon problem, perhaps better called the causality problem, and another is the flatness problem or the entropy problem.

Horizon problem

Let us first consider the horizon problem. The big-bang theory assumes a homogeneous and isotropic universe on large scales. So the metric is assumed to be in the form

$$ds^2 = -dt^2 + a^2(t)d\sigma_{(3)}^2 , \tag{4.19}$$

where $d\sigma_{(3)}^2$ is the 3-metric of a constant-curvature space with K being the curvature, $^{(3)}R^{ij}{}_{km} = K(\delta_k^i \delta_m^j - \delta_m^i \delta_k^j)$. A coordinate system that spans $d\sigma_{(3)}^2$ is said to be comoving because an observer staying at a fixed point on the 3-space is comoving with the expansion of the universe. In this spacetime, the time–time component of the Einstein equations, the Friedmann equation, is

$$H^2 = \frac{\rho}{3M_{Pl}^2} - \frac{K}{a^2} ; \quad H \equiv \frac{\dot{a}}{a} , \tag{4.20}$$

where $M_{Pl}^2 = (8\pi G)^{-1}$ in the units $\hbar = c = 1$, and the trace of the space–space components of the Einstein equations gives

$$\frac{\ddot{a}}{a} = -\frac{\rho + 3P}{3M_{Pl}^2} , \tag{4.21}$$

where ρ is the energy density and P is the pressure of the universe. This latter equation shows that the expansion of the universe is always decelerating as long as $\rho + 3P > 0$. This result may be regarded as a consequence of the attractive nature of the gravitational force.

For simplicity, if we assume a simple equation of state $P/\rho = w = $ constant and $K = 0$, one finds

$$a \propto t^n ; \quad n = \frac{2}{3(1+w)} < 1 \quad \text{for } w > -\frac{1}{3} . \tag{4.22}$$

In particular, during most stages of the early universe, the radiation-domination, $w = 1/3$, is a very good approximation, while the matter-domination, $w = 0$, is a good approximation from the time of baryon–photon decoupling until quite recently. The scale factor behaves as $t^{1/2}$ at the radiation-dominated stage and as $t^{2/3}$ at the matter-dominated stage.

To explain the horizon problem, let us introduce the conformal time $d\eta = dt/a(t)$, and rewrite the metric as

$$ds^2 = a^2(\eta)d\hat{s}^2 ; \quad d\hat{s}^2 = -d\eta^2 + d\sigma_{(3)}^2 . \tag{4.23}$$

Since the conformal transformation of the metric does not change the causal structure, we can discuss the causal structure of the universe with the static metric $d\hat{s}^2$ perfectly well. Now if $w > -1/3$ or $\rho + 3P > 0$, the conformal time is finite in the past because we have

$$\eta = \int_0^t \frac{dt'}{a(t')} \propto \int_0^t \frac{dt'}{t'^n} ; \quad n = \frac{2}{3(1+w)} , \tag{4.24}$$

and the integral is finite for $n < 1$, i.e., for $w > -1/3$. This implies that the size of the future light cone emanating from a point at the beginning of the universe when $\eta = 0$ will cover only a finite fraction of spacetime at any finite time. Conversely, the past light cone emanating from us cannot cover the whole spacelike surface at the beginning of the universe. This is the horizon problem.

Observationally, the horizon problem exhibits itself most clearly in CMB. Since the comoving distance traveled by light is equal to the corresponding conformal time interval, the comoving radius of a causally connected region from the beginning of the universe on the LSS is equal to its conformal time η_{LSS}. From the fact that the LSS is located at redshift $z \sim 10^3$ and the universe has been approximately matter-dominated since then, one finds that each causally connected region covers only a tiny fraction (about 10^{-4}) of the sky (see Fig. 4.1).

The solution is clear: The horizon problem disappears if the conformal time is infinite in the past. Or the problem can be effectively solved if the beginning of the universe $\eta =$

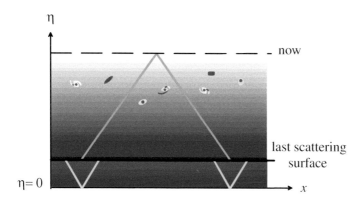

Figure 4.1 Horizon problem. The conformal time of the last scattering surface η_{LSS} from $\eta = 0$ is about 1/30 of that of today, η_0. Reprinted from ref. [46] with kind permission from Springer Science+Business Media B.V.

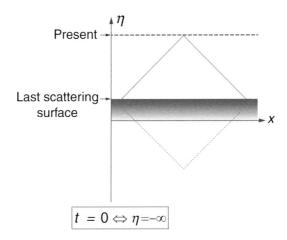

Figure 4.2 The solution to the horizon problem. If the conformal time η at the beginning of the universe is sent to $-\infty$, or to a value sufficiently large and negative, the entire observable universe is within a causally connected region. Reprinted from ref. [46] with kind permission from Springer Science+Business Media B.V.

0 is extended sufficiently back in time so that the future light cone emanating from the beginning of the universe covers the whole visible universe, which is the comoving volume covered by the cross section of the past light cone from the present with the LSS. Since the comoving radius of the visible universe on the LSS is $\eta_0 - \eta_{LSS}$, where η_0 is the conformal time today, the problem is solved if $\eta_{LSS} > \eta_0 - \eta_{LSS}$ (see Fig. 4.2). In Einstein gravity, this means that the equation of state must be $w < -1/3$ or equivalently the expansion of the universe must be accelerating ($\ddot{a} > 0$) for a sufficient lapse of time in the very early universe.

Here we should note that solving the horizon problem does *not* explain the homogeneity and isotropy of the universe. As is clear from the above argument, we had to assume the homogeneity and isotropy of the universe to pose the horizon problem. Explaining the homogeneity and isotropy of the universe in terms of some causal mechanism is a different issue. This point is very often misunderstood in the literature.

Flatness problem

Again we assume a spatially homogeneous and isotropic universe, Eq. (4.19). The Friedmann equation (4.20) tells us that the curvature term K/a^2 is completely negligible in the early universe when it is radiation-dominated, $\rho \propto a^{-4}$. Thus if we consider the initial condition of the universe at an early epoch, the spatial curvature term must be fine-tuned to an extremely small value compared to the energy density. This is the flatness problem. Conversely, if the curvature term was of the same order of magnitude as the density at some epoch in the early universe, the universe must have either collapsed (if $K > 0$) or become completely empty (if $K < 0$) by now. In other words, the flatness problem is equivalent to the oldness problem, that why the universe could have survived for such a long time as 10 Gigayears, which is far greater than any other meaningful timescale in physics.

Alternatively, since the radiation dominates the entropy of the universe, the problem may be rephrased as the existence of huge entropy within the curvature radius of the universe,

$$S = T^3 \left(\frac{a}{\sqrt{|K|}} \right)^3 \approx T_0^3 \left(\frac{a_0}{\sqrt{|K|}} \right)^3 > T_0^3 H_0^3 \approx 10^{87} , \qquad (4.25)$$

where $T_0 \approx 2.7$ K is the CMB temperature today [1] and $H_0 \approx 72$ km/s/Mpc is the Hubble constant [2]. Hence the flatness problem may be called the entropy problem.

It is then apparent that the flatness problem may be solved if there was huge entropy production at a sufficiently early stage of the universe before it collapes or becomes completely empty.

Solution to horizon and flatness problems

A simple and perhaps the best solution to the horizon and flatness problems is given by the inflationary universe [3, 4]. Let us assume that the universe was dominated by a spatially homogeneous scalar field. For a minimally coupled canonical scalar field ϕ, we have

$$\rho = \frac{1}{2}\dot{\phi}^2 + V(\phi), \quad P = \frac{1}{2}\dot{\phi}^2 - V(\phi), \qquad (4.26)$$

so $\rho + 3P = 2(\dot{\phi}^2 - V(\phi))$. Hence, if $\dot{\phi}^2 < V(\phi)$, we may have accelerated expansion. In particular, if the energy density is dominated by the potential energy, $\dot{\phi}^2 \ll V(\phi)$, the motion of the scalar field can be ignored within a few expansion times $\sim H^{-1}$, and the universe expands almost exponentially,

$$H^2 \approx \frac{\rho}{3M_{Pl}^2} \approx \text{constant.} \qquad (4.27)$$

Then the curvature term K/a^2 becomes completely negligible even if it existed in the beginning.

Thus if the universe is dominated by the potential energy, and the potential energy is converted to radiation after a sufficient duration of such a stage, a huge entropy is produced and the horizon and flatness problems are solved simultaneously.

4.2.2 Inflation

Old inflation

Originally the inflationary universe was proposed in the context of grand unified theories. After the successful unification of the electromagnetic and weak interactions by SU(2)×U(1) gauge theory with spontaneous symmetry breaking through the Higgs mechanism, a lot of effort was devoted to also unifying the strong interaction with the now unified electro-weak interaction, called the grand unification. From an estimate of the energy-scale dependence of the coupling constants, the grand unification is expected to occur at around $E \sim 10^{15}$ GeV, and again the Higgs mechanism is responsible for spontaneous symmetry breaking.

Without fine-tuning, one would expect that the symmetry breaking would be a first-order phase transition. Namely, the scalar field ϕ responsible for symmetry breaking has a potential barrier, and the universe was in the meta-stable state, say at $\phi = 0$, before the transition. The potential energy $V(0)$ at the meta-stable state is positive if the vacuum energy of the symmetry-breaking state is non-negative (which should be the case because the vacuum energy of the universe today is non-negative). Inserting $\phi = 0$ and $V = V(0) =$ const. in Eq. (4.26), one immediately finds $P = -\rho$. Thus the universe is dominated by a pure cosmological constant, and it undergoes exponential expansion. This is called the old inflationary scenario.

The old scenario realizes inflation successfully. However, unfortunately it was almost simultaneously realized that the inflationary stage either never ends or ends up with a highly inhomogeneous universe. This was called the graceful exit problem [4]. The reason is fairly simple. Once the universe was stuck at a meta-stable state, any form of the energy density other than the vacuum energy that might have been present prior to inflation dies out quickly. So the transition to a stable vacuum must occur through quantum tunneling, that is, through nucleation of vacuum bubbles by quantum fluctuations.

Nucleated bubbles would expand at the speed of light. So at a state of decelerated expansion, which would be the case for the standard radiation-dominated universe, bubbles would collide with each other and cover the whole three-dimensional space sooner or later, terminating the phase transition through thermalization of the energy stored in the bubble walls.

However, in an exponentially expanding universe, any two points separated by a distance greater than the Hubble length H^{-1} initially will never be in causal contact in the future. This means that even if the bubbles expand at the speed of light, they may not be able to collide with each other if the nucleation rate is less than H^4 per unit volume per unit time. Thus the inflationary stage would never end [5].

On the other hand, if the nucleation rate is greater than H^4, the phase transition occurs too fast, within a Hubble timescale. Thus to begin with, inflation would not be realized. Therefore the only possibility for the realization of inflation for a sufficiently long but finite period of time is the case when the nucleation rate Γ is fine-tuned to a certain value of $O(H^4)$.

However, even in this case the universe would become too inhomogeneous after inflation. The reason is similar to any critical phenomenon in physics. There is a critical value of the nucleation rate $\Gamma_c = O(H^4)$ above which inflation terminates and below which inflation never ends. Since the period of inflation should be long enough, Γ must be fine-tuned to a value extremely close to, but above, Γ_c. This implies that the space would be filled with bubbles of various sizes with an almost self-similar distribution. Namely, the universe would look the same when seen on any length scale. In particular, there would exist bubbles of size comparable to the comoving size of the current Hubble radius. Thus the thermalization of the vacuum energy would never be effective enough to make the universe sufficiently homogeneous and isotropic.

Slow-roll inflation

Soon after it became clear that the old scenario would not work, several authors noticed that if the phase transition was second order and would proceed slowly enough, one could realize inflation successfully. This is called the new inflationary scenario [6, 7]. However, unfortunately it turned out that such a model could not be embedded in a phase transition associated with grand unified theories because of a huge discrepancy in the required magnitude of the coupling constants.

Then it was pointed out by Linde that for a scalar field to evolve slowly enough, a second-order phase transition is not necessary [8]. In fact, the only necessary condition is that the potential slope is flat enough in comparison with the Hubble energy scale. In particular, he argued that the universe might have been born out of a chaotic state at the Planck energy scale in which the scalar field (called the "inflaton" field) is randomly distributed. He argued that even for a simple quartic or quadratic potential, if the energy scales of the potential are much smaller than the Planck scale, the probability that a scalar field will have a fairly homogeneous configuration within a Hubble horizon size region is non-negligible, albeit not very high. Then Eq. (4.26) is applicable to such a region, and the inflaton starts to evolve slowly, leading to almost exponential expansion of the universe.

This specific scenario of inflation in which the universe begins with a chaotic state is called chaotic inflation [8]. However, the crucial point of this scenario is the condition that all the energy scales associated with the potential are smaller than the Planck scale during inflation, $V \ll M_{Pl}^4$, $V' \ll M_{Pl}^3$ and $V'' \ll M_{Pl}^2$, independent of the initial condition. The whole class of models in which inflation is realized by a scalar field which slowly rolls down the potential hill is called slow-roll inflation.

Since Linde's discovery of chaotic inflation, numerous models of inflation have been proposed and studied without relying on specific particle-physics models. This could be regarded as unhealthy if the inflaton should play a physical role in particle physics or in unified theories. Nevertheless, it was not so bad or even good from a different perspective: Thanks to all these previous studies of a large variety of models, we now understand fairly clearly the relations between the model parameters and the observable quantities such as the amplitude and shape of the curvature-perturbation spectrum. Thus at least from a purely phenomenological point of view, we now know more or less how to constrain models of inflation. Below we consider slow-roll inflation in more detail.

Assuming the universe is dominated by a spatially homogeneous scalar field ϕ, the field equation and the Friedmann equation are

$$\ddot{\phi} + 3H\dot{\phi} + V'(\phi) = 0, \quad H^2 = \frac{1}{3M_{Pl}^2}\left[\frac{1}{2}\dot{\phi}^2 + V(\phi)\right], \quad (4.28)$$

where we have neglected the curvature term. The standard slow-roll condition consists of two assumptions. One is that $\ddot{\phi}$ is negligible compared to $3H\dot{\phi}$ in the field equation, that is, the equation of motion is friction-dominated. The other is that the kinetic term $\dot{\phi}^2/2$ is negligible compared to the potential term V in the energy density. Under this condition we have

$$\dot{\phi} \approx -\frac{V'(\phi)}{3H} ; \quad H^2 \approx \frac{V}{3M_{Pl}^2} . \tag{4.29}$$

Then the potential energy dominance implies

$$\epsilon \equiv -\frac{\dot{H}}{H^2} = \frac{\frac{3}{2}\dot{\phi}^2}{\frac{1}{2}\dot{\phi}^2 + V} \approx \frac{3\dot{\phi}^2}{2V} \approx \frac{M_{Pl}^2}{2}\frac{V'^2}{V^2} \equiv \epsilon_V \ll 1 , \tag{4.30}$$

that is, the universe is expanding almost exponentially, and the friction-dominated equation of motion $|\ddot{\phi}/(3H\dot{\phi})| \ll 1$ implies

$$\frac{V''}{3H^2} \approx M_{Pl}^2 \frac{V''}{V} \equiv \eta_V \ll 1 . \tag{4.31}$$

The single-field slow-roll inflation satisfies these conditions.

The important property of slow-roll inflation is that Eq. (4.29) is completely integrable since H is a function of ϕ. In particular, there is one-to-one correspondence between ϕ and t. So instead of the cosmic time t we may measure the time in terms of the value of the scalar field.

Here we introduce a quantity which plays a very important role in the dynamics of slow-roll inflation, namely the number of e-folds counted *backward* in time, say from the end of inflation to an epoch during inflation,

$$\frac{a(t_{end})}{a(t)} = \exp[N(t \to t_{end})] \;\rightarrow\; N = N(\phi) = \int_{t(\phi)}^{t_{end}} H \, dt . \tag{4.32}$$

Its important property is that by definition it does not depend on how and when the inflation began. As shown in Fig. 4.3, N is uniquely determined in terms of the value of the scalar field (up to a constant which depends on the choice of an epoch from which N is computed), and one can associate N with the time at which a given comoving wavenumber k crossed the Hubble radius, $k = aH$, at which the value of the scalar field was ϕ_k; $N = N(\phi_k)$. As we shall see below, this turns out to be an essential quantity for the evaluation of the curvature perturbation from inflation.

There are typically two classes of slow-roll inflation; large-field models and small-field models. The large-field models are those in which the inflaton moves in the field space through a distance greater than the Planck scale, $\Delta\phi \gtrsim M_{Pl}$, during the last 50–60 e-foldings of inflation. A typical example is chaotic inflation. For large-field models, the Hubble constant during inflation is comparatively large, $H \gtrsim 10^{-5}M_{Pl}$. This leads to an amplitude of the tensor perturbation detectable in the near future. In this respect, we mention the fact that the BICEP2 team recently announced that they had detected a B-mode CMB spectrum that may be due to the primordial tensor perturbation [44].

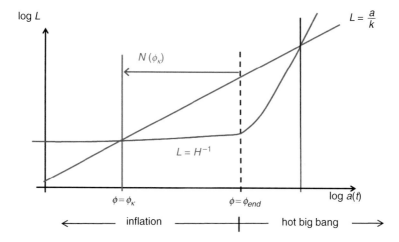

Figure 4.3 The Hubble radius $L = H^{-1}$ and the length scale $L = a/k$ of a comoving wavenumber k in the inflationary cosmology, and the definition of the number of e-folds $N(\phi)$. Reprinted from ref. [46] with kind permission from Springer Science+Business Media B.V.

On the other hand, for small-field models the inflaton moves through a distance much shorter than the Planck scale, $\Delta\phi \ll M_{Pl}$. In this case, detection of the tensor perturbation in the foreseeable future seems impossible.

Before ending this section, let us make a few comments on how inflation ends. As the inflaton evolves, it eventually approaches a minimum of its potential, where the kinetic energy becomes comparable to the potential energy. This is the epoch when the inflaton is no longer slow-rolling and the inflationary stage ends. It is usually assumed that then the inflaton undergoes damped oscillations, converting its potential energy to radiation and thermalizing the universe. This is called reheating of the universe. A variety of reheating mechanisms have been proposed, but they may be mainly classified into two types; perturbative reheating and non-perturbative reheating.

The former is the conventional case when the inflaton couples typically to light fermions and reheating proceeds with the decay of the inflaton field to those relativistic particles [10]. This process is relatively slow, and it can be computed perturbatively. More specifically, denoting the decay rate of the inflaton field by Γ, the damped oscillations last as long as the Hubble expansion rate is larger than the decay rate, $H > \Gamma$. Then, as soon as an epoch when $H = \Gamma$ is reached, the decay occurs almost instantaneously and the universe is thermalized. In this case, the reheating temperature is known to be given by $T_R \sim \sqrt{\Gamma M_{Pl}}$, which is typically much lower than the potential-energy scale at the end of inflation $E_f \sim H_f M_{Pl}$, where H_f is the Hubble scale at the end of slow-roll inflation.

The latter is the case when the inflaton is coupled to another scalar field, say the χ field, with the strength of coupling that gives rise to rapid, resonant particle creation of χ particles [11]. This is called preheating, since it can convert the whole energy stored in the inflaton field to the χ particles within a couple of expansion times. In this case the reheating temperature can be comparable to E_f if the thermalization timescale is sufficiently short.

In both cases, the epoch at which the universe is thermalized may be regarded as the birth of the radiation-dominated, hot big-bang universe. Thus the reheating determines the matter content of the big-bang universe, but to understand it requires an extension of the standard model of particle physics. Lacking any concrete theory or experimental data beyond the standard model, here we will not pursue physics of reheating further.

4.2.3 Cosmological perturbations from inflation

One of the most important predictions of (slow-roll) inflation is the generation of seed cosmological perturbations for the large-scale structure of the universe from the quantum vacuum fluctuations of the inflaton field and the metric [12]. They are called the curvature perturbations since they can be interpreted as perturbations in the spatial curvature on slices of uniform energy density (or comoving slices; see below) on large scales: Roughly speaking they give a spatially dependent curvature term in the Friedmann equation (4.20), $K \to K(x)$.

Another important prediction is the generation of gravitational wave (or tensor) perturbations from the quantum fluctuations of pure gravitational (transverse-traceless) degrees of freedom in the metric [13]. For this, inflation does not have to be caused by a scalar field. It is a very general prediction of an inflationary expanding universe. Hence the detection of the tensor perturbation may be regarded as "the" confirmation of the inflationary universe. As we shall mention below, although it turned out that the amplitude of the tensor perturbation is constrained rather strongly by observation, there is still reasonable hope that it may be detected in the near future.

Curvature perturbation

Let us first consider the curvature perturbation produced from inflation. It arises from the quantum vacuum fluctuations of the inflaton field ϕ. Since a rigorous derivation would take too much space, we leave it to the original literature [14, 15]. Here we give an intuitive, rather hand-waving derivation. We caution that it could well lead to an incorrect result if used blindly.

The vacuum fluctuations of the inflaton field with a comoving wavenumber k are given simply by its positive frequency function, φ_k. Because of the condition $V''/H^2 \ll 1$, on scales $k/a \gg H$, the inflaton field fluctuation behaves like a minimally coupled massless scalar. Hence we have

$$|\langle \delta\phi|\mathbf{k}\rangle|^2 = |\varphi_k|^2, \quad \varphi_k \sim \frac{1}{a^{3/2}\sqrt{2\omega_k}} e^{-i\omega_k t}; \quad \omega_k = \frac{k}{a} \gg H. \qquad (4.33)$$

As the universe expands the physical wavenumber decreases exponentially and becomes smaller than the Hubble parameter, $k/a < H$, or the physical wavelength exceeds the Hubble radius. Then the oscillations of φ_k are frozen. This could be regarded as "classicalization" of the quantum fluctuations. Note that this is merely an interpretation. In a more rigorous sense, freezing of the mode function is a process toward infinite squeezing of the vacuum state.

Setting $a = k/H$ in Eq. (4.33) gives

$$\varphi_k \sim \frac{H}{\sqrt{2k^3}} ; \quad \frac{k}{a} \ll H . \tag{4.34}$$

Therefore the mean square amplitude in unit logarithmic interval of k is

$$\langle \delta\phi^2 \rangle_k \equiv \frac{4\pi k^3}{(2\pi)^3} |\varphi_k|^2 \approx \left(\frac{H}{2\pi} \right)^2_{k/a=H} . \tag{4.35}$$

Inclusion of the non-trivial evolution of the background spacetime and the coupling of the scalar field fluctuation with the metric fluctuation do not change the above estimate if we interpret $\delta\phi$ in the above as evaluated on the flat slicing, that is, on hypersurfaces on which the spatial scalar curvature remains unperturbed.

It is known that the curvature perturbation on the comoving hypersurface $\mathcal{R}_c$ is conserved on superhorizon scales if the perturbation is adiabatic [16]. Furthermore, on superhorizon scales where comoving slices are equivalent to uniform-density slices, it was shown that the conservation of the curvature perturbation holds for a much wider class of gravitational theories not restricted to general relativity [34].

The comoving hypersurface is defined as a surface of uniform ϕ. Then the gauge transformation from the flat slicing to the comoving slicing gives the relation between $\mathcal{R}_c$ and $\delta\phi$,

$$\mathcal{R}_c = -\frac{H}{\dot{\phi}} \delta\phi . \tag{4.36}$$

Since this is conserved for $k/a < H$, the spectrum of the comoving curvature perturbation in unit logarithmic interval of k is given by

$$\mathcal{P}_\mathcal{R}(k) \equiv \frac{4\pi k^3}{(2\pi)^3} |\mathcal{R}_k|^2 \approx \left(\frac{H^2}{2\pi\dot{\phi}} \right)^2_{k/a=H} . \tag{4.37}$$

A rigorous, first-principle derivation of the above result was first done in [14, 15]. Here we emphasize that the above result is valid independent of any subsequent history of the universe. In particular, it is valid for any mechanism of reheating at the end of inflation. This is the unique feature of single-field slow-roll models of inflation.

The important relation of the above result with the number of e-folds was first pointed out in [18]: If we rewrite Eq. (4.32) as

$$N = \int_t^{t_{end}} H \, dt = \int_\phi^{\phi_{end}} \frac{H}{\dot{\phi}} \, d\phi , \tag{4.38}$$

we find

$$\delta N(\phi_k) = \left[\frac{\partial N}{\partial \phi} \delta\phi \right]_{k/a=H} = \left[-\frac{H}{\dot{\phi}} \delta\phi \right]_{k/a=H} = \mathcal{R}_c , \tag{4.39}$$

provided that we identify $\delta\phi$ with the scalar field fluctuation evaluated on the flat hypersurface. This is called the δN formula.

The δN formula implies that we need only knowledge of the background evolution to obtain the power spectrum of the comoving curvature perturbation, once we know the amplitude of the quantum fluctuations of the scalar field at the horizon crossing (i.e. when $k/a = H$). It is quite generally given by $H/(2\pi)$ in slow-roll inflation. With careful geometrical considerations, the δN formula can be extended to general multi-field inflation [19],

$$\mathcal{P}_{\mathcal{R}}(k) = \left(\frac{H}{2\pi}\right)^2 ||\nabla N||^2; \quad ||\nabla N||^2 \equiv G^{ab}(\phi)\frac{\partial N}{\partial\phi^a}\frac{\partial N}{\partial\phi^b}, \quad (4.40)$$

where G^{ab} is the field space metric and it is assumed that the vacuum expectation values are given by

$$\langle\delta\phi^a\,\delta\phi^b\rangle = G^{ab}\left(\frac{H}{2\pi}\right)^2. \quad (4.41)$$

A proof that any perturbation on superhorizon scales can be cast into the form equivalent to the difference between two infinitesimally different background universes was given in [20]. Then the nonlinear generalization of the δN formalism was developed in [22, 33], which will be the topic discussed in Section 4.2.6.

4.2.4 Tensor perturbation

There are not only vacuum fluctuations of the inflaton field but also those of the transverse-traceless part of the metric, $\partial^i h_{ij}^{TT} = \delta^{ij} h_{ij}^{TT} = 0$, that is, the tensor perturbation or gravitational wave degrees of freedom. If we construct the second-order action for h_{ij}^{TT}, we find

$$S \sim \frac{M_{Pl}^2}{8}\int d^4x\sqrt{-g}(\dot{h}_{ij}^{TT})^2 + \cdots. \quad (4.42)$$

To quantize h_{ij}^{TT} it is convenient to normalize the kinetic term to the canonical form. This gives

$$S \sim \frac{1}{2}\int d^4x\sqrt{-g}(\dot{\phi}_{ij})^2 + \cdots; \quad \phi_{ij} \equiv \frac{M_{Pl}}{2}h_{ij}^{TT}. \quad (4.43)$$

If one writes down the field equation for ϕ_{ij}, one finds its mode function ϕ_k obeys exactly the same equation as the one for a minimally coupled massless scalar field,

$$\ddot{\phi}_k + 3H\dot{\phi}_k + \frac{k^2}{a^2}\phi_k = 0. \quad (4.44)$$

Since there are two independent degrees of freedom in ϕ_{ij}, the power spectrum of the tensor perturbation h_{ij}^{TT} is obtained as

$$\mathcal{P}_T(k) = \frac{4}{M_{Pl}} \times 2 \times \frac{4\pi k^3}{(2\pi)^3} |\phi_k|^2 = \frac{8H^2}{(2\pi)^2 M_{Pl}^2} \,. \tag{4.45}$$

Taking the ratio of the tensor spectrum to the curvature perturbation spectrum, we find [19]

$$r \equiv \frac{\mathcal{P}_T}{\mathcal{P}_\mathcal{R}} \leq 8|n_T| = -16\frac{\dot{H}}{H^2} = 16\epsilon \,, \tag{4.46}$$

where n_T is the tensor spectral index, $n_T = d\ln \mathcal{P}_T(k)/d\ln k$, ϵ is the slow-roll parameter introduced in Eq. (4.30), and the equality holds for the case of single-field slow-roll inflation. This is a consistency relation in general slow-roll inflation. As a prototype example, if we consider chaotic inflation [8], which is a typical example of large-field inflation, we expect to have $r \sim 0.1$, which is well within the range of detectability in the near future. In fact, as noted previously, the claimed detection of B-mode polarization in CMB by BICEP2 fits well with the effect of a tensor spectrum from inflation with an amplitude of $r \sim 0.2$ [44].

The important point to be noted is that the existence of the vacuum fluctuations of the tensor part of the metric is a proof of the existence of quantum gravity. These fluctuations exist in any theory of gravity that respects general covariance, apart from possible inessential modifications of the spectrum. Thus a clear detection of the tensor spectrum will be a confirmation of not only the inflationary universe but also quantum gravity.

4.2.5 Deviations from Gaussianity

The standard, single-field slow-roll inflation predicts that the curvature perturbation is a Gaussian random field and it has an almost scale-invariant spectrum. This seems to fit the current observational data quite well [23], but it is quite possible that the actual model will turn out to be non-standard. Maybe it is multi-field, maybe non-slow-roll and/or non-canonical. In such a case, the curvature perturbation may become non-Gaussian. The search for possible non-Gaussian signatures in the primordial curvature perturbation has become one of the important directions in observation in recent years [24].

Here we consider possible origins of non-Gaussianity in the curvature perturbation. Essentially one can classify the origins into three categories: (1) Self-interactions of the inflaton field and/or non-trivial vacua, (2) multi-field dynamics, and (3) nonlinearity in gravity.

The non-Gaussianities of the first category are generated on subhorizon scales during inflation, hence they are of quantum-field-theoretical origin. Those of the second category are usually generated on superhorizon scales either during or after inflation, and they are due to nonlinear coupling of the scalar field to gravity. Since they are generated on superhorizon scales, they are of classical origin. Finally those of the third category are due to nonlinear dynamics in general relativity. Hence they are generated after the scale of interest re-enters the Hubble horizon. Since the last category is not really primordial in nature, let us focus on the first two categories.

Non-Gaussianity from self-interaction/non-trivial vacuum

It is known that conventional self-interactions by the potential are ineffective [25]. This can be seen by considering chaotic inflation, for example. In the simplest case of a quadratic potential, $V = m^2\phi^2/2$, the inflaton is actually a free field apart from the interaction through gravitation perturbations. But the gravitational interaction is Planck-suppressed, i.e., it is always suppressed by a factor $O(M_{Pl}^{-2})$. In the case of a quartic potential, $V = \lambda\phi^4$, it is known that λ should be extremely small, $\lambda \sim 10^{-15}$, in order for it to be consistent with observation.

Thus some kind of unconventional self-interaction is necessary. A popular example is the case of a scalar field with a non-canonical kinetic term such as DBI inflation [26]. In this case the kinetic term takes the form

$$K \sim f^{-1}(\phi)\sqrt{1 - f(\phi)\dot{\phi}^2} \equiv f^{-1}\gamma^{-1}. \qquad (4.47)$$

If we expand this perturbatively,

$$K = K_0 + \delta_1 K + \delta_2 K + \delta_3 K + \cdots, \qquad (4.48)$$

we will find

$$\delta_2 K \propto \gamma^3, \quad \delta_3 K \propto \gamma^{3+2}, \qquad (4.49)$$

since $\delta\gamma = \gamma^3 \, \delta X$, where $X \equiv f\dot{\phi}^2/2$. If we regard the third-order part as the interaction, the above implies that the scalar field fluctuation will be expressed qualitatively as

$$\delta\phi \sim \delta\phi_0 + \gamma^2 \, \delta\phi_0^2 + \cdots, \qquad (4.50)$$

where $\delta\phi_0$ is the free, Gaussian fluctuation. Thus the non-Gaussianity in $\delta\phi$ may become large if γ, which mimics the Lorentz factor, is large [27].

A non-trivial vacuum state is another source of non-Gaussianity. If the universe were a pure de Sitter spacetime, gravitational interaction would be totally negligible in vacuum, except for the effect due to graviton (tensor mode) loops. This may be regarded as due to the maximally symmetric nature of the de Sitter space, $SO(4, 1)$, which has the same number of degrees of symmetry as the Poincaré (Minkowski) symmetry. In slow-roll inflation, the de Sitter symmetry is slightly broken. Nevertheless, the effect induced by this symmetry breaking is small because it is suppressed by the slow-roll parameter $\epsilon = -\dot{H}/H^2$.

However, if the vacuum state does not respect the de Sitter symmetry, there can be a large non-Gaussianity. Such a deviation from the quasi-de Sitter vacuum, usually called the Bunch–Davies vacuum, may occur in various situations, which have been studied e.g. in [28, 29].

Non-Gaussianity from multi-field dynamics

Non-Gaussianity may appear if the energy momentum tensor depends nonlinearly on the scalar field even if the fluctuation of the scalar field itself is Gaussian. This effect is generally important when the fluctuations are on superhorizon scales, i.e., the characteristic

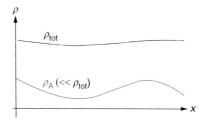

Figure 4.4 An illustration of the energy density configuration in the multi-field case. The density of the A-matter/field ρ_A may vary nonlinearly without significantly affecting the total energy density. Reprinted from ref. [46] with kind permission from Springer Science+Business Media B.V.

wavelength is larger than the Hubble radius. It is small in single-field slow-roll models because the linear approximation is valid to high accuracy [30], generically suppressed by the slow-roll parameter η_V defined in Eq. (4.31).

For multi-field models, however, the contribution to the energy momentum tensor from some of the fields can be highly nonlinear as depicted in Fig. 4.4. The important property of non-Gaussianity in this case is that it is always of the spatially local type. Namely, to second order in nonlinearity, the curvature perturbation will take the form [31],

$$\mathcal{R}_c(x) = \mathcal{R}_{c,0}(x) + \frac{3}{5}f_{NL}^{local}\,\mathcal{R}_{c,0}^2(x)\,, \tag{4.51}$$

where $\mathcal{R}_{c,0}$ is the Gaussian random field and f_{NL}^{local} is a constant representing the amplitude of non-Gaussianity. The factor 3/5 in front of f_{NL}^{local} is due to a historical reason. The reason why it is of local type is simply causality: No information can propagate over a length scale greater than the Hubble horizon scale.

Observationally, this type of non-Gaussianity can be tested by using the so-called squeezed-type templates where one of the wavenumbers, say k_1 in the bispectrum $B(k_1, k_2, k_3)$, is much smaller than the other two, $k_1 \ll k_2 \approx k_3$ [24], and there are a few observational indications that f_{NL}^{local} is actually non-vanishing. For example, the WMAP 7-year data analysis gave a one-sigma bound $11 < f_{NL}^{local} < 53$ (68% CL) [23], although a more recent result by Planck is consistent with the absence of the local-type non-Gaussianity, $-3.1 < f_{NL}^{local} < 8.5$ (68% CL) [32].

4.2.6 δN formalism

As mentioned in Section 4.2.2, the δN formalism is a powerful tool to evaluate the comoving curvature perturbation on superhorizon scales. It then turned out that it can be easily extended to the evaluation of nonlinear, non-Gaussian curvature perturbations [22, 33]. Let us recapitulate its definition and properties:

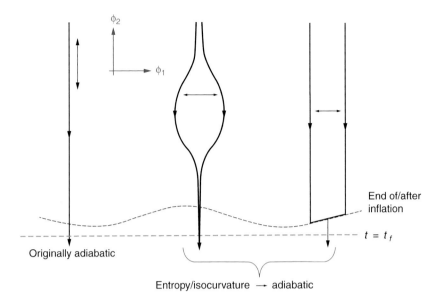

Figure 4.5 Three different types of δN. The field space (ϕ_1,ϕ_2) in the figure represents the degrees of freedom in the initial condition of the universe. The adiabatic limit is defined to be the stage by which all the trajectories converge to a unique one. Reprinted from ref. [46] with kind permission from Springer Science+Business Media B.V.

(1) δN is the perturbation in the number of *e*-folds counted *backward* in time from a fixed final time, say $t = t_f$, to some initial time $t = t_i$.

(2) The final time t_f should be chosen such that the evolution of the universe has become unique by that time, i.e., the universe has reached the adiabatic limit. Then the hyper-surface $t = t_f$ should be identified with a comoving (or uniform-density) slice, and the initial hypersurface $t = t_i$ should be identified with a flat slice.[3]

(3) δN is equal to the conserved (nonlinear) comoving curvature perturbation on super-horizon scales at $t > t_f$.

(4) By definition, it is nonlocal in time. However, because of its purely geometrical defini-tion, it is valid independent of which theory of gravity one considers, provided that the adiabatic limit is reached by $t = t_f$.

There are various kinds of sources that generate δN. They may be classified into three types, as depicted in Fig. 4.5. The left one describes a perturbation along the evolutionary trajectory of the universe. This case is the same as that of single-field slow-roll inflation, in which the comoving curvature perturbation is conserved all the way until it re-enters the horizon. The middle one is the case when a small difference in the initial data develops

---

[3] In linear theory, a flat slice is defined as a hypersurface on which the spatial scalar curvature vanishes [16]. A flat slice generalized to the nonlinear case is defined as a hypersurface on which the determinant of the 3-metric, i.e., the 3-volume element, is homogeneous [33].

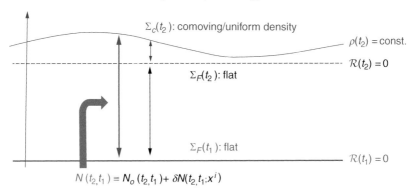

Figure 4.6 Definition of nonlinear δN. It is defined as the perturbation in the number of e-folds from an initial flat slice to a final comoving slice. Reprinted from ref. [46] with kind permission from Springer Science+Business Media B.V.

into a substantial difference in δN. Typically this is realized when there is some instability orthogonal to the trajectory, like the case when the scalar field moves along a ridge. This type of source of δN usually induces a feature in the spectrum and/or bispectrum of the curvature perturbation. The right one represents the case when the perturbation orthogonal to the trajectory does not contribute to the curvature perturbation until or after the end of inflation, but δN is generated due to a sudden transition that brings the universe into an adiabatic stage. Typical examples are curvaton models [33–35] and multi-brid inflation models [36, 37].

Here, for the sake of completeness, let us present the precise definition of the non-linear δN formula. See Fig. 4.6. It is based on the leading-order approximation in the spatial-gradient expansion or the separate-universe approach [33], where spatial derivatives are assumed to be negligible in comparison with time derivatives. At leading order of the spatial-gradient expansion, if we express the spatial volume element as $\sqrt{^{(3)}\gamma} = a^3(t)\exp[3\mathcal{R}(t,x)]$, where $a(t)$ is the scale factor of a fiducial homogeneous and isotropic universe, we easily find that the perturbation in the number of e-folds along a comoving trajectory between two hypersurfaces $t = t_1$ and $t = t_2$ is given by

$$\delta N(t_2, t_1; x^i) = \mathcal{R}(t_2, x^i) - \mathcal{R}(t_1, x^i), \qquad (4.52)$$

where x^i are the comoving coordinates. Here we note that this is purely a geometrical relation. It has nothing to do with any equations of motion.

First we fix the final hypersurface $t = t_2$. It should be taken at the stage when the evolution of the universe has become unique. That is, there exists no isocurvature perturbation any longer that could develop into an adiabatic perturbation at later epochs. Thus the comoving curvature perturbation is conserved at $t > t_2$. In the context of the concordance ΛCDM model of the universe, this corresponds to the final radiation-dominated stage of the universe.

Next we choose the initial slice $t = t_1$. It should be chosen to be flat. Here 'flat' means that the perturbation in the spatial volume element vanishes. Namely, the flat slice is defined as a hypersurface on which $\mathcal{R} = 0$. We note that despite its name, the scalar curvature vanishes only in the linear theory limit: It is non-vanishing in general in the nonlinear case.

Applying the above choice of the initial and final hypersurfaces to Eq. (4.52), it is trivial to see that we have

$$\delta N(t_2, t_1; x^i) = \mathcal{R}_c(t_2, x^i). \tag{4.53}$$

Now by assumption $\mathcal{R}_c$ is conserved at $t > t_2$. So it is the quantity we want to evaluate. This completes the derivation of the nonlinear δN formula.

As mentioned above, since Eq. (4.52) is a pure geometrical relation, so is the nonlinear δN formula (4.53). This is the reason why it can be applied to any theory of gravity as long as it is a geometrical (i.e. general covariant) theory.

Of course, the above definition tells us nothing about how to evaluate it in practice. In this respect, we have a very fortunate situation in the case of inflationary cosmology, in that the evaluation of the quantum fluctuations of the inflaton field, whether it is single- or multi-component, can be most easily done in a gauge in which the time slicing is chosen to be flat [19]. Thus we can choose the initial slice to be an epoch when the scale of our interest has just exited the horizon during inflation. Let the fluctuations of a multi-component scalar field on the flat slice at $t = t_1$ be $\delta\phi^a$. Then, assuming that the values of the scalar field determine the evolution of the universe completely, which is the case for slow-roll inflation, the nonlinear δN can be simply evaluated as

$$\delta N = N(\phi^a + \delta\phi^a) - N(\phi^a), \tag{4.54}$$

where $N(\phi^a)$ is the e-folding number of the fiducial background. In particular, to second order in $\delta\phi^a$, we obtain

$$\mathcal{R}_c = \delta N = \frac{\partial N}{\partial\phi^a}\delta\phi^a + \frac{1}{2}\frac{\partial^2 N}{\partial\phi^a\,\partial\phi^b}\delta\phi^a\,\delta\phi^b + \cdots, \tag{4.55}$$

Comparing this with Eq. (4.51), we see that the curvature perturbation takes a form that is a bit more complicated than the simplest form. Nevertheless, if we consider the bispectrum, i.e., the Fourier component of the three-point function $\langle\mathcal{R}_c(x_1)\mathcal{R}_c(x_2)\mathcal{R}_c(x_3)\rangle$, we find there is a quantity that exactly corresponds to f_{NL}^{local} defined in Eq. (4.51). Namely [22],

$$\frac{3}{5}f_{NL}^{\text{local}} = \frac{G^{ab}G^{cd}N_a N_{bc} N_d}{2\left(||\nabla N||^2\right)^2}; \quad N_a \equiv \frac{\partial N}{\partial\phi^a}, \quad N_{ab} \equiv \frac{\partial^2 N}{\partial\phi^a\,\partial\phi^b}. \tag{4.56}$$

Before concluding this section, we mention the fact that the δN formalism does not require the scalar-field fluctuations to be Gaussian. In fact, except for the last equation in the above, Eq. (4.56), which assumes the Gaussianity of $\delta\phi^a$, the general δN formula (4.54) or its second-order version (4.55) can be used for non-Gaussian $\delta\phi^a$ [38]. Such a case may happen, for example, in multi-field DBI inflation.

4.2.7 Summary

More than three decades have passed since the inflationary universe was first proposed, and there is increasing observational evidence that inflation did take place in the very early universe. Among others, the measured CMB temperature anisotropy is fully consistent with the predictions of inflation that the primordial curvature perturbation spectrum is almost scale-invariant and it is statistically Gaussian.

Inflation also predicts a scale-invariant tensor spectrum, and if the energy scale of inflation is high enough as in the case of chaotic inflation, the tensor–scalar ratio r can be as large as 0.1. As already noted, indeed the recent BICEP2 result strongly indicates such a large value of r [44]. Even if it is not so large, it is argued that a value as small as $r \sim 0.001$ may be detectable in the future [39]. If this is the case, and if the primordial tensor perturbation is clearly detected, it will confirm not only the primordial inflation but also quantum gravity.

Even if the tensor perturbation turns out to be too small to be detected, there may be other interesting signatures of inflation. Non-Gaussianity from inflation is attracting attention as one of those signatures that can distinguish or constrain models of inflation significantly.

We discussed that the origins of primordial non-Gaussianities may be classified into three categories, according to different length scales on which different mechanisms are effective:

(1) Quantum theoretical origin on subhorizon scales during inflation.
(2) Classical nonlinear scalar field dynamics on superhorizon scales during or after inflation.
(3) Nonlinear gravitational dynamics after the horizon re-entry.

In particular we argued that non-Gaussianities in the second case are always of spatially local type. We then mentioned that there are three different kinds of situations in which such local non-Gaussianities can be generated, and described in some detail a very efficient method to compute them, namely, the δN formalism.

Apparently identifying properties of primordial non-Gaussianities in the observational data is extremely important for understanding the physics of the early universe. Here we mentioned only the bispectrum or the three-point function. But higher-order correlation functions may become important as a model discriminator [40–43]. Other types of non-Gaussianity discriminators may also become necessary [44].

What is important is that we are now beginning to test observationally the physics of the very early universe, the physics at an energy scale closer to the Planck scale, at a scale that can never be attained in high-energy accelerator experiments.

Recent observational data released by Planck seem to favor models with non-minimal coupling of gravity to the inflaton [45]. This suggests the inflaton field may be in the gravitational sector, rather than in the matter sector. In other words, modifications of gravity may be essential for the explanation of the primordial inflation. Whether this is really the case or the inflaton can be implemented in an extended model of particle physics is an interesting question to be answered in due course.

Acknowledgments

A substantial part of this chapter is based on a talk given at *Relativity and Gravitation: 100 Years after Einstein in Prague*, June 25–29, 2012, Prague, and published in the proceedings, ref. [46]. This part is reprinted here with kind permission from Springer Science+Business Media B.V.

References

[1] Mather, J. *et al.* 1994. *Astrophys. J.*, **420**, 439–444.
[2] Freedman, W., and Madore, B. 2010. *Ann. Rev. Astron. Astrophys.*, **48**, 673–710.
[3] Sato, K. 1981. *Mon. Not. Roy. Astron. Soc.*, **195**, 467–479.
[4] Guth, A. 1981. *Phys. Rev. D*, **23**, 347–356.
[5] Hawking, S., Moss, I., and Stewart, J. 1982. *Phys. Rev. D*, **26**, 2681.
[6] Linde, A. 1982. *Phys. Lett. B*, **108**, 389–393.
[7] Albrecht, A., and Steinhardt, P. 1982. *Phys. Rev. Lett.*, **48**, 1220–1223.
[8] Linde, A. 1983. *Phys. Lett. B*, **129**, 177–181.
[9] Ade, P., *et al.* [BICEP2 Collaboration]. 2014. *Phys. Rev .Lett.*, **112**, 241101.
[10] Albrecht, A., Steinhardt, P., Turner, M., and Wilczek, F. 1982. *Phys. Rev. Lett.*, **48**, 1437.
[11] Kofman, L., Linde, A., and Starobinsky, A. 1997. *Phys. Rev. D*, **56**, 3258–3295.
[12] Mukhanov, V., and Chibisov, G. 1981. *JETP Lett.*, **33**, 532–535.
[13] Starobinsky, A. 1979. *JETP Lett.*, **30**, 682–685.
[14] Mukhanov, V. 1985. *JETP Lett.*, **41**, 493–496.
[15] Sasaki, M. 1986. *Prog. Theor. Phys.*, **76**, 1036.
[16] Kodama, H., and Sasaki, M. 1984. *Prog. Theor. Phys. Suppl.*, **78**, 1–166.
[17] Wands, D., Malik, K., Lyth, D., and Liddle, A. 2000. *Phys. Rev. D*, **62**, 043527.
[18] Starobinsky, A. 1985. *JETP Lett.*, **42**, 152–155.
[19] Sasaki, M., and Stewart, E. 1996. *Prog. Theor. Phys.*, **95**, 71–78.
[20] Sasaki, M., and Tanaka, T. 1998. *Prog. Theor. Phys.*, **99**, 763–782.
[21] Lyth, D., Malik, K., and Sasaki, M. 2005. *J. Cosmol. Astropart. Phys.*, **0505**, 004.
[22] Lyth, D., and Rodriguez, Y. 2005. *Phys. Rev. Lett.*, **95**, 121302.
[23] Komatsu, E., *et al.* [WMAP Collaboration]. 2011. *Astrophys. J. Suppl.*, **192**, 18.
[24] Komatsu, E. *et al.*, 2009. arXiv:0902.4759.
[25] Maldacena, J. 2003. *J. High Energy Phys.*, **0305**, 013.
[26] Alishahiha, M., Silverstein, E., and Tong, D. 2004. *Phys. Rev. D*, **70**, 123505.
[27] Mizuno, S., Arroja, F., Koyama, K., and Tanaka, T. 2009. *Phys. Rev. D*, **80**, 023530.
[28] Chen, X., Easther, R., and Lim, E. 2008. *J. Cosmol. Astropart. Phys.*, **0804**, 010.
[29] Flauger, R. *et al.* 2010. *J. Cosmol. Astropart. Phys.*, **1006**, 009.
[30] Salopek, D., and Bond, J. 1990. *Phys. Rev. D*, **42**, 3936–3962.
[31] Komatsu, E., and Spergel, D. 2001. *Phys. Rev. D*, **63**, 063002.
[32] Ade, P., *et al.* [Planck Collaboration]. 2014. *Astron. Astrophys.*, **571**, A24.
[33] Lyth, D., and Wands, D. 2002. *Phys. Lett. B*, **524**, 5–14.
[34] Moroi, T., and Takahashi, T. 2001. *Phys. Lett. B*, **522**, 215–221.
[35] Sasaki, M., Valiviita, J., and Wands, D. 2006. *Phys. Rev. D*, **74**, 103003.
[36] Sasaki, M. 2008. *Prog. Theor. Phys.*, **120**, 159–174.
[37] Naruko, A., and Sasaki, M. 2009. *Prog. Theor. Phys.*, **121**, 193–210.
[38] Byrnes, C., Koyama, K., Sasaki, M., and Wands, D. 2007. *J. Cosmol. Astropart. Phys.*, **0711**, 027.
[39] Abazajian, K. *et al.*, 2015. *Astroparticle Physics*, **63**, 55.
[40] Chen, X., Huang, M.-X., and Shiu, G. 2006. *Phys. Rev. D*, **74**, 121301.
[41] Byrnes, C., Sasaki, M., and Wands, D. 2006. *Phys. Rev. D*, **74**, 123519.
[42] Seery, D., and Lidsey, J. 2007. *J. Cosmol. Astropart. Phys.*, **0701**, 008.

[43] Suyama, T., and Yamaguchi, M. 2008. *Phys. Rev. D*, **77**, 023505.

[44] Chen, X. 2010. *Adv. Astron.*, **2010**, 638979.

[45] Ade, P., *et al.* [Planck Collaboration]. 2014. *Astron. Astrophys.*, **571**, A22.

[46] Sasaki, M. 2014. Pages 305–321 of: Bičák, J and Ledvinka, T. (eds) *General Relativity, Cosmology and Astrophysics – Perspectives 100 years after Einstein's stay in Prague*, Fundamental Theories of Physics **177**, Springer Verlag, Heidelberg.

4.3 Precision cosmology
Eiichiro Komatsu

4.3.1 Introduction

"Precision cosmology" is built upon two pillars: precision data and precision theory. It would not be precision cosmology if either was lacking. There are areas in which the data are precise but the physics is not yet well understood (e.g., Type Ia supernovae; small-scale clustering of gas and galaxies), as well as areas in which the physics is well understood but the data are not yet precise enough (e.g., weak gravitational lensing on large scales).

The cosmic microwave background (CMB) offers the best example of precision cosmology to date. The energy spectrum of CMB photons has been measured precisely by the Far Infrared Absolute Spectrophotometer (FIRAS) on NASA's *Cosmic Background Explorer* (*COBE*) satellite [1–3]. The measured spectrum is consistent with the Planck spectrum – a result of precise theoretical understanding of the interaction between photons and matter. The temperature and polarization of the CMB vary across the sky, and this *anisotropy* has been measured precisely by a host of experiments including space-borne experiments such as NASA's *Wilkinson Microwave Anisotropy Probe* (*WMAP*) satellite [4, 5] and ESA's *Planck* satellite [6–8], as well as ground-based experiments such as the *Atacama Cosmology Telescope* (*ACT*) [9–11] and *South Pole Telescope* (*SPT*) [12, 13]. As the amplitude of the anisotropy is of order 10^{-5}, the coupled, linearized Boltzmann–Einstein equations fully describe the physics of temperature and polarization anisotropies of the CMB (see, e.g., [14]). These equations have been solved numerically with 0.1% precision [15–18].

Another good example of precision cosmology is the large-scale structure of the universe, i.e., the distribution of matter in the universe on large scales ($\gg$ 1 Mpc). To be fair, the precision of both data and theory is not yet as good as that of the CMB; however, the situation regarding precision of the data is rapidly improving with the advent of large-scale survey observations of galaxies, covering large ($>$ 1 Gpc^3) comoving volumes. Revolution in the survey volume came with the *Two-degree Field Galaxy Redshift Survey* (*2dFGRS*) [19, 20] and the *Sloan Digital Sky Survey* (*SDSS*) [21–25]. As for theory, while the physics of galaxy clustering on large scales, where clustering of matter is still in the linear regime, is well understood, precision of theory rapidly degrades as we go to smaller spatial scales where clustering becomes non-linear. As the precision of data continues to improve over the next decade, the precision of theory must catch up with it.

In this chapter, we describe the basic physics of the CMB and the main science results from the CMB experiments over the last decade, and discuss the main science goals of precision cosmology with CMB over the next decade.

4.3.2 Spectral distortions

The observed energy spectrum of the CMB is consistent with a Planck spectrum. To quantify "spectral distortions," i.e., possible deviations of the spectrum from a Planck spectrum, we often classify distortions into "μ-type" and "y-type" distortions. Let us define a dimensionless frequency, $x \equiv h\nu/(k_B T_{cmb}) \simeq \nu/(56.78 \text{ GHz})$, where ν denotes frequencies, $T_{cmb} = 2.725$ K (or 2.72548 ± 0.00057 [68% CL]) is the present-day CMB temperature [26], h the Planck constant, and k_B the Boltzmann constant. The distribution function (occupation number) of photons with a Planck spectrum is $f_p(x) \equiv (e^x - 1)^{-1}$. Then the μ- and y-type spectral distortions are defined by

$$f(x) = \frac{1}{e^{x+\mu} - 1}$$

$$= f_p(x) - \mu \frac{e^x}{(e^x - 1)^2} + \cdots \qquad (\mu\text{-type}), \qquad (4.57)$$

$$f(x) = f_p(x) + \frac{y}{x^2} \frac{d}{dx}\left[x^4 \frac{df_p(x)}{dx} \right]$$

$$= f_p(x) + y \frac{xe^x}{(e^x - 1)^2} [x \coth(x/2) - 4] \qquad (y\text{-type}), \qquad (4.58)$$

with $|\mu| \ll 1$ and $|y| \ll 1$ being free parameters constrained by precision measurements of the CMB spectrum. The small distortions, $\Delta f(x) \equiv f(x) - f_p(x)$, can be related to small changes in temperatures via $\Delta f = -x(df_p/dx)(\Delta T/T_{cmb})$. This gives

$$\frac{\Delta T}{T_{cmb}} = -\frac{\mu}{x} \qquad (\mu\text{-type}), \qquad (4.59)$$

$$\frac{\Delta T}{T_{cmb}} = y[x \coth(x/2) - 4] \qquad (y\text{-type}). \qquad (4.60)$$

The asymptotic limits of the y-type distortion are $\Delta T/T_{cmb} \to -2y$ and $+yx$ for $x \ll 1$ and $x \gg 1$, respectively. We observe a temperature *decrement* in $x < 3.830$, a *null* at $x = 3.830$ ($\nu = 217.5$ GHz), and a temperature *increment* in $x > 3.830$.

These parameters can be used to quantify the magnitude of distortions. The current upper bounds on μ and y are $|\mu| < 9 \times 10^{-5}$ and $|y| < 1.5 \times 10^{-5}$ (95% CL), respectively [3]. With the current technology, these limits can be improved by more than three orders of magnitude [27, 28]. But, what is the physics behind these parameters?

The spectrum with a μ-type distortion is simply a Bose–Einstein distribution with a non-zero chemical potential. Therefore, the spectrum with a μ-type distortion is in fact an equilibrium distribution – a distribution obtained in a thermal equilibrium. The chemical potential vanishes (and the spectrum becomes a Planck spectrum) when reactions which do not conserve the number of photons are efficient in thermal equilibrium. Two processes are relevant in the early universe: double Compton scattering ($\gamma + e^- \leftrightarrow \gamma + \gamma + e^-$) [29, 30] and bremsstrahlung ($p + e^- \leftrightarrow p + e^- + \gamma$) [31–33]. Double Compton scattering dominates over bremsstrahlung when the baryon density is lower than $\Omega_B h^2 \approx 0.092$ [34]; thus, double Compton scattering is a dominant process in our universe, which has $\Omega_B h^2 = 0.022$.

The spectrum becomes a Planck distribution as long as the rate of photon production by double Compton scattering at low frequencies (say, $x \approx 10^{-2}$) is greater than the expansion rate of the universe. This occurs when the redshift is greater than $z \approx 2 \times 10^6$. The photons produced at low frequencies are then "Comptonized," i.e., photon energies are re-distributed over all frequencies by Compton scattering, $\gamma + e^- \rightarrow \gamma + e^-$ (which does conserve the number of photons), and the spectrum approaches a Planck distribution at all frequencies (see [35] for the latest calculations).

While production of photons is no longer efficient after $z \approx 2 \times 10^6$, Comptonization is still efficient. The rate of Comptonization is given by $\sigma_T n_e c k_B T_e / (m_e c^2)$ [36], where n_e is the electron number density, σ_T the Thomson scattering cross section, and T_e the electron temperature. Now, suppose that some energy, ΔE, is injected into the plasma after $z \approx 2 \times 10^6$, distorting the spectrum of the CMB away from a Planck spectrum. Compton scattering brings the spectrum back into an equilibrium distribution; however, as the number of photons is conserved, it cannot restore the Planck spectrum. Instead, the spectrum approaches a Bose–Einstein distribution with a non-zero chemical potential [31, 37]. This is the origin of the μ-distortion; namely, it is a consequence of energy injection after $z \approx 2 \times 10^6$. The parameter, μ, is proportional to the amount of injected energy, ΔE, and the relation can be found as follows. The energy density in photons changes from aT^4 to $a\tilde{T}^4(1 - 1.11\mu) = aT^4 + \Delta E/V$ to the first order in μ, where $\tilde{T}$ is a new temperature after energy injection, V a volume, and a a constant. On the other hand, the number density does not change: $b\tilde{T}^3(1 - 1.37\mu) = bT^3$, where b is a constant. Solving these, we find $\mu = 1.4[\Delta E/(aT^4V)] = 1.4(\Delta E/E)$, i.e., μ is 1.4 times the fractional energy injection. More precisely, the μ-type distortion is given by

$$\mu = 1.4 \int_\infty^{5 \times 10^4} dz \, \frac{dQ/dz}{\rho_\gamma(z)} \exp\left[-\left(\frac{z}{2 \times 10^6} \right)^{5/2} \right], \tag{4.61}$$

where dQ/dz is a rate at which energy is injected into the plasma, and $\rho_\gamma(z)$ is the energy density of CMB photons at a given redshift. The rate of Comptonization becomes lower than the expansion rate at $z \approx 5 \times 10^4$, and thus the energy injection after this redshift would not yield a Bose–Einstein distribution. This explains the integration boundary of Eq. (4.61). Spectral distortions caused by energy injection after $z \approx 5 \times 10^4$ would be preserved.

The y-type distortion is created by a gas of hot electrons causing inverse Compton scattering of CMB photons [38, 39], or by a superposition of black-body spectra with different temperatures [40]. For the former case, the parameter y is given by

$$y = \int d\tau \, \frac{k_B(T_e - T_\gamma)}{m_e c^2}, \tag{4.62}$$

where $d\tau = \sigma_T n_e c\, dt$ is a differential optical depth to Thomson scattering, and $T_\gamma = T_{cmb}(1 + z)$ the CMB temperature in the past. This expression is valid when the exchange of energies between photons and electrons is inefficient, $\tau k_B T_e / (m_e c^2) \ll 1$ and $\tau k_B T_\gamma / (m_e c^2) \ll 1$. This effect has been measured routinely toward galaxy clusters, which typically have $y \approx 10^{-4}$ [41]. However, the *average* y-type distortion (averaged over the sky) has not been detected yet. The shock-heated gas in galaxy clusters, groups, galaxies and the intergalactic medium at $z < 5$ would give $y \approx 1.7 \times 10^{-6}$ [42], whereas the photo-heated gas ($T_e \approx 10^4$ K) during the epoch of reionization, $z \approx 10$, would give $y \approx \tau k_B T_e / (m_e c^2) \approx 1.5 \times 10^{-7}$ with $\tau = 0.09$ [5]. These contributions are below the current upper bound on y, but well above the expected sensitivity of future spectroscopic experiments [27, 28].

An example of superposition of black-body spectra with different temperatures is diffusion of photons [43]. When Compton scattering tightly couples photons and electrons, photons cannot propagate freely; instead, photons *diffuse*, and photons having different temperatures mix within a characteristic diffusion scale given by $l_d = \sqrt{N} l_{mfp}$, where $l_{mfp} = (\sigma_T n_e)^{-1}$ is the mean free path of photons, and $N = \sqrt{ct/l_{mfp}}$ is the number of scatterings photons encounter within a given time. This results in a y-type distortion, which is converted into a μ-type distortion with $\mu \approx 10^{-8}$ if diffusion occurs between $z \approx 5 \times 10^4$ and 2×10^6 [44]. One can also think of this problem as energy injection due to dissipation of sound waves [31, 45–47], which occurs via shear viscosity and thermal conduction of the (imperfect) baryon–photon fluid generated by photon diffusion [48–50]. The y-type distortion from energy injection is then given by

$$y = \frac{1}{4} \int_{5 \times 10^4}^{1088} dz\, \frac{dQ/dz}{\rho_\gamma(z)}, \tag{4.63}$$

where $z = 1088$ is the redshift of decoupling of photons from the plasma.

In summary, precision measurements of the CMB spectrum have established the thermal nature of the CMB, thus proving the basic idea of hot Big Bang cosmology. Now, the focus has shifted to measuring small, but expected, distortions of the Planck spectrum. Measuring y and μ, as well as intermediate shapes [30, 44, 51], tells us about a history of energy injection in the universe at both early and late times. These theoretical studies constitute one of the main science goals of future CMB experiments [27, 28].

4.3.3 Temperature anisotropy due to our local motion

Due to the motion of the Earth with respect to the rest frame of the CMB, the observed temperature of the CMB in the sky shows a (predominantly) dipolar pattern. Suppose that the distribution of CMB temperatures in the sky is uniform and is equal to $T_{cmb} = 2.725$ K. Then, the Doppler shift due to the motion of the Earth with respect to the CMB rest frame changes the observed distribution of CMB temperatures toward a given direction, $\hat{n}$, as

$$T_{obs}(\hat{n}) = \frac{T_{cmb}}{\gamma(1 - \hat{n} \cdot \mathbf{v}/c)}, \tag{4.64}$$

where $\gamma \equiv (1 - v^2/c^2)^{-1/2}$ is the relativistic γ factor, and $\mathbf{v}$ is a velocity vector of our motion with respect to the CMB rest frame. Expanding this in terms of v/c, we find $T_{obs}(\hat{n}) = T_{cmb}\left[1 + \hat{n} \cdot \mathbf{v}/c + \mathcal{O}(v^2/c^2)\right]$. The "temperature anisotropy," defined by $\delta T(\hat{n})/T \equiv T_{obs}(\hat{n})/T_{cmb} - 1$, is then given by $\delta T(\hat{n})/T = \hat{n} \cdot \mathbf{v}/c + \mathcal{O}(v^2/c^2)$. Therefore, the leading-order term gives a dipolar pattern ($\ell = 1$ in spherical harmonics expansion) in the sky, while the higher-order terms give ℓth-order multipoles which are suppressed by additional factors of $(v/c)^{\ell-1}$.

Dipole anisotropy was measured in the 1970s [52–55]. A definitive, precise measurement was obtained from the Differential Microwave Radiometer (DMR) on *COBE* [56, 57], and further updated by *WMAP*. The latest determination of the dipole from the *WMAP* seven-year data is $\delta T = 3.355 \pm 0.008$ mK toward $(l, b) = (263°.99 \pm 0°.14, 48°.26 \pm 0°.03)$ (68% CL) in Galactic coordinates [58]. Note that this value is obtained after subtracting the annual motion of the Earth around the Sun (with $v = 30$ km/s). Using the dipole formula above, $\delta T(\hat{n})/T = \hat{n} \cdot \mathbf{v}/c$, we conclude that the Sun moves at 369 km/s with respect to the rest frame of the CMB.

Now, while this is a precise measurement, is it cosmological? Yes. The measured velocity is the sum of the motion of the Sun around the Galactic center; the motion of the Galactic center around the center-of-mass of the Local Group of galaxies; and the motion of the center-of-mass of the Local Group with respect to the rest frame of the CMB. After subtracting the first two components, we obtain the velocity of the center-of-mass of the Local Group with respect to the rest frame of the CMB as $v = 626 \pm 30$ km/s toward $(l, b) = (276° \pm 2°, 30° \pm 2°)$ (68% CL) in Galactic coordinates [59]. This residual velocity must be due to a gravitational pull by objects located at cosmological distances. For example, the Virgo cluster of galaxies at 16.5 Mpc makes a sizable contribution. After subtracting infall into Virgo, we obtain $v = 495 \pm 25$ km/s toward $(l, b) = (275° \pm 2°, 12° \pm 4°)$ (68% CL) in Galactic coordinates [59]. A large velocity still remains after removing the Virgo infall; thus, there must be a gravitational pull caused by a collection of objects at farther distances.

Dipole temperature anisotropy is not the only effect that our local motion can have on the CMB. Equation (4.64) is valid only when the temperature distribution of the CMB in the rest frame is uniform; however, in reality, there is an intrinsic, primordial temperature anisotropy, δT_p. We thus have

$$T_{obs}(\hat{n}) = \frac{T_{cmb} + \delta T_p(\hat{n}')}{\gamma(1 - \hat{n} \cdot \mathbf{v}/c)}, \qquad (4.65)$$

where $\hat{n}'$ is a direction of the CMB photons in the rest frame of CMB, which is different from the observed direction, $\hat{n}$, due to relativistic *aberration*. To the first order in v/c, the relation is given by $\hat{n}' = \hat{n} - \mathbf{d}$, where $\mathbf{d} \equiv \mathbf{v}/c - \hat{n}(\hat{n} \cdot \mathbf{v}/c)$ [60]. Note that $\mathbf{d}$ is perpendicular to the line-of-sight, $\hat{n} \cdot \mathbf{d} = 0$; thus, aberration acts as a deflection of light, just like gravitational lensing, to the first order in v/c. Expanding Eq. (4.65) up to the first order in v/c and extracting the temperature anisotropy, we obtain

$$\frac{\delta T(\hat{n})}{T} = \hat{n} \cdot \mathbf{v}/c + \frac{\delta T_p(\hat{n} - \mathbf{d})}{T}(1 + \hat{n} \cdot \mathbf{v}/c). \tag{4.66}$$

The first term on the right-hand side is the dipole anisotropy described already. The second term is a new piece: it describes aberration and *modulation*. Here, by "modulation" we mean the intrinsic anisotropy being multiplied (modulated) by a spatially-varying function, $1 + \hat{n} \cdot \mathbf{v}/c$. Aberration and modulation change statistical properties of the observed temperature anisotropy. Namely, they create *statistical anisotropy*: even if the intrinsic distribution of temperature anisotropy is statistically isotropic (i.e., the variance of temperature anisotropy is equal in all directions), aberration and modulation make the observed temperature anisotropy statistically anisotropic. An algorithm for extracting such information has been developed [61]. Using this method, the *Planck* team has detected both the aberration and modulation effects, finding a velocity of $v = 384 \pm 78 \pm 115$ km/s (68% CL; statistical and systematic uncertainties) toward the known dipole direction of $(l, b) = (264°, 48°)$ [62], which is in agreement with the latest *WMAP* determination to within the uncertainties.

4.3.4 Primordial temperature anisotropy: scalar modes

The most stunning, successful example of precision cosmology to date is the temperature anisotropy of the CMB beyond dipole anisotropy. To understand temperature anisotropy, we need to understand how light propagates in a clumpy universe, and how it interacts with matter. From now on, we shall set $c \equiv 1$.

The four-momentum of each photon, p^μ, obeys the geodesic equation:

$$\frac{dp^\mu}{d\lambda} + \Gamma^\mu_{\alpha\beta} p^\alpha p^\beta = 0, \tag{4.67}$$

where λ is an affine parameter and $\Gamma^\mu_{\alpha\beta}$ is the affine connection. As for a model of spacetime of a clumpy universe, we use a perturbed flat Friedmann–Lemaître–Robertson–Walker metric given by

$$ds^2 = -(1 + 2\Psi)dt^2 + a^2(t)(1 + 2\Phi)\delta_{ij}\,dx^i\,dx^j, \tag{4.68}$$

in the Newton gauge, where $a(t)$ is the scale factor. We have included only the scalar-mode perturbations: we shall discuss tensor modes in Section 4.3.6. Our convention is such that, in the absence of anisotropic stress in the stress-energy tensor, two perturbations in Eq. (4.68) coincide up to a sign: $\Psi = -\Phi$. Let us write Eq. (4.67) in terms of the magnitude of the three-momentum, $p^2 \equiv g_{ij}p^i p^j$. We obtain, up to the first order in perturbations,

$$\frac{1}{p}\frac{dp}{dt} = -\frac{1}{a}\frac{da}{dt} - \frac{d\Psi}{dt} + \frac{\partial\Psi}{\partial t} - \frac{\partial\Phi}{\partial t}, \tag{4.69}$$

where the total derivative is related to partial derivatives by $d\Psi/dt = \partial\Psi/\partial t + (p^i/p)\partial\Psi/\partial x^i$. The first term on the right-hand side in Eq. (4.69) gives the cosmological redshift, $p \propto 1/a$. The second term gives a redshift (blueshift) as photons climb up (enter into) a gravitational potential. The last term gives an additional redshift/blueshift as Ψ and Φ change while photons propagate through them. The solution of Eq. (4.69) is

$$\ln(ap)_\mathcal{O} = \ln(ap)_\mathcal{E} + (\Psi_\mathcal{E} - \Psi_\mathcal{O}) + \int_{t_\mathcal{E}}^{t_\mathcal{O}} dt\, \frac{\partial}{\partial t}(\Psi - \Phi), \qquad (4.70)$$

where "$\mathcal{O}$" and "$\mathcal{E}$" denote the observed and emitted epochs, respectively. Relating the momentum with temperature anisotropy as $ap \propto aT(1 + \delta T/T)$, Taylor-expanding the logarithm, and adding the Doppler shift we discussed in Section 4.3.3, we obtain

$$\left.\frac{\delta T}{T}\right|_\mathcal{O} = \left.\frac{\delta T}{T}\right|_\mathcal{E} + (\Psi_\mathcal{E} - \Psi_\mathcal{O}) + \int_{t_\mathcal{E}}^{t_\mathcal{O}} dt\, \frac{\partial}{\partial t}(\Psi - \Phi) + \hat{n} \cdot (\mathbf{v}_\mathcal{O} - \mathbf{v}_\mathcal{E}). \qquad (4.71)$$

This result has a simple interpretation.

1. There was an initial temperature anisotropy at the last scattering surface, $\delta T/T|_\mathcal{E}$ (which remains to be calculated), as well as the Doppler shift due to motion of the plasma which last-scattered CMB photons, $-\hat{n} \cdot \mathbf{v}_\mathcal{E}$, where $\mathbf{v}_\mathcal{E}$ is the velocity of the plasma.
2. After the last scattering, photons escape from a potential well, losing energy: $\delta T/T|_\mathcal{E} + \Psi_\mathcal{E} - \hat{n} \cdot \mathbf{v}_\mathcal{E}$.
3. While photons are propagating toward us, photons gain or lose energy depending on how $\Psi - \Phi\ (\approx 2\Psi)$ changes with time, giving $\delta T/T|_\mathcal{E} + \Psi_\mathcal{E} + \int_{t_\mathcal{E}}^{t_\mathcal{O}} dt\, \frac{\partial}{\partial t}(\Psi - \Phi) - \hat{n} \cdot \mathbf{v}_\mathcal{E}$.
4. Finally, photons enter a potential well at our location, $\Psi_\mathcal{O}$, gaining energy. Also, they receive the Doppler shift due to our local motion, giving $\delta T/T|_\mathcal{E} + \Psi_\mathcal{E} - \Psi_\mathcal{O} + \int_{t_\mathcal{E}}^{t_\mathcal{O}} dt\, \frac{\partial}{\partial t}(\Psi - \Phi) + \hat{n} \cdot (\mathbf{v}_\mathcal{O} - \mathbf{v}_\mathcal{E})$.

In particular, $\delta T/T|_\mathcal{E} + \Psi_\mathcal{E} - \Psi_\mathcal{O}$ is known as the Sachs–Wolfe effect, and $\int_{t_\mathcal{E}}^{t_\mathcal{O}} dt\, \frac{\partial}{\partial t}(\Psi - \Phi)$ is known as the integrated Sachs–Wolfe effect. All of these terms were derived by Sachs and Wolfe in 1967 [63].

The integrated Sachs–Wolfe effect vanishes when the universe is completely matter-dominated. Within the standard ΛCDM model, the integrated Sachs–Wolfe effect is important only in two epochs: the energy density of the universe is not completely matter-dominated at the time of decoupling (radiation still contributes a third of the energy density at $z = 1088$); and the energy density is dominated by dark energy at $z < 0.4$. The potentials decay in both epochs, producing an additional temperature anisotropy of $2 \int dt (\partial \Psi/\partial t)$. (The potentials decay inside the horizon in the first epoch, while they decay at all scales in the second epoch.) This effect is quite valuable. The first effect near $z = 1088$ is sensitive to the ratio of the total matter (which includes baryons and dark matter) to the total radiation (which includes photons, neutrinos and any other possible extra radiation species) densities. In the absence of extra radiation species at decoupling, we can measure the total matter density from this "early" integrated Sachs–Wolfe effect [64]. The smaller the total matter density is, the greater the early integrated Sachs–Wolfe effect becomes, enhancing temperature anisotropy at degree scales in the sky. On the other hand, if we fix the matter density from the other measurements (e.g., large-scale structure of the universe), then we can constrain the energy density of extra radiation species [65–67]. The second effect at $z < 0.4$ can be measured by cross-correlating the CMB data with tracers of matter in a low-redshift universe [68]. This "late" integrated Sachs–Wolfe effect is important only on very large scales (more than ten degrees in the sky), and has been detected convincingly [69], providing the direct evidence for dark energy via its influence on the time-dependence of Φ and Ψ.

How do we calculate the initial temperature fluctuation at the last scattering surface, $\delta T/T|_{\mathcal{E}}$? First, we must specify the initial condition for perturbations. This cannot be known a priori without using the observational data. The initial conditions are classified into two possibilities:

- Adiabatic initial condition
- Non-adiabatic initial condition

The current observational data favor the adiabatic initial condition, and we have not yet found any evidence for non-adiabatic initial conditions. Therefore, we shall focus on the adiabatic initial condition. This is the initial condition in which radiation and matter are perturbed in a similar way. More specifically, $\rho_\gamma^{3/4}/\rho_M \propto T^3/\rho_M$ remains unperturbed in adiabatic evolution. (Recall, for example, $\rho_\gamma \propto 1/a^4$ and $\rho_M \propto 1/a^3$ in free expansion.) Therefore, the adiabatic initial condition is given by $\delta T/T = \delta\rho_M/(3\rho_M)$. "Non-adiabatic initial conditions" would have $\delta T/T \neq \delta\rho_M/(3\rho_M)$.

As this is the initial condition, it holds only on very large scales, much larger than the horizon size at the last scattering surface (which would subtend about three degrees in the sky). On such large scales, the density fluctuation during the matter-dominated era is related to the Newtonian potential as $\delta\rho_M/\rho_M = -2\Psi$. This gives, on large scales, the initial temperature fluctuation of $\delta T/T = -\frac{2}{3}\Psi$. Then, the Sachs–Wolfe formula gives

$$\left.\frac{\delta T}{T}\right|_{\odot} = \frac{1}{3}\Psi_{\mathcal{E}} + \cdots \tag{4.72}$$

Therefore, on large scales, an over-density region (i.e., a potential well) appears as a cold spot in the sky. While the temperature at the bottom of the potential well is hotter than the average ($-\frac{2}{3}\Psi$), photons lose more energy (Ψ) as they climb up the potential well, resulting in a cold spot ($-\frac{2}{3}\Psi + \Psi = \frac{1}{3}\Psi$).

On smaller scales (smaller than the horizon size at the decoupling epoch), we must take into account interactions between matter and photons. Photons and electrons interact efficiently via Thomson scattering, and electrons and baryons (protons and helium nuclei) also interact efficiently via Coulomb interaction; thus, we can treat photons and baryons as coupled fluids. These interactions modify the temperature anisotropy away from that given by the initial condition. The relevant equations are continuity and Euler equations for baryons and photons in Fourier space:

$$\dot{\delta}_B = -\frac{k}{a}V_B - 3\dot{\Phi}, \tag{4.73}$$

$$\dot{\delta}_\gamma = -\frac{4}{3}\frac{k}{a}V_\gamma - 4\dot{\Phi}, \tag{4.74}$$

$$\dot{V}_B = -\frac{\dot{a}}{a}V_B + \frac{k}{a}\Psi + \frac{\sigma_T n_e}{R}(V_\gamma - V_B), \tag{4.75}$$

$$\dot{V}_\gamma = \frac{1}{4}\frac{k}{a}\delta_\gamma + \frac{k}{a}\Psi + \sigma_T n_e(V_B - V_\gamma), \tag{4.76}$$

where the dots denote time derivatives, $\delta_B \equiv \delta\rho_B/\rho_B$, $\delta_\gamma \equiv \delta\rho_\gamma/\rho_\gamma$, V_B and V_γ are divergences (divided by k) of baryon and photon velocities relative to the dark matter velocity, respectively, i.e., $\nabla \cdot \mathbf{v} \equiv \int \frac{d^3k}{(2\pi)^3} kVe^{i\mathbf{k}\cdot\mathbf{x}}$ (or $\mathbf{v} = -i(\mathbf{k}/k)Ve^{i\mathbf{k}\cdot\mathbf{x}}$ for a single plane wave), and R is the baryon-to-photon energy-density ratio defined as $R \equiv 3\rho_B/(4\rho_\gamma)$. Here, dark matter enters into the system only via its effects on potentials, Φ and Ψ. We thus have a system of coupled fluids moving in potentials determined primarily by dark matter. Einstein's equations relate the dark matter and baryon densities to Φ and Ψ. From now on, we shall simplify the system by assuming that the universe is matter-dominated at decoupling (see, however, the description of the early integrated Sachs–Wolfe effect above), and set $\Psi = -\Phi$ and $\dot{\Phi} = 0$. Of course, this simplification is not made in the numerical calculations achieving the 0.1% precision [15–18].

We combine these equations using the "tight-coupling approximation," i.e., baryons and photons basically move together. We rewrite the baryon Euler equation as

$$V_B = V_\gamma - \frac{R}{\sigma_T n_e}\left(\dot{V}_B + \frac{\dot{a}}{a}V_B + \frac{k}{a}\Phi\right)$$

$$\approx V_\gamma - \frac{R}{\sigma_T n_e}\left(\dot{V}_\gamma + \frac{\dot{a}}{a}V_\gamma + \frac{k}{a}\Phi\right). \tag{4.77}$$

Here, since we assume that the difference between V_B and V_γ is small, we have replaced V_B with V_γ on the right-hand side. We then use the photon Euler equation to write:

$$V_B = V_\gamma - \frac{R}{\sigma_T n_e}\left[\frac{1}{4}\frac{k}{a}\delta_\gamma + \frac{\dot{a}}{a}V_\gamma + \sigma_T n_e(V_B - V_\gamma)\right]$$

$$\sigma_T n_e(V_B - V_\gamma) = -\frac{R}{1+R}\left[\frac{1}{4}\frac{k}{a}\delta_\gamma + \frac{\dot{a}}{a}V_\gamma\right]. \tag{4.78}$$

Using this in the photon Euler equation, and using the photon continuity equation, $V_\gamma = -\frac{3}{4}\frac{a}{k}\dot{\delta}_\gamma$, we arrive at the following differential equation for the photon energy density:

$$\ddot{\delta}_\gamma + \frac{1+2R}{1+R}\frac{\dot{a}}{a}\dot{\delta}_\gamma + \frac{1}{3(1+R)}\frac{k^2}{a^2}\delta_\gamma = \frac{4}{3}\frac{k^2}{a^2}\Phi. \tag{4.79}$$

This is a wave equation for δ_γ; thus, a coupling between baryons and photons results in acoustic oscillations in the photon density perturbations [49, 70]. Since baryons and photons are coupled, the same oscillations must also be present in the baryon density perturbations as well. Indeed, the acoustic oscillations have been observed both in photons (microwave background) and in the distribution of matter (galaxies).

To obtain a deeper understanding of the structures of the acoustic oscillation, let us focus on the regime where the oscillation frequency is much greater than the expansion rate of the universe. In this case, the wave equation simplifies to

$$\ddot{\delta}_\gamma + \frac{1}{3(1+R)}\frac{k^2}{a^2}\delta_\gamma = \frac{4}{3}\frac{k^2}{a^2}\Phi. \tag{4.80}$$

Since $\dot{\Phi} = 0$ during the matter era, one may rewrite this equation in a suggestive way:

$$\frac{\partial^2}{\partial t^2}\left[\frac{1}{4}\delta_\gamma - (1+R)\Phi\right] + \frac{k^2 c_s^2}{a^2}\left[\frac{1}{4}\delta_\gamma - (1+R)\Phi\right] = 0, \tag{4.81}$$

where c_s is the speed of sound, $c_s^2 \equiv 1/[3(1+R)]$. Note that this speed of sound is smaller than that for the relativistic fluid, $c_s^2 = 1/3$. This is due to the coupling to baryons: the inertia of baryons reduces the speed of sound of the photon–baryon fluid relative to that of the relativistic fluid. The solution to the above wave equation is $\frac{1}{4}\delta_\gamma = (1+R)\Phi + A\cos(kr_s) + B\sin(kr_s)$, where r_s is the "sound horizon" defined by $r_s \equiv \int_0^{t_*} c_s dt/a = 147$ Mpc. The numerical value is obtained for the cosmological parameters best-fit to the *WMAP* data and t_* is the decoupling time. We determine the integration constants, A and B, by noting that, on superhorizon scales, these solutions must match the adiabatic initial condition: $\frac{1}{4}\delta_\gamma = \frac{2}{3}\Phi$. Therefore, we find $A = \frac{2}{3}\Phi - (1+R)\Phi = -\left(\frac{1}{3}+R\right)\Phi$ and $B = 0$. Since $\rho_\gamma \propto T^4$, we can relate $\frac{1}{4}\delta_\gamma$ to $\delta T/T$ as $\frac{1}{4}\delta_\gamma = \delta T/T$. Moreover, since the observed temperature anisotropy is the sum of $\delta T/T$ at the bottom of the potential well and the potential Ψ, we finally obtain the first two terms in the geodesic equation (Eq. (4.69)) in Fourier space as

$$\left.\frac{\delta T}{T}\right|_\mathcal{E} + \Psi_\mathcal{E} = -R\Psi_\mathcal{E} + \left(\frac{1}{3}+R\right)\Psi_\mathcal{E}\cos(kr_s), \tag{4.82}$$

where we have used $\Psi = -\Phi$. What we observe is the power spectrum, which is $(\delta T/T|_\mathcal{E} + \Psi_\mathcal{E})^2$ and has a series of peaks at $k = m\pi/r_s$ with $m = 1, 2, \ldots$. We find that the odd-to-even peak ratios go up as R increases; thus, the odd-to-even peak ratios can be used to determine the baryon density [64]. Together with the early integrated Sachs–Wolfe effect giving the total matter density, we can determine the baryon and dark matter densities separately.

The Doppler shift term, $\hat{n}\cdot\mathbf{v}_\mathcal{E}$, in Eq. (4.69) also plays an important role. As is clear from the photon continuity equation, $V_\gamma = -\frac{3}{4}\frac{a}{k}\dot{\delta}_\gamma$, the photon velocity is given by a time derivative of δ_γ. Therefore, the Doppler shift term is comparable to the density term, with $\cos(kr_s)$ replaced by $c_s\sin(kr_s)$ and without an offset term $R\Psi$. Since $\mathbf{v} = -i(\mathbf{k}/k)V$, we find

$$-\hat{n}\cdot\mathbf{v}_{\gamma,\mathcal{E}} = i\mu(1+3R)\Psi_\mathcal{E}c_s\sin(kr_s), \tag{4.83}$$

where $\mu \equiv \hat{n}\cdot\hat{k}$ is a cosine between the line of sight and the wavenumber vector. As a result, the Doppler shift term fills the power spectrum at the wavenumbers at which the density term vanishes, giving a smoother power spectrum.

Finally, we must mention the effect of damping of acoustic oscillations. The above analysis is based on the tight-coupling approximation. The tight-coupling approximation breaks down at the scales comparable to the photon diffusion scale. We have already discussed photon diffusion in Section 4.3.2; namely, temperature fluctuations are erased within the photon diffusion scale given by $l_d = \sqrt{N} l_{mfp}$, where $l_{mfp} = (\sigma_T n_e)^{-1}$ is the mean free path of photons, and $N = \sqrt{ct/l_{mfp}}$ is the number of scatterings photons encounter within a given time. Temperature fluctuations below this scale are damped exponentially [48–50]. This damping dissipates energy of the acoustic waves into the plasma, creating a y-type distortion, as already discussed in Section 4.3.2.

One last step is required before we compare the above analysis to the observational data. While we have computed $\delta T/T$ in Fourier space, we observe it in our sky, which is a two-dimensional sphere. We thus need to project the Fourier waves onto the sky. To do this, we first decompose the observed temperature anisotropy into spherical harmonics: $\delta T/T|_{\odot}(\hat{n}) = \sum_{\ell m} a_{\ell m} Y_{\ell m}(\hat{n})$. The spherical harmonics coefficients, $a_{\ell m}$, are then related to the Fourier transform of Φ as

$$a_{\ell m} = 4\pi(-i)^{\ell} \int \frac{d^3 \mathbf{k}}{(2\pi)^3} \Phi_{\mathbf{k}} g_{T\ell}(k) Y_{\ell m}^*(\hat{k}), \qquad (4.84)$$

where $g_{T\ell}(k)$ is the so-called "radiation transfer function," which includes both projection of $k \to \ell$ as well as the relation between Φ and temperature fluctuations, such as the physics of acoustic oscillations, the integrated Sachs–Wolfe effect, the Doppler term, photon diffusion damping, etc. For the Sachs–Wolfe effect, we find $\delta T/T|_{\odot}(\hat{n}) = -\Phi(\hat{n}d_A)/3$, where $d_A \simeq 14$ Gpc is the comoving angular diameter distance to the last scattering surface; thus, the radiation transfer function is given by $g_{T\ell} = -j_{\ell}(kd_A)/3$. As this function is sharply peaked at $\ell \approx kd_A$, the $k \to \ell$ projection is simply given by the trigonometric relationship between a half wavelength at the last scattering surface and the angle that it subtends, i.e., $\theta = (\lambda/2)/d_A$ with $\ell = \pi/\theta$ and $\lambda = 2\pi/k$. As the comoving angular diameter distance is given by $d_A = \int_0^{z*} dz/H(z)$ in a flat universe, and $H(z)$ is determined by the *total* energy density of the universe including dark energy, the CMB data are sensitive to the amount of dark energy. (The CMB data are sensitive to dark energy also via the late integrated Sachs–Wolfe effect, as described above.)

Assuming statistical isotropy and homogeneity, the spherical harmonics coefficients and $\Phi_{\mathbf{k}}$ are related to the respective power spectra as $\langle a_{\ell m} a_{\ell' m'}^* \rangle = C_{\ell} \delta_{\ell\ell'} \delta_{mm'}$ and $\langle \Phi_{\mathbf{k}} \Phi_{\mathbf{k}'}^* \rangle = (2\pi)^3 P_{\Phi}(k) \delta^{(3)}(\mathbf{k} - \mathbf{k}')$. The projection, Eq. (4.84), then relates two power spectra as

$$C_{\ell} = \frac{2}{\pi} \int k^2 \, dk \, P_{\Phi}(k) g_{T\ell}^2(k). \qquad (4.85)$$

Now, here is the bottom line. With the advent of sensitive detectors in microwave bands, we have acquired the precision data on C_{ℓ} from the *WMAP* and *Planck* satellites as well as from a host of ground-based and balloon-borne experiments. On the other hand, the linear equations describing the evolution of the photon–baryon system including the influence of dark matter and neutrinos have been solved completely, providing the precision theoretical predictions for $g_{T\ell}(k)$ at the 0.1% level [15–18] as a function of cosmological parameters such as the baryon and dark matter densities. Finally, the theory of cosmic inflation predicts that $P_\Phi(k)$ should obey a power-law spectrum, $P_\Phi(k) \propto k^{n_s-4}$.

If the universe after the photon decoupling epoch at $z = 1088$ is completely transparent to photons, information on the last scattering surface carried by CMB photons is preserved. However, the universe at a low redshift, $z < 6$, is known to be fully ionized [71, 72], and free electrons in this epoch scatter a fraction of CMB photons away from our line of sight. It is likely that this "reionization" of the universe occurred in higher redshifts, say, $z \approx 20$, and it was caused by formation of the first stars in the universe [73]. The fractional amount by which photons are re-scattered is given by $1 - e^{-\tau} \approx \tau$ for $\tau \equiv \int dt\, \sigma_T n_e \ll 1$. While this information is difficult to measure from the temperature data alone, the same scattering produces an additional polarization in CMB photons, which can be used to measure τ precisely ([74]; see Section 4.3.5). The power spectrum of the CMB temperature anisotropy, C_ℓ, is damped uniformly as $C_\ell \to C_\ell e^{-2\tau}$ at roughly $\ell > 20$.

Figure 4.7 is a major milestone in precision cosmology, which shows a truly remarkable agreement between the precision data and the precision theoretical model. A few thousand data points are described by six cosmological parameters: baryon density, $\Omega_B h^2$ (from the odd-to-even peak ratios), dark matter density, $\Omega_{DM} h^2$ (from the first-to-other peak ratios measuring the early integrated Sachs–Wolfe effect), dark energy density, Ω_Λ (from the peak positions measuring d_A), the amplitude and slope of $P_\Phi(k)$ (from the overall amplitude and slope of C_ℓ), and the optical depth to Thomson scattering, τ (from $C_\ell \to C_\ell e^{-2\tau}$ and polarization).

The parameters and associated 68% CL uncertainties are given in Table 4.1. It is clear that the CMB data *alone* are sufficient to determine all of the six parameters of the standard flat ΛCDM model precisely. The parameters from the *WMAP* nine-year data combined with the *ACT* and *SPT* data ("*WMAP+ACT+SPT*") are in good agreement with those from the *Planck* 15.5-month temperature data combined with the *WMAP* nine-year polarization data ("*Planck+WP*") to within the quoted uncertainties. These data convincingly show that the total matter density determined from the early integrated Sachs–Wolfe effect is six times as large as the baryon density determined from the odd-to-even peak ratios, proving the existence of dark matter. They also convincingly show that the matter alone cannot account for the present-day energy density of the universe, erasing any lingering doubts about the existence of dark energy (within the framework of General Relativity). The *Planck* data finally show that, for the first time with the CMB data alone, the spectral tilt, n_s, is less than unity with more than five standard deviations. This measurement has a profound implication for the theory of inflation; in fact, the central value, $n_s = 0.96$, was predicted by the very first model of inflation based upon an R^2 term in the Lagrangian, where R is the Ricci curvature [75, 76]. For cosmologists working on the physics of the early universe, this measurement is perhaps as important as discovery of a Higgs particle.

There is a new twist to this story. While Eq. (4.69) describes changes in the magnitude of photon momentum in a clumpy universe, the *directions* of photon momentum also change due to gravitational lensing. The lensing of the CMB leaves distinct signatures in the statistical properties of CMB temperature and polarization anisotropies [77]. What lensing

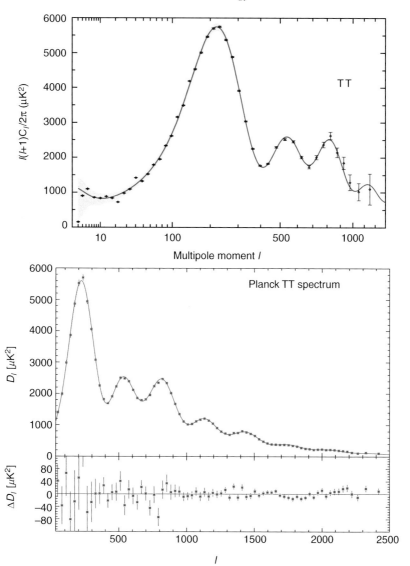

Figure 4.7 Power spectrum of temperature anisotropy of the CMB, $\ell(\ell+1)C_\ell/(2\pi)$, in units of μK^2. (Top) The *WMAP* nine-year data (figure adapted from [4] © AAS. Reproduced with permission). (Bottom) The *Planck* 15.5-month data (figure adapted from [8]. Reproduced with permission © ESo). In both panels the lines show the best-fit models. The bottom panel also shows the difference between the data and the best-fit model. The odd-to-even peak ratios determine the baryon density. The first peak is enhanced with respect to the others by the early integrated Sachs–Wolfe effect (determining the total matter density including baryons and dark matter). Diffusion damping exponentially suppresses the power at high multipoles. The peak positions determine r_s/d_A.

Table 4.1 *The six cosmological parameters of the standard flat ΛCDM model,
determined from the WMAP nine-year temperature and polarization power
spectra combined with the ACT and SPT temperature spectra
("WMAP+ACT+SPT") [5]; the Planck 15.5-month temperature power
spectrum combined with the WMAP nine-year polarization power spectrum
("Planck+WP") [8]; and the Planck 15.5-month temperature power spectrum
combined with the lensing potential power spectrum ("Planck+lensing") [8].*

Parameters	WMAP+ACT+SPT	Planck+WP	Planck+lensing
$\Omega_B h^2$	0.02229 ± 0.00037	0.02205 ± 0.00028	0.02217 ± 0.00033
$\Omega_{DM} h^2$	0.1126 ± 0.0035	0.1199 ± 0.0027	0.1186 ± 0.0031
Ω_Λ	0.728 ± 0.019	$0.685^{+0.018}_{-0.016}$	0.693 ± 0.019
$10^9 \Delta_{\mathcal{R}}^2$ ^{a}	2.167 ± 0.056	$2.196^{+0.051}_{-0.060}$	$2.19^{+0.12}_{-0.14}$
n_s	0.9646 ± 0.0098	0.9603 ± 0.0073	0.9635 ± 0.0094
τ	0.084 ± 0.013	$0.089^{+0.012}_{-0.014}$	0.089 ± 0.032

^{a} This dimensionless amplitude is defined as $\Delta_{\mathcal{R}}^2 \equiv 25k^3 P_\Phi/(18\pi^2)(k = 0.05 \text{ Mpc}^{-1})$.

does is simple: it shifts the arrival directions of CMB photons as $\delta T(\hat{n}) \rightarrow \delta T(\hat{n} + \nabla\phi)$, where ϕ is called the "lensing potential" and is given by

$$\phi(\hat{n}) = \int_0^{d_A} d\chi \, \frac{d_A - \chi}{d_A \chi} [\Phi - \Psi](\hat{n}\chi, \chi), \tag{4.86}$$

in a flat universe. Lensing affects temperature anisotropy in two ways: as lensing effectively changes d_A depending on lines of sight, Fourier modes with a given wavenumber k are projected onto slightly different multipoles, $\ell \approx k d_A$. Lensing thus smears out the acoustic oscillations, reducing the contrasts of peaks and troughs [78]. Lens mapping, $\delta T(\hat{n}) \rightarrow \delta T(\hat{n} + \nabla\phi)$, also breaks the statistical isotropy of temperature anisotropy. An algorithm for measuring this effect exists [79–81]. Both of these effects have been detected with high statistical significance [9, 82–84]. Figure 4.8 shows the measured and predicted angular power spectra of the lensing potential. The prediction comes from the best-fit model to the *Planck* temperature power spectrum shown in Figure 4.7 and the *WMAP* polarization power spectrum at $\ell \leq 23$. The agreement between the data and the model is striking. As most of the CMB lensing is done by the intervening matter at roughly $z < 3$, the agreement shows that we understand the growth history of matter fluctuations from $z = 1088$ to such low redshifts.

The measurement of the CMB lensing has provided an important check of the determination of the optical depth to Thomson scattering, τ, *without using the polarization data*. Free electrons generated by reionization of the universe (done probably by the first generation of stars) scatter a fraction of the CMB photons, reducing the temperature power spectrum as $C_\ell \rightarrow C_\ell e^{-2\tau}$ at roughly $\ell > 20$. As the cosmic variance uncertainties at $\ell < 20$ are large, the overall amplitude of the primordial power spectrum, $P_\Phi(k)$, and

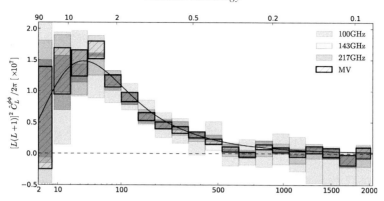

Figure 4.8 Power spectrum of lensing potential measured from the *Planck* 15.5-month temperature data, $[L(L + 1)]^2 C_L^{\phi\phi}/2\pi$ (figure adapted from [84]. Reproduced with permission © ESo) The pink, green, and blue boxes show the measurements from the temperature maps at 100, 143, and 217 GHz, respectively, while the black boxes show the minimum variance estimates combining 143 and 217 GHz. The line shows the prediction from the best-fit model to the temperature power spectrum shown in the bottom panel of Figure 4.7 and the *WMAP* polarization power spectrum at $\ell \leq 23$.

τ are degenerate, and we can measure only a combination, $P_\Phi(k)e^{-2\tau}$. However, the CMB lensing provides an independent measurement of the amplitude of the primordial power spectrum, breaking this degeneracy. As a result, the lensing data shown in Figure 4.8 enable the *Planck* team to determine τ from the temperature data alone. The *Planck* team finds $\tau = 0.089 \pm 0.032$ (68% CL), which is in excellent agreement with $\tau = 0.089^{+0.012}_{-0.014}$ (68% CL) from the *Planck* temperature data combined with the *WMAP* polarization data, but without the lensing information.

4.3.5 Polarization: scalar modes

Thomson scattering creates polarization when CMB photons are last-scattered by free electrons [50, 85–88]. The necessary and sufficient condition for generation of polarization is the existence of quadrupolar temperature anisotropy ($\ell = 2$) around an electron. This implies that polarization is small: in the tight-coupling approximation in which photons and baryons (and electrons) move together, the radiation pattern around an electron is isotropic, and no polarization is produced. Therefore, polarization is produced only when the tight-coupling approximation breaks down. In other words, at the last scattering surface, polarization is significant on the angular scales where the temperature power spectrum damps by photon diffusion.

Polarization (for scalar modes) traces a velocity field of the plasma around gravitational potentials. Suppose that a packet of the plasma is falling into the bottom of the potential well. Due to acceleration, a velocity gradient is generated: the front of the packet falls faster than the back of the packet. Therefore, an electron at the center of the packet observes redshifted photons from both the front and back of the packet, whereas there is no redshift or blueshift from the sides of the packet. This produces a quadrupolar radiation pattern (colder along the motion of the packet and hotter in the perpendicular directions), and the produced polarization is parallel to the motion of the packet. The polarization pattern around a spherically symmetric gravitational potential well is *radial*, and the magnitude of radial polarization is maximal at $2r_s$ (or 1.2 degrees in the sky) from the bottom of the potential well [89]. As the packet approaches the bottom of the potential well, the packet decelerates because of a pressure gradient. In the adiabatic initial condition, the photon density is high at the bottom of the potential well, producing a pressure gradient to decelerate motion of the plasma falling into the potential well. The front of the packet falls slower than the back of the packet. Therefore, an electron at the center of the packet observes blueshifted photons from both the front and back of the packet, whereas there is no redshift or blueshift from the sides of the packet. This produces the opposite quadrupolar radiation pattern (hotter along the motion of the packet and colder in the perpendicular directions), and the produced polarization is *tangential* to the motion of the packet. The magnitude of tangential polarization is maximal at r_s (or 0.6 degrees in the sky) from the bottom of the potential well [89].

These predictions have been confirmed by the *WMAP* polarization data. The bottom panels of Figure 4.9 show the average polarization directions measured around hot and cold

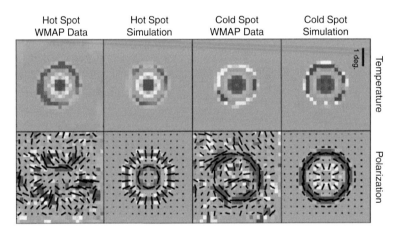

Figure 4.9 Average images of temperature and polarization data. 12387 hot spots and 12628 cold spots are found in the *WMAP* seven-year temperature maps, and the average images of hot and cold spots are shown in the top panels along with the corresponding simulated images. The bottom panels show the average images of the polarization maps around the locations of hot and cold temperature spots, as well as the corresponding simulated images. The size of each image is $5°$ by $5°$. The lines show the polarization directions, and their lengths are proportional to the magnitude of polarization. The colors of the polarization images are chosen such that blue and red show the tangential and radial polarization patterns, respectively. The data show the predicted tangential and radial polarization patterns (E-mode polarization), in excellent agreement with the predictions. The maximum of radial polarization around hot spots occurs at 1.2 degrees from the center, whereas the maximum of tangential polarization around hot spots occurs at 0.6 degrees from the center. Figure adapted from http://wmap.gsfc.nasa.gov/media/101079/index.html (Credit: NASA/WMAP Science Team).

temperature spots. On these angular scales (a few degrees), hot and cold spots correspond to potential wells and hills, respectively. (The high photon energy density at the bottom of the well overcomes the Sachs–Wolfe effect, turning potential wells into hot spots in the sky.) Therefore, we expect each hot spot to come with the radial and tangential polarization patterns at 1.2 and 0.6 degrees from the center, respectively, and each cold spot to come with the opposite patterns. As the magnitude of polarization is small, the *WMAP* cannot detect polarization around each spot; however, by averaging polarization patterns around many spots, we can detect polarization. There are 12387 hot spots and 12628 cold spots outside the Galactic mask in the *WMAP* seven-year temperature map. Averaging the polarization data around these spots, the expected polarization patterns (shown in the "Simulation" columns in Figure 4.9) are clearly detected in the data (shown in the "WMAP Data" columns) [89]. *Planck* detects these patterns with much improved sensitivity, but they are not shown here because the simulation and data look identical [6].

The tangential and radial polarization patterns seen in Figure 4.9 are called the "*E*-mode polarization." There is also the so-called "*B*-mode polarization," whose pattern is tilted by 45 degrees with respect to the *E*-mode pattern. The *B*-mode is a parity-odd pattern whereas the *E*-mode is a parity-even pattern. As we have seen already, the scalar modes (gravitational potentials) create only the *E*-mode polarization. Gravitational lensing by the intervening matter can subsequently convert a fraction of *E*-mode polarization to *B*-mode polarization [90], and this effect has been detected by the *SPT* collaboration [91] and the *Polarbear* collaboration [92–94].

The *B*-mode polarization can also be created by primordial gravitational waves (tensor modes) from inflation [95, 96]. The *BICEP2* collaboration [97] claims to have detected the primordial *B*-mode polarization at 150 GHz [98]; however, detection at other frequencies must be made before interpreting the detected signal as cosmological.

The *E*- and *B*-modes are defined operationally as follows. The polarization field is measured using the so-called Stokes parameters, $Q(\hat{n})$ and $U(\hat{n})$, which represent the directions and magnitudes of polarization as shown in Figure 4.10. As is clear from Figure 4.10, Q and U transform into each other as $Q \rightarrow U \rightarrow -Q \rightarrow -U \rightarrow Q \rightarrow \cdots$ as we rotate coordinates by 45 degrees. We thus use the spin-2 spherical harmonics to expand $Q \pm iU$ as $Q(\hat{n}) \pm iU(\hat{n}) = \sum_{\ell m} {}_{\mp 2}a_{\ell m}{}_{\mp 2}Y_{\ell m}(\hat{n})$. We then define the *E*- and *B*-modes as

$$a_{\ell m}^E \equiv -\frac{{}_2a_{\ell m} + {}_{-2}a_{\ell m}}{2}, \qquad a_{\ell m}^B \equiv -i\frac{{}_2a_{\ell m} - {}_{-2}a_{\ell m}}{2}. \tag{4.87}$$

As temperature spots come with polarization, temperature and polarization are expected to be correlated [99, 100]. Similar to the temperature harmonic coefficients, the *E*-mode polarization coefficients are related to the Fourier transform of Φ as

$$a_{\ell m}^E = 4\pi(-i)^\ell \sqrt{\frac{(\ell+2)!}{(\ell-2)!}} \int \frac{d^3\mathbf{k}}{(2\pi)^3} \Phi_{\mathbf{k}} g_{P\ell}(k) Y_{\ell m}^*(\hat{k}), \tag{4.88}$$

where $g_{P\ell}(k)$ is the "polarization transfer function." As the *E*-mode polarization of scalar modes is produced by the velocity perturbation, the polarization transfer function is

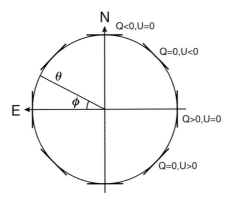

Figure 4.10 Definition of Stokes Q and U. We use Galactic coordinates with north up and east left. Figure adapted from [89] © AAS. Reproduced with permission.

predominantly determined by $\sin(kr_s)$ with $\ell \approx kd_A$ (see Eq. (4.83)), unlike the temperature anisotropy, which is predominantly determined by $\cos(kr_s)$ (see Eq. (4.82)). This implies that the temperature–polarization cross power spectrum, $\langle a_{\ell m} a_{\ell'm'}^{E*} \rangle = C_\ell^{TE} \delta_{\ell\ell'} \delta_{mm'}$, is determined by $\sin(kr_s)\cos(kr_s)$, which changes sign, and the peaks and troughs are shifted with respect to the temperature peaks and troughs (which are determined predominantly by $\cos^2(kr_s)$). Figure 4.11 shows the temperature–E-mode polarization cross power spectra measured from the *WMAP* and *Planck* data. The agreement between the data and the best-fit model, which is found from the temperature power spectrum shown in Figure 4.7 and the *WMAP* polarization power spectrum at $\ell \leq 23$, is striking.

The cross power spectrum offers a powerful, precision test of the standard cosmological model. We fix the basic six cosmological parameters by fitting the temperature power spectrum and the E-mode polarization power spectrum at low multipoles (or the temperature power spectrum and the lensing power spectrum). We can then predict the cross power spectrum *without any more additional free parameters*. The prediction matches with the data at the precision shown in Figure 4.11. This is a great triumph of the standard cosmological model.

The power spectrum of the E-mode polarization is defined by the relation $\langle a_{\ell m}^E a_{\ell'm'}^{E*} \rangle = C_\ell^{EE} \delta_{\ell\ell'} \delta_{mm'}$. As this is determined primarily by $\sin^2(kr_s)$, the peaks and troughs of the E-mode polarization power spectrum occur at the troughs and peaks of the temperature power spectrum, respectively. The E-mode polarization power spectra from the *WMAP*, *Planck*, and a host of ground-based experiments are shown in Figure 4.12. As expected, the E-mode power spectrum rises toward large multipoles, where the temperature power spectrum damps by photon diffusion. (Recall that it is the photon diffusion that creates quadrupolar temperature anisotropy around electrons, hence polarization.) The data and the model are again in excellent agreement.

The measured E-mode power spectrum exhibits a "bump" at low multipoles, $\ell < 10$. This is called the "reionization bump," which is generated by additional Thomson scattering

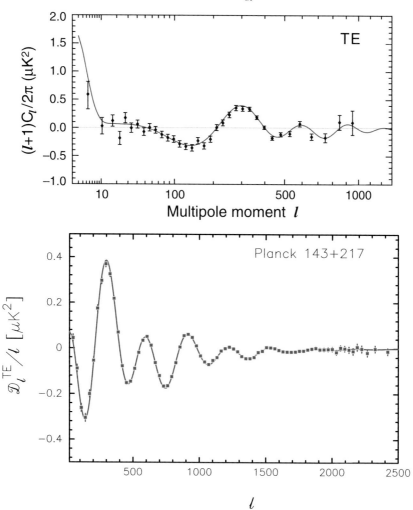

Figure 4.11 Cross power spectrum of temperature and E-mode polarization anisotropies of the CMB, $(\ell + 1)C_\ell^{TE}/(2\pi)$, in units of μK^2. (Top) The *WMAP* nine-year data (figure adapted from [4] © AAS. Reproduced with permission). (Bottom) The *Planck* 15.5-month data (figure adapted from [8]. Reproduced with permission © ESo). In both panels the lines show the best-fit models to the temperature power spectrum shown in Figure 4.7 and the *WMAP* E-mode polarization power spectrum at $\ell \lesssim 23$ (shown in Figure 4.12).

of CMB photons off free electrons produced during the reionization epoch, $z < 20$ [74]. The reason why this signal appears on such large angular scales is simply that this signal is produced at late times, which are closer to us than $z = 1088$ and thus subtend larger angles in the sky. The amplitude of the reionization bump is proportional to τ^2, and thus we can use this measurement to determine the optical depth. The polarization data show that about 9% of photons from $z = 1088$ are re-scattered in a reionized universe (see Table 4.1).

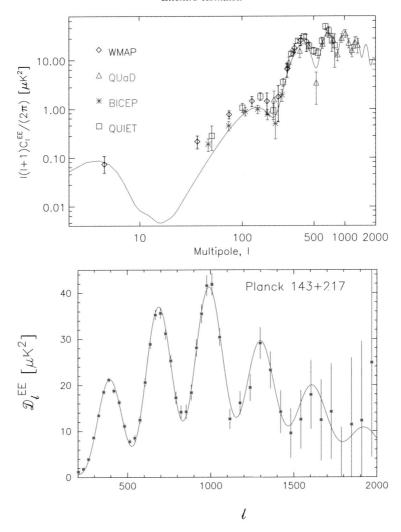

Figure 4.12 Power spectrum of *E*-mode polarization anisotropy of the CMB, $\ell(\ell+1)C_\ell^{EE}/(2\pi)$, in units of $\mu\mathrm{K}^2$. (Top) The *WMAP* nine-year data [4], the full-season *QUaD* data [101], the three-year *BICEP1* data [102], and the *QUIET* 95 GHz data [103]. The *WMAP* data points at $\ell < 10$ provide a measurement of the optical depth, τ. (Bottom) The *Planck* 15.5-month data (figure adapted from [8]. Reproduced with permission © ESo). In both panels the lines show the best-fit models to the temperature power spectra shown in Figure 4.7 and the *WMAP E*-mode polarization power spectrum at $\ell \leq 23$ shown in the top panel.

If we assume that the universe was reionized instantly at a particular redshift, z_{reion}, we find $z_{reion} \simeq 10$. In reality, the reionization is an extended process, and thus we should take this number as the epoch at which the universe is half ionized [104].

Gravitational lensing acts on the Stokes parameters in the same way it acts on temperature: $Q(\hat{n}) \rightarrow Q(\hat{n} + \nabla\phi)$ and $U(\hat{n}) \rightarrow U(\hat{n} + \nabla\phi)$. This transformation converts a fraction (about six percent) of E-mode polarization to B-mode polarization [90]. To the leading order in the lensing potential, the $E \rightarrow B$ conversion is given by (in the absence of primordial B-mode polarization)

$$a_{\ell m}^B = \sum_{LM} \phi_{LM} \sum_{\ell' m'} \tilde{a}_{\ell' m'}^E (-1)^m \begin{pmatrix} \ell & \ell' & L \\ m & -m' & -M \end{pmatrix} W_{\ell \ell' L}, \qquad (4.89)$$

where $\tilde{a}_{\ell m}^E$ is the primordial, unlensed E-mode polarization, and $W_{\ell \ell' L}$ is a known function determining a coupling between ℓ, ℓ', and L [80]. In particular, $W_{\ell \ell' L}$ is non-zero only for $\ell + \ell' + L =$ odd. The Wigner 3-j symbol demands that the angular wavenumbers satisfy the triangular conditions, $|\ell - \ell'| \leq L \leq \ell + \ell'$ etc.

The *SPT* collaboration has recently detected this lensing B-mode with nearly eight standard deviations [91]. They use a cross-correlation technique: they first estimate the lensing potential, ϕ_{LM}, from the distribution of matter, which is traced by dusty galaxies contained in images of the "cosmic infrared background" (CIB) taken by ESA's *Herschel* telescope [105]. They then combine the measured E modes, $a_{\ell m}^E$, and the estimated ϕ_{LM} using Eq. (4.89). The difference between the lensed $a_{\ell m}^E$ that *SPT* measures and the unlensed $\tilde{a}_{\ell m}^E$ yields terms which are of higher order in ϕ_{LM} in Eq. (4.89). By cross-correlating the *predicted* $a_{\ell m}^B$ from Eq. (4.89) and the measured one, the *SPT* detects the expected correlation between the two at nearly eight standard deviations (see Figure 4.13).

The *Polarbear* collaboration also reported a detection of the lensing B-mode in cross-correlation with CIB [92]. More importantly, they have detected the lensing B mode from the CMB data alone using correlations between E and B modes [93]. More recently, they found evidence for the B-mode power spectrum (without reference to E modes or CIB), although the statistical significance of the detection (97.5% CL) is still low [94].

In summary, *all* the effects of the linear scalar-mode perturbations on the CMB predicted by the standard six-parameter flat ΛCDM model, such as the acoustic oscillations in both temperature and E-mode polarization, photon diffusion damping, the effect of dark energy on the late integrated Sachs–Wolfe effect, and gravitational lensing effects on both temperature and polarization (most notably conversion of E modes to B modes), have been measured. The evidence for inflation is strengthened greatly by the discovery of $n_s < 1$ with more than five standard deviations. The future experiments will make these measurements, especially lensing-related measurements, much more precise. Now, what would be the next, *qualitatively new* discoveries in the CMB?

4.3.6 Tensor modes

Inflation predicts the existence of primordial fluctuations not only in scalar modes (gravitational potentials), but also in tensor modes (gravitational waves) [106]. The perturbed metric is given by

$$ds^2 = -dt^2 + a^2(t)(\delta_{ij} + h_{ij})dx^i\, dx^j, \qquad (4.90)$$

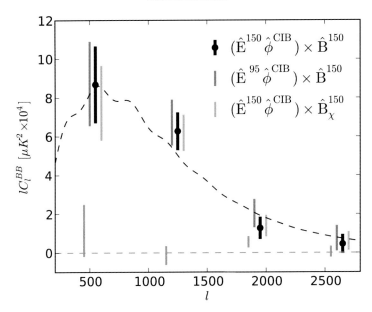

Figure 4.13 Detection of lensing-induced *B*-mode polarization of the CMB by the *SPT* collaboration (figure adapted and reprinted with permission from Fig. 2 of [91]. © 2013 by the American Physical Society). The *B* mode polarization power spectrum, ℓC_ℓ^{BB}, is shown in units of $\mu K^2 \times 10^4$. The measurements are done by cross-correlating the measured *B* modes at 150 GHz with the predicted *B* modes using Eq. (4.89) along with the measured *E* modes and the lensing potential estimated from the CIB. The black and green lines show the measurements using the measured *E* modes at 150 and 95 GHz, respectively. The orange lines show the measurements using alternative decomposition of *E* and *B* modes. The gray lines show a null test. The dashed line shows the prediction of "*Planck*+WP" parameters shown in Table 4.1. Agreement with the data is excellent.

where h_{ij} is a symmetric, traceless and transverse tensor. The tensor perturbation changes distances between test particles while preserving the area, i.e., $\det(\delta_{ij} + h_{ij}) = 1$. For a gravitational wave propagating along the z-axis, we may write

$$h_{ij} = \begin{pmatrix} h_+ & h_\times & 0 \\ h_\times & -h_+ & 0 \\ 0 & 0 & 0 \end{pmatrix}. \tag{4.91}$$

Therefore, a wave would distort space along x and y directions for $h_+ \neq 0$ and $h_\times = 0$, whereas it would distort space along 45 degrees directions for $h_+ = 0$ and $h_\times \neq 0$. Similar to the Stokes parameters, h_+ and $h_\times$ transform to each other as we rotate coordinates by 45 degrees. Thus, gravitational waves are indeed a spin-2 field.

As a gravitational wave propagates, it stretches and contracts space in a quadrupolar way (see Figure 4.14). The stretching along one direction and contraction along another generate redshift in one direction and blueshift in another, creating temperature anisotropy [107–110]. The geodesic equation gives

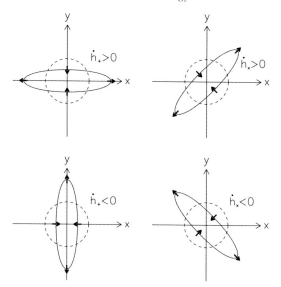

Figure 4.14 How gravitational waves propagating in the z direction stretch and contract space. The stretching and contraction generate redshift and blueshift of photon energies, respectively.

$$\frac{\delta T}{T}\bigg|_O = \frac{\delta T}{T}\bigg|_\mathcal{E} - \frac{1}{2}\hat{n}^i\hat{n}^j\int_{t_\mathcal{E}}^{t_O} dt\,\frac{\partial h_{ij}}{\partial t}. \tag{4.92}$$

As gravitational waves decay inside the horizon, temperature anisotropy from gravitational waves is important only on large scales, roughly at $\ell < 50$. This creates an excess power on large scales, which can be used to constrain the amplitude of primordial gravitational waves.

Let us define the power spectrum of primordial gravitational waves by the equation $\langle h_+(\mathbf{k})h_+^*(\mathbf{k}')\rangle = \langle h_\times(\mathbf{k})h_\times^*(\mathbf{k}')\rangle = (2\pi)^3 P_h(k)\delta^{(3)}(\mathbf{k} - \mathbf{k}')$. We then parametrize the amplitude of the power spectrum of gravitational waves relative to that of the gravitational potential as $r \equiv 4P_h(k_0)/P_\mathcal{R}(k_0)$, where $P_\mathcal{R}(k) = \frac{25}{9}P_\Phi(k)$ and $k_0 = 0.002$ Mpc^{-1}. The current best upper bound on r from the temperature anisotropy is $r < 0.11$ (95% CL) [8]. This is also the best limit that one can ever achieve with the temperature power spectrum alone [67, 111]. To improve upon the limit on r, we must use the polarization data.

Quadrupolar temperature anisotropy produces polarization via Thomson scattering. Unlike the scalar modes producing only E-mode polarization, gravitational waves produce nearly equal amounts of E- and B-mode polarization anisotropies [95, 96]. This is because gravitational waves have two linear polarization states, h_+ and $h_\times$, which are tilted from one another by 45 degrees. As a result, polarization directions produced by these states are also tilted from one another by 45 degrees. Figure 4.15 shows the predicted primordial B-mode spectra with $r = 0.1$, 0.01, and 0.001, along with the predicted lensing B-mode

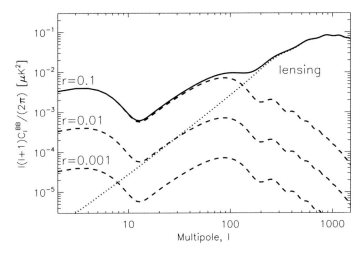

Figure 4.15 Power spectra of B-mode polarization of the CMB, $\ell(\ell + 1)C_\ell^{BB}/(2\pi)$, in units of μK^2. The dashed lines show the spectra from primordial gravitational waves with $r = 0.1, 0.01$, and 0.001. The power at $\ell < 10$ comes from Thomson scattering during the reionization epoch, while that at higher multipoles comes from the last scattering surface at $z = 1088$. The dotted line shows the lensing B-mode power spectrum, and the solid line shows the sum of the lensing B mode and the primordial B mode with $r = 0.1$.

power spectrum. Note how small the amplitudes of B-mode polarization are relative to the measured E-mode polarization from the scalar modes shown in Figure 4.12. The smallness of the B-mode power spectrum poses a serious challenge to the measurement of the primordial B modes: Galactic foreground emission, rather than instrumental noise, will likely be the dominant source of noise in the measurement. A simple analysis shows that removing the effect of foreground down to the level of $r \approx 0.001$ is possible [112], but this is probably the lowest value of r we can realistically achieve in the presence of the Galactic foreground. A host of ground-based and balloon-borne experiments have been proposed and some of them have been taking data already. These experiments are expected to reach $r \approx 0.01$. Going lower in r would require satellite experiments such as those described in [27, 28, 113].

On March 17, 2014, the *BICEP2* collaboration [97] claimed to have detected the primordial B modes [98]. The measurement was done only at a single frequency (150 GHz), and thus it is not possible to draw a firm conclusion about the cosmological origin of the detected signal. Nonetheless, if we interpret the detected signal as cosmological, the inferred tensor-to-scalar ratio is $r \approx 0.1$–0.2. The immediate goal of the current experiments (ground-based, balloon-borne, as well as *Planck*) is to confirm this claim at other frequencies.[4]

---

[4] The subsequent joint analysis performed by the BICEP2 and Planck collaborations did not confirm this. They find that the signal detected by BICEP2 may be due to polarized dust emission in our Galaxy (see ref. [13] in the introduction to this Part). The search for B modes from primordial gravitational waves thus continues.

The future of the CMB lies in the precision measurements and characterizations of the *B*-mode polarization power spectrum. Detection of the signature of *nearly scale-invariant* primordial gravitational waves finally proves that inflation actually occurred in the earliest moment of our universe. The cosmology community endorses this science case as one of the most important goals in the coming decade [114]. This can be achieved by either measuring a bump at $\ell \approx 100$ using ground-based and balloon-borne experiments [115] or a bump at $\ell < 10$ using space-borne experiments [27, 28, 113]. Another subject of focus is the determination of the mass of neutrinos using the power spectrum of the lensing potential measured from the *B*-mode polarization: free-streaming massive neutrinos suppress the growth of density fluctuations, reducing the amplitude of the lensing *B*-mode power spectrum [116]. As the lensing *B*-mode power spectrum peaks on small angular scales, the most efficient way to measure the mass of neutrinos is to use ground-based experiments [117].

In summary, precision cosmology provided by the CMB has transformed our understanding of the universe. We have the standard cosmological model thanks to the experiments and theoretical developments over the last decade. We may witness another revolution in cosmology in the coming decade: observational proof of inflation and the determination of the mass of neutrinos from *B*-mode polarization.

References

[1] Mather, J. C., *et al.* 1990. *Astrophys. J.*, **354**, L37–L40.
[2] Mather, J. C., *et al.* 1994. *Astrophys. J.*, **420**, 439–444.
[3] Fixsen, D. J., *et al.* 1996. *Astrophys. J.*, **473**, 576.
[4] Bennett, C. L., *et al.* [WMAP]. 2013. *Astrophys. J. Suppl.*, **208**, 20.
[5] Hinshaw, G., *et al.* [WMAP Collaboration]. 2013. *Astrophys. J. Suppl.*, **208**, 19.
[6] Ade, P. A. R., *et al.* [Planck Collaboration]. 2014. *Astron. Astrophys.*, **571**, A1.
[7] Ade, P. A. R., *et al.* [Planck collaboration]. 2014. *Astron. Astrophys.*, **571**, A15.
[8] Ade, P. A. R., *et al.* [Planck Collaboration]. 2014. *Astron. Astrophys.*, **571**, A16.
[9] Das, S., *et al.* 2014. *JCAP*, **1404**, 014.
[10] Dunkley, J., *et al.* 2013. *JCAP*, **1307**, 25.
[11] Sievers, J. L., *et al.* 2013. *JCAP*, **1310**, 60.
[12] Story, K. T., *et al.* 2013. *Astrophys. J.*, **779**, 86.
[13] Hou, Z., *et al.* 2014. *Astrophys. J.*, **782**, 74.
[14] Weinberg, S. 2008. *Cosmology*. Oxford: Oxford University Press.
[15] Seljak, U., and Zaldarriaga, M. 1996. *Astrophys. J.*, **469**, 437–444.
[16] Seljak, U., Sugiyama, N., White, M. J., and Zaldarriaga, M. 2003. *Phys. Rev.*, **D68**, 083507.
[17] Lewis, A., Challinor, A., and Lasenby, A. 2000. *Astrophys. J.*, **538**, 473–476.
[18] Blas, D., Lesgourgues, J., and Tram, T. 2011. *JCAP*, **1107**, 034.
[19] Percival, W. J., *et al.* [2dFGRS Collaboration]. 2001. *Mon. Not. Roy. Astron. Soc.*, **327**, 1297.
[20] Cole, S., *et al.* [2dFGRS Collaboration]. 2005. *Mon. Not. Roy. Astron. Soc.*, **362**, 505–534.
[21] Tegmark, M., *et al.* [SDSS Collaboration]. 2004. *Astrophys. J.*, **606**, 702–740.
[22] Eisenstein, D. J., *et al.* [SDSS Collaboration]. 2005. *Astrophys. J.*, **633**, 560–574.
[23] Tegmark, M., *et al.* [SDSS Collaboration]. 2006. *Phys. Rev.*, **D74**, 123507.
[24] Percival, W. J., *et al.* [SDSS Collaboration]. 2010. *Mon. Not. Roy. Astron. Soc.*, **401**, 2148–2168.
[25] Anderson, L., *et al.* 2013. *Mon. Not. Roy. Astron. Soc.*, **427**, 3435–3467.
[26] Fixsen, D. J. 2009. *Astrophys. J.*, **707**, 916–920.
[27] Kogut, A., *et al.* 2011. *JCAP*, **1107**, 025.

[28] André, P., *et al.* 2014. *JCAP*, **1402**, 006.

[29] Danese, L., and de Zotti, G. 1982. *Astron. Astrophys.*, **107**, 39–42.

[30] Burigana, C., Danese, L., and de Zotti, G. 1991. *Astron. Astrophys.*, **246**, 49–58.

[31] Sunyaev, R. A., and Zel'dovich, Ya. 1970. *Astrophys. Space Sci.*, **7**, 20–30.

[32] Illarionov, A. F., and Sunyaev, R. A. 1975. *Sov. Astron.*, **18**, 413–419.

[33] Illarionov, A. F., and Sunyaev, R. A. 1975. *Sov. Astron.*, **18**, 691–699.

[34] Hu, W., and Silk, J. 1993. *Phys. Rev. D*, **48**, 485–502.

[35] Khatri, R., and Sunyaev, R. A. 2012. *JCAP*, **1206**, 038.

[36] Kampaneets, A. S. 1956. *Zh. Eksp. Teor. Fiz.*, **31**, 876.

[37] Weymann, R. 1966. *Astrophys. J.*, **145**, 560–571.

[38] Zel'dovich, Ya. B., and Sunyaev, R. A. 1969. *Astrophys. Space Sci.*, **4**, 301.

[39] Sunyaev, R. A., and Zel'dovich, Ya. B. 1972. *Comments Astrophys. Space Phys.*, **4**, 173–178.

[40] Zel'dovich, Ya. B., Illarionov, A. F., and Sunyaev, R. A. 1972. *Sov. J. Exp. Theor. Phys.*, **35**, 643–648.

[41] Carlstrom, J. E., Holder, G. P., and Reese, E. D. 2002. *Ann. Rev. Astron. Astrophys.*, **40**, 643–680.

[42] Refregier, A., Komatsu, E., Spergel, D. N., and Pen, U.-L. 2000. *Phys. Rev.*, **D61**, 123001.

[43] Khatri, R., Sunyaev, R. A., and Chluba, J. 2012. *Astron. Astrophys.*, **543**, A136.

[44] Chluba, J., and Sunyaev, R. A. 2012. *Mon. Not. Roy. Astron. Soc.*, **419**, 1294–1314.

[45] Sunyaev, R. A., and Zel'dovich, Ya. B. 1970. *Astrophys. Space Sci.*, **9**, 368–382.

[46] Daly, R. A. 1991. *Astrophys. J.*, **371**, 14–28.

[47] Chluba, J., Khatri, R., and Sunyaev, R. A. 2012. *Mon. Not. Roy. Astron. Soc.*, **425**, 1129–1169.

[48] Silk, J. 1968. *Astrophys. J.*, **151**, 459–471.

[49] Peebles, P. J. E., and Yu, J. T. 1970. *Astrophys. J.*, **162**, 815–836.

[50] Kaiser, N. 1983. *Mon. Not. Roy. Astron. Soc.*, **202**, 1169–1180.

[51] Khatri, R., and Sunyaev, R. A. 2012. *JCAP*, **1209**, 016.

[52] Conklin, E. K. 1969. *Nature*, **222**, 971–972.

[53] Henry, P. S. 1971. *Nature*, **231**, 516–518.

[54] Smoot, G. F., Gorenstein, M. V., and Muller, R. A. 1977. *Phys. Rev. Lett.*, **39**, 898.

[55] Cheng, E. S., Saulson, P. R., Wilkinson, D. T., and Corey, B. E. 1979. *Astrophys. J. Lett.*, **232**, L139–L143.

[56] Smoot, G. F., *et al.* 1991. *Astrophys. J. Lett.*, **371**, L1–L5.

[57] Bennett, C. L., *et al.* 1996. *Astrophys. J.*, **464**, L1–L4.

[58] Jarosik, N., *et al.* 2011. *Astrophys. J. Suppl.*, **192**, 14.

[59] Sandage, A., Reindl, B., and Tammann, G. A. 2010. *Astrophys. J.*, **714**, 1441–1459.

[60] Challinor, A., and van Leeuwen, F. 2002. *Phys. Rev.*, **D65**, 103001.

[61] Hanson, D., and Lewis, A. 2009. *Phys. Rev.*, **D80**, 063004.

[62] Aghanim, N., *et al.* [Planck Collaboration]. 2014. *Astron. Astrophys.*, **571**, A27.

[63] Sachs, R. K., and Wolfe, A. M. 1967. *Astrophys. J.*, **147**, 73–90.

[64] Hu, W., and Sugiyama, N. 1995. *Astrophys. J.*, **444**, 489–506.

[65] Hu, W., Scott, D., Sugiyama, N., and White, M. 1995. *Phys. Rev.*, **D52**, 5498–5515.

[66] Hu, W., Eisenstein, D. J., Tegmark, M., and White, M. 1999. *Phys. Rev.*, **D59**, 023512.

[67] Komatsu, E., *et al.* [WMAP Collaboration]. 2009. *Astrophys. J. Suppl.*, **180**, 330–376.

[68] Crittenden, R. G., and Turok, N. 1996. *Phys. Rev. Lett.*, **76**, 575.

[69] Giannantonio, T., Crittenden, R., Nichol, R., and Ross, A. 2012. *Mon. Not. Roy. Astron. Soc.*, **426**, 2581–2599.

[70] Sunyaev, R. A., and Zel'dovich, Ya. 1970. *Astrophys. Space Sci.*, **7**, 3–19.

[71] Gunn, J. E., and Peterson, B. A. 1965. *Astrophys. J.*, **142**, 1633.

[72] Fan, X.-H., *et al.* 2006. *Astron. J.*, **132**, 117–136.

[73] Loeb, A., and Barkana, R. 2001. *Ann. Rev. Astron. Astrophys.*, **39**, 19–66.

[74] Zaldarriaga, M. 1997. *Phys. Rev.*, **D55**, 1822–1829.

[75] Starobinsky, A. A. 1980. *Phys. Lett.*, **B91**, 99–102.

[76] Mukhanov, V. F., and Chibisov, G. V. 1981. *JETP Lett.*, **33**, 532–535.

[77] Lewis, A., and Challinor, A. 2006. *Phys. Reports*, **429**, 1–65.

[78] Seljak, U. 1996. *Astrophys. J.*, **463**, 1.
[79] Hu, W., and Okamoto, T. 2002. *Astrophys. J.*, **574**, 566–574.
[80] Okamoto, T., and Hu, W. 2003. *Phys. Rev.*, **D67**, 083002.
[81] Hirata, C. M., and Seljak, U. 2003. *Phys. Rev.*, **D68**, 083002.
[82] Das, S., *et al.* 2011. *Phys. Rev. Lett.*, **107**, 021301.
[83] van Engelen, A., *et al.* 2012. *Astrophys. J.*, **756**, 142.
[84] Ade, P. A. R., *et al.* [Planck Collaboration]. 2014. *Astron. Astrophys.*, **571**, A17.
[85] Rees, M. J. 1968. *Astrophys. J. Lett.*, **153**, L1.
[86] Basko, M. M., and Polnarev, A. G. 1980. *Mon. Not. Roy. Astron. Soc.*, **191**, 207–215.
[87] Bond, J. R., and Efstathiou, G. 1984. *Astrophys. J.*, **285**, L45–L48.
[88] Bond, J. R., and Efstathiou, G. 1987. *Mon. Not. Roy. Astron. Soc.*, **226**, 655–687.
[89] Komatsu, E., *et al.* [WMAP Collaboration]. 2011. *Astrophys. J. Suppl.*, **192**, 18.
[90] Zaldarriaga, M., and Seljak, U. 1998. *Phys. Rev.*, **D58**, 023003.
[91] Hanson, D., *et al.* [SPTpol Collaboration]. 2013. *Phys. Rev. Lett.*, **111**, 141301.
[92] Ade, P. A. R., *et al.* [POLARBEAR Collaboration]. 2014. *Phys. Rev. Lett.*, **112**, 131302.
[93] Ade, P. A. R., *et al.* [POLARBEAR Collaboration]. 2014. *Phys. Rev. Lett.*, **113**, 021301.
[94] Ade, P. A. R., *et al.* [POLARBEAR Collaboration]. 2014. *Astrophys. J.*, **794**, 1.
[95] Seljak, U., and Zaldarriaga, M. 1997. *Phys. Rev. Lett.*, **78**, 2054–2057.
[96] Kamionkowski, M., Kosowsky, A., and Stebbins, A. 1997. *Phys. Rev. Lett.*, **78**, 2058–2061.
[97] Barkats, D., *et al.* [BICEP1 Collaboration]. 2014. *Astrophys. J.*, **783**, 67.
[98] Ade, P. A. R., *et al.* [BICEP2 Collaboration]. 2014. *Phys. Rev. Lett.*, **112**, 241101.
[99] Coulson, D., Crittenden, R. G., and Turok, N. G. 1994. *Phys. Rev. Lett.*, **73**, 2390–2393.
[100] Crittenden, R. G., Coulson, D., and Turok, N. G. 1995. *Phys. Rev.*, **D52**, 5402–5406.
[101] Brown, M. L., *et al.* [QUaD collaboration]. 2009. *Astrophys. J.*, **705**, 978–999.
[102] Aikin, R., *et al.* [BICEP1 Collaboration]. 2013. arXiv:1310.1422.
[103] Araujo, D., *et al.* [QUIET Collaboration]. 2012. *Astrophys. J.*, **760**, 145.
[104] Zaldarriaga, M. *et al.* 2009. *AIP Conf. Proc.*, **1141**, 179–221.
[105] Holder, G. P., *et al.* 2013. *Astrophys. J.*, **771**, L16.
[106] Starobinsky, A. A. 1979. *JETP Lett.*, **30**, 682–685.
[107] Rubakov, V. A., Sazhin, M. V., and Veryaskin, A. V. 1982. *Phys. Lett.*, **B115**, 189–192.
[108] Fabbri, R., and Pollock, M. 1983. *Phys. Lett.*, **B125**, 445–448.
[109] Abbott, L. F., and Wise, M. B. 1984. *Nucl. Phys.*, **B244**, 541–548.
[110] Starobinsky, A. A. 1985. *Sov. Astron. Lett.*, **11**, 133.
[111] Knox, L. 1995. *Phys. Rev.*, **D52**, 4307–4318.
[112] Katayama, N., and Komatsu, E. 2011. *Astrophys. J.*, **737**, 78.
[113] Hazumi, M., *et al.* 2012. *Society of Photo-Optical Instrumentation Engineers (SPIE) Conference Series*, **8442**, 844219.
[114] Baumann, D., *et al.* [CMBPol Study Team]. 2009. *AIP Conf. Proc.*, **1141**, 10–120.
[115] Abazajian, K. N., *et al.* 2015. *Astroparticle Physics*, **63**, 55–65.
[116] Kaplinghat, M., Knox, L., and Song, Y.-S. 2003. *Phys. Rev. Lett.*, **91**, 241301.
[117] Abazajian, K. N., *et al.* 2015. *Astroparticle Physics*, **63**, 66–80.

4.4 Beyond the standard model
Malcolm A. H. MacCallum

4.4.1 Motivations and possible directions

As described above, perturbed FLRW models fit the observations of the universe very well, even if rather uncomfortably in that they require dark energy, dark matter, and an inflaton field, none of which have so far been observed directly. So why do we want to consider other relativistic models?

Adoption of the cosmological Standard Model as the best model of the cosmos rests on the following (testable) assumptions:

1. We are using the right theory of gravity (GR).
2. Within GR, the only observationally viable global geometry is FLRW.
3. The fits between the model and the data are not significantly affected by consequences of the universe's small-scale inhomogeneities.

It also assumes that the interpretation of the observations is not significantly affected by unknown or neglected astrophysical effects, e.g. possible time-dependent intergalactic absorption of the light from distant supernovae.

The standard model implies that we need an extended understanding of the matter content, involving particles and fields not yet observed terrestrially and not part of the standard model of particle physics[5] (unless, for example, the recently discovered Higgs particle might, with a weakened coupling, act as the inflaton). Modifying one or more of the above assumptions may enable us to avoid the need for one or more of the three types of matter observed only by their astrophysical and cosmological effects.

Changing Assumption 1, replacing GR by some other relativistic theory of gravity, is very widely discussed[6] and may be important. Note that any new theory has to be compatible with special relativity if it is to be observationally viable.

Frameworks organizing gravity theories so that one can make comparisons of whole classes of them with observation, a methodology analogous to the use of the PPN and other frameworks discussed in Chapter 2, are under development, e.g. [1]. However, these do not yet allow us to review the possibilities in any concise way and we will not attempt that here. We only mention that models have been proposed which replace the era before the last scattering surface of the CMB, in the standard model, with some completely different geometry and/or physics. As examples we mention the "pre-big-bang" of [2], the "ekpyrotic" braneworld scenario of [3], and the Conformal Cyclic Cosmology of [4]. At the present time none of these ideas has established itself as a front runner.

It may be worth noting that although the great success of inflation was in predicting an almost flat spectrum of density perturbations, which is transformed by later evolution into the power spectrum observed in the CMB, etc., as described in Komatsu's contribution, the original motivations [5] were different. They were the 'horizon problem' (formulated rather differently from the way described in Sasaki's contribution above), that big-bang models involved causally separated regions which nevertheless showed nearly identical properties, the 'flatness problem' that the best-fit FLRW model had almost the critical density, and the 'monopole problem' of the absence of monopoles. Radical changes from the standard model often entail alternatives to inflation, removing or modifying the need for an inflaton, but do not necessarily provide better answers to these problems.[7]

---

[5] See the contribution of Wands and Maartens to this book for some amplification of this remark.

[6] As shown by M. Fairbairn's count that of the 591 papers submitted to the online arXiv in 2012 discussing dark energy, 287 concerned modified gravity theories. Some of the possibilities are discussed earlier in this Chapter.

[7] I have argued elsewhere [6] that not all of these are real problems and moreover that the first two are not solved, though they are mitigated, by inflation.

Two main types of alternative to assumption 2 have been considered: spatially homogeneous (SH) but anisotropic models, and tractable inhomogeneous models, especially spherically symmetric ones. The motivations are to discover which features are peculiar to the standard model and whether the standard models are the only ones that can fit the data: in particular, inhomogeneous models have been studied as possible alternatives to the introduction of dark energy in the standard model. The most significant of the results of the many studies are summarized below and at greater length in [7–10] and [11].

Inhomogeneous models can also model local features such as galaxies, clusters, and so on in full nonlinearity. The importance of such work in cosmology is shown by the correlations between the masses of supermassive black holes and the properties of their host galaxies, as discussed in Chapter 3. A fuller understanding of black holes' formation, evolution, and interaction with their environments is hence essential to a full understanding of galactic structure and evolution. GR therefore remains crucial to the understanding of galaxies, although many galactic features can be adequately discussed in Newtonian approximations.

The work on non-standard geometries within GR shows that some of the observations which readily fit the standard model, and which could therefore be considered to confirm it, can in fact be equally well fitted by models other than perturbed FLRW. Among these are the apparent isotropy of the universe, the magnitude–redshift relations, and the abundances of the elements. As one example of the extent of such fits one can cite the theorem of [12] for Lemaître–Tolman–Bondi (LTB) models (spherically symmetric dust models) which proves that:

For any given isotropic observations of $\ell(z)$ and $n(z)$, where ℓ is observed luminosity, z is redshift and n is the number count of sources, and any given source evolution with z, there is an LTB model with $\Lambda = 0$ that fits the observations.

Inhomogeneous models even fit some observations better than the standard model [13]. While one may rule out specific inhomogeneous models by a combination of observations or hope to rule out all large-scale inhomogeneities using tests satisfied only by FLRW models [14], Assumption 2 has not yet been fully established observationally.

In support of Assumption 2 one should note the "almost-EGS" theorem[8] of [16], showing that the only models with suitable bounds on derivatives of the covariant multipoles in the distribution of collisionless radiation and in which that radiation is almost isotropic are almost FLRW.

Investigating Assumption 3 requires consideration of the relation between the true lumpy geometry and the global model obtained by smoothing over scales large compared with an individual galaxy.[9]

One aspect is the 'averaging problem' of determining the best-fit smoothed metric. Averaging raises the problem of 'back reaction', arising because the Einstein equations are

---

[8] EGS refers to the exact version of this theorem, due to Ehlers, Geren and Sachs [15].
[9] As Heller has pointed out [17], the issue of the relation between fluid models in the early and the late universe has been little discussed although the averaging is over very different comoving scales – elementary particles in the early universe and galaxies now.

nonlinear. Averaging a product, ab, to get $\langle ab \rangle$ will in general not give the product of the averages, $\langle ab \rangle \neq \langle a \rangle \langle b \rangle$. Similarly the derivative of an average is usually not the average of the derivative, which is relevant both for temporal and for spatial variations. Thus nonlinear equations valid locally will not be valid when quantities are averaged over large scales: the back reaction is the correction one must make because the Einstein tensor of the smoothed metric will not in general be equal to the smoothed Einstein tensor.

The other important aspect is the effect of the small-scale lumpiness on light propagation, which leads to corrections in the relations between model parameters and observation.

As well as changes to the three assumptions above, further extension to or departure from the standard model is provided by the various "multiverse" approaches (see [18]): these may involve regions of our spacetime outside our past light cone, or completely distinct spacetimes. The scientific status of such unobservable universes is debatable and will not be further considered here.

4.4.2 Spatially-homogeneous models

Whereas FLRW models have a 6-dimensional group of spatial symmetries (a G_6, rotations and translations at every point), spatially homogeneous anisotropic models only have an isometry group G_r, $r \leq 4$. Except for the Kantowski–Sachs (K–S) metrics [19], this group always contains a (sub)group G_3 of translations, which can be classified into two main subclasses, A and B, and nine types, Bianchi types I–IX The symmetry group maps the timelike observer world lines orthogonal to the hypersurfaces of homogeneity into one another, and similarly maps invariantly-defined matter worldlines (e.g. those of a perfect fluid) intersecting the hypersurfaces into one another. When the normal and matter worldlines coincide, we say the model is 'orthogonal': otherwise it is 'tilted'. Tilted models are significantly more complicated than orthogonal models.

These models provide fairly tractable solutions of the full nonlinear equations since there is only one essential variable, time, so the equations become ordinary differential equations, but they allow investigation of much more general behaviour than the FLRW models. They can represent anisotropic modes, including rotation and global magnetic fields, which could occur in the real universe (indeed, must do so, if the universe is generic): for these an anisotropic but not necessarily inhomogeneous model is required (see e.g. [20]). They allow new classes of singularities, and modify the implications of nucleosynthesis for baryon density. They may also be good approximations in regions where there is inhomogeneity but spatial gradients are small [21].

In particular, the tilted cases provide the only tractable cosmological solutions we have which involve rotation: rotation is ubiquitous in the universe, and so the vorticity conservation laws suggest there always was and always will be rotation. Thus it is valuable to have solutions in which we can investigate its effects on, for example, nucleosynthesis, and on the CMB, where new classes of anisotropy patterns arise.

There have been many papers analysing the systems of equations governing the K–S and Bianchi models, finding exact solutions of these equations, and discussing physical properties of the models. We mention here only a few important points. For fuller summaries see [7, 9, 10].

At early times in SH models the dynamics is highly anisotropic, and the same is true at late times unless the departure from isotropy is stabilized by $\Lambda > 0$. However, they can be nearly isotropic at intermediate epochs, a behaviour sometimes called 'hesitation dynamics'. In each set of Bianchi models of a type admitting intermediate isotropization, there will be spatially homogeneous models that are linearizations of these Bianchi models about FLRW models. These perturbation modes will occur in any almost-FLRW model that is generic rather than fine-tuned; however, the exact models approximated by these linearizations will be quite unlike FLRW models at very early and very late times. This shows that the almost-FLRW behaviour of the standard model is unstable.

Thus the nature of the initial singularity in SH models can be very different from FLRW models, and SH models have played an important part in the development of our understanding of generic cosmological singularities [21].

Inflation occurs in Bianchi models only if there is not too much anisotropy to begin with [22], and it is not clear that shear and spatial curvature are effectively removed in all inflating cases [23]. Hence, some Bianchi models isotropize due to inflation, but not all, whereas, for example, Bianchi I models can isotropize without inflation, and inflation in them can be anisotropic [24], leading to substantial changes in CMB correlations.

Although the field equations for Bianchi models are relatively simple, and many exact solutions are known, though not for the most complicated cases, the geodesic equations are not so readily solved. Nilsson *et al.* [25] considered the general dynamical system with the geodesic equations included, and pointed out in particular that bounds on shear obtained by considering Bianchi I and V models, which are quite often quoted, are untypical.

Anisotropic universe models give rise to CMB anisotropies. For example many Class B Bianchi models will show a hot-spot and associated spiral pattern in the CMB sky. Some of the most detailed work has involved Class B models where the quadupolar anisotropy found in orthogonal Class A models can be distorted and rotated, leading to hot and cold spots in the sky and contributions to higher multipoles. Bianchi perturbations of FLRW have therefore been considered as models for the observed anomalies in CMB such as the "cold spot".

Note that almost isotropic CMB does not imply, as one might have expected from the almost-EGS theorem [16], that a Bianchi metric is close to FLRW. The reason is that one can have cases where the derivative of the shear is large and oscillatory but the shear itself is small. One can even have models which at a certain instant have exactly isotropic CMB [26]. Pontzen and Challinor [27] have considered linearization of Bianchi models about FLRW models, highlighting "the existence of arbitrarily long near-isotropic epochs in models of general Bianchi type", i.e. the hesitation dynamics. In addition to the effects of the global geometry on the CMB, which largely concern low multipoles, one has to consider whether perturbations will evolve differently and so lead to different small-angular-scale variations: this has not been much explored so far.

4.4.3 Globally inhomogeneous models

Perturbed FLRW models are both inhomogeneous and anisotropic, but they are treated only in perturbation theory. Inhomogeneous models in the fully nonlinear theory have been applied both globally and to model localized inhomogeneities, including, for example, their fully nonlinear effects on observation via lensing. In the global context, issues such as whether inflation could remove inhomogeneity, or whether hierarchical models could fit the data, can be examined: these are essential to judging the robustness of the assumptions of the standard model. In particular, comparison of observations with inhomogeneous models tests the Copernican Principle that we are in no special position in the universe.

Use of inhomogeneous models as global models has in recent times been mainly directed to studying whether they provide adequate explanations for the apparent acceleration explained in the standard model by dark energy. For a recent review see Chapter 15 of [7]. The possibility arises because our observations are of the past light cone and thus cannot readily distinguish between variation in space and variation in time: the inhomogeneous models most used in this context place us in a void, an underdense region.

It turns out that such models can do remarkably well in fitting supernovae, CMB and BAO observations, e.g. fitting magnitude–redshift and/or CMB perturbation data to the same accuracy as the standard model. But they may not fit all details (except perhaps for special choices of initial conditions). To clarify such issues, a number of other tests have been proposed, which may in future provide increasingly stringent constraints, for example the kinematic Sunyaev–Zel'dovich effect [28], which already appears to rule out LTB models [29].

Here are some of the other major points about the use of inhomogeneous models as models of global or individual local structures, taken from the many studies (see Chapters 15, 16 and 19 of [7] for fuller details). In the next section we consider cumulative effects of large numbers of localized lumps.

1. The very early universe may have been significantly different from FLRW models, implying that the initial conditions before inflation may be very anisotropic or inhomogeneous. Inhomogeneous initial conditions can prevent inflation from acting: so the existence of an inflaton would not by itself explain the relatively smooth universe indicated by the CMB. Either one needs fine tuning before inflation or the apparent smoothness does not come from inflation. (A more radical reaction would be to dispense with the inflaton and base the smoothness on earlier initial conditions.)

2. Modelling of the galaxy cluster A2199 [30] showed that velocity perturbations produced density variations more effectively than a pure density perturbation. This conclusion, suggesting that a velocity distribution should also be considered in the standard model, was supported by models of the North and South Galactic Pole voids [31], where the velocity perturbation was the main factor. There they also showed, however, that density and velocity perturbations compatible with the CMB observations could not readily produce the observed density contrasts, within the class of models considered. Faster expansion produced a larger density contrast and nonlinearity can speed up structure

formation. Remaining incompatibilities between CMB measures and observed density contrasts may be resolved by including radiation.

3. Overdensities can grow to underdensities, which does not happen in a linearized picture.

4. The conjecture that universes evolve towards self-similar models, although it is not always correct, has been shown to be a useful guideline characterizing an intriguing part of the dynamics of models of interest, which may help explain some observed structures in an interesting way.

5. Some of the classes considered include exact nonlinear inhomogeneous gravitational waves in an expanding background, whose effects are otherwise known only in the perturbative regime.

As well as general discussions of nonlinear collapse, detailed fits to a number of actual large-scale structures, such as the Local Group, M87, the Great Attractor, and voids with adjacent clusters, have been made. Such models enable inferences about, e.g., the masses of objects. For example, Bolejko and Hellaby [32] used an LTB model for the Shapley concentration and the Great Attractor, and found that "the peculiar velocity maximum near the SC is $\sim$800 km/s inwards, the density between GA and SC must be about $\sim$0.9 times background, the mass of the GA is probably $4-6 \times 10^{15} M_{\odot}$," and "the SC's contribution to the L[ocal] G[roup] motion is negligible". Krasiński and Hellaby [33] modelled M87, a galaxy believed to contain a black hole, by an LTB metric, showing that models with very different black hole ages were indistinguishable observationally. For a full review of such work see [8].

4.4.4 Cumulative effects of small-scale inhomogeneity

By "small-scale" we here mean small compared with the scale of the region visible on the last scattering surface in the standard model: in practice this means we consider the back-reaction effects arising by averaging over regions on the order of, say, 100 Mpc (a scale appropriate for both a fluid approximation and validity of the homegeneity assumption) and the lensing effects on observational relations.

A number of approaches to averaging have been tried (see Chapter 8 of [11] and Chapter 16 of [7]) but none of them is wholly convincing, and in particular most are not covariant. Both of the main difficulties can be considered to come from the absence of a fixed background spacetime, which is of course exactly what we would like to determine. One lies in the definition of the averaging region: if one takes a sphere of radius b in a spacelike surface S agreeing with the rest frame of a given observer in special relativity, Lorentz transformations of it would give regions intersecting the worldline of any other observer crossing S. One could use the cosmic substratum to define an averaging region, which mitigates the problem at the risk of the definitions becoming circular.

The other technical difficulty is that to average one needs to form an integral over an averaging region: this is fine for scalars but for tensors one needs the bitensors associated with the world function [34], based on parallel propagation along geodesics, in order to compare tensors at different points. Even if we could calculate these, they cannot be used to average the metric tensor, since the metric tensor defines the parallel propagation used in this process, and therefore is left invariant by it (since $\nabla_c g_{ab} = 0$). So one has to devise a procedure in which either the field equations are represented only in terms of scalars, possible for example if one takes components relative to a covariantly uniquely defined tetrad, or else bitensors are used only to define averages of quantities other than the metric.

Because there is thus no universally accepted way to calculate the back reaction, there is also no agreement about its magnitude and significance, or even its sign (when formulated as a correction term in the Friedmann equation for the smoothed-out model). For example, one may obtain a term like a positive spatial curvature [35], which could have the dramatic effect of leading to a model that would recollapse when the real universe does not, whereas some earlier papers obtained a term like a negative pressure.

Underlying the study of back reaction is the choice of the background FLRW model to be corrected, the "fitting problem": in particular, the perturbations studied must average out so that the background chosen is unaltered. Most observational studies implicitly use a fitting on the light cone, which could differ from averaging over a spacelike region.

In recent years the most popular approach to averaging has been that of Buchert, see e.g. [36], which averages only scalars and thereby avoids the bitensors problem. It takes into account the nonlinearities in the averages considered. But it does not deal with, for example, the shear and vorticity equations: extensions to do that have used phenomenological assumptions to close the system of equations.

Using explicit inhomogeneous models, rather than approximations or incomplete sets of equations, to compute back reactions runs into other difficulties. For example, "Swiss cheese" models, with spherical regions matched to spheres in an FLRW model, give no back reaction because by definition they exactly match at the junction surfaces.

At present, it seems likely that dynamical back reaction effects make only a small contribution on cosmological scales, but do have to be taken into account in "precision cosmology" (at the order of corrections of a few percent).

Somewhat similar difficulties and variation in results arise in the different attempts to compute the effects of lensing. In the real universe, as pointed out in [37] and [38], observations take place via null geodesics lying in the underdense regions between opaque objects such as stars. Light rays are focused only by the curvature actually inside the beam, not the matter that would be there in a completely uniform model. The effect on observational relations of introducing inhomogeneities into a given background spacetime is twofold: it alters redshifts, and it changes area distances.

The redshift effect can be understood with the following Newtonian analogy. When a void intervenes between the source and the observer, photons drop into a potential well and then climb out, and they exit when the universe is larger than when they went in. If spacetime is static inside the void the redshift changes in and out cancel, but when structure is forming the potential well is changing with time so one gets the Rees–Sciama effect [39]: a change in redshift due to a change in the potential well as the photon traverses it. A nonzero cosmological constant will also lead to such an effect. When the light source is in the void, the photon has to climb out of the void, giving a contribution to observed redshifts.

The area distance depends on the focusing of the beam by the gravitational field. In the light propagation equations, matter contributes locally by its Ricci tensor, which affects the convergence of the rays directly, and non-locally and indirectly by the Weyl tensor it causes. The latter gives the derivative of the shear of the ray congruence and the shear appears quadratically in the equations for the convergence. It is this contribution that we observe in light-bending by the Sun.

Calculations in the standard model use only the smoothed-out Ricci tensor, the Weyl tensor being zero in a strictly FLRW model. Because galaxies are quite diffuse, even if the light passes through a galaxy it would rarely hit a star, and so within a galaxy the averaged Ricci tensor may give a good approximation, which, however, needs justification. Similar remarks apply to the intergalactic medium (except perhaps where there are gas clouds of such a nature that the chance of interaction with an atom or molecule becomes high). On the other hand, calculations for specific models of individual lumps, e.g. a Swiss cheese model, rest on the effect of the Weyl tensor. Since the effects are nonlinear, and hence cannot simply be superposed, one needs to find some way of appropriately combining the influence of regions of varying density encountered by the light.

The most popular is the Dyer–Roeder distance [40, 41], obtained by assuming that only a fraction $\tilde{\alpha}$ of the total mass density is smoothly distributed, i.e. not bound in galaxies, while a fraction $1 - \tilde{\alpha}$ is bound. For matter not passing through bound 'clumps', one replaces the Ricci tensor term in the light propagation equations by $\tilde{\alpha}$ times that term. Thus the Dyer–Roeder proposal is that the main effect of clumpiness is that the light rays by which we observe most distant galaxies pass through less matter than in a corresponding smoothed-out FLRW universe; shear then has a negligible effect, because it is only important for near encounters with isolated masses, when it causes gravitational lensing effects.

This is a good approximation when galaxies are embedded in a fairly uniform intergalactic medium of dark matter, but clearly does not take shear effects and caustics properly into account. If the dark matter is uniform, Dyer–Roeder is good; if dark matter is clustered, it is not so good. But we know that gravitational lensing strong enough to cause significant focusing is a relatively rare effect over the whole sky, suggesting that Dyer–Roeder will be a good approximation.

However, it is not the only approach. One may use stochastic methods [38], or detailed examination of geodesics in universes with lumps, i.e. of lensing effects. The results will of course depend on the statistics of the clumping (see the references in [42]). The over- and under-densities lead to a distribution of magnifications, favouring mild demagnifications but with a long tail of magnifications. Kainulainen and Marra [42] give a general stochastic method for estimating the probability distribution in models where under-densities occupy more volume than over-densities, which might reasonably represent the observed voids and filaments, and show it gives good agreement with Holz and Wald [43]. They conclude that lensing by structures on a scale $10^{15} \, M_\odot/h$ would be important in analyzing the current SNIa data. The residuals in the data already indicate some lensing effect [44].

Specific models of lumps model precisely the difference between Weyl and Ricci focusing of null geodesics: null geodesics in the empty regions are focused only by shear induced by the Weyl tensor. They give results that depend on the choice of modelling, and which vary quite widely. For example, (a) in the work of Kantowski, 'a determination of Ω_0 made by applying the homogeneous distance–redshift relation to SN 1997ap at $z = 0.83$ could be as much as 50% lower than its true value', (b) Biswas and Notari [45] found a negligible integrated effect on area distances in an exact Swiss cheese model with LTB patches in an FLRW background, and (c) Marra *et al.* [46] found the opposite. In [47] they fitted a phenomenological homogeneous model to describe observables in such a Swiss cheese model. Following a fitting procedure based on light-cone averages, they found that the light-cone average of the density as a function of redshift is affected by inhomogeneities because, as the universe evolves, a photon spends more and more time in the (large) voids rather than in the (thin) high-density structures. Although the sole source in the Swiss cheese model is matter, the phenomenological homogeneous model behaves as if it has a dark energy component. However, they find that the holes must have a present size of about 250 Mpc to be able to mimic the concordance model.

As Hanson *et al.* say [48], from the point of view of CMB observations, lensing is a contaminant. As an example of the use of inhomogeneous models, Bolejko [49] considered the case of Swiss cheese models with Szekeres solution interiors, showing that local and uncompensated inhomogeneities can induce temperature fluctuations of amplitude as large as 10^{-3}, and thus can be responsible for the low multipole anomalies observed in the angular CMB power spectrum.

Detailed estimates give an RMS deflection of CMB rays of 2.7 arcmin, with a coherence length of a few degrees (against a 10 arcmin scale of the $\ell = 1000$ modes in the CMB). The lensing effect is detected at high significance levels in the Planck data [50].

Because there are more of them, smaller lenses contribute most, and the peak contribution comes from lenses at redshift $z \sim 2$. As well as the small but non-negligible effect on the CMB power spectrum, this results in the introduction of a vortex-like B-mode polarization. The lensing does not introduce additional polarization: it just realigns existing polarizations. B-mode polarizaton would also arise from gravitational wave modes during inflation. The lensing can be considered as adding non-Gaussianity and anisotropy to the predicted observations. Recent results from the POLARBEAR and BICEP2 collaborations [51, 52] give measurements of both these contributions: the Planck satellite measurements of these effects should in due course provide fuller sky coverage and even greater accuracy.

Conversely, CMB measurements on small angular scales can reconstruct the mass distribution. Recent results [53] show a high correlation between the cosmic infrared background, indicating star formation, and the distribution of (dark) matter shown by lensing.

Overall, it appears the effects of lumps, both by back reaction and by optical effects, do amend the interpretation of cosmological data such as the SN1a m–z relation and the CMB power spectrum, but probably not grossly, e.g. not at the level where they could replace a need for dark energy. However, it is clear that our present models are far from satisfactorily modelling the true nonlinearities.

References

[1] Baker, T., Ferreira, P. G., and Skordis, C. 2013. *Phys. Rev. D*, **87**, 024015.
[2] Gasperini, M., and Veneziano, G. 1993. *Astropart. Phys.*, **1**, 317–339.
[3] Khoury, J., Ovrut, B. A., Steinhardt, P. J., and Turok, N. 2001. *Phys. Rev. D*, **64**, 123522.
[4] Penrose, R. 2006. Before the Big Bang: an outrageous new perspective and its implications. In: *Proc. of EPAC 2006 (Edinburgh, Scotland, 2006)*. Geneva: CERN. http://accelconf.web.cern. ch/AccelConf/e06/PAPERS/THESPA01.PDF.
[5] Guth, A. H. 1981. *Phys. Rev. D*, **23**, 347–356.
[6] MacCallum, M. A. H. 1987. Strengths and weaknesses of cosmological big-bang theory. Pages 121–142 of: Stoeger, W. R. (ed), *Theory and observational limits in cosmology*. Vatican City: Specola Vaticana.
[7] Ellis, G. F. R., Maartens, R., and MacCallum, M. A. H. 2012. *Relativistic cosmology*. Cambridge: Cambridge University Press.
[8] Bolejko, K., Krasiński, A., Hellaby, C., and Célérier, M.-N. 2010. *Structures in the Universe by exact methods: formation, evolution, interactions*. Cambridge: Cambridge University Press.
[9] Coley, A. A. 2003. *Dynamical systems and cosmology*. Astrophysics and Space Science Library, vol. 291. Dordrecht, Boston and London: Kluwer Academic Publishers.
[10] Wainwright, J., and Ellis, G. F. R. 1997. *Dynamical systems in cosmology*. Cambridge: Cambridge University Press.
[11] Krasiński, A. 1997. *Inhomogeneous cosmological models*. Cambridge: Cambridge University Press.
[12] Mustapha, N., Hellaby, C. W., and Ellis, G. F. R. 1999. *Mon. Not. Roy. Astr. Soc.*, **292**, 817–830.
[13] Regis, M., and Clarkson, C. A. 2012. *Class. Quant. Grav.*, **44**, 567–579.
[14] Clarkson, C. A., Bassett, B. A. C. C., and Lu, T. C. 2008. *Phys. Rev. Lett.*, **101**, 011301.
[15] Ehlers, J., Geren, P., and Sachs, R. K. 1968. *J. Math. Phys.*, **9**, 1344–1349.
[16] Stoeger, W. R., Maartens, R., and Ellis, G. F. R. 1995. *Astrophys. J.*, **443**, 1–5.
[17] Heller, M. 1974. *Acta Cosmologica*, **2**, 37–41.
[18] Carr (editor), B. J. 2005. *Universe or Multiverse?* Cambridge: Cambridge University Press.
[19] Kantowski, R., and Sachs, R. K. 1966. *J. Math. Phys.*, **7**, 443–446.
[20] Thorne, K. S. 1967. *Astrophys. J.*, **148**, 51.
[21] Uggla, C. 2013. *Gen. Rel. Grav.*, **45**, 1669–1710.
[22] Rothman, T., and Ellis, G. F. R. 1986. *Phys. Lett. B*, **180**, 19–24.
[23] Raychaudhuri, A. K., and Modak, B. 1988. *Class. Quant. Grav.*, **5**, 225–232.
[24] Gümrükçüoğlu, A. E., Himmetoglu, B., and Peloso, M. 2010. *Phys. Rev. D*, **81**, 063528.
[25] Nilsson, U. S., Uggla, C., Wainwright, J., and Lim, W. C. 1999. *Astrophys. J. Lett.*, **522**, L1–L3.
[26] Lim, W. C., Nilsson, U. S., and Wainwright, J. 2001. *Class. Quant. Grav.*, **18**, 5583–5590.
[27] Pontzen, A., and Challinor, A. 2011. *Class. Quant. Grav.*, **28**, 185007.
[28] Garcia-Bellido, J., and Haugbølle, T. 2008. *J. Cosmol. Astropart. Phys.*, **0804**, 003.
[29] Bull, P., Clifton, T., and Ferreira, P. G. 2012. *Phys. Rev. D*, **85**, 024002.
[30] Krasiński, A., and Hellaby, C. W. 2004. *Phys. Rev. D*, **69**, 023502.
[31] Bolejko, K., Hellaby, C., and Krasiński, A. 2005. *Mon. Not. Roy. Astr. Soc.*, **362**, 213–228.
[32] Bolejko, K., and Hellaby, C. 2008. *Gen. Rel. Grav.*, **40**, 1771–90.
[33] Krasiński, A., and Hellaby, C. W. 2004. *Phys. Rev. D*, **69**, 043502.
[34] Synge, J. L. 1971. *Relativity: the general theory*. Amsterdam: North-Holland. Fourth printing.
[35] Coley, A. A., Pelavas, N., and Zalaletdinov, R. M. 2005. *Phys. Rev. Lett.*, **95**, 151102.
[36] Buchert, T. 2008. *Gen. Rel. Grav.*, **40**, 467–527.
[37] Zel'dovich, Ya. B. 1964. *Sov. Astr.*, **8**, 13.
[38] Bertotti, B. 1966. *Proc. Roy. Soc. Lond. A*, **294**, 195–207.
[39] Rees, M. J., and Sciama, D. W. 1968. *Nature*, **217**, 511.
[40] Dyer, C. C., and Roeder, R. 1974. *Astrophys. J.*, **189**, 167.
[41] Dyer, C. C., and Roeder, R. 1975. *Astrophys. J.*, **196**, 671.
[42] Kainulainen, K., and Marra, V. 2009. *Phys. Rev. D*, **80**, 123020.
[43] Holz, D. E., and Wald, R. M. 1998. *Phys. Rev. D*, **58**, 063501.

[44] Smith, M. *et al.* 2014. *Astrophys. J.*, **780**, 24.

[45] Biswas, T., and Notari, A. 2008. *J. Cosmol. Astropart. Phys.*, **0806**, 021.

[46] Marra, V., Kolb, E. W., Matarrese, S., and Riotto, A. 2007. *Phys. Rev. D*, **76**, 123004.

[47] Marra, V., Kolb, E. W., and Matarrese, S. 2008. *Phys. Rev. D*, **77**, 023003.

[48] Hanson, D., Challinor, A., and Lewis, A. 2010. *Gen. Rel. Grav.*, **42**, 2197–2218. Included in special issue on lensing, ed. P. Jetzer, Y. Mellier and V. Perlick.

[49] Bolejko, K. 2009. *Gen. Rel. Grav.*, **41**, 1737–1755.

[50] Ade, P. A. R. *et al.* 2014. *Astron. Astrophys.*, **571**, A17.

[51] Ade, P. A. R. *et al.* [POLARBEAR Collaboration]. 2014. *Astrophys. J.*, **794**, 171.

[52] Ade, P. A. R. *et al.* [BICEP2 collaboration]. 2014. *Phys. Rev. Lett.*, **112**, 241101.

[53] Ade, P. A. R. *et al.* [Planck collaboration]. 2014. *Astron. Astrophys.*, **571**, A18.

Part Two

New Window on the Universe: Gravitational Waves

Introduction

Gravitational waves provide an opportunity to observe the universe in a completely new way but also give rise to an enormous challenge to take advantage of this opportunity. When Einstein first found wave solutions in linearized general relativity and derived the quadrupole formula, it became clear that a laboratory experiment to produce and detect gravitational waves was impossible, while it was also clear that any gravitational wave signals produced astronomically were too weak to be detected on earth with the instruments available or thought possible at that time. Nearly 100 years later, we are at the confluence of fundamental science and technology that will soon open this new window.

Several lines of development were required to make the search for gravitational waves realistic. Despite the early recognition by Einstein that linearized gravity had wave solutions, the physical reality of gravitational waves remained in dispute for many decades. The reason for this was the absence of formalisms able to separate physical degrees of freedom in the field equations from coordinate (gauge) effects. A well known, striking example was Einstein's conviction that the Einstein–Rosen cylindrical waves [1] were not physical and furthermore that the character of this exact solution proved that there were no physical gravitational waves in the full theory. While Einstein retrieved the correct interpretation in the nick of time [2], the question remained unsettled until correct, gauge-invariant formulations of the problem were developed. The first of these, from Bondi's group [3–5], used the "news function" to demonstrate that, far from the source, one could quantify the energy carried away by gravitational waves. Further developments in understanding equations of motion, gauge freedom, and other methods to identify gravitational waves in the background spacetime led to approximation methods with greater precision and broader application than the original linear waves [6, 7]. In addition, the first half-century of general relativity saw the physically relevant exact solutions of Schwarzschild [8] and, much later, Kerr [9]. In the late 1930s, Oppenheimer and Volkoff [10] and Oppenheimer and Snyder [11] studied spherically symmetric gravitational collapse that predicted the formation of neutron stars and black holes, respectively, even though both categories were considered highly unlikely as real objects for several decades thereafter.

Meanwhile, astronomical technology developed beyond the optical frequency band. Radio astronomy coupled to optical followup led to the discoveries of quasars in 1963 and pulsars in 1967, ushering in the age of relativistic astrophysics – not to mention the discovery of the cosmic microwave background (CMB) in 1965. Suddenly, the universe became much more interesting for general relativity. X-ray detectors aboard rockets found bright point sources outside the solar system, possibly indicating a more violent universe than previously suspected. Supernovae could now be better studied and served as the driver for early general relativistic computer simulations [12]. At approximately the same time as physical systems in the strong-gravity regime became plausible, theoretical developments made it clear that gravitational waves could carry energy and thus could be physically relevant. At some point, improved treatment of the two-body problem in general relativity led to a number of recalculations of Einstein's quadrupole formula, engendering a rather vitriolic and long-reigning dispute over the coefficient. It became clear later (in the 1980s) that consistent approximations yielded Einstein's results [13].

The possibility that detectable gravitational wave sources might exist and the firm theoretical foundation that such waves were a prediction of general relativity inspired Joseph Weber to build a detector to search for them [14]. His first detector, an aluminum cylinder (bar) with piezoelectric readout is described in [15, 16]. He quickly realized that coincident detection – i.e., two or more bars – was an essential ingredient. This first generation of bar detectors reached a sensitivity to gravitational wave strain amplitudes of 10^{-16} by the end of the 1960s [17]. Weber's electrifying claim of the detection of signals associated with the center of our galaxy [16] inspired a worldwide effort, leading to one or more bar detectors operating in China, France, Germany, Italy, Japan, the UK, the US, and the USSR [17, 18]. Unfortunately, even though the subsequent experiments done with this first generation of bars were more sensitive than Weber's, they failed to confirm his detection. In addition to bar development, plans were made to use the earth [19], the moon [20], and a number of spacecraft [21] as gravitational wave detectors. By the first decade of this century, the largest bars had reached a size of 3 meters and a mass of more than 2 tons, were operating at cryogenic temperatures as low as 0.1 K to minimize thermal noise, and had achieved a strain sensitivity of $\approx 10^{-21}/\sqrt{\text{Hz}}$ at 900 Hz. Prototypes of novel designs for resonant-mass detectors were developed in this period, with at least one prototype currently under construction [18].

In the mid 1970s, interferometric detectors were first considered both for the ground and for space. Chapter 5 focuses on highlights of the current status of such instruments. The first prototypes were built by Forward (at Hughes) and later by Weiss (at MIT) and Drever (at Glasgow and Caltech) as well as elsewhere in the world [17]. The laboratory-scale interferometers had sensitivities far worse than those of the contemporaneous bars. However, the interferometers could be scaled up, in contrast to bars (see [22]), but would require facility-class dimensions and funding to achieve sensitivities that were plausible for detection.

We note here that gravitational wave detection pushes precision technology and related theory and thus can yield important scientific spinoffs. An early example is quantum non-demolition. The broad idea is to overcome the standard quantum limit by arranging to decrease the uncertainty in one variable by increasing it in the complementary variable. This was first proposed for bar detectors by Braginsky *et al.* [23] and refined for interferometers by Caves [24]. The concept has been studied extensively in atomic, molecular, and optical physics since then, but is most spectacularly implemented on the kilometer-scale interferometer GEO 600 [25] and was tested successfully on one of the LIGO instruments [26]. It is likely to become a near-term upgrade for the second-generation interferometers Advanced LIGO and Advanced Virgo.

Meanwhile, in the early 1970s, the Uhuru X-ray satellite revealed the X-ray sky for the first time (where brief rocket flights had only given hints). Stellar-mass binaries, including Cyg X-1, a strong black hole candidate, as well as a number of active galactic nuclei, were revealed and could be correlated with radio and optical sources. It became possible to argue that potential gravitational wave sources were abundant.

This time period also saw the first steps toward binary black hole simulations by Smarr and collaborators [27]. However, the landmark discovery of this decade was the binary pulsar PSR1913+16 by Hulse and Taylor [28]. Not only was this the first known system containing two neutron stars, but the orbital parameters made it sufficiently general-relativistic to strongly constrain theories of gravity (see the discussion in Chapter 2). Even more exciting, the period decay due to energy loss by emission of gravitational waves was measurable. Because this system is clean (there are no extraneous effects from tides or matter accretion), it provided incontrovertible experimental evidence of the existence of gravitational waves and also of the correctness of the quadrupole formula (see the discussion in Chapter 6). As a side note, this validation of the quadrupole formula killed off the last vestiges of debate about the coefficient. In addition, the known binary neutron star systems are precursors of gravitational wave sources for ground-based detectors – and, in fact, are the only ones that demonstrably exist. Although supernovae in our galaxy may be detectable sources, the details of nonaxisymmetric collapse are not yet known sufficiently well to predict their strength. What this means is that the orbit decay of binary neutron stars will eventually lead to the final inspiral and merger phase that produces strong gravitational waves at frequencies of up to approximately one kHz (see Chapter 6). Because the time to merger for the known systems is significantly less than the age of the universe, one has reason to hope that such systems in the merger phase occur with sufficient frequency to be a target for human detection if sufficiently sensitive instruments could be built. As more such systems (but still a small number) were discovered and stellar evolution became better understood, event-rate estimates became possible [29]. Chapter 6 focuses on the gravitational wave signatures of these and other proposed sources. To obtain the correct waveform requires either careful (or even rigorous) approximations or robust numerical simulations.

While the best-understood source for ground-based interferometric detectors has been observed electromagnetically in its precursor phase, actual known, named objects, namely white dwarf binaries, are accessible to space-based detectors, which, when constructed and launched, will operate at the much lower frequencies characterizing such systems. Examples are given in Chapter 6.

A byproduct of the expansion of astronomy into non-optical detection was the identi-fication of systems of stellar mass and up to 10^9 $M_\odot$ which could be black holes. Over the years, the black hole hypothesis has become the favored explanation for these systems. Black hole binary systems, if they exist, would be an obvious target for gravitational wave detection since in a clean environment of no accretion discs they would emit no electromagnetic signals. Hawking had proved that a merging binary black hole could emit up to 29% of its mass-energy in gravitational waves [30]. But what was the actual number? To answer this question required numerical simulations in three spatial dimensions (3D). While the 2D (head-on collision) problem was solved reasonably well by Smarr and collaborators [27] in the 1970s, the 3D problem proved astonishingly difficult until it was finally solved in 2005 [31–33] (see Chapter 7).

The growth of relativistic astrophysics made the scientific case for development of gravitational wave detection programs compelling. Although many bar groups left the field after their failure to confirm Weber's claims, several remained and continued to improve their instruments [17]. Nonetheless, though the interferometric prototypes built in the 1970s and 1980s were less sensitive than the bars of their era, it became clear that interferometers were the only way to achieve the needed sensitivity in the frequency range accessible from the ground. Efforts were begun in the US, Germany, UK, France, Italy, and Japan to seek funding to build the kilometer-scale instruments that would be necessary to detect a strong gravitational wave event from the Virgo cluster (see Chapter 5). This distance was chosen as a criterion because it is necessary to probe at least that far to have sufficient likelihood for a detectable event (see Chapter 6). Remarkably, given the novelty and scale (both physically and financially) of the proposed technology, two kilometer-scale projects, LIGO in the US and Virgo, a French/Italian collaboration in Italy, with 4 km and 3 km arm-lengths respectively, were funded and built. Construction took place in the 1990s and data taking by some of the instruments, as well as by the shorter but nearly kilometer-scale GEO 600 and TAMA, began in about 2000 (see Chapter 5).

Meanwhile, building on earlier efforts with data from prototype detectors [34], programs to analyze the data coming from LIGO, Virgo, GEO 600, and TAMA were initiated. The time series $h(t)$ for the gravitational wave strain was analyzed either by matched filtering (comparing to templates) for binary neutron stars or as unmodeled bursts (see Chapters 5 and 6). Note that the former required the substantial computational power provided by advancing technology to become feasible. Toward the end of the 2000s, astrophysically interesting upper limits were obtained, the technological tour-de-force of design sensitivity was achieved, and bar sensitivity was exceeded [35].

In 2010, an upper limit for burst events was published, based on observations by a world-wide network of bars [36]. While the limit itself was not interesting astrophysically, the effort demonstrated the advantages of international collaboration in this field. In addition, the technology to rotate the LSU bar was developed and implemented. This is interesting because a cross-correlation, stochastic measurement with the LIGO Livingston instrument allowed regular reorientation of the bar to impose a periodicity on any signal present since the allowed orientations included a direction where there should be no correlation [37].

Starting in the 1980s, space-based concepts were proposed and evaluated, leading to adoption of LISA (first envisioned by Bender and Faller) as a joint R&D project of NASA and ESA in 1998. The astronomical community became engaged, possibly due to the existence of the white-dwarf-binary calibration sources. Regrettably, in contrast to the ground-based detectors supported through physics programs, the astronomy-centered space programs in Europe and the US never moved LISA into a launchable status. One issue was the alleged riskiness of the technology. While the level of precision needed for LISA is six orders of magnitude less than for, e.g., LIGO due to LISA's proposed arm-lengths of millions of kilometers (see Chapter 5), the picometer level of precision required is still daunting for space-qualified missions and several necessary technologies are new. It is hoped that the often postponed but now firmly scheduled LISA Pathfinder will demonstrate the feasibility of LISA and LISA-like missions (see Chapter 5).

In the past decade or so, increasing interest has arisen in taking advantage of the remarkable timing precision and stability associated with millisecond pulsars to use the monitoring of pulse arrival times to tease out signals from passing gravitational waves [38]. These methods are effective at nanohertz frequencies and are sensitive to signals from supermassive binary black holes and from a stochastic background. In principle, the ability to detect a gravitational wave signal increases with increasing monitoring time (with a power that depends on the type of source). At this time, it is not clear that the systematic errors are completely understood. Nevertheless, there is a significant chance that pulsar timing arrays will detect gravitational waves within this decade (see Chapter 5).

Of course, we should not leave out the imprint of gravitational waves on the cosmic microwave background (see Chapter 3). In particular, at certain angular scales, relic gravitational waves produced in the early universe leave a curl-like tensor imprint, the so-called B-modes, on the cosmic microwave background. If the signal is sufficiently strong, with the ratio of tensor-to-scalar polarization amplitude r large enough to be currently observable, the gravitational waves are likely to have been produced by cosmological graviton production and amplified by inflation. A recent report of such a detection with $r \approx 0.2$ by BICEP2 may have been premature, since the measured effect could be caused by foreground dust or have a significant dust component [39]. Whether or not this possible detection of gravitational waves is confirmed, the discussion centered on the observation emphasizes that direct or indirect detections of gravitational waves reveal information about the universe and its contents that is inaccessible with electromagnetic radiation, cosmic rays, or even neutrinos.

A number of different technologies have been applied to the different programs for gravitational wave detection. These are not in competition but rather are complementary in that they search for gravitational waves in different frequency bands. As with electromagnetic emitters, gravitational waves at different frequencies arise from different sources. In simple terms, the frequency depends inversely on a power of the mass (see Chapter 6). At the highest frequencies of 10 to 10^4 Hz accessible from the ground, the predominant sources are expected to be "chirp signals" from binary neutron stars or binary black holes with masses of order 1–300 $M_\odot$, "monochromatic" signals from non-axisymmetric rotating neutron stars, correlations between the noise from two detectors that characterize a cosmo-logical or astrophysical stochastic background, or unmodeled bursts from supernovae in our galaxy or other sources. The ground-based detectors certainly have the potential to identify the engines of gamma-ray bursts [40]. Future generations may be able to probe the equation of state of nuclear matter, discover intermediate-mass black hole binaries, and reveal as yet unsuspected new phenomena. Space-based observations are sensitive to binary white dwarfs and binary neutron stars in the early inspiral phase and binary black holes in the mass range of 10^5 to 10^7 $M_\odot$. Perhaps more interestingly for general relativity, they can observe the inspiral of a stellar-mass black hole around a supermassive black hole. If the gravitational waves from such sources can be observed with sufficient precision [40], detailed tests of general relativity can be made. One is the definitive "smoking gun" detection showing that the central object is a black hole (see Chapter 6). Another allows one to obtain an interesting bound on the graviton mass through a distortion of the phasing of gravitational waves in a binary black hole inspiral [41]. Finally, pulsar timing arrays can search for gravitational wave signals from supermassive binary black holes that are suspected to exist from their electromagnetic properties as well as ones that are so far unsuspected. Once gravitational waves are detected, it would be natural to expect that the experience of the early radio, X-ray, and γ-ray telescopes will be repeated with the discovery of completely new and unexpected phenomena to, once again, widen our horizons in unforeseen directions.

The next two chapters explore the sources and technologies that comprise gravitational wave science. They emphasize the growth in our knowledge about the sources through observations, theory, and simulations, including the state of our knowledge on their numbers (or number density) in the universe, and their gravitational wave signatures. These are then the targets for decades of technological development for detectors on earth and in space, with the former on the verge of recording their first signals.

When looking back over a century of general relativity, we see that gravitational waves not only represent a future transformative discovery but have also, over the past 30 years or so, changed the nature of the field of gravitational physics. General relativity has gone from a discipline with few practitioners exploring apparently arcane (but often seminal) mathematical properties of the theory to big science with international collaborations yielding publications with nearly 1000 authors whose instruments require an investment on the order of $1 billion (US). A further transformation will occur with the first detections by the second-generation Advanced LIGO and Advanced Virgo instruments, independent of the precise nature of the sources. Unlike the cosmic microwave background measurements or pulsar timing arrays that piggyback on instruments developed for other purposes, the instrumentation developed for gravitational wave detection on the ground and in space represents a completely new way to study the universe. It is likely that the discoveries soon to be made by these advanced detectors will "shake loose" the future of the field by acceler-ating the timeline for the first space-based gravitational wave detectors and by encouraging investment in third-generation instruments such as the proposed Einstein Telescope [42].

References

[1] Einstein, A., and Rosen, N. 1937. *J. Franklin Inst.*, **223**, 43–54.

[2] Kennefick, D. 2005. *Physics Today*, **58**, 43.

[3] Bondi, H. 1957. *Nature*, **179**, 1072–1073.

[4] Pirani, F. A. E. 1957. *Phys. Rev.*, **105**, 1089–1099.

[5] Bondi, H., Pirani, F. A. E., and Robinson, I. 1959. *Proc. Roy. Soc. Lond.*, **A251**, 519.

[6] Isaacson, R. 1968. *Phys. Rev.*, **166**, 1263.

[7] Isaacson, R. 1968. *Phys. Rev.*, **166**, 1272.

[8] Schwarzschild, K. 1916. *Sitzber. Deut. Akad. Wiss. Berlin, Kl. Math-Phys. Tech.*, 189.

[9] Kerr, R. 1963. *Phys. Rev. Lett.*, **11**, 237.

[10] Oppenheimer, J. R., and Volkoff, G. M. 1939. *Phys. Rev.*, **55**, 374.

[11] Oppenheimer, J. R., and Snyder, H. 1939. *Phys. Rev.*, **56**, 455.

[12] May, M. M., and White, R. H. 1966. *Phys. Rev.*, **141**, 1232.

[13] Kennefick, D. 2007. *Traveling at the speed of thought: Einstein and the quest for gravitational waves*. Princeton, NJ: Princeton University Press.

[14] Weber, J., and Wheeler, J. A. 1957. *Rev. Mod. Phys.*, **29**, 509–515.

[15] Weber, J. 1967. *Phys. Rev. Lett.*, **18**, 498.

[16] Weber, J. 1969. *Phys. Rev. Lett.*, **12**, 1320–1324.

[17] Saulson, P. 2005. Receiving gravitational waves. Chapter 9 of: Ashtekar, A. (ed), *100 years of relativity: space-time structure Einstein and beyond*. Singapore: World Scientific.

[18] Aguiar, O. 2011. *Res. Astron. Astrophys.*, **11**, 1–42.

[19] Levine, J., and Stebbins, R. 1972. *Phys. Rev. D*, **6**, 1465–1468.

[20] Giganti, J. J. *et al.* 1977. *Lunar surface gravimeter experiment final report*. See http://www.lpi.usra.edu/lunar/missions/apollo/apollo_17/experiments/lsg/.

[21] Armstrong, J. 2006. *Living Rev. Rel.*, **9**, 1.

[22] Saulson, P. 1994. *Fundamentals of interferometric gravitational wave detectors*. Singapore: World Scientific.

[23] Braginsky, V. B., Vorontsov, Yu. I., and Khalili, F. Ya. 1977. *Sov. Phys. JETP*, **46**, 705.

[24] Caves, C. 1981. *Phys. Rev. D.*, **23**, 1693–1708.

[25] Schnabel, R., *et al.* 2011. *Nature Physics*, **7**, 962–965.

[26] Aasi, J., *et al.* 2013. *Nature Photonics*, **7**, 613–619.

[27] Smarr, L. L. 1979. Gauge conditions, radiation formulae and the two black hole collision. Page 245 of: Smarr, L. L. (ed), *Sources of gravitational radiation*. Cambridge: Cambridge University Press.

[28] Hulse, R. A., and Taylor, J. H. 1975. *Astrophys. J. Lett.*, **195**, L51.

[29] Abadie, J., *et al.* [LIGO Scientific Collaboration]. 2010. *Class. Quant. Grav.*, **27**, 173001.

[30] Hawking, S. 1971. *Phys. Rev. Lett.*, **26**, 1344–1346.

[31] Pretorius, F. 2005. *Phys. Rev. Lett.*, **95**, 121101.

[32] Baker, J. G. *et al.* 2006. *Phys. Rev. Lett.*, **96**, 111102.

[33] Campanelli, M., Lousto, C. O., Marronetti, P., and Zlochower, Y. 2006. *Phys. Rev. Lett.*, **96**, 111101.

[34] Allen, B., *et al.* 1999. *Phys. Rev. Lett.*, **83**, 1498.

[35] Abadie, J., *et al.*, 2012. arXiv:1203.2674.

[36] Astone, P., *et al.* 2010. *Phys. Rev. D*, **82**, 022003.

[37] Abbott, B., *et al.* 2007. *Phys. Rev. D*, **76**, 022001.

[38] Hobbs, G., *et al.* 2010. *Class. Quant. Grav.*, **27**, 084013.

[39] Ade, P. A. R., *et al.* 2014. *Phys. Rev. Lett.*, **112**, 241101.

[40] Sathyaprakash, B. S., and Schutz, B. F. 2009. *Living Rev. Rel.*, **12**, 2.

[41] Will, C. 2014. *Living Rev. Rel.*, **17**, 4.

[42] Punturo, M., *et al.* 2010. *Class. Quant. Grav.*, **27**, 194002.

5

Receiving Gravitational Waves

BEVERLY K. BERGER, KARSTEN DANZMANN, GABRIELA
GONZALEZ, ANDREA LOMMEN, GUIDO MUELLER, ALBRECHT
RÜDIGER AND WILLIAM JOSEPH WEBER

5.1 Introduction

Gravitational waves are a consequence of Einstein's General Theory of Relativity, first presented in 1915 and published in 1916 [1]. Einstein himself linearized his theory and derived wave equations and calculated the gravitational radiation produced by sources in the weak-field, slow-motion limit [2]. As described in the following Chapter, this initial insight has been greatly expanded so that, in general, it is possible to calculate either numerically or analytically the details of the gravitational radiation for a broad range of potential astronomical sources. Much later, in the 1970s, the discovery of the binary neutron star system PSR1913+16 by Hulse and Taylor [3] demonstrated through this natural experiment that gravitational waves carry away energy and angular momentum, causing the neutron star orbit to decay at precisely the predicted rate. Early cosmological gravitational waves imprint a polarization signature in the electromagnetic microwave background that several sensitive instruments may detect. See [4] but also [5] and references therein for further discussion.

These brief remarks gloss over a more complex history where it was unclear whether gravitational waves were real or just gauge artifacts. The theory was finally settled on the side of reality [6]. The standard next step in physics – to build a receiver to directly detect gravitational waves – proved to be extremely challenging. The analog of the Hertz experiment where artificially generated waves are detected within the wave zone will fail because of the undetectably small amplitude (see for example [7]). Astrophysical sources are much stronger but are, of course, more distant. Yet their detection may be possible because gravitational wave receivers respond to amplitude and not to intensity. Nonetheless, the numbers are daunting.

In the early 1960s, J. Weber followed through on a bold vision – that gravitational waves were detectable – by measuring the resonant excitation of acoustic modes in heavy metallic bars, as would be caused by a passing gravitational wave from relatively nearby astrophysical sources [8]. In 1969, he reported coincident signals in two bars separated by hundreds of kilometers that appeared to be correlated with sidereal time [9]. However, this apparent signal was orders of magnitude stronger than might be expected from sources known so far. This announcement inspired a worldwide development of similar bar detectors and improved data analysis, but none were able to confirm the detection [10]. The development of bars with improved sensitivities continued, eventually reaching a length of 3 meters, a mass of more than 2 metric tons, and, to reduce thermal noise, cooled to 0.1 K. See [11] for a recent review of resonant-bar detectors.

Laser interferometry as a way to detect gravitational waves was first discussed by Gertsenshtein and Pustovoit [12] in the early 1960s. Forward [13] built the first prototype and Weiss [14] provided the first detailed study of the limitations of future ground-based laser interferometric detectors. These interferometers have many advantages over bars – their response to gravitational waves covers a wide range of frequencies while the responding masses are mirrors separated by vacuum rather than massive solids. On the other hand, to reach the necessary sensitivity, the separation between the mirrors must be on the scale of kilometers, making the resultant instruments orders of magnitude more expensive. Bender and Faller in the mid 1970s developed the first concepts for an interferometric mission in space [15, 16] where real estate is cheap, but getting there is expensive. In the second half of the 1970s, Sazhin [17] and Detweiler [18] were the first to propose to use pulsars to search for gravitational waves. While two bars (Nautilus and Auriga) operate together with the laser interferometer GEO 600 in Astrowatch mode[1] to detect any nearby supernova [19], most of the current effort to construct gravitational wave receivers focuses on measuring the relative accelerations, or equivalently the tidal deformation, in a constellation of free particles. This is performed both actively – with interferometric measurements between man-made test masses – and passively, in the recorded arrival times of radio bursts from a constellation of pulsars.

Following the discussion in Chapter 3 of [7], we note that a binary star system of masses M_1 and M_2 separated by the distance D and a distance r from us produces, for a reduced quadrupole tensor $I_{\mu\nu}$ evaluated at the retarded time, a gravitational wave amplitude $h_{\mu\nu}$ at the earth

$$h_{\mu\nu} \approx \frac{2G}{rc^4} \ddot{I}_{\mu\nu} \tag{5.1}$$

that can be put in the approximate form depending only on the four length scales in the problem:

$$h \approx \frac{R_1 R_2}{D r}, \tag{5.2}$$

where the R_i are the Schwarzschild radii (R_S) of the two masses, D is their separation, r is the distance to the observer, and $h \equiv |h_{\mu\nu}|$ is the dimensionless strain. If we evaluate this expression for two 1.4 solar mass neutron stars separated by 20 km (approximately two neutron star radii) and located in the Virgo cluster at a distance of about 15 Mpc, we find $h \approx 10^{-21}$, equivalent to the width of an atom over the distance from the earth to the sun.

It must also be noted that the typical gravitational wave frequency of the binary source is twice the orbital frequency, f_{orb}, which may be estimated from Keplerian motion to be

---

[1] Astrowatch-mode refers to a less sensitive detector taking data in case a nearby supernova or other potential source of strong gravitational waves occurs while the more sensitive instruments are unavailable, e.g., due to an ongoing upgrade.

$$f_{\text{orb}} = \sqrt{\frac{GM}{4\pi^2 D^3}} \propto \left(\frac{R_S}{D}\right)^{3/2} \frac{c}{R_S} \qquad (5.3)$$

where R_S/D measures how relativistic the system is while the second factor shows that the gravitational wave frequency scales inversely with the mass of the system. The lessons of Eqs. (5.1) and (5.3) are that, even from very strong sources, the expected signals are tiny and that the categories of sources expected depend on the operating frequencies of the detectors. In addition, likely sources of detectable gravitational waves such as coalescing neutron stars or black holes are extremely rare in an individual galaxy like our own. It is thus important for a gravitational wave detector to be as sensitive as possible to events as distant as possible since typical rates of transient events are proportional to the volume of the universe that is sampled.

In this Chapter, we will discuss three scenarios for receiving gravitational waves: (1) Ground-based, kilometer-scale interferometers. The second generation of this comparatively mature technology is expected to go online in 2015 or 2016 [20, 21]. These instruments operate in the 10–2000 Hz band with greatest sensitivity at around 100 Hz. Signals at these frequencies are generated by stellar mass objects such as neutron-star and black-hole binaries; (2) Space-based, gigameter-scale interferometers target the 10 μHz to 100 mHz range and will be sensitive to mergers between 10^4 to 10^8 solar mass black holes, solar mass black holes falling into massive black holes, and many compact galactic binaries. LISA, a former joint ESA/NASA project, and eLISA, a newly proposed similar design, are the best known concepts [22, 23]; (3) Pulsar timing arrays. This final approach takes advantage of millisecond pulsars – extremely precise clocks placed by nature throughout the Milky Way. The timing of the arrival of the pulses can be affected by a passing gravitational wave. Pulsar timing arrays are already operating and have placed upper limits on gravitational waves from a stochastic background [24] and candidate supermassive black hole binaries [25]. Pulsar timing is sensitive to supermassive black hole binaries in the tens of nanohertz frequency range.

Each of the detector classes will open a new window or – more appropriately – start to listen[2] to its own cosmological concert played by the largest and most compact objects known to mankind. In the following sections, we will discuss the basic detection schemes used by our current and currently planned detectors, review briefly the scientific results obtained by the already operational detectors, and discuss the required advanced detector technologies.

5.2 The basic detection scheme

All direct detection schemes use a concept that was initially invented by Pirani in a thought experiment [6]: Gravitational waves change the proper distance or light travel time between

---

[2] The lengths of gravitational waves compared to source and detector sizes give them many characteristics of sound rather than light.

two or more free falling masses. Because gravitational waves produce strains, the change δL over a distance L scales with the separation:

$$\delta L = \frac{h}{2} L.$$

The above equation underlines the experimental challenge in gravitational wave detection: measuring very small changes in the separations between freely falling test particles. This requires sensitive displacement measurement metrology, whose development has dominated research in this field for decades. It also requires that the freely falling test particles are truly in geodesic motion, as stray accelerations from all non-gravitational forces directly compete with the gravitational wave tidal deformation.

In the long-wavelength limit applicable to interferometric ground-based detectors, $\lambda_{GW} \gg L$, larger distances between the responding test masses will cause larger changes which are easier to detect. Once the distance exceeds half λ_{GW}, the space between the free falling objects will be partly compressed and partly stretched, leading to a cancellation of the overall length change (see, e.g., [7]). The optimum length for gravitational waves in the kilohertz range is on the order of 1000 km while the optimum length for nanohertz waves is in the 10^{14} km range, approximately 10 light years. Just as for electromagnetic radiation detectors, a single type of gravitational wave detector will never be able to span this entire frequency band and cover all known gravitational wave sources.

Practical considerations, both physical and financial, limit ground-based gravitational wave detectors in size to up to a few kilometers, far shorter than the optimum length for their operating frequencies from about 10 Hz to about 10 kHz. All ground-based detectors (working, under construction, or proposed) are much shorter than the gravitational wavelength. With the exception of GEO 600, all detectors use resonant optical cavities to store the light in the arms for times comparable to half of the gravitational wave period to increase the effective detector size. GEO 600 uses signal recycling instead to resonantly enhance the signal inside the interferometer.

The upper limit on the operating frequency coincides with the highest frequencies expected from strong gravitational wave sources while the lower limit is determined by seismic noise (for first-generation instruments) or a combination of radiation pressure noise, suspension thermal noise, and Newtonian (gravity gradient) noise for the second generation or beyond. Currently, two first-generation instruments, LIGO and Virgo, are being upgraded to second-generation Advanced LIGO and Advanced Virgo, with the former planned to be in operation in late 2015 and the latter soon after [26].

The next generation, advanced KAGRA [27] and the Einstein Telescope, ET [28], will push towards lower frequencies and perhaps even go below 1 Hz where gravity-gradient noise (Newtonian noise) caused by changes in the mass distribution surrounding the suspended mirrors will potentially limit all ground-based detectors (see [29] and references therein).

Space-based gravitational wave detectors have the advantages of no exposure to uncontrollable gravity-gradient noise and the potential for a much longer baseline.

LISA and eLISA were optimized to detect mergers between 10^4 to 10^8 solar mass black holes throughout the universe. LISA-type missions use three spacecraft in an equilateral-triangular formation with a (multi-)million kilometer baseline and placed in an earth-like heliocentric orbit, trailing or leading the earth by 10 to 30 degrees. Reference test masses are in free fall inside each of the three spacecraft which create the necessary drag-free environment. LISA Pathfinder [30] is scheduled to be launched in 2015 to test most of the critical technologies including drag-free motion for all LISA-like missions. LISA-type missions are expected to be limited at frequencies below approximately one mHz by stray forces on the test masses and shot noise above a mHz, which is especially detrimental at higher frequencies, where the gravitational wavelengths become smaller than the spacecraft separation.

Over the last decades, many other space-based mission concepts were presented or proposed to different space agencies. The best known are BBO, DECIGO and ALIA, which target the frequency band between Advanced LIGO and eLISA, while Astrod, Lagrange and others aim at frequencies as low as $10\,\mu\text{Hz}$ by using much longer arms reaching and exceeding 1 A.U. All of them use some or all of the same basic LISA technologies while some add features from ground-based detectors such as optical cavities [31–35].

One of the challenges in going to even lower frequencies is to reduce the accelerations from spurious forces. Luckily, nature provides millisecond pulsars, which appear to be massive and isolated enough for one to measure their response to passing gravitational waves at frequencies as low as a nanohertz. In addition, their stable period and pulsar radio signal allows them to act as clocks so that their motions may be monitored. Pulsar timing arrays like NANOGrav, EPTA, and PPTA track the electromagnetic pulses over long periods of time to look for correlations in the pulsar signals [36, 37]. The typical distances between pulsars and the earth as well as each other are of the order of kpc (10^{19} m), significantly larger than the relevant gravitational wavelengths of about 10^{17} m. However, by focusing on a volume of approximately one wavelength centered on the earth, timing residuals from pulsars in a particular direction will be correlated while those in the orthogonal direction will be anti-correlated if a gravitational wave passes by. Monitoring of arrays of pulsars is thus required. The expected modulations of the signal arrival times in pulsar timing are

$$\delta\tau \approx \frac{\delta L}{c} \approx \frac{h}{c}\frac{\lambda_{\text{GW}}}{2} = \frac{h}{2\Omega_{\text{GW}}} \approx 0.1\text{ to }10\,\text{ns}$$

with a period on the order of 10 days to a few years.

These different types of observatories are complementary in that they are sensitive to different wavelength regimes and therefore to different sources. For example, ground-based detectors are sensitive to gravitational waves from binary black holes of up to about 200 solar masses, space-based to about 10^4–10^8 solar masses, and pulsar timing to 10^8–10^{10} solar masses. Additionally, different frequency bands allow different science from the same types of sources at different phases of their evolution. Neutron star binaries, for instance, may be detected by LISA as numerous and monochromatic sources millions of years before their final mergers, allowing a population study in the Milky Way, while the ground-based detectors will observe, out to larger distances, their final inspiral as brief events at audio frequencies, allowing studies of the neutron star equation of state.

5.3 Ground-based detectors

Research with prototype instruments on key technologies throughout the 1970s and 1980s made the construction of kilometer-scale laser interferometric gravitational wave detectors feasible so that the scientific case for providing resources to open a new window on the universe became compelling. By the mid 1990s, the 4 km LIGO instruments were under construction at Livingston, LA and Hanford, WA (also including a third, 2 km detector) in the USA and the 3 km Virgo instrument was under construction at Cascina, Italy. Both these projects were designed with upgrades in mind. In addition, the 600 m GEO 600 near Hannover, Germany and the 300 m TAMA near Tokyo, Japan were also under construction. By the early 2000s, TAMA, GEO 600, and LIGO were operating and taking data with Virgo coming online in 2007.

While the concept of interferometric gravitational wave detectors is simple, the implementation is difficult and complex. A detailed discussion of the issues may be found in [7]. Later reviews may be found in [38, 39]. The level of precision needed to detect gravitational waves imposes daunting restrictions on the tolerances of components and requires elaborate systems to overcome sources of noise. First-generation detectors had to solve the problems of making optical cavities or delay-line systems work to increase the optical storage time, low-noise materials for mirrors with high-reflective, low-loss optical coatings, beyond-the-state-of-the-art laser stability in frequency, power, and spatial mode, the isolation of the test masses from the seismic disturbances in the environment, and the construction of elaborate control systems to maintain the precise alignments and other operating conditions needed for the instruments to function. Figure 5.1 shows the best sensitivity attained for all first-generation detectors. First-generation instruments were limited by photon-counting (shot) noise at high frequencies, seismic disturbances at low frequency, and suspension thermal noise in an intermediate region [7, 38]. Nonetheless, these instruments rose to the challenge such that LIGO and Virgo reached design sensitivity while GEO 600 and TAMA were sufficiently sensitive for important technology development and prototyping and for data taking.

The basic configuration of all ground-based detectors is a Michelson interferometer. A central beam splitter divides a laser field into two parts which propagate in two orthogonal directions. End mirrors placed a distance L away from the beam splitter will retro-reflect the fields. The returning beams recombine at the beam splitter, sending some of the light back towards the laser and some to the second output port, commonly known as the dark port. The power in these two beams depends on the arm length difference or phase difference between the two fields when they recombine. A properly polarized gravitational wave propagating into the plane of the interferometer will modulate the arm length difference with an amplitude $\delta L = hL$.

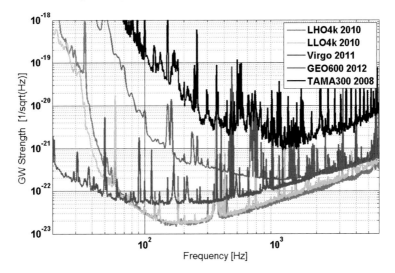

Figure 5.1 Sensitivities of all first-generation ground-based detectors. LIGO is represented by data taken during S6 in 2010 while Virgo is represented by its latest science run during 2011. LIGO's fifth science run (S5) was LIGO's first science run at design sensitivity at frequencies above about 70 Hz. Its final run, S6, used higher laser power and even surpassed the design sensitivity of initial LIGO. GEO 600 is represented by its sensitivity in 2012 during astrowatch when it was the only operating interferometric detector while LIGO and Virgo were being upgraded. Data taken from [26] except for GEO 600 (courtesy of Harald Lück and Hartmut Grote, who also produced the plot).

Over the last 15 years, all existing large detectors have taken data, sometimes for years at a time. The data have been thoroughly analyzed and searched for signatures of signals from different types of sources. These include well modeled signals from inspirals and mergers of compact binary systems containing black holes and/or neutron stars, monochromatic signals from rotating pulsars, and unmodeled signals from brief transients or a stochastic background. None of these searches has yielded a detection but they have nonetheless generated constantly improving upper limits. In addition to the important results which we will briefly summarize here, the developed data analysis methods will be reused and extended for the advanced-detector era. For more on data analysis with the first-generation detectors see the following Chapter and, e.g., [39].

The TAMA detector was the first of the large detectors to take data and analyze them in Aug–Sep 2000. The four LSC detectors (three LIGO detectors and GEO 600) started the first "science run" of the LIGO Scientific Collaboration in Aug–Sep 2001. In the fall of 2005, the three LIGO detectors reached their design sensitivities and began to collect over the following two years to yield one year of triple-coincidence data. The Virgo detector started to participate in this science run in 2007. Following this 5th science run, some of the Advanced LIGO technologies were installed into the two 4 km LIGO detectors, which improved their sensitivity beyond the design sensitivity of initial LIGO. These two LIGO detectors and the Virgo detector started the last coincidence run of the large detectors in 2009. Following a similar upgrade, Virgo started its last science run in 2010. Throughout the years, GEO 600 participated in most science runs. Although it was not competitive in terms of strain sensitivity below a kHz during the later runs, it was the most reliable detector, with duty cycles exceeding 90%.

It should be emphasized that all the first-generation projects have taken and co-analyzed data with at least one other project [40, 41]. An international network of detectors of comparable sensitivity has several advantages, including rejection of apparent events seen by only one instrument – requiring coincident detection can rule out noise (the motivation for the two LIGO sites), and using time-of-flight measurements can provide some information on the location of the source – a key ingredient for coincident electromagnetic detection. A discussion of sky localization with such a network can be found in [42]. LIGO, Virgo, and GEO 600 demonstrated that the goals of the first generation of detectors have been achieved – they can be built and commissioned to reach the expected sensitivity and can be operated stably over long periods of time. Figure 5.1 shows the strain sensitivities of all the initial detectors at their best sensitivity. The achieved strain sensitivity of $2 \times 10^{-23}/\sqrt{\text{Hz}}$ is equivalent to a displacement sensitivity of less than 10^{-19} m$/\sqrt{\text{Hz}}$, more than nine(!) orders of magnitude smaller than the hydrogen atom. These detectors could have seen mergers between 20 solar mass black holes out to 300 Mpc and binary neutron star inspirals in the Virgo cluster of galaxies approximately 15 Mpc away. Unfortunately, none of these events occurred in the accessible volume during the last science runs. Although the detectors did not resolve any gravitational waves, they have produced many astrophysically interesting upper limits on merger rates, on stochastic gravitational wave background radiation, on the eccentricity in the shape of many pulsars, and on the energy released into gravitational waves by supernovae.

Five upper limits stand out and will be described briefly here. More detailed discussion of the results can be found in Chapter 6 of this volume. For the complete list of published results see [43] for those of LIGO and its partners and [44] for additional Virgo publications. The five publications fall into three categories: (1) Beating the spin-down limit for pulsars. Original goals of both LIGO and Virgo were to achieve sufficient sensitivity to either detect or rule out gravitational waves at a level below the energy loss indicated by a slowdown in the pulsar's spin rate. LIGO did this first for the Crab pulsar [45] and Virgo accomplished this later for the Vela pulsar [46]. (2) Beating the big bang nucleosynthesis bound on a stochastic background of gravitational waves [47]. (3) Potentially nearby gamma-ray bursts (GRBs) are not caused by compact binary mergers at the putative distance. Two short, hard GRBs appeared to arise in M31 and M81 respectively, two galaxies close enough that gravitational waves from binary neutron stars, a proposed cause of such bursts, would have been easily detected. None were found, indicating either that the source was far behind the galaxies or that the bursts were produced by something else, such as magnetars [48, 49].

5.3.1 Advanced ground-based detectors

In parallel with operating the first generation of detectors, many groups worldwide began to develop the technologies needed to improve detector sensitivity by another order of

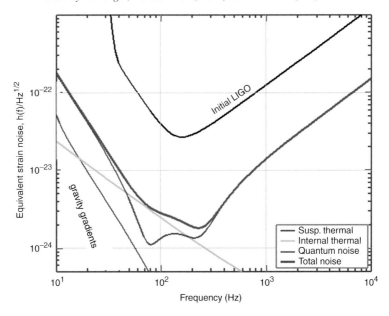

Figure 5.2 The graph shows a typical sensitivity curve for advanced detectors. At low frequencies radiation pressure noise, one of the two quadratures of quantum noise, suspension thermal, and gravity-gradient noise will limit the sensitivity. Coating thermal noise, part of the internal thermal noise, will limit in the mid-frequency range. Shot noise, the second quadrature of quantum noise, will limit at high frequencies. Figure courtesy of the MIT-Caltech LIGO Laboratory.

magnitude and to push towards lower frequencies. These technologies are now used in the second-generation detectors, Advanced LIGO and Advanced Virgo, both well advanced in construction. A typical sensitivity curve for these advanced detectors is shown in Fig. 5.2. The low-frequency range is limited by center-of-mass motion of the mirrors caused by residual accelerations generated by radiation pressure noise, suspension thermal noise, and seismic noise. The mid-frequency range will likely be limited by thermal noise in the optical coatings while the high-frequency range will be limited by sensing noise or shot noise.

At the time of writing, Advanced LIGO and Advanced Virgo were still in their installation phase and had started commissioning some of the subsystems. The experience gained over the 10 years of commissioning and operating the initial detectors should help to accelerate the commissioning phase significantly and the first science run for Advanced LIGO, and possibly for Advanced Virgo, may start as early as late 2015 [50]. The detectors will not have achieved their final sensitivity but are expected to be significantly more sensitive than the initial instruments. It is hoped that before 2018, sufficient sensitivity will have been reached to make the first direct detection of gravitational waves likely. Note that this event will occur approximately 100 years after Einstein's initial prediction of these waves.

An even more ambitious detector is Japan's KAGRA [27]. KAGRA[3] will be the first detector to be built underground in order to minimize environmental noise, anthropogenic noise, and gravitational-gradient noise. KAGRA will have similar sensitivity to Advanced LIGO and Virgo (see Fig. 18 in [38]) and is expected to join them for a science run near the end of this decade. KAGRA plans to employ cryogenics to reduce the temperature of the mirrors and improve its sensitivity. KAGRA can be seen as a generation-2.5 detector to pave the way for potential upgrades of the advanced detectors and for possible third-generation detectors such as the Einstein Telescope [28].

The second generation of ground-based detectors, Advanced LIGO, Advanced Virgo, and KAGRA, will be sensitive in the frequency range between 10 Hz and 10 kHz. The GEO 600 detector has been upgraded to become GEO HF, a high-frequency detector whose goal is to provide additional data at frequencies above several hundred Hz as long as it is competitive with the much longer detectors. While the advanced detectors reuse the site infrastructure and vacuum system of the initial detectors, the internal components are completely upgraded for LIGO and largely upgraded for Virgo. In addition to larger, better mirrors used as test masses, the advanced detectors use significantly increased laser power and, for LIGO, greatly improved seismic isolation and suspension systems. Initial Virgo's goal was to operate at frequencies as low as about 10 Hz with an advanced seismic isolation and suspension system (making it a generation-1.5 instrument). All second-generation detectors will use an additional mirror, the signal recycling mirror, at the output of the Michelson interferometer to extract or coherently enhance the signal inside the interferometer. This technique has already been successfully tested in GEO 600, although not with arm cavities.

Advanced ground-based interferometric gravitational wave detectors are among the most complex scientific instruments ever built. In the following we will describe the core parts of the advanced detectors, discuss the interactions between these parts, outline their fundamental limitations, give examples of how a variety of categories of technical noise sources are suppressed, and finally look into potential future upgrades.

Noise limits of second-generation detectors

Unlike first-generation detectors, which are limited at low frequencies by seismic disturbances, advanced detectors at design sensitivity will be limited by quantum effects over most of their frequency range. These can be understood heuristically as photon-counting uncertainty (shot noise) limiting precision at high frequencies while radiation pressure limits precision at low frequencies. The two are complementary since increasing laser power reduces the influence of shot noise but increases radiation pressure noise. For a time it was not clear if this heuristic argument held up under careful quantum mechanical treatment. Such a treatment, first provided by Caves [51, 52], confirmed this picture. To summarize these more careful arguments, vacuum fluctuations add a random field with an average amplitude of $1/\sqrt{2}$ in natural units ($\hbar\omega = 1$) and an arbitrary phase to the injected

---

[3] KAGRA was formerly called LCGT.

coherent field from the laser. Relevant for us are the vacuum fluctuations that enter the interferometer from the (dark-fringe) output port. They interfere with the incoming laser field and modify the field in each arm.

Constructive interference in one arm requires destructive interference in the other arm and vice versa. As a result, the power in the two arms is only on average identical, while the difference fluctuates with a standard deviation which is proportional to $\sqrt{N}$; a different way to describe this is to assume that the beam splitter randomly transmits or reflects photons with equal probability. This creates an imbalance in the optical power in the two arms which creates an imbalance in the momentum transfer at the test masses or mirrors. The resulting radiation pressure noise is suppressed by the inertia of each test mass with $(4\pi^2 f^2 m)^{-1}$ above the resonance frequency of the pendulum suspension. Radiation pressure noise limits detectors at low frequencies during high-power operations.

Fluctuations in the phase relation between the vacuum fluctuations and the injected field also change the phase of the field in each arm. This differential phase noise is also indistinguishable from a gravitational wave signal and turns into frequency-independent shot noise at the final photodetector. Contrary to the common belief that shot noise decreases with the number of photons, shot noise at the final photodetector increases with $\sqrt{N}$. However, the signal increases with N in an optimized detection scheme and the signal to shot noise ratio increases with $\sqrt{N}$. Shot noise limits all detectors at higher frequencies where all other relevant noise sources have fallen off with some power law. Although shot noise is frequency-independent, it shows up in the sensitivity plot as an inverse transmittance function of an optical cavity which takes into account the response of the interferometer to gravitational waves. Radiation pressure noise and shot noise are two aspects of the same noise, the unified quantum noise, which has its root in the fundamental uncertainties of a coherent field [7, 52].

In the past decade or so, it has become possible to implement vacuum squeezing to reduce shot noise below the standard quantum limit. The first requirement was to learn how to apply squeezing at the frequencies appropriate to gravitational wave detector operation. The breakthroughs were made by [53, 54] and first implemented on GEO 600 [55]. Shortly thereafter it was tested on the LIGO instrument at Hanford [56]. In both cases, it was found possible to apply squeezing at shot-noise-limited frequencies without worsening the performance of the instruments at lower frequencies although neither detector was limited by radiation pressure noise.

Thermal noise can often be a limiting fundamental noise source over some part of a detector's frequency range [7, 57]. The suspension wires, mirror substrates, and mirror coatings are made from atoms which have a finite temperature. The thermal motion of these atoms excites the mechanical eigenmodes of the macroscopic bodies they form. The power spectral density, $S_x(\omega)$, of this excitation can be derived from the fluctuation–dissipation theorem and depends on the susceptibility $\chi(\omega)$ of the mechanical system:

$$S_x(\omega) = \frac{2k_B T}{\omega} \Im\{\chi(\omega)\}.$$

Its imaginary part is proportional to the mechanical loss or dissipation rate of the eigenmode. Materials with a low mechanical loss angle or, equivalently, with a high mechanical Q, will reduce the thermal noise at frequencies different from the resonance frequencies. In general, glasses like fused silica and crystals like sapphire have a higher mechanical Q than metals. This is why the steel wires in the last suspension stages of first-generation detectors have been replaced by fused-silica fibers in the advanced detectors. These are welded to the mirrors and not clamped to minimize the friction at the points of attachment. Suspension thermal noise is not expected to limit the advanced detectors except perhaps during low-power operation, when it could become comparable to radiation pressure noise.

Thermal noise within the mirror substrate can also be a factor through excitation of the material's eigenmodes. These internal modes change the location of the reflecting surface of the mirror while its center of mass is not affected. The mechanical Q of fused silica is high enough that substrate thermal noise will not limit Advanced LIGO or Virgo. However, the dielectric coatings that are required to tailor the reflectivities of the mirror surfaces produce their own thermal noise. These coatings of alternating layers of silicon dioxide and tantalum pentoxide, providing layers of high and low index of refraction, are created by ion beam sputtering. The mechanical Q of these layers is poor compared to the substrate and they act like a thin soft surface from which the light reflects. It is expected that thermal coating noise will limit the advanced detectors in the intermediate frequency range around 100 Hz. Active research programs to develop significantly improved optical coatings have been underway for some time [57].

The facilities themselves have a limit which is set by the residual gas in the vacuum tubes. Fluctuations in the gas density change the index of refraction between the cavity mirrors. Allowed partial pressures range from 10^{-9} mbar for H_2, over 10^{-10} mbar for N_2, to the 10^{-12}–10^{-13} mbar range for more complex hydrocarbons. It is also suspected that hydrocarbons might damage the dielectric coatings when they are exposed to high laser power. Gas molecules also hit and exchange momentum, stochastically, with the mirrors. The resulting Brownian motion of the mirrors is amplified in the Advanced LIGO and Virgo configurations by the small gap between the reaction mass and test mass, which is much smaller than the molecular mean free path but also much smaller than the test mass dimensions (see Figures 5.5 and 5.6). Thus, the same molecule hits the mirror multiple times before it finds its way out of the gap. This leads to correlations in the momentum exchange with the mirror, effectively increasing the pressure fluctuations from this side. The problem is amplified by the fact that many of the installed components are near this gap. Details are discussed in [58]. Rigorous vacuum rules are followed to minimize the outgassing in the central station and the end stations.

The interferometer is operated near a point where it is insensitive in first order to fluctuations in the laser field. However, second-order effects have to be minimized rigorously to reach design sensitivity. One such effect is the coupling of laser beam jitter via a misaligned interferometer into a gravitational wave signal. This coupling puts very stringent requirements on the alignment sensing and control system and on the passive angular stability of all mirrors. A second effect is technical radiation pressure noise. Any imbalance in the optical power stored inside the arms will cause a difference in the radiation pressure on the mirrors. It is assumed that the advanced detectors can balance the power to better than 1% but that would still require that the relative power fluctuations of the laser itself be below $2 \times 10^{-9}/\sqrt{\mathrm{Hz}}$ at the low end of the detection band. In other words, the laser has to be shot-noise-limited for about 300 mW of detected power at 10 Hz. The Advanced LIGO and Advanced Virgo lasers are the only lasers which have ever reached this stability [59].

The core parts of advanced ground-based detectors

The advanced detectors follow similar design philosophies and each core part in Advanced LIGO has its counterpart in Virgo. However, the parts differ in many subtle ways and are sequenced differently in Virgo. In the following, we will focus on the Advanced LIGO design and will only mention the most important differences for Virgo. Figure 5.3 shows the front end of Advanced LIGO. The pre-stabilized laser system (PSL) produces up to

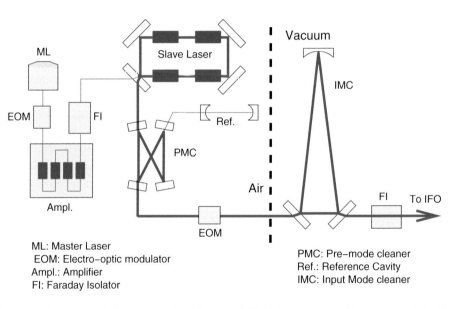

Figure 5.3 Schematic of the Advanced LIGO pre-stabilized laser (PSL) and input optics (IO). The PSL consists of a 2 W master laser, an electro-optical modulator, a 30 W amplifier stage, a high-power Faraday isolator, the 200 W slave laser, and a pre-mode cleaner. The pre-mode cleaner is an optical ring cavity that filters the spatial mode of the laser beam before it is handed over to the input optics subsystem. All lasers are diode pumped solid state lasers which emit at 1064 nm. The PSL also contains a reference cavity to pre-stabilize the laser frequency and a diagnostic breadboard (not shown) to monitor its health internally [60]. The beam is then turned over to the IO, where an EOM provides sidebands to the laser field. The field is then mode-matched into the input mode cleaner and passed through a thermally compensated high-power Faraday isolator before it is injected into the main interferometer. All mode matching, steering, and auxiliary optics have been omitted for clarity.

200 W (20 times the power of the initial LIGO laser) of the most stable single-mode, single-frequency laser light currently available on the planet. It is operated in a climate-controlled, stable environment outside the vacuum system. Once integrated into their respective interferometers, feedback from interferometer signals is used for further stabilization of their amplitudes and frequencies to improve performance by additional orders of magnitude. These laser beams are as close to ideal as is currently possible.

The primary function of the input optics is to condition the beam before it is injected into the main interferometer [61]. One electro-optical modulator (EOM) is located before the vacuum system and is used to modulate the phase of the laser field with multiple frequencies between a few MHz and up to 100 MHz. The generated sidebands are necessary to sense and control the relative positions and orientations of all mirrors that make up the main interferometer. The modulated beam then enters the vacuum system, where a suspended input mode cleaner (IMC) of 30 to 150 meters in length acts as a spatial filter to remove all unwanted higher-order spatial modes of the laser field. The cavity also acts as an intermediate frequency reference to which the laser system is locked. The optical isolation of the laser from the main interferometer is provided by a Faraday isolator (FI). The EOMs and FIs were developed exclusively for the advanced detectors because no commercial devices were able to handle the 200 W of continuous laser power without degrading either the beam profile or the polarization of the field [62, 63].

Figure 5.4 shows the main interferometer with the output mode cleaner and the final photodetector. The core parts in the main interferometer are the beam splitter, the input test masses (ITMs) near the beam splitter and the end test masses (ETMs) at the far ends of the vacuum tubes. Within each arm, the ITMs and ETMs form the long optical cavities. The test masses themselves are typically 40 kg, 30 cm-diameter cylinders of ultra-pure fused silica. They are suspended by multi-stage pendulum systems, which will be discussed in the next section. Dielectric coatings on these mirrors provide the design transmissivity of between 1% and 2% for the ITMs and around one ppm for the ETMs. The beam splitter splits the beam equally at the 45° angle of incidence. The beam splitter substrates are typically thinner than the ITMs or ETMs in order to reduce the optical path in the substrate to avoid distortion of the beam at higher laser powers. They also have a larger diameter to reduce diffraction losses.

The power recycling (PR) and signal recycling (SR) mirrors complete the optical design. As its name suggests, the power recycling mirror returns "waste" light back to the interferometer to enhance the power of the injected field by a factor of 30 to 50. The signal recycling mirror creates an optical cavity to resonantly amplify the signal [65]. The initial LIGO and Virgo detectors implemented only power recycling using a single mirror between the beam splitter and the input optics. Signal recycling without arm cavities has been used in GEO 600 from its very start, and will be employed with arm cavities in the advanced detectors. Advanced LIGO moved the beam-expanding telescope from the input optics and the beam-reducing telescope from the output optics where they had been in initial LIGO and Virgo into the recycling cavities to better contain the spatial mode of the laser fields inside the recycling cavities [66].

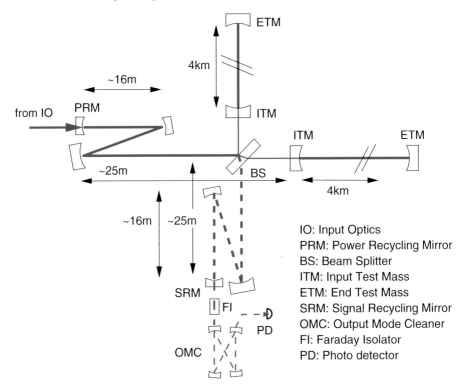

Figure 5.4 Schematic of the Advanced LIGO main interferometer and output optics. The light enters from the left through the power recycling mirror. The following two mirrors expand the beam from about 2 mm to 6 cm Gaussian beam radius before the beam splitter sends the beam into the two 4 km-long arm cavities. The output port is dominated by the signal recycling cavity, which consists of a beam-reducing telescope and the signal recycling mirror. The Faraday isolator separates the output mode cleaner from the main interferometer. The photodetector monitors the power of the output laser field, which is directly proportional to any differential length change in the interferometer. See [64] for more details.

The output optics system conditions the signal field before its detection by the final photodetector (PD). In the initial observatories, a simple beam-reducing telescope focused the beam onto a fast photodetector. The photo-current was then demodulated at one of the sideband phase modulation frequencies to generate the strain signal. This RF-sensing scheme has been replaced by a DC-sensing scheme to improve the signal to shot noise ratio. Figure 5.4 shows a DC-sensing scheme. The signal is first passed through a Faraday isolator (FI). The output mode cleaner (OMC), a 20 to 30 cm-long compact optical cavity, removes all unwanted light such as the RF-sidebands and the higher-order spatial modes, which do not contribute to the signal but increase the shot noise. The PD detects the changes in the output power. These changes are directly proportional to differential length changes in the arms of the two long cavities in the IFO. For additional discussion of past and current sensing schemes see [67, 68].

Length and alignment control of optical components

However, the design of the optical layout is only part of the story. The main difficulty in the operation of sensitive gravitational wave detectors is that all signals depend on the position of every mirror so that the trick is to find a scheme that minimizes the couplings between the different degrees of freedom. The optical subsystems are connected to each other not only by the laser beam but also by the nested feedback loops needed to control the entire instrument. Both cavity length and mirror alignment must be sensed and controlled with extreme precision.

The task of the length sensing and control system is to keep all test masses and the beam splitter within a few femtometers and the recycling cavity mirrors within a few picometers of their nominal positions. This requires control of five longitudinal degrees of freedom: the two arm cavity lengths, the short Michelson interferometer formed between the input test masses and the beam splitter, and the power and the signal recycling cavity lengths. All these degrees of freedom are sensed using a complex phase modulation and demodulation scheme [69] based on [70, 71] and references therein.

Any misaligned mirror makes the interferometer more sensitive to transverse laser-beam motions or beam jitter. This jitter can be described as the excitation of a first-order Hermite–Gauss mode that propagates with the fundamental mode with frequency offset by the jitter frequency. A misaligned mirror in an interferometer scatters light from the first-order Hermite–Gauss mode into the fundamental mode which creates a signal that is indistinguishable from the one generated by a gravitational wave [72]. This is an example of a second-order noise source where one in-band noise source, here the in-band beam jitter, scales with a static or root-mean-square (rms) imperfection, here the rms or static misalignment inside the main interferometer. The requirements on beam jitter and static or rms misalignment can be traded against each other to some degree but are difficult to meet in all cases. Typical requirements for the static or rms misalignment of the test masses are at the femtoradian level while the recycling mirrors have to be aligned at the picoradian level. The alignment is monitored with a technique called wavefront sensing. It uses the reverse scatter process where a misaligned mirror scatters light from the fundamental mode into the first-order Hermite–Gauss mode. The amplitude of this field can be sensed against a reference field using a split photodetector. Again, the challenge is to find signals that allow disentanglement of all misalignments so that each mirror may be individually controlled [73]. The development and debugging of the length and alignment sensing and control system forms a major part of the commissioning effort needed to bring the interferometers to their best sensitivity.

Another part of this effort addresses the mode matching between the different parts of the interferometer. All current detectors plan to operate with the fundamental or TEM_{00} Gauss mode throughout the interferometer. This requires that the beam sizes and radii of curvature of the spatial eigenmodes on the mirror surfaces between the different optical cavities be matched to each other. Any deviations in the radii of curvature of illuminated mirrors or in the distances between these mirrors or any thermal or other lensing produced within the interferometer will lead to mismatches between the spatial eigenmodes. These mismatches reduce the power in the arm cavities, decrease the efficiency of the signal extraction, and increase stray light and the amount of light that contributes only to shot noise and not to the signal. Very tight tolerances are imposed on the radii of curvature of all mirrors, on the surface error, and on the absorption to ensure proper mode matching. For example, the advanced LIGO end test masses have to have a radius of curvature of $2250\,m \pm 10\,m$. The uncertainty corresponds to an uncertainty in the sagitta (depth of the lens surface) over the 12 cm beam profile of 4 nm. The surface error in the central 160 mm has to be below 0.4 nm over all spatial frequencies [64].

The power inside the arm cavities in the advanced detectors might reach 800 kW during operation. Even the sub-ppm absorption in the dielectric coatings and sub-ppm/cm absorption in the substrates will be large enough to deform the mirror surfaces, to create a thermal lens inside the substrates and to degrade the spatial eigenmode of the interferometer by mixing the fundamental mode with the first-order Laguerre–Gauss mode [74, 75]. Several techniques to characterize the mode mismatch have been developed and are used in advanced detectors, including wavefront sensors [76], Hartmann detectors [77], and phase cameras [78]. Once the thermal distortions have been measured, they will be compensated by *fighting fire with fire*. Ring heaters are applied to the barrels of the most exposed substrates, namely the test masses, to reduce the substrate thermal lens and control the radii of curvature on the high-reflectivity (HR) side of the test masses while a CO_2 laser is used to create a negative lens in a compensation plate next to each ITM, the most crucial of the optics, to compensate any remaining substrate thermal lens inside the ITM [79].

The final but important challenge we will address is stray light. Between 10^{20} and 10^{21} photons will be sent into the interferometer each second. Any surface imperfection in the optics will cause some of these photons to scatter out of the main laser beam. Some of these will be phase modulated when they reflect off a slightly shaking surface like the walls of the vacuum tubes or the frames of the suspension systems. Note that *shaking* in this context describes vibrations at the nanometer or even picometer level in the detection band. Some of those photons will find their way back into the main interferometer beam. A single photon frequency-shifted into the shot-noise-limited frequency region will in principle be detectable and can limit the sensitivity. All major components inside the interferometer and most of the input optic components have near-Ångström surface roughness on all length scales between the optical wavelength and the beam size to minimize the scattered light. In addition, a sophisticated baffle system has been installed to catch and absorb as much of the residual scattered light as possible before it reenters the interferometer.

All these techniques will allow advanced detectors to measure length changes as small as $10^{-20}\,m/\sqrt{Hz}$. However, the ability to sense displacements that small is necessary but not sufficient to measure gravitational waves. It is also necessary that the objects between which we measure, the test-mass mirrors, follow their geodesics as rigorously as possible. It is the task of the seismic isolation and suspension systems to ensure this.

Seismic isolation and suspension systems

One of the most challenging problems of ground-based detectors is to isolate the test masses from any ground motion. For science operations, all optical cavities and the short Michelson interferometer must have the proper length to allow the laser field to build up in and only in the right places. This requires the distances between all mirrors to be controlled to a small fraction of the optical wavelength. Typical requirements for the relative rms motions of the mirrors are 10^{-15} m in the control band below 10 Hz and 10^{-19}–10^{-20} m/$\sqrt{\text{Hz}}$ in the detection band above 10 Hz.

Seismic noise is dominated by the microseismic peak at 0.14 Hz caused primarily by ocean waves [80] and then falls off with f^{-2} to f^{-3} above that frequency. At the locations of the LIGO and Virgo detectors, the amplitude of the seismic noise is typically ten orders of magnitude too high at all frequencies below about 20 Hz. Local anthropogenic noise from vehicular traffic or even people walking near the instruments can be reduced by controlling access to the sites and minimizing the use of heavy equipment during science runs. On the other hand, off-site traffic on nearby highways or trains cannot be controlled. The building foundations on which the vacuum tanks rest are isolated from the rest of the building foundations to decouple them from wind, but this isolation is far from perfect. The role of the seismic isolation and suspension systems is to isolate the mirrors from all these spurious accelerations while at the same time providing a means to actively control the position and orientation of each mirror.

Advanced Virgo and Advanced LIGO differ in their suspension systems. It is worthwhile to examine both these approaches in detail. The Virgo group started to develop their superattenuator shown in Fig. 5.5 in the 1990s. The superattenuator was used for initial Virgo and measurements of its transfer function show that it provides enough isolation to be used for Advanced Virgo [81]. The superattenuator is attached to the ground via three 6 m-long aluminum legs. These rest in steel rods that provide an elastic restoring force to the legs. The upper ends of the aluminium legs are connected via a top ring that supports the suspension system. This design acts as an inverted pendulum in which the gravitational force works against the elastic restoring force to lower the resonance frequency of the structure to about 30 mHz [82].

The suspension system consists of a series of pendulums where each stage provides an attenuation of f_{res}^2/f^2 above its resonance frequency. The entire system has a height of 8 m such that the resonance frequency of each stage is below 2 Hz. The intermediate masses are seismic filters which also provide vertical isolation in the form of blade springs from which each following mass is suspended with steel wires. The vertical isolation is necessary because the surface of the mirror will not be perfectly parallel to the gravitational force for various reasons, including the curvature of the earth. This causes local vertical motion of the mirror to look like an apparent displacement of the mirror surface and could limit the sensitivity of the detector.

A marionette is suspended from the last seismic filter, for historical reasons called *Filter 7*. The rotational degrees of freedom of this marionette are controlled with coil/magnet actuators. In initial Virgo, the test mass and a reaction mass were both suspended from the marionette using metal wires. Four magnets were glued on the test mass and inserted into corresponding coils on the reaction mass to actuate on the test mass (see the OSEM discussion below) [83]. For Advanced Virgo, the steel wires will be replaced by fused-silica wires to reduce the thermal suspension noise. This system of

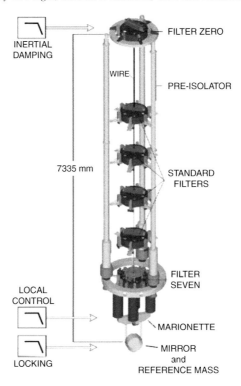

Figure 5.5 The Virgo superattenuator is a multi-stage suspension system which provides at least fourteen orders of magnitude of isolation at all frequencies above 10 Hz. From SAFE "Super Attenuator Facility at EGO," Proposal, VIR-047A-07, 2007, http://origin-ars.els-cdn.com/content/image/1-s2.0-S0927650503002603-gr2.jpg and Science Direct. Courtesy of the Virgo Collaboration.

pendulums and blade springs suppresses the horizontal motion of the test mass above about 10 Hz by fourteen orders of magnitude, sufficient to reach the Advanced Virgo design sensitivity.

Advanced LIGO uses a different design to isolate their test masses from ground motion. The scientific payload in each vacuum chamber is mounted to an intra-vacuum seismic isolator (ISI) platform. Each of these platforms is supported by four hydraulic external pre-isolators (HEPI). Each HEPI uses a continuous laminar flow of hydraulic fluid to apply differential pressure on an actuator plate that controls the height of a corner of the platform. The differential pressure can be adjusted with valves that are arranged in a Wheatstone-bridge-like assembly to reduce the sensitivity to pressure fluctuations [64, 85]. Two different types of ISI platforms exist. Less critical optical components like the recycling mirrors and the input and output optics components are placed on Horizontal Access Module (HAM)-ISI tables. The four test masses and the beamsplitter are suspended from ISIs of the Beam Splitter Chambers (BSCs), which are suspended by HEPIs about 8 m above ground.

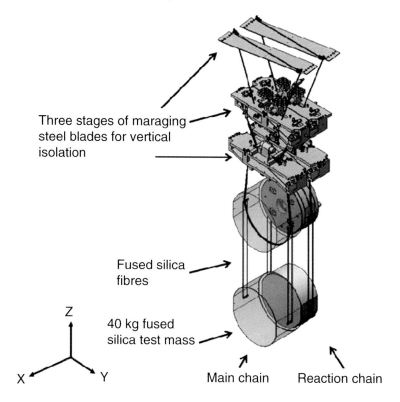

Three stages of maraging steel blades for vertical isolation

Fused silica fibres

Z

40 kg fused silica test mass

X   Y

Main chain   Reaction chain

Figure 5.6 The Advanced LIGO suspension system featuring a main and a reaction chain with four pendulums for horizontal isolation. The top mass and the first two suspended stages include blade springs for vertical isolation [84]. © IOP Publishing. Reproduced with permission. All rights reserved.

Each ISI platform [86, 87] consists of a base and two stages, which are suspended in series via blades and flexure rods in all degrees of freedom, to isolate the last stage from the base motion. Coarse and fine electromagnetic actuators are used between the base and stage one and between stage one and stage two, respectively, to position and align the last stage. The ISI also uses capacitive sensors and seismometers to measure and suppress the residual motion by feeding back to these actuators.

An Advanced LIGO suspension is shown in Fig. 5.6. It uses two quadruple suspension chains for the test mass and the reaction mass [64, 84]. The top blades are mounted to the ISI platform (not shown). The second and third stages consist of a mechanical system which includes additional blade springs. The penultimate masses in both chains are supported by steel wires which hang from the tips of the blade springs. The penultimate mass in the reaction chain is made from aluminum and holds four optical sensors and electro-magnetic actuators (OSEM). These OSEMs consist of a coil, an LED, and a photo-receiver. The penultimate mass in the main chain is made from fused silica. Four magnets are glued onto its surface. Each magnet fits into an opening inside one of the OSEMs and its shadow blocks part of the light between the LED and the photo-receiver. This shadow sensor allows measurement of the relative motion of the two masses. The coil allows generation of a magnetic field that pushes or pulls on the magnet and allows alignment of the penultimate mass and damping of all residual motions. Additional OSEMs have been added to the upper two stages, providing additional opportunities to control the position and orientation within the suspension itself. These OSEMs are used for low-frequency actuation and to damp all motions on the lower resonance frequencies of the mechanical springs and pendulums.

The 40 kg, 30 cm-diameter fused-silica test mass at the bottom of the main chain is welded to the penultimate mass via a fused-silica fiber. The high mechanical Q of this suspension reduces thermal noise in the measurement band. The reaction mass is suspended by steel wires from the penultimate mass in the reaction chain; the requirements for the reaction chain are relaxed compared to the main chain. The reaction masses at the ETMs are partly gold-coated. These gold coatings form electrodes that are used for electrostatic actuation of the test mass itself while all other actuations are applied at upper stages and are filtered by the suspension system.

Ground-based gravitational wave detectors are very complex instruments with highly interconnected subsystems. The number of degrees of freedom that have to be controlled, the precision required to optimize the detectors, and the number of technical noise sources that have to be minimized are mind boggling. These challenges guide the design, manufacturing, and assembly of the detectors and then dominate the commissioning process. These instruments are technical marvels that combine the best in laser technology, optical technology, quantum optics, mechanics, vacuum technology and low-noise electronics.

5.3.2 The future of ground-based gravitational wave detectors

The upcoming science runs of the advanced detectors will last from months to years and will be interrupted by periods of time devoted to improvements of the instruments' performance. The initial interruptions are likely to focus on the reduction of technical noise sources like beam jitter and laser intensity noise or the elimination of scattered light, all of which must be addressed in order to reach the fundamental noise sources, quantum noise and thermal coating noise, at all frequencies. Later interruptions will be dedicated to upgrades intended to lower the fundamental noise sources. One of the most obvious upgrades of the advanced detector is the injection of squeezed light through the dark port. Squeezing uses a non-linear process to correlate the vacuum fluctuations at $\pm f$ around the carrier frequency of the main laser field. These correlations can be used to lower the phase fluctuations and shot noise at the expense of increased amplitude fluctuations and radiation pressure or lower the radiation pressure noise at the expense of increased shot noise [7, 26]. GEO 600 has paved the way for squeezing by routinely operating with phase-squeezed light since 2011 [55, 88], while a test run at the LIGO Hanford site has shown that phase squeezing is also compatible with arm cavities [56]. Both experiments lowered the shot noise by 2 to 3 dB and improved the high-frequency performance. None of the detectors was limited by radiation pressure noise at that time and the injection of squeezed light did not affect the low-frequency performance. Advanced detectors will be limited by radiation pressure noise, and any improvement at low frequencies requires amplitude squeezing instead of phase squeezing, and this would then increase the shot noise.

Ideally, the injected squeezed field in advanced detectors should be amplitude squeezed at low frequencies and phase squeezed at high frequencies. This can be achieved with a filter cavity that changes the phase of the squeezing angle as a function of frequency and thus allows, in principle, tailoring the squeezing angle to lower the quantum noise at all frequencies [89]. The installation of such a filter cavity in the output optics is one of the most likely future upgrades of the advanced detectors. The improvements in sensitivity will be limited by losses in the squeezing system. These losses allow unsqueezed vacuum to enter the interferometer and degrade the correlation between the two frequency components. The losses include absorption in the non-linear crystal, losses in the required Faraday isolators, mode-matching losses between the squeezed field, the filter cavity, the main interferometer and the output mode cleaner, losses inside all cavities and the imperfect quantum efficiency of the photodetector. All these losses can easily add up to 10% of the injected field. This would reduce a 20 dB squeezing amplitude to about 10 dB. Even with these losses, the range would improve by about a factor of 3 at all frequencies limited by quantum noise, or by a factor of 27(!) in the volume of the universe that the detectors can observe for a given astrophysical source. However, squeezing would not help against thermal noise or other noise sources.

The fundamental processes behind mechanical losses in bulk materials are not well understood and are even less understood in sputtered optical coatings and fibers. The current designs and approaches to lower thermal noise are to a large degree based on empirical data. It is known that crystals like sapphire have a higher Q than some amorphous glasses like fused silica, which still outperform metals. Other material parameters such as optical losses, thermo-optical coefficients, thermal conductivity, and thermal expansion coefficients also impact the selection and design of the optics and last suspension stage. This has led to the current choices of fused-silica mirrors with silica and tantala coatings suspended by fused-silica fibers. However, several ideas have been proposed to reduce thermal noise beyond the current levels. The brute-force method is to cool the test masses. The reduction in thermal energy reduces the noise with $\sqrt{T}$, but a larger gain might come from the temperature dependence of many material parameters, including the mechanical loss angle [90]. However, cooling a coated mirror into the range below 100 K will increase the stress in the optical coatings or require better matching of the thermal expansion coefficients of the coatings to the substrates. This further restricts the choice of coatings. Another issue is the cooling process itself. The high laser power that reflects from the cold mirror surface will heat up the central part. This heat has to be removed efficiently without bypassing the suspension system. These are just a few of the challenges associated with cryogenic detectors.

A second approach is to focus on new coating materials. Crystalline coatings have generated very promising initial results in small optical cavities, but it is not obvious if these coatings can meet all requirements and can be produced on a sufficiently large scale to be suitable for large optics in advanced gravitational wave detectors. KAGRA plans to install cryogenic suspensions near the end of this decade and has already tested a cryogenic prototype [91]. Advanced LIGO and Virgo have as yet no plans in this direction.

A few things (nearly) always help to improve sensitivity, such as increasing mass and length. Longer and taller suspension systems have lower resonance frequencies to allow measurement at lower frequencies. Larger masses generally move less when they are excited by external non-gravitational forces and designs with test masses of up to a few hundred kilograms have been proposed. Longer arms will improve the strain sensitivity directly as δL increases but also indirectly as they allow support for wider optical beams to better average over coating thermal noise. The Einstein Telescope (ET) is an ambitious concept to determine if a ground-based telescope could improve upon the second-generation advanced detectors by a further order of magnitude, including extending to lower frequency. A reference design has been developed [28] that calls for 10 km-long arm cavities in a triangular configuration. Note that there is an arm-length beyond which the radius of curvature of the earth will pitch the mirrors with respect to each other, making the non-free-fall vertical motion of the mirrors turn into length changes. The goal of ET and other third-generation designs is to improve the sensitivity in the current detection band and to push to lower frequencies where more mergers of massive black holes are expected to emit. At these lower frequencies, detectors encounter limits caused by changes in local mass distributions that change the local gravitational field. This gravity-gradient or Newtonian noise is caused by seismic waves which lift or lower many hundred tons of rock, soil, sand or water, by changes in the air pressure, by anthropogenic noise caused by moving vehicles and by many other effects that are impossible to eliminate since they are genuinely gravitational. KAGRA will be built into a tunnel system at the Kamioka mine and will be the first underground detector [27]. Tunnels have the advantage that access can be controlled and anthropogenic noise can be minimized. Underground detectors also have the advantage that the highly incoherent surface seismic waves disappear. The longer coherence length of the seismic waves in a homogeneous environment might allow measurement of the seismic spectrum and the use of adaptive filters to subtract it from the gravitational wave signal to allow detection of gravitational waves at frequencies as low as 1 Hz.

This does not mean that the current sites will be abandoned anytime soon. They might not be ideal for low-frequency detectors approaching 1 Hz. These would operate with modest laser powers, massive mirrors, and still-to-be-developed seismic isolation and suspension systems, and might need to be longer and underground. However, the existing infrastructure would still allow the operation of high-frequency detectors that use very high laser powers, less massive test masses, and, possibly, the current seismic isolation and suspension systems. Future ground-based laser-interferometric gravitational wave detectors could be optimized for specific frequency bands and would not necessarily be co-located, thus emulating optical telescopes that also cover only a small fraction of the electromagnetic spectrum and operate mostly independently of each other. While it is possible that these future-generation instruments will look completely different from the currently envisioned instruments, the first direct detection of gravitational waves with the advanced detectors will create the needed impetus for their development by beginning a new era in astronomy and astrophysics.

5.4 Space-based gravitational wave detectors

5.4.1 Introduction

As is true for electromagnetic signals, not all gravitational wavelengths can be detected from the ground. Gravity-gradient noise imposes a fundamental limit (at least for current or envisioned technologies) below about 1 Hz. However, binary black hole inspirals including black holes of up to 10^8 solar masses and compact binary white dwarfs or neutron stars far from merger are expected to generate gravitational waves in the 10^{-5}–0.1 Hz band. To explore the gravitational wave sky at these frequencies requires sensitive gravitational wave detectors in the benign environment of space.

While several space-based gravitational wave detection concepts have been proposed over the years, this section will focus on LISA as the prototype for the class of designs most likely to be launched first. Decades of research and development went into the LISA mission concept, and Europe has dedicated several hundred million Euros to testing most of the LISA technology in the upcoming LISA Pathfinder (LPF) mission [30]. Bender and Faller developed the initial concept which eventually became LISA. The earliest history is described in [92]. Eventually (see [93] for the history up to 1998), in 1998, LISA became a joint project of ESA and NASA. By 2011, most of the technical issues had been solved either by demonstrated performances in tabletop experiments (see also below), with a detailed project design [22], or will be demonstrated in LPF. Unfortunately, at that point, NASA and ESA ended the LISA project because of NASA's other financial commitments, which prohibited a timely implementation of LISA. The European eLISA consortium formed and developed the two-arm e(volved)LISA to fit into an ESA-only mission budget. In late 2013, ESA selected the gravitational universe as the science theme for its L3 mission which is scheduled to launch in 2034 [23]. eLISA is the leading candidate to address this science theme. In the meantime, NASA has studied several mission concepts and favors SGO-mid, an eLISA-like mission with three arms [94, 95]. The sources and science goals for LISA-like missions like eLISA and SGO-mid are described in several reviews [38, 96] and reports [22, 23] and in the following Chapter.

LISA's design calls for 5 million kilometer arm length between the three spacecraft that form its triangular configuration; eLISA and SGO-mid reduced the arm length to 1 million kilometers. This configuration will be placed into an earth-trailing or -leading heliocentric orbit. Each spacecraft contains one or two small test masses (TM) in a drag-free configuration. The proposed implementation of interferometry over million-kilometer distances and designing test masses which respond only to gravity will be described in the following sections. A typical sensitivity plot (here for eLISA) is shown in Fig. 5.7. The low-frequency range is limited by acceleration noise while sensing noise provides the limit

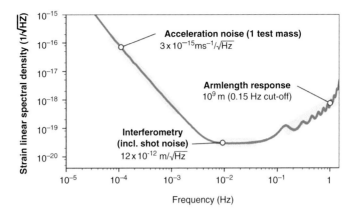

Figure 5.7 As for ground-based detectors, the sensitivity of space-based detectors is often expressed as a linear spectral density averaged over all sky positions and polarization angles. The eLISA sensitivity curve is typical for all LISA-like missions. Credit: eLISA White Paper.

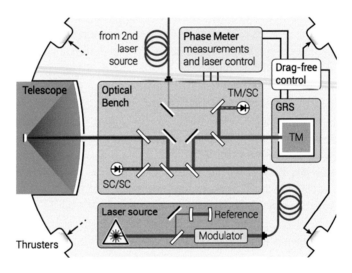

Figure 5.8 A simplified drawing of a scientific payload unit of a LISA-like mission. TM: Test mass, GRS: Gravitational reference sensor. Credit: eLISA White Paper.

above a few mHz. At higher frequencies, the dimensions of the interferometer exceed the wavelengths of the gravitational waves, with a corresponding reduction in sensitivity.

The scientific payloads of LISA-like missions are made up of individual units [97]. A simplified drawing is shown in Fig. 5.8. Each unit contains a free-falling test mass inside a gravitational reference sensor. Each unit also includes a laser system that emits a few watts of laser power in the near infrared and an optical bench to condition the light and form optical beat notes between the local laser and lasers from other units. The phase evolution of each of these beat signals is measured with a phase meter. Last but not least, each unit uses a telescope of 20 to 40 cm in diameter to send and receive the laser beams that are exchanged between the spacecraft. A two-arm mission such as eLISA would require four of these units on three different spacecraft, while a three-arm mission like LISA or SGO-mid would use six units, two for each spacecraft.

5.4.2 Interferometry over millions of kilometers

The interferometric displacement sensitivity of about $10\,\mathrm{pm}/\sqrt{\mathrm{Hz}}$ at a mHz pales compared to the sensitivity of ground-based detectors which is about nine orders of magnitude better in their sweet spot around one hundred Hz. However, while ground-based interferometers measure the displacement between essentially fixed test masses, space-based detectors measure displacements in a highly dynamic interferometer over millions of kilometers. The LISA configuration is not a rigid formation, rather each spacecraft and proof mass is subject to the local gravitational field in its own heliocentric orbit. The different orbits cause variations in the arm lengths on the order of 1% of the total length, relative velocities of up to $\pm 10\,\mathrm{m/s}$, and angular changes in the triangular formation of around 1 degree during the first few years of the mission. These variations grow over time due to the earth's gravitational potential and limit the lifetime of LISA-like missions as stationkeeping is not part of these concepts. The original LISA orbits, $20°$ trailing or leading the earth, allow a mission lifetime of around 8 years, while the drift-away orbits of eLISA limit the lifetime to perhaps five years; the final limit depends on the fine tuning of the orbits and the ability of the interferometry to cope with the large arm-length differences, larger angles, and also larger differential velocities between the three spacecraft.

The interferometric determination of differential changes in the separation of distant test masses in the interferometer arms must be constructed from a combination of signals from three different types of interferometers: (1) The *local interferometer* measures displacements within each unit. More precisely, it measures the distance between the proof mass and a fiducial on the optical bench. (2) The *back-link* monitors the relative phase noise between the two lasers of each local unit. In most designs, the two lasers will be phase-locked to each other. Note that a two-arm mission has only one back-link in the central spacecraft, while a three-arm mission requires a back-link in each spacecraft. (3) The *long-arm interferometer* measures changes in the distance between the fiducials on the optical benches located on opposing spacecraft. It monitors the phase evolution of laser beat signals (SC/SC in Fig. 5.8) on both ends of each arm. The laser on the far spacecraft acts as a transponder while the phase measurement on the far spacecraft measures the noise of the transponder.

The signals of each long-arm interferometer can be combined with the signals from two local interferometers to eliminate the transponder noise and the motion of the fiducials due to noise in the spacecraft position control. This combination is a measure of the changes in the distance between the two proof masses in units of the wavelength of the first laser field. The identical linear combination from the second unit measures the second arm in units of the wavelength of the second laser. The sum of the two signals measures the changes in the average arm length in units of the average wavelength. (This signal is used in ground-based detectors to stabilize the laser frequency to the common arm length, and a modified version can be used in space as well [98, 99].) The difference of the two signals measures changes in the differential arm length in units of the average wavelength. This is the signal we are after. However, it is compromised by changes in the wavelength of the two lasers or simply by laser frequency noise. The back-link signal can be used to eliminate the differential laser frequency noise. The one remaining issue is the typical contribution of the common laser frequency noise in a non-equal-arm Michelson interferometer which scales with the arm-length difference.

It was realized by Tinto *et al.* [100] that the continuous measurement of the phase evolution of the beat signals at various locations within the constellation allows formation of linear combinations between time-shifted phase measurements to create an artificial equal-arm Michelson interferometer. This time-delay interferometry (TDI) can then be used in post-processing to clean up the data streams and eliminate laser frequency noise [101–103].

Several instrumental noise sources limit the sensitivity. The fundamental limit for the interferometry is shot noise. The received power from the far spacecraft will be drastically limited by beam divergence and collecting area from the telescopes with apertures sized to fit aboard a reasonable launch vehicle. This power is expected to be below 1 nW. There are additional technical noise sources. Among the most prominent are phasemeter noise due to timing jitter between the clock and the data acquisition system, clock noise from relative timing jitter among the clocks on the three spacecraft, optical path length changes due to temperature changes on each optical bench, and scattered light which can cause phase noise. Each of these noise sources has generated requirements on the system components to keep them at a tolerable level. See [22] for the most recent list of requirements for LISA, [104] for the current status of the research to meet them, and [98, 103, 105, 106] and references therein for some specific results related to the interferometry.

5.4.3 Achieving free-fall at a level below femto-g

eLISA, or any LISA-like mission, will require that the reference test masses be in free-fall to within a residual spurious acceleration noise of

$$S_a^{1/2} < 3 \times 10^{-15} \, \text{m/s}^2/\text{Hz}^{1/2} \tag{5.4}$$

at frequencies below several mHz. This value is similar to that needed for the advanced terrestrial gravitational wave detectors, but in a frequency band four to six decades lower. This challenging goal dictates a series of features in the design of the system that provides the geodesic reference test bodies.

Space clearly allows a much longer baseline than is possible on earth and eliminates gravity-gradient noise. Spacecraft, such as, most recently, CASSINI [107], have been used as geodesic reference particles for important general-relativity tests. However, they are not "geodesic enough" for eLISA, as the fluctuating force from solar radiation pressure will produce acceleration noise several orders of magnitude too high. Additionally, measuring the center of mass of a realistic spacecraft at the 10 pm level is problematic. For this reason, the fiducial test particles should be geometrically simple test masses that are shielded by co-orbiting spacecraft. Various space gravity missions use an accelerometer with an inertial reference test mass that is electro-statically forced to follow the spacecraft. The recorded applied force is used to correct the observed spacecraft orbit for its non-inertial accelerations in post-processing. However, imposing a test mass acceleration for a typical spacecraft size and mass experiencing radiation pressure from the sun would require electrostatic actuator stability at a prohibitively small level. The solution is to invert this forcing scheme with *drag-free control*, where the spacecraft is actively controlled with micro-Newton thrusters to stay centered on the reference test mass. The test mass itself will not be exposed to any applied forces along the interferometry axes critical to the gravitational wave measurement. Such drag-free control requires the additional complication of a micro-propulsion system, which has already been successfully implemented with Gravity Probe B [108].

The residual acceleration noise for our geodesic reference test mass, with mass m, inside a drag-free satellite can be expressed as an acceleration

$$a_n = \frac{F_n}{m} - \omega_p^2 \, \Delta x. \tag{5.5}$$

The first term accounts for any fluctuating forces acting on the test mass even when it is in its nominally centered position relative to the spacecraft. The second term describes the coupling of the "noisy" spacecraft position, with residual jitter Δx around the test mass, via any static parasitic force gradient or "stiffness" $m\,\omega_p^2$. Satellite control at the several-nm/Hz$^{1/2}$ level requires that this force gradient be smaller than a μN/m, though this noise contribution could be substantially reduced by subtraction of the force calculated from the interferometric measurement of Δx and a calibration of the spacecraft force gradient.

The remaining limit to the test mass's geodesic purity is a long list of fluctuating stray forces, analyzed in various references [109–111] and including Brownian motion due to residual gas, stray electrostatics including charge accumulation, thermal gradients, fluctuating magnetic fields, spacecraft self-gravity, and force cross-talk enabled by the control system.

Proving it: demonstrating free-fall purity on the ground and in orbit

Space-based gravitational wave detection requires a large improvement over the current state-of-the-art in sensitivity to differential acceleration. Experimental verification is thus a must before flying any LISA-like mission. The capacitive readout system and fluctuations in surface forces have been and continue to be studied in torsion pendulums. In these experiments a hollow test mass is suspended inside a prototype gravitational reference sensor (GRS). Such measurements have already established that force noise from thermal gradients [112], from molecular impact Brownian noise [113], from the interaction between noisy test mass charge and stray electrostatic fields [114], and from GRS-related

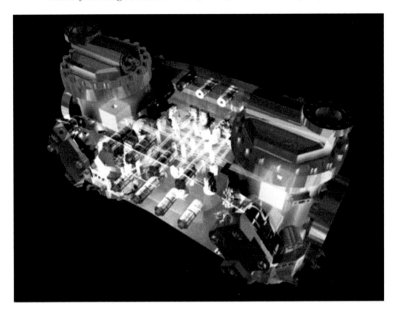

Figure 5.9 Schematic of the LISA Pathfinder payload from [116], including the two test masses housed in their respective GRS and, between these, a zerodur optical bench for the interferometric measurement of the relative test mass displacement. © IOP Publishing. Reproduced with permission.

force gradients [115], can all fit inside a typical LISA noise budget down to the 0.1 mHz frequency band.

However, some tests can only be made in space. LISA Pathfinder (LPF) will take a LISA arm, squeeze it into a single spacecraft and measure the relative acceleration of two free-falling test masses interferometrically. The 40 cm baseline is far too short to detect any gravitational waves, but will test the test-mass acceleration noise and the local test-mass-to-spacecraft interferometry at levels approaching the required sensitivity using hardware that could be inherited by eLISA or any other LISA-like mission. A schematic of LPF is shown in Fig. 5.9.

On the ground, statistical measurements of the torsion pendulum torque noise floor, with an LPF-like test mass inside a prototype LPF GRS connected to its sensing and actuation electronics, allow placement of a global upper limit on stray accelerations from surface force noise at the level of 40 fm/s^2/Hz$^{1/2}$ at 1 mHz [117]. These surface forces are notoriously difficult to calculate from first principles, but the measured performance is roughly within one order of magnitude above the eLISA goals, which are typical for LISA-like missions. These tests form a solid experimental foundation for the more sensitive and more complete LPF space flight test.

LPF itself is slated for launch in the second half of 2015 into an orbit around L1, the first Lagrange point. All key payload subsystems have been delivered at the time of writing and will be integrated with the spacecraft in the next months. In addition to the principal acceleration noise measurement, LPF will isolate many known force disturbances to help build a physical, experimentally anchored model for the limits within which we can achieve geodesic motion with reference test masses. Thus, LPF represents a cornerstone in experimental gravitation metrology, and a stepping stone towards gravitational wave astrophysics from space. Additional details of the design of LPF may be found in [30].

5.4.4 Prospects for a space-based mission

The uniquely powerful science case for gravitational wave observations in the mHz frequency range has been confirmed in many reviews by science and space agency panels in Europe and the United States. Mature mission concepts for LISA and LISA-like missions are based on 20 years of research and development. The main question then is when such a mission might fly.

LISA and later SGO-mid were proposed as three-arm missions where each of the three spacecraft contains two of the individual spacecraft units described previously. One of the cost-reducing measures in going from LISA to eLISA was the change to a two-arm instrument where only the mother spacecraft contains two units and each of the two daughter spacecraft at the far ends contains one unit [23]. A recent study at the behest of NASA [95] compared the scientific reach to cost for several proposed space-based gravitational wave detector designs. To interpret the conclusions of this study, it is instructive to compare the major differences between LISA and eLISA. The latter has shorter arm lengths by a factor of 5, leading to some decrease of sensitivity. It also has a different orbit, leading to a shorter mission lifetime. The most drastic difference, however, is the reduction to two interferometer arms rather than three. The scheme allows measurement of only one polarization at a time and not both, thus losing the ability to analyze the data as redundant interferometers (including one *virtual* configuration that is insensitive to gravitational waves and thus could be used to eliminate systematic noise). In addition, the risk of failure increases because redundancy is lost – a three-armed cluster can function as two-armed if necessary. These are all factors which have to be taken into account during mission selection.

In an ideal world, one would wish to fly a mission with the scientific reach of LISA. Next best would be eLISA with three arms (comparable to SGO-mid). One can be hopeful that direct detection of gravitational waves by ground-based instruments, expected very soon, will not only increase the scientific pressure for exploration of the mHz frequency band but also accelerate the timeline and increase the funding for space-based detectors. Under that scenario, the long-term prospects for a future space-based mission are excellent. LISA has been one of the highest-ranked missions at NASA and ESA for many years. A successful Pathfinder mission in the 2015/16 time frame would address the technical risks that were cited to delay a start of LISA in the past. The time scale to completely develop and launch a mission of the size of LISA, a flagship or cornerstone or L-class mission in NASA- or ESA-jargon, from mission concept selection to launch is typically ten years. Agency budget profiles could allow mission selection to occur near the end of this decade or in the early 2020s, leading to a launch of the space-based gravitational wave observatory in the 2030 timeframe.

5.5 Pulsar timing

5.5.1 Introduction

An array of pulsars functions as a galactic-scale array of clocks. A passing gravitational wave will modulate the arrival times of the pulses from the pulsars. Thus the ensemble of pulsars, a pulsar timing array (PTA), is a gravitational wave detector. The scheme is similar to that of ground-based interferometric techniques where the aim is to detect small phase differences between two electromagnetic signals as they travel down orthogonal detector arms. In the case of pulsar timing, however, the many "arms" of the detector are hundreds of parsecs long, the distance from the earth to the pulsar, and instead of detecting phase differences between two arms, PTAs aim at detecting phase discrepancies between the expected and observed pulsar arrival times.

Pulsar timing is most sensitive to gravitational waves at frequencies of ten to hundreds of nanohertz, corresponding to wave periods near one year. Possible sources of gravitational waves in this range include supermassive black hole binaries, relic gravitational waves, and cosmic strings [118]. There are more than 2000 known pulsars in the galaxy but fewer than 300 of those are millisecond pulsars [119]. Of those, PTAs are currently using the 40 that exhibit the greatest regularity of pulse arrival times. These pulsars' arrival times deviate from predictive models by less than 1 μs [120].

5.5.2 Radio telescopes

The instrumentation that is needed for this experiment includes large radio telescopes. In order to achieve sub-microsecond accuracy in measuring arrival times in pulsars, a large collecting area is needed. A 100-m class telescope is essentially the minimum.

We note that pulsar timing does not require the fine angular resolution offered by, for example, the Square Kilometer Array (SKA), planned for 2025. Rather, the most important commodity is collecting area. Large single-dish telescopes like the Arecibo 300-m telescope in Puerto Rico and the Green Bank 100-m telescope in West Virginia offer a single pixel, and that is sufficient for pulsar timing. Nonetheless, SKA and similar facilities are expected to contribute significantly to pulsar timing by discovering new millisecond pulsars, by allowing greatly improved timing resolution, and by providing accurate distances to pulsars [121, 122].

There are three major collaborations conducting this work. The North American Nanohertz Observatory of Gravitational Waves (NANOGrav[4]) is composed of researchers

---

[4] See, for instance, http://www.nanograv.org.

from the United States and Canada, and its two main telescopes are the Arecibo and Green Bank telescopes mentioned above. The European Pulsar Timing Array (EPTA[5]) is a collaboration of researchers from Europe. They use five 100-m class telescopes: the 76-m Lovell telescope in England, the 94-m Nancay Radio Telescope in France, the 100-m Effelsberg telescope in Germany, the 94-m Westerbork Synthesis Radio Telescope in the Netherlands and the soon to be completed 64-m Sardinia Radio Telescope in Italy. The Parkes Pulsar Timing Array (PPTA[6]) is an Australian collaboration that uses the Parkes 64-m telescope. The International Pulsar Timing Array (IPTA[7]) is composed of these three collaborations [123].

In a typical pulsar, the signal from one pulse is far below the radiometer noise, and the signal must be built up over many turns of the pulsar in order to be measurable. This has two important ramifications. First, "the" time of arrival (TOA) of a pulse is based on averaging many pulses (10,000 or so) right around that pulse. Second, in order to perform this averaging, a timing "model" of the pulsar must be known to great enough accuracy *a priori* so that the folding of the signal from the telescope can be done well.

Once a sufficient signal has been built up, usually over many minutes of observing, a TOA is measured by comparing the accumulated signal to a *template*. The comparison yields an offset between the template and the data. The TOA is this offset plus the recorded start-time of the scan. Once many TOAs have been accumulated, usually over years of observing, a model of the pulsar can be fit to the set of TOAs. Timing models include period, period derivative, right ascension, declination and proper motion of the pulsar, binary parameters if the pulsar is in a binary, perhaps parallax, and perhaps some post-Keplerian binary parameters if the number of degrees of freedom is sufficient. After finding the best-fit model, pulsar timers create a set of *residuals*, the difference between the model and the data. It is in this set of residuals that we look for gravitational waves. (See Fig. 5.10 for an example of residuals adapted from [24]).[8]

5.5.3 Detection principle

As described above, the ensemble of pulsars in our galaxy can be thought of as a system of celestial clocks, and detecting coordinated modulations in the arrival times of these pulsars is tantamount to detecting gravitational waves. However, there are other effects that can modulate pulse arrival times and a PTA must distinguish between these and a real gravitational wave signal. We discuss these in more detail below, but note that gravitational waves will exhibit a telltale *quadrupolar signature* that will distinguish them. We explain this feature by showing the response of pulsar timing to gravitational waves next. The key

---

[5] http://www.epta.eu.org.
[6] http://www.atnf.csiro.au/research/pulsar/ppta/.
[7] http://www.ipta4gw.org.
[8] Note that not all detection methods use the post-fit residuals to search for gravitational waves. Some schemes look in the TOAs [124] and some researchers toy with the idea of looking at the original folded profiles. For the purpose of discussion, however, it is easiest conceptually to think about looking in the residuals.

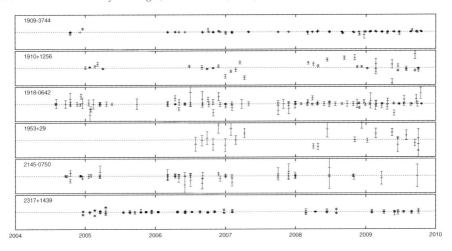

Figure 5.10 Overview of timing residuals for representative NANOGrav sources. The gap in 2007 was due to an extended maintenance period at both the Arecibo and Green Bank telescopes. The full scale of the vertical axis is 10 μs in all cases. Adapted from [24]. © AAS. Reproduced with permission.

to the definitive signature is in the different response the gravitational waves will produce in each of the pulsars in the array.

We begin by expressing the response τ_{GW}, which can be thought of as the residual. For a TT-gauge gravitational wave metric perturbation with form $h_+e_+ + h_\times e_\times$:

$$\tau_{GW}(\hat{k}, t)_j = F^+(\hat{k}, \hat{n}_j)\, g_+(t, L_j, \hat{k} \cdot \hat{n}_j) + F^\times(\hat{k}, \hat{n}_j)\, g_\times(t, L_j, \hat{k} \cdot \hat{n}_j). \tag{5.6}$$

$\hat{n}_j$ is a unit vector pointing to pulsar j, L_j is the distance to that pulsar. $F^{+/\times}$ are geometric functions of $\hat{k}$ and $\hat{n}$ which we omit here for brevity but can be found in [125]. Functions g_+ and $g_\times$ are integrals of h_+ and $h_\times$ as follows [126]:

$$g_{+/\times}(t, L_j, \hat{k}_j \cdot \hat{n}_j) = \int_0^{L_j} h_{+/\times}\left(t - \left(1 + \hat{k} \cdot \hat{n}_j\right)\left(L_j - \lambda\right)\right) d\lambda. \tag{5.7}$$

Note that we are using geometrized units, where $c = G = 1$. Following [126], we assume that a function f exists for which

$$\frac{df_{+/\times}(u)}{du} = h_{+/\times}(u). \tag{5.8}$$

For a plane wave we can then perform the integral to obtain

$$g_{+/\times}(t, L_j, \hat{k}_j \cdot \hat{n}_j) = \frac{f_{+/\times}}{1 + \hat{k} \cdot \hat{n}_j} - \frac{f_{+/\times}\left(t - \left(1 + \hat{k} \cdot \hat{n}_j\right)L_j\right)}{1 + \hat{k} \cdot \hat{n}_j}. \tag{5.9}$$

The first term is the so-called *earth term*, and the second the *pulsar term* [25]. The pulsar term is delayed from the earth term by $(1 + \hat{k} \cdot \hat{n}_j)L_j$, which amounts to thousands of years in

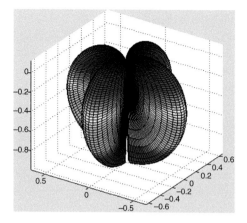

Figure 5.11 Absolute value of the pulsar response to a linearly polarized gravitational wave traveling in the $+z$ direction. Figure courtesy of William A. Coles.

most cases. Thus, a gravitational wave will modulate its signals in a coherent fashion, i.e. the modulation of signals in pulsars at different positions in the sky will be related. More specifically, the signals from two different pulsars could be up to 50% correlated with each other [126].

The response given by Eq. (5.9) has a particular shape in the sky, i.e., for a particular gravitational wave traveling in direction $\hat{k}$ the residuals will have a particular pattern of being *early* and *late*. See Fig. 5.11 for a visual representation of the earth term. Consider the denominator of the equation: $\left(1 + \hat{k} \cdot \hat{n}_j\right)$. When the gravitational wave propagation vector $\hat{k}$ is parallel to the direction to the pulsar $\hat{n}_j$ the denominator is maximum (and is equal to 2) and the response is a minimum. This corresponds to the situation where the pulsar and the gravitational wave source are in opposite parts of the sky. Conversely, when the gravitational wave propagation vector $\hat{k}$ is anti-parallel to the direction to the pulsar $\hat{n}_j$, the denominator is zero. The rationality of the function is saved in this situation by the fact that, in the numerator, the *delay* of the pulsar term is zero, i.e. the earth and pulsar terms cancel out and the numerator is also zero. This corresponds to a perfect alignment between the electromagnetic wave and the gravitational wave, or, in other words, the electromagnetic wave "surfing" on the gravitational wave. This corresponds to the pulsar term exactly canceling out the earth term at all times, and therefore to zero response in the pulsar. However, just away from this null in the response (e.g. the pulsar being one gravitational wavelength away from the line connecting the gravitational wave and the earth) is the maximum response. For a given single sinusoidal source, and a given pulsar, the earth and pulsar terms will either constructively or destructively interfere, depending on the distance of the pulsar. Unfortunately, there is no way at present to determine the distance to the pulsar precisely enough to know *a priori* how these terms will interfere. All single-source detection schemes must therefore account for this. We will discuss these later.

The 'sweet spot' of PTAs near year-long periods   Pulsar timing is generally thought to be most sensitive near periods of a year [24]. Nyquist sets the upper-frequency bound: observing usually occurs at a cadence of once every 3 or 4 weeks, so 600 nHz is the high-frequency cutoff until observing becomes more frequent. The lower-frequency bound is set by the dataspan of observations. Some pulsar data sets are 20 years long, so the low-frequency cutoff is 0.05 yr^{-1}, or about a nanoHertz. The low-frequency end of this range (closer to 1 nHz) has generally been considered the sweet spot of pulsar timing for two reasons: (1) pulsar timing's most impressive claims come from long-term timing rather than short-term timing, i.e. pulsar timing rivals atomic clock precision on long time scales [127], but at short time scales some pulse-phase jitter exists and the atomic clocks are better. So the longest wavelengths represent the most stability in pulsar timing. (2) The amplitude of the sources is largest at the low frequencies. As more realistic source models are considered the location of the sweet spot is the subject of some controversy. Sesana *et al.* [128] have made it clear that, at the high-frequency end of the pulsar timing band, the spectrum is likely to be dominated by a few bright sources. There are some advantages to detecting single sources rather than a stochastic background, so the first source detected is not necessarily going to be at the low-frequency end of the pulsar-timing spectrum.

5.5.4 Data analysis

How will the gravitational wave signals be extracted from the data? For example, a gravitational wave signal from a 3-year-period circular binary black hole would appear as a 1.5-year sinusoid in the pulsar timing residuals. The amplitude of the sinusoid depends on the mass, period, and orientation of the system with respect to the line-of-sight. Some strong signals would be found in a simple periodogram of the data, while others would be found via matched filtering. Additionally, the amplitude depends upon whether the earth and pulsar terms interfere constructively or destructively. This in turn depends upon the distance to the pulsar, which cannot be known well enough *a priori* to determine the phase of the pulsar term. Corbin and Cornish [129] demonstrate that in some circumstances one can determine the distance to the pulsar using the gravitational wave data. Various single-source detection algorithms account for uncertainty in the pulsar distance in different ways [126, 129–132].

Pulsar timing can extract the waveform [126] and localize the gravitational wave source. Several authors have looked at source localization [126, 129, 131] using a variety of methods. Strong burst sources can be localized well to less than 1 square degree. For continuous wave sources it is more difficult because one cannot ignore the pulsar term. Without knowing the pulsar distances, it is not clear that one can do better than 40 square degrees, but there is hope for obtaining the pulsar distances either by mitigating the red noise in pulsars (see the section on detector characterization below) so that the distance can be found from the chirp signal [129], or by obtaining the distances by other means such as timing or VLBI parallax [131]. One interesting thing to note is that if one had an eccentric black hole binary instead of a circular one, it would be a repetitive burst source, and the localization would be much easier.

Stochastic background detection: Instead of one source at a time, pulsar timing may detect an ensemble of sources, e.g., the combined effect of 10,000 black hole binaries distributed around the universe. This signal amounts to a *crinkling* of spacetime, and cannot be detected by looking for periodic signals. Instead, methods for detecting the background amount to exploiting correlations between pulsars. Hellings and Downs [133] pointed out that a plot of the correlation between residuals signals in pairs of the pulsars versus the angle between the pair shows a particular characteristic shape, now called the *Hellings and Downs curve*. For pulsars angularly close to each other on the sky the correlation is high on average, and for pulsars separated by 80 or 90 degrees the correlation is actually negative. Some methods developed for stochastic background detection actually attempt to detect the Hellings and Downs curve [24, 134], whereas other methods use the Hellings and Downs curve as input and determine whether the correlations that are found are consistent with gravitational waves [36].

The absorption of gravitational wave signals by the pulsar model fitting process: One complexity that arises in this analysis is the absorption of gravitational wave signals during the pulsar model fitting procedure. The fitting procedure described on the previous page does not necessarily reveal the intrinsic parameters of the pulsar, and has the potential to actually absorb gravitational waves into pulsar timing model fitting parameters. For example, consider a 2-year-period circular binary black hole impacting pulsar timing residuals by adding a 1-year sinusoid to the TOAs. The 1-year sinusoid would alter the fitted values of the right ascension and declination of the pulsar, unbeknownst to the user.[9] A more widespread effect comes from the fit to the period and the period derivative, which effectively absorbs any quadratic signal in the residuals. Though not many sources are expected to be specifically quadratic in nature, the red spectrum of the gravitational wave background means that for any finite observation of the background, the imprint on the pulsar timing signals is likely to look quadratic. Another way of saying this is that the longest-period signals (of order the length of the dataspan) are likely to be absorbed by this fit to period and period derivative. For more information see [135].

What else can mimic a gravitational wave signal in the data? There are three effects that will produce deviations in pulsar timing that will be correlated among pulsars [123]: terrestrial clock errors, errors in the solar system ephemerides, and gravitational waves. Each of the three has a different spatial shape, and so can be distinguished. Terrestrial clock errors will produce a monopolar signature, ephemerides errors will produce a dipolar signature, and gravitational waves will produce a quadrupolar signature. We describe each below.

The time standard used by pulsar timing is UTC, kept by the Bureau Internationale des Poids et Mesures (BIPM). This time is a weighted average of the time kept by a collection of cesium clocks and is accurate to one part in 10^{15} [123] on time scales of months. This is a bit better than, but of the same order as, pulsar timing, so it is reasonable to ask if an error in BIPM would mimic a gravitational wave in its effect on pulsar timing. In fact, if a clock

---

[9] The right ascension and declination are used in the pulsar timing model to account for earth's rotation about the sun, so an offset in right ascension or declination produces a 1-year sinusoid in the pulsar residuals.

error existed, it would have the same effect on all pulsars in the array, simultaneously, i.e., it would be monopolar.

In accounting for the motion of the earth (and therefore the radio telescopes) around the sun, pulsar timers use the solar system ephemerides provided by the Jet Propulsion Laboratory. If the mass of Jupiter used to create the model is off by a small fraction (one part in 10^6), the effect on the residuals will be of order 1 μs [136]. This would affect all the residuals simultaneously, but it would have a different sign in different pulsars. Jupiter's mass error would cause an error in the estimated position of the solar system barycenter, so pulsars that were in the direction of the shift would have falsely reduced residuals, and pulsars in the opposite direction would have falsely increased residuals. Thus the spatial signature would be dipolar and would be distinguishable from gravitational waves.

The signature from gravitational waves in pulsar timing is not strictly quadrupolar, but it is often referred to as such because the gravitational waves themselves exhibit a quadrupolar pattern. Once the integral described previously has been performed, however, some of the quadrupolar symmetry is broken. The shape of the resulting signature is shown in Fig. 5.11.

Detecting alternative theories of gravity: In general relativity, there are two transverse gravitational wave polarization modes. However, alternative theories of gravity can include up to 6 gravitational wave polarization modes. Lee *et al.* [137] find that if these extra polarizations exist, they could be detected in 5 years with a PTA of 40 to 60 pulsars. Chamberlin and Siemens [138] show that sensitivity to the vector and scalar-longitudinal modes can increase dramatically for pulsar pairs with small angular separations. For example, the J1853+1303/J1857+0943 pulsar pair, with an angular separation of about 3°, is about 10,000 times more sensitive to a longitudinal component of the stochastic background than to the transverse components. Detection of an extra gravitational wave polarization could provide the first evidence for the violation of general relativity.

Also, in general relativity, gravitational waves travel at the speed of light, and the graviton is therefore massless. However, in some alternative theories of gravity the graviton has mass, and the gravitational wave is therefore dispersive. Lee *et al.* [139] estimate that it should be possible to detect this dispersion using PTAs. In particular they conclude that massless gravitons can be distinguished from gravitons heavier than 3×10^{-22} eV with 5 years of bi-weekly observations of 60 pulsars with a pulsar rms timing accuracy of 100 ns. This is not as good as limits placed currently by [140] using galaxy cluster observations and even stronger but model-dependent bounds (see the discussion in [141]), but the two methods are independent and pulsar timing would represent an important substantiation of the cluster results.

5.5.5 When will pulsar timing detect gravitational waves?

Simulations show that detection of gravitational waves using PTAs is possible within a decade, and could occur as early as 2016 [142]. Current sensitivity limits reach optimistic predictions of the amplitude of the stochastic gravitational wave background. In other words, we are just entering a sensitivity range that could produce a detection of the stochastic background. How quickly our sensitivity improves depends upon the number M of pulsars added to the array, the cadence, c, at which those pulsars are observed, and the rms of the timing residuals σ of those pulsars. It also depends upon the amplitude A of the stochastic background; the dependence on the aforementioned parameters is different depending upon the strength of the gravitational wave signal relative to the noise. Siemens *et al.* [142] derive scaling relations for the signal-to-noise ratio in the weak- and intermediate-signal regimes. Based on reasonable estimates of the stochastic background of gravitational waves due to supermassive binary black holes, we are already in the intermediate-signal regime for a few pulsars, and after 5 to 10 years we will be in the intermediate-signal regime for most pulsars. For the weak-signal regime the sensitivity scales as:

$$\langle \rho \rangle \propto Mc \frac{A^2}{\sigma^2} T^{\beta}, \tag{5.10}$$

where $\beta = 13/3$ for a background due to supermassive binary black holes. After enough time passes and the amplitude of the lowest frequency of the stochastic background exceeds the level of the white noise, the intermediate-signal regime scaling laws apply:

$$\langle \rho \rangle \propto M \left(\frac{A}{\sigma \sqrt{c}} \right)^{1/\beta} T^{1/2}. \tag{5.11}$$

These relations are summarized for various numbers of pulsars and various amplitudes of the background in Fig. 5.12 [142].

For more details on these calculations please see [142]. Interestingly, as time goes by, the number of pulsars retains its importance, i.e., the sensitivity continues to scale linearly with the number of pulsars, but the dependence on the timespan and cadence weakens considerably. This transition is caused by the random nature of the gravitational wave signal

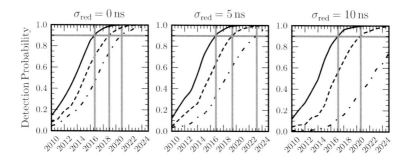

Figure 5.12 Detection probability versus time in years for the NANOGrav pulsar timing array. The dash–dot, dashed and solid lines show different amplitudes of the background: $A = 5.6 \times 10^{-16}$, $A = 1 \times 10^{-15}$, and $A = 2 \times 10^{-15}$, respectively. Siemens *et al.* [142] used the combined-epoch root-mean-squares of the 17 pulsars in Fig. 5.10 [24], started timing a total of 35 pulsars in 2013, and added 3 new pulsars per year with an epoch-combined rms of 200 ns prior to hardware improvements. Reproduced, with permission, from [142]. © IOP Publishing. All rights reserved.

becoming the dominant contribution to the rms of the pulsars. This source of scatter does not get reduced via added integration time, but only by the addition of more pulsars to the array. See [143] for discussion of some additional constraints on the signals potentially discoverable by PTAs.

5.5.6 Concluding remarks on pulsar timing

PTAs are robust, tunable, and pointable gravitational wave detectors, capable of measuring characteristics of single sources of gravitational waves in addition to those of a stochastic background. PTAs are estimated to reach detection sensitivity within the next decade, possibly as soon as 2016, depending on the amplitude of the background and the amount of red noise in the timing data.

5.6 Summary

We have come a long way since Einstein presented his General Theory of Relativity in 1915 and published his derivation of gravitational waves in 1918. One hundred years later we are on the verge of gravitational wave astronomy becoming a reality! We can expect that in a few years, after the first detections, multi-wavelength gravitational wave astronomy will observe signals at frequencies ranging from the inverse of the age of the universe to many kHz. Future detectors will observe high event rates at high signal-to-noise ratio and will become part of mainstream astronomy.

With gravitational waves we will be able to survey the entire universe. They will allow us to investigate the formation and evolution of binary systems of neutron stars and stellar-mass black holes not only in the Milky Way, but out to redshifts on the order of 10. Low-frequency detectors will have the sensitivity to observe supermassive black hole mergers out to redshifts of 20 and beyond, if they exist, to study the history and evolution of the universe. Coalescing binaries of black holes or neutron stars can serve as standard candles for absolute luminosity distance. If a redshift can be obtained through electromagnetic observations, or with gravitational waves alone through statistical reasoning or through the mass–redshift degeneracy breaking for neutron star binary observations, this will give us a model-independent determination of the distance–redshift relation, the cosmic expansion history, and the dark energy equation of state. High-precision observations of inspirals will test gravity in the dynamical strong-field regime, and probe the early universe at the TeV energy scale. Gravitational wave astronomy will play a unique role in the scientific landscape of the 2020s and 2030s. (For a detailed discussion of gravitational wave sources and their science, see Chapter 6 in this volume.)

Ground-based laser interferometric detectors are now relying on well-developed technology. The second-generation detectors Advanced LIGO and Advanced Virgo are expected to start routine observations of signals in the frequency band from a few hertz to many kilohertz by the end of this decade, with an amplitude sensitivity and distance reach ten times those of the first generation. By that time, GEO 600 will leave Astrowatch mode and serve as an advanced interferometry testbed. The even more advanced KAGRA detector will join the network in the early 2020s, as may LIGO India, and increase the sensitivity and directionality of the international network considerably. Even beyond this, the prospects for the future are bright. Planning for a third-generation detector in Europe, the Einstein Telescope (ET), is well advanced and a conceptual design is available. Similar investigations in the US are underway. Once the second-generation detectors are routinely observing signals, we can expect a lot of interest in this field, and the preparation of the third generation will begin in earnest, aiming at another factor of ten improvement in the sensitivity and distance reach.

At these sensitivities, the event rates will become comparable to event rates in electromagnetic astronomy such as rates of supernovae or signals from gamma ray detectors. Most gravitational wave signals will also allow precise determination of source parameters such as masses and spins of merging objects well beyond anything previously known in astronomy.

But the third generation will not be the end either. We now understand that the so-called Standard Quantum Limit is really not a limit at all for the laser interferometric detection of gravitational waves, and can be circumvented by non-classical quantum optical arrangements. These are already being tested in the laboratory and will soon be employed in meter-scale prototype interferometers. The other major limiting noise source, thermal noise, will probably in the future be mitigated not just by cooling but also by new types of crystalline mirror coatings or even coating-free mirrors, possibly combined with displacement-noise-free readout methods. All in all, there appear to be no fundamental limits for the sensitivity of ground-based gravitational wave detectors in sight that cannot be solved by technical ingenuity. That means several orders of magnitude in sensitivity enhancement are imaginable on the ground for frequencies between a hertz and many kilohertz. The only remaining barrier on the ground, for which there is no remedy in sight, seems to be gravity-gradient noise at low frequencies. So, this low-frequency range below 1 Hz will probably always remain the realm of space technologies.

The very-lowest-frequency gravitational waves, at frequencies around the inverse of the age of the universe, originating from primordial processes like inflation, will be detectable by the imprint they have left on the cosmic microwave background (CMB) in the form of tensor modes visible through B-mode polarization patterns. While a detector in space might help to observe these modes, it may well be that ground-based microwave telescopes may see them first, even though not over the full sky. Very little is known about the strength of these modes from other gravitational wave observation, because the details of the frequency spectrum of these primordial signals depend in a complicated way on what happened during inflation. While it is most likely that the high-frequency spectrum of these modes will be slightly falling in energy density and the ground-based detectors will have little chance to see them, there are also other models that even lead to a rise. At the time of writing of this review, there is evidence of a possible observation of a B-mode polarization signature in the CMB, but the discussion is ongoing and the results are not clear.

Gravitational waves at nanohertz to microhertz frequencies, originating from supermassive black hole mergers, will be detectable by pulsar timing arrays (PTAs) in the not too distant future. Pulsar timing is almost equivalent to multi-arm laser interferometry, but the light sources are the pulsars themselves, the arms are much longer and signal-to-noise ratio is lower. The performance of pulsar timing is dependent on the timing performance of the employed 100 m-class radio telescopes on the ground. At current noise levels and source strength estimates, a first stochastic signal might become detectable as early as 2016, but in any case is likely in the next decade before 2024.

The frequency range from a fraction of a millihertz up to a hertz is probably the most important of all, because it contains both the most secure and the most fascinating sources of the gravitational wave sky, i.e. the known galactic binary stars as guaranteed calibration sources, and the coalescing massive binary black holes. The latter will be visible throughout the entire universe as far as the beginning of their existence. This frequency range is the realm of space-based laser interferometers, generically called LISA, or LISA-like missions. They are employing laser interferometry between free-flying test masses in drag-free spacecraft in heliocentric orbits with million-kilometer arms. LISA has been under study now for over twenty years. The technology is well understood in the laboratory and a space mission (LISA Pathfinder) testing the LISA noise models will be launched by ESA in 2015. LPF is almost complete as far as the production of payload space hardware, satellites and integration is concerned. The science theme *The Gravitational Universe* has just been selected for the L3 large-mission launch slot of the European Space Agency ESA, with international contributions limited to 20% of the mission. The ESA L3 launch is currently scheduled for 2034, if it remains an ESA-dominated mission. But it is entirely possible that in the end LISA will be a truly international mission with contributions from ESA, NASA, China and Japan, in an as yet to be discussed scenario. This would be the ultimate demonstration that gravitational wave research is a glowing example of international collaboration.

The future is bright for gravitational wave astronomy. One hundred years after Einstein's General Relativity, we are about to add another sense to our perception of the universe. Across the whole gravitational wave spectrum, ranging from frequencies around the inverse of the age of the universe up to many kHz, experimental efforts are under way to finally listen to the dark side of the universe.

Acknowledgments

The authors acknowledge the help of Harald Lück and Hartmut Grote for the preparation of Fig. 5.1. We also acknowledge the very useful comments we received from many members of the LIGO and Virgo Science Collaborations, the eLISA consortium, and the NanoGrav collaboration during the preparation of the manuscript.

B.K.B., K.D., G.G., G.M., and A.R. gratefully acknowledge the support of the United States National Science Foundation for the construction and operation of the LIGO Laboratory, the Science and Technology Facilities Council of the United Kingdom, the Max

Planck Society, the Volkswagen Foundation and the State of Niedersachsen/Germany for support of the construction and operation of the GEO 600 detector. K.D., W.J.W., and G.M. gratefully acknowledge the support of the European Space Agency, the Deutsches Zentrum für Luft- und Raumfahrt (DLR), the Agenzia Spaziale Italiana (ASI), and the National Aeronautics and Space Administration for the support of the LISA Pathfinder and the technology development for LISA and eLISA. W.J.W. acknowledges the continuous support for the LPF R&D by the INFN. A.N.L. gratefully acknowledges the support of the United States National Science Foundation for the NanoGrav collaboration. In addition, G.G. has received partial support from NSF grants PHY-1205882 and PHY-0905184, G.M. has received partial support from NSF grants PHY-1306594 and PHY-0969935 and NASA grant NNX12AE97G. A.N.L. has received partial support from NSF grants IIA-0968296 and AST-0748580.

References

[1] Einstein, A. 1916. *Annalen der Physik*, **49**, 769–823.
[2] Einstein, A. 1918. *Sitzungsber. Preuss. Akad. Wiss.*, 154.
[3] Hulse, R. A., and Taylor, J. H. 1975. *Astrophys. J. Lett.*, **195**, L51.
[4] Ade, P. A. R., *et al.* 2014. *Phys. Rev. Lett.*, **112**, 241101.
[5] Ade, P. A. R., *et al.* 2014. A joint analysis of BICEP2/Keck Array and Planck. arXiv: 1502.00612 [Astro-ph.CO].
[6] Pirani, F. 1956. *Acta Phys. Polon.*, **15**, 389.
[7] Saulson, P. 1994. *Fundamentals of interferometric gravitational wave detectors*. Singapore: World Scientific.
[8] Weber, J. 1960. *Phys. Rev.*, **117**, 306–313.
[9] Weber, J. 1969. *Phys. Rev. Lett.*, **12**, 1320–1324.
[10] Kafka, R., and Schnupp, L. 1978. *Astron. Astrophys.*, **70**, 97–103.
[11] Aguiar, O. 2011. *Res. Astron. Astrophys.*, **11**, 1–42.
[12] Gertsenshtein, M., and Pustovoit, V. 1963. *Sov. Phys. JETP*, **16**, 433–435.
[13] Moss, G. E., Miller, L. R., and Forward, R. L. 1971. *Appl. Optics*, **10**, 2495–2498.
[14] Weiss, R. 1972. *Quarterly Progress Report of the Research Laboratory of Electronics of the Massachusetts Institute of Technology*, **105**, 54–76.
[15] Faller, J. E., and Bender, P. L. 1984. A possible laser gravitational-wave experiment in space. Pages 689–690 of: *Precision measurements and fundamental constants II*. NBS Special Publication 617.
[16] Faller, J. E. *et al.* 1985. Space antenna for gravitational wave astronomy. In: *Proceedings of the Colloquium on Kilometric Optical Arrays in Space*. ESA report SP-226.
[17] Sazhin, M. 1978. *Soviet Astr.*, **22**, 36.
[18] Detweiler, S. 1979. *Astrophys. J.*, **234**, 1100.
[19] Astone, P., *et al.* 2013. *Phys. Rev. D*, **86**, 082002.
[20] Harry, G., *et al.* 2010. *Class. Quant. Grav.*, **27**, 084006.
[21] Degallaix, J., *et al.* 2013. *Astronomical Society of the Pacific Conference Series*, **467**, 151.
[22] Danzmann, K., *et al.* 2011. *LISA, Unveiling a hidden Universe*. ESA/SRE Report, see http://www.rssd.esa.int/index.php?project=LISA&page=LISA_doc.
[23] Danzmann, K., *et al.* 2013. *The gravitational Universe: the eLISA whitepaper*. ESA Report, see https://www.elisascience.org/whitepaper/.
[24] Demorest, P., *et al.* 2013. *Astrophys. J.*, **762**, 94.
[25] Jenet, F.A., Lommen, A., Larson, S.L., and Wen, L. 2004. *Astrophys. J.*, **606**, 799.
[26] Adhikari, R. 2014. *Rev. Mod. Phys.*, **86**, 121.

[27] Aso, Y., *et al.* 2013. *Phys. Rev. D*, **88**, 043007.
[28] Punturo, M., *et al.* 2010. *Class. Quant. Grav.*, **27**, 194002.
[29] Driggers, J. C., Harms, J., and Adhikari, R. X. 2012. *Phys. Rev. D*, **86**, 102001.
[30] Antonucci, F., *et al.* 2012. *Class. Quant. Grav.*, **29**, 124014.
[31] Cutler, C., and Holz, D.E. 2009. *Phys. Rev. D*, **80**, 104009.
[32] Kawamura, S., *et al.* 2008. *J. Phys.: Conf. Ser.*, **122**, 012006.
[33] Bender, P. L., Begelmann, M. C., and Gair, J. R. 2013. *Class. Quant. Grav.*, **30**, 165017.
[34] Burston, R., Gizon, L., Appourchaux, T., and Ni, W.-T. 2008. *J. Phys.: Conf. Ser.*, **118**, 012043.
[35] McKenzie, K., *et al.* 2011. *Lagrange: a space-based gravitational-wave detector with geometric suppression of spacecraft noise*. NASA-Document-URS225851.
[36] van Haasteren, R., *et al.* 2011. *Mon. Not. R. Astr. Soc.*, **414**, 3117.
[37] Yardley, D., *et al.* 2011. *Mon. Not. R. Astr. Soc.*, **407**, 1777.
[38] Pitkin, M., Reid, S., Rowan, S., and Hough, J. 2011. *Living Rev. Rel.*, **14**, 5.
[39] Abbott, B., *et al.* 2009. *Rep. Prog. Phys.*, **72**, 076901.
[40] Abbott, B., *et al.* 2005. *Phys. Rev. D*, **72**, 122004.
[41] Abadie, J., *et al.* 2012. *Sensitivity achieved by the LIGO and Virgo gravitational wave detectors during LIGO's sixth and Virgo's second and third science runs*. arXiv:1203.2674 [gr-qc].
[42] Fairhurst, S. 2011. *Class. Quant. Grav.*, **28**, 105021.
[43] LIGO Scientific Collaboration and others. 2014. *LSC observational and instrument papers*. https://www.lsc-group.phys.uwm.edu/ppcomm/Papers.html.
[44] Virgo Collaboration and others. 2014. *Virgo publication list*. https://www.ego-gw.it/editorialboard/list.aspx.
[45] Abbott, B., *et al.* B. 2008. *Astrophys. J. Lett.*, **683**, 45.
[46] Abadie, J., *et al.* 2011. *Astrophys. J.*, **737**, 93
[47] Abadie, J., *et al.* 2009. *Nature*, **460**, 990–994.
[48] Abbott, B., *et al.* 2008. *Astrophys. J.*, **681**, 1419.
[49] Abbott, B., *et al.* 2012. *Astrophys. J.*, **755**, 2.
[50] Aasi, J., *et al.* 2014. *Prospects for localization of gravitational wave transients by the Advanced LIGO and Advanced Virgo observatories*. arXiv:1304.0670 [gr-qc].
[51] Caves, C. 1980. *Phys. Rev. Lett.*, **45**, 75.
[52] Caves, C. 1981. *Phys. Rev. D*, **23**, 1693–1708.
[53] McKenzie, K. *et al.* 2004. *Phys. Rev. Lett.*, **93**, 161105.
[54] Vahlbruch, H., *et al.* 2008. *Phys. Rev. Lett.*, **100**, 033602.
[55] Schnabel, R., *et al.* 2011. *Nature Physics*, **7**, 962–965.
[56] Aasi, J., *et al.* 2013. *Nature Photonics*, **7**, 613–619.
[57] Harry, G. M., Bodiya, T., and DeSalvo, R. (eds). 2012. *Optical coatings and thermal noise in precision measurement*. Cambridge: Cambridge University Press.
[58] Dolesi, R., *et al.* 2011. *Phys. Rev. D*, **84**, 063007.
[59] Willke, B., *et al.* 2006. *J. Phys.: Conf. Ser.*, **32**, 270–275.
[60] Kwee, P., *et al.* 2012. *Optics Express*, **20**, 10617.
[61] Rong, H., *et al.* 2002. Characterization of the LIGO input optics system. Pages 1837–1838 of: Gurzadyan, V. G., Jantzen, R. T., and Ruffini, R. (eds), *Proceedings of the Ninth Marcel Grossmann Meeting on General Relativity*. Singapore: World Scientific.
[62] Palashov, P., *et al.* 2012. *JOSA B*, **29**, 1784– 1792.
[63] Dooley, K., *et al.* 2012. *Rev. Sci. Instrum.*, **83**, 033109.
[64] Advanced LIGO Team. 2011. *Advanced LIGO reference design*. https://dcc.ligo.org/public/0001/M060056/002/AdL-reference-design-v2.pdf.
[65] Grote, H., *et al.* 2004. *Class. Quant. Grav.*, **21**, S473.
[66] Arain, M., *et al.* 2008. *Optics Express*, **16**, 10018–10032.
[67] Ward, R., *et al.* 2008. *Class. Quant. Grav.*, **25**, 114030.
[68] Fricke, T., *et al.* 2012. *Class. Quant. Grav.*, **29**, 065005.

[69] Abbott, R., *et al.* 2010. *T1000298: advanced LIGO length sensing and control final design.* LIGO Document Control Center, https://dcc.ligo.org/cgi-bin/DocDB/ShowDocument?. submit=Number&docid=T1000298&version=.

[70] Black, E. 2001. *Amer. J. Phys.*, **69**, 79–87.

[71] Strain, K., *et al.* 2003. *Appl. Optics*, **42**, 1244–1256.

[72] Mueller, G. 2005. *Optics Express*, **18**, 7118–7132.

[73] Barsotti, L., *et al.* 2010. *Class. Quant. Grav.*, **27**, 084026.

[74] Hello, P., *et al.* 1990. *French J. Phys.*, **51**, 2243–2261.

[75] Winkler, W., *et al.* 1991. *Phys. Rev. A*, **44**, 7022–7036.

[76] Mueller, G., *et al.* 2000. *Optics Lett.*, **25**, 266–268.

[77] Brooks, A., *et al.* 2005. *Gen. Rel. Grav.*, **37**, 1575–1580.

[78] Goda, K., *et al.* 2004. *Optics Lett.*, **29**, 1452.

[79] Lawrence, R., *et al.* 2004. *Optics Lett.*, **27**, 2635–2637.

[80] Raab, F., and Coyne, D. 1997. *Effect of microseismic noise on a LIGO interferometer.* LIGO-T960187-01, http://www.ligo-wa.caltech.edu/ligo_science/museism.pdf.

[81] Acernese, F., *et al.* 2010. *Astroparticle Phys.*, **33**, 182–189.

[82] Losurdo, G., *et al.* 1999. *Rev. Sci. Instrum.*, **70**, 2507–2515.

[83] Ballardin, G., *et al.* 2001. *Rev. Sci. Instrum.*, **72**, 3643.

[84] Aston, S., *et al.* 2012. *Class. Quant. Grav.*, **29**, 235004.

[85] Abbott, R., *et al.* 2004. *Class. Quant. Grav.*, **21**, S915.

[86] Abbott, R., *et al.* 2002. *Class. Quant. Grav.*, **19**, 1591–1597.

[87] Matichard, F., *et al.* 2010. *Prototyping, testing and performance of the two-stage seismic isolation system for advanced LIGO gravitational wave detectors.* American Society for Precision Engineering, Spring 2010 meeting.

[88] Grote, H., *et al.* 2013. *Phys. Rev. Lett.*, **110**, 181101.

[89] Kimble, H., *et al.* 2001. *Phys. Rev. D*, **65**, 022002.

[90] Yamamoto, K., *et al.* 2006. *Phys. Rev. D*, **74**, 022002.

[91] Agatsuma, K., *et al.* 2010. *Class. Quant. Grav.*, **27**, 084022.

[92] Bartusiak, M. 2000. *Einstein's unfinished symphony.* Washington: Joseph Henry Press.

[93] Danzmann, K., *et al.* 1998. *LISA: Laser Interferometer Space Antenna.* Pre-Phase-A Report, NASA, http://lisa.nasa.gov/Documentation/ppa2.08.pdf.

[94] Livas, J. 2013. *Status of space-based gravitational-wave observatories (SGOs).* https://conferences.lbl.gov/getFile.py/access?contribId=237&sessionId=30&resId=0& materialId=slides&confId=36.

[95] Weiss, R., *et al.* 2012. *Gravitational-wave mission concept study final report.* NASA Report, http://pcos.gsfc.nasa.gov/physpag/GW_Study_Rev3_Aug2012-Final.pdf.

[96] Sathyaprakash, B. S., and Schutz, B. F. 2009. *Living Rev. Rel.*, **12**, 2.

[97] Amaro-Seoane, P., *et al.* 2012. *eLISA: astrophysics and cosmology in the millihertz regime.* arXiv:1201.3621 [astro-ph.CO].

[98] Sutton, A., and Shaddock, D. 2008. *Phys. Rev. D*, **78**, 082001.

[99] Livas, J., *et al.* 2009. *Class. Quant. Grav.*, **26**, 094016.

[100] Tinto, M., *et al.* 1999. *Phys. Rev. D*, **59**, 102003.

[101] Shaddock, D., *et al.* 2004. *Phys. Rev. D*, **70**, 081101.

[102] Mitryk, S., *et al.* 2010. *Class. Quant. Grav.*, **27**, 084012.

[103] Mitryk, S., *et al.* 2012. *Phys. Rev. D*, **86**, 122006.

[104] Vitale, S. 2014. *Space-borne gravitational wave observatories.* arXiv:1404:3136 [gr-qc].

[105] Sutton, A., *et al.* 2010. *Optics Express*, **18**, 20759–20773. http://dx.doi.org/10.1364/OE.18.020759.

[106] Sweeney, D., *et al.* 2012. *Optics Express*, **20**, 25603.

[107] Armstrong, J. W., Iess, L., Tortora, P., and Bertotti, B. 2003. *Astrophys. J.*, **599**, 806–813.

[108] Everitt, C., *et al.* 2011. *Phys. Rev. Lett.*, **106**, 221101.

[109] Schumaker, B. 2003. *Class. Quant. Grav.*, **20**, S239–S254.

[110] Stebbins, R., *et al.* 2004. *Class. Quant. Grav.*, **21**, S653–S660.

[111] Antonucci, F., *et al.* 2011. *Class. Quant. Grav.*, **28**, 094002.

[112] Carbone, L., *et al.* 2007. *Phys. Rev. D.*, **76**, 102003.
[113] Cavalleri, A., *et al.* 2009. *Phys. Rev. Lett.*, **103**, 140601.
[114] Antonucci, F., *et al.* 2012. *Phys. Rev. Lett.*, **108**, 181101.
[115] Cavalleri, A., *et al.* 2009. *Class. Quant. Grav.*, **26**, 094012.
[116] Armano, M., *et al.* 2009. *Class. Quant. Grav.*, **26**, 094001.
[117] Cavalleri, A., *et al.* 2009. *Class. Quant. Grav.*, **26**, 094017.
[118] Jenet, F., *et al.* 2006. *Astrophys. J.*, **653**, 1571.
[119] Stovall, K., Lorimer, D. R., and Lynch, R. S. 2013. *Searching for millisecond pulsars: surveys, techniques and prospects.* arXiv:1308.4612 [astro-ph.GA].
[120] Hobbs, G., *et al.* 2010. *Class. Quant. Grav.*, **27**, 084013.
[121] Liu, K. *et al.* 2011. *Mon. Not. R. Astr. Soc.*, **417**, 2916–2926.
[122] Smits, R. *et al.* 2011. *Astron. Astrophys.*, **528**, A108.
[123] Manchester, R., *et al.* 2013. *Publ. Astron. Soc. Australia*, **30**, 17.
[124] van Haasteren, R., and Levin, Y. 2013. *Mon. Not. R. Astr. Soc.*, **428**, 1147.
[125] Burt, B. J., Lommen, A. N., and Finn, L. S. 2011. *Astrophys. J.*, **730**, 17.
[126] Finn, L. S., and Lommen, A. N. 2010. *Astrophys. J.*, **718**, 1400.
[127] Splaver, E. M. *et al.* 2005. *Astrophys. J.*, **620**, 405.
[128] Sesana, A., Vecchio, A., and Volonteri, M. 2009. *Mon. Not. R. Astr. Soc.*, **394**, 2255.
[129] Corbin, V., and Cornish, N. J. 2010. *Pulsar timing array observations of massive black hole binaries.* arXiv:1008.1782 [astro-ph.HE].
[130] Yardley, D., *et al.* 2010. *Mon. Not. R. Astr. Soc.*, **407**, 669.
[131] Lee, K., *et al.* 2011. *Mon. Not. R. Astr. Soc.*, **414**, 3251.
[132] Ellis, J., Jenet, F., and McLaughlin, M. 2012. *Astrophys. J.*, **753**, 96.
[133] Hellings, R. W., and Downs, G. S. 1983. *Astrophys. J. Lett.*, **265**, L39.
[134] Jenet, F. A., Creighton, T., and Lommen, A. 2005. *Astrophys. J. Lett.*, **627**, L125.
[135] Lommen, A., and Demorest, P. 2013. *Class. Quant. Grav.*, **30**, 224001.
[136] Champion, D., *et al.* 2010. *Astrophys. J.*, **720**, L201.
[137] Lee, K., Jenet, F. A., and Price, R. H. 2008. *Astrophys. J.*, **685**, 1304.
[138] Chamberlin, S. J., and Siemens, X. 2012. *Phys. Rev. D*, **85**, 082001.
[139] Lee, K. *et al.* 2010. *Astrophys. J.*, **722**, 1589.
[140] Goldhaber, A. S., and Nieto, M. M. 1974. *Phys. Rev. D*, **9**, 1119.
[141] Goldhaber, A. S., and Nieto, M. M. 2010. *Rev. Mod. Phys.*, **82**, 939.
[142] Siemens, X., Ellis, J., Jenet, F., and Romano, J. D. 2013. *Class. Quant. Grav.*, **30**, 224015.
[143] Cutler, C. *et al.* 2014. *Phys. Rev. D*, **89**, 042003.

6

Sources of Gravitational Waves: Theory and Observations

ALESSANDRA BUONANNO AND B. S. SATHYAPRAKASH

6.1 Historical perspective

James Clerk Maxwell discovered in 1865 that electromagnetic phenomena satisfied wave equations and found that the velocity of these waves in vacuum was numerically the same as the speed of light [1]. Maxwell was puzzled at this coincidence between the speed of light and his theoretical prediction for the speed of electromagnetic phenomena and proposed that "light is electromagnetic disturbance propagated through the field according to electromagnetic laws" [1].

Because any theory of gravitation consistent with special relativity cannot be an action-at-a-distance theory, in many ways, Maxwell's theory, being the first relativistic physical theory, implied the existence of gravitational waves (GWs) in general relativity (GR). Indeed, years before Einstein derived the wave equation in the linearised version of his field equations and discussed the generation of GWs as one of the first consequences of his new theory of gravity [2, 3], Henri Poincaré proposed the existence of *les ondes gravifiques* purely based on consistency of gravity with special relativity [4]. However, for many years GWs caused much controversy and a lot of doubt was cast on their existence [5–8]. The year 1959 was, in many ways, the turning point – it was the year of publication of a seminal paper by Bondi, Pirani and Robinson [6] on the exact plane wave solution with cylindrical symmetry and the energy carried by the waves [9]. This paper proved that wave solutions exist not just in the weak-field approximation and that GWs in GR carry energy and angular momentum away from their sources. These results cleared the way for Joseph Weber [10] to start pioneering experimental efforts. The discovery of the Hulse–Taylor binary [11], a system of two neutron stars in orbit around each other, led to the first observational evidence for the existence of gravitational radiation [12]. The loss of energy and angular momentum to GWs causes the two stars in this system to slowly spiral in towards each other. Since 1974, a few other pulsar binaries have been discovered. For the most relativistic binary, the observed rate of change of the period agrees with the GR prediction to better than 0.03% [13].

Once the most obvious theoretical impediments in defining gravitational radiation were tackled and observational evidence of the existence of the radiation was firmly established, serious research on the modeling of astrophysical and cosmological sources of GWs began, alongside experimental and data-analysis efforts to detect GWs. This chapter examines the last thirty years of endeavours that have brought the field to the dawn of the first direct detections of GWs, hopefully marking the hundredth anniversary of the first articles by Einstein on gravitational radiation [2, 3].

Impressive theoretical advances and a few breakthroughs have occurred since the publication of the two notable reviews: *Gravitational radiation* in 1982 [14] and *Three hundred years of gravitation* in 1987 [15]. Since then, a network of ground-based, GW laser interferometers has been built and has taken data.[1] Sophisticated and robust analytical techniques have been developed to predict highly-accurate gravitational waveforms emitted by compact-object binary systems (compact binaries, for short) during the inspiral, but also plunge, merger and ringdown stages (see Sections 6.2 and 6.3). After many years of attempts, today numerical-relativity (NR) simulations[2] are routinely employed to predict merger waveforms and validate analytical models of binary systems composed of black holes (BHs) and/or neutron stars (NSs). Simulations of binary systems containing NSs are becoming more realistic, and robust connections to astrophysical, observable phenomena (i.e., electromagnetic counterparts) are under development (see Section 6.3). The internal structure of NSs has still remained a puzzle, but there are hints of superfluidity of neutrons in the core. Neutron-star normal modes are now understood in far more detail, and a new relativistic instability, that could potentially explain NSs in X-ray binaries, has been discovered (see Section 6.4). Supernova simulations are getting more sophisticated and are able to include a variety of micro- and macro-physics (full GR, neutrino transport, weak interaction physics, magnetohydrodynamics), but not all in a generic 3-dimensional simulation with realistic equations of state. Many challenges remain, including understanding the basic supernova mechanism of core collapse and bounce (see Section 6.4). Thirty years ago, only a few rough predictions of GW signals from the primordial dark age of the Universe existed. Today we know a plethora of physical mechanisms in the early Universe that could generate GWs (see Section 6.5). Because of the weakness of GW signals, scientists working closely with the experiments have established strong collaborations with theorists, astrophysicists and cosmologists, so that searches for GWs are fully optimised (see Sections 6.3, 6.4 and 6.5). Concurrently with the construction of initial interferometers, geometrical approaches to optimizing data analysis were developed, which now form the backbone of all GW data-analysis quests. As we shall discuss, searches with initial LIGO and Virgo detectors have already produced astrophysically and cosmologically significant upper limits. Pulsar timing arrays (PTAs) have reached unprecedented sensitivity levels and could detect a stochastic background from a population of supermassive black holes over the next five years.

Although the progress has been tremendous and the field has reached an unprecedented degree of maturity, several challenges still need to be tackled to take full advantage of the discovery potential of ground- and space-based detectors, and pulsar timing arrays. Part of this chapter is also devoted to those challenges.

---

[1] See Chapter 5 in this volume.
[2] See also Chapter 7 in this volume.

Later in this chapter we will discuss sources that can be observed in different types of detectors. These include ground-based interferometers such as initial and Advanced LIGO (iLIGO and aLIGO), Virgo and Advanced Virgo (AdV), KAGRA and Einstein Telescope (ET) (see Fig. 6.6), space-based detectors LISA and eLISA (see Fig. 6.7) and PTAs and the Square Kilometre Array (SKA).[3]

Lastly, while this Chapter was being finalised, the BICEP2 experiment claimed to have observed the polarisation of the cosmic microwave background (CMB) photons (the so-called B-modes), caused by primordial GWs [17]. If confirmed, this result will constitute a landmark discovery in cosmology and GW science, enabling us to probe epochs very close to the Big Bang. However, at this stage it is not clear if the observed signal is truly primordial in nature and not due to astrophysical foregrounds and synchrotron emission by intervening dust [18, 19]. Measuring the imprint of primordial GWs on the CMB is different from, but equally relevant to, the detection of GWs with ground- and space-based detectors or PTAs. Indeed, detectors like BICEP2 infer the presence of GWs through their interaction with the CMB radiation at the time the latter was produced. They do not detect GWs passing by the detector on the Earth today. LIGO, Virgo, KAGRA, eLISA, PTA, etc., will probe contemporary GWs, yielding spectral and sky position data, and a plethora of new and unique information about our Universe and its contents. The focus of this Chapter is on the new window on the Universe that those detectors will enable us to open.

6.2 Analytical approximation methods

Progress over the past three decades: There is no doubt that the field of analytical relativity has matured considerably and has made tremendous progress since the notable Les Houches school on *Gravitational radiation* in 1982 [14]. During the last thirty years there have been significant advances on the problems emphasised in the historic discussion organised and moderated by A. Ashtekar at the end of the school. The validity of the quadrupole formula for gravitational radiation far away from a binary source was questioned by some of the participants at the Les Houches school, appealing to the ongoing debate at that time [20–27] relating to the difficulties in describing nonlinearities in GR within a precise mathematical framework. At the Les Houches school, the Paris group (Damour, Deruelle, · · ·) proposed a consistent framework in terms of which to formulate questions about the two-body dynamics and GW emission. The Paris group was motivated by the impressive observational work related to the discovery of the Hulse–Taylor binary pulsar [11] and wanted to develop a mathematical framework that could match the standards set by the ever more accurate pulsar-timing data. Eventually, as we shall discuss below, a precise mathematical framework was also needed for *direct* detection of gravitational waves on the ground, because in this case nonlinearities play a role that is much more crucial than in binary pulsar observations.

---

[3] ET is a third-generation detector concept being studied in Europe whose conceptual design study was completed in 2011 [16]. The European Space Agency has selected GW Observatory as the science theme for the 3rd large mission (L3) in its future science program, scheduled for a launch in 2034. eLISA is the current straw-man design for L3.

Between 1980 and 1992 important theoretical foundations in gravitational radiation and post-Newtonian (PN) theory were established by a number of researchers [28–41]. However, during those years, the analytical work on the two-body problem was considered mostly academic. It was not clear how relevant it would be to push calculations beyond the quadrupole formula for the direct observation of GWs. The first important turning point was in 1993 when Cutler *et al.* [42] pointed out the importance of computing the GW phasing beyond the leading order. Many crucial developments took place in the subsequent years [43–58]. The second important turning point, which brought theory and observations closer, occurred in the mid and late 1990s when the construction of LIGO, Virgo, GEO600 and TAMA 300 detectors started [59]. The TAMA 300 and LIGO detectors took the first data in 1999 [60] and 2002 [61], respectively. (The first ever coincident operation of a pair of interferometers was between the Glasgow 10 m and Garching 30 m prototypes [62].) The third turning point happened in the late 1990s and early 2000s when, pressed by the construction of GW interferometers, the analytical effective-one-body (EOB) approach [63, 64] made a bold prediction for the late inspiral, merger and ringdown waveform emitted by comparable-mass binary BHs. The EOB formalism builds on PN and perturbation theory results, and it is guided by the notion that non-perturbative effects can be captured analytically if the key ingredients that enter the two-body dynamics and GW emission are properly resummed about the (exact) test-particle limit results. Moreover, in the early 2000s, a pragmatic, numerical and analytical, hybrid approach aimed at predicting the plunge and merger waveform was bravely carried out [65, 66]. This approach, called the Lazarus project, consisted of evolving the binary system in full numerical relativity (NR) for less than an orbit just prior to merger before stopping the evolution, extracting the spacetime metric from the results of the simulation of a deformed BH, and using perturbation theory calculations to complete the evolution during ringdown. The fourth relevant turning point occurred in 2005, when, after more than thirty years of attempts, the first numerical-relativity simulations of binary BHs at last unveiled the merger waveforms [67–69]. Since then, synergies and interplays between different analytical and numerical techniques to solve the two-body problem in GR have grown considerably. A few paradigms were broken, in particular the nature of the binary BH merger waveform, which turned out to be much simpler than what most people had expected or predicted. Finally, recent years have seen remarkable interactions between GW data analysts, astrophysicists and theorists to construct templates to be used for the searches, making analytical relativity a crucial research area for experiments that will soon revolutionise our understanding of the Universe.

In the rest of this section we shall discuss the main approximation methods that have been developed to study the two-body problem in GR, highlighting some of the key theoretical ideas that have marked the last thirty years. As we shall see, in all the approximation schemes one needs to develop a conceptual and consistent framework and solve difficult technical problems. Since Einstein did not conceive the theory of GR starting from approximations to it, we shall see in relation to GWs what we gain and what we lose from the original theory when investigating it through approximation methods.

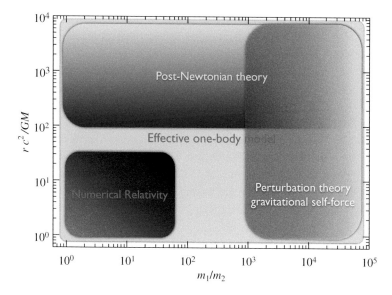

Figure 6.1 Current range of validity of the main analytical and numerical methods to solve the two-body problem.

Physical scales and methods in the two-body problem: Three main methods have been proposed to tackle the two-body problem in GR: the PN approach, perturbation theory (and the gravitational self-force formalism), and NR. The first step towards setting up those approximation methods consists in identifying some small (dimensionless) parameters. In Fig. 6.1 we show the range of validity of each method using the parameters $rc^2/(GM)$ and $m_1/m_2 \geq 1$, where m_1 and m_2 are the binary's component masses, $M = m_1 + m_2$ is its total mass and r is the separation between the two bodies. If we consider astrophysical sources that are held together by gravitational interactions, then, as a consequence of the virial theorem, $v^2/c^2 \sim GM/(rc^2)$, where v is the characteristic velocity of the bodies in the binary.

The PN formalism expands the dynamics and gravitational waveforms in powers of v/c. It is valid for any mass ratio but, in principle, only for slow motion, which, for self-gravitating objects, also implies large separations. Perturbation theory is suitable to describe the motion and radiation of a small body moving around a large body. It expands the Einstein equations around the BH metric in powers of the mass ratio m_2/m_1. At leading order the small-body moves along geodesics of the background spacetime and can reach any speed $v \lesssim c$. When taking into account the back-reaction of the gravitational field of the small body on its motion, one needs to develop a consistent framework free of divergences. This is done within the gravitational self-force (GSF) formalism. Finally, NR solves the Einstein equations on a computer. In principle, it could be used for any mass ratio, binary separation and velocity. However, the computational cost and the requirements on the

accuracy of the numerical solutions limit its range of validity. For the past thirty years
the two-body problem has been tackled keeping in mind that each method has a domain of
validity displayed in Fig. 6.1. As we shall see below, by proceeding without blinkers, recent
work at the interfaces between the different methods has demonstrated that the limits of
validity of those approaches are more blurred than expected – for example PN calculations
can be pushed into the mildly relativistic regime $v/c \lesssim 0.1$ and GSF predictions could
be used also for intermediate-mass (or perhaps even comparable-mass) binary systems.
Moreover, an analytical formalism, namely, the EOB approach, exists that can incorporate
the results of the different methods in such a way as to span the entire parameter space and
provide highly-accurate templates to search for BH binaries in GW data.

Because of space limitations, the presentation of the different approximation methods
will be sketchy and incomplete. The reader is referred to the original articles, reviews
[70–75] and books [76–79] for more details.

6.2.1 Post-Newtonian formalism

The Einstein field equations $R_{\alpha\beta} - g_{\alpha\beta} R/2 = 8\pi G T_{\alpha\beta}/c^4$ can be recast in a convenient form
by introducing the field $h^{\alpha\beta} = \sqrt{-g}\, g^{\alpha\beta} - \eta^{\alpha\beta}$, which is a measure of the deviation of the
background from the Minkowski metric $\eta_{\alpha\beta}$, and imposing the harmonic gauge condition
$\partial_\beta h^{\alpha\beta} = 0$ [76],

$$\Box h^{\alpha\beta} = \frac{16\pi G}{c^4}|g|T^{\alpha\beta} + \Lambda^{\alpha\beta} \equiv \frac{16\pi G}{c^4}\tau^{\alpha\beta}, \tag{6.1}$$

where $\Box$ is the D'Alembertian operator in flat spacetime, $g \equiv \det(g_{\alpha\beta})$, $T^{\alpha\beta}$ is the matter
stress-energy tensor and $\Lambda^{\alpha\beta}$ depends on non-linear terms in $h^{\mu\nu}$ and $g^{\mu\nu}$ and their derivatives. By imposing no-incoming-radiation boundary conditions, one can formally solve
Eq. (6.1) in terms of retarded Green functions

$$h^{\alpha\beta}(t,r) = \frac{16\pi G}{c^4}\Box_{\text{ret}}^{-1}\tau^{\alpha\beta} = -\frac{4G}{c^4}\int \frac{\tau^{\alpha\beta}(t-|r-r'|/c, r')}{|r-r'|}\,\mathrm{d}^3 r'. \tag{6.2}$$

Limiting to leading order in G and considering $r \equiv |r| \gg d$ (i.e., the field point is at a
far greater distance compared to the size d of the source), we can expand the integrand in
powers of $1/r$ and find at leading order

$$h^{\alpha\beta}(t,r) = -\frac{4G}{c^4 r}\int T^{\alpha\beta}(t - r/c + n\cdot r'/c, r')\mathrm{d}^3 r', \tag{6.3}$$

where $n = r/r$. Let us assume that the source is a PN source (i.e., it is slowly moving,
weakly stressed and weakly self-gravitating). This means that $|T^{0i}/T^{00}| \sim \sqrt{|T^{ij}/T^{00}|} \sim$
$\sqrt{|U/c^2|} \ll 1$, where U is the source's Newtonian potential. It is customary to indicate the
magnitude of the above small quantities with a small parameter ϵ, which is essentially v/c,
where v is the characteristic, internal velocity of the source. Assuming that the source's size
is d and it oscillates at frequency ω, the characteristic speed of the source is $v \sim \omega d$. From
analogy to the electromagnetic case, we expect $\lambda_{\text{GW}} \sim (c/v)d$. For slow motion $v/c \ll 1$,

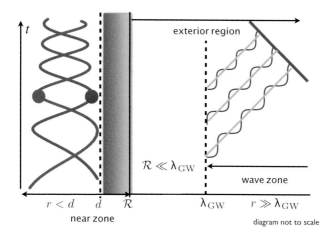

Figure 6.2 Schematic diagram of an inspiraling binary showing various scales used in the PN formalism.

thus $\lambda_{\text{GW}} \gg d$ and the source is located well within one wavelength. Historically, the region at a distance $r \ll \lambda_{\text{GW}}$ from the source, extending to $\mathcal{R}$ with $\mathcal{R} \ll \lambda_{\text{GW}}$, has been denoted the *near zone*, whereas the region which extends to $r \gg \lambda_{\text{GW}}$ is denoted the *wave zone* (see Fig. 6.2).

Using the conservation of the energy–momentum tensor at linear order in G, that is $\partial_\alpha T^{\alpha\beta} = 0$, and expanding the integral (6.3) in powers of v/c, one can obtain the gravitational field at linear order in G as a function of the derivatives of the source multipole moments [28]. As originally derived by Einstein [3] and then by Landau and Lifshitz, at lowest order in the wave-generation formalism, the gravitational field in the transverse-traceless (TT) gauge and in a suitable radiative coordinate system $X^\mu = (cT, X)$ reads ("far-field quadrupole formula")

$$h_{ij}^{\text{TT}} = \frac{2G}{c^4 R} \sum_{k,l} \mathcal{P}_{ijkl}(N) \left[\frac{d^2}{dT^2} Q_{kl} \left(T - \frac{R}{c} \right) + \mathcal{O} \left(\frac{1}{c} \right) \right] + \mathcal{O} \left(\frac{1}{R^2} \right), \qquad (6.4)$$

where $R = \sqrt{\sum_i X_i^2}$ is the distance to the source, $N = X/R$ is the unit vector from the source to the observer, and $\mathcal{P}_{ijkl} = \mathcal{P}_{ik} \mathcal{P}_{jl} - \mathcal{P}_{ij} \mathcal{P}_{kl}/2$ is the TT projection operator, where $\mathcal{P}_{ij} = \delta_{ij} - N_i N_j$ is the operator that projects onto the plane orthogonal to N. The radiative coordinate system X^μ can be related to the source-rooted coordinate system $x^\mu = (ct, x)$ [71]. The source quadrupole moment at Newtonian order is

$$Q_{ij}(t) = \int_{\text{source}} d^3 x' \, \rho(t, x') \left(x_i' x_j' - \frac{1}{3} \delta_{ij} x'^2 \right), \qquad (6.5)$$

where ρ is the Newtonian mass density. The gravitational field (6.4) is generally referred to as Newtonian because the evolution of the quadrupole moment of the source is computed using Newton's law of gravity.

Several methods have been proposed for going beyond the leading-order result to solve Eq. (6.2) approximately. In the near zone the solution $h_{\mu\nu}$ can be written in terms of instantaneous potentials, while in the wave zone retardation effects must be taken into account. Note that at higher orders gravitational waves can themselves act as sources; so $\tau^{\alpha\beta}$ in Eq. (6.2) contains both compact and non-compact support terms.

The multipolar post-Minkowskian–post-Newtonian formalism: Early on Epstein and Wagoner [80], and Thorne [28] proposed a PN extension of the Landau–Lifshitz derivation and computed the $\mathcal{O}(v^2/c^2)$ corrections to the quadrupole formula for binary systems. However, in their approach the formal application of the PN expansion leads to divergent integrals. Building on pioneering work by Bonnor *et al.* [81, 82] and Thorne [28], Blanchet and Damour [32, 37, 83] introduced a GW generation formalism in which the corrections to the leading quadrupolar formalism are obtained in a mathematically well-defined way. In this approach the wave-zone expansion, in which the external vacuum metric is expanded as a multipolar post-Minkowskian (MPM) series (i.e., a non-linear expansion in G), is matched to the near zone expansion, in which a PN expansion (i.e., a non-linear expansion in $1/c$) is applied, the coefficients being in the form of a multipole expansion. The multipolar expansions employ a multipole decomposition in irreducible representations of the rotation group. The gravitational field all over space can be obtained by matching the near-zone to the wave-zone fields and the matching can be accomplished at all orders, as shown by Blanchet and collaborators [84–87]. We shall denote this approach the MPM-PN formalism.

Wave-zone multipolar post-Minkowskian approximation: In the multipolar post-Minkowskian expansion, as one moves away from a weakly self-gravitating source, the spacetime quickly approaches the Minkowskian spacetime. At a distance r from the source with mass M, the deviations from the metric scale like r_s/r with $r_s = 2GM/c^2$. Thus, in the region $d < r < +\infty$, one can solve the vacuum Einstein equations iteratively in powers of G. Indeed, one sets $\sqrt{-g}g^{\alpha\beta} = \eta^{\alpha\beta} + Gh_1^{\alpha\beta} + G^2 h_2^{\alpha\beta} + \cdots$, that is $h^{\alpha\beta} = \sum_{n=1} G^n h_n^{\alpha\beta}$, and substitutes this expression into the Einstein equations (6.1) in vacuum, obtaining

$$\Box h^{\alpha\beta} = \Lambda^{\alpha\beta} = N^{\alpha\beta}(h,h) + M^{\alpha\beta}(h,h,h) + \mathcal{O}(h^4)\,, \tag{6.6}$$

and equates terms of the same order in G.

At linear order in G, the most general homogeneous solution satisfying the harmonic gauge can be written in terms of symmetric trace-free tensors (STFs), which are a complete basis of the rotation group's representation, that is $h_1^{\alpha\beta} = \sum_{\ell=0} \partial_L [K_L^{\alpha\beta}(t-r/c)/r]$, where $L = i_1, \ldots, i_\ell$ denotes a multi-index composed of ℓ STF indices $i_1, \ldots, i_\ell$, ranging from 1 to 3. More explicitly, one finds that the most general solution satisfying the gauge condition is given by $h_1^{\alpha\beta} = k_1^{\alpha\beta} + \partial^\alpha \varphi_1^\beta + \partial^\beta \varphi_1^\alpha - \eta^{\alpha\beta} \partial_\rho \varphi_1^\rho$, where $k_1^{\alpha\beta}$ and φ_1^α can be expressed in terms of two *source* multipole moments, $\{I_L, J_L\}$, that encode properties of the source and four multipole functions, $\{W_L, X_L, Y_L, Z_L\}$, that parameterise gauge transformations. Eventually, these six moments can be reduced to two gauge-inequivalent canonical ones $\{M_L, S_L\}$. At quadratic order in G, one needs to invert the equation $\Box h_2^{\alpha\beta} = N^{\alpha\beta}(h_1, h_1)$.

Because $h_1^{\alpha\beta}$ is known only in the exterior region ($r > d$), one cannot employ retarded or advanced Green functions, for which it is necessary to know the solution everywhere in space. However, at each finite PN order only a finite number of multipole moments contribute. So, one applies a multipolar post-Minkowskian expansion outside the source for $d/r < 1$ and introduces a regularisation to extend $N^{\alpha\beta}(h_1, h_1)$ at the origin. Blanchet and Damour proposed that at each order G^n the function $\Lambda_n^{\alpha\beta}$ be multiplied by r^B, B being a complex number whose real part is positive and sufficiently large, so that $r^B \Lambda_n^{\alpha\beta}$ is regular at the origin. Then, the solution of $\Box h_n^{\alpha\beta} = r^B \Lambda_n^{\alpha\beta}$ is obtained by using retarded Green functions, analytic continuation in the complex B-plane and extracting the coefficient of the zeroth power of r ($B = 0$), that is $u_n^{\alpha\beta} = \text{FP}_{B=0}\left\{\Box_{\text{ret}}^{-1}[r^B \Lambda_n^{\alpha\beta}]\right\}$, where FP stands for finite part [32]. The most general solution is obtained by adding to the inhomogeneous solution $u_n^{\alpha\beta}$ the homogeneous solution $v_n^{\alpha\beta}$, such that the harmonic gauge condition $\partial_\alpha h_n^{\alpha\beta} = 0$ is satisfied. Thus, $h_n^{\alpha\beta} = u_n^{\alpha\beta} + v_n^{\alpha\beta}$. The solution $h_n^{\alpha\beta}$ depends on the canonical multipole moments M_L and S_L, which at this stage do not know anything about the matter source. They only parameterise the most general solution of the Einstein equations in vacuum.

Near-zone post-Newtonian approximation: In the near zone one wants to obtain the solution of Eq. (6.2) in a multipolar PN expansion. Expanding $h^{\alpha\beta}$ and $\tau^{\alpha\beta}$ in powers of $1/c$, that is $h^{\alpha\beta} = \sum_{n=2} {}^{(n)}h^{\alpha\beta}/c^n$ and $\tau^{\alpha\beta} = \sum_{n=-2} {}^{(n)}\tau^{\alpha\beta}/c^n$, substituting them into the Einstein equations (6.1) and equating terms of the same order in $1/c$, one finds the relation [78] $\nabla^2[{}^{(n)}h^{\alpha\beta}] = 16\pi G[{}^{(n-4)}\tau^{\alpha\beta}] + \partial_t^2[{}^{(n-2)}h^{\alpha\beta}] \equiv {}^{(n)}f^{\alpha\beta}$. If one were solving the above differential equation via Poisson integrals, the presence of non-compact-support terms at some order in the PN expansion would prevent the integral from converging at spatial infinity. The convergence problem becomes even more severe when the multipolar expansion $1/|r - r'| = 1/r + r \cdot r'/r^3 + \cdots$ is applied. These were the problems that affected the original method of Epstein, Wagoner and Thorne. They were overcome by applying a more sophisticated mathematical method to invert the Laplacian and find the correct solution. This method is a variant of the analytic continuation technique developed by Blanchet and Damour in the wave zone, and it was carried out in [84–88]. Basically, one multiplies the function ${}^{(n)}f^{\alpha\beta}$ by r^B, where B is a negative real number whose modulus is sufficiently large that the integral is regular at spatial infinity. Then, the inhomogeneous solution is derived by analytic continuation in the complex B-plane and extracting the coefficient of the pole at $B = 0$, that is ${}^{(n)}u^{\alpha\beta} = \text{FP}_{B=0}\left\{(\nabla^2)^{-1}[{}^{(n)}f^{\alpha\beta} r^B]\right\}$. The most general solution is obtained by adding the homogeneous solution ${}^{(n)}v^{\alpha\beta}$, that is regular at the origin $r = 0$, to the inhomogeneous solution, that is ${}^{(n)}h^{\alpha\beta} = {}^{(n)}u^{\alpha\beta} + {}^{(n)}v^{\alpha\beta}$. As derived in [84–89], the solution in the near zone that matches the external field and satisfies correct boundary conditions at infinity involves a specific homogenous solution which can be expressed in terms of STF tensors as $\sum_{\ell=0} \partial_L[F_L^{\alpha\beta}(t - r/c)/r - F_L^{\alpha\beta}(t + r/c)/r]$ and it is fixed by matching it to the post-Minkowskian solution. In the region $d < r < \mathcal{R}$ the multipolar PN and post-Minkowskian series are both valid and one can resort to the standard method of matched asymptotic expansion to relate them, obtaining the solution over all space, $0 < r < +\infty$. In particular, the matching allows expression of the canonical multipole moments M_L and S_L (or the source multipole moments $\{I_L, J_L, W_L, X_L, Y_L, Z_L\}$)

in terms of integrals that extend over the matter and gravitational fields described by the PN-expanded $\tau^{\alpha\beta}$.

Direct integration of the relaxed Einstein equations: A different formalism that also cures the convergence issues that plagued previous brute-force, slow-motion approaches to gravitational radiation from isolated sources, was developed by Will, Wiseman and collaborators [52, 90, 91]. It is called the Direct Integration of the Relaxed Einstein Equation (DIRE). It differs from the MPM-PM approach in the definition of the source multipole moments. In both formalisms, the moments are generated by the PN expansion of $\tau^{\alpha\beta}$ in Eq. (6.2). However, in the DIRE formalism they are defined by compact-support integrals terminating at the radius $\mathcal{R}$ enclosing the near zone, while in the MPM-PN approach, as we have discussed, the moments are defined by integrals covering the entire space and are regularised using the finite-part procedure.

Gravitational waveform in the wave zone: When neglecting terms of order $1/R^2$ or higher, the general expression of the TT waveform that goes beyond the leading-order term (6.4) reads

$$h_{ij}^{\mathrm{TT}} = \frac{4G}{c^2 R} \sum_{k,q} \mathcal{P}_{ijkq}(N) \sum_{\ell=2}^{+\infty} \frac{1}{c^\ell \ell!} \left[N_{L-2} U_{kqL-2} - \frac{2\ell\, N_{mL-2}\, \varepsilon_{mn(k}\, V_{q)nL-2}}{c(\ell+1)} \right], \quad (6.7)$$

where the integer ℓ refers to the multipolar order, ε_{ijk} is the Levi-Civita antisymmetric symbol, parentheses denote symmetrisation and U_L and V_L are the multipole moments at infinity (called *radiative* multipole moments), which are functions of the retarded time $T - R/c$. The radiative multipoles U_L and V_L can be expressed in terms of the canonical multipole moments M_L and S_L as $U_L(T) = d^\ell M_L/dT^\ell + \mathcal{F}_L[M(T'), S(T')]$ and $V_L(T) = d^\ell S_L/dT^\ell + \mathcal{G}_L[M(T'), S(T')]$ where $\mathcal{F}_L$ and $\mathcal{G}_L$ are multi-linear retarded functionals of the full past behaviour, with $T' < T$. For example, the radiative mass-type multipole reads

$$U_L = \frac{d^\ell M_L}{dT^\ell} + \frac{2GM}{c^3} \int_0^\infty d\tau\, M_L^{(\ell+2)}(T_R - \tau) \left[\log\left(\frac{c\tau}{2r_0}\right) + \kappa_\ell \right] + \mathcal{O}\left(\frac{1}{c^5}\right), \quad (6.8)$$

where $T_R = T - R/c$, r_0 is an arbitrary length scale and κ_ℓ are constants. The term at order $1/c^3$ in the equation above describes the effect of back scattering of the gravitational waves on the Schwarzschild-like curvature associated with the total mass M of the source, the so-called *tail terms* [35, 92]. At order $1/c^5$, one has another hereditary term called the *memory term*, which is generated by non-linear interactions between multipole moments [40]. At order $1/c^6$, tails back scatter again with the Schwarzschild-like curvature generating *tail-of-tail terms* [93].

Gravitational-wave flux at infinity: The GW energy flux (or luminosity) $\mathcal{L}$ can be expressed in terms of the radiative multipole moments. It reads:

$$\mathcal{L} = \sum_{\ell=2} \frac{G}{c^{2\ell+1}} \left[\frac{(\ell+1)(\ell+2)}{(\ell-1)\ell\ell!(2\ell+1)!!} \left(\frac{dU_L}{dT}\right)^2 \right.$$

$$\left. + \frac{4\ell(\ell+2)}{c^2(\ell-1)(\ell+1)!(2\ell+1)!!} \left(\frac{dV_L}{dT}\right)^2 \right]. \quad (6.9)$$

and at leading order, using $U_{ij} = d^2 Q_{ij}/dT^2 + \mathcal{O}(1/c^3)$, the luminosity reduces to the famous "Einstein quadrupole formula"

$$\mathcal{L} = \frac{G}{5c^5} \left[\frac{d^3 Q_{ij}}{dT^3} \frac{d^3 Q_{ij}}{dT^3} + \mathcal{O}\left(\frac{1}{c^2}\right) \right]. \tag{6.10}$$

Gravitational radiation reaction: The gravitational radiation acts back on the motion of the binary through a radiation-reaction force. To have an explicit temporal representation of the waveform h_{ij}^{TT} and the fluxes, one needs to solve the problem of motion of the source, including radiation-reaction effects. The first relativistic terms in the equations of motion, at the 1PN order, were derived by Lorentz and Droste [94]. Then Einstein, Infeld and Hoffmann obtained the full 1PN corrections using the surface-integral method [95], in which the equations of motions are deduced from the vacuum field equations and they are valid for any compact object (NS, BH, etc.). Petrova [96], Fock [97] and Papapetrou [98] also obtained the equations of motion for the centres of extended bodies at 1PN order. Kimura, Ohta and collaborators introduced the ADM Hamiltonian formalism for doing PN computations [99, 100] and started the computation of the equations of motion for non-spinning bodies at 2PN order [101, 102]. The equations of motion for non-spinning point masses through 2.5PN order in harmonic coordinates were obtained by Damour and Deruelle [29, 31], who built on the non-linear iteration of the metric proposed by Bel *et al.* [103]. The gravitational radiation-reaction force in the equations of motion at 2.5PN order made it possible to unambiguously test general relativity through the observation of the secular acceleration in the orbital motion of binary pulsars [104]. Quite importantly, because of the *effacement principle*, the 2.5PN equations of motion are independent of the internal structure of the bodies [105]. In fact, the latter effect appears only at 5PN order for compact bodies. The 2.5PN equations of motion for point masses were also derived using extended compact objects by Kopeikin [106]. Itoh, Futamase and Asada [107] also computed the equations of motion through 2.5PN order, using a variant of the surface-integral method. In Table 6.1 we summarise the current status of the computation of the two-body equations of motion.

Advances in regularisation method: In the MPM-PN approach the two bodies are treated as point particles using delta functions. As a consequence, singularities appear when computing the near-zone metric and the equations of motion, because the gravitational field needs to be computed at the location of the particles. Thus, it is necessary to introduce a regularisation. Until the early 2000s the Hadamard regularisation was employed [108], but it does not provide unambiguous results at 3PN order, and so dimensional regularisation, which is a well-known regularisation scheme in particle physics, was adopted [109].

Canonical Hamiltonian approach: Another analytical approach that has been very effective in computing the two-body equations of motion at high PN orders is the canonical Hamiltonian approach. The canonical Hamiltonian formulation of GR was developed in 1958–1963 by Dirac [110–112], Arnowitt, Deser, Misner (ADM) [113, 114], and Schwinger [115], with important contributions by DeWitt [116], and Regge and

Teitelboim [117] in the 1960s and 1970s. The original motivation of the formulation was the quantisation of GR.

ADM introduced suitable coordinate conditions that allow solution of the field and constraint equations for the metric coefficients such that they become Minkowskian asymptotically. By adopting such a coordinate system and imposing the constraint equations, one obtains the *reduced* Hamiltonian $H_{\text{reduced}} = E[h_{ij}^{\text{TT}}, \pi^{ij\,\text{TT}}, x_A, p_A]$, which contains the full information for the dynamical evolution of the canonical field variables h_{ij}^{TT} and $\pi^{ij\,\text{TT}}$ and the canonical particle variables x_A and p_A. The reduced Hamiltonian has been explicitly computed by solving the constraint equations in a PN expansion and by adopting a suitable regularisation procedure. It contains a matter piece, an interaction piece that yields the radiation-reaction force, and a radiation piece. The energy and angular momentum losses can be computed through surface integrals in the wave zone. The ADM Hamiltonian has been successfully extended to gravitationally interacting spinning particles using a tetrad generalisation of the ADM canonical formalism. This allowed the computation of spin–orbit and spin–spin couplings at quite high PN orders. So far, the ADM Hamiltonian approach has been developed mostly for the conservative dynamics through high PN orders [30, 48, 49, 99, 100, 109, 118–124].

Effective-field theory approach: The two-body equations of motion and gravitational radiation have also been computed using Feynman diagrams, in GR by Bertotti and Plebanski [125], Hari Dass and Soni [126], and in scalar–tensor theories by Damour and Esposito-Farese [127]. An important turning point occurred in 2006, when Goldberger and Rothstein proposed a more systematic use of Feynman diagrams within effective field theory (EFT) to describe non-relativistic extended objects coupled to gravity [128]. As discussed above, there are three relevant scales in the two-body problem: r_s, the internal structure scale or the size of the compact object; r, the orbital separation, and $\lambda_{\text{GW}} \sim r(c/v)$, the radiation wavelength, where v is the typical velocity of the body in the binary. To carry out calculations in a systematic manner at high orders in v, Goldberger and Rothstein took advantage of the separation among those scales to set up a tower of effective field theories that account for effects at each scale [129].

When using the EFT approach to describe the object's size r_s, one does not need to resort to a specific model of the short-scale physics to resolve the point-particle singularity. Instead, one integrates out the internal structure of the object by matching onto an effective theory that captures the relevant degrees of freedom. Thus, one systematically parameterises the ignorance of the internal structure by building an effective point-particle Lagrangian that includes the most general set of operators consistent with the symmetry of GR (i.e., with general coordinate invariance). The operators in the point-particle Lagrangian have coefficients which encapsulate the properties of the internal structure of the extended objects. Moreover, short-distance divergences can be regularised and renormalised using standard methods in quantum field theory. Given a model for the internal structure, the values of the coefficients in the point-particle Lagrangian can be adjusted by a short-distance matching calculation so that they reproduce the observables of the isolated object. The resulting effective point-particle Lagrangian correctly describes length scales all the way to the orbital separation r. Then, to describe the binary problem at the scale r, it would be necessary to go beyond the point-particle effective Lagrangian and integrate out all modes of the graviton with wavelengths between the scales r_s and r, so that one obtains an effective Lagrangian of composite particles interacting with long-wavelength modes of the gravitational field.

So far, the EFT approach has recovered and confirmed the PN results previously obtained with the MPM-PN, DIRE and ADM canonical Hamiltonian methods for the two-body equations of motion up to 3.5PN order when spins are neglected [130, 131]. When spin effects are included, the EFT approach has extended the knowledge of the conservative dynamics and multipole moments to high PN orders [132–143].

Summary of results and final remarks: In Table 6.1 we summarise the impressive current status of PN calculations for the conservative dynamics, equations of motion and waveforms, for compact objects with and without intrinsic rotation (or spin) and tidal effects. We indicate by $n/2$-PN the PN term of formal order $\mathcal{O}(1/c^n)$ relative to the leading non-spinning term.[4] (For PN results of highly eccentric systems the reader may consult [71] and references therein.)

In summary, during the last thirty years there has been tremendous progress in the PN computation of the two-body equations of motion and gravitational radiation. Results obtained using different analytical techniques (i.e., point particles described by Dirac delta-functions, surface-integral methods, post-Minkowskian and PN expansions, canonical Hamiltonian formalism, and the EFT approach) have been compared with each other and agree. We stressed at the beginning of Section 6.2 that Einstein did not conceive the theory of GR starting from approximations to it. In spite of this, the PN approximation to the two-body dynamics is able to capture all relevant features of the weak-field, slow-motion dynamics and it can approximate quite well the theory of GR as long as the motion of comparable-mass compact objects is not highly relativistic and PN corrections at the highest order currently known are included. Future work at the interface between numerical and analytical relativity will be able to determine more precisely the PN region of accuracy, but this will imply doing very long NR simulations (i.e., simulations over hundreds of orbits) for generic binary configurations.

As we shall discuss in Section 6.3.3, current PN results allow us to compute the GW phasing with sufficient accuracy to detect quasi-circular, neutron-star inspirals and extract physical parameters if neutron stars carry mild spins. Moreover, the knowledge of higher-order PN calculations has made it possible to test the reliability of the PN expansion as the two bodies approach each other and also understand the practicability of extending those calculations at any PN order. It turns out that as one approaches the last stages of inspiral crucial quantities that enter the computation of the waveforms start being very sensitive to the PN truncation error, leading to unreliable results. As we shall see below, by wisely

---

[4] The spin of a rotating body is on the order $S \sim mlv_{\text{rot}}$, where m and l denote the mass and typical size of the body, respectively, and where v_{rot} represents the velocity of the body's surface. Here, we consider bodies which are both compact, $l \sim Gm/c^2$, and maximally rotating, $v_{\text{rot}} \sim c$. For such objects the magnitude of the spin is roughly $S \sim Gm^2/c$.

Table 6.1 *State-of-the-art of* PN *calculations for compact binaries with comparable masses. We list main references that contributed to the current accuracy. Unless otherwise specified, n/2-PN refers to the* PN *term of formal order* $\mathcal{O}(1/c^n)$ *relative to the leading non-spinning term.*

	No Spin	Spin-Linear	Spin-Squared	Tidal
Conservative Dynamics	4PN[a] [119, 120, 131] [124, 156–162]	3.5PN [50, 52, 139] [138, 163–167]	3PN [50, 52, 136] [135, 168–170]	7PN[b] [153–155]
Energy Flux at Infinity	3.5PN [93, 171, 172]	4PN [173–176]	2PN [51, 52, 177–179]	6PN [180]
RR Force	4.5PN [35, 91, 181, 182, 506]	4PN [183–185]	4.5PN [186]	6PN [153]
Waveform Phase[c]	3.5PN [187]	4PN [173, 175, 176]	2PN [52, 177–179, 188]	6PN [180, 189]
Waveform Amplitude[e]	3PN[d] [190–193]	2PN [188, 195]	2PN [51, 52, 188, 195]	6PN [154, 180]
BH Horizon Energy Flux[g]	5PN [196]	3.5PN [197, 198]	4PN[f] [197, 198]	– –

[a] Partial higher-order PN terms in the two-body energy for circular orbits have been computed both analytically and numerically [144–149]. The work of Bini and Damour built on [150–152].
[b] 2PN tidal effects in the conservative dynamics are known explicitly only for circular orbits.
[c] We refer to quasi-circular orbits only.
[d] The -2 spin-weighted $(2, 2)$ mode is known through 3.5PN order [194].
[e] We refer to quasi-circular orbits only.
[f] Spin couplings beyond the squared ones have also been computed.
[g] We count the PN order with respect to the leading-order luminosity at infinity. BH horizon flux terms start at 4PN and 2.5PN orders in the non-spinning and spinning case, respectively.

combining different analytical techniques one can avoid such shortcomings and further improve the accuracy of the two-body dynamics and GW emission up to coalescence.

6.2.2 Perturbation theory and gravitational self force

Extreme-mass-ratio inspirals composed of a stellar-mass compact object orbiting a supermassive or massive BH are promising sources for space-based and (future) ground-based detectors. The orbits are expected to be highly eccentric, non-equatorial and relativistic.

To detect such binary systems and extract the strong-field information encoded in the space-time around the larger body, one needs to model very accurately the equations of motion of the smaller body orbiting the BH and develop a consistent, appropriate wave-generation formalism. The PN framework, which is limited to slow velocities, is not suitable in this case. One needs to include relativistic effects at all orders and expand the field equations in the binary mass ratio (see Fig. 6.1). Henceforth, we let m denote the mass of the small object, M the mass of the central object and $q = m/M$ the mass ratio. Typically, extreme-mass-ratio binaries have $q \lesssim 10^{-5}$. If $g_{\mu\nu}$ is the metric of the background spacetime, the perturbation produced by the particle is $h_{\mu\nu} = \mathrm{g}_{\mu\nu} - g_{\mu\nu}$, where $\mathrm{g}_{\mu\nu}$ is the metric of the perturbed spacetime. The metric perturbation can be written as $h_{\alpha\beta} = \sum_{n\geq1} h_{\alpha\beta}^{(n)}$ with $h_{\alpha\beta}^{(n)} \propto q^n$. In the limiting case of a very small test mass orbiting a heavy central mass, one works at first order in q and obtains equations for the linear perturbations of the background geometry roughly of the kind $\Box_g h_{\alpha\beta}^{(1)} \sim T_{\alpha\beta}[z] \sim \mathcal{O}(m)$, where $T_{\alpha\beta}$ is the energy–momentum tensor of the small body and z^μ its worldline. [Essentially one obtains Eq. (6.1) with $\eta_{\mu\nu}$ replaced by the background metric $g_{\mu\nu}$ and the higher-order terms $\Lambda_{\mu\nu}$ neglected.] Those equations were derived in the 1950s and 1970s for the metric perturbations by Regge–Wheeler and Zerilli (RWZ) [199, 200] in the Schwarzschild case, and for the curvature perturbations by Teukolsky [201] in the Kerr case. Using suitable gauges, those equations (or variants of them [202]) can be integrated analytically, for quasi-circular orbits, by PN expansion in powers of v/c, v being the velocity of the small body, obtaining the gravitational radiation and luminosity at very high PN orders. (Strictly speaking, one computes analytically the Green function associated with the master equations.) In Table 6.2 we summarise the current status of PN calculations in BH perturbation theory. Furthermore, at leading order in the computation of the radiation field, one can assume that the small test mass moves along an adiabatic sequence of geodesics of the fixed background spacetime and compute the

Table 6.2 *State-of-the-art of* PN *calculations in BH perturbation theory (i.e., for an extreme-mass-ratio compact binary) in the case of quasi-circular orbits. We list main references that contributed to the current accuracy. Unless otherwise specified, n/2-PN refers to the* PN *term of formal order* $\mathcal{O}(1/c^n)$ *relative to the leading non-spinning term.*

	No Spin	Spin-Linear	Spin-Squared
Energy Flux	22PN	4PN	4PN
at Infinity	[53, 57, 219–223]	[55, 56, 58]	[58, 224]
BH Horizon	6PN	6.5PN	6.5PN[a]
Flux[b]	[150, 225, 226]	[150, 226, 227]	[150, 226, 227]

[a] Spin couplings beyond the squared ones are also present.

[b] We count the PN order with respect to the leading-order luminosity at infinity. BH horizon flux terms start at 4PN and 2.5PN orders in the non-spinning and spinning case, respectively.

gravitational radiation numerically solving the RWZ and Teukolsky equations [53, 54, 203–205]. Much progress has been made in the last twenty years to evolve those equations in a robust, accurate and fast way [206–213], and compute the gravitational waveform $h_{\alpha\beta}^{(1)}$ in the wave zone. Today, time-domain RWZ and Teukolsky equations can compute not only the waveform emitted during the very long inspiral stage, but also the plunge, merger and ringdown stages [213–218].

The perturbation sourced by the small test mass not only produces outgoing radiation in the wave zone that removes energy and angular momentum from the particle, but also produces a field in the near zone that acts on the test mass (i.e., the GSF) and gradually diverts it from its geodesic motion. Besides conservative terms, the GSF contains dissipative contributions that are responsible for the radiation-reaction force. The finite-mass corrections to the orbital motion due to the GSF are important for detection of extreme-mass-ratio binaries and extraction of parameters. Although at any given time the GSF yields fractional corrections to the motion of the small body on the order of $q \ll 1$, these corrections accumulate over the very large number of cycles ($\sim 1/q$), thus producing effects that cannot be neglected.

The computation of the GSF is not an easy task because the field generated by the particle's motion diverges on the particle's worldline. Indeed, the gravitational field is infinite at the particle's position. Thus, one first needs to isolate the field's singular part. Quite interestingly, a careful analysis shows that the singular piece does not affect the motion of the particle, but only contributes to the particle's inertia and renormalises its mass. The regular field is solely responsible for the GSF.

The case of a point electric charge moving in flat spacetime is well understood and dates back to work by Lorentz, Abrahams, Poincaré and Dirac [228]. The extension to curved spacetime has not been straightforward [229–231]. The proper definitions of the singular and regular Green functions from the Hadamard elementary functions for the wave equation in curved spacetime were obtained only in 2003 by Detweiler and Whiting [232]. In curved spacetime the GSF is non-local in time. It is given by a tail integral describing radiation that is first emitted by the particle and then comes back to the particle after interacting with the spacetime curvature. Because the regular field that is fully responsible for the GSF satisfies a homogeneous wave equation, it is a free radiation field that interacts with the particle and carries information about its past. We now follow [74] and sketch the derivation of the GSF equation at first order.

The gravitational perturbations produced by a point particle of mass m can be described by the trace-reversed potentials $\gamma_{\mu\nu} = h_{\mu\nu} - (1/2)(g^{\rho\lambda}h_{\rho\lambda})g_{\mu\nu}$. Imposing the Lorenz gauge $\gamma^{\mu\nu}{}_{;\mu} = 0$, one finds that the trace-reversed potentials satisfy the equation $\Box\gamma^{\alpha\beta} + 2R_\mu{}^\alpha{}_\nu{}^\beta\gamma^{\mu\nu} = -16\pi G T^{\alpha\beta}/c^4$, where covariant differentiation uses the background metric $g_{\mu\nu}$, $\Box = g^{\mu\nu}\nabla_\mu\nabla_\nu$, $T^{\alpha\beta}$ being the point mass's energy–momentum tensor. The solutions for the potentials are obtained in terms of retarded Green functions $G_{+\,\mu\nu}^{\alpha\beta}$ as $\gamma^{\alpha\beta}(x) = 4m\int_\gamma G_{+\,\mu\nu}^{\alpha\beta}(x,z)u^\mu u^\nu\,d\tau$, where the integral is done on the body worldline and $u^\mu = dz^\mu/d\tau$. The perturbations $h_{\mu\nu}$ are derived by inverting the equation $\gamma_{\mu\nu} = h_{\mu\nu} - (1/2)(g^{\rho\lambda}h_{\rho\lambda})g_{\mu\nu}$. Furthermore, the equations of motion of the small body are obtained: (i) imposing that the body follows geodesics in the metric $g_{\mu\nu}$ of the perturbed spacetime, (ii) removing the singular part $h^S_{\mu\nu}$ from the retarded perturbation and (iii) postulating that it is the regular part $h^R_{\mu\nu}$ that acts on the small body. They read

$$a^\mu = -\frac{1}{2}(g^{\mu\nu} + u^\mu u^\nu)(2h^{\text{tail}}_{\nu\lambda\rho} - h^{\text{tail}}_{\lambda\rho\nu})u^\lambda u^\rho, \tag{6.11}$$

with

$$h^{\text{tail}}_{\mu\nu\lambda} = 4m \int_{-\infty}^{\tau^-} \nabla_\lambda \left(G_{+\mu\nu\mu'\nu'} - \frac{1}{2}g_{\mu\nu}G_{+\ \ \rho\mu'\nu'}^{\ \ \rho} \right)(z(\tau),z(\tau'))u^{\mu'}u^{\nu'}\,d\tau', \tag{6.12}$$

where $\tau^- = \tau - \epsilon$ is introduced to avoid the singular behaviour when $\tau' = \tau$, $z(\tau)$ is the current position of the particle; all tensors with unprimed indices are evaluated at the current position, while tensors with primed indices are evaluated at prior positions $z(\tau')$. Finally, on the particle worldline the regular field is $h^R_{\mu\nu;\lambda} = -4m(u_{(\mu}R_{\nu)\rho\lambda\xi} + u_\lambda R_{\mu\rho\nu\xi})u^\rho u^\xi + h^{\text{tail}}_{\mu\nu\lambda}$.

The equation of motion (6.11) was first derived in 1996 by Mino, Sasaki and Tanaka [230], and then by Quinn and Wald [231]. It is known as the MiSaTaQuWa equation of motion. It is important to notice that whereas in the original derivation the MiSaTaQuWa equation appears as the geodesic equation in the metric $g_{\mu\nu} + h^{\text{tail}}_{\mu\nu}$, in the interpretation by Detweiler and Whiting, it is a geodesic equation in the (physical) metric $g_{\mu\nu} + h^R_{\mu\nu}$, which is regular on the worldline of the body and satisfies the Einstein equations in vacuum. The derivation in [230, 231] is limited to point masses, but Gralla and Wald [233] and Pound [234] demonstrated that the MiSaTaQuWa equation applies to any compact object of arbitrary internal structure. The MiSaTaQuWa equation of motion is not gauge-invariant [235] and relies on the Lorenz gauge condition. To obtain physically meaningful results, one needs to combine the MiSaTaQuWa equation of motion with the metric perturbations $h_{\mu\nu}$ to obtain gauge-invariant quantities that can be related to physical observables.

Although considerable progress has been made in the last several years to develop methods to calculate the metric perturbation and GSF at first order [235, 236], the majority of the work has focused on computing the GSF on a particle that moves on a *fixed* worldline of the background spacetime – for example, for a static particle [237], radial [238], circular [239, 240] and eccentric [241, 242] geodesics in the Schwarzschild spacetime. Methods to compute the GSF on a particle orbiting a Kerr BH have been proposed (e.g., see [243]) and actual implementations are underway. Recently, Warburton *et al.* [244] carried out the first calculation, in Schwarzschild spacetime, that takes into account changes in the particle's worldline as the GSF acts on the particle. It is important to stress that it is computationally very intensive to integrate the Einstein equations for very long inspiraling orbits. For this reason approximation methods have also been developed. Assuming that secular effects associated with the GSF accumulate on time scales much longer than the orbital period, one can employ the adiabatic approximation [245–249], which uses a field that is sourced by a geodesic and neglects periodic effects and the conservative portion of the GSF. For some choices of the binary parameters the adiabatic approximation can be sufficiently accurate for detection, but it has been shown [250] to be generically inaccurate for extracting binary parameters.

The GSF program is not yet complete. To obtain sufficiently accurate templates produced by extreme-mass-ratio binaries, the GSF needs to be computed at first order in q, but the gravitational energy flux at second order in q. This implies that the metric perturbations need to be computed at second order. The formalism to derive metric perturbations at second order has been developed [251–253]; calculations are underway and might be completed in a few years. Nevertheless, steady advances in the knowledge of the GSF have already been used to derive interesting, physical effects and higher-order PN terms, as we now discuss.

Barack and Sago [241] combined the conservative pieces of the GSF and metric perturbations to calculate the frequency shift in the innermost, stable circular orbit that originates from the GSF of the small body in Schwarzschild spacetime (see [254] for the extension to the Kerr spacetime). In [255] a similar shift was computed in the rate of periastron advance for eccentric orbits. In 2008 Detweiler pointed out [240] that the time component of the velocity vector, u^t, of a small test mass in the Schwarzschild spacetime is gauge-invariant with respect to transformations that preserve the helical symmetry of the perturbed spacetime. The inverse of u^t is an observable, as it is the gravitational redshift experienced by photons emitted by the orbiting body and observed at a large distance on the orbital axis. The redshift has been computed numerically through a GSF calculation as a function of the (gauge-invariant) orbital frequency and it has been compared to analytical predictions [144, 148, 149] and used to extract as yet unknown higher-order PN terms beyond the test-particle limit [144, 145, 147, 256, 257]. The latter was possible through the first law of binary BH dynamics [145, 257, 258]. Quite importantly, the redshift factor has been shown to be simply related to the binding energy and angular momentum of circular-orbit binaries [257]. Thus, the knowledge of the redshift can be employed to compute relativistic effects linear in q in the (specific) binding energy and angular momentum [146, 259, 260]. Other dynamical invariants have also been derived [261–263].

Lastly, in the absence of GSF results at second order in q and of comparisons to NR simulations for intermediate-mass-ratio binaries, it is difficult to assess the region covered by GSF results in Fig. 6.1. It is generally believed that the knowledge of relativistic effects at first order in q in the conservative dynamics and second order in q in the dissipative sector would be able to describe only waveforms from extreme-mass ratio inspirals having $q \lesssim 10^{-5}$. However, results at the interface between GSF, PN theory and NR are suggesting that leading-order GSF results may have a much larger range of validity including intermediate-mass-ratio binaries and perhaps even comparable-mass binaries when q is replaced by the symmetric mass ratio $mM/(m+M)^2$ [264–266]. Approximations to GR continue to be surprisingly successful.

6.2.3 The effective-one-body formalism

In 1998–2000, motivated by the construction of LIGO and Virgo detectors and the absence of merger waveforms for comparable-mass binary BHs from NR, an analytical approach that combines the PN expansion and perturbation theory, known as the effective-one-body

(EOB) approach [63, 64], was introduced. This novel approach was aimed at modeling analytically both the motion and the radiation of coalescing binary systems over the entire binary evolution (i.e., from the inspiral to the plunge, then the merger and the final ring-down). Several predictions [64, 267] of the EOB approach have been broadly confirmed by the results of NR simulations. These include: (i) the blurred, adiabatic transition from the inspiral to a plunge, which is merely a continuation of the inspiral, (ii) the extremely short merger phase, (iii) the simplicity of the merger waveform (i.e., the absence of high-frequency features in it, with the burst of radiation produced at the merger being filtered by the potential barrier surrounding the newborn BH), (iv) estimates of the radiated energy during the last stages of inspiral, merger and ringdown (0.6% to 5% of the binary total mass depending on BH spin magnitude) and spin of the final BH (e.g., $0.8M_{\mathrm{BH}}^2$ for an equal-mass binary, M_{BH} being the final BH mass), and (v) prediction that a Kerr BH will promptly form at merger even when BHs carry spin close to extremal. Soon after its inception, the EOB model was extended to include leading-order spin effects [268] and higher-order PN terms that meanwhile became available [269].

We now describe how the EOB formalism is able, in principle, to predict the full wave-form emitted by coalescing binary systems using the best information available from ana-lytical relativity. In Section 6.3.3, we shall show that the EOB approach can be made highly accurate by calibrating it to NR simulations, so that it can be used for detection and parameter estimation by ground- and space-based GW detectors.

There are three key ingredients that enter the EOB approach: (i) the conservative, two-body dynamics (or Hamiltonian), (ii) the radiation-reaction force and (iii) the gravitational waveform emitted during inspiral, merger and ringdown. In building these ingredients the EOB formalism relies on the assumption that the comparable-mass case is a smooth deformation of the test-particle limit. Moreover, each ingredient is crafted through a resum-mation of the PN expansion to incorporate non-perturbative and strong-field effects that are lost when the dynamics and the waveforms are Taylor-expanded in PN orders. The construction of the three ingredients leverages previous results. The finding of the EOB Hamiltonian was inspired by results in the 1970s aimed at describing the binding energy of a two-body system composed of comparable-mass, charged particles interacting electro-magnetically [270]. The resummation of the radiation-reaction force was initially inspired by the Padé resummation of the energy flux [271]. The description of the merger–ringdown waveform was inspired by results in the 1970s on the radial infall of a test particle in a Schwarzschild BH [272], where it was found that the direct gravitational radiation from the particle is strongly filtered by the potential barrier once the test particle is inside it. The construction of the EOB merger–ringdown waveform was also inspired by results in the close-limit approximation [273], in which one switches from the two-body to the one-body description close to the peak of the BH potential barrier. The recent description of the EOB inspiral–plunge waveform [274] was inspired by the multiplicative (or factorised) structure of the waveform in the test-particle limit case. We now review the three basic ingredients.

In the physical, *real* description, the centre-of-mass conservative dynamics of two parti-cles of masses m_1 and m_2 and spins S_1 and S_2 is described by the PN-expanded Hamiltonian

$H_{\mathrm{PN}}(\boldsymbol{Q},\boldsymbol{P},\boldsymbol{S}_1,\boldsymbol{S}_2)$, where $\boldsymbol{Q}$ and $\boldsymbol{P}$ are the relative position and momentum vectors. The basic idea of the EOB approach is to construct an auxiliary, *effective* description of the real conservative dynamics in which an effective particle of mass $\mu = m_1\, m_2/(m_1 + m_2)$ and effective spin $\boldsymbol{S}_*(\boldsymbol{S}_1,\boldsymbol{S}_2)$ moves in a deformed, Kerr-like geometry $g_{\mu\nu}^{\mathrm{eff}}(M,\boldsymbol{S}_{\mathrm{Kerr}};\nu)$, with mass $M = m_1 + m_2$ and spin $\boldsymbol{S}_{\mathrm{Kerr}}(\boldsymbol{S}_1,\boldsymbol{S}_2)$, such that the effective conservative dynamics is equivalent (when expanded in powers of $1/c$) to the original, PN-expanded dynamics. The deformation parameter is the symmetric mass ratio $\nu = \mu/M$, ranging from $\nu = 0$ (test-particle limit) to $\nu = 1/4$ (equal masses). Exactly solving the problem of a spinning, effective particle in this deformed, Kerr-like geometry amounts to introducing a particular *non-perturbative* method for resumming the PN-expanded equations of motion.

As done originally in [63], even if the problem is purely classical, it is quite instructive to obtain such a mapping between the real and effective dynamics by thinking quantum mechanically. For simplicity, let us restrict ourselves to non-spinning particles. Instead of considering the classical Hamiltonians $H_{\mathrm{real}}(\boldsymbol{Q},\boldsymbol{P})$ and $H_{\mathrm{eff}}(\boldsymbol{q},\boldsymbol{p})$ and their bounded orbits, we consider the energy levels $E_{\mathrm{real}}(N_{\mathrm{real}},J_{\mathrm{real}})$ and $E_{\mathrm{eff}}(N_{\mathrm{eff}},J_{\mathrm{eff}})$ of the quantum bounded states associated with the Hamiltonian operators. The energy levels depend on the principal quantum number N and the total-angular-momentum quantum number J, and they can be computed in a gauge-invariant manner within the Hamilton–Jacobi formalism, where N and J correspond to classical action variables. We sketch in Fig. 6.3 the real and effective descriptions. Because in quantum mechanics the action variables are quantised in integers, it is most natural to map the real and effective descriptions, requiring that the quantum numbers be the same (i.e., imposing $N_{\mathrm{real}} = N_{\mathrm{eff}}$ and $J_{\mathrm{real}} = J_{\mathrm{eff}}$), but allowing the energy axis to change. Doing the mapping explicitly, Buonanno and Damour [63] found the following, simple relation between the real and effective non-relativistic energies

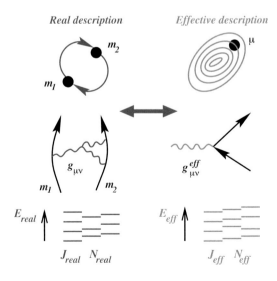

Figure 6.3 The real and effective descriptions in the EOB formalism.

($E_{\text{real}}^{\text{nr}} \equiv E_{\text{real}} - Mc^2$ and $E_{\text{eff}}^{\text{nr}} \equiv E_{\text{eff}} - \mu c^2$): $E_{\text{eff}}^{\text{nr}} = E_{\text{real}}^{\text{nr}}[1 + (v/2)(E_{\text{eff}}^{\text{nr}}/(\mu c^2))]$, which can be re-written as

$$\frac{E_{\text{eff}}}{\mu c^2} = \frac{E_{\text{real}}^2 - m_1^2 c^4 - m_2^2 c^4}{2 m_1 m_2 c^4}. \tag{6.13}$$

Remarkably, the relation (6.13) coincides with the one found in quantum electrodynamics [270], resumming part of the Feynman diagrams when mapping the one-body relativistic Balmer formula onto the two-body one, which describes charged particles of comparable masses interacting electromagnetically (e.g., positronium). (The mapping between the effective and real Hamiltonians can be also obtained through a suitable canonical transformation [63].) The improved resummed or EOB Hamiltonian, obtained by inverting the expression Eq. (6.13), reads [63]

$$H_{\text{EOB}} = Mc^2 \left[\sqrt{1 + 2v \left(\frac{H_{\text{eff}}}{\mu c^2} - 1 \right)} - 1 \right], \tag{6.14}$$

with

$$H_{\text{eff}}(\boldsymbol{r}, \boldsymbol{p}) = \mu c^2 \sqrt{A(r) \left[1 + \frac{p^2}{\mu^2 c^2} + \left(B(r)^{-1} - 1 \right) \frac{(\boldsymbol{n} \cdot \boldsymbol{p})^2}{\mu^2 c^2} + Q_4(\boldsymbol{p}) \right]}, \tag{6.15}$$

where $A(r)$ and $B(r)$ are the radial potentials of the effective metric $ds_{\text{eff}}^2 = -A(r)c^2 \, dt^2 + B(r)dr^2 + r^2 \, d\Omega^2$ and $Q_4(\boldsymbol{p})$ is a non-geodesic term quartic in the linear momentum that appears at 3PN order [269]. The metric potentials differ from the Schwarzschild ones by terms proportional to v. They can be computed in a PN series by matching the effective and real dynamics, thus $A_k(r) = \sum_{i=0}^{k+1} a_i(v)/r^i$ and $B_k(r) = \sum_{i=0}^{k} b_i(v)/r^i$. The EOB Hamilton equations read

$$\frac{d\boldsymbol{r}}{dt} = \frac{\partial H_{\text{real}}}{\partial \boldsymbol{p}}, \qquad \frac{d\boldsymbol{p}}{dt} = -\frac{\partial H_{\text{real}}}{\partial \boldsymbol{r}} + \mathcal{F}, \tag{6.16}$$

where $\mathcal{F}$ denotes the radiation-reaction force that can be expressed, assuming the energy balance equation and quasi-circular orbits, in terms of the GW energy flux at infinity [64, 271] and through the BH horizons [275, 276]. Using -2 spin-weighted spherical harmonics $_{-2}Y_{\ell m}(\theta, \phi)$, the gravitational polarisations can be written as $h_+(\theta, \phi; t) - ih_\times(\theta, \phi; t) = \sum_{\ell, m} {}_{-2}Y_{\ell m}(\theta, \phi) h_{\ell m}(t)$. The most recent description of the EOB inspiral–plunge modes proposed in [214, 274] read (symbolically)

$$h_{\ell m}^{\text{insp-plunge}}(t) = h_{\ell m}^{(N, \epsilon)} \, \hat{S}_{\text{eff}}^{(\epsilon)} \, T_{\ell m} \, e^{i\delta_{\ell m}} f_{\ell m} \, N_{\ell m}, \tag{6.17}$$

where the term $h_{\ell m}^{(N, \epsilon)}$ is the leading Newtonian mode, ϵ denotes the parity of the mode, the factor $T_{\ell m}$ resums the leading-order logarithms of tail effects, the term $e^{i\delta_{\ell m}}$ is a phase correction due to sub-leading-order logarithms, while the function $f_{\ell m}$ collects the remaining PN terms. Finally, the term $N_{\ell m}$ is a non-quasi-circular correction that models deviations from quasi-circular motion [277], which is assumed when deriving all the other factors in Eq. (6.17).

Inspired by results in the 1970s [272], the EOB approach assumes that the merger is very short in time, although broad in frequency, and builds the merger–ringdown signal by attaching a superposition of quasi-normal modes (QNMs) [64] to the plunge phase of the signal. Following the close-limit result [273], in a first approximation the plunge and QNM signals are matched at the light ring (i.e., at the unstable photon circular orbit), where the peak of the potential barrier around the newborn BH is located. Thus, the EOB merger–ringdown waveform is built as a linear superposition of QNMs of the final Kerr BH [64, 278]:

$$h_{\ell m}^{\text{merger-RD}}(t) = \sum_{n=0}^{N-1} A_{\ell mn} e^{-i\sigma_{\ell mn}(t - t_{\text{match}}^{\ell m})}, \qquad (6.18)$$

where N is the number of overtones [279, 280], $A_{\ell mn}$ is the complex amplitude of the nth overtone, and $\sigma_{\ell mn} = \omega_{\ell mn} - i/\tau_{\ell mn}$ is the complex frequency of this overtone with positive (real) frequency $\omega_{\ell mn}$ and decay time $\tau_{\ell mn}$. The complex QNM frequencies are known functions of the mass and spin of the final Kerr BH.

Finally, the full inspiral–plunge–merger–ringdown EOB waveform is obtained by joining the inspiral–plunge waveform $h_{\ell m}^{\text{inspiral-plunge}}(t)$ and the merger–ringdown waveform $h_{\ell m}^{\text{merger-RD}}(t)$ at the matching time $t_{\text{match}}^{\ell m}$ as [64]

$$h_{\ell m}^{\text{EOB}}(t) = h_{\ell m}^{\text{inspiral-plunge}}(t)\,\theta(t_{\text{match}}^{\ell m} - t) + h_{\ell m}^{\text{merger-RD}}(t)\,\theta(t - t_{\text{match}}^{\ell m}), \qquad (6.19)$$

where $\theta(t)$ is the Heaviside step function. For $t > t_{\text{match}}$ the GW emission is no longer driven by the orbital motion, but by the ringing of spacetime itself and the production of QNMs.

In Fig. 6.4 we show in the top panel the first, full EOB waveform for a non-spinning, equal-mass binary BH obtained in [64] (see [267] for first, full EOB spinning, precessing waveforms). In the lower panel we show the EOB GW and twice orbital-angular frequencies, the former flattening at late times at the least damped QNM frequency of the newborn BH, the latter having a peak around the EOB light ring. During the inspiral and plunge stages the GW emission is driven by the motion of the effective particle. As the effective particle passes through the EOB light ring, the direct GW emission from the effective particle is filtered by the potential barrier around the newborn BH, and the GW radiation is driven by the perturbed spacetime geometry through the emission of QNMs. Soon after the NR breakthrough, Buonanno *et al.* [278] compared EOB to NR waveforms, finding very reasonable agreement for the late inspiral, plunge, merger and ringdown stages. In particular, the EOB light ring was found to be located very close to the peak of the NR waveform and close to the location of the common apparent horizon, supporting the idea that GW radiation is well described by a superposition of QNMs as the two BHs merge. In Section 6.3.3 we shall discuss more recent, sophisticated comparisons and also calibrations of EOB waveforms. The EOB inspiraling dynamics has been compared directly to that produced in NR simulations through the computation of the periastron advance and the binding energy. The agreement is remarkable [264, 281, 282], even when no information from NR is used to improve the EOB model.

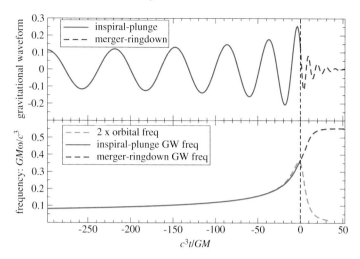

Figure 6.4 Gravitational waveform (upper panel), GW and twice orbital-angular frequencies (lower panel) from inspiral, plunge, merger and ringdown stages of a non-spinning, equal-mass binary BH as predicted in the EOB approach [63, 64].

The possibility of analytically modeling the merger waveform in the EOB approach stems from the waveform's simplicity. Does the simplicity imply that nonlinearities of GR do not play an important role? Not at all. Comparisons between numerical and analytical PN and EOB waveforms during the last 15–20 orbits of evolution have demonstrated that the best agreement with NR results is obtained when corrections up to the highest PN order available today are included. Thus, as expected, non-linear effects are present and dominant in the strong-field regime. The waveform simplicity is the result of (i) the presence of only one characteristic scale close to merger, when radiation-reaction, orbital and spin-precession time scales become of the same order of magnitude, (ii) the formation of a potential barrier around the newborn BH filtering the direct radiation from the merger *burst*, and (iii) the highly dissipative nature of disturbances in the BH spacetime because of QNMs.

The EOB conservative dynamics and waveforms have been extended to spinning BHs in [268, 283–288] and [289], respectively. In particular, motivated by the construction of an EOB Hamiltonian for spinning systems, that reduces to the Hamiltonian of a spinning particle in the extreme-mass-ratio limit, Barausse *et al.* [284] worked out, for the first time, the Hamiltonian of a spinning particle in curved spacetime at all orders in PN theory and linear in the particle's spin. The EOB approach has also been extended to NS binary systems, incorporating tidal effects in the dynamics and waveforms [154, 155].

To gain more insight and improve the transition from merger to ringdown [213, 214, 216–218] combined the EOB approch with numerical studies in BH perturbation theory. Concretely, they used the EOB formalism to compute the trajectory followed by an object spiraling and plunging into a much larger BH, and then used that trajectory in the source term of either the time-domain RWZ [199, 200] or the Teukolsky equation [201]. Solving those equations is significantly less expensive than evolving a BH binary in full numerical relativity. The possibility of using the test-particle limit to infer crucial information about the merger waveform of bodies of comparable masses follows from the universality of the merger process throughout the binary parameter space. The EOB approach has also been employed to generate quasi-circular, equatorial, very long, inspiraling waveforms in the extreme-mass-ratio limit, with accuracy comparable to those produced by the Teukolsky-equation code [290, 291]. Finally, the EOB formalism has been improved by taking advantage of important developments in the GSF formalism and its interface with PN theory [257]. In particular, using those results, the potentials entering the EOB metric have been derived at PN orders higher than those previously known [146, 161, 259, 292].

6.3 Compact-object binaries

Binary systems of compact objects[5] are the prime target for observation of almost all GW detectors. Loss of energy and angular momentum to GWs causes the companion stars of a binary system to spiral in toward each other, making the system more relativistic, in turn leading to a greater luminosity and a faster rate of inspiral. A vast majority of astronomical binaries are non-relativistic, thus they have negligible luminosity in GWs. Neutron stars and BHs are the most compact objects in the Universe, with orbital velocities that can reach close to the speed of light; therefore, the most luminous sources are also the most compact and strongly-gravitating systems. Indeed, when BH binaries merge $v/c \sim 1/\sqrt{2}$ (see, e.g., Eq. (3.2) of [293]) and luminosities could reach the phenomenal levels of $\sim 4 \times 10^{50}\,\mathrm{W} \sim 10^{24}\,L_\odot$, independent of the total mass of the binary.

6.3.1 Frequency–mass diagram

In Fig. 6.5 we plot several characteristic frequencies of equal-mass systems on (quasi-circular) orbits and illustrate the time scales over which they evolve. The frequency range 1 nHz–100 nHz is targeted by pulsar timing arrays (PTAs), 30 μHz–1 Hz by space-based interferometers and 1 Hz–10 kHz by ground-based or underground detectors. The dotted lines are frequencies starting from which a binary would last for 1 min, 1 day and 3 years until merger. As a binary evolves, its orbital frequency changes. For equal-mass sources that begin above the "30 yr (1 yr, 1 day) to chirp" line, it will be possible to measure their chirp mass and luminosity distance after a 30-year (1-yr, 1-day) observational period; for systems that begin below the "30 yr (1 yr, 1 day) to chirp" line, it will not be possible to infer the two quantities even after 30 years (respectively, 1 yr and 1 day).

During the final stages of the evolution, the component stars of a binary are no longer able to stay on stable orbits. They plunge towards each other when the orbital frequency is larger than a certain value, which we assume, for simplicity, to be that corresponding to the last stable orbit (LSO) of a Schwarzschild BH, i.e. $f_{\mathrm{merge}} \simeq c^3/(6^{3/2}\pi GM) \simeq (M/(10\,M_\odot))^{-1}\,440\,\mathrm{Hz}$. A binary's GW luminosity reaches its peak soon after it reaches

---

[5] The compactness of a body of mass M and size r_s is defined as $\mathcal{C} \equiv GM/(c^2 r_s)$. For BHs $\mathcal{C}_{\mathrm{BH}} = 0.5$, for NSs, depending on the EoS, $\mathcal{C}_{\mathrm{NS}} \sim 0.2$–$0.4$, while for the Sun $\mathcal{C}_\odot \ll 1$.

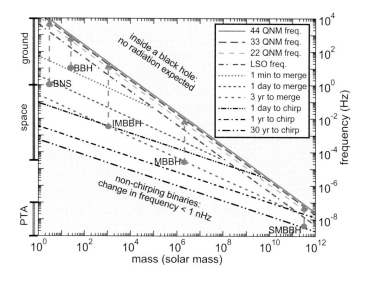

Figure 6.5 Frequency–mass diagram for equal-mass compact binaries.

the LSO. The merged object in all cases, except for very-low-total-mass NS binaries, is a highly deformed BH that quickly settles down to a quiescent Kerr state, emitting a characteristic spectrum of damped sinusoidal GWs – the quasi-normal modes (QNMs) [294]. The complex frequencies of the QNMs depend on the mass and spin of the final BH. Figure 6.5 shows, for a non-spinning BH, the frequencies of the dominant quadrupole mode (i.e., $\ell = 2, m = 2$ mode labeled '22 QNM freq.') and two of the higher-order modes (labelled "33 QNM freq." and "44 QNM freq.", see [279] for QNM frequencies) that are expected to carry a significant amount of energy. Although BHs have an infinite sequence of modes of higher frequencies, numerical simulations of BH mergers reveal that they are devoid of any appreciable energy in modes with $\ell > 4$ [295] and so we do not expect sources to radiate significantly in the top shaded region.

6.3.2 Zoo of compact-object binaries

Compact binaries occur in a very large range of masses and mass ratios. In Fig. 6.5 we show a few examples, but only for equal-mass, non spinning systems: (i) a NS binary of total mass $3\,M_\odot$ (BNS) that would be visible for about 15 minutes from 10 Hz in aLIGO/AdV and a few days from 1 Hz in ET, and it would chirp up in far less than a day from 1 Hz, (ii) a $20\,M_\odot$ BH binary (BBH) that lasts for almost 40 seconds from 10 Hz in aLIGO/AdV and $\sim$ 5 hours from 1 Hz in ET, (iii) a $10^3\,M_\odot$ intermediate-mass BH binary (IMBBH) that would chirp up in just one day from 3 mHz, but takes 3 years to merge, sweeping the bands of both eLISA and ground-based detectors, (iv) a $2 \times 10^6\,M_\odot$ massive BH binary (MBBH) in the eLISA band that would chirp up in less than a year, but takes 3 years to merge, starting at 30 μHz and (v) a $3 \times 10^{11}\,M_\odot$ supermassive BH binary in the PTA band that takes hundreds of years to merge, but would chirp up in just 30 years.

Stellar-mass binaries: Binaries of stellar-mass compact objects could contain two NSs, a NS and a BH or two BHs. Advanced LIGO and Virgo are well positioned to observe all such systems. In the case of a binary composed of two NSs the merger dynamics can be quite complex and depends on the binary's total mass M, the mass of the final remnant M_f and the maximum NS mass M_{max}^{NS} allowed by the (unknown) NS EoS. For the majority of mergers, the final remnant is expected to be a BH with or without an accretion disk, except on rare occasions when $M_f < M_{max}^{NS}$, in which case the final remnant can be a NS [296–298]. A BH with an accretion disk might promptly form if $M_{max}^{NS} < M \lesssim 3 M_\odot$ and the component masses are different from each other. However, if $M_{max}^{NS} < M \lesssim 3 M_\odot$ and the component masses are similar, a transient object called a *hypermassive* NS may form [297, 299]. The hypermassive NS is expected to be a non-axisymmetric ellipsoid supported against collapse by a combination of thermal pressure and differential rotation [298] and can delay BH formation for 1 ms to 1 s [298, 300]. This phase could witness quite a lot of rotational energy emitted as GWs, with a spectrum that is characteristic of the NS EoS [301, 302]. A BH without an accretion disk is not a very likely outcome, but it can happen if $M \gtrsim 3 M_\odot$ [296, 297, 303]. The hypermassive NS phase and BHs with accretion disks could both be accompanied by significant emission of electromagnetic radiation [304].

In the case of NS–BH binaries and binary BHs, the merger essentially produces a highly deformed BH that quickly settles down to a quiescent Kerr state by emitting QNM radiation. In the case of NS–BH mergers with mild mass ratios (say, $m_{NS}/m_{BH} \gtrsim 1/3$ for a non-spinning BH), the NS is tidally disrupted before the LSO and forms an accretion disk, which can generate electromagnetic signals; for smaller mass ratios the NS directly plunges into the BH without forming an accretion disk [305–307].

It is apparent from Fig. 6.5 that every binary in the band of ground-based detectors will merge within just a few days. How many such mergers might we expect each year within a given volume of the Universe? From the small number of observed pulsar binaries it is not possible to reliably estimate the merger rate. The estimated median rate is about one event per year in 100 Mpc^3, but it could be a factor of 100 smaller or a factor of 10 greater due mainly to uncertainties in the distance to radio pulsars, their radio luminosity function, the opening angle of the radio beam and the incompleteness of radio surveys [308]. Rate predictions based on the evolution of populations of massive-star binaries (called *population synthesis*) are also highly uncertain due to poor knowledge of the relevant astrophysics (e.g., supernova kick velocities and stellar metallicity). The upshot of these predictions is that aLIGO, with a horizon distance[6] of ~ 450 Mpc for binary NS mergers, might observe between 0.4 and 400 mergers per year [308].

There is currently no evidence for compact binaries in which one or both of the components is a BH. Population synthesis models predict a median rate of 5 binary BH

---

[6] The *horizon distance* of a detector is defined as the distance at which a face-on binary located directly above the plane of the detector produces an SNR of 8. The *reach* of a detector is the distance at which a randomly oriented and located source produces the same SNR; the reach of a detector is a factor of 2.26 smaller than its horizon distance [309]. A detector has roughly 50% efficiency, i.e. is able to see half of all sources, within its reach [309].

mergers and 30 NS–BH mergers per Gpc^3/year; also in this case the uncertainties are large. Advanced LIGO, with a horizon distance of 2.2 Gpc and 930 Mpc for these sources, could detect 10 and 20 mergers per year, respectively [308]. Metallicity of stars plays a key role in the evolution of massive stars. Black holes could be far more common in the Universe for low metallicities because stars would lose far less of their mass by stellar wind due to lower opacities, leading to more massive remnants towards the end of main sequence evolution. Expected binary BH detection rates in aLIGO for low metallicities are a factor of 10 larger [310] and binary NS rates a factor of 10 smaller. More recently, it has been noted that high-mass X-ray binaries, such as IC10 X-1 and NGC300 X-1, could be progenitors of BH binaries, in which case their merger rate could be far higher [311].

Supermassive and intermediate-mass black-hole binaries: There is growing evidence that certain galactic nuclei contain binary supermassive BHs [312] and eLISA would observe binary mergers if their total mass is in the range 10^4–$10^7 M_\odot$ (see Fig. 6.5). The merger rate of binaries of interest to eLISA is highly uncertain. This is because there are only a handful of such candidate binaries that would merge within the Hubble time [313]. Detailed modeling of these systems is very difficult due to there being many unknown astrophysical parameters, including the mass function and spin distributions of massive BHs and the process by which they grow [314]. eLISA could observe ~ 10–100 mergers per year, depending on the model that is used for the formation and growth of massive BHs [315, 316]. PTAs are expected to detect in five years or more the background produced by a population of $> 10^7 M_\odot$ supermassive BH binaries [317]; while they are not likely to observe mergers, they could detect individual systems if binaries of appropriate masses exist at redshifts $z \sim 0.1$–1, with orbital periods of ~ 1–30 years [318].

Although there is strong observational evidence for the existence of stellar-mass (5–30 $M_\odot$) and supermassive ($\gtrsim 10^6 M_\odot$) BHs, little is known about BHs of intermediate mass $\sim 10^2$–$10^5 M_\odot$, not to mention their binaries. However, there are hints that certain ultra-luminous X-ray sources (e.g., HLX-1 in ESO 243-49 [319]) might host intermediate-mass BHs. If a population of such objects exists and they grow by merger of smaller BHs, then, depending on their masses, ET and eLISA will be able to detect them (see Fig. 6.5). Their merger rates are highly speculative and range from 10 to 100 per year [316, 320, 321].

Extreme-mass-ratio binaries: When one of the companion masses is far smaller than the other (e.g., a $10 M_\odot$ BH orbiting a $10^6 M_\odot$ BH), we have the problem of a test body in *near*-geodesic motion in BH geometry. Such binaries are called *extreme-mass-ratio binaries,* as the mass ratio could be stupendously small, $\sim 10^{-6}$–10^{-4}. eLISA would be best suited to observe the inspiral of stellar- and intermediate-mass BHs into massive 10^4–$10^7 M_\odot$ BHs (see Fig. 6.5). Supermassive BHs in galactic nuclei are believed to grow by the infall of stellar mass and intermediate-mass BHs. Such events could be observed by eLISA at cosmological distances. For instance, the inspiral of a $10 M_\odot$ BH into a $10^6 M_\odot$ supermassive BH at 1 Gpc would be visible in eLISA. The rates in this case too are highly uncertain and range from a few to several hundreds per year [316, 322, 323].

Figures 6.6 and 6.7 plot the characteristic amplitude $h_c \equiv \sqrt{f}|H(f)|$, where $H(f)$ is the Fourier transform of the signal, for several non-spinning, equal-mass binaries with random

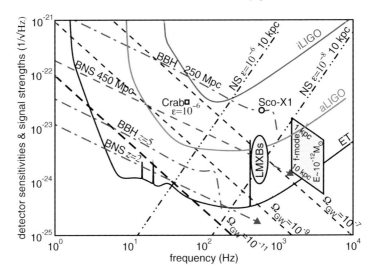

Figure 6.6 Sources of GWs for the ground-based detectors iLIGO, aLIGO and ET. For continuous waves and stochastic backgrounds the characteristic amplitudes of the signals are plotted assuming an integration time of 1 year.

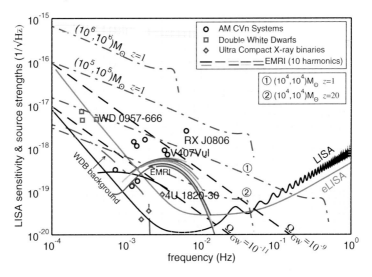

Figure 6.7 Sources of GWs for the space-borne detectors LISA and eLISA. For double white dwarfs, AM CVn systems, X-ray binaries and the stochastic background, the characteristic amplitude is computed assuming an integration time of 1 year.

orientation and sky position; for BH–BH binaries we use the inspiral–merger–ringdown signal, for BNS systems we have plotted only the inspiral part of the signal terminated at the LSO frequency because the spectrum of the merger signal is known only numerically and it varies greatly, depending on the microphysics and NS EoS. Also plotted are the detector noise amplitude spectra of three generations of ground-based interferometers in Fig. 6.6 and two versions of LISA in Fig. 6.7.

Black-hole binaries of mass 50–2000 $M_\odot$ can be detected by aLIGO/AdV at redshifts $z \sim 0.3$–1.4. The largest confirmed BH mass in the stellar range is $\sim 15\,M_\odot$ [324], but there are hints of even heavier BHs of 23–34 $M_\odot$ [325]. Theoretically, low-metallicity massive stars could lead to BHs of 50 $M_\odot$ or higher [310]. For binary systems of total mass 50–100 $M_\odot$, detectors are sensitive to the final moments of merger, when the strong-field dynamics dominates the evolution. Comprehensive studies have shown that, depending on the mass ratio, full inspiral–merger–ringdown waveforms should be used as matched filters when $M \gtrsim 10$–15 $M_\odot$, if one wishes to have a loss in detection rate of no more than 10% [293]. Space-based detectors would observe the merger dynamics from $\sim 10^5$–$10^7\,M_\odot$ binaries, with SNRs ~ 300 for sources at $z \sim 3$ [316, 326]. These SNRs are so large that templates would need to be improved beyond their current status, so as not to bias the estimation of the system's masses and its position on the sky.

6.3.3 Interface between theory and observations

A search for GWs from sources with known amplitude and phase evolution can make use of *matched filtering*, which involves cross-correlating the detector output with a copy of our best guess of the expected signal called a *template*. If the template matches the signal well, then the correlation between the noisy signal and the template builds up with time, giving rise, on average, to a positive output. Matched filtering, however, is very sensitive to the signal's phase evolution; even tiny phasing errors in the template can destroy the cross correlation. It is critical to have accurate templates so that the SNR lost due to incorrect templates is negligible.

For low-mass, inspiraling binary systems, carrying very mild spins, i.e., for NS–NS binaries, any 3.5PN approximant is accurate enough for detection [293], a remarkable result of the PN formalism. Until what time in the binary evolution (or, equivalently, for which total mass of the binary) is it safe to employ PN approximants in GW searches? Is there one particular PN approximant that is more accurate than others for any mass ratio and spin? Such questions remained unsolved for many years and were among the motivations of the EOB formalism in late 1990s. To cope with those uncertainties, during the few years before the NR breakthrough, *detection-template families* were developed [277, 327–329] and some of them were used in LIGO searches [330, 331]. To incorporate possible systematics present in PN approximants, those template families either extended the binary parameter space to unphysical regions or incorporated higher-order physical effects, so that they could reach higher overlaps with both PN approximants and EOB waveforms. Eventually, after the NR breakthrough in 2005, PN approximants started to be compared to highly accurate NR waveforms [332, 333] and also to EOB waveforms calibrated to NR waveforms [293]. It was found that for $M \gtrsim 10$–15 $M_\odot$ non-quasi-circular effects cannot be neglected, and templates that include inspiral, merger and ringdown should be employed to avoid a large loss in the detection rate [293]. It was also found that PN approximants did not perform very well for large mass ratios, i.e., for NS–BH binaries or IMBHs. This is because in the PN approach, exact, known results in the test-particle limit are expanded in a PN series, washing out crucial non-perturbative information – a drawback that was another motivation for the EOB formalism.

In recent years, a variety of studies have been carried out at the interface between analytical and numerical relativity. The results have indicated that the best way to provide accurate templates for a successful detection and extraction of binary parameters is to combine the knowledge from all the available methods: PN, GSF and NR. One could try to directly combine PN-computed waveforms with NR waveforms, thus building a hybrid waveform. However, if the goal is to produce *highly* accurate templates, this method would still require high computational cost, because the different PN approximants agree sufficiently well with each other *only* at large separations, thus the hybridisation should start hundreds of GW cycles before merger [334–336]. An alternative avenue is provided by the EOB approach.

Analytical vis-à-vis numerical relativity: As the earliest comparisons with NR waveforms demonstrated [278, 337] (cf. Section 6.2.3), the EOB formalism is able to describe waveforms emitted during the inspiral, plunge, merger and ringdown stages using *only* analytical information. Those first comparisons employed the 3.5PN EOB dynamics and leading-order PN waveforms. Subsequent studies carried out with highly accurate NR waveforms revealed the necessity of including higher-order PN terms in the EOB dynamics, energy flux and waveforms if the goal is to develop highly accurate templates for aLIGO/AdV searches. As a consequence, higher-order PN terms (in particular, the test-particle limit terms) are included in the gravitational modes $h_{\ell m}$ [274, 289, 338]. Since PN corrections are not yet fully known in the two-body dynamics, higher-order PN terms are included in the EOB dynamics with arbitrary coefficients [295, 338–346], which are then calibrated by minimising the phase and amplitude difference between EOB and NR waveforms aligned at low frequency. Those coefficients have been denoted *adjustable or flexible parameters*. In particular, EOB non-spinning waveforms (including the first four subdominant modes) have been developed with any mass ratio and shown to be indistinguishable from highly accurate NR waveforms with mass ratios 1–6 up to SNRs of ~ 50 [347]. Note, however, that current NR waveforms cover the full detector bandwidth only for binaries with total mass larger than $M \gtrsim 100 M_\odot$, thus those results are not yet conclusive. EOB waveforms are also stable with respect to the length of the numerical waveforms [336]. EOB waveforms for non-precessing systems with any mass ratio and spin have also been developed and calibrated to existing, highly accurate numerical waveforms, which, however, do not yet span the overall parameter space [346]. EOB waveforms for precessing systems can be built from those for non-precessing ones [348]; they capture remarkably well the spin-induced modulations in the long inspiral of NR waveforms and will be calibrated and improved in the near future. In Fig. 6.8 we show the agreement between state-of-the-art EOB [276] and NR waveforms for an equal-mass BH–BH binary with both spins aligned with the orbital angular momentum and quasi-extremal (top panel), and a single-spin binary BH with mass ratio 5, precessing with mild spin magnitude (bottom panel).

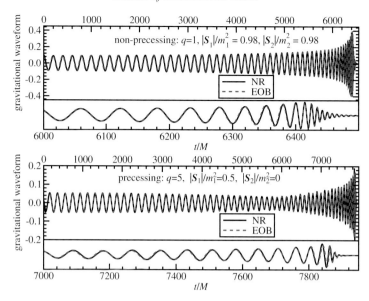

Figure 6.8 State-of-the-art comparison [346] between (calibrated) EOB and NR waveforms for quasi-extremal, non-precessing spins (top panel) and precessing spins (bottom panel); lower parts show the final few cycles. (The x-axis should be c^3t/GM and spins should be multiplied by c/G to get appropriate physical units.)

Starting with [337, 349], a more phenomenological avenue has also been followed to produce inspiral–merger–ringdown waveforms. In this case, the original motivation was to provide aLIGO/AdV detectors with inspiral, merger and ringdown waveforms that could be computed efficiently during searches and used to detect high-mass coalescing compact binaries. The full waveforms are constructed by first matching inspiral PN templates and NR waveforms in either the time or frequency domain, and then fitting this hybrid waveform in the frequency domain to a stationary-phase-approximation-based template, augmented by a Lorentzian function for the ringdown stage. As NR waveforms started spanning larger regions of the parameter space, the phenomenological waveforms have been improved [350], and extended to non-precessing [351] and precessing [352] binary BHs.

Today, highly accurate NR waveforms, having several tens (~ 70) of GW cycles and generic BH-spin orientations, can be produced by the pseudo-spectral Einstein code (SpEC) of the Simulating eXtreme Spacetime (SXS) collaboration [353]. Although such waveforms do not span the entire parameter space and are not long enough to be employed as search templates, they do allow testing of the stability of calibrated analytical waveforms with respect to the length of the simulations, improvement of the accuracy of analytical templates and also discrimination between different PN/EOB approximants. We expect that longer and more accurate numerical waveforms will be produced even more efficiently in the near future. Furthermore, the production of many short numerical waveforms with finite-difference codes [354] will continue to help with extracting interesting information about the characteristics of the merger signal. Eventually, all those advances will reduce systematics in the analytical waveforms, so that they can be used not only to observe GWs with advanced detectors, but also, in the future, by space-based detectors to extract binary parameters and to test general relativity at high SNRs ($\sim 10^3$).

6.3.4 Results from LIGO and Virgo

Data from several science runs of iLIGO and Virgo have been analysed to search for compact binary coalescences. No GW signals were found, but the results were used to set upper limits on merger rates and exclusion distances to short, hard, gamma-ray bursts (GRBs). The searches have used PN templates for "low-mass" binaries [355] with total mass $< 25\,M_\odot$ and component masses $> 1\,M_\odot$ and EOB templates calibrated to NR waveforms [339] for "high-mass" binaries [356] with total mass in the range $[25, 100]\,M_\odot$ and mass ratio in the range $1 \leq m_1/m_2 \leq 6$. The horizon distance to low-mass systems during iLIGO Science Run S6 and Virgo Science Run VSR3 was 40 Mpc, 80 Mpc and 90 Mpc, for binaries with masses $(1.35 + 1.35)\,M_\odot$, $(1.35 + 5.0)\,M_\odot$ and $(5.0 + 5.0)\,M_\odot$, respectively. The corresponding upper limits, in units of $\mathrm{Mpc}^{-3}\,\mathrm{Myr}^{-1}$, were 130, 31 and 6.4 [355]. These limits were derived for binaries with non-spinning components; if spins are included the upper limits are 15% higher for binaries containing one or more BHs. The horizon distance to high-mass binaries [356] ranged from 230 Mpc for a $(14 + 14)\,M_\odot$ binary to nearly 600 Mpc for a $(50 + 50)\,M_\odot$ binary; the corresponding upper limits, in units of $\mathrm{Mpc}^{-3}\,\mathrm{Myr}^{-1}$, were 0.87 and 0.07. Additionally, searches for intermediate-mass BHs with component masses $50\,M_\odot$ to $350\,M_\odot$ were carried out [357] using an excess-power algorithm in the time–frequency plane, setting upper limits in the range 0.14–13 $\mathrm{Mpc}^{-3}\,\mathrm{Myr}^{-1}$ for equal-mass binaries.

The upper limits from the low-mass searches are roughly two orders of magnitude away from the expected "realistic" merger rates [308]. Advanced detectors, at their design sensitivity [358], will improve the range by a factor of ~ 10, and increase the search volume by a factor of 1000. We can, therefore, expect the network of aLIGO/AdV/KAGRA to be making detections once they reach their design sensitivities. The current plan [358] is to collect data intermittently for periods of 3 to 6 months each year as the detectors' performances are improved from their initial sensitivity in late 2015 to their design sensitivity by the end of the decade. There is a fair chance that the first detections might happen around 2017, when the range for binary NSs is expected to be ~ 100 Mpc (or a horizon distance of 250 Mpc). Initial detections, with two or three detectors, might only have a moderate SNR (~ 12–15) and it might only be possible to localise events to within hundreds of square degrees. The addition of KAGRA and a new detector in India [359] will help improve the angular resolution of the network to tens of square degrees and facilitate easier EM follow-up of mergers. The analysis methods deployed in the searches have proven their ability to make use of the predicted waveforms to identify events and measure their parameters and compute the false-alarm probability and statistical significance of detected events. The best example of this is the GW100916 event (popularly called the *Big Dog* event) that was secretly injected into the iLIGO–Virgo data streams as part of the *Detection Challenge* (http://www.ligo.org/news/blind-injection.php). It was successfully identified by the on-line and off-line analysis pipelines, attributing a very high significance to the detected event [355].

Finally, new collaborative efforts have been established to coordinate activities between GW data analysts (or astronomers), numerical relativists and analytical relativists which have enabled testing of data analysis pipelines, production of a variety of NR simulations of binary BHs, and building of more robust models for use in searches and in extracting astrophysical information from the data [360–363].

6.3.5 Science targets and challenges

Gravitational waves from compact sources will unravel many unsolved problems in astronomy, fundamental physics and cosmology. In this section we will discuss science targeted by GW observations and the challenges that must be addressed in achieving those targets.

Science targets

Formation and evolution of stellar-mass compact binaries: Coalescing compact binaries form from massive stars. Their formation involves a number of stellar processes. These include the evolution of the two stars through the main sequence; gravitational collapse; supernova of the more massive star and associated kick that could disrupt the binary; evolution of the binary through the common envelope phase, wherein the larger giant star transfers mass to its compact companion; the second supernova and associated kick [364]. The parameters of the final compact binary remnant depend on all these factors, as well as on the metallicity, chemical composition, masses and spins of the progenitors and the initial separation of the stars [310]. Understanding the formation and evolution of these compact objects is an open problem in astrophysics as many of the mechanisms mentioned above are poorly understood, not accessible from observations and difficult to model.

Different evolutionary models of compact binaries predict different coalescence rates for the three types of compact binary mergers. By determining the rates for these different populations it will be possible to discriminate amongst the many competing models that are currently prevalent. Determining the mass function and spin distributions (i.e., spin magnitudes and relative orientations) and mass ratios of companion stars will add more discriminatory power to single out the correct model that describes the formation and evolutionary mechanisms [365]. Some models also predict a gap in the largest NS mass allowed by the EoS and the smallest mass of a BH formed by stellar evolution. Gravitational-wave observations could help verify the existence of such a mass gap [366].

Gamma-ray bursts: Observing GWs in coincidence with GRBs will have a tremendous impact on understanding the progenitors of GRBs and how they are powered. Moreover, coincident observation of GRBs and GWs will help identify the host galaxy of a GW source and measure its redshift. If progenitors of short GRBs (shGRBs) are binary NS mergers [367], then this would help measure both the luminosity distance and the redshift to the source, *without* the use of the cosmic distance ladder [368, 369]. Clearly, such observations will have a great potential for precision cosmology [370–372].

If binary NS mergers are progenitors of shGRBs then in addition to beamed emission of gamma-rays, they could also emit isotropic radiation at optical and infra-red wavelengths. The neutron-rich material that is ejected in the process of merger could produce heavy nuclei that decay by r-process radioactivity, powering a transient optical and infra-red source called a *kilonova* [304]. Follow-up observations with the Hubble Space Telescope of the shGRB 130603B at a redshift of $z = 0.356$ provided the first evidence of a kilonova [373, 374], with an apparent magnitude of 25.8. At 200 Mpc, the distance reach of ground-based advanced detectors to binary NS mergers, such transients would have an apparent magnitude of about 23 and be observable by some of the ground-based optical and near-IR telescopes, but sky localisation of GW events will be a challenge.

Cosmology: Compact binary sources are quite unique for cosmology as they are standard candles [368]: for binaries that chirp up during the course of observation one can measure the source's luminosity distance from GW observations alone. Weak gravitational lensing would bias distance measurements of individual sources [375], but a large population of events, as might be expected in the case of ET, can average out lensing biases [371]. If the host galaxy of a merger event is identified and its redshift measured, then we can use a population of binary coalescence events to infer cosmological parameters. In fact, GW observations might also measure the redshift through galaxy clustering using wide-field galaxy surveys [368, 369]. Moreover, tidal effects can be used to determine the source's redshift, provided the NS EoS is known [376].

There is strong observational evidence that galactic centres, including the Milky Way [377], host supermassive BHs of 10^6–$10^{12}\,M_\odot$. When and how did such BHs form? Did the BHs precede the galaxies or did they form after the galaxies were assembled? What were their initial masses and how did they grow? These are among the most pressing unsolved questions in cosmology. Gravitational-wave observations by PTAs, eLISA and ground-based interferometers will together cover the entire spectrum of BH binaries up to $z \sim 5$–20, redshifts so large that the Universe was in its infancy assembling the first stars and galaxies.

Neutron-star equation of state: By determining the EoS of NSs we can infer the composition and structure of NS cores, which has remained largely unknown nearly half-a-century after the discovery of the first pulsar, although astronomical observations have begun to indicate hints of neutron superfluidity in the core [378, 379]. Tidal interaction in compact binaries, where one or both of the companions is a NS, depends on the EoS. Tidal effects are imprinted in the inspiral phasing starting at 5PN order beyond the leading term, and in the merger and post-merger dynamical phases, as well. Advanced LIGO/AdV could distinguish between extreme models of equations of state by observing ~ 25 NS binary inspirals [380], while ET should measure the EoS quite accurately with a single loud event [381].

Testing gravitational dynamics: Signals from coalescing compact binaries can be used to probe the strong-field dynamics of gravity and as such facilitate tests of GR and its alternatives. Proposed tests either assume that GR is correct and look for small deviations from GR [382, 383], or begin with an alternative theory of gravity and determine the degree to which observations favour the alternative [384, 385]. Orbits of small BHs plunging into massive BHs are very complex and capture the non-linear dynamics of gravity. By observing the emitted radiation one can map out the geometry of the massive objects that reside in galactic nuclei and check if it agrees with the Kerr geometry or if BHs have extra "hair" [386–388]. An alternative approach checks for consistency of GWs from QNMs emitted during the ringdown phase of a BH binary merger [389].

Challenges

Critical problems in numerical relativity: It is important to continue to test the accuracy of analytical inspiral–merger–ringdown waveforms by comparing them to long NR wave-forms and to characterise any systematic biases in parameter estimation that might result due to inaccurate modeling. If BHs carry large spins, waveforms emitted by NS–BH and BH–BH binaries with mass ratios $\gtrsim 4$ will have modulations due to spin-induced preces-sion, which accumulate mostly during the long inspiral. Thus, more comprehensive studies are needed to understand the dynamics as a function of mass ratio, BH spins and EoSs of NSs. Moreover, at present, accurate NR waveforms span the entire aLIGO/AdV bandwidth only if their total mass is larger than $\sim 100\,M_\odot$ and their mass ratio is $\lesssim 10$. Longer wave-forms for larger mass ratios and generic spin configurations would be highly challenging, but they will be invaluable in validating analytical models, even if they sparsely sample the full parameter space.

Binary neutron star and NS–BH merger simulations that use realistic EoSs and include neutrino transport and other microphysics will be necessary to extract the best science from the data. Such simulations will also provide more accurate templates for the merger and bar-mode instability phases to infer redshifts from GW observations alone. Analytical and semi-analytical models of NS–NS and NS–BH mergers will be crucial for parameter estimation and Bayesian hypothesis testing. This will be a huge challenge for the bar-mode instability regime, where the signal does not seem to have any phase coherence (see, e.g., simulations in [301, 302]). Even so, accurate modeling of the expected spectrum or time–frequency content of the signal can be useful in understanding the physics of dynamical instabilities. Several related questions remain open, such as the role of large NS magnetic fields and spins in the merger dynamics of binary NSs and NS–BHs [304].

Critical problems in analytical relativity: As discussed in Section 6.2.1, advances in PN theory over the last thirty years will enable the detection of BNS inspirals (if NSs carry mild spins) with negligible loss in detection rates and in extraction of binary parameters. When spins are present, the PN phasing and amplitude are not known as accurately as in the non-spinning case, causing PN approximants to differ substantially even during the inspiral phase [390]. Thus, spin couplings through at least 4.5PN order beyond the leading order are required in the conservative dynamics and gravitational flux at infinity, and through a similar PN order in the BH-absorbed horizon flux. Once available, spin couplings will also be employed in the EOB formalism to further improve inspiral–merger–ringdown waveforms. Considering that today the two-body, non-spinning conservative dynamics is known at 4PN order, it will be relevant to derive the energy flux at 4PN order to allow the computation of the phase evolution at 4PN order, further validating analytical templates against NR waveforms.

Recent results that take advantage of GSF calculations have allowed the computation of new terms in the conservative dynamics at PN orders higher than 4PN. However, if we consider that to fully complete the computation of the dynamics at 3PN and 4PN orders, it was necessary to overcome novel, specific subtleties that appeared at those PN orders (notably, at 3PN order one had to replace the Hadamard regularisation with dimensional regularisation and at 4PN order it was necessary to introduce a non-local-in-time action to properly include tail effects), it is difficult to imagine that, in the future, PN calculations could be systematically and automatically extended to an *arbitrarily high* PN order by simply using algebraic computer programs. This limitation does not depend on the particular technique that is used (MPM-PN or DIRE formalisms, EFT or Hamiltonian canonical formalism), but it seems to be simply a consequence of quibbles and complexities of the nonlinearities of GR. It is worth noticing that at present it does not also seem necessary to have Taylor-expanded PN results at arbitrarily high PN orders to detect the signals and to extract the best science. A combination of PN, perturbative, and NR results and resummation techniques can effectively work around the problem.

As discussed in Section 6.2.2, to obtain sufficiently accurate templates for extreme-mass-ratio binaries, the metric perturbations need to be computed at second order. The formalism has been developed, calculations are underway and should be completed in the next few years.

Refining analytic source models and waveforms: Current searches in ground-based detectors assume that compact binaries are on quasi-circular orbits when they enter the detector sensitivity band (see, e.g., [356]). This is very likely a good approximation for binaries formed in fields as radiation back reaction circularises a binary much faster than the orbit decays [391, 392]. However, eccentric binaries may form through the Kozai mechanism or dynamical capture in dense stellar environments [393–396] and in this case the eccentricity can be large at merger. Eccentric-binary event rates are very uncertain. Eccentric-binary waveforms are not required for detection in aLIGO/AdV/KAGRA, unless the eccentricity at 20 Hz is larger than 0.1. Faithful models that take into account eccentricity would be needed to detect GWs from highly eccentric binaries or to accurately estimate parameters of mildly eccentric binaries.

Except for [61], past analysis pipelines have mostly searched for binaries composed of non-spinning objects in the two-dimensional space of component masses [356]. It is important to explore the relevance of including spins for detection. Several studies have indicated that due to degeneracies in the binary parameter space, template families containing a single effective spin could be sufficient to detect waveforms emitted by double-spin binary systems [328, 329, 397–399]. Moreover, precession-induced modulations can be incorporated in template waveforms in an efficient way, reducing also the dimensionality of the parameter space [328, 329, 348, 366, 400]. However, we still lack a comprehensive study that spans the entire parameter space and determines, after taking into account how the improvements in sensitivity can be negated by increases in false-alarm probability, where in the parameter space each category of searches – single- and/or double-spin, non-precessing or precessing – is needed (for some first steps in this direction see [401, 402]). Furthermore, studies are needed on the systematic biases in the estimation of parameters due to the use of incomplete waveform models, especially when spin-induced modulations in NS–BH and BH–BH systems are included. At present EOB waveforms for spinning BH binaries are computationally expensive to generate (although far faster than doing NR simulations) and hence not suitable for use in Markov-Chain Monte Carlo-based parameter estimation methods. Quite importantly, accelerated waveform-generation techniques that use reduced-order algorithms or singular-value decomposition techniques have been proposed to address such problems [403–407].

Finally, using PN results and NR simulations, analytical templates that extend up to merger and include tidal effects are under development [408, 409]. They will be needed to extract the best information on tidal effects and EoS in NS–NS and NS–BH coalescences. Including tidal effects in *point-particle* templates calibrated to highly accurate BH–BH waveforms from NR (e.g., EOB waveforms) is the best way of controlling systematic errors due to lack of knowledge of higher-order point-particle terms in the PN expansion. In fact, employing inspiraling, point-particle PN templates augmented by tidal effects leads to large systematic biases and limits the extraction of tidal information [410, 411].

Lastly, all sources of systematic effects in GR waveforms need to be under control if one wants to measure possible deviations from GR. Current waveforms are likely not to satisfy this requirement in the majority of the parameter space for ground-based and especially space-based detectors. A comprehensive study using accurate waveforms from alternative theories of gravity for the inspiral, plunge and merger stages is needed to understand the reduction in systematic errors in GR waveforms that is necessary to make deviations from GR detectable.

Synergy between EM and GW observations: Following up GW events using EM telescopes, and likewise analysing GW data at the time of EM transients, will be invaluable in enhancing the scientific returns of observations. Since the EM sky is full of transient events, understanding which EM transients to follow up in GW data is important as otherwise coincidences will lose significance. While many EM transients are easily identifiable, challenges remain in unraveling the nature of astronomical transients [412]. A study of the fraction of EM transients that might look like GW progenitors, and hence contribute to false coincidences, is desirable.

EM followups of GW events rely on accurate estimation of source position [413]. What are the methods by which we might be able to improve sky localisation? For example, subdominant harmonics in systems with large mass ratios [414, 415] and binaries with rapidly spinning components [416] could enhance sky resolution and galaxy surveys could help target specific sky patches [417]. A proper understanding of biases in the estimation of sky position due to inaccurate waveform models or the use of galaxy catalogues is necessary.

Pulsar timing arrays are approaching astrophysically relevant sensitivity levels and setting limits on the detectability of supermassive BH binaries [418]. SKA could observe continuous waves from an isolated supermassive BH binary, if one exists in the relevant frequency range, within $z \lesssim 1$ [419]. The challenge here would be to control systematics in pulsar timing noise and to discover a large number (~ 100) of stable (timing noise $\lesssim 20$ ns) millisecond pulsars.

6.4 Isolated compact objects

Over the past 30 years, astronomy has made great strides in observing compact objects and their environments and equally raised new puzzles about the interior structure of NSs. In particular, X- and gamma-ray observations have identified new potential sources of GWs. On the theoretical front, there is now a vast amount of literature aiming to understand the structure and composition of NS cores and observational signatures expected of them. Supernova simulations have become more sophisticated, yielding predictions of GW amplitudes that are far more pessimistic than those of three decades ago. On the other hand, challenges remain in producing realistic simulations that resolve all the relevant scales and include all the macro- and micro-physics. Gravitational-wave observations with initial interferometers at design sensitivity have broken new ground, setting the best upper limits on the strength of GWs from known pulsars. These observations are already constraining theoretical models and advanced interferometers will greatly improve upon them. In this section we will take a census of the most important GW sources of isolated compact objects, the current observational status and science targets and challenges for future observations.

6.4.1 A menagerie of neutron-star sources

Neutron stars in isolation with a time-varying quadrupole moment are potential sources of GWs. The birth of a NS in a supernova, a non-axisymmetric spinning NS, a NS accreting from a companion in a low-mass X-ray binary, differentially rotating NSs, could all produce GWs.

For an isolated body the energy available for radiation is in the form of its gravitational binding energy, rotational energy or energy stored in its magnetic field. Most of the available energy might be emitted in a burst of GWs, resulting in a source with a large amplitude, or else the energy might leak out slowly over a long period of time, giving a continuous, but low-amplitude, source of radiation.

Supernovae: Neutron stars are born in the aftermath of the gravitational collapse of a massive star of ~ 8–$100 \, M_\odot$ or when the core of a white dwarf becomes more massive than the Chandrasekhar limit of $1.4 \, M_\odot$. Supernovae were the prime targets for the first

GW detectors and they are still among the most important sources. The Galactic supernova rate is uncertain and is thought to be 0.01–0.1 per year, but the rate within about 5 Mpc could be one per few years [420].

In supernovae, GWs are emitted at the expense of the gravitational binding energy. The time scale over which the radiation is emitted is the dynamical free-fall time $\tau_{FF} \sim \sqrt{4\pi/G\rho_{NS}}$, where $\rho_{NS} \sim 5 \times 10^{17}$ kg m^{-3} is the mean density of a NS.[7] Thus the time scale for collapse is $\tau_{FF} \sim 2$ ms and the frequency of GWs would be $f \sim \tau_{FF}^{-1} \sim 500$ Hz. Numerical simulations also reveal that the time-domain waveform is a short burst and the energy in the burst is spread over a frequency range of 200 Hz to 1 kHz, with the peak of the radiation at $f_{peak} \simeq 500$ Hz [421, 422]. If a fraction $\epsilon \sim 10^{-8}$ of the rest-mass energy of the star is converted to GWs then the characteristic amplitude of the signal would be $h_c \sim 2 \times 10^{-22}$ Hz$^{-1/2}$. From Fig. 6.6 we see that a Galactic supernova from a random sky position would be easily observable in aLIGO, producing an SNR[8] of $\sim \sqrt{2/5}\, h_c/\sqrt{S_h(f_{peak})} \sim 30$. At 4 Mpc, the distance at which the rate could be one per few years, the characteristic amplitude would be $h_c \sim 5 \times 10^{-24}$ Hz$^{-1/2}$, which would be observable in ET with a similar SNR if $\epsilon \sim 10^{-6}$.

Spinning neutron stars: A NS that is perfectly spherically symmetric or spinning about its symmetry axis emits no radiation since its quadrupole moment would not vary with time. Non-axisymmetric NSs would produce radiation at twice the spin frequency. The amplitude of GWs for a NS at a distance R is [78] $h_0 = 4\pi^2 G\epsilon\, I_{zz} f^2/(c^4 R)$, where f is the frequency of GWs and $\epsilon \equiv (I_{xx} - I_{yy})/I_{zz}$ is the NS ellipticity given in terms of the principal moments of inertia with respect to the rotation axis, I_{xx}, I_{yy} and I_{zz}. Typical NS moments of inertia are $I \sim 3 \times 10^{38}$ kg m^2, so a NS at 10 kpc, spinning at 50 Hz (GW frequency of 100 Hz) and an ellipticity of 10^{-6}, has an amplitude of $h_0 \simeq 3 \times 10^{-27}$.

Since GWs from spinning NSs are essentially continuous waves (CWs), Fourier transforming the signal would focus all its power into one frequency bin. The SNR grows as the square-root of the integration period. Thus, the characteristic strain amplitude h_c of a signal integrated over a time T is $h_c = h_0\sqrt{T}$, which for the example considered above is $h_c \sim 1.7 \times 10^{-23}$ Hz$^{-1/2}$, for $T = 1$ yr. For the Crab (B0531+21), the youngest known pulsar, at a distance of 2 kpc and spin frequency of 30 Hz, $h_c \sim 3 \times 10^{-23}$ Hz$^{-1/2}$, for the same ellipticity. Figure 6.6 plots the characteristic amplitude as a function of GW frequency for two choices of ellipticities, $\epsilon = 10^{-6}$ and 10^{-8}, for NSs at 10 kpc, located and oriented randomly with respect to the detector, for an integration time of 1 year. Neutron stars of spin frequencies in the range 20–100 Hz (GW frequencies of 40–800 Hz) would be accessible to advanced detectors if their ellipticities are $\epsilon \gtrsim 1.6 \times 10^{-5}(f/100\,\text{Hz})^{-2}(R/10\,\text{kpc})$.

Nearly 2000 pulsars are currently known[9] and it is estimated that our galaxy is host to $\sim 10^9$ NSs. The fraction of NSs accessible to the gravitational window is uncertain.

---

[7] The density of the pre-collapse star is not relevant as most of the energy in GWs is emitted in the final moments of the collapse and core bounce.

[8] Laser interferometers have the best response to burst sources that occur directly above their plane, but for an event at a random position on the sky and for waves of arbitrary polarisation the response and the SNR are a factor of $\sqrt{2/5}$ smaller.

[9] ATNF pulsar catalogue: http://www.atnf.csiro.au/people/pulsar/psrcat/.

The biggest uncertainty is the ellipticity that can be sustained in a NS, with largest estimates of $\epsilon \sim 10^{-4}$ [423], but more typically $\epsilon \sim 10^{-6}$ or smaller [424]. Statistical arguments suggest that a NS with ellipticity $\epsilon = 10^{-6}$ could be close enough to have an amplitude of $h_{\max} \simeq 1.6 \times 10^{-24}$ in the frequency range 250–680 Hz [425].

Figure 6.7 shows the expected amplitudes for several known AM CVn systems, white dwarf binaries, and X-ray binaries in the eLISA band. The latter sources are often referred to as *calibration sources,* because eLISA should see them at these amplitudes.

Pulsar glitches and magnetar flares: Radio pulsars have very stable spins and their periods (P) change very slowly over time. Their small spin-down rate ($\dot{P} \lesssim 10^{-12}$) is occasionally marked by a sudden increase in angular frequency Ω, an event that is called a *glitch* [426]. To date more than 300 glitches have been observed in about 100 pulsars [427]. Vela (B0833–45) is a nearby ($R \sim 300\,\mathrm{pc}$) pulsar in which 16 glitches have been observed since its discovery in 1969. The magnitude of a glitch is measured in terms of the fractional change in the angular velocity, which is found to be in the range $\Delta\Omega/\Omega \sim 10^{-5}$–$10^{-11}$. Some time after a glitch, the pulsar returns to its regular spin-down evolution. Pulsar glitches are not the only transient phenomena observed in NSs. Sources of giant X- and gamma-ray flashes are thought to arise in highly magnetised NSs, called *magnetars,* with B-fields $\sim 10^{15}$–10^{16} Gauss. The source of high-energy radiation is believed to be powered by the decay of the magnetic field associated with stellar quakes [428].

Pulsar glitches and magnetar flares could excite a spectrum of normal-mode oscillations of the ultra-dense NS core, the characteristic mode frequencies varying over a range of 1.5–6 kHz and damping times $\tau \sim$ ms, depending on the mode in question and the NS EoS [429]. The energy in normal modes could be emitted as a narrow-band burst of exponentially damped sinusoidal GWs. Figure 6.6 shows plausible characteristic amplitudes produced by normal modes of energy $10^{-12}\,M_\odot$, for mode frequencies in the range of 1.5–4 kHz and NS distances in the range 1 kpc to 10 kpc. Third-generation detectors like ET should be able to detect such amplitudes in coincidence with radio observations.

Low-mass X-ray binaries: Low-mass X-ray binaries (LMXBs) are accreting NSs or BHs that emit bursts of X-ray flashes lasting for about 10 s and repeat once every few hours or days, with millisecond oscillations in burst intensity [430]. X-ray bursts are believed to be caused by thermonuclear burning of infalling matter, while oscillations are suspected to be caused by the NS spin. About 100 galactic LMXBs are known to date as well as many extra-galactic ones. Inferred spin frequencies of NSs in LMXBs seem to have an upper limit of about 700 Hz [430], although this is nowhere close to the value at which centrifugal break-up would limit the star's spin frequency. It has been proposed that GWs might be responsible for limiting the spin frequencies of NSs in LMXBs [431–433].

The expected characteristic amplitude of gravitational radiation is shown in Fig. 6.6 for the well-known LMXB Sco X-1 and for the known galactic population of LMXBs. Advanced detectors could detect Sco X-1 if it is losing all of its accreted angular momentum to GWs (which is unlikely to be the case), while ET targets the full galactic population [434].

6.4.2 Results from LIGO and Virgo

Searches for burst signals: Searches for bursts of GWs essentially fall into one of two classes: all sky, blind searches and astrophysically triggered searches. In the first approach there is no a priori information about what to look for and when. The goal of this approach is to detect radiation from unmodeled, or poorly modeled, transient sources, as well as hitherto unknown sources, that last for less than ~ 1 second. Since no assumption about the nature of GWs is made, this approach has the greatest serendipitous discovery potential. The search algorithm uses wavelet transforms to look for excess power [435] and the sensitivity of the search is characterised in terms of the root-sum-square strain amplitude h_{rss} of the signal.[10] Analysing data from the various science runs has determined that the rate of strong GW bursts (i.e., bursts with $h_{\mathrm{rss}} > 10^{-19}\,\mathrm{Hz}^{-1/2}$ in the frequency region from 70 Hz to 3 kHz) reaching the Earth is less than 1.3 events per year at 90% confidence [436]. For hypothesised standard-candle sources that emit 1 $M_\odot$ equivalent of energy in GWs as sine-Gaussian waveforms, the inferred rate density of events in the local Universe, in units of Mpc^{-3} yr^{-1}, is less than 10^{-6} in the frequency range 100–200 Hz and less than 10^{-2} in the frequency range 1–2 kHz. Alternatively, generic burst sources within 10 kpc emitted less than $\simeq 2 \times 10^{-8}\, M_\odot$ in GWs in the frequency range 100–200 Hz; that limit increases to $\simeq 10^{-5}\, M_\odot$ at 1 kHz [436].

In the second approach, analysis is carried out around the time of an astrophysical transient, such as a supernova or a magnetar flare. Knowledge of the epoch and sky position of the event helps reduce the amount of data that needs to be searched for, which in turn decreases the false-alarm probability and improves the search sensitivity. Searches have been carried out at the time of pulsar glitches [437], magnetar flares [438] and GRBs [439]. Of particular significance is the search for GWs around the time of GRB070201 [440]. The event in this case was a shGRB that is believed to have followed either from giant quakes in highly magnetised NSs or from merging binary NSs. The location of GRB070201 coincides with the spiral arms of the Andromeda galaxy (M31) at 780 kpc. LIGO detectors, which were taking data at the time of this event, would have quite easily detected signals from a merging NS binary at this distance, but not bursts associated with a magnetar flare. The analysis found no plausible GW candidates within a 180 s window around the time of the GRB and in particular excluded binary NS–NS and NS–BH mergers at M31 with more than 99% confidence [440]. The analysis also concluded that isotropic energy in GWs from the source, if it were at M31, was most likely less than $4.4 \times 10^{-4}\, M_\odot$, lending support for the possibility that this was the first Soft Gamma Repeater flare observed outside the Milky Way.

More recently, searches have also been performed around the times of 128 long GRBs and 26 shGRBs [439]; no GW candidates of any significance were found, which meant that the bursts could not have occurred closer than a certain distance determined by the horizon distance of the detectors in the direction of the GRBs. The maximum exclusion

---

[10] The strain amplitude h_{rss} is defined as $h_{\mathrm{rss}} = \sqrt{\int \left[\, |h_+(t)|^2 + |h_\times(t)|^2 \,\right] dt}$.

distance for the population of shGRBs was 80 Mpc, which is not surprising since the closest known shGRB is at a distance of $\sim$ 500 Mpc. However, extrapolating current results to advanced detectors makes it seem quite plausible that GWs coincident with GRBs could be detected within 2.5–5 years of observing [439, 441] or, if none are detected, could yield upper bounds on the number of GRBs arising from merging binaries.

Searches for continuous waves: Most CW signals are monochromatic in the rest frame of their sources. Their detection is complicated by the fact that the signal received at the detector is modulated due to the Earth's motion. Because of Doppler modulation in frequency, the spectral lines of fixed-frequency sources spread power into many Fourier bins about some mean frequency. Although the modulation of the signal makes the search prohibitively expensive, imprinted in the modulation is the source's position on the sky; it will be possible to resolve the source's location subject to the Rayleigh criterion, $\delta\theta = 2\pi\lambda/L$, where $\delta\theta$ is the angular resolution, λ is the wavelength of the radiation and $L = 2\,\mathrm{AU}$ is the baseline for an observation period of 1 year. At a frequency of 100 Hz, $\delta\theta \sim 2''$. In the case of CW sources, two different types of searches have been performed: searches for known pulsars (with precisely known sky position and frequency evolution) and blind searches (sources with unknown sky position and spin frequency).

Searches for CW signals from known, isolated pulsars, not being limited by computational resources, have achieved the best possible sensitivity [442]. In particular, upper limits on the strength of GWs from the Crab and Vela pulsars have now been constrained by these observations to be well below the level expected from the observed rate at which these pulsars are spinning down. The loss in energy to GWs from the Crab pulsar is less than 1% of the rotational energy lost due to the observed spin down [442]; the corresponding number for Vela is 10% [437]. A search for CW signals in the frequency range of 100–300 Hz from the compact central object, believed to be a NS, in the supernova remnant Cassiopea A at 3 kpc, has set the best upper limit on the strain amplitude, $\sim 3 \times 10^{-24}$, and on the equatorial ellipticity, 0.4–4×10^{-4}, as well as setting the first ever limit on the amplitude of r-modes in this young NS [443].

Given the possibility that the strongest CW sources may be electromagnetically quiet or remain undiscovered, an all sky, all frequency search for such unknown sources is very important, though computationally formidable. Clever and computationally efficient algorithms and distributed volunteer-computing Einstein@Home [444] have made the searches ever more sensitive, and have been successful in discovering new radio pulsars in old radio data (see, e.g., [445, 446]). A blind GW search using Einstein@Home excluded signals in the 50 Hz to 1.2 kHz band, with upper limits on strain amplitudes $\sim 10^{-24}$–10^{-23} depending on the frequency of the source. For example, strain amplitudes greater than 7.6×10^{-25} were excluded at 152.5 Hz (the frequency where the LIGO S5 run had the best sensitivity), over a 0.5 Hz-wide band [447]. This means there are no NSs at this frequency within 4 kpc and spinning down faster than 2 nHz s^{-1} that have ellipticities greater than 2×10^{-4}. Targeted searches for sources within 8 pc of the galactic centre Sag A*, in the frequency range of 78–496 Hz, and with maximum spin-down rates of $\sim -8 \times 10^{-8}$ Hz s^{-1}, have achieved the best sensitivities for blind searches, ruling out NSs with GW amplitudes larger than $\sim 3 \times 10^{-25}$ at frequencies around 150 Hz in this region of the sky (for details and caveats see [448]).

Advanced detectors will beat the spin-down limit of several pulsars [442]. For the fastest pulsars in the frequency range 200–400 Hz, advanced detectors will reach ellipticity limits of $\sim 10^{-8}$ (or a differential radius of 100 microns in 10 km!), significantly below the spin-down limits; ET will be sensitive to ellipticities as low as 10^{-9} [16].

6.4.3 Science targets and challenges

Neutron stars are the most compact objects with matter known today. They have strong surface gravity that is responsible for very intense sources of X-rays and gamma-rays. Their dense cores could be superfluid and might consist of hyperons, quark–gluon plasma or other exotica [430] and are, therefore, laboratories of high-energy nuclear physics. Observing a representative sample of the galactic population of NSs could transform astrophysical studies of compact objects, but there are still some challenges in theoretical modeling of NSs and analysis of data.

Science targets

Physics of low-mass X-ray binaries: Detecting GWs from LMXBs should help us to understand the mechanism limiting spin frequencies in LMXBs. The centrifugal breakup of NS spins for most EoS is ~ 1500 Hz, far greater than the maximum spin of ~ 700 Hz inferred from X-ray observations [449]. It has, therefore, been a puzzle as to why NS spin frequencies are stalled. One reason for this could be that some mechanism operating in the NS is emitting GWs and the resulting loss in angular momentum explains why NSs cannot be spun up beyond a certain frequency. The exact mechanism causing the emission of GWs can account for this amplitude if NSs can support an effective ellipticity of $\epsilon \sim 10^{-8}$. This ellipticity could be produced by a time-varying, accretion-induced quadrupole moment [432], by relativistic instabilities (e.g. r-modes) [450], or by large toroidal magnetic fields [451]. Targeted observations of known LMXBs could confirm or rule out astrophysical models of such systems.

Understanding supernovae: Supernovae produce the Universe's dust and some of its heavy elements; their cores are laboratories of complex physical phenomena requiring general relativity, nuclear physics, magneto-hydrodynamics, neutrino viscosity and transport, and turbulence to model them. Much of the physics of supernovae is poorly understood: How non-axisymmetric is the collapse? How much energy is converted to GWs and over what time scale? What causes shock revival in supernovae that form a NS: neutrino, acoustic and/or magneto-rotational mechanisms? Depending on the supernova mechanism, the predicted energy in GWs from supernovae varies by large factors (see, e.g., [452, 453]), indicating the complexity of the problem in numerical simulations. Until we know the mechanism that revives the stalled shock it will not be possible to correctly predict the amplitude of the emitted gravitational radiation or its time–frequency structure.

Gravitational-wave observations could provide some of the clues for solving these questions [422, 454]. Moreover, GWs could also be produced by neutrino emission during the supernova explosion and the signal spectrum could extend down to $\sim$ 10 Hz [455]. More realistic studies are needed to quantify this signal.

Testing neutron-star models: Models of NSs are mostly able to compute their maximum ellipticity by subjecting the crust to breaking strains with predicted ellipticities ranging from $\epsilon \sim 10^{-4}$ (for exotic EoS) [423] to 10^{-7} for conventional crustal shear [456]. Large toroidal magnetic fields of order 10^{15} G could sustain ellipticities of order 10^{-6} [457] and accretion along magnetic fields might produce similar, or a factor of 10 larger, deformations [458]. The large range in possible ellipticities shows that GW observations could have a potentially high impact and science return in this area. Confirmed detections of NSs with known distances will severely constrain models of the crustal strengths, a catalogue of CW sources would help us understand the galactic supernova rate and their demographics could lead to insights on the evolutionary scenarios of compact objects.

Challenges

Interfacing theory with searches: Models that can accurately predict the spectrum and complex mode frequencies will be very useful in searches for GW signals at the time of glitches in pulsars and flares in magnetars [429]. In fact, robust predictions can also help to tune detectors to a narrower band, where signals are expected, with greater sensitivity. To take advantage of such techniques, which will become feasible in the era of routine observations, models would need to become realistic and reliable.

Supernova simulation is one area where a breakthrough in understanding the core bounce that produces the explosion could be critical to produce reliable models. At present, it is not clear if models will ever be able to produce waveforms that can be deployed as matched filters. A catalogue of predicted waveforms is routinely used to calibrate the sensitivity of a search (see, e.g., [436]). Accurate models of the frequency range and spectral features of the emitted radiation will obviously aid in better quantification of the search sensitivity.

The problem of blind searches: Looking for CWs in GW data is a computationally formidable problem [459–461]. Blind searches have to deal with many search parameters, such as the sky position of the source, its spin frequency and one or more derivatives of the spin frequency [460]. The number of floating-point operations grows as the 5th power of the integration time T for a blind search with unknown sky position, unknown spin frequency and one spin-down parameter [462]. Most blind searches are able to coherently integrate the data for about a few hours to days (depending on the number of spin-down parameters searched for) [447, 448] and the sensitivity of searches will always be limited by the available computing resources. Algorithms that can integrate for longer periods are desirable, as are multi-step hierarchical searches that could achieve optimal sensitivity given the computational power.

6.5 Gravitational radiation from the early Universe

During the past 30 years, several new predictions for GW signals from the primordial Universe have been made, greatly stimulated by the construction and operation of the first GW detectors and the planning of future experiments. The epoch of big-bang nucleosynthesis (BBN), when light elements first formed, is the earliest epoch of the Universe that we understand today with any confidence. The Universe was only a second old at this epoch, it was radiation dominated and had a temperature of ~ 1 MeV. In contrast, the Universe was much older (age of $\sim 10^5$–10^6 years) and cooler (temperature of ~ 1 eV), when the CMB radiation, measured today with amazing accuracy, was emitted. It is expected that the Universe is filled with cosmic neutrinos produced when the Universe's temperature was 1 MeV, but this has not been observed, yet. In fact, no primordial background, of radiation or particles, produced before the epoch of CMB has ever been detected.

6.5.1 Primordial sources and expected strengths

Gravitational waves emitted prior to BBN in the so-called *dark age* would travel unscathed, due to their weak interaction with matter, and provide us with a view of the Universe at that time.

A rapidly varying gravitational field during inflation can produe a stochastic background of GWs by parametric "amplification" of quantum, vacuum fluctuations [464, 465]. Today this background would span the frequency range of 10^{-16}–10^{10} Hz, which covers the frequency band of current and future detectors on the ground and in space (see Figs. 6.6 and 6.7). This is the same mechanism that is believed to have produced the scalar density perturbations that led to the formation of large scale structures in the Universe. Single-field, slow-roll models of inflation predict that the background slightly decreases as the frequency increases, $\Omega_{GW} = \Omega_0 (f/f_{eq})^{n_T}$, for $f > f_{eq}$ with $n_T \lesssim 0$, while it rises as a power law, $\Omega_{GW} = \Omega_0 (f/f_{eq})^{-2}$, for $f < f_{eq}$ where $\Omega_{GW}(f) = d\rho_{GW}(f)/d \log f/\rho_C$ [466], with $\rho_C = 3H_0^2/(8\pi G)$ and H_0 the present value of the Hubble parameter. The transition frequency $f_{eq} \simeq 10^{-16}$ Hz corresponds to the Hubble radius at the time of matter–radiation equality, redshifted to the current epoch. The value of Ω_0 is not known, but the current upper limit on the tensor-to-scalar ratio from the CMB [468] implies $\Omega_0 \lesssim 10^{-15}$. A cosmological GW background would leave an imprint in the CMB polarisation map [469, 470] and, as mentioned in Section 6.1, the BICEP2 [17] experiment has claimed a detection of this signature. However, further scrutiny suggests that BICEP2 result cannot be excluded from being of astrophysical origin [18, 19]. Concurrently with the construction of ground-based detectors and the planning of the next generation of experiments, studies of the GW background from inflation have been refined and several physical effects that may impact the high-frequency portion of the spectrum have been predicted [471–474].

The preheating phase, which occurs at the end of inflation, is a highly non-thermal phase that creates transient density inhomogeneities with time-varying mass multipoles, which would generate a stochastic background of GWs [475]. Symmetry-breaking phase transitions or the ending stages of brane inflation might witness the creation of cosmic (super)strings [476–478]. Due to their large tension these strings undergo relativistic oscillations and thereby produce GWs, causing them to shrink in size and disappear. However, they could be constantly replaced by smaller loops that break off from loops of size larger than the Hubble radius. In this way, a network of cosmic (super)strings could generate a stochastic GW background but could also produce bursts of gravitational radiation when cusps and kinks form along strings [479–483]. Likewise a strong first order phase transition could create bubbles of true vacuum which collide with each other and produce a GW background [484, 485]. For more details on the mechanisms responsible for gravitational radiation in the early Universe, we refer the reader to the reviews [466, 467, 486, 487] and references therein.

6.5.2 Results from LIGO and Virgo

Stochastic signals are a type of continuous waves, but with two important differences. In general they do not arrive from any particular direction and, by definition, have no predictable phase evolution. Therefore, conventional matched filtering would not work and sliding data of one detector relative to another (to account for the difference in arrival time) has no particular advantage. Even so, data from one detector could serve as a "template" to detect the same stochastic signal present in another detector. If the detectors are located next to each other and have the same orientation then a simple cross-correlation of their outputs weighted by their noise spectral densities would result in the optimal SNR [488]. The problem with two nearby detectors is that they will have a common noise background that would contaminate the correlated output. By placing detectors far apart, one could mitigate the effect of common noise, but in that case wavelengths smaller than the distance between the detectors will not all be coherent in the two detectors, which effectively reduces the sensitivity bandwidth. This is appropriately taken into account in the cross-correlation statistic of a pair of detectors by using what is called the *overlap reduction function* [489], which is a function of frequency that accounts for the lack of coherence in stochastic signals in detectors of different orientations that are separated by a given distance [463]. Additionally, since the template is essentially noisy, the amplitude SNR grows as the *fourth-root* of the product of the effective bandwidth Δf of the detector and the duration T over which the data is integrated [488]. In the case of PTAs, the detection technique is similar; instead of a pair of detectors one constructs the correlation between the timing residuals of many stable millisecond pulsars.

The energy density in GWs is related to the strain power spectrum $S_{GW}(f)$ by [463] $S_{GW}(f) = 3H_0^2 \, \Omega_{GW}(f)/(10\pi^2 f^3)$. The *characteristic amplitude* h_c of a stochastic background, i.e. the strain amplitude produced by a background after integrating for a time T over a bandwidth Δf, is given by $h_c^2 = \sqrt{T \, \Delta f} \, S_{GW}$. In Figs. 6.6 and 6.7 we plot (with dotted lines) $h_c(f)$ for several values of Ω_{GW} assumed to be independent of f, setting $T = 1$ yr and $\Delta f = 100$ Hz. Due to the overlap reduction function, the SNR is built up mostly from the low-frequency part of the signal such that $\lambda_{GW} \gtrsim d/2$, where d is the distance between detectors. A stochastic signal would be detectable if it stays above the noise curve roughly over a frequency band $\Delta f \simeq f$. Advanced detectors should detect $\Omega_{GW} \geq 10^{-9}$ at tens of Hz, while ET, due to its much improved low-frequency sensitivity and collocated detectors, could detect $\Omega_{GW} \sim 10^{-11}$, and eLISA could detect $\Omega_{GW} \geq 10^{-12}$ at mHz frequencies. For an integration time of $T = 5$ yr and bandwidth of $\Delta f = 6$ nHz, a stochastic background with $\Omega_{GW} = 2.5 \times 10^{-10}$ would be detectable by a PTA with an SNR of 5, assuming 20 millisecond pulsars that have an rms stability of 100 ns [490, 491]. SKA will improve the sensitivity of PTAs by three orders of magnitude to $\Omega_{GW} \sim$ few $\times 10^{-13}$ [492].

Gravitational radiation from the early Universe does not merely generate stochastic backgrounds: bursts from cusps and kinks along cosmic (super)strings produce power-law signals in the frequency domain [482] that can be searched for using matched-filtering techniques. Data from several science runs of iLIGO and Virgo detectors have been analysed to search for signals from the early Universe. In those science runs, the detectors' sensitivity has passed the BBN bound [493, 494] in the frequency band around 100 Hz, $\Omega_{GW} < 6.9 \times 10^{-6}$, but not yet the CMB bound [495]. The results have started to exclude regions of the parameter space of expected signals from cosmic (super)strings [494, 496, 497], and have constrained the equation of state of the Universe during the dark age [494, 498]. Moreover, pulsar timing observations have set physically meaningful upper limits for the supermassive BH binary background ($\Omega_{GW} < 1.3 \times 10^{-9}$ at 2.8 nHz) [499] and cosmic (super)strings [500].

6.5.3 Science targets and challenges

Stochastic GW signals might carry a signature of unexplored physics in the energy range $\sim 10^9$ GeV to $\sim 10^{16}$ GeV. The detection of GWs from the dark age could therefore be revolutionary. The spectrum of the detected radiation could reveal phase transitions that might have occurred in the Universe's early history, unearth exotic remnants like cosmic (super)strings, and prove that a cosmic inflationary phase existed and that gravity can be reconciled with quantum mechanics. No other observation can ever take us closer to the origin of our Universe and hence the science potential of discovering primordial GWs will be immense. However, the challenges in this area are equally formidable.

As mentioned before, one of the biggest problems in identifying a stochastic GW background is how to disentangle it from the environmental and instrumental noise backgrounds. At present, data from two or more detectors are cross correlated to see if there is any statistical excess. If the detectors are geographically widely separated then one could reasonably hope that the environmental noise backgrounds in different instruments do not correlate, although correlations could exist due to large-scale magnetic fields, cosmic rays and anthropogenic noise and the like. When the number of detectors grows, noise correlations decrease. However, because of the overlap reduction function, the sensitivity to a stochastic background diminishes quickly with geographically separated detectors. Due to environmental and instrumental noise, searching for stochastic backgrounds in collocated detectors like ET or eLISA will be a real challenge.

In the case of PTAs, the problem is less severe as one is looking for correlations in the residuals of the arrival times of radio pulses from an array of millisecond pulsars, after subtracting the model of the pulsar from the original data. In principle, any systematics in timing residuals could be mitigated by integrating the correlation over long periods, but the problem here is that the time scale for integration tends to be large (i.e., tens of years).

In the case of the GW background from inflation, CMB bounds on inflationary potentials are not very informative on the value of Ω_0 except for placing an upper limit. In the absence of probes from epochs prior to BBN, it is very hard to infer the equation of state of the Universe between the end of inflation and the epoch when the radiation era started. Thus, it is difficult to predict the spectral slope of the relic GW background from BBN (when the Universe was certainly radiation-dominated) to the scales where PTA, space-based and ground-based detectors are sensitive [498]. It is customary and (perhaps) natural to assume that the slope is the same over the huge range of frequency, spanning twenty orders of magnitude, and that it can be determined by CMB observations. However, as we look backward past BBN, a stiff energy component might overtake radiation as the dominant component in the cosmic energy budget [501–505], without coming into conflict with any current observational constraints. The detection of the B-mode polarisation in the CMB will certainly have a huge impact, determining Ω_0, but we will still not know whether the slope remains the same for twenty orders of magnitude.

Predictions for the GW background from preheating at the end of inflation [475, 487] lie typically (except for some choices of parameters in hybrid inflation) in the MHz frequency range where no GW detectors exist or are currently planned. Experimental proposals would need to be made, but since, when holding Ω_{GW} fixed, the noise spectral density decreases as the frequency increases, the requirements on the detector sensitivity would be very hard to achieve. Finally, depending on string parameters, both the stochastic background and single powerful bursts from cusps and kinks of cosmic (super)strings could be observed by PTAs, eLISA and the aLIGO/AdV/KAGRA network. More robust predictions of (super)string loop sizes (large versus small loop sizes at birth) will be important to restrict regions of parameter space that are searched over (see [487] and references therein).

Acknowledgements

We are greateful to Luc Blanchet and Eric Poisson for carefully reading and providing comments on the manuscript. We have benefited from useful discussions with Leor Barack (who supplied a Mathematica code to compute EMRI spectra in Fig. 6.7), Chris Messenger, Bernard Schutz, Riccardo Sturani and Patrick Sutton. A.B. acknowledges partial support from NSF Grant No. PHY-1208881 and NASA Grant NNX09AI81G. B.S.S. acknowledges support from STFC (UK) grant ST/L000962/1, ST/L000342/1 and ST/J000345/1.

References

[1] Maxwell, J. 1865. *Phil. Trans. R. Soc. Lond.*, **155**, 459–512.
[2] Einstein, A. 1916. *Sitzungsber. Preuss. Akad. Wiss.*, 688–696.

[3] Einstein, A. 1918. *Sitzungsber. Preuss. Akad. Wiss.*, 154–167.
[4] Poincaré, H. 1905. *Comptes Rendus Acad. Sci. Paris*, **140**, 1504–1508.
[5] Einstein, A., and Rosen, N. 1937. *J. Franklin Inst.*, **223**, 43–54.
[6] Bondi, H., Pirani, F., and Robinson, I. 1959. *Proc. Roy. Soc. Lond.*, **A251**, 519.
[7] Kennefick, D. 2007. *Traveling at the speed of thought: Einstein and the quest for gravitational waves*. Princeton, NJ: Princeton University Press.
[8] Saulson, P. R. 2011. *Gen. Rel. Grav.*, **43**, 3289–3299.
[9] Bondi, H., van der Burg, M., and Metzner, A. 1962. *Proc. Roy. Soc. Lond.*, **A269**, 21–52.
[10] Weber, J. 1960. *Phys. Rev.*, **117**, 306.
[11] Hulse, R. A., and Taylor, J. H. 1975. *Astrophys. J.*, **195**, L51–L53.
[12] Weisberg, J. M., Nice, D. J., and Taylor, J. H. 2010. *Astrophys. J.*, **722**, 1030–1034.
[13] Kramer, M. 2013. Probing gravitation with pulsars. Pages 19–26 of: van Leeuwen, J. (ed), *Neutron stars and pulsars (IAU291)*. Cambridge: Cambridge University Press.
[14] Deruelle, N., and Piran, T. (eds). 1982. *Gravitational radiation (Les Houches Winter School, 1982)*. Amsterdam: North Holland.
[15] Hawking, S. W., and Israel, W. (eds). 1987. *Three hundred years of gravitation*. Cambridge: Cambridge University Press.
[16] Abernathy, M., *et al.* 2011. *Einstein gravitational wave telescope: conceptual design study*. Available from European Gravitational Observatory, document number ET-0106A-10. https://tds.ego-gw.it/itf/tds/index.php?callContent=2&callCode=8709.
[17] Ade, P., *et al.* 2014. *BICEP I: detection of B-mode polarization at degree angular scales*. arXiv:1403.3985.
[18] Mortonson, M. J., and Seljak, U. 2014. *A joint analysis of Planck and BICEP2 B modes including dust polarization uncertainty*. arXiv:1405.5857.
[19] Flauger, R., Hill, J. C., and Spergel, D. N. 2014. *Toward an understanding of foreground emission in the BICEP2 region*. arXiv:1405.7351.
[20] Wagoner, R., and Will, C. 1976. *Astrophys. J.*, **210**, 764–775.
[21] Ehlers, J., Rosenblum, A., Goldberg, J., and Havas, P. 1976. *Astrophys. J.*, **208**, L77–L81.
[22] Ehlers, J. 1980. Isolated systems in general relativity. Pages 279–294 of: Ehlers, J., Perry, J. J., and Walker, M. (eds), *Proceedings of the Niath Texas Symposium on Relativistic Astrophysics*.
[23] Walker, M., and Will, C. 1980. *Phys. Rev. Lett.*, **45**, 1741–1744.
[24] Walker, M., and Will, C. M. 1980. *Astrophys. J.*, **242**, L129–L133.
[25] Damour, T., and Deruelle, N. 1981. *Phys. Lett.*, **A87**, 81.
[26] Futamase, T., and Schutz, B. F. 1983. *Phys. Rev. D*, **28**, 2363–2372.
[27] Futamase, T. 1983. *Phys. Rev. D*, **28**, 2373–2381.
[28] Thorne, K. S. 1980. *Rev. Mod. Phys.*, **52**, 299–339.
[29] Damour, T., and Deruelle, N. 1985. *Ann. Inst. H. Poincaré*, **44**, 107–132.
[30] Damour, T., and Schäfer, G. 1985. *Gen. Rel. Grav.*, **17**, 879.
[31] Damour, T., and Deruelle, N. 1986. *Ann. Inst. H. Poincaré Phys. Théor.*, **44**, 263–292.
[32] Blanchet, L., and Damour, T. 1986. *Phil. Trans. Roy. Soc. Lond. A*, **320**, 379.
[33] Blanchet, L. 1987. *Proc. R. Soc. A*, **409**, 383–399.
[34] Damour, T., and Schäfer, G. 1988. *Nuovo Cim.*, **B101**, 127.
[35] Blanchet, L., and Damour, T. 1988. *Phys. Rev. D*, **37**, 1410–1435.
[36] Blanchet, L., and Schäfer, G. 1989. *Mon. Not. Roy. Astron. Soc.*, **239**, 845–867.
[37] Blanchet, L., and Damour, T. 1989. *Ann. Inst. H. Poincaré Phys. Théor.*, **50**, 377–408.
[38] Damour, T., and Iyer, B. R. 1991. *Ann. Inst. Henri Poincaré, A*, **54**, 115–164.
[39] Lincoln, C., and Will, C. 1990. *Phys. Rev. D*, **42**, 1123–1143.
[40] Blanchet, L., and Damour, T. 1992. *Phys. Rev. D*, **46**, 4304–4319.
[41] Thorne, K. S. 1992. *Phys. Rev. D*, **45**, 520–524.
[42] Cutler, C. *et al.* 1993. *Phys. Rev. Lett.*, **70**, 2984–2987.
[43] Blanchet, L., and Schäfer, G. 1993. *Class. Quant. Grav.*, **10**, 2699–2721.
[44] Iyer, B. R., and Will, C. 1993. *Phys. Rev. Lett.*, **70**, 113–116.
[45] Blanchet, L., Damour, T., and Iyer, B. R. 1995. *Phys. Rev. D*, **51**, 5360–5386.
[46] Blanchet, L. *et al.* 1995. *Phys. Rev. Lett.*, **74**, 3515–3518.

[47] Blanchet, L., Iyer, B. R., Will, C. M., and Wiseman, A. G. 1996. *Class. Quant. Grav.*, **13**, 575–584.

[48] Jaranowski, P., and Schäfer, G. 1997. *Phys. Rev. D*, **55**, 4712–4722.

[49] Jaranowski, P., and Schäfer, G. 1998. *Phys. Rev. D*, **57**, 7274–7291.

[50] Kidder, L. E., Will, C. M., and Wiseman, A. G. 1993. *Phys. Rev. D*, **47**, 4183–4187.

[51] Kidder, L. E. 1995. *Phys. Rev. D*, **52**, 821–847.

[52] Will, C. M., and Wiseman, A. G. 1996. *Phys. Rev. D*, **54**, 4813–4848.

[53] Poisson, E. 1993. *Phys. Rev. D*, **47**, 1497–1510.

[54] Tagoshi, H., and Nakamura, T. 1994. *Phys. Rev. D*, **49**, 4016–4022.

[55] Poisson, E. 1993. *Phys. Rev. D*, **48**, 1860–1863.

[56] Shibata, M., Sasaki, M., Tagoshi, H., and Tanaka, T. 1995. *Phys. Rev. D*, **51**, 1646–1663.

[57] Tagoshi, H., and Sasaki, M. 1994. *Prog. Theor. Phys.*, **92**, 745–772.

[58] Tagoshi, H., Shibata, M., Tanaka, T., and Sasaki, M. 1996. *Phys. Rev. D*, **54**, 1439–1459.

[59] Abramovici, A. *et al.* 1992. *Science*, **256**, 325–333.

[60] Tagoshi, H., *et al.* [TAMA Collaboration]. 2001. *Phys. Rev. D*, **63**, 062001.

[61] Abbott, B., *et al.* [LIGO Scientific Collaboration]. 2004. *Phys. Rev. D*, **69**, 122001.

[62] Nicholson, D. *et al.* 1996. *Phys. Lett.*, **A218**, 175–180.

[63] Buonanno, A., and Damour, T. 1999. *Phys. Rev. D*, **59**, 084006.

[64] Buonanno, A., and Damour, T. 2000. *Phys. Rev. D*, **62**, 064015.

[65] Baker, J. G., Campanelli, M., Lousto, C., and Takahashi, R. 2002. *Phys. Rev. D*, **65**, 124012.

[66] Baker, J. G. *et al.* 2001. *Phys. Rev. Lett.*, **87**, 121103.

[67] Pretorius, F. 2005. *Phys. Rev. Lett.*, **95**, 121101.

[68] Campanelli, M., Lousto, C. O., Marronetti, P., and Zlochower, Y. 2006. *Phys. Rev. Lett.*, **96**, 111101.

[69] Baker, J. G. *et al.* 2006. *Phys. Rev. Lett.*, **96**, 111102.

[70] Sasaki, M., and Tagoshi, H. 2003. *Living Rev. Rel.*, **6**, 6.

[71] Blanchet, L. 2006. *Living Rev. Rel.*, **9**, 4.

[72] Futamase, T., and Itoh, Y. 2007. *Living Rev. Rel.*, **10**, 2.

[73] Barack, L. 2009. *Class. Quant. Grav.*, **26**, 213001.

[74] Poisson, E., Pound, A., and Vega, I. 2011. *Living Rev. Rel.*, **14**, 7.

[75] Damour, T. 2013. *The general relativistic two body problem.* arXiv:1312.3505.

[76] Landau, L. D., and Lifshitz, E. M. 1962. *Classical theory of fields.* Second edn. Reading, MA: Addison Wesley.

[77] Schutz, B. 2009. *A first course in general relativity.* Cambridge Cambridge University Press.

[78] Maggiore, M. 2008. *Gravitational waves – volume 1.* First edn. New York, NY: Oxford University Press.

[79] Blanchet, L., Spallicci, A., and Whiting, B. (eds). 2011. *Mass and motion in general relativity. Berlin*: Springer.

[80] Epstein, R., and Wagoner, R. V. 1975. *Astrophys. J.*, **197**, 717–723.

[81] Bonnor, W. B., and Rotenberg, M. A. 1961. *Roy. Soc. Lond. Proc. Series A*, **265**, 109–116.

[82] Bonnor, W. B., and Rotenberg, M. A. 1966. *Roy. Soc. Lond. Proc. Series A*, **289**, 247–274.

[83] Blanchet, L. 1987. *Proc. R. Soc. Lond. A*, **409**, 383.

[84] Blanchet, L. 1995. *Phys. Rev. D*, **51**, 2559–2583.

[85] Blanchet, L. 1998. *Class. Quant. Grav.*, **15**, 1971–1999.

[86] Poujade, O., and Blanchet, L. 2002. *Phys. Rev. D*, **65**, 124020.

[87] Blanchet, L., Faye, G., and Nissanke, S. 2005. *Phys. Rev. D*, **72**, 044024.

[88] Blanchet, L. 1996. *Phys. Rev. D*, **54**, 1417–1438.

[89] Blanchet, L. 1993. *Phys. Rev. D*, **47**, 4392–4420.

[90] Pati, M., and Will, C. 2000. *Phys. Rev. D*, **62**, 124015.

[91] Pati, M. E., and Will, C. M. 2002. *Phys. Rev. D*, **65**, 104008.

[92] Wiseman, A. 1993. *Phys. Rev. D*, **48**, 4757–4770.

[93] Blanchet, L. 1998. *Class. Quant. Grav.*, **15**, 113–141.

[94] Lorentz, H., and Droste, J. 1917. De beweging van een stelsel lichamen onder den invloved van hunne onderlinge aantrekking, behandeld folgens de theoric van Einstein. I. *Kon. Akad. Wetensch.*, **26**, 392–403. English translation as "The motion of a system of bodies under the influence of their mutual attraction, according to Einstein's theory" in Zeeman, P., and Fokker, A. (eds), *H. A. Lorentz: collected papers*, Vol. 5, pp. 330–355. The Hague: Nijhoff (1937).

[95] Einstein, A., Infeld, L., and Hoffmann, B. 1938. *Annals Math.*, **39**, 65–100.

[96] Petrova, N. 1949. *JETP*, **19**, 989–999.

[97] Fock, V. 1939. *J. Phys.*, **1**, 81–116.

[98] Papapetrou, A. 1951. *Proc. Phys. Soc. A*, **64**, 57–75.

[99] Kimura, T. 1961. *Prog. Theor. Phys.*, **26**.

[100] Ohta, T., Okamura, H., Kimura, T., and Hiida, K. 1974. *Prog. Theor. Phys.*, **51**, 1598–1612.

[101] Ohta, T., Okamura, H., Kimura, T., and Hiida, K. 1973. *Prog. Theor. Phys.*, **50**, 492–514.

[102] Ohta, T., Okamura, H., Kimura, T., and Hiida, K. 1974. *Prog. Theor. Phys.*, **51**, 1220–1238.

[103] Bel, L. *et al.* 1981. *General Relativity and Gravitation*, **13**, 963–1004.

[104] Damour, T., and Taylor, J. H. 1992. *Phys. Rev. D*, **45**, 1840–1868.

[105] Damour, T. 1987. The problem of motion in Newtonian and Einsteinian gravity. Chapter 6, pages 128–198 of: Hawking, S. W., and Israel, W. (eds), *Three hundred years of gravitation*. Cambridge: Cambridge University Press.

[106] Kopeikin, S. M. 1985. *Sov. Astron.*, **29**, 516–524.

[107] Itoh, Y., Futamase, T., and Asada, H. 2001. *Phys. Rev. D*, **63**, 064038.

[108] Blanchet, L., and Faye, G. 2000. *J. Math. Phys.*, **41**, 7675–7714.

[109] Damour, T., Jaranowski, P., and Schäfer, G. 2001. *Phys. Lett.*, **B513**, 147–155.

[110] Dirac, P. A. M. 1958. *Roy. Soc. Lond. Proc. A*, **246**, 333–343.

[111] Dirac, P. A. 1959. *Phys. Rev.*, **114**, 924–930.

[112] Dirac, P. A. 1959. *Phys. Rev. Lett.*, **2**, 368–371.

[113] Arnowitt, R., Deser, S., and Misner, C. W. 1960. *Phys. Rev.*, **117**, 1595–1602.

[114] Arnowitt, R., Deser, S., and Misner, C. W. 1960. *J. Math. Phys.*, **1**, 434–439.

[115] Schwinger, J. 1963. *Phys. Rev.*, **130**, 1253–1258.

[116] DeWitt, B. S. 1967. *Phys. Rev.*, **160**, 1113–1148.

[117] Regge, T., and Teitelboim, C. 1974. *Annals of Physics*, **88**, 286–318.

[118] Schäfer, G. 1985. *Annals of Physics*, **161**, 81–100.

[119] Jaranowski, P., and Schäfer, G. 1999. *Phys. Rev. D*, **60**, 124003.

[120] Damour, T., Jaranowski, P., and Schäfer, G. 2000. *Phys. Rev. D*, **62**, 021501.

[121] Damour, T., Jaranowski, P., and Schäfer, G. 2001. *Phys. Rev. D*, **63**, 044021.

[122] Ledvinka, T., Schäfer, G., and Bičák, J. 2008. *Phys. Rev. Lett.*, **100**, 251101.

[123] Jaranowski, P., and Schäfer, G. 2012. *Phys. Rev. D*, **86**, 061503.

[124] Jaranowski, P., and Schäfer, G. 2013. *Phys. Rev. D*, **87**, 081503.

[125] Bertotti, B., and Plebanski, J. 1960. *Annals of Physics*, **11**, 169–200.

[126] Hari Dass, N., and Soni, V. 1982. *J. Phys.*, **A15**, 473.

[127] Damour, T., and Esposito-Farese, G. 1996. *Phys. Rev. D*, **53**, 5541–5578.

[128] Goldberger, W. D., and Rothstein, I. Z. 2006. *Phys. Rev. D*, **73**, 104029.

[129] Goldberger, W. D., and Rothstein, I. Z. 2006. *Gen. Rel. Grav.*, **38**, 1537–1546.

[130] Gilmore, J. B., and Ross, A. 2008. *Phys. Rev. D*, **78**, 124021.

[131] Foffa, S., and Sturani, R. 2011. *Phys. Rev. D*, **84**, 044031.

[132] Porto, R. A. 2006. *Phys. Rev. D*, **73**, 104031.

[133] Porto, R. A., and Rothstein, I. Z. 2006. *Phys. Rev. Lett.*, **97**, 021101.

[134] Kol, B., and Smolkin, M. 2008. *Class. Quant. Grav.*, **25**, 145011.

[135] Porto, R. A., and Rothstein, I. Z. 2008. *Phys. Rev. D*, **78**, 044013.

[136] Porto, R. A., and Rothstein, I. Z. 2008. *Phys. Rev. D*, **78**, 044012.

[137] Porto, R. A., Ross, A., and Rothstein, I. Z. 2011. *JCAP*, **1103**, 009.

[138] Porto, R. A. 2010. *Class. Quant. Grav.*, **27**, 205001.

[139] Levi, M. 2010. *Phys. Rev. D*, **82**, 104004.

[140] Levi, M. 2012. *Phys. Rev. D*, **85**, 064043.

[141] Hergt, S., Steinhoff, J., and Schäfer, G. 2012. *Annals of Physics*, **327**, 1494–1537.

[142] Hergt, S., Steinhoff, J., and Schäfer, G. 2014. *J. Phys.: Conf. Ser.*, **484**, 012018.
[143] Porto, R. A., Ross, A., and Rothstein, I. Z. 2012. *JCAP*, **1209**, 028.
[144] Blanchet, L., Detweiler, S. L., Le Tiec, A., and Whiting, B. F. 2010. *Phys. Rev. D*, **81**, 084033.
[145] Blanchet, L., Buonanno, A., and Le Tiec, A. 2013. *Phys. Rev. D*, **87**, 024030.
[146] Le Tiec, A., Barausse, E., and Buonanno, A. 2012. *Phys. Rev. Lett.*, **108**, 131103.
[147] Shah, A. G., Friedman, J. L., and Whiting, B. F. 2014. *Phys. Rev.*, **D89**, 064042.
[148] Blanchet, L., Faye, G., and Whiting, B. F. 2014. *Phys. Rev.*, **D89**, 064026.
[149] Bini, D., and Damour, T. 2013. *High-order post-Newtonian contributions to the two-body gravitational interaction potential from analytical gravitational self-force calculations.* arXiv:1312.2503.
[150] Mano, S., Suzuki, H., and Takasugi, E. 1996. *Prog. Theor. Phys.*, **95**, 1079–1096.
[151] Mano, S., and Takasugi, E. 1997. *Prog. Theor. Phys.*, **97**, 213–232.
[152] Mano, S., Suzuki, H., and Takasugi, E. 1996. *Prog. Theor. Phys.*, **96**, 549–566.
[153] Vines, J. E., and Flanagan, E. E. 2013. *Phys. Rev. D*, **88**, 024046.
[154] Damour, T., and Nagar, A. 2010. *Phys. Rev. D*, **81**, 084016.
[155] Bini, D., Damour, T., and Faye, G. 2012. *Phys. Rev. D*, **85**, 124034.
[156] Blanchet, L., and Faye, G. 2001. *Phys. Rev. D*, **63**, 062005.
[157] de Andrade, V. C., Blanchet, L., and Faye, G. 2001. *Class. Quant. Grav.*, **18**, 753–778.
[158] Blanchet, L., Damour, T., and Esposito-Farese, G. 2004. *Phys. Rev. D*, **69**, 124007.
[159] Itoh, Y., and Futamase, T. 2003. *Phys. Rev. D*, **68**, 121501.
[160] Foffa, S., and Sturani, R. 2013. *Phys. Rev. D*, **87**, 064011.
[161] Bini, D., and Damour, T. 2013. *Phys. Rev.*, **D87**, 121501.
[162] Damour, T., Jaranowski, P., and Schäfer, G. 2014. *Phys. Rev. D*, **89**, 064058.
[163] Tagoshi, H., Ohashi, A., and Owen, B. J. 2001. *Phys. Rev. D*, **63**, 044006.
[164] Faye, G., Blanchet, L., and Buonanno, A. 2006. *Phys. Rev. D*, **74**, 104033.
[165] Damour, T., Jaranowski, P., and Schäfer, G. 2008. *Phys. Rev. D*, **77**, 064032.
[166] Hartung, J., and Steinhoff, J. 2011. *Annalen der Physik*, **523**, 783–790.
[167] Marsat, S., Bohé, A., Faye, G., and Blanchet, L. 2013. *Class. Quant. Grav.*, **30**, 055007.
[168] Levi, M. 2010. *Phys. Rev. D*, **82**, 064029.
[169] Steinhoff, J., Hergt, S., and Schäfer, G. 2008. *Phys. Rev. D*, **77**, 081501.
[170] Steinhoff, J., Hergt, S., and Schäfer, G. 2008. *Phys. Rev. D*, **78**, 101503.
[171] Blanchet, L., Damour, T., Esposito-Farese, G., and Iyer, B. R. 2004. *Phys. Rev. Lett.*, **93**, 091101.
[172] Blanchet, L., Damour, T., Esposito-Farese, G., and Iyer, B. R. 2005. *Phys. Rev. D*, **71**, 124004.
[173] Blanchet, L., Buonanno, A., and Faye, G. 2006. *Phys. Rev. D*, **74**, 104034.
[174] Blanchet, L., Buonanno, A., and Faye, G. 2011. *Phys. Rev. D*, **84**, 064041.
[175] Bohé, A., Marsat, S., and Blanchet, L. 2013. *Class. Quant. Grav.*, **30**, 135009.
[176] Marsat, S., Bohé, A., Blanchet, L., and Buonanno, A. 2014. *Class. Quant. Grav.*, **31**, 025023.
[177] Gergely, L. A. 2000. *Phys. Rev. D*, **61**, 024035.
[178] Gergely, L. A. 2000. *Phys. Rev. D*, **62**, 024007.
[179] Mikoczi, B., Vasuth, M., and Gergely, L. A. 2005. *Phys. Rev. D*, **71**, 124043.
[180] Vines, J., Flanagan, E. E., and Hinderer, T. 2011. *Phys. Rev. D*, **83**, 084051.
[181] Konigsdorffer, C., Faye, G., and Schäfer, G. 2003. *Phys. Rev. D*, **68**, 044004.
[182] Nissanke, S., and Blanchet, L. 2005. *Class. Quant. Grav.*, **22**, 1007–1032.
[183] Will, C. M. 2005. *Phys. Rev. D*, **71**, 084027.
[184] Zeng, J., and Will, C. M. 2007. *Gen. Rel. Grav.*, **39**, 1661–1673.
[185] Wang, H., Steinhoff, J., Zeng, J., and Schäfer, G. 2011. *Phys. Rev. D*, **84**, 124005.
[186] Wang, H., and Will, C. M. 2007. *Phys. Rev. D*, **75**, 064017.
[187] Blanchet, L., Faye, G., Iyer, B. R., and Joguet, B. 2002. *Phys. Rev. D*, **65**, 061501.
[188] Arun, K. G., Buonanno, A., Faye, G., and Ochsner, E. 2009. *Phys. Rev. D*, **79**, 104023.
[189] Flanagan, E. E., and Hinderer, T. 2008. *Phys. Rev. D*, **77**, 021502.
[190] Blanchet, L., Iyer, B. R., Will, C. M., and Wiseman, A. G. 1996. *Class. Quant. Grav.*, **13**, 575–584.

[191] Arun, K., Blanchet, L., Iyer, B. R., and Qusailah, M. S. 2004. *Class. Quant. Grav.*, **21**, 3771–3802.
[192] Kidder, L. E. 2008. *Phys. Rev. D*, **77**, 044016.
[193] Blanchet, L., Faye, G., Iyer, B. R., and Sinha, S. 2008. *Class. Quant. Grav.*, **25**, 165003.
[194] Faye, G., Marsat, S., Blanchet, L., and Iyer, B. R. 2012. *Class. Quant. Grav.*, **29**, 175004.
[195] Buonanno, A., Faye, G., and Hinderer, T. 2013. *Phys. Rev. D*, **87**, 044009.
[196] Taylor, S., and Poisson, E. 2008. *Phys. Rev. D*, **78**, 084016.
[197] Alvi, K. 2001. *Phys. Rev. D*, **64**, 104020.
[198] Chatziioannou, K., Poisson, E., and Yunes, N. 2013. *Phys. Rev. D*, **87**, 044022.
[199] Regge, T., and Wheeler, J. A. 1957. *Phys. Rev.*, **108**, 1063–1069.
[200] Zerilli, F. 1970. *Phys. Rev. D*, **2**, 2141–2160.
[201] Teukolsky, S. A. 1973. *Astrophys. J.*, **185**, 635–647.
[202] Sasaki, M., and Nakamura, T. 1982. *Prog. Theor. Phys.*, **67**, 1788.
[203] Cutler, C., Poisson, E., Sussman, G., and Finn, L. 1993. *Phys. Rev. D*, **47**, 1511.
[204] Apostolatos, T., Kennefick, D., Poisson, E., and Ori, A. 1993. *Phys. Rev. D*, **47**, 5376–5388.
[205] Cutler, C., Kennefick, D., and Poisson, E. 1994. *Phys. Rev. D*, **50**, 3816.
[206] Hughes, S. A. 2000. *Phys. Rev. D*, **61**, 084004.
[207] Hughes, S. A. 2001. *Phys. Rev. D*, **64**, 064004.
[208] Fujita, R., and Tagoshi, H. 2004. *Prog. Theor. Phys.*, **112**, 415–450.
[209] Sundararajan, P., Khanna, G., and Hughes, S. A. 2007. *Phys. Rev. D*, **76**, 104005.
[210] Sundararajan, P., Khanna, G., Hughes, S. A., and Drasco, S. 2008. *Phys. Rev. D*, **78**, 024022.
[211] Fujita, R., and Tagoshi, H. 2005. *Prog. Theor. Phys.*, **113**, 1165–1182.
[212] Zenginoglu, A., and Khanna, G. 2011. *Phys. Rev. X*, **1**, 021017.
[213] Bernuzzi, S., Nagar, A., and Zenginoglu, A. 2011. *Phys. Rev. D*, **84**, 084026.
[214] Damour, T., and Nagar, A. 2007. *Phys. Rev. D*, **76**, 064028.
[215] Sundararajan, P., Khanna, G., and Hughes, S. A. 2010. *Phys. Rev. D*, **81**, 104009.
[216] Bernuzzi, S., and Nagar, A. 2010. *Phys. Rev. D*, **81**, 084056.
[217] Bernuzzi, S., Nagar, A., and Zenginoglu, A. 2011. *Phys. Rev. D*, **83**, 064010.
[218] Barausse, E. *et al.* 2012. *Phys. Rev. D*, **85**, 024046.
[219] Sasaki, M. 1994. *Prog. Theor. Phys.*, **92**, 17–36.
[220] Tanaka, T., Tagoshi, H., and Sasaki, M. 1996. *Prog. Theor. Phys.*, **96**, 1087–1101.
[221] Fujita, R., and Iyer, B. 2010. *Phys. Rev. D*, **82**, 044051.
[222] Fujita, R. 2012. *Prog. Theor. Phys.*, **127**, 583–590.
[223] Fujita, R. 2012. *Prog. Theor. Phys.*, **128**, 971–992.
[224] Tanaka, T., Mino, Y., Sasaki, M., and Shibata, M. 1996. *Phys. Rev. D*, **54**, 3762.
[225] Poisson, E., and Sasaki, M. 1995. *Phys. Rev. D*, **51**, 5753–5767.
[226] Tagoshi, H., Mano, S., and Takasugi, E. 1997. *Prog. Theor. Phys.*, **98**, 829–850.
[227] Mino, Y. *et al.* 1997. *Prog. Theor. Phys. Suppl.*, **128**, 1–121.
[228] Dirac, P. A. M. 1938. *Roy. Soc. Lond. Proc. Series A*, **167**, 148–169.
[229] DeWitt, B. S., and Brehme, R. W. 1960. *Annals of Physics*, **9**, 220–259.
[230] Mino, Y., Sasaki, M., and Tanaka, T. 1997. *Phys. Rev. D*, **55**, 3457–3476.
[231] Quinn, T. C., and Wald, R. M. 1997. *Phys. Rev. D*, **56**, 3381–3394.
[232] Detweiler, S. L., and Whiting, B. F. 2003. *Phys. Rev. D*, **67**, 024025.
[233] Gralla, S. E., and Wald, R. M. 2008. *Class. Quant. Grav.*, **25**, 205009.
[234] Pound, A. 2010. *Phys. Rev. D*, **81**, 024023.
[235] Barack, L., and Ori, A. 2001. *Phys. Rev. D*, **64**, 124003.
[236] Detweiler, S., Messaritaki, E., and Whiting, B. 2003. *Phys. Rev. D*, **67**, 104016.
[237] Keidl, T. S., Friedman, J. L., and Wiseman, A. G. 2007. *Phys. Rev. D*, **75**, 124009.
[238] Barack, L., and Lousto, C. O. 2002. *Phys. Rev. D*, **66**, 061502.
[239] Barack, L., and Sago, N. 2007. *Phys. Rev. D*, **75**, 064021.
[240] Detweiler, S. L. 2008. *Phys. Rev. D*, **77**, 124026.
[241] Barack, L., and Sago, N. 2009. *Phys. Rev. Lett.*, **102**, 191101.
[242] Barack, L., and Sago, N. 2010. *Phys. Rev. D*, **81**, 084021.
[243] Barack, L., and Ori, A. 2003. *Phys. Rev. Lett.*, **90**, 111101.

[244] Warburton, N. *et al.* 2012. *Phys. Rev. D*, **85**, 061501.
[245] Mino, Y. 2003. *Phys. Rev. D*, **67**, 084027.
[246] Sago, N. *et al.* 2006. *Prog. Theor. Phys.*, **115**, 873–907.
[247] Sago, N., Tanaka, T., Hikida, W., and Nakano, H. 2005. *Prog. Theor. Phys.*, **114**, 509–514.
[248] Hughes, S. A., Drasco, S., Flanagan, E. E., and Franklin, J. 2005. *Phys. Rev. Lett.*, **94**, 221101.
[249] Drasco, S., and Hughes, S. A. 2006. *Phys. Rev. D*, **73**, 024027.
[250] Hinderer, T., and Flanagan, E. E. 2008. *Phys. Rev. D*, **78**, 064028.
[251] Rosenthal, E. 2006. *Phys. Rev. D*, **74**, 084018.
[252] Gralla, S. E. 2012. *Phys. Rev. D*, **85**, 124011.
[253] Pound, A. 2012. *Phys. Rev. Lett.*, **109**, 051101.
[254] Isoyama, S. *et al.* 2014. *Gravitational self-force correction to the innermost stable circular equatorial orbit of a Kerr black hole.* arXiv:1404.6133.
[255] Barack, L., Damour, T., and Sago, N. 2010. *Phys. Rev. D*, **82**, 084036.
[256] Blanchet, L., Detweiler, S. L., Le Tiec, A., and Whiting, B. F. 2010. *Phys. Rev. D*, **81**, 064004.
[257] Le Tiec, A., Blanchet, L., and Whiting, B. F. 2012. *Phys. Rev.*, **D85**, 064039.
[258] Friedman, J. L., Uryu, K., and Shibata, M. 2002. *Phys. Rev. D*, **65**, 064035.
[259] Barausse, E., Buonanno, A., and Le Tiec, A. 2012. *Phys. Rev. D*, **85**, 064010.
[260] Akcay, S., Barack, L., Damour, T., and Sago, N. 2012. *Phys. Rev. D*, **86**, 104041.
[261] Barack, L., and Sago, N. 2011. *Phys. Rev. D*, **83**, 084023.
[262] Shah, A. G., Friedman, J. L., and Keidl, T. S. 2012. *Phys. Rev. D*, **86**, 084059.
[263] Dolan, S. R. *et al.* 2014. *Phys. Rev. D*, 064011.
[264] Le Tiec, A. *et al.* 2011. *Phys. Rev. Lett.*, **107**, 141101.
[265] Le Tiec, A. *et al.* 2013. *Phys. Rev. D*, **88**, 124027.
[266] Nagar, A. 2013. *Phys. Rev. D*, **88**, 121501.
[267] Buonanno, A., Chen, Y., and Damour, T. 2006. *Phys. Rev. D*, **74**, 104005.
[268] Damour, T. 2001. *Phys. Rev. D*, **64**, 124013.
[269] Damour, T., Jaranowski, P., and Schäfer, G. 2000. *Phys. Rev. D*, **62**, 084011.
[270] Brezin, E., Itzykson, C., and Zinn-Justin, J. 1970. *Phys. Rev. D*, **1**, 2349–2355.
[271] Damour, T., Iyer, B. R., and Sathyaprakash, B. 1998. *Phys. Rev. D*, **57**, 885–907.
[272] Davis, M., Ruffini, R., and Tiomno, J. 1972. *Phys. Rev. D*, **5**, 2932–2935.
[273] Price, R. H., and Pullin, J. 1994. *Phys. Rev. Lett.*, **72**, 3297–3300.
[274] Damour, T., Iyer, B. R., and Nagar, A. 2009. *Phys. Rev. D*, **79**, 064004.
[275] Nagar, A., and Akcay, S. 2012. *Phys. Rev. D*, **85**, 044025.
[276] Taracchini, A., Buonanno, A., Hughes, S. A., and Khanna, G. 2013. *Phys. Rev. D*, **88**, 044001.
[277] Damour, T., Iyer, B. R., Jaranowski, P., and Sathyaprakash, B. 2003. *Phys. Rev. D*, **67**, 064028.
[278] Buonanno, A., Cook, G. B., and Pretorius, F. 2007. *Phys. Rev. D*, **75**, 124018.
[279] Berti, E., Cardoso, V., and Will, C. M. 2006. *Phys. Rev. D*, **73**, 064030.
[280] Berti, E., Cardoso, V., and Starinets, A. 2009. *Class. Quant. Grav.*, **26**, 163001.
[281] Hinderer, T. *et al.* 2013. *Phys. Rev. D*, **88**, 084005.
[282] Damour, T., Nagar, A., Pollney, D., and Reisswig, C. 2012. *Phys. Rev. Lett.*, **108**, 131101.
[283] Damour, T., Jaranowski, P., and Schäfer, G. 2008. *Phys. Rev. D*, **78**, 024009.
[284] Barausse, E., Racine, E., and Buonanno, A. 2009. *Phys. Rev. D*, **80**, 104025.
[285] Barausse, E., and Buonanno, A. 2010. *Phys. Rev. D*, **81**, 084024.
[286] Nagar, A. 2011. *Phys. Rev. D*, **84**, 084028.
[287] Barausse, E., and Buonanno, A. 2011. *Phys. Rev. D*, **84**, 104027.
[288] Balmelli, S., and Jetzer, P. 2013. *Phys. Rev. D*, **87**, 124036.
[289] Pan, Y. *et al.* 2011. *Phys. Rev. D*, **83**, 064003.
[290] Yunes, N. *et al.* 2010. *Phys. Rev. Lett.*, **104**, 091102.
[291] Yunes, N., *et al.* 2011. *Phys. Rev. D*, **83**, 044044.
[292] Damour, T. 2010. *Phys. Rev. D*, **81**, 024017.
[293] Buonanno, A. *et al.* 2009. *Phys. Rev. D*, **80**, 084043.
[294] Berti, E., Cardoso, V., and Starinets, A. O. 2009. *Class. Quant. Grav.*, **26**, 163001.
[295] Pan, Y. *et al.* 2011. *Phys. Rev. D*, **84**, 124052.
[296] Shibata, M., and Taniguchi, K. 2006. *Phys. Rev. D*, **73**, 064027.

[297] Baiotti, L., Giacomazzo, B., and Rezzolla, L. 2008. *Phys. Rev. D*, **78**, 084033.
[298] Duez, M. D. 2010. *Class. Quant. Grav.*, **27**, 114002.
[299] Shibata, M., Taniguchi, K., and Uryu, K. 2005. *Phys. Rev. D*, **71**, 084021.
[300] Bartos, I., Brady, P., and Marka, S. 2013. *Class. Quant. Grav.*, **30**, 123001.
[301] Shibata, M., Karino, S., and Eriguchi, Y. 2003. *Mon. Not. Roy. Astron. Soc.*, **343**, 619.
[302] Baiotti, L., De Pietri, R., Manca, G. M., and Rezzolla, L. 2007. *Phys. Rev. D*, **75**, 044023.
[303] Shibata, M., Taniguchi, K., and Uryu, K. 2003. *Phys. Rev. D*, **68**, 084020.
[304] Rosswog, S. 2011. *Proceedings of Science*, 11th Symposium on Nuclei in the Cosmos, NIC XI, Heidelberg, 032.
[305] Shibata, M., and Taniguchi, K. 2011. *Living Rev. Rel.*, **14**, 6.
[306] Foucart, F. 2012. *Phys. Rev. D*, **86**, 124007.
[307] Foucart, F. *et al.* 2013. *Phys. Rev. D*, **87**, 084006.
[308] Abadie, J., *et al.* [LIGO Scientific]. 2010. *Class. Quant. Grav.*, **27**, 173001.
[309] Finn, L. S., and Chernoff, D. F. 1993. *Phys. Rev. D*, **47**, 2198–2219.
[310] Belczynski, K. *et al.* 2010. *Astrophys. J. Lett.*, **715**, L138–L141.
[311] Bulik, T., Belczynski, K., and Prestwich, A. 2011. *Astrophys. J.*, **730**, 140.
[312] Komossa, S. 2003. *AIP Conf. Proc.*, **686**, 161–174.
[313] Dotti, M., Sesana, A., and Decarli, R. 2012. *Adv. Astron.*, **2012**, 940568.
[314] Arun, K. G., *et al.* 2009. *Class. Quant. Grav.*, **26**, 094027.
[315] Sesana, A., Gair, J., Berti, E., and Volonteri, M. 2011. *Phys. Rev. D*, **83**, 044036.
[316] Seoane, P. A., *et al.* 2013. *The gravitational universe*. arXiv:1305.5720.
[317] Siemens, X., Ellis, J., Jenet, F., and Romano, J. D. 2013. *Class. Quant. Grav.*, **30**, 224015.
[318] Sesana, A. 2013. *Braz. J. Phys.*, **43**, 314–319.
[319] Farrell, S. A. *et al.* 2009. *Nature*, **460**, 73–75.
[320] Gair, J., Mandel, I., Miller, M., and Volonteri, M. 2011. *Gen. Rel. Grav.*, **43**, 485–518.
[321] Amaro-Seoane, P., and Santamaria, L. 2010. *Astrophys. J.*, **722**, 1197–1206.
[322] Sesana, A., Gair, J., Mandel, I., and Vecchio, A. 2009. *Astrophys. J. Lett.*, **698**, L129–L132.
[323] Gair, J. R. 2009. *Class. Quant. Grav.*, **26**, 094034.
[324] Orosz, J. A. *et al.* 2007. *Nature*, **449**, 872–875.
[325] Prestwich, A. *et al.* 2007. *Astrophys. J.*, **669**, L21–L24.
[326] Amaro-Seoane, P. *et al.* 2012. *eLISA: astrophysics and cosmology in the millihertz regime*. arXiv:1201.3621.
[327] Buonanno, A., Chen, Y., and Vallisneri, M. 2003. *Phys. Rev. D*, **67**, 024016.
[328] Buonanno, A., Chen, Y., and Vallisneri, M. 2003. *Phys. Rev. D*, **67**, 104025.
[329] Buonanno, A. *et al.* 2005. *Phys. Rev. D*, **72**, 084027.
[330] Abbott, B., *et al.* [LIGO Scientific Collaboration]. 2006. *Phys. Rev. D*, **73**, 062001.
[331] Abbott, B., *et al.* [LIGO Scientific]. 2008. *Phys. Rev. D*, **78**, 042002.
[332] Boyle, M. *et al.* 2007. *Phys. Rev. D*, **76**, 124038.
[333] Lovelace, G., Boyle, M., Scheel, M. A., and Szilágyi, B. 2012. *Class. Quant. Grav.*, **29**, 045003.
[334] Damour, T., Nagar, A., and Trias, M. 2011. *Phys. Rev. D*, **83**, 024006.
[335] MacDonald, I. *et al.* 2013. *Phys. Rev. D*, **87**, 024009.
[336] Pan, Y. *et al.* 2014. *Phys. Rev. D*, **89**, 06150(R).
[337] Pan, Y. *et al.* 2008. *Phys. Rev. D*, **77**, 024014.
[338] Taracchini, A. *et al.* 2012. *Phys. Rev. D*, **83**, 104034.
[339] Buonanno, A. *et al.* 2007. *Phys. Rev. D*, **76**, 104049.
[340] Damour, T. *et al.* 2008. *Phys. Rev. D*, **77**, 084017.
[341] Damour, T. *et al.* 2008. *Phys. Rev. D*, **78**, 044039.
[342] Buonanno, A. *et al.* 2009. *Phys. Rev. D*, **79**, 124028.
[343] Damour, T., and Nagar, A. 2009. *Phys. Rev. D*, **79**, 081503.
[344] Pan, Y. *et al.* 2010. *Phys. Rev. D*, **81**, 084041.
[345] Damour, T., Nagar, A., and Bernuzzi, S. 2013. *Phys. Rev. D*, **87**, 084035.
[346] Taracchini, A. *et al.* 2014. *Phys. Rev. D*, **89**, 061502(R).
[347] Littenberg, T. B., Baker, J. G., Buonanno, A., and Kelly, B. J. 2013. *Phys. Rev. D*, **87**, 104003.

[348] Pan, Y. *et al.* 2014. *Phys. Rev. D*, **89**, 084006.
[349] Ajith, P., *et al.* 2007. *Class. Quant. Grav.*, **24**, S689–S699.
[350] Santamaría, L., *et al.* 2010. *Phys. Rev. D*, **82**, 064016.
[351] Ajith, P. *et al.* 2011. *Phys. Rev. Lett.*, **106**, 241101.
[352] Hannam, M. *et al.* 2013. *Twist and shout: a simple model of complete precessing black-hole-binary gravitational waveforms.* arXiv:1308.3271.
[353] Mroue, A. H. *et al.* 2013. *Phys. Rev. Lett.*, **111**, 241104.
[354] Pekowsky, L., O'Shaughnessy, R., Healy, J., and Shoemaker, D. 2013. *Phys. Rev. D*, **88**, 024040.
[355] Abadie, J., *et al.* 2012. *Phys. Rev. D*, **85**, 082002.
[356] Aasi, J., *et al.* 2013. *Phys. Rev. D*, **87**, 022002.
[357] Abadie, J., *et al.* 2012. *Phys. Rev. D*, **85**, 102004.
[358] Aasi, J., *et al.* 2013. *Prospects for localization of gravitational wave transients by the Advanced LIGO and Advanced Virgo observatories.* arXiv:1304.0670.
[359] Iyer, B. *et al.* 2012. https://dcc.ligo.org/LIGO-M1100296.
[360] Aylott, B. *et al.* 2009. *Class. Quant. Grav.*, **26**, 165008.
[361] Ajith, P. *et al.* 2012. *The NINJA-2 catalog of hybrid post-Newtonian/numerical-relativity waveforms for non-precessing black-hole binaries.* arXiv:1201.5319.
[362] Hinder, I. *et al.* 2014. *Class. Quant. Grav.*, **31**, 025012.
[363] Aasi, J., *et al.* 2014. *The NINJA-2 project: detecting and characterizing gravitational waveforms modelled using numerical binary black hole simulations.* arXiv:1401.0939.
[364] Postnov, K., and Yungelson, L. 2005. *Living Rev. Rel.*, **9**, 6.
[365] O'Shaughnessy, R., Kim, C., Kalogera, V., and Belczynski, K. 2008. *Astrophys. J.*, **672**, 479–488.
[366] Hannam, M. *et al.* 2013. *Astrophys. J.*, **766**, L14.
[367] Fong, W., and Berger, E. 2013. *Astrophys. J.*, **776**, 18.
[368] Schutz, B. F. 1986. *Nature*, **323**, 310.
[369] Del Pozzo, W. 2012. *Phys. Rev. D*, **86**, 043011.
[370] Dalal, N., Holz, D. E., Hughes, S. A., and Jain, B. 2006. *Phys. Rev. D*, **74**, 063006.
[371] Sathyaprakash, B., Schutz, B., and Van Den Broeck, C. 2010. *Class. Quant. Grav.*, **27**, 215006.
[372] Nissanke, S. *et al.* 2010. *Astrophys. J.*, **725**, 496–514.
[373] Berger, E., Fong, W., and Chornock, R. 2013. *Astrophys. J. Lett.*, **774**, L23.
[374] Tanvir, N. R. *et al.* 2013. *Nature*, **500**, 547–549.
[375] Holz, D. E., and Hughes, S. A. 2005. *Astrophys. J.*, **629**, 15–22.
[376] Messenger, C., and Read, J. 2012. *Phys. Rev. Lett.*, **108**, 091101.
[377] Ghez, A. *et al.* 2005. *Astrophys. J.*, **620**, 744–757.
[378] Shternin, P. S. *et al.* 2011. *Mon. Not. Roy. Astron. Soc.*, **412**, L108–L112.
[379] Page, D., Prakash, M., Lattimer, J. M., and Steiner, A. W. 2011. *Phys. Rev. Lett.*, **106**, 081101.
[380] Del Pozzo, W. *et al.* 2013. *Phys. Rev. Lett.*, **111**, 071101.
[381] Markakis, C. *et al.* 2009. *J. Phys.: Conf. Ser.*, **189**, 012024.
[382] Mishra, C. K., Arun, K., Iyer, B. R., and Sathyaprakash, B. 2010. *Phys. Rev. D*, **82**, 064010.
[383] Li, T. G. F., *et al.* 2012. *Phys. Rev. D*, **85**, 082003.
[384] Will, C. M. 2006. *Living Rev. Rel.*, **9**, 3.
[385] Yunes, N., and Siemens, X. 2013. *Living Rev. Rel.*, **16**, 9.
[386] Ryan, F. D. 1997. *Phys. Rev. D*, **56**, 1845.
[387] Barack, L., and Cutler, C. 2007. *Phys. Rev. D*, **75**, 042003.
[388] Gair, J., Vallisneri, M., Larson, S., and Baker, J. 2013. *Living Rev. Rel.*, **16**, 7.
[389] Gossan, S., Veitch, J., and Sathyaprakash, B. 2012. *Phys. Rev. D*, **85**, 124056.
[390] Nitz, A. H. *et al.* 2013. *Phys. Rev. , **D88**, 124039.
[391] Peters, P. C., and Mathews, J. 1963. *Phys. Rev.*, **131**, 435–440.
[392] Kowalska, I. *et al.* 2011. *Astron. Astrophys.*, **527**, A70.
[393] Miller, M., and Hamilton, D. 2002. *Astrophys. J.*, **576**, 894–898.
[394] Wen, L. 2003. *Astrophys. J.*, **598**, 419–430.

[395] Kocsis, B., and Levin, J. 2012. *Phys. Rev. D*, **85**, 123005.
[396] Naoz, S., Kocsis, B., Loeb, A., and Yunes, N. 2013. *Astrophys. J.*, **773**, 187.
[397] Pan, Y., Buonanno, A., Chen, Y., and Vallisneri, M. 2004. *Phys. Rev. D*, **69**, 104017.
[398] Buonanno, A., Chen, Y., Pan, Y., and Vallisneri, M. 2004. *Phys. Rev. D*, **70**, 104003.
[399] Ajith, P. 2011. *Phys. Rev. D*, **84**, 084037.
[400] Schmidt, P., Hannam, M., and Husa, S. 2012. *Phys. Rev. D*, **86**, 104063.
[401] Harry, I., and Fairhurst, S. 2011. *Class. Quant. Grav.*, **28**, 134008.
[402] Privitera, S. *et al.* 2014. *Phys. Rev. D*, **89**, 024003.
[403] Cannon, K. *et al.* 2010. *Phys. Rev. D*, **82**, 044025.
[404] Cannon, K., Hanna, C., and Keppel, D. 2011. *Phys. Rev. D*, **84**, 084003.
[405] Field, S. E. *et al.* 2011. *Phys. Rev. Lett.*, **106**, 221102.
[406] Field, S. E. *et al.* 2013. *Fast prediction and evaluation of gravitational waveforms using surrogate models.* arXiv:1308.3565.
[407] Pürrer, M. 2014. *Frequency domain reduced order models for gravitational waves from aligned-spin black-hole binaries.* arXiv:1402.4146.
[408] Bernuzzi, S., Nagar, A., Thierfelder, M., and Brugmann, B. 2012. *Phys. Rev. D*, **86**, 044030.
[409] Damour, T., Nagar, A., and Villain, L. 2012. *Phys. Rev. D*, **85**, 123007.
[410] Favata, M. 2014. *Phys. Rev. Lett.*, **112**, 101101.
[411] Yagi, K., and Yunes, N. 2013. *Phys. Rev.*, **D89**, 021303.
[412] Djorgovski, S., *et al.* 2011. *Towards an automated classification of transient events in synoptic sky surveys.* arXiv:1110.4655.
[413] Fairhurst, S. 2011. *Class. Quant. Grav.*, **28**, 105021.
[414] Van Den Broeck, C., and Sengupta, A. 2007. *Class. Quant. Grav.*, **24**, 1089–1113.
[415] Capano, C., Pan, Y., and Buonanno, A. 2014. *Phys. Rev. D*, **89**, 102003.
[416] Raymond, V. *et al.* 2009. *Class. Quant. Grav.*, **26**, 114007.
[417] Hanna, C., Mandel, I., and Vousden, W. 2014. *Astrophys. J.*, **784**, 8.
[418] Arzoumanian, Z., *et al.* 2014. *NANOGrav limits on gravitational waves from individual supermassive black hole binaries in circular orbits.* arXiv:1404.1267.
[419] Burke-Spolaor, S. 2013. *Class. Quant. Grav.*, **30**, 224013.
[420] Ando, S., Beacom, F., and Yüksel, H. 2005. *Phys. Rev. Lett.*, **95**, 171101.
[421] Dimmelmeier, H., Ott, C. D., Marek, A., and Janka, H.-T. 2008. *Phys. Rev. D*, **78**, 064056.
[422] Ott, C. D. 2009. *Class. Quant. Grav.*, **26**, 204015.
[423] Owen, B. J. 2005. *Phys. Rev. Lett.*, **95**, 211101.
[424] Andersson, N. *et al.* 2011. *Gen. Rel. Grav.*, **43**, 409–436.
[425] Knispel, B., and Allen, B. 2008. *Phys. Rev. D*, **78**, 044031.
[426] Chamel, N., and Haensel, P. 2008. *Living Rev. Rel.*, **11**.
[427] Espinoza, C. M., Lyne, A. G., Stappers, B. W., and Kramer, M. 2011. *Mon. Not. Roy. Astr. Soc.*, **414**, 1679–1704.
[428] Kaspi, V. M. 2010. *Proc. Nat. Acad. Sci.*, **107**, 7147–7152.
[429] Andersson, N., and Kokkotas, K. 1998. *Mon. Not. Roy. Astron. Soc.*, **299**, 1059–1068.
[430] Chakrabarty, D. 2005. Millisecond pulsars in X-ray binaries. Page 279 of: Rasio, F. A., and Stairs, I. H. (eds), *Binary radio pulsars*. Astronomical Society of the Pacific Conference Series, vol. 328.
[431] Wagoner, R. V. 1984. *Astrophys. J.*, **278**, 345–348.
[432] Bildsten, L. 1998. *Astrophys. J. Lett.*, **501**, L89.
[433] Chakrabarty, D. *et al.* 2003. *Nature*, **424**, 42–44.
[434] Watts, A. L., Krishnan, B., Bildsten, L., and Schutz, B. F. 2008. *Mon. Not. Roy. Astr. Soc.*, **389**, 839–868.
[435] Klimenko, S., Yakushin, I., Mercer, A., and Mitselmakher, G. 2008. *Class. Quant. Grav.*, **25**, 114029.
[436] Abadie, J., *et al.* 2012. *Phys. Rev. D*, **85**, 122007.
[437] Abadie, J., *et al.* [LIGO Scientific Collaboration, Virgo Collaboration]. 2011. *Astrophys. J.*, **737**, 93.
[438] Abadie, J., *et al.* 2011. *Astrophys. J. Lett.*, **734**, L35.

[439] Abadie, J., *et al.* 2012. *Astrophys. J.*, **760**, 12.
[440] Abbott, B., *et al.* 2008. *Astrophys. J.*, **681**, 1419–1430.
[441] Metzger, B. D., and Berger, E. 2012. *Astrophys. J.*, **746**, 48.
[442] Aasi, J., *et al.* [The LIGO Scientific Collaboration]. 2014. *Astrophys. J.*, **785**, 119.
[443] Abadie, J., *et al.* [LIGO Scientific]. 2010. *Astrophys. J.*, **722**, 1504–1513.
[444] Abbott, B., *et al.* [LIGO Scientific Collaboration]. 2009. *Phys. Rev. D*, **79**, 022001.
[445] Knispel, B. *et al.* 2013. *Astrophys. J.*, **774**, 93.
[446] Allen, B. *et al.* 2013. *Astrophys. J.*, **773**, 91.
[447] Aasi, J., *et al.* 2013. *Phys. Rev. D*, **87**, 042001.
[448] Aasi, J., *et al.* 2013. *Phys. Rev. D*, **88**, 102002.
[449] Benacquista, M. J., and Downing, J. M. 2013. *Living Rev. Rel.*, **16**, 4.
[450] Andersson, N., Kokkotas, K. D., and Stergioulas, N. 1999. *Astrophys. J.*, **516**, 307–314.
[451] Cutler, C. 2002. *Phys. Rev. D*, **66**, 084025.
[452] Muller, B., Janka, H.-T., and Dimmelmeier, H. 2010. *Astrophys. J. Suppl.*, **189**, 104–133.
[453] Ott, C. D. *et al.* 2013. *Astrophys. J.*, **768**, 115.
[454] Logue, J. *et al.* 2012. *Phys. Rev. D*, **86**, 044023.
[455] Mueller, E. *et al.* 2004. *Astrophys. J.*, **603**, 221–230.
[456] Ushomirsky, G., Cutler, C., and Bildsten, L. 2000. *Mon. Not. R. Astron. Soc.*, **319**, 902.
[457] Cutler, C. 2002. *Phys. Rev. D*, **66**, 084025.
[458] Payne, D., Melatos, A., and Phinney, E. 2003. Gravitational waves from an accreting neutron star with a magnetic mountain. Pages 92–95 of: Centrella, J. (ed), *Astrophysics of gravitational wave sources*. AIP Conference Proceedings, vol. 686. Melville, NY: American Institute of Physics.
[459] Schutz, B. F. 1989. Data analysis requirements of networks of detectors. Page 315 of: Schutz, B. F. (ed), *NATO ASIC Proc. 253: Gravitational wave data analysis*.
[460] Brady, P. R., Creighton, T., Cutler, C., and Schutz, B. F. 1998. *Phys. Rev. D*, **57**, 2101–2116.
[461] Jaranowski, P., Krolak, A., and Schutz, B. F. 1998. *Phys. Rev. D*, **58**, 063001.
[462] Creighton, J., and Anderson, W. 2011. *Gravitational-wave physics and astronomy*. Weinheim: Wiley-VCH.
[463] Allen, B., and Romano, J. D. 1999. *Phys. Rev. D*, **59**, 102001.
[464] Starobinsky, A. A. 1979. *JETP Lett.*, **30**, 682–685.
[465] Grishchuk, L. P. 1975. *Sov. Phys. JETP*, **40**, 409–415.
[466] Allen, B. 1988. *Phys. Rev. D*, **37**, 2078–2085.
[467] Maggiore, M. 2000. *Phys. Rep.*, **331**, 283.
[468] Ade, P., *et al.* 2013. *Planck 2013 results. XVI. Cosmological parameters*. arXiv:1303.5076.
[469] Kamionkowski, M., Kosowsky, A., and Stebbins, A. 1997. *Phys. Rev. D*, **55**, 7368–7388.
[470] Seljak, U., and Zaldarriaga, M. 1997. *Phys. Rev. Lett.*, **78**, 2054–2057.
[471] Weinberg, S. 2004. *Phys. Rev. D*, **69**, 023503.
[472] Pritchard, J. R., and Kamionkowski, M. 2005. *Annals of Physics*, **318**, 2–36.
[473] Smith, T. L., Kamionkowski, M., and Cooray, A. 2006. *Phys. Rev. D*, **73**, 023504.
[474] Boyle, L. A., and Steinhardt, P. J. 2008. *Phys. Rev. D*, **77**, 063504.
[475] Khlebnikov, S. Y., and Tkachev, I. I. 1997. *Phys. Rev. D*, **56**, 653–660.
[476] Kibble, T. 1976. *J. Phys.*, **A9**, 1387–1398.
[477] Vilenkin, A. 1985. *Phys. Rept.*, **121**, 263.
[478] Sarangi, S., and Tye, S. H. 2002. *Phys. Lett.*, **B536**, 185–192.
[479] Berezinsky, V., Hnatyk, B., and Vilenkin, A. 2000. *Superconducting cosmic strings as gamma-ray burst engines*. astro-ph/0001213.
[480] Damour, T., and Vilenkin, A. 2000. *Phys. Rev. Lett.*, **85**, 3761–3764.
[481] Copeland, E. J., Myers, R. C., and Polchinski, J. 2004. *JHEP*, **06**, 013.
[482] Damour, T., and Vilenkin, A. 2005. *Phys. Rev. D*, **71**, 063510.
[483] Olmez, S., Mandic, V., and Siemens, X. 2010. *Phys. Rev. D*, **81**, 104028.
[484] Turner, M. S., and Wilczek, F. 1990. *Phys. Rev. Lett.*, **65**, 3080–3083.
[485] Kamionkowski, M., Kosowsky, A., and Turner, M. 1994. *Phys. Rev. D*, **49**, 2837–2851.

[486] Buonanno, A. 2003. *TASI lectures on gravitational waves from the early Universe.* arXiv:0303085.

[487] Binetruy, P., Bohé, A., Caprini, C., and Dufaux, J.-F. 2012. *JCAP*, **1206**, 027.

[488] Thorne, K. 1987. Gravitational radiation. Chapter 9, pages 330–458 of: Hawking, S. W., and Israel, W. (eds), *Three hundred years of gravitation.* Cambridge: Cambridge University Press.

[489] Flanagan, E. E. 1993. *Phys. Rev. D*, **48**, 2389–2407.

[490] Sesana, A., Vecchio, A., and Colacino, C. N. 2008. *Mon. Not. Roy. Astr. Soc.*, **390**, 192–209.

[491] Sesana, A. 2013. *Mon. Not. Roy. Astr. Soc.*, **433**, L1–L5.

[492] Kramer, M. 2004. *Fundamental physics with the SKA: strong-field tests of gravity using pulsars and black holes.* astro-ph/0409020.

[493] Copi, C. J., Schramm, D. N., and Turner, M. S. 1997. *Phys. Rev. D*, **55**, 3389–3393.

[494] Abbott, B. P., *et al.* [LIGO Scientific]. 2009. *Nature*, **460**, 990.

[495] Smith, T., Pierpaoli, E., and Kamionkowski, M. 2006. *Phys. Rev. Lett.*, **97**, 021301.

[496] Abbott, B. P., *et al.* [LIGO Scientific]. 2009. *Phys. Rev. D*, **80**, 062002.

[497] Aasi, J., *et al.* 2014. *Phys. Rev. Lett.*, **112**, 131101.

[498] Boyle, L. A., and Buonanno, A. 2008. *Phys. Rev. D*, **78**, 043531.

[499] Shannon, R. *et al.* 2013. *Science*, **342**, 334–337.

[500] Jenet, F. A. *et al.* 2006. *Astrophys. J.*, **653**, 1571–1576.

[501] Peebles, P., and Vilenkin, A. 1999. *Phys. Rev. D*, **59**, 063505.

[502] Giovannini, M. 1998. *Phys. Rev. D*, **58**, 083504.

[503] Giovannini, M. 1999. *Phys. Rev. D*, **60**, 123511.

[504] Riazuelo, A., and Uzan, J.-P. 2000. *Phys. Rev. D*, **62**, 083506.

[505] Tashiro, H., Chiba, T., and Sasaki, M. 2004. *Class. Quant. Grav.*, **21**, 1761–1772.

[506] Gopakumar, A., and Iyer, B. R. 1997. *Phys. Rev. D*, **56**, 7708–7731.

Part Three

Gravity is Geometry, after all

Introduction

Einstein's general relativity is a mathematically beautiful application of geometric ideas to gravitational physics. Motion is determined by geodesics in spacetime, tidal effects between physical bodies can be read directly from the curvature of that spacetime, and the curvature is closely tied to matter and its motion in spacetime. When proposed in 1915, general relativity was a completely new way to think about physical phenomena, based on the geometry of curved spacetimes that was largely unknown to physicists.

While the geometric nature of Einstein's theory is beautiful and conceptually simple, the fundamental working structure of the theory as a system of partial differential equations (PDEs) is much more complex. Einstein's equations are not easily categorized as wave-like or potential-like or heat-like, and they are pervasively nonlinear. Hence, despite the great interest in general relativity, mathematical progress in studying Einstein's equations (beyond the discovery of a small collection of explicit solutions with lots of symmetry) was quite slow for a number of years.

This changed significantly in the 1950s with the appearance of Yvonne Choquet-Bruhat's proof that the Einstein equations can be treated as a well-posed Cauchy problem [1]. The long-term effects of this work have been profound: Mathematically, it has led to the present status of Einstein's equations as one of the most interesting and important systems in PDE theory and in geometrical analysis. Physically, the well-posedness of the Cauchy problem for the Einstein equations has led directly to our present ability to numerically simulate (with remarkable accuracy) solutions of these equations which model a wide range of novel phenomena in the strong-field regime.

The Cauchy formulation of general relativity splits the problem of solving Einstein's equations, and studying the behavior of these solutions, into two equally important tasks: First, one finds an initial data set – a "snapshot" of the gravitational field and its rate of change – which satisfies the *Einstein constraint equations*, which are essentially four of the ten Einstein field equations. Then, using the rest of the equations, one evolves the gravitational fields forward and backward in time, thereby obtaining the spacetime and its geometry. The first of these tasks, working with initial data and the constraint equations, is the focus of Chapter 8. The second task, involving evolution, is the subject of Chapter 9. Chapter 7 discusses the ideas and methods which are used to carry out both of these tasks numerically, as well as some of the novel insights that have been obtained from this numerical work.

It is not surprising for the field equations of a physical theory to include constraint equations. Maxwell's equations, for example, include the constraints $\nabla \cdot B = 0$ and $\nabla \cdot E = 4\pi\rho$. However, the Einstein constraints (see Chapter 8) are much harder to handle than their Maxwellian counterparts, so considerable mathematical research has gone into working with them over the years.

The approach which has been most successfully used over the years for constructing and analyzing solutions of the constraint equations is the *conformal method*. Based on the work of Lichnerowicz [2] and York [3], the conformal method splits the initial data into "seed data", which can be freely chosen, and "determined data", which one obtains if possible by solving the constraint equations with the seed data specified (see Chapter 8 for details). The goal of the conformal method is two-fold: to obtain an effective parametrization of the "degrees of freedom" of the gravitational field, and to construct initial data sets which incorporate the physics of interest.

For initial data with constant mean curvature (CMC), the conformal method works very well. This is true for data on compact manifolds, for asymptotically Euclidean (AE) data, and for asymptotically hyperbolic (AH) data. It also works very well for data sets with sufficiently small mean curvature (near-CMC), so long as there are no conformal Killing fields present. However, more generally, much less is known. There are special classes of non-CMC seed data for which the conformal method has been shown to work [4], but there are others for which it appears to behave quite badly [5]. While to date these latter classes are quite restricted, the signs of trouble are clear: It may well be that the conformal method is effective for CMC and near-CMC initial data, but is ineffective more generally.

Are there modifications of the conformal method which might get around these looming difficulties? We first note that while the *conformal thin sandwich method* [6] is a popular alternative to the conformal method among numerical relativists, Maxwell has shown [7] that these two methods are mathematically equivalent; consequently they succeed or fail together. Thus, one expects that more substantial modifications may be necessary to obtain a method which works for all solutions of the constraint equations. Maxwell [8] has proposed one possible direction for such modifications, in terms of his "drift" vector treatment of the mean curvature in the seed data, but much more work is needed to determine if this or any other modification is likely to be successful.

While the conformal method has to date been the dominant tool used for constructing and analyzing solutions of the constraint equations, there are other approaches available. Most notable are the *gluing techniques*, which are designed to combine two or more known solutions of the constraints into a single one. Roughly speaking, there are two types of gluing techniques which have proven to be effective. *Connected sum* gluing starts with a pair of smooth solutions of the constraints, chooses a point in each, and produces a new solution of the constraints on the connected sum manifold[1] which is identical to the original

---

[1] Roughly speaking, the connected sum of a pair of manifolds is obtained by removing a neighborhood of a point from each, and then adding a tube which connects the manifolds where the neighborhoods have been removed.

solutions away from the chosen gluing points. This gluing technique [9, 10] works for any pair of initial data sets – compact, AE, or AH – so long as a certain non-degeneracy condition (see [10]) holds at the gluing points. It also works for most coupled-in matter source fields. It has been used to prove a number of interesting results, including that there exist maximally-extended globally hyperbolic solutions of the constraints containing *no* CMC Cauchy surfaces.

The other type of gluing was introduced in 2000 in the landmark work of Corvino [11], in which it is shown that for any time-symmetric initial data set which solves the constraints and is asymptotically Euclidean in a suitable sense, and for any bounded open interior region W, there is a new solution of the constraints which is identical to the original in W, and is identical to a subset of the Schwarzschild initial data outside of some bounded set containing W. The surprise here is that quite general interior gravitational configurations can be smoothly glued to a Schwarzschild exterior. Equally surprisingly, Corvino and Schoen show in [12] (see Chapter 8) that this result extends to non-time-symmetric initial data, with the exterior region being Kerr rather than Schwarzschild. Based on this work, one can show that there is a large class of asymptotically flat solutions of the Einstein equations which admit the conformal compactification (and the corresponding "Scri" structure) of the sort proposed by Penrose for studying asymptotically flat solutions.

Very recently, a further surprising variant of Corvino–Schoen-type gluing has been developed by Carlotto and Schoen. They show [13] that, just as one can construct a solution of the constraints by gluing an arbitrary interior region across an annular transition region to an exterior Schwarzschild or Kerr region, one can also do so by gluing an arbitrary solid conical region (stretching out to spatial infinity) across a transition region to a region of Minkowski initial data with a conical region removed. Moreover, one can do this without significantly changing the ADM mass. Both the Carlotto–Schoen conical gluing and the Corvino–Schoen annular gluing [14] can be used to produce N-body initial data sets of specific design which satisfy the constraints. The evolutions of such N-body initial data sets have not yet been studied.

In addition to the conformal method and the gluing techniques, there is one other procedure which has been used to produce solutions of the constraints. Designed to construct solutions with fixed interior boundaries, Bartnik's "quasispherical method" [15] chooses coordinates in such a way that one of the constraints takes the form of a parabolic (heat-type) equation with the radial coordinate as "time". Specifying data on the interior boundary, one "evolves" outward towards $r \to \infty$. Originally proposed as a tool for studying the Bartnik quasilocal mass, the quasispherical method has been used as well for constructing dynamical horizon initial data sets [16]. It would be useful to explore possible further applications.

The development of procedures for constructing solutions of the constraint equations is only one part of the broad scale of research focused on initial data sets and the constraints. Another very important area aims to understand the energy, momentum and angular momentum of isolated gravitational systems, both for the system as a whole and for specified subregions. Such notions, well understood in Newtonian physics, are not obvious in general relativity, especially for non-isolated subregions. Globally, for a data set which is asymptotically Euclidean, the ADM formalism [17] provides definitions which are correct from the Hamiltonian perspective, and are theoretically useful. In particular, the global ADM energy E_{ADM} and the global ADM momentum P_{ADM} are both crucial for the statement and proof of two of the major results of mathematical relativity: the *Positive Mass Theorem* and the *Penrose Inequality Theorem*. As discussed in Chapter 8, the first of these states that for every AE initial data set satisfying the constraints, the corresponding E_{ADM} and P_{ADM} satisfy the inequality $E_{ADM} \geq |P_{ADM}|$, with equality holding if and only if the initial data set generates Minkowski spacetime. The Penrose Inequality Theorem, which thus far has been proven only for time-symmetric initial data sets, states that for time-symmetric AE solutions of the constraints the ADM mass $\mu_{ADM} := \sqrt{E_{ADM}^2 - |P_{ADM}|^2}$ satisfies the inequality $\mu_{ADM} \geq \sqrt{A/(16\pi)}$, where A is the surface area of the outermost horizon in the data set.

The proofs of both of these theorems – the first by Schoen and Yau [18] and then independently by Witten [19] during the late 1970s, the second by Huisken and Ilmanen [20] and then independently by Bray [21] at the turn of the millennium – were major triumphs of geometric analysis. To a large extent, they showed mathematicians working in geometric analysis that general relativity could be a fertile source of mathematically deep – and solvable – problems. Indeed, formulating and proving a version of the Penrose Inequality Conjecture for non-time-symmetric initial data is an outstanding open problem in mathematical relativity and geometric analysis.

While the definitions, major results, and major open questions regarding global quantities like E_{ADM} are agreed upon by most relativists, much less is settled for the localized quantities. There is agreement regarding the properties that a "quasilocal mass" should have. Yet, as discussed in Chapter 8, there is a wide variety of possible definitions, including the Brown–York mass, the Liu–Yau mass, the Wang–Yau mass, the Hawking mass, and the Bartnik mass. In addition, there are definitions of quasilocal mass relying on isolated horizons and on dynamical horizons. All have their virtues and their difficulties, and their studies motivate a variety of challenging problems in geometric analysis.

One of the geometric structures which arose as a useful tool for the Schoen–Yau proof of the Positive Mass Theorem is the marginally outer trapped surface or MOTS. In a spacetime, an embedded two-dimensional spacelike surface Ξ is a MOTS if one of the two families of future-directed null geodesics orthogonal to Ξ has vanishing expansion everywhere along Ξ. Such behavior is closely tied to that of an apparent horizon, and consequently (through Hawking–Penrose-type singularity theorems) to the development of a black hole. One also notes that if Ξ is contained in a time-symmetric initial data set, then it must be a minimal surface in that Riemannian manifold.

Given the strong tie between MOTS and black holes, and between MOTS and minimal surfaces, it is not surprising that the analysis of MOTS – conditions for their existence in a given spacetime, conditions for their stability, restrictions on their allowed geometry – has proven to be very rich. Chapter 8 covers these topics extensively, including an outline of a very nice MOTS-based simplification of the proof of the Positive Mass Theorem [22].

From a physical point of view, the analysis of initial data sets and solutions of the Einstein constraint equations is all about specifying physically interesting initial configurations of gravitational systems, understanding the properties of these systems, and identifying the degrees of freedom of the gravitational field if possible. Correspondingly, from a physical point of view the analysis of the evolution equations is all about determining how (according to general relativity) such initial configurations evolve in time: Do they form black holes? Do they form singularities? If they produce gravitational radiation, what forms might the radiation take for different initial configurations?

Choquet-Bruhat's well-posedness theorem shows that (in suitable coordinates) Einstein's field equations constitute a (nonlinear) hyperbolic PDE system. However, there are two very important ways in which the Einstein system differs from other geometrically-based hyperbolic PDE systems such as the Yang–Mills equations or wave maps on Minkowski spacetime. First, while the Yang–Mills and wave-map examples have a fixed background spacetime which can be used to define long-time existence, singularities, and energy functionals unambiguously, with Einstein's theory the spacetime evolves along with the solution, so these concepts are much harder to pin down. Second, since general relativity uses solutions of Einstein's equations to model (classical) physical phenomena, there is physical interest in individual solutions and their evolution; for Yang–Mills and wave maps this is not the case.

These two distinguishing characteristics of Einstein's equations have had two important consequences. The lack of a fixed background has made it difficult until recently to use some of the standard ideas and tools of nonlinear hyperbolic PDE analysis. The physical interest in individual solutions has made the pursuit of numerical simulations of solutions a very important part of general relativity since the 1970s.

While there was little progress in either the numerical or the hyperbolic PDE analysis of Einstein's equations in the two decades following Choquet-Bruhat's theorem, important insights were obtained regarding singularities (see [23]). It was known from the study of explicit solutions that singularities (in the sense of unbounded curvature) *can* develop, but it was not known if this was just an artifact of the symmetries of solutions such as Friedmann–Lemaître–Robertson–Walker (FLRW) and Schwarzschild. The Hawking–Penrose "singularity theorems", which were proven using the properties of geodesic congruences in curved spacetimes, did show that if one identifies geodesic incompleteness with singularities, then singularities occur in large classes of solutions. However, in making this identification, one obscures the issue of the physical meaning of a singularity occurring in a given spacetime. For example, in Taub–NUT spacetimes, geodesic incompleteness marks the breakdown of causality (with closed timelike paths appearing), while in FLRW spacetimes, geodesic incompleteness occurs because of unbounded tidal curvature.

Restated as the *Strong Cosmic Censorship* (SCC) conjecture, this issue regarding the generic nature of solutions of Einstein's equations which are geodesically incomplete is one of the outstanding questions of mathematical relativity. Proposed during the late 1960s by Penrose, the SCC conjecture can be stated as follows: *Among all sets of constraint-*

satisfying initial data whose maximal developments[2] *are geodesically incomplete, all but a set of measure zero have unbounded curvature, and cannot be extended (across a "Cauchy horizon") as non-globally hyperbolic solutions.* In fact there *are* many families of globally hyperbolic solutions which are geodesically incomplete, have bounded curvature, and can be extended across Cauchy horizons [26]; this is fully consistent with SCC, since its claim is that *generic* geodesically incomplete solutions have unbounded curvature and cannot be extended; SCC allows *some* solutions to behave otherwise.

As discussed in Chapter 9, Strong Cosmic Censorship has been carefully formulated and proven for a number of restricted families of solutions, such as the T^3 Gowdy solutions [27]. Whether or not it holds generally is an open question.

Also open and also one of the central questions of mathematical relativity is the *Weak Cosmic Censorship* (WCC) conjecture. Despite its name, the WCC conjecture (also due to Penrose) neither implies nor is implied by SCC. It can be stated as follows: *Among all sets of constraint-satisfying initial data sets whose maximal developments are asymptotically flat and contain a curvature singularity, in all but a set of measure zero there is an event horizon which shields the singular region from observation by asymptotic observers.* In other words, WCC conjectures that in solutions of Einstein's equations, gravitational collapse generically leads to the formation of a black hole. As with SCC, genericity is a key part of the conjecture: There *are* gravitationally collapsing solutions with naked singularities, but WCC conjectures that this does not generically occur. Restricted families of solutions for which WCC has been proven are discussed in Chapter 9.

Once it has been determined that a nonlinear hyperbolic PDE system has solutions which evolve from regular initial data and become singular in finite time, one of the main questions becomes the *stability* of solutions. That is, fixing a particular solution Ψ, is it true that the evolution of every solution with initial data close to that of Ψ must evolve essentially as does Ψ? In addition to its mathematical interest, stability is important physically, since Ψ is useful in modeling physical systems only if it is stable in this sense.

For Einstein's equations, stability is not at all obvious. For example, considering Minkowski spacetime, does one expect very small gravitational perturbations of flat data to disperse (stability), or to concentrate and form a black hole (instability)?

The Christodoulou–Klainerman proof of the stability of Minkowski spacetime [28] is one of the iconic results in mathematical relativity. Besides resolving an important question concerning Einstein's theory, this proof showed that many of the difficulties impeding the analysis of Einstein's equations as a nonlinear PDE system could be overcome. It showed that coordinate freedom could be controlled using null foliations, and it showed that the Bel–Robinson tensor could be used to construct an effective energy functional for Einstein's theory (see details in Chapter 9). Many of the ideas developed in this work were used subsequently by Christodoulou [29] to prove that certain ("pulse") classes of regular initial

---

[2] As proven by Choquet-Bruhat and Geroch [24], for every initial data set which satisfies the constraints, there is a unique (up to diffeomeorphism) globally hyperbolic spacetime which is evolved (via Einstein's equations) from that initial data, and which extends all other globally hyperbolic solutions evolved from the same data. This spacetime is called the *maximal development* of that data set. A recent new proof [25] of this result avoids the use of Zorn's Lemma.

data necessarily evolve into spacetimes with trapped surfaces and therefore very likely into black holes.

Minkowski spacetime is not the only solution of Einstein's equations which has been shown to be stable. Almost ten years before the Christodoulou–Klainerman work, Friedrich proved [30] that the De Sitter spacetime, which is a solution of the Einstein equations with positive cosmological constant, is stable. To do this, he relied on his conformal reformulation of the field equations (based on the higher-order Bianchi identities) and the conformal properties of the De Sitter spacetime, thereby avoiding some of the problems encountered in proving long-time existence for solutions near Minkowski spacetime. Stability has also been proven for a number of solutions which have a steady or accelerating rate of expansion to the future. As discussed in Chapter 9, Andersson and Moncrief [31] have shown that the Milne spacetimes[3] are stable. In this work, a coordinate choice which results in an elliptic-hyperbolic form for the evolution equations plays a crucial role. In more recent work, Ringström [32, 33] has examined solutions of certain Einstein–scalar-field theories which are characterized by accelerated rates of expansion and shown that these solutions are stable. Followup work [34, 35] suggests that his techniques are robust for expanding solutions.

One of the most challenging open questions in general relativity is whether the Kerr solutions are stable. This is a mathematically very rich problem, which has motivated a significant amount of research (some discussed in Chapter 9). Since the Kerr solutions are believed by many to accurately model the final state resulting from the gravitational collapse of large-mass stars, it is very important physically to determine if the Kerr solutions are stable. We note that general perturbations of Schwarzschild solutions are *not* expected to evolve to Schwarzschild solutions; rather they are expected generally to evolve to Kerr solutions.

Stability and cosmic censorship are not the only questions arising in the study of the long-time behavior of solutions of Einstein's equations. One would like to explore the long-time, asymptotic behavior of families of solutions which are not necessarily small perturbations of known solutions like Minkowski, Kerr, Milne, or FLRW. While much less is known about how to do this, two very active areas of research that are discussed in Chapter 9 could be very helpful. In Section 9.2, efforts to reduce as much as possible the regularity (differentiability) needed for well-posedness are discussed. Such efforts, which have steadily progressed from Choquet-Bruhat's original theorem (requiring smooth data) to the recent remarkable L^2 bounded curvature version [36], are important for long-time evolution studies because they establish crucial break-down and continuation criteria for evolving solutions. That is, if one can prove well-posedness for data with bounded curvature, then one knows that evolution continues so long as the curvature indeed remains bounded. Combining results of this sort with the control of certain energy functionals (designed to control curvature) would lead to long-time existence.

---

[3] The Milne spacetimes are obtained by topologically compactifying the constant-mean-curvature hyperboloids in the future lightcone of a point in Minkowski spacetime. Together these CMC hypersurfaces foliate the spacetime, which is flat but is geodesically incomplete to the past. To the future, the CMC hypersurfaces have a steady rate of expansion.

A scenario for defining and establishing control of energy functionals is outlined in Section 9.3. For solutions involving negative scalar curvature metrics, one can choose certain gauges and coordinates, and carry out a reduction[4] which results in a Hamiltonian formulation with a well-defined, monotonic Hamiltonian (energy) functional. Combining this reduction with lessons learned in studying long-time existence for the Yang–Mills equations [37] could lead to very interesting insights into the dynamics of Einstein's equations.

In mathematical analyses of PDE systems, the emphasis is usually on understanding the properties and the behavior of large sets of solutions. For a PDE system used to model physics, one is interested in individual solutions as well. Numerical simulation of solutions is crucial for such studies; consequently much effort has gone into developing numerical relativity since the early 1970s.

Although the techniques used for carrying out numerical simulations are very different from those used in performing mathematical analysis, the same basic issues – controlling diffeomorphisms, choosing gauges and coordinates, distinguishing physical effects from coordinate effects, dealing with a non-fixed background spacetime – have caused difficulties in both enterprises. Consequently, advances in one of these enterprises have often been very helpful in the other. A key example of this is the mathematical development of generalized harmonic coordinates as a tool for proving well-posedness, followed by the crucial role played by this coordinate choice in carrying out the first stable numerical simulation of a binary black hole coalescence [38], in turn followed by the use of generalized harmonic coordinates in proving the stability of solutions with accelerated expansion [32].

Throughout most of its history, from the 1960s to the present, numerical relativity has been primarily focused on simulating the coalescence of compact binaries, and determining the consequent generation of gravitational radiation as well as the properties of the remaining object. This is largely because of the central role such collisions are expected to play in generating radiation observable by LIGO, Virgo and other detectors, and the need to understand the detailed features of this radiation if it is to be detected in practice. Despite significant investments in both human effort and resources, it was a long trail from the earliest efforts to simulate black hole collisions by Hahn and Lindquist in 1964 [39] and the early development work of Smarr and Eppley during the 1970s, to the first stable simulations of these collisions by Pretorius [38] and independently by groups led by Campanelli [40] and by Centrella [41] in 2005, culminating in the current systematic production of simulations of a large parameter space of collisions of binary systems, including neutron stars as well as black holes. Chapter 7 presents much of the story of this difficult (and adventurous) journey.

In addition to its role in simulating sources of gravitational radiation, numerical relativity has been very useful for other explorations of general relativity. Perhaps most noteworthy is the discovery in such simulations by Choptuik [42] of *critical phenomena* in gravitational collapse. Critical phenomena are found by considering the evolutions of each of a

---

[4] Such a reduction effectively solves some of the constraints and implements the coordinate and gauge conditions in such a way that the evolution projects down to a phase space with fewer dynamical variables. See Section 9.3 for details.

one-parameter (λ) family of initial data sets, with those solutions evolving from data with large λ collapsing to form black holes, and those solutions evolving from data with small λ dispersing. At the transition value λ_c between data leading to black holes and data leading to dispersion, the evolved solution has been found to have special features including certain types of (continuous or discrete) time symmetry described by scaling laws. Critical behavior has been observed (via simulations) to occur for essentially all such choices of one-parameter families of data ("universality") and has been found to occur for Einstein's equations coupled to a very wide variety of source fields. Details are presented in Chapter 7. Interestingly, while the evidence for critical behavior from numerical simulations is overwhelming, there has been no mathematical proof of its existence.

Also noteworthy is the significant role which has been played by numerical simulations in the study of the behavior of the gravitational field near the singularity in families of solutions defined by T^2 or $U(1)$ spatial isometry (including the Gowdy solutions). These simulations, initiated by Berger and Moncrief [43], indicated the prevalence of "AVTD" behavior in Gowdy solutions as well as in polarized T^2 symmetric and polarized $U(1)$ symmetric solutions. That is, as one evolves toward the singularity in these solutions (at least in certain chosen coordinate systems), spatial derivative terms in the evolution equations appear to be dominated by time derivative terms; consequently, an observer following a coordinate path sees the fields evolve more and more like a Kasner solution, with each observer seeing an independent Kasner. This somewhat peculiar behavior was first described during the 1960s by Lifshitz and Khalatnikov, who predicted that it should characterize large classes of solutions. In later years, with Belinskii [44], they amended their prediction to claim that solutions with singularities would generically exhibit what is now known as "BKL" or "Mixmaster" behavior: each observer would see an endless sequence of Kasner-like evolutions, punctuated by "bounces" leading from one Kasner era to another. While it was widely believed during the 1980s that neither AVTD nor BKL behavior was likely to be seen, the numerical work by Berger, Moncrief, Garfinkle, Weaver, Isenberg, and others during the 1990s has supported the contention that AVTD and BKL behavior are there, but with the added feature of "spikes". Indeed, it has now been proven by Ringström [45] that T^3 Gowdy solutions *are* generically AVTD; this demonstration is a crucial step in his proof that Strong Cosmic Censorship holds for these Gowdy solutions.

Chapter 7 discusses a number of other important numerical relativity projects which have indicated surprising physical phenomena in general relativity. Most notable is the apparent instability of anti-De Sitter spacetime which has been discovered numerically [46]. This result is very surprising for two reasons: (i) it contrasts with the proven stability of Minkowski spacetime and De Sitter spacetime, and (ii) it has been shown that anti-De Sitter spacetime is linearly stable.[5] Also very interesting are the studies of the complex dynamics of perturbations of black strings in $4 + 1$ dimensions [47]. In both of these cases, mathematical questions have motivated the studies, and the numerical work has provided very strong indications of unambiguous answers to these questions.

---

[5] A solution is linearly stable if the corresponding linearized Einstein equation operator has negative eigenvalues.

One area not discussed in these chapters in which the partnership of numerical and mathematical work has been very useful is the study of the strong-field regime during the formation and coalescence of black holes. Mathematically, the definition of black holes has traditionally been associated with event horizons, the future boundaries of the causal past of future null infinity. They separate the spacetime region from which light can escape to infinity from those regions from which light can *never* do so. Unfortunately, an immediate consequence is that the notion of an event horizon is not only extremely global but it is also teleological; one has to know the entire spacetime before one can locate it. In particular, an event horizon can grow in regions which have weak – or even zero – spacetime curvature, *in anticipation* of a future gravitational collapse. Therefore, in the study of gravitational collapse or dynamics of compact objects, one cannot locate event horizons *during* evolution. In practice, numerical simulations have used "apparent horizons" on the chosen family of Cauchy hypersurfaces to indicate incipient black holes because neither of the two light fronts emanating from such surfaces is expanding; light is trapped at least instantaneously. However, this notion is tied to the choice of Cauchy hypersurfaces. To get around this issue, one can consider spacetime world tubes of apparent horizons, known as *dynamical horizons* if they evolve, and *isolated horizons* if they do not [48–51]. As black hole markers, these structures have the advantage that they are determined by the local spacetime geometry in their immediate vicinity; they do not require knowledge of the full spacetime and in particular they do not exist in a flat spacetime. Dynamical horizons and isolated horizons have been studied both analytically and numerically. From these studies, tools have been developed to extract physical information, such as the mass, angular momentum, and (source) multipole moments of black holes, even in dynamical regions, in a coordinate-invariant fashion. These structures have also made it possible to meaningfully compare results of simulations that use different initial data, coordinates and foliations in the strong-field region. Finally, they provide analytical tools to probe *how* black holes approach their final, universal equilibrium state, represented by the geometry of the Kerr horizon, and if the process has universal features [52, 53]. While these developments were inspired by simulations, they have also had interesting applications on the analytical side, resulting in significant generalizations of the laws of black hole mechanics [54, 55] to fully dynamical situations [50], enhanced understanding of black hole evaporation [56] and improved calculations of horizon entropy using quantum geometry, as summarized in Chapter 11.

One very elusive issue for which dynamical horizons could be helpful is to find a comprehensive definition of black holes which makes sense in dynamical spacetimes, as well as in spacetimes which are not asymptotically flat [57]. One problem with using dynamical horizons, however, is that they are generally not unique [58]; consequently they are not good candidates for canonically locating the surface of a black hole.

Finally, we note one other very elusive physical issue which calls for both numerical and mathematical studies, likely in partnership, concerning gravitational waves. As explained in Part II, although Einstein analyzed gravitational waves in the linearized approximation and derived the quadrupole formula soon after his discovery of the field equations, there was considerable confusion over the next three decades regarding the physical reality of gravitational waves because it was difficult to disentangle physical effects from coordinate artifacts. The situation was clarified only in the 1960s when the notion of *Bondi news* was shown to provide a coordinate-invariant characterization of gravitational radiation [59]. That framework continues to provide the conceptual foundations of the gravitational wave theory discussed in Chapter 6. However, it assumes a zero cosmological constant; $\Lambda = 0$. This is a problem, because cosmological observations have established that $\Lambda > 0$ and, surprisingly, we do not yet have a satisfactory analog of the Bondi news, or viable expressions of energy, momentum and angular momentum carried by gravitational waves in full, nonlinear general relativity for $\Lambda > 0$. Indeed, even the asymptotic symmetries that are needed to introduce these notions do not extend to the $\Lambda > 0$ case. In cosmological applications, it has been sufficient just to use linearized gravitational waves. Similarly, for isolated systems such as compact binaries or collapsing stars, working with $\Lambda = 0$ should be an excellent approximation in most calculations since the observed value of Λ is so small. However, as of now, we do not have any theory for $\Lambda > 0$, whose predictions are to be approximated by the $\Lambda = 0$ calculations. Only when we have this theory and a detailed understanding of how Λ affects physical quantities can we be confident that the current $\Lambda = 0$ calculations do approximate reality sufficiently well. Moreover, there may well be subtle effects – similar to the Christodoulou memory effect in the $\Lambda = 0$ case – that have eluded us and which will be uncovered only when the $\Lambda > 0$ case is well understood.

For a number of years following its pristine birth one hundred years ago, general relativity was all too often admired and adulated rather than tested and put to work. Just as this has now changed in the experimental and observational realm, as reported in Parts One and Two, we see here in Part Three that this has dramatically changed in the mathematical and numerical realm as well. Einstein's theory is now recognized as a fresh source of interesting and challenging mathematical problems, and the successful solution of some of these problems has attracted increasing numbers of mathematicians with innovative skills. In turn, with more powerful analytical tools available, relativists have been able to explore the physical implications of general relativity more effectively. The intriguing mix of beautiful mathematics and profound physics, always an enticing feature of general relativity, is now a central reason for expecting rapid progress both in its use as an effective tool for studying gravitational physics and in its role as a challenging field of geometric analysis.

References

[1] Y. Choquet-Bruhat, *Acta Math.* **88** 141–225 (1952).
[2] A. Lichnerowicz, *J. Math. Pures Appl.* **23** 37–63 (1944).
[3] J. York, *J. Math. Phys.* **13** 125–130 (1972).
[4] M. Holst, G. Nagy and G. Tsogtgerel, *Commun. Math. Phys.* **288** 549–613 (2009).
[5] D. Maxwell, *Commun. Math. Phys.* **302** 697–736 (2011).
[6] J. York, *Phys. Rev. Lett.* **82** 1350–1353 (1999).
[7] D. Maxwell, unpub. gr-qc 1402.5585.
[8] D. Maxwell, unpub. gr-qc 1407.1467.
[9] J. Isenberg, R. Mazzeo and D. Pollack, *Commun. Math. Phys.* **231** 529–568 (2002).
[10] P. Chruściel, J. Isenberg and D. Pollack, *Commun. Math. Phys.* **257** 29–42 (2005).
[11] J. Corvino, *Commun. Math. Phys.* **214** 137–189 (2000).
[12] J. Corvino and R. Schoen, *J. Diff. Geom.* **73** 185–217 (2006).

[13] A. Carlotto and R. Schoen, unpub. math 1407.4766.
[14] P. Chruściel, J. Corvino and J. Isenberg, *Commun. Math. Phys.* **304** 637–647 (2011).
[15] R. Bartnik, *J. Diff. Geom.* **37** 31–71, (1993).
[16] R. Bartnik and J. Isenberg *Class. Quant. Grav.* **23** 2559–2569 (2006).
[17] R. Arnowitt, S. Deser and C. Misner, *Phys. Rev.* **122** 997–1006 (1961).
[18] R. Schoen and S.-T. Yau, *Commun. Math. Phys.* **65** 45–76 (1979).
[19] E. Witten, *Commun. Math. Phys.* **80** 381–402 (1981)
[20] G. Huisken and T. Ilmanen, *J. Diff. Geom.* **59** 353–437 (2001).
[21] H. Bray, *J. Diff. Geom.* **59** 177–267 (2001).
[22] M. Eichmair, L.-H. Huang, D. Lee and R. Schoen, unpub. math 1110.2087.
[23] R. Geroch, *J. Math. Phys.* **9** 450–465 (1968).
[24] Y. Choquet-Bruhat and R. Geroch, *Commun. Math. Phys.* **14** 329–335 (1968).
[25] J. Sbierski, unpub. gr-qc 1309.7591.
[26] V. Moncrief, *Phys. Rev. D* **23** 312–315 (1981).
[27] H. Ringström, *Ann. Math.* **170** 1181–1240 (2009).
[28] D. Christodoulou and S. Klainerman, *The global nonlinear stability of Minkowski space*, Princeton U. Press (1994).
[29] D. Christodoulou, *The formation of black holes in general relativity*, EMS (2009).
[30] H. Friedrich, *J. Geom. Phys.* **3** 101–117 (1986).
[31] L. Andersson and V. Moncrief, in *Cauchy problem in general relativity*, ed. P. Chruściel and H. Friedrich, Birkhäuser (2004).
[32] H. Ringström, *Invent. Math.* **173** 123–208 (2008).
[33] H. Ringström, *Commun. Math. Phys.* **290** 155–218 (2009).
[34] C. Svedberg, *Ann. Inst. Henri Poincaré* **12** 849–917 (2011).
[35] X. Luo and J. Isenberg, *Ann. Phys.* **334** 420–454 (2013).
[36] S. Klainerman, I. Rodnianski and J. Szeftel, unpub. gr-qc 1204.1772.
[37] D. Eardley and V. Moncrief, *Commun. Math. Phys.* **83**, 171–212 (1982).
[38] F. Pretorius, *Phys. Rev. Lett.* **95** 121101 (2005).
[39] S. Hahn and R. Lindquist, *Ann. Phys.* **29** 304–331 (1964).
[40] M. Campanelli *et al.*, *Phys. Rev. Lett.* **96** 111101 (2006).
[41] J. Baker *et al.*, *Phys. Rev. Lett.* **96** 111102 (2006).
[42] M. Choptuik, *Phys. Rev. Lett.* **70** 9–12 (1994).
[43] B. Berger and V. Moncrief, *Phys. Rev. D* **48**, 4676–4687 (1993).
[44] V. Belinskii, E. Lifshitz, and I. Khalatnikov, *Zh. Eksp. Teor. Fiz.* **62**, 1606–1613 (1972).
[45] H. Ringström, *Ann. Math.* **170** 1181–1240 (2009).
[46] P. Bizon, *Gen. Rel. Grav.* **46** 1724 (2014).
[47] L. Lehner and F. Pretorius, *Phys. Rev. Lett.* **105** 101102 (2010).
[48] S. A. Hayward, unpub. `gr-qc 0008071`
[49] A. Ashtekar *et al.*, *Phys. Rev. Lett.* **85**, 3564–3567 (2000).
[50] A. Ashtekar and B. Krishnan, *Living Rev. Rel.* **7** 10 (2004).
[51] E. Gourgoulhon, J. L. Jaramillo, *Phys. Rept.* **423** 159–294 (2006).
[52] A. Ashtekar and M. Campiglia, *Phys. Rev. D* **88**, 064045 (2013)
[53] R. Owen, *Phys. Rev. D* **80**, 084012 (2009).
[54] J. M. Bardeen, B. Carter and S. W. Hawking, *Commun. Math. Phys.* **31**, 161–170 (1973)
[55] *Black holes, Les Houches 1972*, C. DeWitt and B. DeWitt eds (Gordon and Breach, New York, 1973)
[56] A. Ashtekar, F. Pretorius and F. M. Ramazanoğlu, *Phys. Rev. Lett.* **106** 161303 (2011); *Phys. Rev. D* **83**, 044040 (2011).
[57] J. M. M Senovilla, In *Relativity and gravitation: 100 years after Einstein in Prague*, J. Bičák and T. Ledvinka, eds (Springer, Berlin, 2014)
[58] A. Ashtekar and G. J. Galloway, *Adv. Theor. Math. Phys.* **9** 1–30 (2005).
[59] *Proceedings on theory of gravitation*, I. Infeld ed (Gauthier Villars, Paris and PWN, Warsaw (1964)).

7

Probing Strong-Field Gravity
Through Numerical Simulations

MATTHEW W. CHOPTUIK, LUIS LEHNER AND FRANS
PRETORIUS

This chapter describes what has been learned about the dynamical, strong-field regime of general relativity via numerical methods. There is no rigorous way to identify this regime, in particular since notions of energies, velocities, length and timescales are observer-dependent at best, and at worst are not well-defined locally or even globally. Loosely speaking, however, dynamical strong-field phenomena exhibit the following properties: there is at least one region of spacetime of characteristic size R containing energy E where the compactness $2GE/(c^4R)$ approaches unity, local velocities approach the speed of light c, and luminosities (of gravitational or matter fields) can approach the Planck luminosity c^5/G. A less physical characterization, though one better suited to classifying solutions, involves spacetimes where even in "well-adapted" coordinates the non-linearities of the field equations are strongly manifest. In many of the cases where these conditions are met, numerical methods are the only option available to solve the Einstein field equations, and such scenarios are the subject of this chapter.

Mirroring trends in the growth and efficacy of computation, numerical solutions have had greatest impact on the field in the decades following the 1987 volume [1] celebrating the 300th anniversary of Newton's *Principia*. However, several pioneering studies laying the foundation for subsequent advances were undertaken before this, and they are briefly reviewed in Section 7.1 below. Though this review focuses on the physics that has been gleaned from computational solutions, there are some unique challenges in numerical evolution of the Einstein equations; these as well as the basic computational strategies that are currently dominant in numerical relativity are discussed in Section 7.2. As important as computational science has become in uncovering details of solutions too complex to model analytically, it is a rare moment when qualitatively new physics is uncovered. The standout example in general relativity is the discovery of critical phenomena in gravitation collapse (Section 7.3.1); another noteworthy example is the formation of so-called spikes in the approach to cosmological singularities (Section 7.3.7). A significant motivation for obtaining solutions in the dynamical strong field has been to support the upcoming field of gravitational wave astronomy, which requires predictions of emitted waveforms for optimal detection and parameter extraction. The expected primary sources are compact-object mergers, where numerical methods are crucial in the modeling of the final stages of the events. Binary black hole systems are discussed in Section 7.3.2, black hole–neutron star and binary neutron stars systems in Section 7.3.3. Though not of astrophysical or experimental relevance – barring the existence of an unexpectedly small Planck energy scale – the ultra-relativistic limit of the two-body problem is of considerable theoretical interest, and this is discussed in Section 7.3.5. Spurred by the gauge-gravity dualities of string theory, the study of higher-dimensional gravity has been very active in the past decade; related numerical discoveries are presented in Section 7.3.6. Some miscellaneous topics are mentioned in Section 7.3.8, and we conclude the review in Section 7.4 with a discussion of open problems for the coming years.

Regarding notation, for the most part we report results in geometric units where Newton's constant G and the speed of light c are set to unity, though for clarity some expressions will explicitly include these constants. In referring to the dimensionality of a manifold, metric or tensor field, we will use lower case "d" for spacetime dimensions, and upper case "D" for purely spacelike dimensions; e.g., "4d" refers to $3 + 1$ spacetime dimensions (this latter "$n + 1$" form we will also use), and "3D" means three spatial dimensions.

7.1 Historical perspective

The similarly oriented book released in 1987 [1] gave a snapshot of the various interesting subjects and problems in gravitational research. However, there was no chapter on numerical solution of the Einstein equations, even though the subfield of numerical relativity had been in active development for over a decade by then. The discipline was still coming into its own, and the breadth and scope of works within its purview was still limited. Nevertheless, these incipient studies did provide a hint of developments to come as the know-how, computational resources and experience improved. It is thus important to set some perspective by describing a subset of works leading to the current status of the field.

The particular topics we review later in this chapter are weighted towards developments that have occurred within the past decade or two. This is natural as numerical relativity has been a rapidly growing field during this time. However, as mentioned, the foundations for building a mature field were initiated before, and here we briefly discuss, loosely organized by subject, some of these more important early results. Unfortunately, due to space limitations we cannot mention all the relevant works, nor discuss those we do mention in any detail. Also, we do not include results, in particular the more recent ones, that are discussed elsewhere in this book.

Binary black hole mergers. The first attempt at a numerical solution of the binary black hole merger problem was made by Hahn and Lindquist [2] in 1964. At that time the term "black hole" had not yet been coined, and the full significance of the problem, in particular with regard to gravitational wave emission and black hole mergers in the universe, was not recognized. Using Gaussian normal coordinates, Misner's "wormhole" initial data [3], representing two black holes initially at rest, was evolved until $t \approx m/2$ (with $m = \sqrt{A/(16\pi)}$, A being the area of each throat). At that point numerical errors had grown too large to warrant further evolution, but it was nonetheless possible to measure the mutual attraction between the holes, and the fact that the throats were beginning to pinch off. Smarr [4, 5] and Eppley [6] independently revisited the head-on collision problem a decade later, now with a profound new understanding of black holes gained in the preceding years, both from theory and from observations suggesting that they likely exist in the universe. These works used the same initial data as Hahn and Lindquist, Čadež coordinates to simultaneously conform to the throats and approach the usual spherical polar coordinates at large distances [7], and maximal slicing. The culmination of these studies showed that the collision emitted radiation of order 0.1% of the total mass, and that the waveform was very similar to that computed from a perturbative calculation (the first indications of the "relative simplicity" of black hole merger waveforms discussed in Section 7.3.2).

In anticipation of construction of the LIGO gravitational wave detectors, and the recognized need for waveform models to enable detection, the head-on collision calculations were reinitiated by the NCSA group in the early 1990s [8]. The new simulations offered improved treatment of the Čadež coordinate singularity and radiation extraction, but essentially confirmed previous results. With hindsight, it is amusing that in [5] the status of this field was summarized as "The two black hole collision problem has been largely completed." This was the prevailing opinion through the mid 1990s, with the consensus being that the most significant impediment to solving the full 3D merger problem was simply lack of available computational power. This turned out not to be the case, and a tremendous effort by the community was expended in going from the first short-lived grazing-collision simulations reported in 1997 [9] to the breakthroughs in 2005 [10–12] that facilitated stable evolution of the full problem and the impressive results that have followed (see [13] for more discussion of this development). As briefly discussed in Section 7.2, some of the key stumbling blocks were related to the underlying mathematical character of the Einstein field equations and the existence of geometric singularities inside black holes. This is not to say that limited computational power was not an issue; in fact it did hamper the effort to rapidly find solutions to the more fundamental problems, as numerous attempts to isolate and solve issues in a symmetry-reduced (or similar) setting that could be tackled more quickly with available computational resources failed when carried over to the full problem.

Gravitational collapse. Numerical studies of the gravitational collapse of stars began with the work of May and White [14], who looked at the collapse of ideal fluid spheres with a Γ-law equation of state (specifically $\Gamma = 5/3$). They found, depending on the initial conditions, that collapse would continue to black hole formation, or halt and then bounce (a necessary condition for an eventual supernova). The "second generation" of codes was developed over the next couple of decades, with the pioneering efforts of, among others, Wilson [15, 16], Shapiro and Teukolsky [17], Stark and Piran [18], Nakamura [19, 20] and Evans [21]. Advances included evolution of axisymmetric models to study the effects of rotation and asymmetries, solution of the hydrodynamic equations written in conservative form, improvement in the handling of axis coordinate singularities, development of moving-mesh methods, incorporation of effects of neutrino emission, and exploration of a variety of slicing and spatial coordinate conditions. The more recent studies of stellar collapse are reviewed in Section 7.3.4.

Although many studies of gravitational collapse are motivated by application to the wide variety of observed phenomena attributed to stellar collapse, there has been considerable work on more theoretical scenarios, in particular critical collapse, reviewed in Section 7.3.1. A notable work we mention here is the evolution of collapsing, axisymmetric configurations of collisionless matter by Shapiro and Teukolsky [22]. In all cases the formation of a geometric singularity was observed, but, intriguingly, for sufficiently prolate distributions no apparent horizon was found when the time-slice ran into the singularity. This could be a slicing issue in that a horizon could still form at a later time; however, the threshold of prolateness above which no horizons were found is consistent with the hoop conjecture [23], suggesting these cases are examples of the violation of the weak cosmic censorship conjecture in asymptotically flat spacetimes. Another work of theoretical interest that had significant impact on the foundations of the field was the numerical study of cylindrical gravitational wave spacetimes by Piran [24] (here "collapse" of non-linear waves always leads to naked singularities, though the spacetime is not asymptotically flat). In particular, the modern notions of free vs constrained evolution were introduced, and the utility of using coordinate conditions to modify the structure of the numerical scheme was demonstrated.

Binary neutron star, black hole/neutron star mergers. Unlike the binary black hole problem which featured extensive early development around the head-on collision case, relatively little work on the full general-relativistic modeling of binary neutron star or black hole/neutron star mergers, head-on or otherwise, was undertaken until the early 2000s, as reviewed in Section 7.3.3. A notable exception is the head-on collision study done by Wilson in the late 1970s [16]; he found (by applying the quadrupole formula to the matter dynamics) that, similar to the black hole case, $\sim 0.1\%$ of the rest mass of the spacetime is emitted in gravitational waves.

Initial data. The numerical initial data problem in general relativity deserves an entire chapter by itself, and unfortunately we are unable to devote space to it here (for reviews see [25–27], and the books mentioned below). We would, however, be remiss not to mention the York formalism for the construction of initial data [28, 29]. This has become the standard method for producing generic initial data for a wide range of problems. Moreover, it provides the framework in which modern ADM-based [30] Cauchy evolution schemes are written, in particular being the starting point to develop the now commonly employed BSSN formalism discussed in Section 7.2. A few other notable initial-data-related works include the Bowen–York closed-form solutions to the momentum constraints for black hole initial data [31], the "puncture" initial data [32] (which has had significant influence beyond initial data, leading to the stable evolution of black hole spacetimes without the need for excision) and the use of apparent horizons to provide boundary conditions and implement singularity excision [33].

Miscellaneous. We conclude this historical review by listing a few other developments of import to the growth of the field in the 1980s and 1990s.

- *Cosmology.* We discuss work related to cosmological singularities in Section 7.3.7, and numerical studies of cosmic bubble collisions and local inhomogeneities in cosmology in Section 7.3.8, but mention here that much pioneering work on numerical cosmologies and spacetimes with related symmetries began with the works of Centrella, Anninos, Wilson, Kurki-Suonio, Laguna and Matzner, Berger and Moncrief [34–38]

- *Boson stars.* The numerical study of these self-gravitating soliton-like configurations of scalar fields has a long history that begins with calculations in the late 1960s by Kaup [39] and by Ruffini and Bonnazola [40], who found spherically-symmetric, static solutions in the Einstein–Klein–Gordon model. A major resurgence of interest in the subject was sparked by Colpi, Shapiro and Wassermann's discovery that the addition of a non-linear self-interaction could lead to boson star masses that, in contrast to those originally constructed, were in an astrophysically interesting range [41]. Much subsequent work investigating a wide variety of types of boson stars and related objects has been carried out since, and we touch on some representative calculations in Sections 7.3.1 and 7.3.5. Once more, however, space limitations preclude a thorough coverage of this topic and we direct the interested reader to reviews such as [42].

- *Excision.* The first successful simulation incorporating the use of black hole excision to eliminate geometric singularities from the computational domain was presented by Seidel and Suen [43].

- *Hyperbolic evolution schemes.* One of the influential efforts predating the wave of activity searching for stable hyperbolic evolution schemes discussed in Section 7.2 was the formulation of Bona and Masso [44].

- *The Grand Challenge (1993–1998).* This large-scale NSF-funded project, aimed at solving the black hole inspiral and merger problem, involved essentially all US-based numerical relativists and, crucially, many computer scientists. The notable results culminating from this effort include propagation of a single Schwarzschild black hole through a 3D mesh [45], early efforts in refined gravitational wave extraction methods and improved outer boundary conditions [46], and the development of a characteristic code that could stably evolve even highly perturbed single black hole spacetimes [47].

7.2 Numerical relativity: current state of the art

Before beginning our review of the important physics garnered from numerical solutions of the Einstein field equations over the past few decades, we describe some of the key insights obtained along both formal and numerical fronts that have made these ventures possible. Of course, it is impossible to exhaustively cover all of them; we thus choose particularly relevant ones that have had a strong influence on the field.

7.2.1 Mathematical formalism

Any numerical study of a dynamical process requires solving a suitably formulated initial value (or initial-boundary-value) problem. That is, given a set of evolution equations, together with a specified state of the system at an initial time, its future evolution can be obtained by a numerical integration. At face value any covariant theory is at odds with this requirement, unless a suitable 'time' foliation of the spacetime is introduced. In the case of the Einstein equations, projections (tangential and normal to each leaf of the foliation) provide a natural hierarchy of evolution and constraint equations. The latter are tied to the fact that the Einstein equations are overdetermined with respect to the physical degrees of freedom, and allow one to chose different combinations of equations to solve for the spacetime to the future of the initial hypersurface. As a result, one can distinguish *free-evolution* approaches – where only evolution equations are employed to this effect – from *constrained (partially constrained)* approaches where (some of the) constraints are used to solve for a subset of variables [24]. We also note that the freedom in choosing a foliation gives rise to Cauchy (or $D + 1$), characteristic or hyperboloidal formulations, in the case of spacelike, null or spacelike-but-asymptotically-null foliations respectively. For a review of related numerical approaches, see e.g., [48]; a more pedagogical exposition of the basic concepts can be found in recent textbooks on the subject [49–52].

On the Cauchy front, early efforts employed the so-called York-ADM formulation [28] (a reformulation of the standard Hamiltonian-based ADM approach). This formulation is geometrically appealing in that it provides evolution equations for the intrinsic and extrinsic curvatures of the foliation. However, beyond spacetimes where symmetries allow for reducing the dimensionality of the problem, numerical evolution with the York-ADM method exhibits instabilities. In the early 2000s [53], it was recognized that such a formulation is only weakly hyperbolic, implying that at the analytical level the system of PDEs lacks properties required to achieve robust numerical implementation [54]. A flurry of activity in the following years provided much insight on how to deal with this issue and construct (desirable) symmetric or strongly hyperbolic formulations by suitable modifications of the equations (primarily via the addition of new constraints and the use of appropriate coordinate conditions – see e.g. [55–57] for reviews on the subject). It turns out that arbitrarily many formulations could be defined with these desirable properties and all are, of course, equivalent at the analytical level.

At the discrete level, however, this is not the case. In particular in free-evolution schemes it is challenging to control the magnitude of the truncation error associated with the constraint equations (which are not explicitly imposed except at the initial time). Such errors compound at different rates in different formulations, and the practical physical time needed in order to achieve accurate results in simulations thus varies significantly. Two formulations have been shown empirically to display the robustness needed to construct a large class of 4-dimensional spacetimes (without symmetries) that are of particular relevance to the contemporary astrophysical and theoretical physics problems discussed here. These are the generalized harmonic evolution with constraint damping [10] (closely related to the Z4 formalism, see e.g. [58]), and the BSSN (or BSSNOK) approach [59–61].

Harmonic coordinates have a history even older than the field equations themselves – having been used by Einstein in his search for a relativistic theory of gravity as early as 1912 [62] – and have played an important role in many key discoveries of properties of the field equations since (see the introduction in [63]). Harmonic coordinates can be defined as the requirement that each spacetime coordinate obeys a homogeneous scalar wave equation. Enforcing this at the level of the Einstein equations converts the latter to a form that is manifestly symmetric hyperbolic. This desirable property is maintained if freely specifiable functions are added as source terms in the wave equations, resulting in *generalized* harmonic coordinates. In principle, the source functions allow arbitrary gauges to be implemented within a harmonic evolution scheme [64]. Constraint damping terms [65] are further added to tame what otherwise would result in exponential growth of truncation error.[1] This formulation has also been particularly useful in finding stationary black hole solutions in higher dimensions where, due to the stationary nature of the spacetime, the coordinate freedom can be exploited to define a convenient, strictly elliptic problem [68]. The BSSN formulation is an extension of the York-ADM approach which introduces several additional (constrained) variables to remove particular offending terms from the equations, and this, together with a judicious choice of coordinates, ensures strong hyperbolicity of the underlying equations.

When black holes are evolved, the geometric singularities inside the horizons need to be dealt with in some manner to avoid numerical problems with infinities. One such method is excision [33], where an inner *excision boundary* is placed inside each apparent horizon to remove the singular region from the computational domain. Due to the causal structure of the spacetime, the characteristics of the evolution equations ostensibly all point out of the computational domain at the excision boundary, and no boundary conditions are placed there. Excision is commonly used in harmonic evolution schemes. For evolution within the BSSN approach, an alternative *moving puncture* method has proven successful [11, 12]. This is an extension of puncture initial data, where the puncture point inside the horizon that formally represents spatial infinity on the other side of a "wormhole" is now evolved in time. With the typical gauges employed during evolution the interior geometry evolves to a so-called "trumpet" slice, where the puncture asymptotes to the future timelike infinity of the other universe [69, 70]. Effectively then, the puncture also excises the singularity from the computational domain.

On the characteristic front, the structure of the equations is significantly different from the Cauchy problem, as the foliation is defined using characteristic surfaces. The system of equations displays a natural hierarchy of evolution equations, constraints, and a set of hypersurface equations for variables that asymptotically are intimately connected to the physical degrees of freedom of the theory [71]. Beyond spherically symmetric applications, numerical codes employing this formulation show a remarkable degree of robustness, and can stably evolve highly distorted single black hole spacetimes [47, 72]. However, the rigidity in the choice of coordinates (being tied to characteristics) implies that difficulties arise when caustics and crossovers develop. For astrophysical purposes, the main role of the characteristic approach has been to provide a clean gravitational wave extraction procedure [73, 74] and to study isolated black holes. Outside of the astrophysical domain, it has

---

[1] The idea of constraint damping via the addition of terms to the evolution equations which vanish if the constraint equations hold true dates back to several years earlier for the Einstein equations [66], and since that time this idea has also effectively been applied to other systems of PDEs that have internal constraints, in particular Maxwell's equations [67].

been convenient in studies of black hole interiors (e.g. [75]) and has become the predominant approach used to exploit the AdS/CFT (Anti-de Sitter/Conformal Field Theory) duality of string theory (see [76] and references therein). Here gravity in asymptotically AdS spacetimes is used to study field-theory problems, some examples of which are described in Section 7.3.6.

Finally, the hyperboloidal formulation [77] adopts a Cauchy approach in a conformally related spacetime where the physical spacetime is recovered as a subset of a larger one. This allows studying, within a single framework, both the local and the global structure of spacetime (as the larger manifold covers spacelike, null and timelike infinity in a natural way). However, it has received considerably less attention than the other approaches, though some interesting first steps have been carried out (e.g. [78, 79]).

7.2.2 Numerical methods

The subject of numerical analysis as it pertains to solutions of problems in applied mathematics is, of course, vast. Even when restricting to what is relevant for numerical relativity applications, the breadth of methods employed is considerable and, naturally, depends on the particular goal one has in mind. From constructing initial data and evolving the solution to the future of an initial hypersurface, to extracting physical information from the numerical results, an abundance of different techniques and methods have been used. Here we provide some brief comments with the aim of imparting a basic understanding of the available options.

Any numerical implementation ultimately renders the problem of interest into an algebraic problem for a discrete number of variables that describe the sought-after solution. For gravitational studies, this involves devising approximate methods to solve the relevant partial differential equations. Such methods can be conveniently visualized as providing a way to discretize the underlying variables that describe the problem as well as the spatial derivatives within the equations, and providing a recipe to advance the solution in discrete time.

The technique most commonly used is the *finite difference* (FD) method, with the solution represented by its value at discrete grid-points covering the manifold of interest. Discrete spatial derivatives are defined through suitable Taylor expansions, which can provide high-order-accurate approximations for smooth solutions. Further refinements can be achieved through the use of discrete derivative approximations that satisfy summation by parts. This property is a direct analog of integration-by-parts that is often exploited to obtain estimates controlling the behavior of general solutions at the analytical level [54, 55]. What has proven especially useful for many problems is the adoption of *adaptive mesh refinement* (AMR) to efficiently resolve a large range of relevant spatio-temporal lengthscales, and without *a priori* knowledge of the development of small-scale features [80–82]. Another discretization approach used is the *pseudo-spectral* method [83, 84]: here the solution is expanded in terms of a suitably chosen basis (e.g. Chebyshev or Fourier), which also provides a simple way to compute spatial derivatives. The coefficients of the expansion provide the sought-after solution. Pseudo-spectral methods provide a highly efficient way to achieve highly accurate results for smooth solutions. As with AMR for FD, adaptive, multi-domain decomposition methods can be employed for efficiently resolving the length scales in the problem (see e.g. [85]).

Advancing the solution in time requires integrating the discrete values of the FD solution, or spectral coefficients, via suitable operators. One common approach is the method of lines, where, having discretized the spatial part of the problem, accurate methods devised for ordinary differential equations are used to integrate the variables in time.

The above discretization approaches are also well-suited to evolving additional fields coupled to gravity that are smooth and, in particular, do not develop discontinuities (such as scalar or electromagnetic fields). However, when matter sources such as neutron stars are incorporated, the equations of general-relativistic hydrodynamics (or magnetohydrodynamics) must also be solved. These can be expressed in a way fully consistent with the approaches employed to integrate Einstein equations (for a review on this topic see [86]). Nevertheless, as the solution to the hydrodynamic equations can induce discontinuities even for smooth initial conditions, *finite-volume* methods [87] are most often adopted as they are especially suited to accurately handle such features.

7.3 Strong-field gravity

As mentioned, numerical simulations are often the only way to gain insights into the behavior of gravity in the strongly non-linear regime. This arises naturally in gravitational collapse, systems involving black holes and neutron stars, ultra-relativistic collisions and cosmology. Here questions range from fundamental explorations of the theory itself, to the resolution of questions of astrophysical relevance, such as the characteristics of gravitational wave signals produced in compact binary mergers, or the effect of highly dynamical and strong gravitational fields on matter/gas/plasma and their role in powering spectacular phenomena like gamma-ray bursts. Complex numerical simulations have been developed over the past few decades to start answering these long-standing questions, and have produced results that often raise new questions. In what follows, we discuss some of the more important findings and open questions, organizing the presentation by subject area. These examples are necessarily limited in scope and presentational depth, but serve as illustrations of the breadth of problems addressed with simulations. For a complementary, recent review article on numerical relativity and its applications, see [88].

7.3.1 Critical phenomena in gravitational collapse

Overview

Gravitational collapse, including the process of black hole formation, is one of the hallmarks of general relativity. As has already been noted in Section 7.1, although simulations play an increasingly dominant role in advancing our understanding of collapse scenarios that we believe play out in the universe, they also provide the means to perform detailed studies of more fundamental aspects of the process. Albeit reflective of more than a little theorist's conceit, we can view computer programs as numerical laboratories which – paralleling real experiments in non-linear science – are endowed with one or more control parameters that are varied in order to unearth and elucidate the phenomenology exhibited by the setup.

Critical phenomena are concerned with families of solutions to the coupled Einstein-matter equations (including the vacuum case), where a continuous parameter p labels the family members, and it is assumed that the spacetimes are constructed dynamically – usually via simulation – starting from prescribed initial data that depends on p. The initial data typically represents some bounded distribution of initially imploding matter and p is chosen to control the maximal strength of the gravitational interaction that ensues. For p sufficiently small, gravity remains weak during the evolution, and the spacetime is regular everywhere (if the matter is massless radiation, for example, the radiation will disperse to infinity, leaving flat spacetime in its wake). For p sufficiently large, gravity becomes strong enough to trap some of the matter in a black hole, with mass M_{BH}, and within which a singularity forms. For some critical value $p^\star$ lying between the very-weak and very-strong limits, the solution corresponds to the threshold of black hole formation, and is known as a critical spacetime for the given model. Collectively, the properties of these special configurations, as well as the features associated with the spacetimes close in solution space to the precisely critical solution, comprise what is meant by critical behaviour. Evidence to date suggests that virtually any collapse model that admits black hole formation will contain critical solutions.

It transpires that M_{BH} can be formally viewed as an order parameter, in the statistical mechanical sense, and most of the critical solutions identified to date can be sorted into two basic classes based on the behaviour of M_{BH} at threshold. Specifically, solution-space behaviour corresponding to both first- and second-order phase transitions is seen, defining what are called Type I and Type II critical solutions, respectively. Thus, in the Type I case M_{BH} is *finite* at the threshold, so $M_{BH}(p)$ exhibits a gap (jump) at $p = p^\star$. Conversely, in the Type II instance M_{BH} becomes *infinitesimal* as $p \to p^\star$ from above, and there is no gap. It is also crucial to observe that the precisely critical solution for $p = p^\star$ does *not* contain a black hole.

There are three key features associated with both types of black hole critical phenomena: universality, symmetries and scaling.

Concerning the first property, most, if not all, Type II solutions, and some Type I solutions, exhibit a type of universality in the sense that one finds the same critical configuration through numerical experimentation as sketched above, irrespective of the specific way p parameterizes the initial data. This implies a certain type of uniqueness, or at least isolation, of the critical spacetimes in solution space, analogous to the uniqueness of the Schwarzschild solution as the endpoint of black hole formation in spherical symmetry.

Secondly, Type I critical solutions generically possess a time-translational symmetry, which is either continuous (so the solution is static or stationary) or discrete (so the solution is periodic). In the discrete case, the oscillation frequency forms part of the precise description of the critical solution, and is usually determined by an eigenvalue problem. Type II critical solutions, on the other hand, typically have a scale, or homothetic, symmetry and are therefore scale-invariant. Once again, this symmetry can be either continuous or discrete (CSS/DSS for continuously/discretely self-similar). Continuously self-similar solutions have long been studied in relativity as well as in many other areas of science, frequently arising in situations where the underlying physics has no intrinsic length scale. On the other hand, it is safe to say that the observation of discrete self-similarity in the earliest numerical calculations of critical collapse came as a complete surprise. For DSS solutions, the analogue of the frequency of periodic Type I configurations is known as the echoing exponent, Δ. When expressed in coordinates adapted to the self-similarity, a DSS solution is oscillatory with period Δ; each complete oscillation represents a shrinking of the scale of the dynamics by a factor of e^{Δ}. In relation to the collapse process, self-similar behaviour of either type is particularly interesting because it means that at criticality the strong-field regime propagates to arbitrarily small spatiotemporal scales. Indeed, as the self-similar solution "focuses in" to an accumulation event at the center of the collapse, curvature quantities grow without bound, but with no formation of an event horizon. Thus, Type II critical solutions possess naked singularities and have significant relevance to the issue of weak cosmic censorship. However, in terms of their origin from collapse, it is imperative to note that these naked singularities are produced from (infinite) fine tuning of the initial data, so they are therefore not generic with respect to initial conditions.

The third characteristic of critical collapse is the emergence of scaling laws. Empirically, these are determined by studying the behaviour of certain physical quantities as a function of the family parameter p as $p \rightarrow p^{\star}$. For Type I behaviour a typical scaling law measures the time interval, τ, during which the dynamically-evolving configuration is close to the precisely critical solution. One finds

$$\tau \sim -\sigma \ln|p - p^{\star}| \qquad (7.1)$$

where σ is called the time-scaling exponent. In Type II collapse the black hole mass scales according to

$$M_{\text{BH}} \sim C|p - p^{\star}|^{\gamma} \qquad (7.2)$$

where γ is known as the mass-scaling exponent, and C is a family-dependent constant. For both types of behaviour, if the critical solution is universal with respect to the initial data, then so is the corresponding scaling exponent. Again, this is the case for all known Type II solutions, but not so for most Type I transitions where, as discussed in more detail below, a particular critical configuration is typically one member of an entire branch of unstable solutions.

The scaling laws can be understood in terms of perturbation theory. The key observation [89–91] is that the appearance of critical solutions through a fine-tuning process – wherein one of two distinct end states characterizes the long-time dynamics – suggests that they have a *single* unstable perturbative mode. The inverse of the Lyapunov exponent associated with that mode is then precisely the scaling exponent. Furthermore, leading subdominant modes can give rise to additional scaling laws, involving for example charge and angular momentum [92].

The scaling relations (7.1) and (7.2) underscore the fact that near criticality, there is exponentially sensitive dependence on initial conditions, and that irrespective of the original choice of initial data parametrization, it is the transformed quantity $\ln |p - p^\star|$ which is most natural in describing the phenomenology. In simulations one wants to compute with $|p - p^\star|$ as small as possible in order to most accurately determine the threshold solutions and their associated exponents. This is especially true for the Type II case since the asymptotically flat boundary conditions that are normally adopted are incompatible with self-similarity, so one relies on calculations which probe scales as small as possible to ensure that boundary effects are minimized. In practice, investigation of the critical regime is ultimately limited by the fact that p can only be fine-tuned to machine precision. For spherical and axisymmetric calculations, this can be accomplished with current technologies, but, for the Type II scenarios, only if the numerical algorithm provides sufficient spatiotemporal dynamic range using a technique such as adaptive mesh refinement (AMR) [80]. Indeed, at this point it should be emphasized that almost all known critical solutions have been computed in models that impose symmetry restrictions. Spherical symmetry has been most commonly adopted. There have been some axisymmetric calculations, but very few fully 3D studies.

As already mentioned, the appearance of critical solutions seems completely generic, irrespective of the matter content of the model. However, details of the phenomenology are dependent on a number of factors including the following: the type of matter and the specifics of any self-interaction terms, the imposed symmetries, the spacetime dimensionality and the asymptotic boundary conditions. The remainder of this section is devoted to a summary of a necessarily incomplete selection of the many numerical studies performed to date, organized by the type of matter employed. In order to highlight the state of the art in the subject, there is some bias towards more recent calculations, and an attempt has been made to impart some sense of the wide variety of scenarios that have been explored. Those interested in more information are directed to the excellent comprehensive reviews of the subject [92, 93].

Scalar fields

Critical collapse was first studied in the model of a spherically symmetric minimally coupled massless scalar field [94]. Using several families of initial data, a single Type II DSS solution with $\gamma \approx 0.37$ and $\Delta \approx 3.44$ was found (see Fig. 7.1). Due to the extreme dynamical range required to fully resolve the critical solution, use of adaptive mesh refinement was crucial. Remarkably, the mass-scaling relation (7.2) provided a good fit for M_{BH} even when $|p - p^\star|$ was large enough that the final black hole contained most of the total mass of the spacetime. The main results of [94] have been confirmed many times since using a variety of different algorithms and coordinate systems. Assuming the existence of a spacetime with a discrete homotheticity, certain regularity conditions and a tailored numerical approach, the critical solution and associated exponents were computed to very high accuracy in [95].

Figure 7.1 Type II discretely self-similar critical solution computed from the collapse of a spherically symmetric distribution of massless scalar field. The figure shows the late time configuration of the scalar field from a marginally subcritical evolution where the family parameter has been tuned to approximately a part in 10^{15}. The radial coordinate is logarithmic, making the discretely self-similar (echoing) nature of the solution evident: each successive echo represents a change in scale of $e^\Delta \approx 31$. The data were generated using the axisymmetric code described in [98].

Additionally, analysis of the implications of the discrete self-similarity led to the prediction and observation of a modulation of the mass-scaling law (7.2) with period $\Delta/(2\gamma)$ [95, 96]. Finally, an important analysis in [97] found that all non-spherical modes of the critical solution decay, strongly suggesting that the same threshold configuration would appear if the symmetry restriction were relaxed.

Critical solutions from axisymmetric massless scalar collapse using multiple initial data families were constructed in [98]. For the most part the threshold configurations could be described as the spherical solution plus perturbations (measured values for the scaling exponents were $\gamma \approx 0.28$–0.41 and $\Delta \approx 2.9$–3.5), but there were also indications of a single asymmetric mode which, as it grew, produced two separated regions (on axis) within which the solution locally resembled the spherical one. This observation is in conflict with [97], but the accuracy of the results was insufficient to convincingly demonstrate that the growth was genuine and not a reflection of limitations in the simulations. Adaptive mesh refinement was again crucial.

Very recently, a study of massless scalar collapse using a fully 3D code has been carried out [99] and, in fact, represents the first calculations of Type II general-relativistic critical phenomena in 4 spacetime dimensions without symmetry restrictions. Four initial data families defining a spherical matter distribution deformed to varying degrees with a Y_{21} spherical harmonic anisotropy were considered. Even though AMR was used, compute-time limitations kept the tuning of $p/p^\star$ to about a part in 10^4. Nonetheless, evidence for the emergence of the spherically symmetric critical solution with $\gamma \approx 0.37$–0.38 was found. There were also preliminary indications of echoing with $\Delta \approx 3.1$–3.3.

The massless scalar model has no intrinsic length scale so, in retrospect, the appearance of a Type II solution at threshold is natural. Introduction of a mass, μ, breaks scale invariance and, as shown in [100], complicates the picture of criticality. For initial data with a length scale λ the massless behaviour is recovered when $\lambda\mu \ll 1$. However, for $\lambda\mu \gtrsim 1$, a Type I transition is seen, with a critical solution which is one of the periodic, starlike configurations (oscillons) admitted by the model and constructed in [101]. As with relativistic perfect-fluid stars, the oscillons comprise a one-parameter family that can be labeled by the central density. As the central density increases the stellar mass also increases, but only up to a point, whereafter dynamical instability sets in and the stars reside on the so-called unstable branch – it is precisely one of these unstable solutions that sits at the Type I transition. This latter type of behaviour was also observed in [102] using a massive complex scalar field whose static solutions, known as boson stars, also have stable and unstable branches. In this instance stable stars were driven to a Type I threshold via an imploding pulse of massless scalar field, whose overall amplitude was used as the tuning parameter.

Investigation of circularly symmetric massless scalar collapse in $2 + 1$ AdS spacetime [103, 104] represents one of the few instances where critical behaviour in a non-asymptotically flat setting has been seen (but also see the discussion of the turbulent instability of $3 + 1$ AdS [105] in Section 7.3.8). Evidence for a Type II transition with a CSS solution was found – with a mass-scaling exponent $\gamma \approx 1.2$ – but a thorough understanding of the picture of criticality here is still lacking. In particular, an analytic CSS solution that shows good agreement with the numerical results has been found [106], but seems to have additional unstable modes. Its existence also seems paradoxical in the sense that, heuristically, the cosmological constant should be irrelevant on the small scales pertinent to scale-invariance, yet is essential in the construction of the solution.

Vacuum

Historically, the second example of black hole critical phenomena discovered was in the collapse of pure gravitational waves [107] in axisymmetry. The study employed one family of initial data representing initially incoming pulses of gravitational radiation with quadrupolar angular dependence, and with an overall amplitude factor serving as the control parameter. Evidence for a Type II transition was found, with a discretely self-similar critical solution that was centred in the collapsing energy. The computations yielded an estimated mass-scaling exponent $\gamma \approx 0.37$ and an echoing factor $\Delta \approx 0.6$. The calculations did not use AMR. However, due to the use of spherical polar coordinates, increased central resolution could be achieved with a moving-mesh technique. Nonetheless, the dynamic range of the code was very limited relative to that used in [94], so it was quite fortuitous that Δ in this case was quite small.

It is truly remarkable that in the two decades that have elapsed since the publication of [107], and despite several additional assaults on the problem and a vast increase in the available amount of computer resources, little progress has been made in reproducing and extending these early results. One notable exception is [108] in which the collapse of axisymmetric Brill waves was studied, using several different families of data with varying degrees of anisotropy. Once more, evidence for a Type II transition was found in all of the experiments, with a scaling exponent γ – measured in this instance through the scaling of a curvature invariant in subcritical collapse [109] – in the range 0.37–0.4. However, in stark contrast to the observations in [107], most of the computed critical solutions showed accumulation on rings at finite distances from the origin, rather than at the origin itself. Additionally, indications of echoing were seen, but with an estimated $\Delta \approx 1.1$ significantly different from that reported in [107]. Development of a more complete understanding of the critical behaviour of collapsing gravitational waves, both in axisymmetry and in the full 3D case, remains one of the most important unresolved issues in this field.

In $D + 1$ dimensions, with D even, application of a co-homogeneity two symmetry reduction to the vacuum Einstein equations yields a set of wave equations dependent only on a single radial dimension. In contrast to those resulting from a spherically symmetric reduction, these equations admit asymptotically-flat, radiative solutions [110–113]. For $D = 4$, and adopting the so-called biaxial ansatz, Type II DSS behaviour was found, with $\Delta \approx 0.47$ and $\gamma \approx 0.33$ [110]. Analogous results were found for $D = 8$, where $\Delta \approx 0.78$ and $\gamma \approx 1.64$ [112]. The more general triaxial ansatz for $D = 4$ was considered in [111]. Here, the biaxial critical solution still appears at threshold. However, due to a discrete symmetry in the model, the critical surface actually contains three copies of the configuration. As well, on the boundaries of the basins of attraction of these copies, a different DSS solution with two unstable modes was predicted and computed using a two-parameter tuning process. Additional numerical experiments have shown that the critical-surface boundaries have a fractal structure [113].

Fluids

Studies of critical behaviour with perfect fluid sources have been extremely important in the development of the subject, not least since it was in this context that understanding of the phenomena in terms of unstable perturbative modes was developed. The first calculations focused on spherically symmetric simulations with a fluid equation of state (EOS), $P = k\rho$, where P and ρ are the fluid pressure and energy density, respectively, and with the specific choice $k = 1/3$ (radiation fluid) [89]. A continuously self-similar critical solution was found with a mass-scaling exponent $\gamma \approx 0.36$. In addition, the critical solution was computed independently by adopting a self-similar ansatz, and was shown to be in excellent agreement with the simulation results, and it was suggested that a perturbation analysis could be used to at least approximately compute γ. Such an analysis was carried out in [90], where both the critical solution and its linear perturbations were determined, and it was shown that there was a single unstable mode whose inverse Lyapunov exponent yielded the same value of γ as seen in the simulations. Interestingly, at this time the values of γ that had emerged from the three models for which threshold solutions had been identified were numerically the same to the estimated level of numerical accuracy, suggesting that the mass-scaling exponent might be universal across all matter models. However, the results of [91] (performed at the same time as [90]), where critical solutions and their perturbative modes were determined via the self-similar ansatz for many values of the EOS parameter k in the range 0.01–0.888, showed definitively that γ was in general model-dependent. A more extensive perturbation analysis [93, 114] suggested that the spherical solutions will appear at threshold when spherical symmetry is relaxed only for values of k in the range $1/9 < k \lesssim 0.49$; for other values of k, additional unstable modes were found. These conclusions have yet to be verified through simulations, and it will be very interesting to do so.

The $P = k\rho$ EOS is scale-invariant (and in fact is the only EOS compatible with self-similarity [115]) so Type II critical behaviour is expected. For more general equations of state, including the commonly adopted ideal gas law, intrinsic length scales appear and, as anticipated from the massive-scalar-field studies [100], the critical phenomenology becomes richer. In particular, the expectation that unstable stars can appear as Type I critical solutions was confirmed in [116] using the ideal gas EOS and the same type of experiments as performed in [102]. Type II behaviour with this EOS also appears when the fluid internal density of the configuration is much larger than the rest-energy density [117–119], in which case the EOS limits to the scale-free equation, and the measured mass-scaling exponents agree with those computed from a scale-invariant ansatz.

A possible cosmological application of Type II fluid collapse was posited in [120], where it was argued that the mass-scaling relation should apply to the formation of primordial black holes, since the exponential decay of the scale of density fluctuations entering the horizon at any epoch provides an intrinsic fine-tuning mechanism. This leads to a modification of the usual mass function for the primordial black holes, which incorporates the prediction that holes of sub-horizon scale could form at all times.

Over the past few years, significant progress has been made in extending the investigations of Type I critical behaviour with fluids to the axisymmetric [121–124] and 3D [125] arenas. Almost all studies have adopted a stiff ($k = 1$) ideal gas EOS (with static or stationary solutions interpreted as neutron stars), and the work reported in [125] also incorporated rotational and magnetic effects. In [121] a Type I transition was observed in the head-on collisions of two neutron stars where several different tuning parameters, including the stellar mass and the index k, were employed. Clear evidence of lifetime scaling for subcritical evolutions was seen. It was also suggested that the change in the EOS that occurs as a real post-collision remnant cools could provide a natural tuning mechanism, so that if the cooling was sufficiently slow, the critical solution might have astrophysical relevance. Further simulations of head-on collisions [122, 124] have corroborated these findings, and it was demonstrated in [122] that the end state of the marginally subcritical collision was well-described by a perturbed star on the stable branch. Intriguingly, the lifetime scaling measured in [122] exhibits a periodic modulation of σ – analogous to that seen in the mass-scaling exponent for DSS Type II transitions – that has yet to be explained. The fact that stars on an unstable branch can be identified as Type I solutions has also been demonstrated in a more direct fashion, through the use of initial data families where the tuning parameter perturbs (or effectively perturbs) a star known or suspected to be one-mode unstable. This strategy was employed in [123] to demonstrate the criticality of an unstable spherical configuration, with an accurate computation of σ. Finally, in [125] evidence for the threshold nature of rotating unstable stars – both non-magnetized and magnetized – with preliminary evidence of lifetime scaling was reported. This last study, along with [99], provides a tantalizing glimpse of what lies in store for this field as symmetry restrictions are relaxed and the physical realism of models is enhanced.

Other types of matter

Spherical collapse of an $SU(2)$ Yang–Mills field within the magnetic ansatz was studied in [126], and was the first model where both Type I and Type II behaviour was observed. Here the $n = 1$ member of the Bartnik and McKinnon countable sequence of static configurations [127], which had previously been shown to have one unstable mode, is the attractor for the Type I transition, while a DSS solution with $\Delta = 0.74$ and $\gamma = 0.20$ was also found. The model exhibits another transition, strictly in the black-hole sector of solution space, where colored black holes arise at the threshold and where M_{BH} has a gap as one tunes across it [128].

Spherically symmetric self-gravitating σ-models (wave-maps), which typically incorporate dimensionless tunable coupling constants, have been shown to display especially rich critical phenomenology. Notably, transitions between CSS and DSS Type II behaviour as the coupling is varied have been seen in both the 2-dimensional non-linear model [129] and the $SU(2)$ case [130]. The transition in the latter instance is particularly interesting, displaying behaviour where near-critical evolutions approach and depart from a CSS solution episodically.

Finally, Type I critical behaviour has been seen in the collapse of collisionless matter in spherical symmetry – with or without a particle mass – where the threshold solutions are static [131–133] and appear to exhibit the expected properties of Type I solutions, including lifetime scaling. In the massless case it has been argued that there should be *no* one-mode unstable solutions [134], and this apparent contradiction with the numerical results remains another unsolved puzzle.

7.3.2 Binary black hole mergers

The non-linear nature of general relativity has several interesting consequences for how it describes particles and the gravitational interaction between them. First, technical caveats aside, the simplest possible solution describing the geometry of an idealized point-like distribution of chargeless, spinning matter is a Kerr black hole. From an external observer's perspective there is thus *no* geometrical realization of a point-like structure, as the event horizon prevents length scales smaller than the energy (in geometric units) of the black hole from being probed. Second, there is no analogue of a Newtonian potential that can be superposed to come up with a simple description of the interaction of two black holes. In consequence, a detailed understanding of one of the most basic interactions in gravity, the two-body problem, requires numerical solution of the field equations. On the other hand, thanks to the "no-hair" properties of black holes, the merger of two Kerr black holes is expected to describe the merger of all astrophysical black holes essentially exactly, the only idealization being that the presence of surrounding matter is ignored.

The discussion in the previous paragraph assumes many properties of solutions to the field equations not yet proven with mathematical rigor. Chief among them are that weak cosmic censorship holds in these scenarios, and that any black hole that forms in our universe (specifically here via the merger of two black holes, but implicitly also by processes that led to the initial black holes) evolves to a geometry locally describable by a unique member of the Kerr family (again modulo perturbations from the exterior universe). Other than intrinsic theoretical interest to understand merger geometries, finding numerical solutions for specific examples can provide strong evidence for these assumptions. However, the most pressing reason to study the binary black hole problem in recent years has been to support the effort to observe the universe in the gravitational wave spectrum. As discussed in more detail elsewhere in this volume (see Chapter 6), theoretical models of expected waveforms are necessary for successful detection and to decipher the properties of sources. A host of tools have been developed to tackle this problem for black hole mergers, including post-Newtonian expansions, black hole perturbation theory, the effective-one-body (EOB) approach, and the geodesic self-force problem applicable to extreme-mass-ratio mergers. For comparable-mass mergers, perturbative methods break down near coalescence, and this is where numerical relativity contributes most to the problem. The rest of this section is devoted to an overview of what has been learned about these final stages of the merger from numerical solutions, restricting to the four-spacetime-dimension case. For more detailed reviews see [13, 135].

One of the results that was immediately obvious from the first full merger simulations of equal-mass, non-spinning black holes [10–12], and since then for the large swath of parameter space simulated (see for example [136]), is the relative simplicity of the structure in the emitted waves during the transition from inspiral to ringdown (see the left panel of Fig. 7.2). This is the regime of evolution where the strongest-field dynamics is manifest, and the perturbative approaches applicable before and after should be least reliable. Certainly the perturbative inspiral calculations do break down evolving forwards to merger, and similarly for extending the quasi-normal ringdown backwards to this time. However, there is no significant intermediate regime of dynamics between the two, and with guidance from the numerical simulations, the perturbative waveforms can be stitched together with relatively simple matching conditions (this is a rapidly advancing subfield; see [137–139] for a few recent examples at the time this chapter was written).

With regard to the issues of theoretical interest discussed above, no simulation has shown a violation of cosmic censorship, and the final state, to within the accuracy of the simulations and the level that researchers have scrutinized the geometry, is a member of the Kerr family. Moreover, though it is unlikely that the quasi-normal mode spectrum of Kerr is able to describe all possible perturbations, in cases studied to date the post-merger waveforms can indeed be well approximated as sums of quasi-normal modes. Of course,

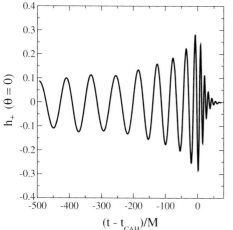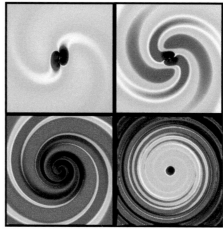

Figure 7.2 Depictions of the gravitational waves emitted during the merger of two equal-mass (approximately) non-spinning black holes [140]. Left: The plus-polarized component h_+ of the wave measured along the axis perpendicular to the orbital plane. t_{CAH} on the horizontal axis is the time a common apparent horizon is first detected. Right: A color-map of the real component of the Newman–Penrose scalar Ψ_4 (proportional to the second time derivative of h_+ far from the BH) multiplied by r along a slice through the orbital plane (green is 0, toward violet (red) positive (negative)). From top left to bottom right the time $(t - t_{CAH})/M$ of each panel is approximately $-150, -75, 0, 75$.

here we have a rather restrictive class of astrophysically minded "initial conditions" for the perturbed Kerr black hole formed by the merger of two black holes. We note that single black holes perturbed by gravitational waves have also been studied numerically beyond the linear regime, and similar conclusions hold [141, 142].

Some of the more important numbers that have been provided by numerical simulations include the total energy and angular momentum radiated during merger (and consequently the final mass and spin of the remnant black hole), the spectra of quasi-normal modes excited, and the recoil, or "kick" velocity of the final black hole to balance net linear momentum radiated. It is beyond the scope of this chapter to list all these numbers. However, in brief, for a baseline reference, it has been found that two equal-mass, non-spinning black holes beginning on a zero-eccentricity orbit at "infinite" separation radiate $\sim 4.8\%$ of the net gravitational energy during inspiral, merger and ringdown, ultimately becoming a Kerr black hole with dimensionless spin parameter $a \sim 0.69$ (due to the symmetry of this system, there is zero recoil). The waveform spectrum is dominated by the quadrupole mode in a spin-weight-2 spheroidal harmonic mode decomposition; the next-to-leading order is the octupole mode, which is strongly subdominant, though it briefly grows to an amplitude around 1/5th that of the quadrupole mode near merger [140] (the energy of a mode scales as its amplitude squared). Changing the mass ratio decreases the energy radiated by roughly the square of the symmetric mass ratio η, the final black hole spin drops linearly with η, new multipole moments in the waveform are introduced (reflecting the quadrupole moment of the effective energy distribution of the two-particle source) and such changes can result in recoil velocities as high as ~ 175 km/s [143–146]. Introducing spin for the initial black holes can alter the radiated energies by up to a factor of roughly 2 (higher for spins aligned with the orbital angular momenta, lower otherwise) [147], can increase (decrease) the final spin for initial spin aligned (anti-aligned) with the orbital angular momentum (the largest aligned-spin case simulated to date begins with equal initial spins of $a \sim 0.97$, merging to a black hole with $a \sim 0.95$ [148]), can result in precession of the orbital plane which correspondingly modulates the multipole structure of the waveform observed along a given line of sight [149–151], and perhaps most remarkably can produce recoil velocities of several thousand km/s for appropriately aligned high-magnitude spins [152, 153]. Figure 7.3 illustrates some of the results obtained for equal-mass, fast-spinning binary black holes.

There are many astrophysical consequences of large recoil velocities, in particular for supermassive black hole mergers; we briefly mention a few here, together with some broader consequences of mergers on surrounding matter (for recent more detailed reviews see [155, 156]). First, the velocities for near-equal-mass, high-spin mergers are large enough to significantly displace the remnant from the galactic core, or for the highest velocities even eject the black hole from the host galaxy altogether. This may be in some tension with observations that seem to suggest that all sufficiently massive galaxies harbour supermassive black holes. If the system has a circumbinary accretion disk, the recoil would carry the inner part of the disk with it, and this could be observable in Doppler-displaced emission lines relative to the galactic rest frame [157]. The near-impulsive perturbation to the gravitational potential in the outer parts of the accretion disk could lead to the formation of strong shocks, producing observable electromagnetic emission on timescales of a month

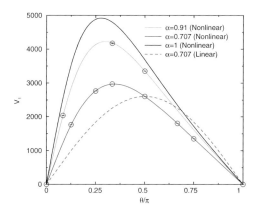

Figure 7.3 Recoil velocities from equal-mass, spinning binary black hole merger simulations (circles) together with analytical fitting functions. Each black hole has the same spin magnitude α, with equal but opposite components of the spin vector within the orbital plane, and θ is the initial angle between each spin vector and the orbital angular momentum. The dashed line corresponds to a fitting formula that depends linearly on the spins, while solid lines add non-linear spin contributions. (Reprinted, with permission, from Fig. 2 of [154]. © 2011 by the American Physical Society.)

to a year afterwards [158]. (Note that regardless of the recoil, the entire accretion disk will experience an impulsive change in potential due to the near-instantaneous loss of energy from gravitational wave emission at merger, also producing electromagnetic emission post-merger [159]). Earlier studies have suggested that prior to merger the accretion rate, and hence the luminosity of the nucleus, would be low as the relatively slow migration of the inner edge of the accretion disk decouples from the rapidly shrinking orbit of the binary. Post merger then, AGN-like emission could be re-ignited once the inner edge of the disk reaches the new innermost stable circular orbit (ISCO) of the remnant black hole. This will be displaced from the galactic center if a large recoil occurred, and could be observable in nearby galaxies (see for example [160]). However, more recent simulations of circumbinary disks using ideal magnetohydrodynamics for the matter show that complete decoupling does not occur, and relatively high accretion rates can be maintained all the way to merger [161, 162]. (The left panel of Fig. 7.4 illustrates a binary black hole system accreting surrounding gas.) The binary orbit can cause a modulation in the induced luminosity of the system, which may be observable. A displaced central black hole will also have its loss-cone refilled, increasing the frequency of close encounters with stars and their subsequent tidal disruption by the black hole, with rates as high as 0.1/yr; the disruption could produce observable electromagnetic emission [163]. Yet another exciting

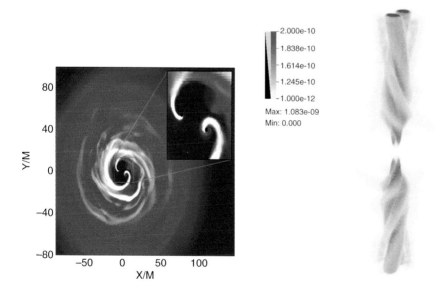

Figure 7.4 Left: Rest-mass density induced by a supermassive black hole binary interacting with a magnetized disk prior to when the binary "decouples" from the disk, namely when the gravitational wave backreaction timescale becomes smaller than the viscous timescale (reprinted, with permission, from Fig. 2 of [162]. ©2012 by the American Physical Society). Right: Poynting flux produced by the interaction of an orbiting binary black hole interacting with a surrounding magnetosphere. The "braided" jet structure is induced by the orbital motion of the black holes (from [164] © AAAS).

prospect for an electromagnetic counterpart is an analog of the standard Blandford–Znajek mechanism (to extract rotational energy from a spinning black hole) induced by a tightening binary within a circumbinary disk. In particular, numerical simulations have uncovered that binary black holes can extract both rotational and translational kinetic energies when there is surrounding plasma [164]. Not only can this power strong dual Poynting jets (emanating from each black hole), but also the jets will increase in strength until merger, making them indirect "spacetime tracers" – the right panel of Fig. 7.4 depicts the resulting "braided" structure of the Poynting flux.

As a final comment we note that the majority of work, both numerical and analytic, has been devoted to studying zero-eccentricity mergers, due to the prevailing view that these will dominate event rates. However, there are binary formation mechanisms that can produce high-eccentricity mergers (see the discussion in [165] for an overview and references). One of the interesting results from the handful of studies including large eccentricity performed to date [166–168] is that zoom-whirl orbital dynamics is possible for comparable-mass binaries. In the test-particle limit, zoom-whirl orbits are perturbations of the class of unstable circular geodesics that exist within the ISCO; further, they exhibit extreme sensitivity to initial conditions where sufficiently fine-tuned data can exhibit an arbitrary number of near-circular "whirls" at periapse for a fixed-eccentricity geodesic. Away from the test-particle limit gravitational wave emission adds dissipation to the system, though what the simulations show is that even in the comparable-mass limit the dissipation is not strong enough to eradicate zoom-whirl dynamics, but merely limits how long it can persist.

7.3.3 Black hole–neutron star/binary neutron star mergers

Non-vacuum compact binary systems – i.e., those involving at least one neutron star – are also the subject of intense scrutiny. These systems produce powerful gravitational waves and likely also lead to intense neutrino and electromagnetic emission that could be detected by transient surveys or by dedicated follow up by the astronomical community. In particular they are posited to be the progenitors of short gamma-ray bursts (sGRBs) and a host of other transient phenomena [169, 170]. Signals from these systems can thus carry a wealth of information about gravity, the behavior of matter at nuclear densities, and binary populations and their environments. The challenge for simulations is to obtain predictions to confront with observations.

Relative to the two-black-hole case, the most obvious complication in the simulation of binaries with neutron stars is the need to include non-gravitational physics. The simplest relativistic model of a neutron star couples relativistic hydrodynamics to the Einstein equations and, using a simplified equation of state (EOS), the first successful simulations of binary neutron star mergers within this framework were presented in [171, 172]. Since the time of those studies, the community has made steady progress in exploring the full parameter space relevant to astrophysical mergers, while simultaneously increasing the fidelity of the matter modeling through inclusion of the electromagnetic interaction, neutrino and radiation transport, nuclear reactions, and other physics. A crucial unknown here is the EOS that describes matter at nuclear densities: it plays a leading role in the phenomenology of the system as, for a given stellar mass, it regulates the star's radius, affects its response to tidal forces, and affects its ability to resist collapse to a black hole when it accretes matter (or collides with another star). Given the difficulty of first-principles calculations or probing similar conditions in laboratories, detailed knowledge of the nuclear density EOS is likely to come only through astronomical observations, and prospects for doing this through gravitational waves are particularly exciting – see for example [173, 174].

While the pericenter is large these systems evolve much like black hole binaries. The orbit shrinks due to the emission of gravitational radiation, with the internal details of the stars playing essentially no role. However, finite-body effects become important as the orbit tightens. In the remainder of this paragraph we focus on binary neutron stars, returning to black hole–neutron star systems in the following paragraph. Tidal forces deform both stars (which can even induce crust-shattering [175]), leaving subtle imprints in the ensuing gravitational waves. This behavior intensifies until the point of merger, when the local velocities reach a sizable fraction of the speed of light, ending in a violent collision that ejects neutron-rich matter due to shock heating and extreme tidal forces. Figure 7.5 (left panel) illustrates waveforms obtained in an equal-mass binary neutron star system for

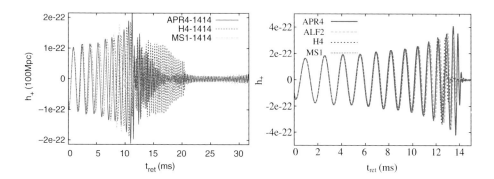

Figure 7.5 Examples of the "plus" polarization component of gravitational waves from binary neutron star mergers, measured 100 Mpc from the source along the direction of the orbital angular momentum. The different curves correspond to different choices of the EOS of the neutron star matter, labeled APR4, ALF2, H4 and MS1. For a $1.4 M_\odot$ neutron star, the APR4, ALF2, H4, MS1 EOS give radii of 11.1, 12.4, 13.6, 14.4 km respectively. Left: Mergers of an equal-mass binary neutron star system (with $m_1 = m_2 = 1.4 M_\odot$). A hypermassive neutron star (HMNS) is formed at merger, but how long it survives before collapse to a black hole strongly depends on the EOS. The H4 case collapses to a black hole ≈ 10 ms after merger; the APR and MS1 cases have not yet collapsed $\simeq 35$ ms after merger when the simulations where stopped (the MS1 EOS allows a maximum total mass of $2.8 M_\odot$, so this remnant may be stable). The striking difference in gravitational wave signatures is self-evident (reprinted with permission from Fig. 9 of [189]. ©2013 by the American Physical Society). Right: Emission from black hole–neutron star mergers, with $m_{\rm BH} = 4.05 M_\odot, m_{\rm NS} = 1.35 M_\odot$. Variation with EOS is primarily due to coalescence taking place earlier for neutron stars with larger radii (from [190]).

different EOS models, demonstrating how significantly this can affect the behavior. In general terms, for the typically expected neutron star masses of 1.2–$1.8\,M_\odot$ the merger yields a hot, differentially rotating, hypermassive neutron star (HMNS). Such an object will promptly collapse to a black hole if the total binary mass is above 2.6–$2.8\,M_\odot$, depending on the stiffness of the EOS. Otherwise, a delayed collapse takes place as the star is initially supported by differential rotation and thermal pressure. During this stage, the HMNS rotates and emits gravitational waves with frequencies in the range $2 \lesssim f \lesssim 4\,\mathrm{kHz}$, with a characteristic frequency proportional (and relatively close) to the Keplerian velocity $(M_{\mathrm{HMNS}}/R_{\mathrm{HMNS}}^3)^{1/2}$ (e.g., [176]). On a scale of tens of milliseconds, however, such support diminishes due to gravitational radiation, angular momentum transport via hydrodynamical and electromagnetic effects, and cooling due to neutrino emission (these effects have just begun to be studied, e.g. [177–179]). The black holes that form in both prompt or delayed cases are (reasonably) well-described by a Kerr solution with a spin parameter $J/M^2 \lesssim 0.8$, surrounded by left-over material, much of which is bound (e.g., [180]) and can form an accretion disk with a mass on the order of $\simeq 0$–$0.3\,M_\odot$. The amount depends on the EOS, mass ratio, and electromagnetic fields (though this latter effect is still largely unexplored) and is intuitively larger for longer-lived HMNS as more angular momentum is transferred outwards to the envelope. Importantly, this is enough material to form a sufficiently massive disk as called for in models of sGRBs. Some material will be ejected (again, the amount depending upon various parameters) and will decompress to form heavy elements through the r-process; if these merger events are frequent this could account for a significant fraction, if not the majority, of such elements in the Universe. Subsequent decay of the more radioactive isotopes could lead to a so-called kilo- or macronova (reports of the afterglow of the recent sGRB 130603B are consistent with this [181, 182]). Observation of these signatures together with gravitational wave observations will allow us to make contact between simulations and the birth of a black hole. However, gravitational waves emitted during the HMNS and collapse stages have a higher frequency than those reachable by LIGO/VIRGO/KAGRA, and it will take third-generation facilities to detect them. Nevertheless, up to the frequencies that existing (and near-future) detectors can probe, subtle differences in the gravitational waveforms should allow us to constrain the radius of the neutron stars to within 10% [183, 184]. Simulations are further probing possible counterpart signals from neutrino production [185] and electromagnetic interactions [186–188].

Black hole–neutron star binaries display even more complex merger dynamics. Indeed, at an intuitive level one expects significant differences to arise depending on whether the tidal radius R_T ($\propto R_{\mathrm{NS}}(3M_{\mathrm{BH}}/M_{\mathrm{NS}})^{1/3}$) lies inside or outside the black hole's inner most stable circular orbit radius (R_{ISCO}), which ranges from M_{BH} to $9M_{\mathrm{BH}}$ for a prograde to retrograde orbit about a maximally spinning black hole. This is clearly borne out in simulations exploring a range of mass ratios and black hole spins, showing markedly different behavior in the ensuing dynamics and gravitational waves produced. Qualitatively, for sufficiently high spins and/or sufficiently low mass ratios, the star significantly disrupts instead of plunging into the black hole. As a result, gravitational waves promptly "shut-off" at a frequency related to the star's EOS. Figure 7.5 (right panel) illustrates this for different EOS models in a 3 : 1 mass ratio black hole–neutron star system. When disruption occurs during the merger, a significant amount of material, in the range 0.01–$0.3\,M_\odot$, can remain outside R_{ISCO}. This material will be on trajectories having a range of eccentricities, with the fraction that is bound falling back to accrete onto the black hole at a rate governed by the familiar law $\dot{M} \propto t^{-5/3}$ [191, 192]. The details, however, depend on many factors, including spin–orbit precession as illustrated in Fig. 7.6. The matter that is ejected ($\lesssim 0.05\,M_\odot$) can be have speeds up to $\simeq 0.2c$ [190, 193]. This, together with the amount of likely accretion, is in the range assumed by models predicting that black hole–neutron star mergers can power sGRBs, kilonovae and related electromagnetic counterparts. Consequently, a similar array of electromagnetic signatures and r-process elements could result as with binary neutron star mergers, and the gravitational wave signals could be ideal to differentiate between them. For the subset of black hole–neutron star mergers where $R_T \lesssim R_{\mathrm{ISCO}}$, the star plunges into the black hole with little or no material left behind, and the resulting gravitational wave signal will be much like that of a binary black hole system with the same binary parameters. Counterparts such as sGRBs or kilonova requiring significant accretion disks or unbound matter are therefore not favored for this sub-class of binary. Nevertheless, interesting electromagnetic precursors could be induced by magnetosphere–black hole interactions prior to merger (e.g. [194–196]).

Figure 7.6 General-relativistic hydrodynamic simulations of the merger of a $9.8\,M_\odot, a = 0.9$ black hole with a $1.4\,M_\odot$ neutron star. (Reprinted, with permission, from Fig. 5 of [192]. ©2013 by the American Physical Society.) The top two panels are from a case where the spin and orbital angular momentum vectors are aligned; the bottom two where the initial (~ 9 orbits before merger) misalignment is $40°$. The left two panels are at a time when half the material has been absorbed by the black hole, showing matter densities above $\sim 6 \times 10^{10}$ g/cm^3; the right two are 5 ms later, showing densities above $\sim 6 \times 10^9$ g/cm^3, and the facing quadrant has been cut from the top-right rendering. These results illustrate the profound effect spin-induced precession can have on the matter disruption and subsequent accretion.

An important observation is that black hole–neutron star systems are, in all likelihood, more massive than binary neutron star systems. Therefore, the wave frequency peaks at lower frequencies than binary neutron star mergers, offering better prospects for observing non-linear effects by near-future detectors. Indeed, since the characteristics of gravitational waves depend on masses, spins and the EOS, black hole–neutron star systems provide perhaps the best prospects for extracting key physical information about neutron stars [174, 197]. To date the majority of simulations have focused on a black hole spin aligned with the orbital angular momentum, with the exception of [192], which showed that the above conclusions hold qualitatively even with inclinations $\lesssim 30°$ of the spin axis away from alignment. For larger inclinations, a smaller fraction of the disrupted material forms a disk on a short timescale following merger, while a larger fraction follows an eccentric trajectory and returns to interact with the black hole on longer timescales.

As in the binary black hole case, incipient efforts are examining encounters with high initial orbital eccentricity in non-vacuum binaries (e.g. [198, 199]). Qualitatively, much of the same phenomenology of outcomes can occur as with quasi-circular inspirals (except that now zoom-whirl orbital dynamics is also possible), though the details can be drastically different. For example, in high-eccentricity encounters of a neutron star with a black hole, tidal disruption can occur for higher-mass-ratio systems and smaller black hole spin, as the effective inner most stable orbit is closer to the black hole for eccentric orbits. There can also be multiple, partial disruptions on each of the last several periapse passages, ejecting larger amounts of material and leaving behind more massive accretion disks than otherwise possible. On close periapse passages (even without disruption) f-modes can be impulsively excited in the star, or both stars in a binary neutron star encounter. These modes are too low-amplitude/high-frequency to be directly observed with the current generation of ground-based gravitational wave detectors, though they may indirectly be measured in the leading-order part of the waveform, since from the perspective of the binary the f-modes are a new channel of energy dissipation. The impulsive tidal interaction may also cause crust-shattering [200], leading to electromagnetic emission similar to the resonant-excitation-induced shattering in quasi-circular inspirals [175].

For merger simulations involving neutron stars, the current frontier is to add more matter physics to the models (resistive magnetohydrodynamics, radiation and neutrino physics, multi-component fluids, "realistic" high-temperature equations of state, etc.). Given the many orders of magnitude of spatial and temporal scales involved, as well as the complexity of the microphysics, it will likely be several years before both realistic models and the computational power necessary to simulate them accurately are available. Due to space constraints we will not list all the directions currently being pursued, referring the reader to recent reviews in [201–203].

7.3.4 Gravitational collapse to a neutron star or black hole

Considerable efforts have been undertaken to study gravitational collapse to a neutron star or a black hole, in particular within the context of core-collapse supernovae. Here, stars with masses in the range $10\,M_\odot \lesssim M \lesssim 100\,M_\odot$ at zero-age main sequence form cores which

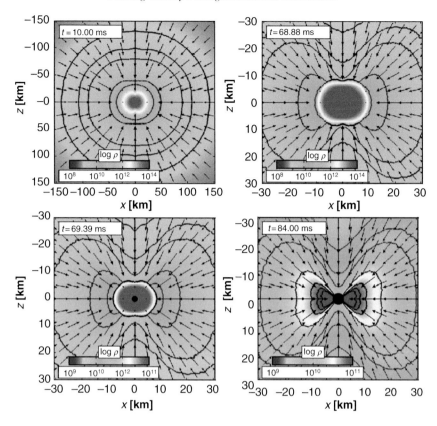

Figure 7.7 Density colormaps of the meridional plane of a collapsing $75\,M_\odot$ star superposed with velocity vectors at various times after bounce (and with different spatial ranges to zoom-in on particularly relevant behavior). The collapse first forms a proto-neutron star, which later collapses to a black hole (shown in the bottom panels). (Reprinted from [212], Fig. 3, with permission. ©2011 by the American Physical Society.)

can exceed the Chandrasekhar mass and become gravitationally unstable. This leads to collapse which compresses the inner core to nuclear densities, at which point the full consequences of general relativity must be accounted for. Depending upon the mass of the core, it can "bounce" or collapse to a black hole. Figure 7.7 displays representative snapshots of the behavior of a collapsing $75\,M_\odot$ star at different times. The collapse forms a proto-neutron star which later collapses to a black hole. In the case of a bounce, an outward propagating shock wave is launched which collides with still infalling material and stalls. Observations of core-collapse supernovae imply that some mechanism is capable of reviving the shock, which is then able to plow through the stellar envelope and blow up the star. This process is extremely energetic, releasing energies on the order of 10^{53} erg, the majority of which is emitted in neutrinos. For several decades now, the primary motivation driving theoretical and numerical studies has been to understand what process (or combination of processes) mediates such revival, and how (for a recent review see [204]). Several suspects have been identified: heating by neutrinos, (multidimensional) hydrodynamical instabilities, magnetic fields and nuclear burning (see e.g. [205, 206]). With the very disparate time and space scales involved, a multitude of physically relevant effects to consider, and the intrinsic cost to accurately model them (e.g., radiation transport is a 7-dimensional problem), progress has been slow. Moreover, electromagnetic observations do not provide much guidance to constrain possible mechanisms, as they can not peer deep into the central engine. On the other hand, observations of gravitational waves and neutrinos have the potential to do so, provided the explosion is sufficiently close to us. Thus, in addition to exploring mechanisms capable of reviving the stalled shock, simulations have also concentrated on predicting specific gravitational wave and neutrino signatures.

Modeling gravity using full general relativity has only recently been undertaken [207], though prior to this some of the more relevant relativistic effects were incorporated (e.g. [208–211]). While the full resolution of the problem is still likely years ahead, interesting insights into fundamental questions and observational prospects have been garnered. For example, simulations have shown that in rotating core-collapse scenarios, gravitational waves can be produced and their characteristics are strongly dependent on properties of the collapse: the precollapse central angular velocity, the development of non-axisymmetric rotational instabilities, postbounce convective overturn, the standing accretion shock instability (SASI), proto-neutron star pulsations, etc. If a black hole forms, gravitational wave emission is mainly determined by the quasi-normal modes of the newly formed black hole. The typical frequencies of gravitational radiation can lie in the range $\simeq 100$–1500 Hz, and so are potential sources for advanced earth-based gravitational wave detectors (though the amplitudes are sufficiently small that it would need to be a galactic event). As mentioned, the characteristics of these waveforms depend on the details of the collapse, and hence could allow us to distinguish the mechanism inducing the explosion. Neutrino signals have also been calculated, revealing possible correlations between oscillations of gravitational waves and variations in neutrino luminosities. However, current estimates suggest neutrino detections would be difficult for events taking place farther away than at kpc distances [207].

7.3.5 Ultra-relativistic collisions

Some of the early interest in the ultra-relativistic collision problem stemmed from investigations by Penrose [213] into its relevance to questions of weak cosmic censorship. It was known that collisions of gravitational waves with planar symmetry in 4d lead to the formation of naked singularities regardless of how "weak" the initial curvature is. This is not considered a serious counter-example to weak cosmic censorship as the spacetime is not asymptotically flat, nor for that matter are there black hole solutions with planar symmetry in 4d vacuum Einstein gravity (with zero cosmological constant), so in a sense the question of censorship is not particularly meaningful here. However, taking the infinite-boost limit of the Schwarzschild metric (scaling the rest mass m to zero as the boost $\gamma \to \infty$

while the energy $E = m\gamma$ remains finite [214]) results in the Lorentz contraction of the curvature to a plane-fronted gravitational shock wave, with Minkowski spacetime on either side. One can then consider what happens when two such shock waves, traveling in opposite directions, collide. Given the resemblance between the two scenarios, the infinitely boosted black hole collision is a natural place to test weak cosmic censorship, especially since the geometry approaches Minkowski spacetime transverse to the center of each shock sufficiently rapidly to remove the trivial objections to the plane-symmetric gravitational wave collisions. Penrose found a trapped surface in a zero-impact parameter, infinite-γ black hole collision, and, although the metric to the causal future of the collision is unknown, this is a good indication that weak cosmic censorship does holds here.

More recently two additional lines of research have come to the fore motivating the study of ultra-relativistic collision geometries. The first is a consequence of the observation that if extra spatial dimensions exist, then the true Planck scale could be much different from the effective 4-dimensional scale one would otherwise expect [215, 216]. In particular, a "natural" solution to the hierarchy problem results if the Planck energy is on the order of a TeV. If that is the case it was conjectured that particle collisions at the Large Hadron Collider (LHC) and in cosmic-ray collisions with the earth with center-of-mass energies above this could result in black hole formation [217, 218].[2] The conjecture is essentially based on two premises: that Thorne's hoop conjecture [222] can be applied to the collision to deduce whether the purely classical gravitational interaction between particles will cause a black hole to form, and if so, that the quantum interactions are sufficiently "local" to not alter this conclusion (until Hawking evaporation becomes significant). The second motivation comes from applications of the AdS/CFT correspondence of string theory to attempt to explain the formation and early-time dynamics (before hadronization) of the quark–gluon plasma formed in relativistic heavy-ion collisions (RHICs) (see Section 7.3.6 for more on this). Here, the gravitational dual to a heavy-ion collision is conjectured to be an ultra-relativistic black hole collision in the bulk asymptotically AdS spacetime.

The first 4d ultra-relativistic head-on black hole collision simulations (up to $\gamma \approx 3$) were carried out in [223], followed by several studies with general impact parameters [224, 225], including the effects of black hole spin [226, 227], and collisions in higher dimensions [228]. A wealth of interesting results have emerged, a select few of which we briefly summarize here. Outcomes of most interest to LHC searches include the critical impact parameters for black hole formation, and the energy and angular momentum lost to gravitational waves as a function of the impact parameter. This determines the formation cross-section and initial spectrum of black hole masses that will subsequently Hawking evaporate. Extrapolated results from 4d head-on collisions give $14 \pm 3\%$ energy emitted in gravitational waves, roughly half the Penrose trapped-surface calculation, though consistent with the 16% obtained using perturbative analytic methods [229]. As the impact parameter increases, the radiated energy and now angular momentum increases, though the former

---

[2] To date, the LHC has not seen evidence for black hole formation in searches of collisions with center-of-mass energies up to 8 TeV [219, 220]; likewise, no signs of black hole formation have yet been observed in cosmic-ray collisions [221], the most energetic of which can have much larger center of mass energies.

is still less than trapped-surface estimates [230]. Qualitative features of the spectrum of emitted waves can be understood by appealing to the analytic zero-frequency and point-particle-limit calculations [231]. The largest gravitational wave fluxes arise near the threshold impact parameter. Here, in the 4d case, the binary exhibits behavior akin to zoom-whirl dynamics of black hole geodesics, though not in 5d (presumably due to the stronger effective gravitational potential, related to the fact that there are no stable circular orbits about Myers–Perry black holes in dimensions greater than 4) [225, 228].

Because of the zoom-whirl behavior in 4d, it was argued in [166] that as a consequence of tuning to the threshold in the large-γ limit essentially all the initial kinetic energy of the black holes would be converted and radiated out as gravitational waves. However, the results presented in [227] show this is likely not true, due to what *appears* to be strong self-absorption of the emitted gravitational energy by the black holes. These simulations only went to $\gamma \approx 2.5$; however, if they in fact provide a decent approximation of the large-γ limit, then one concludes that as much as half the kinetic energy could be converted to rest-mass energy in the black holes (the rest to gravitational waves), *even* in close scattering encounters (which is in fact consistent with a perturbative calculation in the extreme-mass-ratio limit [232]). The surprising consequence of this is that two "microscopic" black holes each of rest-mass m scattered off one another with $\gamma \gg 1$ and a finely tuned impact parameter could grow to two "macroscopic" black holes moving apart sub-relativistically, each with rest-mass $\sim m\gamma/2$.

A further intriguing result for the 5d case presented in [228] is that for a small range of impact parameters near threshold, curvature invariants grow rapidly at the center of mass shortly after what appears to be a scattering event, though the code crashed before the final outcome could be determined. No encompassing apparent horizon is detected then, which could simply be because of the nature of the time coordinate employed, or could be due to a naked singularity that is forming. If the former, this would be a new outcome to the black hole scattering problem in 5d, namely three black holes; if the latter, this would be another example (in addition to the Gregory–Laflamme instability of black strings, and possible prolate dust collapse [22]) showing a violation of weak cosmic censorship.

The high-speed limit has also shed some light on the mechanism responsible for large recoil velocities seen in merger simulations of astrophysically relevant inspirals with certain spin configurations (see the discussion in Section 7.3.2). In particular, there have been suggestions that the large recoils require the formation of a common horizon to effect the gravitational wave emission of "field momentum" associated with what are otherwise purely kinematical properties of the orbit; in reaction the merger remnant receives a kick in a direction that conserves linear momentum [233]. However, in high-speed merger simulations with similar black hole spin setups, even in scattering cases where a common horizon does *not* form, large recoils are observed [226]. This is consistent with the heuristic explanation of the superkicks presented in [13] as arising from frame-dragging-induced Doppler boosting of the radiation emitted by the binary motion.

The first ultra-relativistic collision simulations of "solitons" (non-singular compact distributions of matter) were carried out in [234], consisting of the head-on collision of two boson stars each with compactness $2M/R \approx 1/20$, and center-of-mass boosts up to $\gamma = 4$

($v \approx 0.968$). The main goal of the study was to test the hoop-conjecture arguments for black hole formation; hence the use of boson stars as model particles given that their self-interaction is weak compared to gravity in this limit. Black hole formation was observed above a critical boost $\gamma_c \approx 3$, roughly one third the value predicted by the hoop conjecture. Similar results were later obtained using an ideal fluid (fermion) star as the model particle [235, 236]. The study in [235] used less compact stars that pushed the critical boost to $\gamma_c \approx 8.5$, but again found this to be a similar factor less than the hoop-conjecture estimate. It was argued that the lower thresholds are due to the compression of one particle by gravitational focusing of the near-shock geometry of the other particle, and vice-versa. This conclusion was anticipated by a geodesic model of black hole formation presented in [237]. It is remarkable that such a simple model, and for that matter the trapped-surface calculations as well, predicts the qualitative properties of what is ostensibly the regime where the most dynamical, non-linear aspects of the Einstein equations are manifest. On the other hand, a recent calculation of the gravitational self-force using effective-field-theory techniques in the large-γ limit shows that many simplifications arise here; in particular the non-linear interactions coming from gravitational bulk vertices are suppressed by factors of $1/\gamma^4$ [238] (see also [239]). Aside from giving strong evidence that the hoop conjecture is applicable to the classical collision problem, these studies also support the expectation that the outcomes of sufficiently supercritical $\gamma > \gamma_c$ collisions are insensitive to the details of the particle self-interactions. This is essential for black hole formation in super-Planck particle collisions to be a robust conclusion, despite the lack of detailed calculations in a full quantum (gravity) theory. This also justifies the use of black holes as model particles, which from a classical gravity perspective is (in theory) a simpler problem to simulate, due to the absence of matter.

The motivation and applications of the AdS/CFT correspondence in string theory are discussed below in Section 7.3.6; here we briefly comment on what RHIC-motivated studies have taught us about ultra-relativistic collisions. The relevant spacetime for this problem is 5d AdS, and in particular the Poincaré wedge, as its boundary is conformal to 4d Minkowski spacetime. Solving the full Einstein equations in 5 spacetime dimensions without symmetries and resolving the geometry dual to highly boosted concentrations of energy would be an extremely challenging computation to perform. To date then, existing studies (see [76] for a review) have made simplifying approximations: each particle is modeled as a finite-width gravitational wave with planar symmetry transverse to the collision axis.[3] This effectively reduces the numerical evolution to $2 + 1$ dimensions, and characteristic approaches have proven highly successful for this problem. Though the topology and asymptotics are quite different from the 4d asymptotically flat case, there is some similarity. Most relevant to this discussion is that the infinite-boost limit is similar to the Penrose/Aichelburg–Sexl superposed shock-wave construction; in both cases trapped surfaces can be found [240], yet the full solution to the causal future of the collision is unknown. The numerics have solved the finite-width planar-collision problem, showing that a black hole (with planar topology) does form in this case, and resolving the spacetime

---

[3] Note added in proof: A first simulation without symmetry assumptions in 5d AdS has recently been presented in [313].

to the future of the shock. In particular, post-collision along the future lightcone of the collision, the amplitude of the shock, as projected onto the Minkowski boundary, decays as a power law in time; within the lightcone, after a time roughly consistent with inferred thermalization times in RHIC experiments, the near-boundary metric fluctuations transform to a state that can be characterized as an expanding, cooling hydrodynamic flow [76, 241]. For further details on numerical relativity applications in the realm of high energy see [242].

7.3.6 Gravity in $d \neq 4$

Beyond ultra-relativistic collisions, numerical relativity has also been crucial in exploring the behavior of gravity in both stationary and time-dependent scenarios beyond $d = 4$. There are several motivations for doing so. On one end there is the desire to understand gravity at a fundamental level by contrasting known behavior in $d = 4$ to what arises in different dimensions. Higher dimensions are required by string theory, and this has inspired many speculative theories: for example TeV scale gravity/braneworlds [216, 243], some models of inflation [244] and modern cyclic models [245] of the universe. Lower dimensions have also been used to provide a simpler setting to gain intuition about quantum gravity (e.g., in $(2 + 1)$- [246] and in $(1 + 1)$-dimensional dilaton gravity [247]). At the other end, compelling practical reasons are provided by the role gravity may play in understanding phenomena described by field theories through holography [248, 249].

For more information, readers are directed to the recent book [250]. Here, for brevity we mainly focus on time-dependent problems, though we briefly review stationary solutions; in particular, those that are relevant to existing or future dynamical studies.

Black holes in dimensions $d > 4$

Understanding the landscape of stationary solutions with event horizons has been the focus of considerable effort [251, 252]. This work has illustrated how much richer the space of stationary black object solutions in higher dimensions is compared to the $d = 4$ case. A case in point is the broader class of topological structures allowed, which includes hyperspherical black holes, black rings and a combination of these latter two giving "black Saturns", black strings, black branes, etc. Interestingly, several topologically distinct solutions can have the same asymptotic charges, showing some degree of non-uniqueness of black hole solutions in higher dimensions. However, a particularly intriguing conjecture is that uniqueness can be restored by the additional requirement of *stability*. This possibility is implied by the fact that linear perturbations of many of these solutions are unstable. As a further contrast with $d = 4$ stationary black holes, there is no "Kerr-like" bound for spinning black holes in $d \geq 5$ as they can have arbitrarily large angular momenta. Again, related to the uniqueness issue, these ultra-spinning black holes are unstable [253]. Numerical solutions are required to understand the non-linear dynamics of unstable black objects, and to date this has been achieved only for black strings in $d = 5$ [254] and ultra-spinning black holes in $d \in 5..8$ [255, 256].

Black strings are black hole solutions extended along a trivial (optionally) compactified extra dimension. For simplicity, and because it is the one studied numerically, we restrict the discussion to $d = 5$, and so the static black string is given by the $d = 4$ Schwarzschild solution crossed with a circle of (asymptotic) length L. Gregory and Laflamme showed that linearized perturbations of such a black string admit exponentially growing modes above some critical L/M (with M the mass per unit length of the black string) [257]. Further, thermodynamical arguments suggested that above this ratio the entropically preferred solution would be a $d = 5$ Schwarzschild–Tangherlini black hole. Thus, it appeared possible that the effect of these growing modes would be to eventually cause the black string to pinch-off and give rise to a spherical configuration. Naturally, if that happened, weak cosmic censorship would be violated, indicating yet again that gravity in $d = 4$ is rather special.

To understand the dynamical behavior of the solution, a full non-linear – and so necessarily numerical – analysis is required. Such a study was presented in [254], and revealed that the instability unfolds in a self-similar fashion, where the black string horizon at any given time could be seen as thin strings connected by hyperspherical black holes of different radii (see Fig. 7.8). As the evolution proceeds, pieces of the string shrink further while others give rise to spherical black hole bulges, and the horizon develops a fractal structure. Interestingly, such behavior is reminiscent of the one displayed by a thin column of fluid through the Rayleigh–Plateau instability (see [258]). In the case of the black string, extrapolating the numerical results shows that the ever-thinning string regions eventually reach zero size, revealing a massless naked singularity in *finite time*. Thus, perturbed black strings

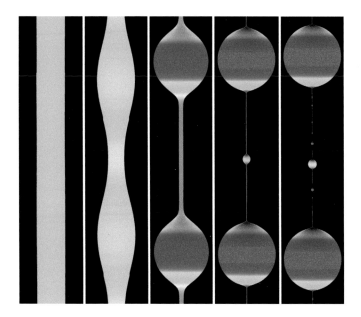

Figure 7.8 A sequence of snapshots showing the evolution (left to right) of an unstable black string; see [254] for details.

do provide a counter-example to the weak cosmic censorship conjecture, though in $d = 5$. In still higher dimensions, the outcome is expected to be qualitatively similar up to a critical dimension beyond which stable, non-uniform black string states are entropically favored. Perturbative analysis indicates that the critical dimension is $d = 13$ [259], though recent work making use of a local Penrose inequality suggests it may be as low as $d = 11$ [260].

This result has application beyond black string spacetimes, as many of the higher-dimensional black hole solutions have a near-horizon geometry that can be mapped, in appropriate regions, to black strings. For instance, ultra-spinning black holes satisfy the Gregory–Laflamme instability condition around the polar region [253]. Such black holes are thus expected to develop growing deformations about the poles of the horizon when perturbed. These can be both axisymmetric modes that would evolve toward axially "pinched" or ring-like configurations [261], and also non-axisymmetric modes. The latter, however, would induce a time-varying quadrupole moment that would radiate angular momentum, allowing for the possibility that gravitational wave emission could regulate the instability, in particular if the non-axisymmetric modes are the dominant unstable ones. Numerical simulations for systems with angular momentum mildly higher than the critical value show this is precisely the case [255, 256], where a "bar-mode" develops that radiates angular momentum until the black hole settles down to a subcritical, stable state. For larger initial spins it has been speculated that non-axisymmetric modes can grow more rapidly than gravitational wave emission can reduce the spin to subcritical, and the horizon might then fragment into multiple pieces [253, 256]. These cases have yet to be explored beyond the linear level.

As a last example we mention that solutions describing large black holes in Randall–Sundrum models were numerically constructed in [262], disproving a conjecture that such solutions could not exist [263, 264]. Moreover, the particular Ricci flow method employed to obtain the solutions argues implicitly in favour of their stability.

AdS/CFT duality applications

The AdS/CFT correspondence [248, 249] provides a remarkable framework to study certain strongly coupled gauge theories in d dimensions by mapping to weakly coupled gravitational systems in $d + 1$ dimensions. A large body of work has been built since the introduction of this correspondence; we will not review it here. Rather, we concentrate on a handful of applications where numerical simulations have been crucial to the understanding of the gravitational aspects of the problems. The relevant spacetimes typically involve black holes, and are asymptotically AdS, the latter property creating delicate issues on both analytical and numerical fronts. This is in part due to the timelike nature of the AdS boundary, with the consequence that the correct specification of boundary data (in addition to the initial configuration) is crucial for a well-defined evolution that can be mapped to the CFT description of the problem. The boundary conditions can be derived from the limiting behavior of the Einstein equations approaching the boundary, together with constraints from imposing that the spacetime approaches AdS at the appropriate rate in a suitable gauge.

Many interesting applications have been pursued using holography, spurred by work beginning soon after the formulation of the correspondence indicating the rather spectacular breadth of possible applications to finite-temperature field theory (see [265] for a review). Some highlights include that the hydrodynamic behavior of field theory is captured by correlation functions in the low-momentum limit, that hydrodynamic modes in relevant field theory states correspond to low-lying quasi-normal modes of an AdS black brane solution, and that for such states there is a universal viscosity-to-entropy-density ratio $\eta/s = 1/(4\pi)$ for a broad class of theories with gravitational duals. The value of η/s is remarkably close to that inferred from hydrodynamic models of the quark–gluon plasma (QGP) formed in relativistic heavy-ion collisions, and this observation has led to the new approach of using AdS/CFT to try to understand the QGP (for a review see [266]). Though the $N = 4$ SYM theory of the duality is not deconfined QCD, there are sufficient similarities that one might hope the former can give insights into aspects of the problem not easily calculable via traditional techniques (perturbative Feynman diagrams and lattice QCD). For example, using AdS/CFT purely gravitational studies can be used to estimate the thermalization timescales post-collision, and the subsequent evolution of the expanding plasma to the point of hadronization. As mentioned above in Section 7.3.5, a series of groundbreaking works [267–269] studied the behavior of the spacetime when two gravitational shock waves collide. This is expected to offer a decent approximation to the dynamics along the beam axis in central (head-on) collisions. The results for the thermalization time (one definition of which is the time after collision when the boundary stress tensor of the CFT is well approximated by the leading-order terms in the hydrodynamic expansion) are broadly consistent with the times inferred from experiment. Moreover, the subsequent hydrodynamic flow exhibits a form of boost-invariance similar to the predicted Bjorken flow [270, 271], the difference being characterizable as a modest dependence of the energy scale of the flow on rapidity [76, 241]. Figure 7.9 illustrates the energy density measured at the AdS boundary from simulations on the gravitational side; the top image is from a shock collision simulation, and the bottom is the relaxation of a highly perturbed black hole. Soon after the collision (left) and from the beginning of ringdown (right), a hydrodynamical description on the field theory side matches the observed near-boundary metric behavior to an excellent degree.

Another front where the duality is being exploited is to understand the behavior of a system in the ground state of a given Hamiltonian when a "quenched interaction" is introduced. Here the response of an initial thermal equilibrium state of the theory under rapid variations of suitable operators can be studied using the correspondence. As the behavior on the gravitational side is governed by the dynamics of an appropriately perturbed black hole, a universal response is uncovered which, on the CFT side, means the system responds in a way that depends only on the conformal dimension of the quench operator in the vicinity of the ultraviolet fixed point of the theory [272–274].

As a last example we mention an application of the duality in the opposite direction: using knowledge of the behavior on the field-theory side to discover and analyze novel features on the gravitational side. It is well known that field theories at sufficiently high

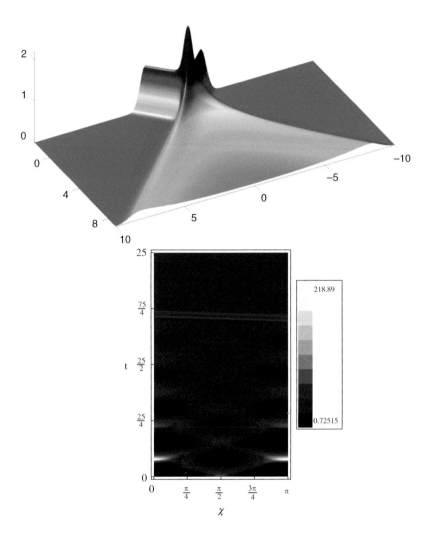

Figure 7.9   Top: Energy density in planar shock collisions as a function of time t and longitudinal position z. The shocks approach each other along the z axis and collide at $t = 0, z = 0$. The collision produces "debris" that fills the forward lightcone (from [76]). Bottom: Depiction of the energy density of a 4d boundary flow dual to the evolution of a highly perturbed 5d black hole in asymptotically global AdS spacetime (the radius of the black hole settles to 5 in geometric units, where the AdS length scale is $L = 1$) (from [279]). The boundary has topology $\mathbb{R} \times S^3$, and χ is an angular coordinate; hence the image represents an initial high-density (hence, pressure) enhancement on the equator ($\chi = \pi/2$) that propagates back and forth between the equator and the poles ($\chi = 0, \pi$). This result is from a pure 5d vacuum gravity simulation, yet the projected boundary dynamics matches that of a relativistic conformal fluid to within better than 1%, even in the early stages when the perturbation is highly non-linear.

energies admit a hydrodynamical description; this motivated works that established a duality between gravity and hydrodynamics for relativistic, conformal fluids. Specifically, it was shown that in the limit of long-wavelength perturbations of black holes the Einstein equations projected onto the AdS boundary reduce to the familiar relativistic hydrodynamics for a viscous fluid (e.g. [275]). Numerical work has demonstrated that the hydrodynamic description matches the behavior of full, non-linear solutions of the Einstein equations surprisingly well in many situations, as mentioned above, with RHIC applications (and see Fig. 7.9). Taken in the other direction then, this duality implies that phenomena familiar in hydrodynamics should arise in gravity. In particular, motivated by this, arguments were presented that gravity could exhibit turbulent dynamics, with a direct energy cascade in $4 + 1$ dimensions and the opposite in $3 + 1$ [276]. Furthermore, in $(3 + 1)$-dimensional gravity a quasi-conserved quantity should arise that is related to the conservation of entropy in hydrodynamics [277]. These observations have recently been demonstrated in ground-breaking numerical simulations of perturbed black branes [278] (see Fig. 7.10), showing that the horizon geometry reflects the turbulent behavior of the boundary projection, and develops a fractal-like structure over the corresponding range of lengthscales.

7.3.7 Singularities

Numerical simulations have played a significant role in analyzing the nature of singularities, in particular those which are often called "cosmological" due to the spacetimes having compact spatial topology (for a thorough review, see [281]). One of the long-standing questions has been the generic nature of singularities; i.e., what is the geometry

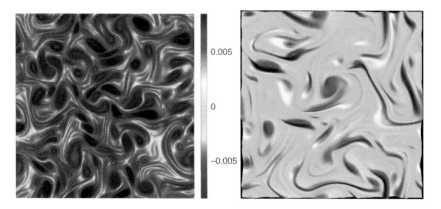

Figure 7.10 Left: Vorticity of gravitational perturbations of a planar black hole as obtained through a $3 + 1$ simulation of Einstein equations in AdS (from [76]). Right: Vorticity of a hydrodynamical field obtained in a $2 + 1$ viscous hydrodynamic simulation with a background fluid configuration dual to a planar black hole (from [280]). Exploiting the fluid/gravity duality allows for constructing the full metric of the dual $3 + 1$ spacetime to excellent accuracy.

of a spacetime approaching a singularity if no symmetries are presumed? For a vacuum spacetime, much of the research was inspired by the early work of Belinskii, Lifschitz and Khalatnikov (BKL, [282]), who conjectured that the generic singularity is spacelike, local, and oscillatory. The "local" part of the conjecture is that, in an appropriate gauge, the spatial gradients in the field equations become irrelevant compared to the temporal gradients, and hence the dynamics at any spatial point reduces to a set of ordinary differential equations in time. The oscillatory (or "mixmaster") aspect then describes the dynamics of one of these points, claiming that the solution consists of an infinite, chaotic sequence of transitions between epochs, and in each epoch the geometry is well-described by one member of the Kasner family of geometries. A Kasner geometry is a homogeneous but anisotropic solution to the field equations consisting of two contracting and one expanding spatial direction (in the approach to the singularity). Several objections were raised to the BKL conjecture, in particular that the assumptions they employed restricted their conclusions to local aspects of homogeneous cosmologies, and hence had little bearing on the generic, global properties of the spacetimes [283]. Numerical simulations have been key in resolving these disputes, gathering evidence in favour of the BKL conjecture [38, 284, 285], though discovering a surprising caveat in the process. This discovery was of so-called "spikes" that develop at isolated regions in the geometry [38]. A spike is a small-lengthscale feature where the spatial gradients are *not* small, and hence are important in governing the local dynamics of the geometry. Spikes seem to undergo oscillatory transitions similar to the mixmaster behavior of non-spike worldlines [286]. However, since (in the approach to the singularity) they shrink rapidly with time, even numerical simulations imposing planar symmetry (so $(1+1)$-dimensional evolution) have not been able to follow their dynamics for long enough to conclusively demonstrate this. Due to these resolution challenges spikes have not been studied in scenarios with less symmetry, and so whether spike-like features exist beyond co-dimension 1 is also not known.

An important point to make with regard to the above discussion of genericity and singularities is that it strictly applies only to these so-called cosmological singularities, and not *necessarily* to those formed in gravitational collapse to black holes. There is some expectation that local properties of the singularities should be the same whether in a cosmological or black hole setting (indeed, the interior geometry of Schwarzschild is locally Kasner). However, the interior (Cauchy) horizons of rotating and charged black holes develop into null singularities when perturbed, and have very different structure from the spacelike singularities in cosmology [287]. There are also arguments that null singularities are "as generic" as spacelike singularities [288], and may also be relevant in a cosmological setting [289].

7.3.8 Miscellaneous

Here we briefly discuss two miscellaneous topics where numerics have played an important role, and do not naturally fit into the main topic sections above.

Stability of AdS

As opposed to Minkowski and de Sitter spacetimes where global non-linear stability with respect to small perturbations has been established [290, 291], the related question has yet to be resolved in AdS. A key difference between AdS and the other two spacetimes is the fact that infinity is timelike, and acts like a confining boundary; namely the future lightcone from any event on an interior timelike observer's worldline will reach the boundary and return to intersect the worldline a finite proper time later. Ground-breaking numerical and analytical work [105] studied the spherically symmetric Einstein–Klein–Gordon system in asymptotically AdS spacetime, and uncovered that a black hole eventually forms from an arbitrarily small initial perturbation. This can heuristically be understood as a direct result of the confining property of AdS – energy cannot dissipate away, and as a consequence of non-linear interaction, eventually a configuration will develop in which the central density becomes sufficiently large to cause gravitational collapse. A more quantitative explanation was given in [105], where through a resonance mechanism there is a secular transfer of energy from large to small scales, ending when a black hole forms. Furthermore, at the threshold of black hole formation, the spacetime behaves self-similarly, and the solution corresponds to the one seen in the asymptotically flat case [94] (as expected, since the AdS scale is irrelevant for a small black hole). Related work has argued that this behavior should still be present in the absence of symmetries, and also when only gravitational perturbations are considered [292]. While these studies suggested AdS is unstable to arbitrarily small, generic perturbations, more recent follow-up work has demonstrated the existence of large classes of initial data that are stable [293–295]. Applying these results to the AdS/CFT correspondence, given that black hole formation is synonymous with thermalization, this implies (perhaps unsurprisingly) that there are large classes of states in the dual CFT that do not thermalize.

Formation and evaporation of CGHS black holes

Two-dimensional dilaton models of black hole evaporation were a popular subject of research a couple of decades ago, and though much was learned about the quantum nature of black holes from them, one could argue that no consensus results were obtained regarding the final near-Planck stages of evaporation (whether the black hole evaporates completely, or if there is a remnant, or a naked singularity, or a baby universe, etc.), or whether information is lost. One such popular model is that of Callan, Giddings, Harvey and Strominger (CGHS) [296]. Though extensively studied before, many interesting quantitative and qualitative features of solutions to the semi-classical CGHS equations of motion were missed until a recent numerical study [297–299]. One of the more interesting results revealed here is that there are two distinct classes of solutions: those that can be identified as microscopic (with initial masses of the order of the Planck mass or less) and those that are macroscopic (with initial masses a few times the Planck mass or larger). Remarkably, for macroscopic cases, after a brief transient, the evaporating spacetime and Hawking flux asymptote to a universal solution, irrespective of details of the matter distribution that formed the black hole. Evaporation continues until the dynamical horizon shrinks to zero area, whence it encounters a singularity of the semi-classical equations (though this singularity is weaker than that arising in the classical solution). The future Cauchy horizon of this singularity is regular, in contrast with earlier suggestions that it would propagate to infinity in a "thunderbolt". An improvement to the Bondi mass of the spacetime proposed in [300] shows that there is still on the order of a Planck mass "remnant" in the singularity, though this would presumably be resolved with higher-order quantum corrections.

This behavior is very different from that of black holes initially formed with only of the order of the Planck mass (the microscopic branch). Earlier studies had missed this distinction, and focused all attention on the physically less relevant microscopic solutions. The macroscopic branch also turns out to be quite challenging to solve numerically, where scales of the order M_{Planck} in the initial vacuum state are exponentially "inflated" to scales of order $e^{M/M_{\text{Planck}}}$ in the outgoing Hawking flux (a manifestation of the red-shift of outgoing radiation in black hole spacetimes, though in the evaporating case this red-shift remains finite, since what was a null event horizon becomes instead a timelike dynamical horizon).

One unusual aspect of 2d dilaton gravity that prevents straightforward application of 2d results to the more relevant 4d case is that in the former case there are two distinct, causally disconnected null infinities ("left" and "right"). This effectively disassociates the quantum state of the "ingoing" (right-to-left moving quanta, say) matter that forms the black hole from the "outgoing" (left to right) vacuum that becomes the Hawking flux. The semi-classical results together with the arguments presented in [300] suggest that the evolution of this vacuum sector is unitary, though little information about the infalling matter is retrievable from the Hawking flux. There is also no sign of any "firewall" [301] along the dynamical horizon at the semi-classical level.

Cosmic bubble collisions

Within the eternal-inflation paradigm, our observable universe is contained in one of many bubbles formed from an inflating metastable vacuum [302]. Collisions between bubbles can potentially leave a detectable imprint on the cosmic microwave background radiation (see reviews [303, 304]). While this scenario was initially studied through phenomenological models, recent works have concentrated on providing a quantitative connection between particular scalar-field models giving rise to eternal inflation and the detailed signatures imprinted on the CMB. To this end, the intrinsically non-linear nature of the bubbles and their collisions has been studied numerically within full general relativity [305, 306]. Simulations have revealed, in particular, the following: (i) the energy released in the collision of identical vacuum bubbles goes mostly into the formation of localized field configurations such as oscillons; (ii) the structure of the potential considered is the dominant factor determining the immediate outcome of a collision; and (iii) slow-roll inflation can occur to the future of a collision. Interestingly, these studies indicate that the signature in the CMB is well-described by a set of four phenomenological parameters whose values can be only probabilistically determined.

Inhomogeneity in cosmology

The majority of applications of general relativity to cosmology over the past decades have utilized analytical methods. For observational cosmology this is because the observed homogeneity and isotropy of the universe implies that its large scale structure is well-described by known exact solutions (the Friedmann–Robertson–Walker–Lemaître metrics) with deviations from the FRWL solutions small and hence amenable to treatment by perturbation theory. There has however been some concern, in particular in light of the discovery of the present-day accelerated expansion of the universe, that large-scale inhomogeneities such as filaments and voids, or small-scale non-linear inhomogeneities such as stars, can alter the assumptions made to study the largest-scale dynamics of the cosmos (see [307] for a recent review). Some of these questions can be addressed by numerical solutions within full general relativity, in particular whether local, non-linear inhomogeneities can affect the overall expansion compared to a homogeneous universe with the same average stress-energy context (though "averaging" is itself an issue of some delicacy here). Only recently have simulations of such scenarios been considered [308–310]. In the latter two studies, universes with a positive cosmological constant and filled with a periodic lattice of black holes (thus the most extreme example of non-linearity possible in general relativity) were evolved. The results were that the effective expansion rate was consistent with that of an equivalent homogeneous dust-filled universe.

7.4 Unsolved problems

The aforementioned list of studies, while impressive in its own right, describes only a portion of the interesting phenomena where numerical relativity can shed light on important questions, as well as open up new research directions from them. The following is a (necessarily incomplete) list of such questions.

- *Strongly gravitating/highly dynamical scenarios and astrophysics.* While it is clear that simulations have played a key role in uncovering the behavior and characteristics of gravitational wave emission from compact binaries in astrophysical settings, much work, and many opportunities, remain. Indeed, even in the case of binary black holes where "only" the Einstein equations in vacuum are required, higher mass ratios and/or nearly-maximal spinning configurations have proven difficult and costly. The possible existence of intermediate-mass black holes strongly motivates understanding the former class of binaries. Non-vacuum systems require a more complex description due to the additional, and often involved, matter physics. The rewards for this complexity are that now in addition to gravitational waves, electromagnetic and/or neutrino emission become possible, with the consequence that the opportunity for simulations to make contact with observation is extremely rich. The overarching goals of such simulations are to obtain first-principles descriptions of the detailed observational signatures across the range of emission channels the binaries might produce. The challenge to do so comes largely from the disparate time/length scales introduced by a plethora of physical processes, and by the complexity of the microphysics. Unlike the Einstein equations, to make simulations of realistic matter tractable invariably requires simplified models of the fundamental equations. There is much opportunity here for synergy among the relevant communities: numerical relativity, gravitational wave observation, theoretical and observational astronomy, and nuclear physics.

- *Fundamental questions.* Numerical simulations will continue to play a key role in exploring questions about the fundamental nature of Einstein gravity. There is no shortage of tantalizing questions remaining to be explored, including the non-linear development of superradiant and other black hole instabilities in four- and higher-dimensional spacetimes (see [311] for a first such study in 4d), the nature of generic singularities inside rotating black holes (c.f. the "mass inflation" phenomenon [287]), the dynamics of near-extremal black holes, cosmological domain wall and gravitational shock-wave collisions without symmetry assumptions, critical collapse without symmetries, black hole collisions and other dynamical non-linear interactions in asymptotically AdS spacetimes (in particular with AdS/CFT applications in mind), tests of the limits of the hoop and cosmic censorship conjectures, the possible development – and consequences – of turbulence in gravity and fractal horizon structures, etc. (some of these are discussed further below). If past discoveries, such as critical phenomena and the "turbulent" instability in AdS spacetimes, are any indication, many surprises await. Furthermore, these might have counterparts in other physical systems, and important insights might be gained in both directions based on analogies between different physical systems, and based on similarities and differences between the mathematical models of these systems.

- *Critical collapse.* As reviewed above, the vast majority of studies of the threshold of gravitational collapse have been carried out in spherical symmetry. That the original study of the axisymmetric gravitational wave critical solution has defied attempts at a detailed solution for almost twenty years now hints at a very interesting and rich geometric structure awaiting discovery. For axisymmetric scalar field collapse, the inconsistency between perturbative results suggesting that all non-spherical perturbations decay and a numerical study that hinted at a second, "focusing" instability remains to be resolved. Collapse without any symmetry assumptions is essentially uncharted territory.

- *Black hole instabilities in higher dimensional, asymptotically flat spacetimes.* A number of black holes in higher dimensions have been argued to be unstable, in particular by making connection to Gregory–Laflamme-type instabilities. These arguments stem from the realization that the near-horizon geometry in (at least portions of) these black hole spacetimes can be mapped to unstable black string solutions, and so should display related phenomenology. This is the case for ultra-spinning black holes, black rings, black Saturns, etc. (e.g. [253]). Whether in all cases these black holes yield the rich behavior observed in perturbed, unstable black strings is yet unknown. For instance, in rapidly spinning black holes the instability induces a non-trivial, time-dependent quadrupole that radiates angular momentum that could shut off the instability. This has already been observed in numerical simulations of Myers–Perry black holes in 6–8 dimensions, though only for cases with relatively mild angular momenta [256]. For sufficiently large angular momentum (recall that there is no upper bound in higher dimensions) the time scale of the Gregory–Laflamme instability is shorter than the expected gravitational wave emission time required to reduce the spin by enough to stabilize the system. A related problem is to consider highly prolate,"cigar-shaped" black holes in higher dimensions. Certainly, barring small-scale length-wise perturbations, such a black hole would tend to become spherical on a time scale of order equal to the light crossing time τ_{CT} of the black hole. However, such horizons that are sufficiently thin should locally be Gregory–Laflamme unstable on a time scale much shorter than τ_{CT}. Both of the aforementioned problems appear tractable in the near future.

- *Gravitational behavior in $d \neq 4$ and holography.* As discussed above, holography has opened the door for numerical relativity to be exploited in problems outside the gravitational arena. Indeed, applications relevant to quark–gluon plasmas, condensed matter physics and quantum quenches (processes in which the physical couplings of a quantum system are abruptly changed) have recently been undertaken. While there is already an impressive body of work in this context, it is important to point out that most studies in this field have been in non-dynamical settings, and existing dynamical studies have assumed symmetries to yield a tractable computational problem. As a consequence, current results have certain limitations to the applicability and generality of the physics that can be drawn from them. This leaves much room for novel future work.

- *High-speed/soliton collisions.* Many questions remain in this topic. For soliton collisions, the nature of the black hole formation threshold solution is unknown; possibilities include a "universal" gravitational critical solution irrespective of the nature of the matter, or alternatively the critical solution of the matter field that the soliton is composed of. In the infinite-boost limit, the geometry to the causal future of the shockwave collision is unknown. Very few studies of finite-boost, higher-dimensional collisions relevant to super-Planck-scale particle collisions have been conducted. In particular, only trivial topologies without brane tension have been considered, and charge has been ignored, which could be important at LHC energies. The intriguing suggestion of naked-singularity formation in grazing 5d collisions shown in [228] needs further investigation. A detailed study of the radiation emitted in large-impact-parameter encounters in 4d would allow comparison with the effective-field-theory calculations that suggest the problem simplifies in this limit [238, 239] (the difficulty with such a study is that the black holes lose little energy in the encounter, and hence numerical evolution over a very long time will be required to allow the gravitational waves to get sufficiently far ahead of the black holes, unless novel gravitational-wave-extraction methods are developed). For high-speed-collision applications to heavy-ion collisions via AdS/CFT, future work includes relaxing symmetries to model non-central collisions, and introducing refinements to allow the dual CFT to better approximate QCD (for example, trying to model effects of confinement with additional matter fields, or via dynamics in the S^5 manifold of $AdS_5 \times S^5$ that are usually assumed to be trivial).

- *Alternative theories of gravity.* Numerical relativity has also recently ventured into studying astrophysical binary systems within alternative gravity theories. Incipient investigations within scalar–tensor theories have uncovered an unexpected dynamical scalarization phenomena driven by the dynamics of binary neutron stars [312]. This phenomenon has a significant impact on the orbiting behavior with clear consequences for the gravitational wave signals from the system. In all likelihood this behavior is only a token of the rich phenomenology awaiting to be discovered upon closer inspection of relevant theories, and can have astrophysical/observational consequences.

• *Supernova.* Most core-collapse supernova simulations to date have not incorporated full general relativity. Given that the problem of the explosion mechanism(s) is still unsolved and likely to be highly sensitive to the underlying physics, making the codes fully relativistic is another crucial step in the direction of more realistic modeling of the physics of this highly complex problem.

As is clear from this list, there is no shortage of interesting applications for numerical studies. With the rapid development of numerical relativity over the past decades and its expansion to fields outside of pure classical general relativity, it is impossible to tell what a future review might have in store. At the same time, however, it is safe to predict that many exciting results will fill its pages!

Acknowledgements

This work was supported in part by CIFAR; NSERC Discovery Grants (MWC and LL); NSF grants PHY-1065710, PHY1305682 and the Simons Foundation (FP). Research at Perimeter Institute is supported by the Government of Canada through Industry Canada and by the Province of Ontario through the Ministry of Research and Innovation.

References

[1] Israel, W., and Hawking, S. W. 1987. *Three hundred years of gravitation.* Cambridge: Cambridge University Press.

[2] Hahn, S. G., and Lindquist, R. W. 1964. *Ann. Phys.,* **29**, 304–331.

[3] Misner, C. W. 1960. *Phys. Rev.,* **118**, 1110–1111.

[4] Smarr, L. L. 1975. *The structure of general relativity with a numerical illustration: the collision of two black holes.* Ph.D. thesis, University of Texas, Austin, Austin, Texas.

[5] Smarr, L. L. 1979. Basic concepts in finite differencing of partial differential equations. Page 139 of: Smarr, L. L. (ed), *Sources of gravitational radiation.* Cambridge: Cambridge University Press.

[6] Eppley, K. 1975. *The numerical evolution of the collision of two black holes.* Ph.D. thesis, Princeton University, Princeton, New Jersey.

[7] Smarr, L., *et al.* 1976. *Phys. Rev. D,* **14**, 2443–2452.

[8] Anninos, P., *et al.* 1993. *Phys. Rev. Lett.,* **71**, 2851–2854.

[9] Brügmann, B. 1999. *Int. J. Mod. Phys. D,* **8**, 85–100.

[10] Pretorius, F. 2005. *Phys. Rev. Lett.,* **95**, 121101.

[11] Campanelli, M., *et al.* 2006. *Phys. Rev. Lett.,* **96**, 111101.

[12] Baker, J. G., *et al.* 2006. *Phys. Rev. Lett.,* **96**, 111102.

[13] Pretorius, F. 2009. Binary black hole coalescence. Pages 305–369 of: Colpi, M. *et al.* (eds), *Physics of relativistic objects in compact binaries: from birth to coalescence.* Heidelberg: Springer.

[14] May, M. M., and White, R. H. 1966. *Phys. Rev.*, **141**, 1232–1241.

[15] Wilson, J. R. 1971. *Astrophys. J.*, **163**, 209.

[16] Wilson, J. R. 1979. A numerical method for relativistic hydrodynamics. Pages 423–445 of: Smarr, L. L. (ed), *Sources of gravitational radiation.* Cambridge: Cambridge University Press.

[17] Shapiro, S. L., and Teukolsky, S. A. 1980. *Astrophys. J.*, **235**, 199–215.

[18] Stark, R. F., and Piran, T. 1985. *Phys. Rev. Lett.*, **55**, 891–894. Erratum: ibid. **56**, 97 (1986).

[19] Nakamura, T. 1981. *Prog. Theor. Phys.*, **65**, 1876–1890.

[20] Nakamura, T. 1983. *Prog. Theor. Phys.*, **70**, 1144–1147.

[21] Evans, C. R. 1986. An approach for calculating axisymmetric gravitational collapse. Pages 3–39 of: Centrella, J. M. (ed), *Dynamical spacetimes and numerical relativity.* Cambridge: Cambridge University Press.

[22] Shapiro, S. L., and Teukolsky, S. A. 1991. *Phys. Rev. Lett.*, **66**, 994–997.

[23] Thorne, K. S. 1972. Nonspherical gravitational collapse: a short review. Page 231 of: Klauder, J. (ed), *Magic without magic: John Archibald Wheeler.* San Francisco, CA: Freeman.

[24] Piran, T. 1980. *J. Comp. Phys.*, **35**, 254–283.

[25] Cook, G. B. 2000. *Living Rev. Rel.*, **3**.

[26] Gourgoulhon, E. 2007. *J. Phys.: Conf. Ser.*, **91**, 012001.

[27] Pfeiffer, H. P. 2005. *Initial data for black hole evolutions.* Ph.D. thesis, Cornell University, Ithaca, New York.

[28] York Jr., J. W. 1979. Kinematics and dynamics of general relativity. Pages 83–126 of: Smarr, L. L. (ed), *Sources of gravitational radiation.* Cambridge: Cambridge University Press.

[29] York Jr., J. W., and Piran, T. 1982. The initial value problem and beyond. Pages 147–176 of: Matzner, R. A., and Shepley, L. C. (eds), *Spacetime and geometry: the Alfred Schild lectures.* Austin, TX: University of Texas Press.

[30] Arnowitt, R., Deser, S., and Misner, C. W. 1962. The dynamics of general relativity. Pages 227–265 of: Witten, L. (ed), *Gravitation: an introduction to current research.* New York: Wiley.

[31] Bowen, J. M., and York Jr., J. W. 1980. *Phys. Rev. D*, **21**, 2047–2056.

[32] Brandt, S. R., and Brügmann, B. 1997. *Phys. Rev. Lett.*, **78**, 3606–3609.

[33] Thornburg, J. 1987. *Class. Quant. Grav.*, **4**, 1119–1131.

[34] Centrella, J. M. 1980. *Phys. Rev. D*, **21**, 2776–2784.

[35] Centrella, J. M., and Wilson, J. R. 1984. *Astrophys. J. Suppl. Ser.*, **54**, 229–249.

[36] Anninos, P., Centrella, J. M., and Matzner, R. A. 1991. *Phys. Rev. D*, **43**, 1808.

[37] Kurki-Suonio, H., Laguna, P., and Matzner, R. A. 1993. *Phys. Rev. D*, **48**, 3611–3624.

[38] Berger, B. K., and Moncrief, V. 1993. *Phys. Rev. D*, **48**, 4676–4687.

[39] Kaup, D. J. 1968. *Phys. Rev.*, **172**, 1331–1342.

[40] Ruffini, R., and Bonazzola, S. 1969. *Phys. Rev.*, **187**, 1767–1783.

[41] Colpi, M., Shapiro, S. L., and Wasserman, I. 1986. *Phys. Rev. Lett.*, **57**, 2485–2488.

[42] Liebling, S. L., and Palenzuela, C. 2012. *Living Rev. Rel.*, **15**, 6.

[43] Seidel, E., and Suen, W.-M. 1992. *Phys. Rev. Lett.*, **69**, 1845–1848.

[44] Bona, C., and Massó, J. 1993. A vacuum fully relativistic 3D numerical code. Pages 258–264 of: d'Inverno, R. A. (ed), *Approaches to numerical relativity.* Cambridge: Cambridge University Press.

[45] Cook, G. B., *et al.* 1998. *Phys. Rev. Lett.*, **80**, 2512–2516.

[46] Abrahams, A. M., *et al.* 1998. *Phys. Rev. Lett.*, **80**, 1812–1815.

[47] Gómez, R., *et al.* 1998. *Phys. Rev. Lett.*, **80**, 3915–3918.

[48] Lehner, L. 2001. *Class. Quant. Grav.*, **18**, R25–R86.

[49] Bona, C., and Palenzuela, C. (eds). 2005. *Elements of numerical relativity.* Lecture Notes in Physics, vol. 673. Berlin/Heidelberg: Springer.

[50] Alcubierre, M. 2008. *Introduction to 3+1 numerical relativity.* Oxford: Oxford University Press.

[51] Baumgarte, T. W., and Shapiro, S. L. 2010. *Numerical relativity: solving Einstein's equations on the computer*. Cambridge: Cambridge University Press.

[52] Gourgoulhon, E. (ed). 2012. *3+1 Formalism in general relativity*. Lecture Notes in Physics, vol. 846. Berlin: Springer.

[53] Kreiss, H.-O., and Ortiz, O. E. 2002. Some mathematical and numerical questions connected with first and second order time-dependent systems of partial differential equations. Pages 359–370 of: Frauendiener, J., and Friedrich, H. (eds), *The conformal structure of space-time*. Lecture Notes in Physics, vol. 604. Berlin: Springer.

[54] Gustafsson, B., Kreiss, H.-O., and Oliger, J. 1995. *Time dependent problems and difference methods*. New York: Wiley.

[55] Sarbach, O., and Tiglio, M. 2012. *Living Rev. Rel.*, **15**, 9.

[56] Friedrich, H., and Rendall, A. D. 2000. The Cauchy problem for the Einstein equations. Pages 127–223 of: Schmidt, B. G. (ed), *Einstein's field equations and their physical Lecture Notes in Physics*, vol. 540. Berlin: Springer.

[57] Reula, O. A. 1998. *Living Rev. Rel.*, **1**.

[58] Alic, D., *et al.* 2012. *Phys. Rev. D*, **85**, 064040.

[59] Nakamura, T., Oohara, K.-I., and Kojima, Y. 1987. *Prog. Theor. Phys. Suppl.*, **90**, 1–218.

[60] Shibata, M., and Nakamura, T. 1995. *Phys. Rev. D*, **52**, 5428–5444.

[61] Baumgarte, T. W., and Shapiro, S. L. 1998. *Phys. Rev. D*, **59**, 024007.

[62] Renn, J., and Sauer, T. 1999. Heuristics and mathematical representation in Einstein's search for a gravitational field equation. Page 87 of: Goenner, H., Renn, J., Ritter, J., and Sauer, T. (eds), *The expanding worlds of general relativity*. Basel: Birkhäuser.

[63] Lindblom, L., *et al.* 2006. *Class. Quant. Grav.*, **23**, S447–S462.

[64] Garfinkle, D. 2002. *Phys. Rev. D*, **65**, 044029.

[65] Gundlach, C., *et al.* 2005. *Class. Quant. Grav.*, **22**, 3767–3774.

[66] Brodbeck, O., *et al.* 1999. *J. Math. Phys.*, **40**, 909–923.

[67] Palenzuela, C., Lehner, L., and Yoshida, S. 2010. *Phys. Rev. D*, **81**, 084007.

[68] Headrick, M., Kitchen, S., and Wiseman, T. 2010. *Class. Quant. Grav.*, **27**, 035002.

[69] Hannam, M. D. *et al.* 2007. *Phys. Rev. Lett.*, **99**, 241102.

[70] Hannam, M. D. *et al.* 2008. *Phys. Rev. D*, **78**, 064020.

[71] Winicour, J. 1998. *Living Rev. Rel.*, **1**.

[72] Gómez, R., *et al.* 1998. *Phys. Rev. D*, **57**, 4778–4788.

[73] Bishop, N. T., *et al.* 1997. *Phys. Rev. D*, **54**, 6153–6165.

[74] Reisswig, C., *et al.* 2010. *Class. Quant. Grav.*, **27**, 075014.

[75] Brady, P. R., and Smith, J. D. 1995. *Phys. Rev. Lett.*, **75**, 1256–1259.

[76] Chesler, P. M., and Yaffe, L. G., 2013. arXiv:1309.1439.

[77] Friedrich, H. 2002. Conformal Einstein evolution. Pages 1–50 of: Frauendiener, J., and Friedrich, H. (eds), *The conformal structure of space-time*. Lecture Notes in Physics, vol. 604. Berlin: Springer.

[78] Frauendiener, J. 1998. *Phys. Rev. D*, **58**, 064002.

[79] Husa, S. 2002. Problems and successes in the numerical approach to the conformal field equations. Pages 239–260 of: Frauendiener, J., and Friedrich, H. (eds), *The conformal structure of space-time*. Lecture Notes in Physics, vol. 604. Berlin: Springer.

[80] Berger, M. J., and Oliger, J. 1984. *J. Comput. Phys.*, **53**, 484.

[81] Choptuik, M. W. 1989. Experiences with an adaptive mesh refinement algorithm in numerical relativity. In: Evans, C. R., Finn, L. S., and Hobill, D. W. (eds), *Frontiers in numerical relativity*. Cambridge: Cambridge University Press.

[82] Lehner, L., Liebling, S. L., and Reula, O. A. 2006. *Class. Quant. Grav.*, **23**, S421–S446.

[83] Boyd, J. P. 1989. *Chebyshev and Fourier spectral methods*. New York: Springer-Verlag.

[84] Grandclement, P., and Novak, J. 2009. *Living Rev. Rel.*, **12**.

[85] Szilágyi, B., Lindblom, L., and Scheel, M. A. 2009. *Phys. Rev. D*, **80**, 124010.

[86] Font, J. A. 2008. *Living Rev. Rel.*, **11**, 7.

[87] LeVeque, R. J. 1992. *Numerical methods for conservation laws*. Basel: Birkhäuser Verlag.

[88] Cardoso, V., Gualtieri, L., Herdeiro, C., and Sperhake, U. 2014. *Exploring new physics frontiers through numerical relativity*. To appear in Living Reviews in Relativity.
[89] Evans, C. R., and Coleman, J. S. 1994. *Phys. Rev. Lett.*, **72**, 1782–1785.
[90] Koike, T., Hara, T., and Adachi, S. 1995. *Phys. Rev. Lett.*, **74**, 5170–5173.
[91] Maison, D. 1996. *Phys. Lett.*, **B366**, 82–84.
[92] Gundlach, C., and Martin-Garcia, J. M. 2007. *Living Rev. Rel.*, **10**, 5.
[93] Gundlach, C. 1998. *Adv. Theor. Math. Phys.*, **2**, 1–49.
[94] Choptuik, M. W. 1993. *Phys. Rev. Lett.*, **70**, 9–12.
[95] Gundlach, C. 1997. *Phys. Rev. D*, **55**, 695–713.
[96] Hod, S., and Piran, T. 1997. *Phys. Rev. D*, **55**, 3485–3496.
[97] Martin-Garcia, J. M., and Gundlach, C. 1999. *Phys. Rev. D*, **59**, 064031.
[98] Choptuik, M. W., Hirschmann, E. W., Liebling, S. L., and Pretorius, F. 2003. *Phys. Rev. D*, **68**, 044007.
[99] Healy, J., and Laguna, P., 2013. arXiv:1310.1955.
[100] Brady, P. R., Chambers, C. M., and Goncalves, S. M. 1997. *Phys. Rev. D*, **56**, 6057–6061.
[101] Seidel, E., and Suen, W. 1991. *Phys. Rev. Lett.*, **66**, 1659–1662.
[102] Hawley, S. H., and Choptuik, M. W. 2000. *Phys. Rev. D*, **62**, 104024.
[103] Husain, V., and Olivier, M. 2001. *Class. Quant. Grav.*, **18**, L1–L10.
[104] Pretorius, F., and Choptuik, M. W. 2000. *Phys. Rev. D*, **62**, 124012.
[105] Bizon, P., and Rostworowski, A. 2011. *Phys. Rev. Lett.*, **107**, 031102.
[106] Garfinkle, D. 2001. *Phys. Rev. D*, **63**, 044007.
[107] Abrahams, A. M., and Evans, C. R. 1993. *Phys. Rev. Lett.*, **70**, 2980–2983.
[108] Sorkin, E. 2011. *Class. Quant. Grav.*, **28**, 025011.
[109] Garfinkle, D., and Duncan, G. C. 1998. *Phys. Rev. D*, **58**, 064024.
[110] Bizon, P., Chmaj, T., and Schmidt, B. G. 2005. *Phys. Rev. Lett.*, **95**, 071102.
[111] Bizon, P., Chmaj, T., and Schmidt, B. G. 2006. *Phys. Rev. Lett.*, **97**, 131101.
[112] Bizon, P., *et al.* 2005. *Phys. Rev. D*, **72**, 121502.
[113] Szybka, S. J., and Chmaj, T. 2008. *Phys. Rev. Lett.*, **100**, 101102.
[114] Gundlach, C. 1998. *Phys. Rev. D*, **57**, 7080–7083.
[115] Cahill, M. E., and Taub, A. H. 1971. *Commun. Math. Phys.*, **21**, 1–40.
[116] Noble, S. C. 2003. *A numerical study of relativistic fluid collapse*. Ph.D. thesis, The University of British Columbia, Vancouver, British Columbia. [arXiv:gr-qc/0310116].
[117] Neilsen, D. W., and Choptuik, M. W. 2000. *Class. Quant. Grav.*, **17**, 761–782.
[118] Noble, S. C., and Choptuik, M. W. 2008. *Phys. Rev. D*, **78**, 064059.
[119] Novak, J. 2001. *Astron. Astrophys.*, **376**, 606–613.
[120] Niemeyer, J. C., and Jedamzik, K. 1998. *Phys. Rev. Lett.*, **80**, 5481–5484.
[121] Jin, K.-J., and Suen, W.-M. 2007. *Phys. Rev. Lett.*, **98**, 131101.
[122] Kellermann, T., Rezzolla, L., and Radice, D. 2010. *Class. Quant. Grav.*, **27**, 235016.
[123] Radice, D., Rezzolla, L., and Kellermann, T. 2010. *Class. Quant. Grav.*, **27**, 235015.
[124] Wan, M.-B. 2011. *Class. Quant. Grav.*, **28**, 155002.
[125] Liebling, S. L., *et al.* 2010. *Phys. Rev. D*, **81**, 124023.
[126] Choptuik, M. W., Chmaj, T., and Bizon, P. 1996. *Phys. Rev. Lett.*, **77**, 424–427.
[127] Bartnik, R., and McKinnon, J. 1988. *Phys. Rev. Lett.*, **61**, 141–144.
[128] Choptuik, M. W., Hirschmann, E. W., and Marsa, R. L. 1999. *Phys. Rev. D*, **60**, 124011.
[129] Liebling, S. L., and Choptuik, M. W. 1996. *Phys. Rev. Lett.*, **77**, 1424–1427.
[130] Lechner, C., *et al.* 2002. *Phys. Rev. D*, **65**, 081501.
[131] Andreasson, H., and Rein, G. 2006. *Class. Quant. Grav.*, **23**, 3659–3678.
[132] Olabarrieta, I., and Choptuik, M. W. 2002. *Phys. Rev. D*, **65**, 024007.
[133] Rein, G., Rendall, A. D., and Schaeffer, J. 1998. *Phys. Rev. D*, **58**, 044007.
[134] Martin-Garcia, J. M., and Gundlach, C. 2002. *Phys. Rev. D*, **65**, 084026.
[135] Centrella, J., *et al.* 2010. *Ann. Rev. Nucl. Particle Sci.*, **60**, 75–100.
[136] Hinder, I., *et al.* 2013. *Class. Quant. Grav.*, **31**, 025012.
[137] Ajith, P., *et al.* 2011. *Phys. Rev. Lett.*, **106**, 241101.
[138] Pan, Y., *et al.*, 2013. arXiv:1307.6232.

[139] Damour, T., Nagar, A., and Bernuzzi, S. 2013. *Phys. Rev. D*, **87**, 084035.
[140] Buonanno, A., Cook, G. B., and Pretorius, F. 2007. *Phys. Rev. D*, **75**, 124018.
[141] Bishop, N. T., *et al.* 1997. *Phys. Rev. D*, **56**, 6298–6309.
[142] Chu, T., Pfeiffer, H. P., and Cohen, M. I. 2011. *Phys. Rev. D*, **83**, 104018.
[143] Gonzalez, J. A. *et al.* 2007. *Phys. Rev. Lett.*, **98**, 091101.
[144] Baker, J. G., *et al.* 2006. *Astrophys. J.*, **653**, L93–L96.
[145] Herrmann, F., *et al.* 2007. *Class. Quant. Grav.*, **24**, 33.
[146] Berti, E. *et al.* 2007. *Phys. Rev. D*, **76**, 064034.
[147] Campanelli, M., Lousto, C., and Zlochower, Y. 2006. *Phys. Rev. D*, **74**, 041501.
[148] Hemberger, D. A., *et al.* 2013. *Phys. Rev. D*, **88**, 064014.
[149] Schmidt, P., *et al.* 2011. *Phys. Rev. D*, **84**, 024046.
[150] Boyle, M., Owen, R., and Pfeiffer, H. P. 2011. *Phys. Rev. D*, **84**, 124011.
[151] O'Shaughnessy, R., *et al.* 2011. *Phys. Rev. D*, **84**, 124002.
[152] Campanelli, M., *et al.* 2007. *Astrophys. J. Lett.*, **659**, L5–L8.
[153] González, J. A., *et al.* 2007. *Phys. Rev. Lett.*, **98**, 231101.
[154] Lousto, C. O., and Zlochower, Y. 2011. *Phys. Rev. Lett.*, **107**, 231102.
[155] Komossa, S. 2012. *Adv. Astron.*, **2012**.
[156] Schnittman, J. D. 2013. *Class. Quant. Grav.*, **30**, 244007.
[157] Komossa, S., Zhou, H., and Lu, H. 2008. *Astrophys. J.*, **678**, L81–L84.
[158] Lippai, Z., Frei, Z., and Haiman, Z. 2008. *Astrophys. J.*, **676**, L5–L8.
[159] Milosavljevic, M., and Phinney, E. 2005. *Astrophys. J.*, **622**, L93–L96.
[160] Loeb, A. 2007. *Phys. Rev. Lett.*, **99**, 041103.
[161] Noble, S. C., *et al.* 2012. *Astrophys. J.*, **755**, 51.
[162] Farris, B. D., *et al.* 2012. *Phys. Rev. Lett.*, **109**, 221102.
[163] Stone, N., and Loeb, A. 2011. *Mon. Not. Roy. Astr. Soc.*, **412**, 75–80.
[164] Palenzuela, C., Lehner, L., and Leibling, S. L. 2010. *Science*, **329**, 927–930.
[165] East, W. E., McWilliams, S. T., Levin, J., and Pretorius, F. 2013. *Phys. Rev. D*, **87**, 043004.
[166] Pretorius, F., and Khurana, D. 2007. *Class. Quant. Grav.*, **24**, S83–S108.
[167] Healy, J., Levin, J., and Shoemaker, D. 2009. *Phys. Rev. Lett.*, **103**, 131101.
[168] Gold, R., and Bruegmann, B. 2013. *Phys. Rev. D*, **88**, 064051.
[169] Metzger, B., and Berger, E. 2012. *Astrophys. J.*, **746**, 48.
[170] Piran, T., Nakar, E., and Rosswog, S. 2013. *Mon. Not. R. Astron. Soc.*, **430**, 2121–2136.
[171] Shibata, M., and Uryu, K. 2000. *Phys. Rev. D*, **61**, 064001.
[172] Nakamura, T., and Oohara, K.-I. 1999. *A way to 3D numerical relativity – coalescing binary neutron stars.* arXiv:gr-qc/9812054.
[173] Read, J. S. *et al.* 2013. *Phys. Rev. D*, **88**, 044042.
[174] Lackey, B. D., *et al.*, 2013. arXiv:1303.6298.
[175] Tsang, D., *et al.* 2012. *Phys. Rev. Lett.*, **108**, 011102.
[176] Hotokezaka, K., *et al.* 2011. *Phys. Rev. D*, **83**, 124008.
[177] Anderson, M., *et al.* 2008. *Phys. Rev. Lett.*, **100**, 191101.
[178] Sekiguchi, Y., *et al.*, 2012. arXiv:1206.5927.
[179] Kaplan, J., *et al.* 2013. *Phys. Rev. D*, **88**, 064009.
[180] Rezzolla, L. *et al.* 2010. *Class. Quant. Grav.*, **27**, 114105.
[181] Tanvir, N. R., *et al.* 2013. *Nature*, **500**, 547–549.
[182] Berger, E., Fong, W., and Chornock, R. 2013. *Astrophys. J. Lett.*, **774**, L23.
[183] Hinderer, T., *et al.* 2010. *Phys. Rev. D*, **81**, 123016.
[184] Markakis, C., *et al.* 2009. *J. Phys.: Conf. Ser.*, **189**, 012024.
[185] Sekiguchi, Y., *et al.* 2011. *Phys. Rev. Lett.*, **107**, 051102.
[186] Lehner, L., *et al.* 2012. *Phys. Rev. D*, **86**, 104035.
[187] Kyutoku, K., Ioka, K., and Shibata, M. 2014. *Mon. Not. Roy. Astron. Soc.*, **437**, L6. [arXiv:1209.5747].
[188] Palenzuela, C., *et al.* 2013. *Phys. Rev. Lett.*, **111**, 061105.
[189] Hotokezaka, K., *et al.* 2013. *Phys. Rev. D*, **88**, 044026.
[190] Kyutoku, K., Ioka, K., and Shibata, M. 2013. *Phys. Rev. D*, **88**, 041503.

[191] Chawla, S., *et al.* 2010. *Phys. Rev. Lett.*, **105**, 111101.
[192] Foucart, F., *et al.* 2013. *Phys. Rev. D*, **87**, 084006.
[193] Foucart, F. 2012. *Phys. Rev. D*, **86**, 124007.
[194] Hansen, B. M., and Lyutikov, M. 2001. *Mon. Not. Roy. Astron. Soc.*, **322**, 695.
[195] McWilliams, S. T., and Levin, J. 2011. *Astrophys. J.*, **742**, 90.
[196] Paschalidis, V., Etienne, Z. B., and Shapiro, S. L. R. *Phys. Rev. D*, **88**, 021504.
[197] Lackey, B. D., *et al.* 2012. *Phys. Rev. D*, **85**, 044061.
[198] Stephens, B. C., East, W. E., and Pretorius, F. 2011. *Astrophys. J. Lett.*, **737**, L5.
[199] Gold, R., *et al.* 2012. *Phys. Rev. D*, **86**, 121501.
[200] Tsang, D. 2013. *Astrophys. J.*, **777**, 103.
[201] Duez, M. D. 2010. *Class. Quant. Grav.*, **27**, 114002.
[202] Pfeiffer, H. P. 2012. *Class. Quant. Grav.*, **29**, 124004.
[203] Faber, J. A., and Rasio, F. A. 2012. *Living Rev. Rel.*, **15**, 8.
[204] Ott, C. D. 2009. *Class. and Quant. Grav.*, **26**, 063001.
[205] Janka, H.-T., *et al.* 2007. *Phys. Reports*, **442**, 38–74.
[206] Burrows, A., *et al.* 2007. *Phys. Reports*, **442**, 23–37.
[207] Ott, C. D., *et al.* 2013. *Astrophys. J.*, **768**, 115.
[208] Dimmelmeier, H., Font, J. A., and Müller, E. 2002. *Astron. Astrophys.*, **388**, 917–935.
[209] Obergaulinger, M., *et al.* 2006. *Astron. Astrophys.*, **457**, 209–222.
[210] Müller, B., Janka, H.-T., and Dimmelmeier, H. 2010. *Astrophys. J. Suppl. Ser.*, **189**, 104–133.
[211] Wongwathanarat, A., Janka, H.-T., and Mueller, E. 2013. *Astron. Astrophys.*, **552**, A126.
[212] Ott, C. D., *et al.* 2011. *Phys. Rev. Lett.*, **106**, 161103.
[213] Penrose, R. 1966. General relativistic energy flux and elementary optics. In: Hoffman, B. (ed), *Perspectives in geometry and relativity*. Indiana University Press.
[214] Aichelburg, P., and Sexl, R. 1971. *Gen. Rel. Grav.*, **2**, 303–312.
[215] Arkani-Hamed, N., Dimopoulos, S., and Dvali, G. 1998. *Phys. Lett. B*, **429**, 263–272.
[216] Randall, L., and Sundrum, R. 1999. *Phys. Rev. Lett.*, **83**, 3370–3373.
[217] Giddings, S. B., and Thomas, S. D. 2002. *Phys. Rev. D*, **65**, 056010.
[218] Feng, J. L., and Shapere, A. D. 2002. *Phys. Rev. Lett.*, **88**, 021303.
[219] Chatrchyan, S., *et al.* [CMS Collaboration]. 2013. *JHEP*, **1307**, 178.
[220] Aad, G., *et al.* [ATLAS Collaboration]. 2013. *Phys. Rev. D*, **88**, 072001.
[221] de los Heros, C. 2007. *ArXiv Astrophysics e-prints*.
[222] Thorne, K. S. 1972. Nonspherical gravitational collapse – a short review. Page 231 of: Klauder, J. (ed), *Magic without magic: John Archibald Wheeler*. San Francisco, CA: Freeman.
[223] Sperhake, U. *et al.* 2008. *Phys. Rev. Lett.*, **101**, 161101.
[224] Shibata, M., Okawa, H., and Yamamoto, T. 2008. *Phys. Rev. D*, **78**, 101501.
[225] Sperhake, U., *et al.* 2009. *Phys. Rev. Lett.*, **103**, 131102.
[226] Sperhake, U., *et al.* 2011. *Phys. Rev. D*, **83**, 024037.
[227] Sperhake, U., Berti, E., Cardoso, V., and Pretorius, F. 2013. *Phys. Rev. Lett.*, **111**, 041101.
[228] Okawa, H., Nakao, K.-i., and Shibata, M. 2011. *Phys. Rev. D*, **83**, 121501.
[229] D'eath, P. D., and Payne, P. N. 1992. *Phys. Rev. D*, **46**, 694–701.
[230] Eardley, D. M., and Giddings, S. B. 2002. *Phys. Rev. D*, **66**, 044011.
[231] Berti, E. *et al.* 2010. *Phys. Rev. D*, **81**, 104048.
[232] Gundlach, C., *et al.* 2012. *Phys. Rev. D*, **86**, 084022.
[233] Gralla, S. E., Harte, A. I., and Wald, R. M. 2010. *Phys. Rev. D*, **81**, 104012.
[234] Choptuik, M. W., and Pretorius, F. 2010. *Phys. Rev. Lett.*, **104**, 111101.
[235] East, W. E., and Pretorius, F. 2013. *Phys. Rev. Lett.*, **110**, 101101.
[236] Rezzolla, L., and Takami, K. 2013. *Class. Quant. Grav.*, **30**, 012001.
[237] Kaloper, N., and Terning, J. 2008. *Int. J. Mod. Phys.*, **D17**, 665–672.
[238] Galley, C. R., and Porto, R. A. 2013. *JHEP*, **1311**, 096.
[239] Gal'tsov, D., Spirin, P., and Tomaras, T. N. 2013. *JHEP*, **1301**, 087.
[240] Grumiller, D., and Romatschke, P. 2008. *JHEP*, **0808**, 027.
[241] Casalderrey-Solana, J., Heller, M. P., Mateos, D., and van der Schee, W. 2013. *Phys. Rev. Lett.*, **111**, 181601.

[242] Cardoso, V. *et al.* 2012. *Class. Quant. Grav.*, **29**, 244001. [arXiv:1201.5118].
[243] Antoniadis, I., *et al.* 1998. *Phys. Lett. B*, **436**, 257–263.
[244] Baumann, D., 2009. arXiv:0907.5424.
[245] Lehners, J.-L. 2008. *Phys. Rept.*, **465**, 223–263.
[246] Carlip, S. 2005. *Living Rev. Rel.*, **8**, 1.
[247] Gegenberg, J., and Kunstatter, G., 2009. arXiv:0902.0292.
[248] Maldacena, J. M. 1998. *Adv. Theor. Math. Phys.*, **2**, 231–252.
[249] Aharony, O., *et al.* 2000. *Phys. Rept.*, **323**, 183–386.
[250] Horowitz, G. T., 2012. *Black holes in higher dimensions*. Cambridge: Cambrdige University Press.
[251] Emparan, R., and Reall, H. S. 2008. *Living Rev. Rel.*, **11**, 6.
[252] Reall, H. S., 2012. arXiv:1210.1402.
[253] Emparan, R., and Myers, R. C. 2003. *JHEP*, **0309**, 025.
[254] Lehner, L., and Pretorius, F. 2010. *Phys. Rev. Lett.*, **105**, 101102.
[255] Shibata, M., and Yoshino, H. 2010. *Phys. Rev. D*, **81**, 021501.
[256] Shibata, M., and Yoshino, H. 2010. *Phys. Rev. D*, **81**, 104035.
[257] Gregory, R., and Laflamme, R. 1993. *Phys. Rev. Lett.*, **70**, 2837–2840.
[258] Eggers, J. 1993. *Phys. Rev. Lett*, **71**, 3458.
[259] Sorkin, E. 2004. *Phys. Rev. Lett.*, **93**, 031601.
[260] Figueras, P., Murata, K., and Reall, H. S. 2012. *JHEP*, **1211**, 071.
[261] Dias, O. J., *et al.* 2009. *Phys. Rev. D*, **80**, 111701.
[262] Figueras, P., and Wiseman, T. 2011. *Phys. Rev. Lett.*, **107**, 081101.
[263] Tanaka, T. 2003. *Prog. Theor. Phys. Suppl.*, **148**, 307–316.
[264] Emparan, R., Fabbri, A., and Kaloper, N. 2002. *JHEP*, **0208**, 043.
[265] Son, D. T., and Starinets, A. O. 2007. *Ann. Rev. Nucl. Part. Sci.*, **57**, 95–118.
[266] DeWolfe, O., *et al.*, 2013. arXiv:1304.7794.
[267] Chesler, P. M., and Yaffe, L. G. 2009. *Phys. Rev. Lett.*, **102**, 211601.
[268] Chesler, P. M., and Teaney, D., 2011. arXiv:1112.6196.
[269] Chesler, P. M., and Yaffe, L. G. 2011. *Phys. Rev. Lett.*, **106**, 021601.
[270] Bjorken, J. 1983. *Phys. Rev. D*, **27**, 140–151.
[271] Luzum, M., and Romatschke, P. 2008. *Phys. Rev. C*, **78**, 034915.
[272] Buchel, A., Lehner, L., and Myers, R. C. 2012. *JHEP*, **1208**, 049.
[273] Buchel, A., *et al.* 2013. *JHEP*, **1305**, 067.
[274] Buchel, A., Myers, R. C., and van Niekerk, A. 2013. *Phys. Rev. Lett.*, **111**, 201602.
[275] Hubeny, V. E., Minwalla, S., and Rangamani, M., 2011. arXiv:1107.5780.
[276] Van Raamsdonk, M. 2008. *JHEP*, **0805**, 106.
[277] Carrasco, F., *et al.* 2012. *Phys. Rev. D*, **86**, 126006.
[278] Adams, A., Chesler, P. M., and Liu, H., 2013. arXiv:1307.7267.
[279] Bantilan, H., Pretorius, F., and Gubser, S. S. 2012. *Phys. Rev. D*, **85**, 084038.
[280] Green, S. R., Carrasco, F., and Lehner, L., 2013. arXiv:1309.7940.
[281] Berger, B. K. 2002. *Living Rev. Rel.*, **5**, 1.
[282] Belinskii, V., Lifshitz, E., and Khalatnikov, I. 1972. *Zh. Eksp. Teor. Fiz.*, **62**, 1606–1613.
[283] Barrow, J. D., and Tipler, F. J. 1979. *Phys. Rept.*, **56**, 371–402.
[284] Berger, B., *et al.* 1998. *Mod. Phys. Lett.*, **A13**, 1565–1574.
[285] Garfinkle, D. 2004. *Phys. Rev. Lett.*, **93**, 161101.
[286] Lim, W. C., *et al.* 2009. *Phys. Rev. D*, **79**, 123526.
[287] Poisson, E., and Israel, W. 1990. *Phys. Rev. D*, **41**, 1796–1809.
[288] Ori, A., and Flanagan, É. É. 1996. *Phys. Rev. D*, **53**, 1754.
[289] Dafermos, M., 2012. arXiv:1201.1797.
[290] Christodoulou, D., and Klainerman, S. 1993. *The global nonlinear stability of the Minkowski space*. Princeton, NJ: Princeton University Press.
[291] Friedrich, H. 1986. *J. Geom. Phys.*, **3**, 101–117.
[292] Dias, O. J., Horowitz, G. T., and Santos, J. E. 2012. *Class. Quant. Grav.*, **29**, 194002.
[293] Buchel, A., Liebling, S. L., and Lehner, L. 2013. *Phys. Rev. D*, **87**, 123006.

[294] Dias, O. J., *et al.* 2012. *Class. Quant. Grav.*, **29**, 235019.
[295] Maliborski, M., and Rostworowski, A., 2013. arXiv:1303.3186.
[296] Callan, C. G., *et al.* 1992. *Phys. Rev. D*, **45**, 1005–1009.
[297] Ashtekar, A., Pretorius, F., and Ramazanoğlu, F. M. 2011. *Phys. Rev. Lett.*, **106**, 161303.
[298] Ashtekar, A., Pretorius, F., and Ramazanoğlu, F. M. 2011. *Phys. Rev. D*, **83**, 044040.
[299] Ramazanoğlu, F. M., and Pretorius, F. 2010. *Class. Quant. Grav.*, **27**, 245027.
[300] Ashtekar, A., Taveras, V., and Varadarajan, M. 2008. *Phys. Rev. Lett.*, **100**, 211302.
[301] Almheiri, A., *et al.* 2013. *JHEP*, **1302**, 062.
[302] Guth, A. H. 2007. *J. Phys.*, **A40**, 6811–6826.
[303] Aguirre, A., and Johnson, M. C. 2011. *Rept. Prog. Phys.*, **74**, 074901.
[304] Kleban, M. 2011. *Class. Quant. Grav.*, **28**, 204008.
[305] Johnson, M. C., Peiris, H. V., and Lehner, L. 2012. *Phys. Rev. D*, **85**, 083516.
[306] Wainwright, C. L., *et al.*, 2013. arXiv:1312.1357.
[307] Buchert, T., and Räsänen, S. 2012. *Ann. Rev. Nucl. Particle Sci.*, **62**, 57–79.
[308] Zhao, X., and Mathews, G. J. 2011. *Phys. Rev. D*, **83**, 023524.
[309] Yoo, C.-M., Okawa, H., and Nakao, K.-i. 2013. *Phys. Rev. Lett.*, **111**, 161102.
[310] Yoo, C.-M., and Okawa, H. 2014. *ArXiv e-prints*.
[311] East, W. E., Ramazanoğlu, F. M., and Pretorius, F. 2014. *Phys. Rev. D*, **89**, 061503.
[312] Barausse, E., *et al.* 2013. *Phys. Rev. D*, **87**, 081506.
[313] Chesler, P. M., and Yaffe, L. G. 2015. arXiv:1501.04644.

8

Initial Data and the Einstein Constraint Equations

GREGORY J. GALLOWAY, PENGZI MIAO AND RICHARD SCHOEN

8.1 Introduction

Solutions of the Einstein equations evolve from initial data given on a three-dimensional manifold M. The initial position and velocity of the gravitational field are given by a Riemannian metric g and a symmetric $(0, 2)$ tensor K. The metric g will be the metric induced on M as a spacelike hypersurface in the spacetime S which evolves from the data, and the tensor K will be the second fundamental form of M in S. Thus an initial data set is given by a triple (M, g, K). There is currently interest in higher-dimensional gravity in the physics community, so when convenient we will discuss initial data on an n-dimensional manifold M^n which will evolve to an $(n + 1)$-dimensional spacetime S^{n+1} $(n \geq 3)$.

A basic fact of life for the Einstein equations is that the initial data g and K cannot be freely specified, but must satisfy a system of $n + 1$ nonlinear partial differential equations. These are called the constraint equations, and Section 8.2 deals with recent progress on solving this set of equations. On the one hand the constraint equations present a complication in the study of the initial value problem since it is a difficult (and as yet unsolved) problem to fully analyze their solutions. On the other hand, it is because of the constraint equations that physical notions of energy and momentum can be defined. It is also because of them that geometric and topological restrictions hold in certain cases on the initial manifold M, and for black holes in Σ.

We do not have the space here to give a comprehensive survey of the initial value problem, so instead we have focused on several questions on which there has been recent progress and which are currently active areas of investigation. We have chosen to give brief outlines of the main ideas involved in the study of these specific questions rather than to attempt to touch on all aspects of the field. For more detailed information on various aspects of the subject we refer the reader to the recent surveys of the subject by Corvino and Pollack [38] and by Bartnik and Isenberg [15] as well as the influential survey by Choquet-Bruhat and York [31].

In Section 8.2 we begin with a brief introduction to the constraint equations. We then introduce the conformal method for solving the equations and we describe recent work on solving these equations on compact manifolds M. We then introduce asymptotically flat initial data and discuss the asymptotic behavior of solutions. We explain several theorems which show that one can approximate general solutions by solutions which have very nice asymptotics.

In Section 8.3 we discuss notions of energy and momentum for the gravitational field. We begin by recalling the total ADM energy and momentum for an isolated system and the positive mass theorem. We then give an account of several quasi-local notions provided by the Hamilton–Jacobi approach, with an emphasis on the recently proposed Wang–Yau mass. In connection with the Penrose inequality, we also discuss monotonicity results for the Hawking mass along geometric deformations in spacetimes. We end this section by describing the Bartnik mass and related questions.

In Section 8.4 we discuss the theory of marginally outer trapped surfaces (MOTS), which arise in connection with black hole formation. We describe the basic existence theorems which have been developed in recent years. We also introduce the important notion of stability for MOTS, which is responsible for the geometric and topological restrictions on black holes. We then describe recent work on topological censorship, the restrictions on the topology of MOTS-free initial data sets, and the use of MOTS to prove the spacetime positive mass theorem.

8.2 The Einstein constraint equations

The Einstein equations in an $(n + 1)$-dimensional spacetime Σ may be written

$$Ric(\mathfrak{g}) - \frac{1}{2} R(\mathfrak{g})\mathfrak{g} = T,$$

where $\mathfrak{g}$ is the gravitational field, a Lorentz signature metric on $\mathcal{S}$, and T is the stress-energy tensor of any matter fields which are present. This is a system of $(n + 1)(n + 2)/2$ equations which determine the evolution of $\mathfrak{g}$ and the matter fields from initial data posed on a spacelike hypersurface M^n. The initial data for $\mathfrak{g}$ consist of the induced metric g gotten by restricting $\mathfrak{g}$ to M and the second fundamental form K of M in $\mathcal{S}$. The triple (M, g, K) will be referred to as an initial data set. In order to have a well-posed Cauchy problem it is also necessary to specify initial conditions for the matter fields which are present.

A unique feature of the Einstein equations is that $n + 1$ of the equations may be expressed entirely in terms of the initial data. To see this we consider a Lorentz frame $e_0, e_1, \ldots, e_n$ defined along M and adapted to M in the sense that e_0 is the timelike unit normal vector field to M and $e_1, \ldots, e_n$ are an orthonormal basis for the tangent space to M at each point. In this frame the Einstein equations may be written

$$R_{ab} - \frac{1}{2} R\mathfrak{g}_{ab} = T_{ab}, \ 0 \le a, b \le n.$$

The $n + 1$ equations gotten by setting $a = 0$ and $0 \le b \le n$ may be written in terms of the initial data by using the Gauss and Codazzi equations from differential geometry. These equations are called the *constraint equations* and they may be written

$$\frac{1}{2}(R(g) - \|K\|^2 + Tr(K)^2) = \mu \tag{8.1}$$

$$div(K - Tr(K)g) = J, \tag{8.2}$$

where $\mu = T_{00}$ is the energy density of the matter as observed by an observer moving in the e_0 direction and $J_i = T_{0i}$, $i = 1, \ldots, n$ is the observed momentum density of the matter. The expressions on the left side depend only on g and K, while the expressions on the right involve the initial data for the matter fields. When matter fields are present we will always assume the *dominant energy condition* $\mu \geq \|J\|$.

In the brief survey given in this section we will consider solvability of the constraint equations only for the vacuum Einstein equations. These are the equations

$$Ric(\mathfrak{g}) = 0,$$

and the vacuum constraint equations become

$$R(g) - \|K\|^2 + Tr(K)^2 = 0 \tag{8.3}$$

$$div(K - Tr(K)g) = 0. \tag{8.4}$$

In addition to the initial conditions, one needs to specify boundary conditions in order to have a global (in space) well-posed Cauchy problem. In this survey we will consider only two types of boundary conditions. First we will describe recent progress on solving the vacuum constraint equations on compact manifolds M; this is the case of cosmological spacetimes. Secondly we will describe localization methods for the constraint equations which are most relevant to the case of asymptotically flat initial data; this is the case of isolated gravitating systems.

8.2.1 The conformal method

Notice that the system of vacuum constraint equations gives us $n + 1$ equations for the unknowns g, K, which consist of $n(n + 1)$ local functions. Thus it is an underdetermined system of equations and one would expect to be able to specify $(n - 1)(n + 1)$ functions freely and solve the equations for the remaining $n + 1$ unknowns. The most common way to do this is the conformal method, which goes back to Lichnerowicz [70] and Choquet-Bruhat and York [31]. In this method, one chooses free data consisting of a conformal class of Riemannian metrics specified by a chosen metric γ, a divergence-free and trace-free symmetric $(0, 2)$ tensor σ, and a function τ. We will describe below how to choose such tensors σ. We then seek a solution of the vacuum constraint equations of the form

$$g = u^{N-2}\gamma, \ K = \frac{\tau}{n}u^{N-2}\gamma + u^{-2}(\sigma + LW)$$

where u is a positive function, W is a one-form, $N = 2n/(n - 2)$, and L is the conformal Killing operator (with respect to γ) which takes one-forms to trace-free symmetric $(0, 2)$ tensors

$$LW = \mathscr{L}_W \gamma - \frac{2}{n} div(W)\gamma,$$

where $\mathscr{L}$ is the Lie derivative. Thus the data we specify is the conformal class of the metric g, the mean curvature $\tau = Tr_g(K)$, and the conformal class of the trace-free part of K. The unknowns are now u, W, which consist locally of $n + 1$ functions, so the constraint system becomes a determined system. Using the conformal covariance property of the divergence operator on trace-free tensors k

$$div_g(u^{-2}k) = u^{-N} \, div_\gamma(k),$$

we may write the vacuum constraint equations as

$$\frac{4(n-1)}{n-2}\Delta u + Ru = -\frac{n-1}{n}\tau^2 u^{N-1} + \|\sigma + LW\|^2 u^{-N-1} \tag{8.5}$$

$$-\frac{1}{2}L^* LW = \frac{n-1}{n} u^N \, d\tau, \tag{8.6}$$

where L^* is the formal adjoint of L and all operators and norms are taken with respect to the background metric γ.

We now digress to discuss L^* in more detail and we explain how to construct tensors σ which are trace-free and divergence-free. First we consider L to be an operator from one-forms to symmetric trace-free $(0, 2)$ tensors and we see that for any smooth trace-free symmetric $(0, 2)$ tensor k and one-form W we have

$$\int_M \langle LW, k \rangle \, dv = 2 \int_M W_{i:j}k^{ij} \, dv = -2 \int_M \langle W, div(k) \rangle \, dv,$$

where the colon indicates the spatial covariant derivative. Therefore we have $L^*(k) = -2 \, div(k)$. Now L is an elliptic operator and since M is compact, L becomes a Fredholm operator between spaces of tensors with appropriate topologies. It follows that the space of trace-free symmetric $(0, 2)$ tensors has an orthogonal direct sum decomposition as the image of L direct sum the kernel of L^*. This means that, for a fixed metric, we can take any trace-free symmetric k and write it uniquely as $k = LW + \sigma$, where W is a one-form and σ is divergence-free. Because of elliptic regularity theory, the components LW and σ in this decomposition are as smooth as k. This gives a description of the σ which are required for the conformal method.

In the special case that τ is taken to be a constant function, we are looking for solutions of the vacuum constraint equations which have constant mean curvature (CMC) τ, and in this case we can take $W = 0$ and the equations (8.5), (8.6) reduce to the scalar equation

$$\frac{4(n-1)}{n-2}\Delta u + Ru = -\frac{n-1}{n}\tau^2 u^{N-1} + \|\sigma\|^2 u^{-N-1}.$$

In this case the work of Isenberg [63], following earlier important work by Lichnerowicz, Choquet-Bruhat, and York (see [63] for references) completely settles the existence problem.

When τ is not constant there is much less known about solvability of the equations. For near-CMC data there are results by Isenberg and Moncrief [64] and by Allen, Clausen, and Isenberg [1]. More recently there have been results by Holst, Nagy, and Tsogtgerel [60] and by Maxwell [76, 77] which allow general mean curvature τ, but instead assume that σ is small. There are also some non-CMC cases for which non-existence is known (see Isenberg and Ó Murchadha [66]).

There is a recent general result concerning solvability of (8.5), (8.6) which does not require any smallness conditions. This is due to Dahl, Gicquaud, and Humbert [40]. The proof is in the general spirit of the analysis of blow-up phenomena for solutions of nonlinear equations in that it considers a regularized problem which is always solvable and studies how solutions can blow up when the regularization is removed.

The authors of [40] make the following three assumptions:

1. the function τ is either strictly positive or strictly negative;
2. the manifold (M, γ) has no conformal Killing vector fields (that is, L is injective);
3. if the Yamabe invariant of γ is non-negative, then σ is not identically zero.

In this case the regularized problem is gotten by replacing (8.6) by the equation

$$-\frac{1}{2}L^*LW = \frac{n-1}{n}u^{N-\epsilon}\,d\tau, \qquad (8.7)$$

where ϵ is a small positive number. Using ideas of Maxwell [76] the authors prove the following:

Proposition 8.2.1 *([40]) Under the assumptions (1), (2), and (3) and appropriate regularity assumptions, the regularized system (8.5), (8.7) has a solution for any $\epsilon > 0$.*

The authors now consider the limit of the solutions of the regularized problem as ϵ goes to zero. They discover a 'blow-up' equation

$$-\frac{1}{2}L^*LW = \alpha_0\sqrt{\frac{n-1}{n}}\|LW\|\frac{d\tau}{\tau} \qquad (8.8)$$

where $\alpha_0 \in (0, 1]$. The main theorem is now the following.

Theorem 8.2.2 *([40]) Under the assumptions (1), (2), and (3) above and appropriate regularity assumptions, one or both of the following assertions hold:*

 (i) *The system (8.5), (8.6) has a solution (u, W). Furthermore the set of solutions is compact in an appropriate topology.*
(ii) *There exists a nontrivial solution W of (8.8) for some $\alpha_0 \in (0, 1]$.*

While this theorem is not an unqualified existence theorem, it gives a general criterion for existence (namely that there be no non-zero solution of (8.8)) which can be checked in some cases. In particular, the authors show that it implies and improves most of the known existence results which have been obtained. We now describe some of the consequences.

They first observe that the set of metrics γ and mean curvature functions τ for which (8.8) has no solution is open.

Proposition 8.2.3 *([40]) The set of metrics γ and functions τ for which (8.8) has no solution is open in the C^1 topology.*

It follows that if condition (ii) of Theorem 8.2.2 does *not* hold for γ and τ, then it also does not hold for metrics and functions which are sufficiently close in the C^1 topology. In particular, condition (i) must hold and we have existence for the constraint system (8.5), (8.6) for such nearby data and any choice of σ.

There are two conditions under which the authors are able to show that (8.8) has no solution.

Proposition 8.2.4 *([40]) Assume conditions (1), (2), and (3) and appropriate regularity assumptions. If $Ric(\gamma) \leq -\lambda\gamma$ for some positive constant λ and*

$$\max_M \left\| \frac{d\tau}{\tau} \right\| < \sqrt{\frac{n}{2(n-1)}}\lambda,$$

then (8.8) has no nontrivial solution and hence (8.5), (8.6) have at least one solution.

Note that by Bochner's theorem the condition of negative Ricci curvature implies that there are no non-vanishing conformal Killing vector fields, so this condition implies a strong injectivity condition on L. This condition on the Ricci curvature can be replaced by an injectivity bound on the operator L. Precisely, consider the constant C_γ given by

$$C_\gamma = \inf \frac{(\int_M \|LV\|^2 \, dv_\gamma)^{1/2}}{(\int_M \|V\|^N \, dv_\gamma)^{1/N}},$$

where the infimum is taken over nontrivial one-forms V. The Sobolev inequality and elliptic estimates imply that C_γ is positive if L is injective. The following result gives another criterion for existence of solutions of the constraint equations.

Proposition 8.2.5 *([40]) Assume conditions (1), (2), and (3) and appropriate regularity. If*

$$\left\| \frac{d\tau}{\tau} \right\| < \frac{1}{2}\sqrt{\frac{n}{n-1}} \, C_\gamma,$$

then the constraint system (8.5), (8.6) has at least one solution and the space of solutions is compact in an appropriate topology.

It would be reasonable to expect that the existence of a nonzero solution of (8.8) would be a non-generic situation so that the set of (γ, τ) for which there is no such solution would be open and dense. The authors of [40] prove a density result of this type with a very weak topology (C^0) on the space of metrics.

Proposition 8.2.6 *([40]) Given a nonzero function τ, let $\mathcal{R}(M, \tau)$ be the set of metrics γ for which there is no nonzero solution of (8.8) for the pair (γ, τ). Then the set $\mathcal{R}(M, \tau)$ is dense with respect to the C^0 topology in the space of all smooth metrics on M.*

One might hope that the blow-up equation (8.8) never has a nonzero solution and thus we would always have existence for the constraint system. The authors show that this is not true by constructing a metric γ and positive function τ on S^n for which (8.8) has a nonzero solution for some $\alpha_0 \in (0, 1]$. This suggests that the best one can hope for is an existence theorem under generic conditions for the non-CMC conformal method. Of course it would be nice to be able to give a clear description of a general set of sufficient conditions for solvability.

Finally we make a brief comment about solving the constraint equations for asymptotically flat initial data. The conformal method may again be used to do this, and the method works with great success in the case of maximal, $\tau = 0$, data (see Cantor [25]). The general case is also largely unknown for asymptotically flat data. Analogues of the results of [40] which have been described above are currently being explored for the asymptotically flat case. The problem is complicated by the need to solve the corresponding equations with solutions which decay at infinity at an appropriate rate. In the next section we will consider the asymptotically flat case in more detail and discuss the asymptotic behavior which one can impose on solutions.

8.2.2 The asymptotically flat case

An asymptotically flat initial data set is given by (M, g, K), where each end of M is diffeomorphic to $\mathcal{R}^n$ minus a ball, and such that on each end coordinates $x^1, \ldots, x^n$ can be introduced so that the data fall off to the trivial data $g_{ij} = \delta_{ij}$ and $K = 0$. More precisely we seek solutions which satisfy

$$g_{ij} = \delta_{ij} + O_m(|x|^{-p}), \quad K_{ij} = O_{m-1}(|x|^{-p-1}), \tag{8.9}$$

where $m \geq 2$ is an integer and p is a positive real number. Here the notation that a tensor field h is $O_k(|x|^{-q})$ means

$$|\partial^\alpha h(x)| \leq c|x|^{-q-|\alpha|}$$

for any multi-index α with $0 \leq |\alpha| \leq k$, where we calculate the derivatives of the components of h with respect to the Euclidean metric. In other words it means that all derivatives up to order k fall off correspondingly faster (as they would if there were an expansion in inverted spherical harmonics near infinity). It turns out that the ADM energy and linear momentum (see Section 8.3) exist provided $p > (n-2)/2$, so we will always make that assumption. Furthermore, if $p > n-2$ the energy would be zero and hence the data trivial (by the positive energy theorem), so we will always assume that $(n-2)/2 < p \leq n-2$.

To understand the asymptotic behavior which we can hope to achieve we consider the simplest nontrivial exact solution, the n-dimensional Schwarzschild solution. It has initial data given in conformally flat form on $M = \mathcal{R}^n \setminus \{0\}$ by

$$g_{ij} = \left(1 + \frac{E}{2|x|^{n-2}}\right)^{\frac{4}{n-2}} \delta_{ij}, \quad K_{ij} = 0.$$

For $E > 0$ this metric satisfies $R(g) = 0$ and has two asymptotically flat ends, one for $|x|$ large and the other for $|x|$ near 0 (we can see this by changing coordinates to inverted coordinates $y = c^2 x/|x|^2$, where $c = (E/2)^{1/(n-2)}$).

We first discuss the construction of time-symmetric initial data; that is, $K = 0$. This is then the study of Riemannian manifolds which are asymptotically flat and have vanishing scalar curvature. The simplest way to construct a general family of solutions of this type on $M = \mathcal{R}^n$ is to consider a metric γ which is a small perturbation of the Euclidean metric, say $\gamma_{ij} = \delta_{ij} + h_{ij}$, where h is compactly supported and small. One may then use the conformal method to find a solution of the form $g = u^{4/(n-2)}\gamma$, $K = 0$. The equation (8.5) becomes the linear equation

$$\frac{4(n-1)}{n-2}\Delta_\gamma u + R(\gamma)u = 0.$$

When $R(\gamma)$ is sufficiently small this has a unique solution satisfying $u > 0$ and $u(x) \to 1$ as $|x| \to \infty$ (see for example [25]). Since $R(\gamma)$ is compactly supported it follows that u is a Euclidean harmonic function for $|x|$ large and therefore it has an expansion in inverted spherical harmonics of the form

$$u(x) = 1 + \frac{E}{2|x|^{n-2}} + O_m(|x|^{1-n})$$

for any positive integer m. Thus we see that solutions of this type are Schwarzschild to leading order. We will refer to such a solution as being conformally flat near infinity.

A more sophisticated use of the conformal method was made by Schoen and Yau [93] as a step to simplify the asymptotic behavior in the proof of the positive energy theorem. They showed the following.

Theorem 8.2.7 *([93]) Among all time-symmetric solutions of the vacuum constraint equations satisfying the asymptotic conditions (8.9) with $(n-2)/2 < p \le n-2$ and $m \ge 2$, those which are conformally flat near infinity are dense in an appropriate topology in which the ADM energy is continuous. In particular, there is a dense set of solutions of the time-symmetric vacuum constraint equations which are Schwarzschild to leading order.*

The question of whether the asymptotic behavior can be improved to make a solution exactly equal to a Schwarzschild solution near infinity was taken up by Corvino [36]. To explain his theorem, we note that if asymptotically flat coordinates are chosen then there is an $(n + 1)$-parameter family of Schwarzschild metrics (parametrized by a point $C \in \mathcal{R}^n$ and by the mass parameter E) given by

$$g_{E,C} = \left(1 + \frac{E}{2|x - C|^{n-2}}\right)^{\frac{4}{n-2}} \delta.$$

If a solution is identical to one of the metrics $g_{E,C}$ outside a compact set, we refer to such a metric as being Schwarzschild near infinity and we may now state the theorem.

Theorem 8.2.8 *([36]) Assume that g satisfies (8.9) with $p = n - 2$ and $m \geq 4$. There is a sequence of metrics $g^{(i)}$ which are Schwarzschild near infinity and which agree with g on larger and larger balls such that the $g^{(i)}$ converge to g in an appropriate norm in which the energy is continuous. In particular we have $E(g^{(i)}) \to E(g)$.*

We remark that in [36] it is assumed that g is Schwarzschild to leading order at infinity and that it has a well defined center of mass C. Under these conditions it is also shown that $C^{(i)} \to C$. The weaker asymptotic condition of Theorem 8.2.8 was handled in a recent paper of Chruściel, Corvino, and Isenberg [32].

The proof of Theorem 8.2.8 is much more sophisticated than that of Theorem 8.2.7 and we give an outline here before discussing the spacetime version of the theorem (i.e., the version for data which are not time-symmetric) and applications. The first step of the proof is to choose a large radius ρ and perform a cutoff of g to the metric $g_{E,C}$ in the annulus $A_\rho := B_{2\rho} \setminus B_\rho$. This is done by choosing a smooth cutoff function $\chi(|x|)$ which is equal to 1 inside B_ρ and equal to 0 outside $B_{2\rho}$ and considering the metric $\tilde{g}_{E,C} = \chi g + (1 - \chi) g_{E,C}$. Now the scalar curvature $\tilde{R}_{E,C}$ of $\tilde{g}_{E,C}$ is zero inside B_ρ and outside $B_{2\rho}$. Moreover, we have

$$\tilde{R}_{E,C} = O_{m-2}(\rho^{-n}) \text{ in } A_\rho.$$

The idea now is to treat $\tilde{g}_{E,C}$ as an approximate solution and attempt to correct it by a small perturbation; that is, we seek a small symmetric $(0,2)$ tensor h with support in the annulus A_ρ such that

$$R(\tilde{g}_{E,C} + h) \equiv 0.$$

There are two fundamental issues to be encountered in trying to do this: (1) the tensor h must be smooth with compact support in A_ρ, and (2) the choices of E and C must be made carefully in order for this to work.

Concerning issue (1), we point out that standard elliptic methods cannot produce solutions h which vanish to high order at the boundary of a domain. In our case the equation we are solving is underdetermined and this makes it possible. We can rewrite our equation in the form

$$R(\tilde{g}_{E,C} + h) = \tilde{R}_{E,C} + \tilde{\mathcal{L}}(h) + Q(h) = 0$$

where $\tilde{\mathcal{L}}$ is a linear operator and $Q(h)$ a quadratic error term. The operator $\tilde{\mathcal{L}}$ is close to the linearized scalar curvature operator $\mathcal{L}$ at the Euclidean metric. This operator is

$$\mathcal{L}(h) = div(div\ h) - \Delta(Tr(h)).$$

We will need to prove surjectivity of this operator in suitable spaces, so in order to do that we compute the adjoint operator $\mathcal{L}^*$ and consider injectivity questions for it. We see that

$$\mathcal{L}^*(f) = Hess(f) - \Delta(f)\ \delta.$$

We observe that in a connected open subset of $\mathcal{R}^n$ the operator $\mathcal{L}^*$ has an $(n + 1)$-dimensional kernel $\mathcal{K}$ consisting of the affine functions $a + b \cdot x$, where $a \in \mathcal{R}$ and $b \in \mathcal{R}^n$.

This is because if $\mathcal{L}^*(f) = 0$, then $\Delta(f) = 0$ follows from taking the trace, and therefore *Hess*$(f) = 0$. Problem (1) is now handled by proving a very strong injectivity statement for $\mathcal{L}^*$ called the "basic estimate",

$$\|f\|_2 \leq c\|\mathcal{L}^*(f)\|_2,$$

which holds provided that f is approximately orthogonal to the kernel $\mathcal{K}$; here the norms $\|\cdot\|_2$ are $\mathcal{L}^2$ norms on A_ρ. The key point in this estimate is that it is a global estimate which does not require a boundary condition on f. This can be used to prove injectivity in weighted L^2 spaces L^2_w, where $w = \varphi^{2N}$ for φ a smooth positive function in A_ρ which is equal to the distance to the boundary near ∂A_ρ, and where N will be chosen large. Thus the norm is

$$\|f\|_{2,w}^2 = \int_{A_\rho} f^2 w \, dx.$$

We then have the injectivity estimate

$$\|f\|_{2,w} \leq c\|L^*(f)\|_{2,w},$$

provided that f is almost orthogonal to $\mathcal{K}$. Since the dual space of L^2_w is the space $L^2_{w^{-1}}$, this estimate implies surjectivity of L in such spaces. Note that finiteness of the norm $\|f\|_{2,w^{-1}}$ requires f to decay rapidly at the boundary ∂A_ρ.

The requirement in the basic estimate that f be approximately orthogonal to $\mathcal{K}$ leads to obstructions for solving the constraint equations. We can approximate the space $\mathcal{K}$ by a space $\mathcal{K}_0$ of functions with compact support in A_ρ. The theory we have described then implies that for any choice of E and C we can find a small h so that the metric $\hat{g}_{E,C} := \tilde{g}_{E,C} + h$ has scalar curvature $R_{E,C} \in \mathcal{K}_0$.

Finally we come to issue (2), which is the question of how to adjust the parameters E and C to make the argument work. The point is that the functions in $\mathcal{K}_0$ are parametrized by $n + 1$ numbers $a \in \mathcal{R}$ and $b \in \mathcal{R}^n$. The solution process we have described then gives us a map from $\mathcal{R}^{n+1}$ to $\mathcal{R}^{n+1}$. The remainder of the argument involves showing that 0 lies in the image of this map, so that there is a choice of E and C which leads to a metric $\hat{g}_{E,C}$ with scalar curvature zero. Roughly speaking the variation of the energy parameter E primarily affects the scalar parameter a, and variation of the center-of-mass parameter C primarily affects the vector parameter b. This metric then solves our problem for a given large ρ. The approximating sequence is obtained by doing this process for a sequence $\rho_i \to \infty$ to obtain the sequence $g^{(i)}$.

In the case $n = 3$, a spacetime version of these results was developed by Corvino and Schoen [39] and by Chruściel and Delay [34]. For general data (M, g, K) with appropriate asymptotics (so that angular momentum and center of mass can be defined), these authors prove the following result.

Theorem 8.2.9 *([34, 39]) Given a vacuum initial data set (M, g, K) with appropriate asymptotic behavior, there is a sequence of vacuum initial data $(g^{(i)}, K^{(i)})$ on M which agree with (g, K) in larger and larger balls, which are identical with a slice of a Kerr* *spacetime near infinity, and which converges to (g, K) in an appropriate topology in which the energy, center of mass, linear momentum, and angular momentum are continuous.*

The idea of the proof of this theorem is similar to that of Theorem 8.2.8, but the proof is technically more difficult. In this case the operator is replaced by the linearized constraint operator at the trivial data. This is the operator

$$T(h, k) = (\mathcal{L}(h), div(h)),$$

where $\mathcal{L}$ is the linearized scalar curvature operator. The adjoint operator is then

$$T^*(f, W) = \left(\mathcal{L}^*(f), -\frac{1}{2}\mathscr{L}_X\delta \right),$$

where $\mathscr{L}$ is the Lie derivative. In particular the kernel of T^* on any connected open subset of $\mathcal{R}^3$ is given by pairs $(a + b \cdot x, Y)$, where Y is a Killing vector field of $\mathcal{R}^3$. Thus the kernel is 10-dimensional, so we must consider a 10-parameter family of approximate solutions with effective parameters in order to do the gluing. Such a 10-parameter family is given by slices in Kerr spacetimes relative to fixed coordinates where the energy, center of mass, linear momentum, and angular momentum are allowed to vary. As in the proof of Theorem 8.2.8, it must be shown that there is a choice of an approximate solution from this 10-parameter family which leads to a solution of the vacuum constraint equations.

This construction has various applications since it allows one to trivialize the geometry of the spacetime in a full neighborhood of spatial infinity. One dramatic application is to justify the notion of *asymptotically simple* spacetimes proposed by Penrose. These are spacetimes for which the conformal compactification near null infinity is a smooth manifold with boundary. It was not previously known whether one could expect nontrivial examples of such asymptotically simple spacetimes to exist. An application of this work is that such spacetimes exist in abundance and can be expected to be dense in the full space of solutions. For the construction of asymptotically simple spacetimes, see Chruściel and Delay [33] and Corvino [37].

There is another application of Theorem 8.2.8 which involves showing that a minimizer for the Bartnik quasi-local mass is a static metric. This application is described in the next section of this chapter.

We close this section by describing a spacetime (i.e., non-time-symmetric) version of Theorem 8.2.7. We must ask the question: What is the natural spacetime version of conformally flat asymptotics? The basic feature of time-symmetric data that is important in the application to the positive energy theorem is that the constraint equation (in this case $R = 0$) has solutions defined by functions which are harmonic to leading order. This allows us to show that these solutions are Schwarzschild to leading order. It turns out that the following asymptotic form, called *harmonic asymptotics*, plays a similar role in the case of general data

$$g = u^{\frac{4}{n-2}}\delta, \quad K_{ij} - Tr(K)\delta_{ij} = u^2(\mathcal{L}X)_{ij} = u^2(\partial_i X_j + \partial_j X_i),$$

where u is a positive function which is asymptotic to 1 at infinity, X is a one-form which decays at infinity, and $\mathscr{L}$ denotes the Lie derivative of the Euclidean metric δ with respect to X. If such asymptotics can be achieved, then the constraint equations near infinity become the equations (computed with respect to δ)

$$8\Delta u = u\left(-|\mathscr{L}X|^2 + (1/2)(Tr(\mathscr{L}X))^2\right)$$

$$\Delta X_i + 4u^{-1}u_j(\mathscr{L}X)_i^j - 2u^{-1}u_i\, Tr(\mathscr{L}X) = 0,$$

where the second equation is written with respect to a Euclidean basis. Note that these are harmonic to leading order and hence expansions can be derived. It turns out that the important asymptotic quantities – the energy, the linear momentum, the center of mass, and the angular momentum – can all be read off the asymptotic expansions of u and X. For example we have

$$u(x) = 1 + \frac{E}{2|x|^{n-2}} + \frac{n-2}{2}\frac{E}{|x|^{-n}}(C \cdot x) + O_2(|x|^{-n}),$$

where E is the energy and C the center of mass. The expansion for X contains the linear and angular momentum; for example (see [44], Lemma 5)

$$X_i(x) = -\frac{n-1}{n-2}P_i|x|^{2-n} + O_2(|x|^{1-n}),$$

where $P = (P_1, \ldots, P_n)$ is the ADM linear momentum vector. The result that solutions of the vacuum constraint equations in harmonic asymptotics are dense is given by the following density theorem of Corvino and Schoen [39]:

Theorem 8.2.10 *([39]) The set of solutions of the vacuum constraint equations with harmonic asymptotics is dense (for a suitable weighted Sobolev topology) in the space of all solutions. This topology is such that the energy and linear momentum are continuous.*

This density result was extended to matter solutions satisfying the dominant energy condition by Eichmair, Huang, Lee, and Schoen [44] in their treatment of the spacetime positive mass theorem. This proof, which involves the use of asymptotically planar MOTS, is described in the last section of this chapter.

8.3 Quasi-local mass

In an isolated system where gravitation is weak near infinity, there is a notion of total mass at spatial infinity, defined by Arnowitt, Deser and Misner who applied a Hamilton–Jacobi analysis of the Einstein–Hilbert action to the system ([7, 8]). In terms of an asymptotically flat initial data set[1] (M, g, K), the total energy of such a system is

$$E = \lim_{r \to \infty} \frac{1}{16\pi} \int_{S_r} (g_{ij,i} - g_{ii,j})dS^j,$$

---

[1] In this section we restrict our attention to 3-dimensional initial data sets and $(3 + 1)$-dimensional spacetimes. Apart from the multiplicative constants, the expressions for E and P_i remain the same in higher dimensions; cf., [44].

where S_r is the coordinate sphere of coordinate radius r (and where the indices i as well as the indices j are summed from 1 to 3); and the total linear momentum is $P = (P_1, P_2, P_3)$, where

$$P_i = \lim_{r \to \infty} \frac{1}{8\pi} \int_{S_r} [K_{ij} - (\operatorname{tr} K) g_{ij}] dS^j.$$

Assuming the dominant energy condition is satisfied, one expects that (E, P_1, P_2, P_3) is a future-directed causal vector so that the total mass $\mathfrak{m}$ of the system can be defined as

$$\mathfrak{m} = \sqrt{E^2 - |P|^2}.$$

This is the content of the positive mass theorem, which was first proved by Schoen and Yau [91, 94] via a geometric method using minimal hypersurfaces and solutions to Jang's equation [67], and later by Witten [103] using spinors.

Positive Mass Theorem *Let (M, g, K) be a 3-dimensional, complete, asymptotically flat initial data set satisfying the dominant energy condition. Then $E \geq |P|$, and $E = 0$ if and only if (M, g, K) can be isometrically embedded into the Minkowski spacetime.*

In [16] Beig and Chruściel showed that the equality case $E = |P|$ in the above theorem occurs only when $E = 0$. Hence, the rigidity part of the positive mass theorem can be rephrased as $\mathfrak{m} = 0$ if and only if the spacetime is flat along M.

Given the positive mass theorem, a naturally related question is whether the total mass of an isolated system can be consistently found by computing contributions from extended bodies in the system. More precisely, if Ω is a compact spacelike hypersurface with boundary $\partial\Omega$ in a spacetime, what constitutes a suitable notion of mass in Ω counting the effect of both gravitation and matter distribution? This is known as the quasi-local mass problem in general relativity (cf. [88]).

In special relativity, with gravity not present, the energy integral of the stress-energy tensor (with respect to an observer) on Ω depends only on the boundary surface $\partial\Omega$. Thus, based on conservation-law ideas, one expects in general relativity that an appropriate notion of quasi-local mass should also depend on the geometry of $\partial\Omega$ alone.

Compared to the definition of the total mass for an isolated system, an appropriate notion of quasi-local mass $\mathfrak{m}(\cdot)$ has the advantage of being applicable to finitely extended bodies in a non-isolated system where gravitation can be strong. Some basic properties that one would like $\mathfrak{m}(\cdot)$ to satisfy include the following:

- Positivity: $\mathfrak{m}(\partial\Omega) \geq 0$ assuming that the dominant energy condition holds.
- Vanishing property: $\mathfrak{m}(\partial\Omega) = 0$ if the spacetime is flat along Ω.
- Asymptotic to the total mass: $\lim_{k\to\infty} \mathfrak{m}(\partial\Omega_k) = \mathfrak{m}$ if $\{\Omega_k\}$ forms an exhaustion of an asymptotically flat initial data set (M, g, K) whose total mass is $\mathfrak{m}$.

There have been various approaches towards defining quasi-local mass (cf. [97] and the references therein). Among them, the definitions proposed by Brown and York [22, 23] and Wang and Yau [100, 101] are based on a Hamiltonian formulation; the Hawking mass [56] is motivated by the investigation of gravitational radiation; and the Bartnik mass [11] is an analogue of the electrostatic capacity of a conducting body. Below we provide an account of these definitions, focusing on their mathematical aspects and on related open questions.

8.3.1 Quasi-local mass via Hamilton–Jacobi approach

In what follows, Σ always denotes a closed, connected, spacelike 2-surface in a $(3+1)$-dimensional spacetime $(\mathcal{S}, \mathfrak{g})$ satisfying the dominant energy condition. The Lorentz metric $\mathfrak{g}$ is also often denoted by $\langle \cdot, \cdot \rangle$. Under suitable orientation assumptions (for instance Σ bounding a spacelike hypersurface), we assume that a continuous choice of an outward spacelike direction can be made in $(T\Sigma)^\perp$, the normal bundle of Σ.

As noted above, expressions for the total energy-momentum of an isolated system were obtained (by ADM) via a Hamilton–Jacobi analysis of the Einstein–Hilbert action to the system. Applying similar analyses to the time history of a compact spacelike hypersurface bounded by Σ (cf. [22, 23, 58, 69]), one obtains a surface integral on Σ, known as the surface Hamiltonian, which we describe below.

Surface Hamiltonian

Suppose Σ is a boundary component of a compact spacelike hypersurface Ω. Along Σ, let n be the future timelike unit normal to Ω and let ν be the outward unit normal to Σ in Ω. Given a future timelike unit vector field T along Σ, decompose $T = Nn + X$, where N is the lapse function and X is the shift vector tangent to Ω. The surface Hamiltonian in [58] is

$$\mathcal{H}(T, n) = -\frac{1}{8\pi} \int_\Sigma [NH - (K - (\operatorname{tr} K)g)(X, \nu)] \, d\sigma. \qquad (8.10)$$

Here H is the mean curvature of Σ in Ω with respect to ν, K is the second fundamental form of Ω with respect to n, g is the induced Riemannian metric on Ω, $\operatorname{tr} K$ is the trace of K with respect to g and $d\sigma$ is the area form on Σ. Observe that

$$(K - (\operatorname{tr} K)g)(X, \nu) = -(\operatorname{tr}_\Sigma K)\langle X, \nu \rangle + K(T^\|, \nu)$$

where $\operatorname{tr}_\Sigma K$ is the trace of K restricted to Σ and $T^\|$ is the projection of X (hence T) to the tangent space of Σ. In terms of the vectors

$$\boldsymbol{H} = (\operatorname{tr}_\Sigma K)n - H\nu, \quad \boldsymbol{J} = -(\operatorname{tr}_\Sigma K)\nu + Hn,$$

where $\boldsymbol{H}$ is the mean curvature vector of Σ in $\mathcal{S}$ and $\boldsymbol{J}$ is the dual of $\boldsymbol{H}$ in $(T\Sigma)^\perp$ obtained by reflecting $\boldsymbol{H}$ across the inward future null direction, one can rewrite $\mathcal{H}(T, n)$ as

$$\mathcal{H}(T, n) = \frac{1}{8\pi} \int_\Sigma \langle \boldsymbol{J}, T \rangle - \langle \overline{\nabla}_{T^\|} \nu, n \rangle \, d\sigma, \qquad (8.11)$$

where $\overline{\nabla}$ is the connection in $\mathcal{S}$. Expression (8.11) indicates that $\mathcal{H}(T, n)$ depends on n only through the connection 1-form $\alpha_\nu(\cdot) = \langle \overline{\nabla}_{(\cdot)} \nu, n \rangle$ on Σ.

To define quasi-local energy and mass using this approach, one needs to choose a reference surface Hamiltonian $\mathcal{H}_0(T_0, n_0)$, which is often computed along the image of an isometric embedding of Σ into some reference space. Once a choice of $\mathcal{H}_0(T_0, n_0)$ has been made, the corresponding quasi-local energy is defined as the difference

$$\mathcal{H}(T, n) - \mathcal{H}_0(T_0, n_0). \tag{8.12}$$

Brown–York mass

In [22, 23], Brown and York defined a quasi-local mass by taking the reference to be an isometric embedding $\iota : \Sigma \to \mathcal{R}^3$, where $\mathcal{R}^3$ is viewed as a time-symmetric slice in $\mathcal{R}^{3,1}$. The existence of such an embedding is usually guaranteed by an assumption that Σ has positive Gaussian curvature [83, 89], in which case $\iota(\Sigma)$ is a convex surface in $\mathcal{R}^3$. Choose $T_0 = n_0$ to be the future timelike unit normal to $\mathcal{R}^3$ in $\mathcal{R}^{3,1}$; one then obtains a reference surface Hamiltonian

$$\mathcal{H}_0(T_0, n_0) = -\frac{1}{8\pi} \int_\Sigma H_0 \, d\sigma, \tag{8.13}$$

where H_0 denotes the mean curvature of $\iota(\Sigma)$ in $\mathcal{R}^3$ and is treated as a function on Σ by identifying Σ with $\iota(\Sigma)$. In $\mathcal{S}$, suppose Σ is a boundary component of a spacelike hypersurface Ω. If we choose $T = n$ to be the future timelike unit normal to Ω (i.e., $N = 1$ and $X = 0$), then the corresponding physical surface Hamiltonian is

$$\mathcal{H}(T, n) = -\frac{1}{8\pi} \int_\Sigma H \, d\sigma.$$

The Brown–York mass of Σ in Ω is then defined as

$$\mathfrak{m}_{BY}(\Sigma; \Omega) = \frac{1}{8\pi} \int_\Sigma (H_0 - H) \, d\sigma. \tag{8.14}$$

Since the reference in the definition of $\mathfrak{m}_{BY}(\Sigma; \Omega)$ is an embedding into a time-symmetric slice in $\mathcal{R}^{3,1}$, one expects that $\mathfrak{m}_{BY}(\Sigma; \Omega)$ has good properties when Ω is also time-symmetric in $\mathcal{S}$. In [95], Shi and Tam proved that, if Ω is time-symmetric, then $\mathfrak{m}_{BY}(\Sigma; \Omega) \geq 0$ and $\mathfrak{m}_{BY}(\Sigma; \Omega) = 0$ if and only if Ω is isometric to a domain in $\mathcal{R}^3$.

By definition, $\mathfrak{m}_{BY}(\Sigma; \Omega)$ depends on both Σ and Ω. In particular, $\mathfrak{m}_{BY}(\Sigma; \Omega)$ can be negative even for certain spacelike hypersurfaces Ω in Minkowski spacetime $\mathcal{R}^{3,1}$. For instance, if Ω is a geodesic ball in $\mathbb{H}^3 = \{(t, x_1, x_2, x_3) \in \mathcal{R}^{3,1} \mid t^2 - |x|^2 = 1, \, t > 0\}$, then $\mathfrak{m}_{BY}(\partial\Omega; \Omega) < 0$.

Liu–Yau mass

A generalization of the Brown–York mass was given by Liu and Yau in [71] (see also Booth and Mann [24], Epp [47], and Kijowski [69]). In this definition, the reference is still an isometric embedding of Σ into $\mathcal{R}^3$; hence the reference surface Hamiltonian $\mathcal{H}_0(T_0, n_0)$ remains the same as in (8.13). The difference is to choose $T = n = \mathbf{J}/|\mathbf{H}|$ in the physical surface Hamiltonian $\mathcal{H}(T, n)$, which yields

$$\mathcal{H}(T, n) = -\frac{1}{8\pi} \int_\Sigma |\mathbf{H}| \, d\sigma.$$

Under the assumptions that Σ has positive Gaussian curvature and $\boldsymbol{H}$ is inward spacelike, the Liu–Yau mass of Σ is defined as

$$\mathfrak{m}_{LY}(\Sigma) = \frac{1}{8\pi} \int_{\Sigma} (H_0 - |\boldsymbol{H}|)\, d\sigma. \tag{8.15}$$

Clearly $\mathfrak{m}_{LY}(\Sigma)$ depends only on Σ. If Σ bounds a time-symmetric Ω, then $\mathfrak{m}_{LY}(\Sigma) = \mathfrak{m}_{BY}(\Sigma; \Omega)$. In general, it is proved by Liu and Yau in [71, 72] that $\mathfrak{m}_{LY}(\Sigma) \geq 0$ as long as Σ bounds a compact spacelike hypersurface Ω; moreover, $\mathfrak{m}_{LY}(\Sigma) = 0$ only if $\mathcal{S}$ is flat along Ω.

In [84], Ó Murchadha, Szabados and Tod found that there exist closed spacelike 2-surfaces, lying on the light cone of the Minkowski spacetime $\mathcal{R}^{3,1}$, whose Liu–Yau mass is strictly positive. Later, Miao, Shi and Tam proved in [79] that, for any $\Sigma \subset \mathcal{R}^{3,1}$ whose Liu–Yau mass is defined, $\mathfrak{m}_{LY}(\Sigma) > 0$ unless Σ lies in a spacelike hyperplane.

Wang–Yau mass

In [100, 101], Wang and Yau introduced a new definition of quasi-local mass using isometric embeddings of Σ into $\mathcal{R}^{3,1}$ as references.

Given an isometric embedding $\iota : \Sigma \to \mathcal{R}^{3,1}$ and a future timelike unit vector $T_0 \in \mathcal{R}^{3,1}$, one defines an associated reference surface Hamiltonian $\mathcal{H}_0(T_0, n_0)$ by decomposing $T_0 = T_0^{\parallel} + T_0^{\perp}$, where $T_0^{\parallel}$ is tangent to $\iota(\Sigma)$ and $T_0^{\perp}$ is normal to $\iota(\Sigma)$, and choosing $n_0 = N^{-1} T_0^{\perp}$ where $N = (1 + |T_0^{\parallel}|^2)^{\frac{1}{2}}$. It follows from (8.11) that

$$\mathcal{H}(T_0, n_0) = \frac{1}{8\pi} \int_{\Sigma} \left[(1 + |T_0^{\parallel}|^2)^{\frac{1}{2}} \langle H_0, v_0 \rangle - \langle \nabla_{T_0^{\parallel}}^{\mathcal{R}^{3,1}} v_0, n_0 \rangle \right] d\sigma, \tag{8.16}$$

where Σ is identified with $\iota(\Sigma)$, $\nabla^{\mathcal{R}^{3,1}}$ is the connection in $\mathcal{R}^{3,1}$ and v_0 is the unique outward spacelike unit vector in the normal bundle of $\iota(\Sigma)$ with $v_0 \perp n_0$. Here the outward spacelike direction in the normal bundle of $\iota(\Sigma)$ is specified by the direction of $-H_0$, where H_0 is the mean curvature vector of $\iota(\Sigma)$ and is assumed to be spacelike. By abuse of notation, $\langle \cdot, \cdot \rangle$ denotes the metric product on $\mathcal{R}^{3,1}$.

There is a good physical and geometric reason to make the above choice of n_0 and v_0. Consider the time history $\mathcal{P}$ of $\iota(\Sigma)$ with respect to the time direction specified by T_0 in $\mathcal{R}^{3,1}$. Locally around any $p \in \iota(\Sigma)$, $\mathcal{P}$ is a 3-dimensional hypersurface obtained by translating $\iota(\Sigma)$ along $-T_0$. By definition, n_0 is tangent to $\mathcal{P}$ and v_0 is normal to $\mathcal{P}$ at p. In particular, if $\hat{\Sigma}$ is the image of $\iota(\Sigma)$ near p under the projection onto the orthogonal complement of T_0, then v_0 is also a unit normal vector to $\hat{\Sigma}$.

Having picked T_0 and n_0, one chooses a corresponding pair of T and n along Σ in $\mathcal{S}$ as follows. Supposing that the mean curvature vector $\boldsymbol{H}$ of Σ in $\mathcal{S}$ is spacelike, define T and n by requiring $T = T_0^{\parallel} + Nn$ and

$$\langle \boldsymbol{H}, T \rangle = \langle \boldsymbol{H_0}, T_0 \rangle. \tag{8.17}$$

Condition (8.17) is imposed so that the expansion of Σ along T in $\mathcal{S}$ is the same as the expansion of $\iota(\Sigma)$ along T_0 in $\mathcal{R}^{3,1}$. With such a choice of T and n, the physical surface Hamiltonian is given by

$$\mathcal{H}(T,n) = \frac{1}{8\pi} \int_{\Sigma} \left[(1 + |T_0^{\parallel}|^2)^{\frac{1}{2}} \langle \boldsymbol{H}, \nu \rangle - \langle \overline{\nabla}_{T_0^{\parallel}} \nu, n \rangle \right] d\sigma, \tag{8.18}$$

where ν is the unique outward spacelike unit vector in $(T\Sigma)^{\perp}$ with $\nu \perp n$. (As in the case of $\iota(\Sigma)$, here the outward spacelike direction in $(T\Sigma)^{\perp}$ is specified by $-\boldsymbol{H}$.)

In terms of $\tau = -\langle \iota, T_0 \rangle$, which is the time function on Σ associated with ι and T_0, one obtains from (8.16), (8.17) and (8.18) that

$$\mathcal{H}_0(T_0, n_0) = -\frac{1}{8\pi} \left\{ \int_{\Sigma} \sqrt{(\Delta_{\Sigma}\tau)^2 + |\boldsymbol{H}_0|^2(1 + |\nabla_{\Sigma}\tau|^2)} \, d\sigma \right.$$
$$\left. + \int_{\Sigma} (\Delta_{\Sigma}\tau)\theta_0 \, d\sigma - \int_{\Sigma} \left\langle \nabla_{\nabla_{\Sigma}\tau}^{\mathcal{R}^{3,1}} \frac{\boldsymbol{J}_0}{|\boldsymbol{H}_0|}, \frac{\boldsymbol{H}_0}{|\boldsymbol{H}_0|} \right\rangle d\sigma \right\}$$

and

$$\mathcal{H}(T, n) = -\frac{1}{8\pi} \left\{ \int_{\Sigma} \sqrt{(\Delta_{\Sigma}\tau)^2 + |\boldsymbol{H}|^2(1 + |\nabla_{\Sigma}\tau|^2)} \, d\sigma \right.$$
$$\left. + \int_{\Sigma} (\Delta_{\Sigma}\tau)\theta \, d\sigma - \int_{\Sigma} \left\langle \overline{\nabla}_{\nabla_{\Sigma}\tau} \frac{\boldsymbol{J}}{|\boldsymbol{H}|}, \frac{\boldsymbol{H}}{|\boldsymbol{H}|} \right\rangle d\sigma \right\},$$

where Δ_{Σ} denotes the Laplacian and ∇_{Σ} denotes the gradient on Σ, where $\boldsymbol{J}_0$ and $\boldsymbol{J}$ are the duals of $\boldsymbol{H}_0$ and $\boldsymbol{H}$ respectively, and where $\theta_0 := \sinh^{-1}\left(\frac{-\Delta_{\Sigma}\tau}{\sqrt{1 + |\nabla_{\Sigma}\tau|^2}|\boldsymbol{H}_0|} \right)$ and $\theta :=$ $\sinh^{-1}\left(\frac{-\Delta_{\Sigma}\tau}{\sqrt{1 + |\nabla_{\Sigma}\tau|^2}|\boldsymbol{H}|} \right)$. When $\iota(\Sigma)$ indeed is the graph of τ over some closed 2-surface $\hat{\Sigma}$ in the orthogonal complement of T_0, one checks that $\mathcal{H}_0(T_0, n_0)$ is equal to

$$\mathcal{H}_0(T_0, n_0) = -\frac{1}{8\pi} \int_{\hat{\Sigma}} \hat{H} \, d\hat{\sigma}, \tag{8.19}$$

where $\hat{H}$ is the mean curvature of $\hat{\Sigma}$ in $\mathcal{R}^3$. The fact that such an embedding exists for a suitable τ was proved in [101]:

Theorem on Isometric Embedding into $\mathcal{R}^{3,1}$ *If τ is a function on Σ such that $\sigma + d\tau \otimes d\tau$ has positive Gaussian curvature, where σ is the given Riemannian metric on Σ, then there exists a unique isometric embedding of Σ into $\mathcal{R}^{3,1}$ such that Σ is the graph of $t = \tau$ over a convex surface $\hat{\Sigma}$ in $\mathcal{R}^3 = \{t = 0\}$.*

Following (8.12), the Wang–Yau quasi-local energy of Σ with respect to ι and T_0 is given by

$$E_{WY}(\Sigma, \iota, T_0) = \mathcal{H}(T, n) - \mathcal{H}_0(T_0, n_0).$$

Viewed as a functional of τ, $E_{WY}(\Sigma, \iota, T_0) = E_{WY}(\Sigma, \tau)$ takes the form of

$$E_{WY}(\Sigma, \tau) = \frac{1}{8\pi} \int_{\hat{\Sigma}} \hat{H} \, d\hat{\sigma} - \frac{1}{8\pi} \left\{ \int_{\Sigma} \sqrt{(\Delta_{\Sigma}\tau)^2 + |\boldsymbol{H}|^2(1 + |\nabla_{\Sigma}\tau|^2)} \, d\sigma \right.$$
$$\left. + \int_{\Sigma} (\Delta_{\Sigma}\tau)\theta \, d\sigma - \int_{\Sigma} \left\langle \overline{\nabla}_{\nabla_{\Sigma}\tau} \frac{\boldsymbol{J}}{|\boldsymbol{H}|}, \frac{\boldsymbol{H}}{|\boldsymbol{H}|} \right\rangle d\sigma \right\}. \tag{8.20}$$

The Wang–Yau quasi-local mass is defined as

$$\mathfrak{m}_{WY}(\Sigma) = \inf E_{WY}(\Sigma, \tau), \tag{8.21}$$

where the infimum is taken over all admissible functions τ (cf. [100, 101] for the precise meaning of *admissibility*). Wang and Yau proved in [100, 101] that $\mathfrak{m}_{WY}(\Sigma) \geq 0$ and $\mathfrak{m}_{WY}(\Sigma) = 0$ if $\Sigma \subset \mathcal{R}^{3,1}$. In [27, 102], Chen, Wang and Yau showed that $\mathfrak{m}_{WY}(\cdot)$ approaches the ADM mass and Bondi mass at spatial and null infinity respectively.

Related questions

The Euler–Lagrange equation of the quasi-local energy functional $E_{WY}(\Sigma, \tau)$ was derived in [101]. It is a fourth-order equation for the function τ, given by

$$0 = -\left(\hat{H}\hat{\sigma}^{ab} - \hat{\sigma}^{ac}\hat{\sigma}^{bd}\hat{h}_{cd}\right)\frac{(\nabla^2_\Sigma \tau)_{ab}}{\sqrt{1 + |\nabla_\Sigma \tau|^2}}$$

$$+ \operatorname{div}_\Sigma \left(\frac{\nabla_\Sigma \tau}{\sqrt{1 + |\nabla_\Sigma \tau|^2}} \cosh\theta |H| - \nabla_\Sigma \theta - V\right), \tag{8.22}$$

where $\hat{H}$, $\hat{h}$ are the mean curvature and the second fundamental form respectively of the convex surface $\hat{\Sigma}$ in $\mathcal{R}^3$ which has the induced metric $\hat{\sigma} = \sigma + d\tau \otimes d\tau$, and where ∇^2_Σ is the Hessian on Σ, $\theta = \sinh^{-1}\left(\frac{-\Delta_\Sigma \tau}{\sqrt{1 + |\nabla_\Sigma \tau|^2}|H|}\right)$, and V is the vector field on Σ dual to the 1-form $\left\langle \nabla_{(\cdot)} \frac{J}{|H|}, \frac{H}{|H|}\right\rangle$.

A fundamental question regarding the Wang–Yau mass is whether there exists a solution τ_0 to (8.22) that minimizes $E_{WY}(\Sigma, \tau)$, hence realizes $\mathfrak{m}_{WY}(\Sigma)$. In the basic case when Σ bounds a time-symmetric hypersurface Ω in $\mathcal{S}$, one wonders if the global minimum of $E_{WY}(\Sigma, \tau)$ occurs at $\tau_0 = 0$; i.e., if

$$\mathfrak{m}_{WY}(\Sigma) = \mathfrak{m}_{BY}(\Sigma; \Omega).$$

The second variation of $E_{WY}(\Sigma, \tau)$ at $\tau = 0$ (cf. [80, 81]) is given by

$$\delta^2 E_{WY}(\Sigma, \cdot)|_{\tau=0}(\eta) = \frac{1}{8\pi}\int_\Sigma \left[\frac{(\Delta_\Sigma \eta)^2}{H} + (H_0 - H)|\nabla_\Sigma \eta|^2 - \Pi_0(\nabla_\Sigma \eta, \nabla_\Sigma \eta)\right]d\sigma,$$

where Π_0 is the second fundamental form of Σ when it is isometrically embedded in $\mathcal{R}^3$. It is natural to ask if

$$\delta^2 E_{WY}(\Sigma, \cdot)|_{\tau=0}(\eta) \geq 0, \ \forall \ \eta \in W^{2,2}(\Sigma). \tag{8.23}$$

Note that this inequality involves only the induced metric on Σ and the mean curvature function H of Σ in Ω. Therefore, if (8.23) is true, it would provide a necessary condition for a function H on a 2-sphere Σ to be the mean curvature of Σ in some compact Riemannian 3-manifold of non-negative scalar curvature which has Σ as its boundary. In [80], it was shown that such a necessary condition is different from that provided by Shi and Tam in [95].

In the general spacetime setting, assuming $\Sigma \subset \mathcal{S}$ satisfies the property that the vector field V in (8.22) is divergence-free, i.e. $\operatorname{div}_\Sigma V = 0$, Chen, Wang and Yau obtained various results concerning the minimizing property of the critical point $\tau_0 = 0$ in [28]. In [27], Chen, Wang and Yau considered the above question near the spatial or null infinity of an asymptotically flat spacetime and proved that a series solution exists for (8.22) and the solution minimizes $E_{WY}(\Sigma, \cdot)$ locally.

There is another intriguing question related to the reference surface Hamiltonian $\mathcal{H}_0(T_0, n_0)$ used in the definition of Wang–Yau mass. It follows from (8.11) and (8.19) that

$$\frac{1}{8\pi} \int_\Sigma \langle J_0, T_0 \rangle - \langle \nabla^{\mathcal{R}^{3,1}}_{T_0^\parallel} v_0, n_0 \rangle d\sigma = -\frac{1}{8\pi} \int_{\hat{\Sigma}} \hat{H} \, d\hat{\sigma} \tag{8.24}$$

for a closed, spacelike 2-surface $\Sigma \subset \mathcal{R}^{3,1}$ which is a graph over some convex surface $\hat{\Sigma}$ in the orthogonal complement of a constant timelike unit vector T_0. This combined with the classical Minkowski inequality for convex surfaces in $\mathcal{R}^3$ implies

$$-\frac{1}{8\pi} \int_\Sigma \langle J_0, T_0 \rangle + \langle \nabla^{\mathcal{R}^{3,1}}_{T_0^\parallel} v_0, n_0 \rangle d\sigma \geq \sqrt{\frac{|\Sigma|}{4\pi}}, \tag{8.25}$$

where $|\Sigma|$ denotes the area of Σ. Expression (8.25) closely resembles a conjectured inequality

$$-\frac{1}{8\pi} \int_\Sigma \langle J_0, T_0 \rangle \geq \sqrt{\frac{|\Sigma|}{4\pi}} \tag{8.26}$$

proposed by Penrose in [87] for past null convex surfaces in $\mathcal{R}^{3,1}$ and by Gibbons in [54] for general surfaces in $\mathcal{R}^{3,1}$ (also see Mars [74] and Mars and Soria [75]). In [21], Brendle, Hung and Wang proved a sharp Minkowski inequality in the hyperbolic 3-space $\mathbb{H}^3$, which was later shown by Wang in [99] to be equivalent to (8.26) for surfaces $\Sigma \subset \mathbb{H}^3 \subset \mathcal{R}^{3,1}$. Recently, Brendle and Wang proved in [20] that a Penrose–Gibbons type inequality, similar to (8.26), holds for a large class of surfaces in the Schwarzschild spacetime.

We end this subsection by noting some very recent progress [29, 30] towards defining other quasi-local conserved quantities. Via the optimal isometric embeddings used in the definition of $\mathfrak{m}_{WY}(\cdot)$, Chen, Wang and Yau in [29, 30] proposed new definitions of quasi-local angular momentum and quasi-local center of mass.

8.3.2 Hawking mass

Motivated by his investigation of outgoing gravitational radiation, Hawking [56] gave the following definition of quasi-local mass:

$$\mathfrak{m}_H(\Sigma) = \sqrt{\frac{|\Sigma|}{16\pi}} \left(1 - \frac{1}{16\pi} \int_\Sigma |H|^2 \, d\sigma \right), \tag{8.27}$$

where H is the mean curvature vector of Σ in the spacetime.

An important feature of $\mathfrak{m}_H(\cdot)$ is that it varies monotonically under various geometric deformations. This feature is best illustrated in the time-symmetric case in which $\mathfrak{m}_H(\cdot)$ played a successful role in one of the proofs of the Riemannian Penrose inequality. Supposing that Σ lies in a time-symmetric hypersurface M with an induced Riemannian metric g, Geroch [53] discovered that $\mathfrak{m}_H(\cdot)$ is monotonically nondecreasing along connected surfaces $\{\Sigma_t\}$ evolving smoothly in (M, g) by the inverse mean curvature flow (IMCF). A smooth family of closed surfaces $\{\Sigma_t\}$ evolving in a Riemannian manifold is said to satisfy IMCF if their velocity is outward-normal with speed equal to the reciprocal of their mean curvature. This connection between $\mathfrak{m}_H(\cdot)$ and IMCF was later explored by Jang and Wald in [68] to argue for the Riemannian Penrose inequality in the case that the flow remains smooth forever. In [62], Huisken and Ilmanen developed a theory of weak solutions to IMCF along which the Geroch monotonicity of $\mathfrak{m}_H(\cdot)$ is maintained. As an application, they were able to prove the Riemannian Penrose inequality so long as the outermost horizon of (M, g) is connected:

Riemannian Penrose Inequality *Let (M, g) be a complete, asymptotically flat 3-manifold with non-negative scalar curvature and total mass $\mathfrak{m}$ whose outermost horizon Σ has total surface area A. Then*

$$\mathfrak{m} \geq \sqrt{\frac{A}{16\pi}},$$

and the equality holds if and only if (M, g) is isometric to the spatial Schwarzschild manifold of mass $\mathfrak{m}$ outside their respective horizons.

We note that the multiple-component case of the above theorem was established by Bray [17] using a different method. For the current developments on the Penrose inequality, readers are referred to the comprehensive article [74] by Mars.

In the spacetime setting, Bray, Hayward, Mars and Simon [18] examined necessary conditions on *uniformly area-expanding flows* of 2-surfaces in an ambient spacetime under which $\mathfrak{m}_H(\cdot)$ is monotone. A smooth family of surfaces $\{\Sigma_t\}$ evolving in a spacetime is said to be uniformly area-expanding provided $\frac{\partial}{\partial t} d\sigma_t = d\sigma_t$, where $d\sigma_t$ is the area form on Σ_t. It follows from the first variation of area that this is equivalent to requiring $\langle \frac{\partial F}{\partial t}, \mathbf{H} \rangle = -1$ where $\frac{\partial F}{\partial t}$ is the velocity of the flow. Assuming $\frac{\partial F}{\partial t}$ is normal to Σ_t and $\mathbf{H}$ is spacelike, one can write such a flow as

$$\frac{\partial F}{\partial t} = -\frac{\mathbf{H}}{\langle \mathbf{H}, \mathbf{H} \rangle} + \beta \frac{\mathbf{J}}{\langle \mathbf{H}, \mathbf{H} \rangle},$$

where β is some function on Σ_t and $\mathbf{J}$ is the dual of $\mathbf{H}$ in the normal bundle of Σ. Building on a general derivative formula of $\mathfrak{m}_H(\cdot)$ obtained by Malec, Mars and Simon in [73], Bray and Jauregui recently proved in [19] that $\mathfrak{m}_H(\cdot)$ is always monotonically nondecreasing along a uniformly area-expanding flow $\{\Sigma_t\}$ if each Σ_t is connected and *time-flat*. As introduced in [19], a spacelike 2-surface Σ is called time-flat if

$$\mathrm{div}_\Sigma \alpha_H = 0, \tag{8.28}$$

where α_H is the connection 1-form on Σ defined by

$$\alpha_H(v) = \left\langle \overline{\nabla}_v \frac{\boldsymbol{J}}{|\boldsymbol{H}|}, \frac{\boldsymbol{H}}{|\boldsymbol{H}|} \right\rangle.$$

It is conjectured in [19] that a closed, time-flat 2-surface in $\mathcal{R}^{3,1}$ must be contained in a spacelike hyperplane.

The fact that the same connection 1-form α_H also appeared in the study of the Wang–Yau mass $\mathfrak{m}_{WY}(\Sigma)$ (cf. (8.20) and (8.22)) suggests that the time-flat condition (8.28) can be understood in terms of Wang–Yau quasi-local energy $E_{WY}(\Sigma, \cdot)$. By (8.22), Σ is time-flat if and only if $\tau_0 = 0$ is a solution to the critical-point equation of $E_{WY}(\Sigma, \cdot)$. Applying results concerning $E_{WY}(\Sigma, \cdot)$ from [28, 81], Chen, M.-T. Wang and Y.-K. Wang recently established the rigidity of time-flat surfaces in $\mathcal{R}^{3,1}$ in several important cases [26].

8.3.3 Bartnik mass

Another important definition of quasi-local mass, which is analogous to the usual electrostatic capacity of a conducting body, was made by Bartnik in [11, 12].

Most features of the Bartnik mass can be seen in the basic case when Σ lies in a time-symmetric hypersurface. Suppose (M, g) is an asymptotically flat, time-symmetric initial data set and the surface $\Sigma = \partial\Omega$ for some bounded region Ω in M. The Bartnik mass of Σ in Ω is defined as

$$\mathfrak{m}_B(\Sigma; \Omega) = \inf\{\mathfrak{m}(\tilde{M}) \mid \tilde{M} \in \mathcal{PM}_o\}, \tag{8.29}$$

where $\mathfrak{m}(\cdot)$ is the total mass functional and $\mathcal{PM}_o$ denotes the set of asymptotically flat 3-manifolds $\tilde{M}$ of non-negative scalar curvature such that $\tilde{M}$ contains Ω isometrically and $\tilde{M}$ has no horizons. In this setting, a horizon simply means a stable minimal 2-sphere, and the no-horizon condition is imposed to prevent $\inf \mathfrak{m}(\tilde{M})$ from being zero trivially. The non-negativity of $\mathfrak{m}_B(\Sigma; \Omega)$ follows directly from this definition and the positive mass theorem. The strict positivity of the Bartnik mass, namely that $\mathfrak{m}_B(\Sigma; \Omega) > 0$ unless Ω is locally flat, was established by Huisken and Ilmanen in [62]. (Huisken and Ilmanen also introduced a variation of the Bartnik mass in [62] by allowing elements in $\mathcal{PM}_o$ to have horizon boundaries.)

Arguing based on the physical expectation that a mass minimizer should have no matter fields and no gravitational dynamics, Bartnik conjectured in [11, 12] that $\mathfrak{m}_B(\Sigma; \Omega)$ is achieved by the total mass of a unique asymptotically flat 3-manifold (M_s, g_s) with boundary Σ, satisfying

$$g_s|_\Sigma = g|_\Sigma, \quad H(\Sigma, M_s) = H(\Sigma, \Omega), \tag{8.30}$$

which is a time-symmetric hypersurface in a static vacuum spacetime. Here $H(\Sigma, M_s)$ and $H(\Sigma, \Omega)$ are the mean curvature of Σ in M_s and Ω respectively. In [36], Corvino proved that a mass minimizer (in a larger class of competitors) is necessarily static. Partial results concerning the existence of a static metric extension were given in [9, 10, 78]. Besides the main conjecture above, the definition of $\mathfrak{m}_B(\Sigma; \Omega)$ also gives rise to a general extension question – given a bounded Ω, how does one construct an asymptotically flat extension $\tilde{M}$ with zero scalar curvature satisfying the boundary condition (8.30)? As the traditional conformal method is limited by the two boundary conditions in (8.30), a parabolic method was implemented by Bartnik in [13] to construct a large class of extensions with prescribed scalar curvature (also see [95, 96]).

When Ω is contained in an arbitrary, asymptotically flat, initial data set (M, g, K), the Bartnik mass of $\Sigma = \partial \Omega$ is defined similarly as

$$\mathfrak{m}_B(\Sigma; \Omega) = \inf\{\mathfrak{m}(\tilde{M}) \mid \tilde{M} \in \mathcal{PMS}_o\}, \tag{8.31}$$

where $\mathfrak{m}(\tilde{M}) = \sqrt{E^2 - |P|^2}$ is the length of the ADM energy-momentum 4-vector of $\tilde{M}$ and $\mathcal{PMS}_o$ is the set of asymptotically flat initial data sets $(\tilde{M}, \tilde{g}, \tilde{K})$ extending (Ω, g, K) as a solution of the constraint equations and satisfying some "no-horizon" condition (see [12] for details). It is conjectured by Bartnik in [11, 12] that $\mathfrak{m}_B(\Sigma; \Omega)$ is achieved by a stationary initial data set (M_S, g_S, K_S) satisfying the boundary conditions

$$g_S|_\Sigma = g|_\Sigma, \ H(\Sigma, M_S) = H(\Sigma, \Omega),$$
$$\alpha_S(\cdot) = \alpha(\cdot), \quad \mathrm{tr}_\Sigma K_S = \mathrm{tr}_\Sigma K. \tag{8.32}$$

Here $\alpha_S(\cdot) = K_S(\cdot, \nu_S), \alpha(\cdot) = K(\cdot, \nu)$ are the corresponding connection 1-forms on Σ with ν_S, ν being the unit normal to Σ in M_S, Ω respectively. So far, little is known about this stationary extension conjecture. In the boundaryless case, the characterization of stationary initial data sets as critical points for $\mathfrak{m}(\cdot)$ was given in [14].

8.4 Marginally outer trapped surfaces

Consider a co-dimension-two spacelike submanifold Σ in a spacetime $(\mathcal{S}, \mathfrak{g})$. Under suitable orientation assumptions, there exist two (globally defined) families of future-directed null geodesics issuing orthogonally from Σ. If one of the families has vanishing expansion along Σ, then Σ is called a marginally outer trapped surface, or MOTS for short.[2] Physically, MOTSs represent an extreme gravitational situation, and their presence in an initial data set has important physical implications for the spacetime that evolves from this data. As shown in [43], a Penrose-type singularity theorem holds for MOTSs. Under appropriate energy conditions, the maximal globally hyperbolic development of an asymptotically flat initial data set containing a MOTS will be null geodesically incomplete. Moreover, if this development is asymptotically flat in an appropriate sense then the presence of a MOTS signals the existence of a black hole (or white hole); cf. [35, 57]. Thus, MOTSs are associated with both singularities and black holes in spacetime.

MOTSs arose in a more purely mathematical context in the work of Schoen and Yau [94] concerning the existence and regularity of solutions to Jang's equation, in connection with their proof of positivity of mass. Quite surprisingly, MOTSs were shown to be obstructions

---

[2] The terminology 'apparent horizon' is also used, but more often this refers to an outermost MOTS.

to global existence. The analysis in [94] has led in recent years to significant advances in our understanding of the geometric analytic properties of MOTSs. In particular, despite the absence of a variational characterization for MOTSs like that for minimal surfaces, MOTSs have been shown to satisfy a number of analogous properties. We shall touch upon some of these developments here; see the excellent survey article [2] for further details.

8.4.1 Basic definitions

Consider an n-dimensional initial data set (M, g, K). As in the introduction, Section 8.1, we may assume that this initial data set is embedded in a spacetime $(\mathcal{S}, \mathfrak{g})$, so that M is a spacelike hypersurface in $\mathcal{S}$, and g and K are the induced metric and second fundamental form, respectively, of M.

Let Σ be a closed (compact without boundary)[3] two-sided hypersurface in M. Then Σ admits a smooth unit normal field ν in M, unique up to sign. By convention, refer to such a choice as outward pointing. Let u be the future-pointing timelike unit vector orthogonal to M. Then $l_+ = u + \nu$ (resp., $l_- = u - \nu$) is a future-directed outward-pointing (resp., future-directed inward-pointing) null normal vector field along Σ.

Associated with l_+ and l_- are the two *null second fundamental forms*, χ_+ and χ_-, respectively, defined as

$$\chi_\pm : T_p\Sigma \times T_p\Sigma \to \mathcal{R}, \qquad \chi_\pm(X, Y) = \mathfrak{g}(\bar{\nabla}_X l_\pm, Y). \tag{8.33}$$

The *null expansion scalars* (or *null mean curvatures*) $\theta_\pm$ of Σ are obtained by tracing $\chi_\pm$ with respect to the induced metric γ on Σ,

$$\theta_\pm = \text{tr}_\gamma \chi_\pm = \gamma^{AB} \chi_{\pm AB} = \text{div}_\Sigma l_\pm. \tag{8.34}$$

Physically, θ_+ (resp., θ_-) measures the divergence of the outgoing (resp., ingoing) light rays emanating from Σ.

The null expansion scalars can be expressed solely in terms of the initial data (M^n, g, K). We have

$$\theta_\pm = \text{tr}_\gamma K \pm H, \tag{8.35}$$

where H is the mean curvature of Σ within M (given by the divergence of ν along Σ). In particular, in the time-symmetric case, i.e., when $K = 0$ (and hence M is totally geodesic in $(\mathcal{S}, \mathfrak{g})$), θ_+ is just the mean curvature of Σ in M.

For round spheres in Euclidean slices of Minkowski space, or, more generally, for large 'radial' spheres in asymptotically flat initial data sets, one has $\theta_- < 0$ and $\theta_+ > 0$ (with the obvious choice of inside and outside). However, in regions of spacetime where the gravitational field is strong, one may have both $\theta_- < 0$ and $\theta_+ < 0$, in which case Σ is called a *trapped surface*. The concept of a trapped surface was introduced by Penrose

---

[3] In Section 8.4.6, we will consider MOTS Σ with boundary.

and plays a key role in the Penrose singularity theorem: Under appropriate curvature and causality conditions, a spacetime containing a trapped surface must be singular, in the sense of being future null geodesically incomplete; cf. [57, 85, 86].

Focusing attention on the outward null normal only, we say that Σ is an *outer trapped surface* if $\theta_+ < 0$. Finally, we define Σ to be a *marginally outer trapped surface (MOTS)* if θ_+ vanishes identically. Note that in the time-symmetric case, a MOTS is simply a minimal hypersurface in M. It is in this sense that MOTSs may be viewed as spacetime analogues of minimal surfaces.

MOTSs arise naturally in a number of situations. For example, cross sections[4] of the event horizon in stationary (i.e., steady-state) black hole spacetimes are MOTSs. This may be understood as follows (see [57] for details). The event horizon is a null hypersurface in spacetime ruled by null geodesics. One may consider the divergence of these null geodesics towards the future. In the steady-state limit, this divergence vanishes. Since these null geodesics are orthogonal to the cross section, the cross section is a MOTS. In dynamical black hole spacetimes, MOTSs are expected to occur inside the event horizon (and, in fact, they cannot occur outside [35, 57]). There are old heuristic arguments for the existence of MOTSs in this case, based on considering the boundary of the *trapped region* inside the event horizon. These heuristic ideas have recently been made rigorous, first by Andersson and Metzger [5] for three-dimensional initial data sets, and then by Eichmair [41, 42] for initial data sets up to dimension seven. These results rely on a basic existence result for MOTSs discussed in Section 8.4.3.

8.4.2 Stability of MOTSs

In [3, 4], Andersson, Mars and Simon introduced a notion of stability for MOTSs. In this case, stability is associated with variations of the null expansion under deformations of a MOTS.

Let Σ be a MOTS in the initial data set (M, g, K) with outward unit normal v. Consider a normal variation of Σ in M, i.e., a variation $t \to \Sigma_t$ of $\Sigma = \Sigma_0$ with variation vector field $V = \frac{\partial}{\partial t}|_{t=0} = \phi v$, $\phi \in C^\infty(\Sigma)$. Let $\theta(t)$ denote the null expansion of Σ_t with respect to $l_t = u + v_t$, where u is the future-directed timelike unit normal to M and v_t is the outer unit normal to Σ_t in M. A computation shows

$$\frac{\partial \theta}{\partial t}\bigg|_{t=0} = L(\phi), \tag{8.36}$$

where $L : C^\infty(\Sigma) \to C^\infty(\Sigma)$ is the operator [4]

$$L(\phi) = -\Delta\phi + 2\langle X, \nabla\phi \rangle + \left(\frac{1}{2}S - (\mu + J(v)) - \frac{1}{2}|\chi_+|^2 + \operatorname{div} X - |X|^2 \right)\phi. \tag{8.37}$$

In the above, the differential operators are defined with respect to the induced metric $\gamma = \langle , \rangle$ on Σ, S is the scalar curvature of Σ, X is the vector field on Σ dual to the one-form

---

[4] By "cross section" we mean the smooth compact intersection of the event horizon with a spacelike hypersurface.

$K(v, \cdot)|_{T\Sigma}$, and μ and J are as in (8.1) and (8.2). In the time-symmetric case, $X \equiv 0$, and L reduces to the classical stability operator of minimal surface theory.

A MOTS Σ is said to be stable (resp., strictly stable) if there exists a normal variation, with variation vector field $V = \phi v$ (with $\phi > 0$) such that $\frac{\partial \theta}{\partial t}\big|_{t=0} \geq 0$ (resp., $\frac{\partial \theta}{\partial t}\big|_{t=0} > 0$). In other words, a MOTS Σ is stable if there exists an outward variation of Σ such that the null expansion is (infinitesimally) nonincreasing. If Σ is strictly stable then there exists a foliation $t \to \Sigma_t$, $t \in (-\epsilon, \epsilon)$ of $\Sigma = \Sigma_0$ such that Σ_t has negative outward null expansion for $t \in (-\epsilon, 0)$, and positive outward null expansion for $t \in (0, \epsilon)$. In particular, by the maximum principle, Σ must be a locally outermost MOTS. Thus, a stable MOTS Σ may be viewed as an *infinitesimally outermost* MOTS.

The MOTS stability operator L is not in general self-adjoint. However, as observed in [4], the Krein–Rutman theorem implies that the eigenvalue $\lambda_1(L)$ of L with the smallest real part, which is referred to as the principal eigenvalue of L, is necessarily real. Moreover, there exists an associated eigenfunction ϕ which is strictly positive. It can then be shown that Σ is stable (resp., strictly stable) if and only if $\lambda_1(L) \geq 0$ (resp., $\lambda_1(L) > 0$).

Let Σ be a stable MOTS in (M, g, K); hence, there exists $\phi > 0$ such that $L(\phi) \geq 0$. As shown in [52], one may derive from this inequality the following *MOTS stability inequality*,

$$\int_\Sigma |\nabla \psi|^2 + \left(\frac{1}{2}S - (\mu + J(v)) - \frac{1}{2}|\chi|^2\right)\psi^2 \geq 0, \qquad (8.38)$$

for any $\psi \in C^1(\Sigma)$. This inequality is remarkably similar to the well-known stability inequality of minimal surface theory, and in fact reduces to the latter in the time-symmetric case.

Stable MOTSs arise in various situations. For example, outermost MOTSs, in the sense defined below, are stable. This includes, in particular, compact cross sections of the event horizon in stationary black hole spacetimes obeying the null energy condition. More generally, the results mentioned above of Andersson and Metzger, and of Eichmair, establish natural criteria for the existence of outermost MOTSs.

8.4.3 An existence result for MOTSs

In dynamical black hole spacetimes, one would expect trapped or outer trapped surfaces to form in the black hole region. But the occurrence of an outer trapped surface in a spacelike hypersurface that obeys a mild asymptotic flatness condition leads to the existence of a MOTS, as is implied by the following basic existence result:

Theorem 8.4.1 *Let W^n be a connected compact manifold-with-boundary in an initial data set (M^n, g, K), $3 \leq n \leq 7$. Suppose, that the boundary of W can be expressed as a disjoint union, $\partial W = \Sigma_{in} \cup \Sigma_{out}$, such that Σ_{in} is outer trapped (i.e., $\theta_+ < 0$ with respect to the null normal pointing into W) and Σ_{out} is outer untrapped (i.e., $\theta_+ > 0$ with respect to the null normal pointing out of W). Then there exists a stable MOTS Σ in W that separates Σ_{in} from Σ_{out}.*

While variational methods are used to prove the existence of minimal surfaces, such methods are not in general available for proving the existence of MOTSs. In fact, the approach that is taken to prove their existence (as originally proposed by the third author) is to induce blow-up of solutions to Jang's equation.

Existence and regularity results for solutions of Jang's equation for 3-dimensional asymptotically flat initial data sets were established by Schoen and Yau [94] in their proof of the positive mass theorem in the general non-time-symmetric case. They interpreted Jang's equation geometrically as a prescribed mean curvature equation of sorts. Given an initial data set (M, g, K), consider graphs of functions $u : M \to \mathcal{R}$ in the initial data set $(\bar{M}, \bar{g}, \bar{K})$ of one higher dimension, where $\bar{M} = M \times \mathcal{R}$, $\bar{g} = g + dt^2$, and $\bar{K}$ is the pullback of K to $\bar{M}$ via projection along the $\mathcal{R}$ factor. Jang's equation may then be written as

$$H(u) + \mathrm{tr}_u \bar{K} = 0, \qquad (8.39)$$

where $H(u)$ is the mean curvature of graph(u) in $(\bar{M}, \bar{g})$ and $\mathrm{tr}_u \bar{K}$ is the partial trace of $\bar{K}$ over the tangent spaces of graph(u). (Note the similarity to the MOTS equation $\theta_+ = 0$; cf., Equation (8.35).) Schoen and Yau established the global existence of solutions of a regularized version of Equation (8.39), and then studied the behavior of the limits of these solutions as the regularization parameter tends to zero. They showed that the only possible obstruction to global existence for Jang's equation is the presence of MOTSs in the initial data, where, in fact, the solution may have cylindrical blow-ups.

The fact that such solutions can blow up in the presence of MOTSs now becomes a feature (rather than a drawback) of Jang's equation, which allows one to establish the existence of MOTSs. Theorem 8.4.1 was first proved in three dimensions by Andersson and Metzger [5], and then proved up to seven dimensions by Eichmair [41] using different methods and a regularity theory based on geometric measure theory. We refer the reader to [2] for an excellent discussion of Jang's equation and a detailed account of its use in proving the existence of MOTSs.

Let Σ be a MOTS in (M, g, K), and let $U \subset M$ be a neighborhood of Σ that is separated by Σ. (Since Σ is two-sided, such a neighborhood always exists.) We say the Σ is outermost in U if there are no outer trapped ($\theta_+ < 0$) or marginally outer trapped ($\theta_+ = 0$) surfaces in U outside of, and homologous to, Σ. One easily sees that outermost MOTSs are necessarily stable. Using the Andersson–Metzger results [5] and those of Eichmair [41, 42], in the setting of Theorem 8.4.1, the MOTS Σ can be constructed so as to be outermost in W.

8.4.4 Black hole topology

The Kerr solution is an asymptotically flat,[5] stationary (i.e., time-independent and rotating) axisymmetric solution of the vacuum Einstein equations. It is determined by two parameters: the mass parameter m and the angular momentum parameter a. When $a = 0$, the Kerr

---

[5] In the spacetime setting this means "asymptotically Minkowskian". One way to make this precise is in terms of *conformal compactification*; see e.g. [57, 98].

solution reduces to the Schwarzschild solution. The Kerr solution contains an event horizon (provided $a < m$), and hence represents a steady-state rotating black hole.

It is a widely held belief that "true" astrophysical black holes "settle down" to a Kerr solution in the steady-state limit. This belief is largely based on classical results of Carter, Hawking and Robinson that establish the uniqueness of the Kerr solution among all asymptotically flat, stationary solutions of the vacuum Einstein equations; cf. [57]. (The proof assumes analyticity, but there has been recent progress in removing this assumption; cf. [6].)

A basic step in the proof of the uniqueness of the Kerr solution is Hawking's theorem on the topology of black holes in $3 + 1$ dimensions.

Theorem 8.4.2 (Hawking's black hole topology theorem) *Suppose* $(\mathcal{S}, \mathfrak{g})$ *is a* $(3 + 1)$-*dimensional asymptotically flat stationary black hole spacetime obeying the dominant energy condition. Then cross sections* Σ *of the event horizon are topologically 2-spheres.*

Hawking's proof is variational in nature. Using the dominant energy condition and the Gauss–Bonnet theorem, he shows that if Σ has genus ≥ 1 then Σ can be deformed outward to an outer trapped surface (the torus is actually borderline for the argument). However, there can be no outer trapped (or even marginally outer trapped) surface outside the event horizon. Such a surface would be visible from 'null infinity', but there are arguments precluding that possibility [35, 98].

String theory and various related developments (e.g., gauge/gravity duality) have generated a great deal of interest in gravity in higher dimensions, and, in particular, in higher-dimensional black holes. In this context, one of the first questions to arise was whether black hole uniqueness holds in higher dimensions.

With impetus coming from the development of string theory, in the 1980s Myers and Perry [82] constructed natural higher-dimensional generalizations of the Kerr solution, which, in particular, have spherical horizon topology. These Myers–Perry black holes painted a picture consistent with the Kerr solution. However, in 2002, Emparan and Reall [46] discovered a remarkable example of a $(4 + 1)$-dimensional asymptotically flat, stationary vacuum black hole spacetime with horizon topology $S^2 \times S^1$, which they dubbed the black ring. Their example demonstrated that black hole uniqueness does not hold in higher dimensions, and that horizon topology need not be spherical. This caused a great surge of activity in the study of higher dimensional black holes; for a detailed account of these developments, see the recent monograph [61].

The question naturally arose as to what horizon topologies are allowed in higher dimensions; what, if any, are the restrictions on the topology? This question was addressed in [52], in which a higher-dimensional generalization of Hawking's topology theorem was obtained. Recall that a manifold Σ is of *positive Yamabe type* if it admits a metric of positive scalar curvature.

Theorem 8.4.3 ([52]) *Let* Σ *be a stable MOTS in an* n-*dimensional,* $n \geq 3$, *initial data set* (M, g, K), *satisfying the dominant energy condition,* $\mu \geq |J|$. *Then* Σ *is of positive Yamabe type, unless* Σ *is Ricci flat,* $\chi_+ = 0$ *and* $\mu + J(\nu) = 0$ *along* Σ.

Thus, apart from these exceptional circumstances, Σ must be of positive Yamabe type. But Σ being of positive Yamabe type implies many well-known restrictions on the topology. We consider here two basic examples, and for simplicity we assume Σ is orientable.

Case 1. dim $\Sigma = 2$ (dim $\Sigma = 3 + 1$). In this case, Σ being of positive Yamabe type means that Σ admits a metric of positive Gaussian curvature. Hence, by the Gauss–Bonnet theorem, Σ is topologically a 2-sphere, and we recover Hawking's theorem.

Case 2. dim $\Sigma = 3$ (dim $\Sigma = 4 + 1$). As shown in [55, 92], if Σ is of positive Yamabe type then it cannot contain any $K(\pi, 1)$ spaces in its prime decomposition. It follows that Σ must be a connected sum of $S^2 \times S^1$ spaces and spherical spaces S^3/Γ (where we have made use of the positive resolution of the Poincaré conjecture). Thus, the basic horizon topologies in the case dim $\Sigma = 3$ are S^3 (realized by the Myers–Perry black holes) and $S^2 \times S^1$ (realized by the black ring).

In the time-symmetric case, Theorem 8.4.3 reduces to a result of Schoen and Yau [92], which is fundamental to their study of manifolds of positive scalar curvature. Analogous to this Riemannian result, Theorem 8.4.3 is a consequence of the stability inequality (8.38). Let λ_1 be the principal eigenvalue of the operator $L_0 = -\triangle + \left(\frac{1}{2}S - (\mu + J(v)) - \frac{1}{2}|\chi|^2\right)$. It follows from (8.38) and the Raleigh formula for self-adjoint operators that $\lambda_1 \geq 0$. Consider on Σ the conformally related metric, $\tilde{\gamma} = \phi^{\frac{2}{n-2}}\gamma$, where γ is the induced metric on Σ and ϕ is a positive eigenfunction corresponding to $\lambda_1(L_0)$. A computation, using $L_0(\psi) = \lambda_1\psi$, shows that the scalar curvature $\tilde{S}$ of the metric $\tilde{\gamma}$ is given by

$$\tilde{S} = \psi^{-\frac{2}{n-2}}\left(2\lambda_1 + 2(\mu + J(v)) + |\chi_+|^2 + \frac{n-1}{n-2}\frac{|\nabla\psi|^2}{\psi^2}\right). \tag{8.40}$$

Since $\mu + J(v) \geq \mu - |J| \geq 0$, we have that $\tilde{S} \geq 0$. By further standard metric deformations, the scalar curvature of Σ can be made strictly positive, unless various quantities vanish identically. (Alternatively, (8.38) implies that the principal eigenvalue of the conformal Laplacian is non-negative, and in the case $n \geq 4$ one can use this to obtain a metric of positive scalar curvature.)

A drawback of Theorem 8.4.3 is that it allows certain possibilities that one would like to rule out. For example the theorem does not rule out the possibility of a vacuum black hole spacetime with toroidal horizon topology. If one assumes that Σ is an outermost MOTS, as defined in Section 8.4.3, then the exceptional case in Theorem 8.4.3 can be eliminated; cf. [50].

We refer the reader to [61, Chapter 7] for further discussion concerning the topology of higher-dimensional black holes.

8.4.5 The topology of initial data sets

In this subsection we consider the topology of asymptotically flat initial data sets. Although there is interest in higher dimensions, here we will restrict our attention to three-dimensional initial data manifolds. More specifically, we explore the relationship between the topology of initial data manifolds and the occurrence of MOTSs in the initial data. There is both a physical and a geometric rationale for considering this. As noted at the beginning of the section, the presence of a MOTS in an initial data manifold has important physical implications for the spacetime that evolves from this data. These physical considerations, taken together with the connections between MOTSs and topology, relate naturally to the notion of topological censorship, as described in detail in [43]. We will briefly address this later in this subsection. On the geometric side, there is a long tradition in Riemannian geometry of using minimal surfaces to study the topology of a Riemannian manifold. In some sense, we are considering here the non-time-symmetric version of this program.

General relativity imposes no a priori restrictions on the topology of asymptotically flat solutions to the Einstein equations. This is made explicit in the following result of Isenberg, Mazzeo and Pollack [65].

Theorem 8.4.4 *Let N be any compact n-dimensional manifold, and $p \in N$ arbitrary. Then $N \setminus \{p\}$ admits an AF initial data set satisfying the vacuum constraint equations.*

However, as shown in [43], for three-dimensional asymptotically flat initial data sets, if the topology is nontrivial, there must be MOTSs present in the initial data. In fact, for these results, a slightly more general notion of MOTS is needed.

Definition 8.4.1 (Immersed MOTS [43]) *Given an initial data set (M, g, K), we say that a subset $\Sigma \subset M$ is an immersed MOTS if there exists a finite cover $\tilde{M}$ of M with covering map $p : \tilde{M} \to M$ and there exists a MOTS $\tilde{\Sigma}$ in $\tilde{M}$ (with respect to the pulled-back data (p^*h, p^*K)) such that $p(\tilde{\Sigma}) = \Sigma$.*

A simple example of an immersed MOTS (that is not a MOTS) occurs in the so-called $\mathcal{RP}^3$ geon; see e.g. [48] for a detailed description. The $\mathbb{RP}^3$ geon is a globally hyperbolic spacetime that is double covered by the extended Schwarzschild spacetime. Its Cauchy surfaces have the topology of $\mathcal{RP}^3$ minus a point. The Cauchy surface that is covered by the $T = 0$ slice in the extended Schwarzschild spacetime has one asymptotically flat end (identical to an end in the Schwarzschild slice), and contains a projective plane Σ that is covered by the unique minimal sphere $\tilde{\Sigma}$ in the Schwarzschild slice. Since Σ is not two-sided, it is not a MOTS. However, since the slice $T = 0$ is totally geodesic, $\tilde{\Sigma}$ is a MOTS, and so Σ is an immersed MOTS.

The following result is obtained in [43]:

Theorem 8.4.5 *Let (M, g, K) be a 3-dimensional AF initial data set. If M is not diffeomorphic to $\mathcal{R}^3$ then M contains an immersed MOTS.*

In the assumption of asymptotic flatness above, more than one end is allowed. Thus, there is a compact set $K \subset M$ such that $M \setminus K$ is diffeomorphic to $\sqcup_{i=1}^{k} E_i$, where each E_i is diffeomorphic to $\mathcal{R}^3 \setminus B_1(\mathbf{0})$, with the data on each end satisfying the limit conditions $g \to \delta$ and $K \to 0$ at suitable rates as $|x| \to \infty$ (cf. Section 8.2.2).

It was noted at the beginning of this section that a Penrose-type singularity theorem holds for MOTSs. This theorem extends easily to immersed MOTSs; cf. [43, Theorem 3.2,

Corollary 3.5]. Thus, Theorem 8.4.5 may be viewed as an 'initial data singularity theorem' (induced by topology): If the topology of M^3 is not trivial then M^3 contains an immersed MOTS, and we expect the spacetime that evolves from this data to be singular in the sense of being null geodesically incomplete.

It is interesting to note that Theorem 8.4.5 does not require the dominant energy condition, or any other energy condition. In this sense, Theorem 8.4.5 is a pure topological result: MOTSs detect nontrivial topology. In fact, this theorem may be viewed as a non-time-symmetric version of a theorem of Meeks, Simon and Yau (1982), which implies (without any curvature assumption) that an asymptotically flat 3-manifold that is not diffeomorphic to $\mathcal{R}^3$ contains an embedded stable minimal sphere or projective plane.

The proof of Theorem 8.4.5 relies in an essential way on geometrization. It is a consequence of geometrization that the fundamental group of a compact 3-manifold N^3 is *residually finite*. In fact this result was known since the 1980s, modulo geometrization, from work by Hempel [59]. Residual finiteness guarantees that if $\pi_1(N) \neq 0$, then N admits a finite nontrivial cover, even if $\pi_1(N)$ is infinite.

We give an outline of the proof of Theorem 8.4.5. In particular, we argue the contrapositive: Suppose that M does not contain any immersed MOTSs.

Step 1: M has only one end. If not, truncate the ends of M to obtain a compact body W with boundary, whose boundary components correspond to the ends of M. Label one of the boundary components Σ_{in} and label the union of the others Σ_{out}. By taking the truncations sufficiently far out, the asymptotic flatness guarantees that the barrier conditions in Theorem 8.4.1 are satisfied, and hence W contains a MOTS, contrary to assumption.

Step 2: M is orientable. If not, the orientable double cover $\tilde{M}$ has two asymptotically flat ends, and hence, as in Step 1, contains a MOTS Σ. The projection of Σ under the covering map is then an immersed MOTS in M, contrary to assumption.

Step 3: Hence, we can write M as $M = \mathcal{R}^3 \# N$, where N is a compact orientable 3-manifold. We argue that N must be simply connected. If not, by residual finiteness, N admits a nontrivial finite cover. It follows that M admits a finite nontrivial cover $\tilde{M}$. But then $\tilde{M}$ has more than one (but finitely many) asymptotically flat ends. As in Step 1, $\tilde{M}$ contains a MOTS, and so M contains an immersed MOTS, contrary to assumption.

Step 4: Hence, by Poincaré, N is diffeomorphic to S^3 and thus M is diffeomorphic to $\mathcal{R}^3$, as was to be shown.

We now wish to consider initial data sets with horizons. Let (M^3, g, K) be an asymptotically flat initial data set such that M^3 is a manifold-with-boundary, whose boundary ∂M is a MOTS. Physically, we are to think of M as an asymptotically flat spacelike slice in the domain of outer communications (the DOC – the region outside all black holes and white holes) whose boundary ∂M corresponds to a cross section of the event horizon. At the initial data level, this cross section is represented by a MOTS. Since there can be no MOTS (or immersed MOTS) in the DOC, we will adopt this as an assumption on $M \setminus \partial M$.

The techniques used in the proof of Theorem 8.4.5 carry over, with some minor modifications, to this setting of MOTS boundary; hence one can prove the following in a fairly similar manner [43]:

Theorem 8.4.6 *Let (M, g, K) be a 3-dimensional asymptotically flat initial data set such that M is a manifold-with-boundary, whose boundary ∂M is a compact MOTS (perhaps with multiple components). If there are no immersed MOTS in $M \setminus \partial M$, then M is diffeomorphic to $\mathcal{R}^3$ minus a finite number of open balls.*

Topological censorship has to do with the idea that the DOC should have simple topology, that nontrivial topology (which tends to lead to the formation of singularities), should end up hidden behind the event horizon. There are a number of spacetime results that support the principle of topological censorship; see, e.g., [35, 48, 49, 51]. Theorem 8.4.6 may be viewed as an *initial data version* of topological censorship: M^3, which represents a spacelike slice in the DOC, has topology as simple as possible.

8.4.6 MOTS and the spacetime positive mass theorem

The development of the theory of MOTSs has made it possible to give a direct MOTS-based proof of the general non-time-symmetric version of the positive mass theorem in a manner analogous to Schoen and Yau's [91] original minimal-surface-based proof in the time-symmetric case. An outline of this approach for $n = 3$ was first described in [90]; a complete, detailed proof, as well as an extension to higher dimensions, has now been given in [44]. Here we discuss some of the main ideas.

For this purpose one now works with a compact MOTS Σ with boundary. The notion of stability discussed in Section 8.4.2 carries over in a straightforward manner to this case. One now considers normal variations with the variational vector field $\phi\nu$ for smooth functions ϕ on Σ that vanish on $\partial\Sigma$. Equations (8.36) and (8.37) still hold for all such functions ϕ. Then Σ is stable if and only if $\lambda_1(L) \geq 0$, where $\lambda_1(L)$ is the principal Dirichlet eigenvalue of the MOTS stability operator L (see equation (8.37)). If Σ is stable, it follows that the stability inequality (8.38) holds for all $\psi \in C^1(\Sigma)$ that vanish on $\partial\Sigma$.

The 'Plateau problem for MOTS' has been solved by Eichmair [41]. Eichmair has established the existence of MOTSs with boundary under barrier conditions somewhat similar to Theorem 8.4.1. We paraphrase Theorem 5.1 in [41]: Let W^n be a connected compact manifold with connected boundary in an initial data set (M^n, g, K), $3 \leq n \leq 7$. Let $\Gamma^{n-2} \subset \partial W^n$ be a closed $(n-2)$-dimensional submanifold that separates ∂W^n into two parts, along one of which $\theta^+ > 0$ with respect to the null normal pointing out of W, and along the other $\theta^+ < 0$ with respect to the null normal pointing into W. Then there exists a MOTS Σ with boundary $\partial\Sigma = \Gamma$, and with $\Sigma \setminus \partial\Sigma \subset \text{int } W$. Moreover, as shown in [45, Theorem 2.3], there exists such a MOTS Σ that is stable.

With these preliminaries, we can now describe some elements of the MOTS-based proof of the positive mass theorem presented in [44].

Theorem 8.4.7 *Let (M, g, K) be an n-dimensional, $3 \leq n \leq 7$, asymptotically flat initial data set satisfying the dominant energy condition. Then $E \geq |P|$, where (E, P) is the ADM energy-momentum vector of (M, g, K).*

The proof is an induction on dimension. We shall focus on the $n = 3$ base step of the induction. As in [44], for simplicity we shall assume (M, g, K) has only one end. Let (x^1, x^2, x^3) denote Euclidean coordinates on the end.

Suppose that the mass inequality is violated for (M, g, K); i.e., that $E < |P|$. By the density theorem, [44, Theorem 18] (see the discussion at the end of Section 8.2.2), we may assume without loss of generality that (M, g, K) has harmonic asymptotics and that the dominant energy condition holds strictly: $\mu > |J|$. We may further assume that P points in the $-\partial_3$ direction.

The harmonic asymptotics and the assumption that $E < |P|$ imply that for $\Lambda > 0$ sufficiently large, the plane $x^3 = \Lambda$ has null expansion $\theta_+ > 0$ with respect to the upward pointing normal. In fact, a computation shows that, with these asymptotics, $\theta_+ = 2(|P| - E)\Lambda r^{-3} + O(r^{-3})$ ([44, Lemma 5]). Similarly, for $\Lambda > 0$ sufficiently large, the plane $x^3 = -\Lambda$ has null expansion $\theta_+ < 0$ with respect to the upward-pointing normal. Fix Λ so that both null expansion inequalities hold.

Consider the compact region $W_{\rho,\Lambda}$ bounded by the planes $x^3 = \pm\Lambda$ and the cylinder $C_\rho : (x^1)^2 + (x^2)^2 = \rho^2$. Let $\Gamma_{\rho,0} \subset \partial W_{\rho,\Lambda}$ be the circle obtained as the intersection of C_ρ with the plane $x^3 = 0$. For Λ as chosen, and for sufficiently large ρ, $W_{\rho,\Lambda}$ satisfies the barrier conditions with respect to $\Gamma_{\rho,0}$ of Eichmair's existence result for MOTSs with boundary, as discussed above. (There is a small technicality due to the 'corners' where C_ρ and the planes $x^3 = \pm\Lambda$ meet, but this can be dealt with.) Thus, essentially by [41, Theorem 5.1], there exists a stable MOTS Σ_ρ spanning $\Gamma_{\rho,0}$ and contained between the planes $x^3 = \pm\Lambda$.

Now let $\rho \to \infty$: By [44, Lemma 8], which relies on Eichmair's GMT-based compactness and regularity theory for MOTS [41, 42], there exists a sequence $\rho_j \to \infty$ and an associated sequence of stable MOTSs Σ_j spanning $\Gamma_{\rho_j,0}$, and contained between the planes $x^3 = \pm\Lambda$, such that the sequence Σ_j converges to a smooth complete MOTS Σ_∞ asymptotic to a plane $x^3 = a$. As a limit of stable MOTSs, Σ_∞ satisfies the stability inequality (8.38) for all $\psi \in C^1(\Sigma_\infty)$ with compact support. Setting $\psi = 1$ in (8.38) (which is justified by a careful cutoff argument; see [44, Section 8.4]) and using the strict dominant energy condition gives

$$\int_{\Sigma_\infty} K > 0,\tag{8.41}$$

where K is the Gaussian curvature of Σ_∞. On the other hand, asymptotic estimates and the Gauss–Bonnet formula imply

$$\int_{\Sigma_\infty} K = 2\pi \chi(\Sigma_\infty) - 2\pi \leq 0,\tag{8.42}$$

which yields a contradiction. Thus, when $n = 3$, $E \geq |P|$.

Now consider the induction step. Let (M, g, K) be an n-dimensional, $4 \leq n \leq 7$, asymptotically flat initial data set such that $E < |P|$. The regions $W_{\rho, \Lambda}$ can be defined just as in the case $n = 3$, and, for large Λ and ρ, will have boundaries with the same convexity properties. Hence, by a similar procedure, one can obtain a complete MOTS Σ_∞ as a limit of stable MOTS Σ_j, such that each Σ_j is contained between the planes $x^n = \pm \Lambda$ and has boundary $\partial \Sigma_j = C_{\rho_j} \cap \{x^n = a_j\}$, for some $a_j \in (-\Lambda, \Lambda)$. But now the sequence Σ_j is constructed in such a way as to satisfy a somewhat stronger stability condition. By a delicate 'height-picking' argument (which is more subtle in the MOTS case), one can construct the sequence Σ_j so that Σ_∞ satisfies the MOTS stability inequality not only for normal variations of compact support, but also for variations that are approximately vertical translations near infinity. It is then shown that the induced metric on Σ_∞ can be conformally deformed to an asymptotically flat metric having zero scalar curvature and negative energy. But this contradicts the positive mass theorem in dimension $n - 1$.

Acknowledgments

The work of the first author was partially supported by NSF grant DMS-1313724 and by a grant from the Simons Foundation (Grant No. 63943). The work of the second author was partially supported by Simons Foundation Collaboration Grant for Mathematicians #281105. The work of the third author was partially supported by NSF grant DMS-1105323.

References

[1] P. T. Allen, A. Clausen, and J. Isenberg, Near-constant mean curvature solutions of the Einstein constraint equations with non-negative Yamabe metrics, *Class. Quant. Grav.* **25** (2008), 075009, 15 pp.

[2] L. Andersson, M. Eichmair, and J. Metzger, Jang's equation and its applications to marginally trapped surfaces, in: *Complex analysis and dynamical systems IV: part 2. general relativity, geometry, and PDE*, Contemporary Mathematics, vol. 554, (AMS and Bar-Ilan), 2011.

[3] L. Andersson, M. Mars, and W. Simon, Local existence of dynamical and trapping horizons, *Phys. Rev. Lett.* **95** (2005), 111102.

[4] ———, Stability of marginally outer trapped surfaces and existence of marginally outer trapped tubes, *Adv. Theor. Math. Phys.* **12** (2008), no. 4, 853–888.

[5] L. Andersson and J. Metzger, The area of horizons and the trapped region, *Commun. Math. Phys.* **290** (2009), no. 3, 941–972.

[6] S. Alexakis, A. D. Ionescu, and S. Klainerman, Uniqueness of smooth stationary black holes in vacuum: small perturbations of the Kerr spaces, *Commun. Math. Phys.* **299** (2010), no. 1, 89–127.

[7] R. Arnowitt, S. Deser, and C. W. Misner, Coordinate invariance and energy expressions in general relativity, *Phys. Rev.* **122**, (1961), 997–1006.

[8] ———The dynamics of general relativity, in *Gravitation: an introduction to current research*, 1962, pp. 227–265 Wiley, New York. arXiv:gr-qc/0405109

[9] M. Anderson, Local existence and uniqueness for exterior static vacuum Einstein metrics, arXiv:1308.3642.

[10] M. Anderson, and M. Khuri, On the Bartnik extension problem for the static vacuum Einstein equations, *Class. Quant. Grav.* **30**, (2013), 125005.

[11] R. Bartnik, New definition of quasilocal mass, *Phys. Rev. Lett.*, **62**, (1989), 2346–2348.

[12] _____, Energy in general relativity, Lecture Notes at National Tsing Hua Univeristy, Hsinchu, Taiwan, July 1992.

[13] _____, Quasi-spherical metrics and prescribed scalar curvature, *J. Diff. Geom.* **37** (1993), 31–71.

[14] _____ Phase space for the Einstein equations, *Commun. Anal. Geom.* **13**, (2005), no. 5, 845–885.

[15] R. Bartnik and J. Isenberg, The constraint equations, in *The Einstein equations and the large scale behavior of gravitational fields*, 1–38, Birkhäuser, Basel, 2004.

[16] R. Beig, and P. T. Chruściel, Killing vectors in asymptotically flat space-times. I. Asymptotically translational Killing vectors and the rigid positive energy theorem, *J. Math. Phys.* **37** (1996), no. 4, 1939–1961.

[17] H. Bray, Proof of the Riemannian Penrose inequality using the positive mass theorem, *J. Diff. Geom.* **59** (2001), no. 2, 177–267.

[18] H. Bray, S. Hayward, M. Mars, and W. Simon, Generalized inverse mean curvature flows in space-time, *Commun. Math. Phys.* **272** (2007), no. 1, 119–138.

[19] H. Bray, and J. L. Jauregui, Time flat surfaces and the monotonicity of the spacetime Hawking mass, arXiv:1310.8638.

[20] S. Brendle, and M.-T. Wang, A Gibbons–Penrose inequality for surfaces in Schwarzschild spacetime, to appear in *Commun. Math. Phys.*, arXiv:1303.1863.

[21] S. Brendle, P.-K. Hung, and M.-T. Wang, A Minkowski-type inequality for hypersurfaces in the Anti-deSitter–Schwarzschild manifold, arXiv:1209.0669.

[22] J. D. Brown, and J. W. York, Jr., Quasilocal energy in general relativity, in *Mathematical aspects of classical field theory* (Seattle, WA, 1991), volume 132 of Contemporary Mathematics, pages 129–142. American Mathematical Society, Providence, RI, 1992.

[23] _____, Quasilocal energy and conserved charges derived from the gravitational action, *Phys. Rev. D* (3), **47** (1993), no. 4, 1407–1419.

[24] I. S. Booth, and R. B. Mann, Moving observers, nonorthogonal boundaries, and quasilocal energy, *Phys. Rev. D* (3), **59** (1999), no. 6, 064021.

[25] M. Cantor, The existence of non-trivial asymptotically flat initial data for vacuum spacetimes, *Commun. Math. Phys.* **57** (1977), 83–96.

[26] P.-N. Chen, M.-T. Wang, M.-T. and Y.-K.Wang, Rigidity of time-flat surfaces in the Minkowski spacetime, arXiv:1310.6081.

[27] P.-N. Chen, M.-T. Wang, and S.-T. Yau, Evaluating quasilocal energy and solving optimal embedding equation at null infinity, *Commun. Math. Phys.* **308** (2011), no. 3, 845–863.

[28] _____, Minimizing properties of critical points of quasi-local energy, to appear in *Commun. Math. Phys.*, arXiv:1302.5321.

[29] _____, Quasilocal angular momentum and center of mass in general relativity, arXiv:1312.0990.

[30] _____, Conserved quantities in general relativity: from the quasi-local level to spatial infinity, arXiv:1312.0985.

[31] Y. Choquet-Bruhat and J. W. York, The Cauchy problem, *General relativity and gravitation*, Vol. 1, Plenum, New York, 1980, 99–172.

[32] P. T. Chruściel, J. Corvino, and J. Isenberg, Construction of N-body initial data sets in general relativity, *Commun. Math. Phys.* **304** (2011), 637–647.

[33] P. T. Chruściel and E. Delay, Existence of non-trivial, vacuum, asymptotically simple spacetimes, *Class. Quant. Grav.* **19** (2002), L71–L79.

[34] _____, On mapping properties of the general relativistic constraints operator in weighted function spaces, with applications, *Mém. Soc. Math. Fr.* (N. S.) No. 94 (2003), vi + 103 pp.

[35] P. T. Chruściel, G. J. Galloway, and D. Solis, Topological censorship for Kaluza–Klein space-times, *Ann. Inst. Henri Poincaré* **10** (2009), no. 5, 893–912.

[36] J. Corvino, Scalar curvature deformation and a gluing construction for the Einstein constraint equations, *Commun. Math. Phys.* **214** (2000), 137–189.

[37] ———, On the existence and stability of the Penrose compactification, *Ann. Inst. Henri Poincaré* **8** (2007), 597–620.

[38] J. Corvino and D. Pollack, Scalar curvature and the Einstein constraint equations, in *Surveys in geometric analysis and relativity*, 145–188, Advanced Lectures in Mathathematics (ALM), **20**, International Press, Somerville, MA, 2011.

[39] J. Corvino and R. Schoen, On the asymptotics for the vacuum Einstein constraint equations, *J. Diff. Geom.* **73** (2006), 185–217.

[40] M. Dahl, R. Gicquaud, and E. Humbert, A limit equation associated to the solvability of the vacuum Einstein constraint equations by using the conformal method, *Duke Math. J.* **161** (2012), 2669–2697.

[41] M. Eichmair, The Plateau problem for marginally outer trapped surfaces, *J. Diff. Geom.* **83** (2009), no. 3, 551–583.

[42] ———, Existence, regularity, and properties of generalized apparent horizons, *Commun. Math. Phys.* **294** (2010), no. 3, 745–760.

[43] M. Eichmair, G. J. Galloway, and D. Pollack, Topological censorship from the initial data point of view, *J. Diff. Geom.* **95** (2013), no. 3, 389–405.

[44] M. Eichmair, L.-H. Huang, D. A. Lee, and R. Schoen, The spacetime positive mass theorem in dimensions less than eight, 2011, arXiv:1110.2087v1.

[45] M. Eichmair and J. Metzger, Jenkins–Serrin type results for the Jang equation, 2012, arXiv:1205.4301.

[46] R. Emparan and H. S. Reall, A rotating black ring solution in five dimensions, *Phys. Rev. Lett.* **88** (2002), no. 10, 101101, 4.

[47] R. J. Epp, Angular momentum and an invariant quasilocal energy in general relativity, *Phys. Rev. D* **62** (2000), no. 12. 124108.

[48] J. L. Friedman, K. Schleich, and D. M. Witt, Topological censorship, *Phys. Rev. Lett.* **71** (1993), no. 10, 1486–1489.

[49] G. J. Galloway, On the topology of the domain of outer communication, *Class. Quant. Grav.* **12** (1995), no. 10, L99–L101.

[50] ———, Rigidity of marginally trapped surfaces and the topology of black holes, *Commun. Anal. Geom.* **16** (2008), no. 1, 217–229.

[51] G. J. Galloway, K. Schleich, D. M. Witt, and E. Woolgar, Topological censorship and higher genus black holes, *Phys. Rev. D* (3) **60** (1999), no. 10, 104039, 11.

[52] G. J. Galloway and R. Schoen, A generalization of Hawking's black hole topology theorem to higher dimensions, *Commun. Math. Phys.* **266** (2006), no. 2, 571–576.

[53] R. Geroch, Energy extraction, *Ann. N.Y. Acad. Sci.* **224** (1973), 108–117.

[54] G. W. Gibbons, Collapsing shells and the isoperimetric inequality for black holes, *Class. Quant. Grav.* **14** (1997), 2905–2915.

[55] M. Gromov and H. B. Lawson, Jr., Positive scalar curvature and the Dirac operator on complete Riemannian manifolds, *Inst. Hautes Études Sci. Publ. Math.* (1983), no. 58, 83–196 (1984).

[56] S. W. Hawking, Gravitational radiation in an expanding universe, *J. Math. Phys.* **9** (1968), 598–604.

[57] S. W. Hawking and G. F. R. Ellis, *The large scale structure of space-time*, Cambridge University Press, London, 1973, Cambridge Monographs on Mathematical Physics, No. 1.

[58] S. W. Hawking, and G. T. Horowitz, The gravitational Hamiltonian, action, entropy and surface terms, *Class. Quant. Grav.* **13** (1996) (6), 1487–1498.

[59] J. Hempel, Residual finiteness for 3-manifolds, in *Combinatorial group theory and topology* (Alta, Utah, 1984), *Ann. of Math. Stud.*, vol. 111, Princeton University Press, Princeton, NJ, 1987, pp. 379–396.

[60] M. Holst, G. Nagy, and G. Tsogtgerel, Rough solutions of the Einstein constraints on closed manifolds without near-CMC conditions, *Commun. Math. Phys.* **288** (2009), 547–613.

[61] G. Horowitz (ed.), *Black holes in higher dimensions*, Cambridge University Press, London, 2012.

[62] G. Huisken, and T. Ilmanen, The inverse mean curvature flow and the Riemannian Penrose inequality, *J. Diff. Geom.* **59** (2001), no. 3, 353–437.

[63] J. Isenberg, Constant mean curvature solutions of the Einstein constraint equations on closed manifolds, *Class. Quant. Grav.* **12** (1995), 2249–2274.

[64] J. Isenberg and V. Moncrief, Some results on nonconstant mean curvature solutions of the Einstein constraint equations, in *Physics on manifolds* (Paris 1992), Mathematical Physics Studies, vol. 15, Kluwer, Dordrecht, 1994, 295–302.

[65] J. Isenberg, R. Mazzeo, and D. Pollack, On the topology of vacuum spacetimes, *Ann. Henri Poincaré* **4** (2003), no. 2, 369–383.

[66] J. Isenberg and N. Ó Murchadha, Non-CMC conformal data sets which do not produce solutions of the Einstein constraint equations, *Class. Quant. Grav.* **21** (2004), S233–S241, A spacetime safari: essays in honor of Vincent Moncrief.

[67] P. S. Jang, On the positivity of energy in general relativity, *J. Math. Phys.* **19** (1978), no. 5, 1152–1155.

[68] P. S. Jang, and R. Wald, The positive energy conjecture and the cosmic censor hypothesis, *J. Math. Phys.* **18** (1977), 41–44.

[69] J. Kijowski, A simple derivation of canonical structure and quasi-local Hamiltonians in general relativity, *Gen. Rel. Grav.* **29** (1997), no. 3, 307–343.

[70] A. Lichnerowicz, L'intégration des équations de la gravitation relativiste et le problème des n corps, *J. Math. Pures Appl.* **23** (1944), 37–63.

[71] C.-C. M. Liu, and S.-T. Yau, Positivity of quasilocal mass, *Phys. Rev. Lett.* **90** (2003) no. 23, 231102

[72] ———, Positivity of quasilocal mass II, *J. Amer. Math. Soc.* **19** (2006) 181–204.

[73] E. Malec, M. Mars, and W. Simon, On the Penrose inequality for general horizons, *Phys. Rev. Lett.* **88** (2002), no. 12, 121102.

[74] M. Mars, Present status of the Penrose inequality, *Class. Quant. Grav.* **26** (2009), 193001.

[75] M. Mars, and A. Soria, On the Penrose inequality for dust null shells in the Minkowski spacetime of arbitrary dimension, *Class. Quant. Grav.* **29** (2012), 135005 (2012).

[76] D. Maxwell, A class of solutions of the vacuum Einstein constraint equations with freely specified mean curvature, *Math. Res. Lett.* **16** (2009), 627–645.

[77] ———, A model problem for conformal parametrizations of the Einstein constraint equations, *Commun. Math. Phys.* **302** (2011), 697–736.

[78] P. Miao, On existence of static metric extensions in General Relativity, *Commun. Math. Phys.* **241** (2003), 27–46.

[79] P. Miao, Y.-G. Shi, and L.-F. Tam, On geometric problems related to Brown–York and Liu–Yau quasilocal mass, *Commun. Math. Phys.* **298**, (2010), no. 2, 437–459.

[80] P. Miao, and L.-F. Tam, On second variation of Wang–Yau quasi-local energy, *Ann. Inst. Henri Poincaré*, onlinefirst, (2013), DOI: 10.1007/s00023-013-0279-z

[81] P. Miao, L.-F. Tam, and N.-Q. Xie, Critical points of Wang–Yau quasi-local energy, *Ann. Inst. Henri Poincaré*, **12** (2011), no. 5, 987–1017.

[82] R. C. Myers and M. J. Perry, Black holes in higher-dimensional space-times, *Annals of Physics* **172** (1986), no. 2, 304–347.

[83] L. Nirenberg, The Weyl and Minkowski problems in differential geometry in the large, *Commun. Pure Appl. Math.* **6** (1953), 337–394.

[84] N. Ó Murchadha, L. B. Szabados, and K. P. Tod, Comment on "Positivity of quasilocal mass", *Phys. Rev. Lett.* **92** (2004), 259001.

[85] B. O'Neill, *Semi-Riemannian geometry*, Academic Press, New York, 1983.

[86] R. Penrose, Gravitational collapse and space-time singularities, *Phys. Rev. Lett.* **14** (1965), 57–59.

[87] ———, Naked singularities, *Ann. N.Y. Acad. Sci.* **224** (1973), 125–134

[88] ———, Some unsolved problems in classical general relativity, *Seminar on differential geometry, Ann. of Math. Stud.*, **102** (1982), 631–668, Princeton University Press.

[89] A. V. Pogorelov, Regularity of a convex surface with given Gaussian curvature, *Mat. Sbornik N.S.*, **31**(73), (1952), 88–103.

[90] R. Schoen, Mean curvature in Riemannian geometry and general relativity, *Global theory of minimal surfaces*, Clay Math. Proc., vol. 2, Amer. Math. Soc., Providence, RI, 2005, pp. 113–136.

[91] R. Schoen and S.-T. Yau, On the proof of the positive mass conjecture in general relativity, *Commun. Math. Phys.* **65** (1979), no. 1, 45–76.

[92] ———, On the structure of manifolds with positive scalar curvature, *Manuscripta Math.* **28** (1979), no. 1–3, 159–183.

[93] ———, Positivity of the total mass of a general spacetime, *Phys. Rev. Lett.* **43** (1979), 1457–1459.

[94] ———, Proof of the positive mass theorem. II, *Commun. Math. Phys.* **79** (1981), no. 2, 231–260.

[95] Y.-G. Shi, and L.-F. Tam, Positive mass theorem and the boundary behaviors of compact manifolds with nonnegative scalar curvature, *J. Diff. Geom.* **62** (2002), 79–125.

[96] B. Smith, and G. Weinstein, Quasiconvex foliations and asymptotically flat metrics of nonnegative scalar curvature, *Commun. Anal. Geom.* **12** (2004), no. 3, 511–551.

[97] L. B. Szabados, Quasi-local energy-momentum and angular momentum in GR: a review article, *Living Rev. Rel.*, **12** (2009) no. 4, URL: relativity.livingreviews.org/Articles/lrr-2009-4

[98] R. M. Wald, *General relativity*, University of Chicago Press, Chicago, IL, 1984.

[99] M.-T. Wang, Quasilocal mass and surface Hamiltonian in spacetime, arXiv:1211.1407.

[100] M.-T. Wang, and S.-T. Yau, Quasilocal mass in general relativity, *Phys. Rev. Lett.* 102 (2009), no. 2, no. 021101, 4 pp.

[101] ———, Isometric embeddings into the Minkowski space and new quasi-local mass, *Commun. Math. Phys.* **288** (2009), no. 3, 919–942.

[102] ———, Limit of quasilocal mass at spatial infinity, *Commun. Math. Phys.* **296** (2010), no. 1, 271–283.

[103] E. Witten, A new proof of the positive energy theorem, *Commun. Math. Phys.* **80** (1981), 381–402.

9
Global Behavior of Solutions to Einstein's Equations

9.1 Introduction
James Isenberg

From a purely analytical perspective, Einstein's equations constitute a formidable PDE system. They mix constraint equations with evolution equations, their manifest character (hyperbolic or not) depends on the choice of coordinates, they are defined and studied on a spacetime manifold which is field-dependent and therefore not fixed, and the system is nonlinear in a serious way. These features make it challenging to study Einstein's equations using the analytical techniques and ideas which have been successfully applied to other nonlinear PDE systems. This is especially true of those analyses concerned with global, evolutionary aspects of solutions of Einstein's equations, which are the focus of interest in this chapter.

During the past thirty years, it has become apparent that the most successful way to meet these challenges and understand the behavior of solutions of Einstein's equations is to recognize the fundamental role played by spacetime geometry in general relativity and exploit some of its structures. Indeed, the Christodoulou–Klainerman proof of the stability of Minkowski spacetime [1] provides a good example of this: It relies strongly on the use of spacetime geometric structures such as null foliations, maximal hypersurfaces, "almost Killing fields", and the Bel–Robinson tensor, combined with sophisticated use of standard analytical tools such as the control of energy functionals and hyperbolic radiation estimates. The more recent work of Christodoulou and others, which has discovered sufficient conditions for the formation of trapped surfaces and black holes, also relies on a strong alliance between geometric insight and the mastery of analytical technique.

As we saw in Chapters 3 and 4, gravitational effects play an important role both in astrophysics and cosmology. However, the two areas feature two fairly distinct branches of the mathematical analysis of solutions of Einstein's equations: One works with asymptotically flat solutions to Einstein's equations and often focuses on issues related to black holes, while the other works predominantly with spatially closed solutions and is more focused on the nature of cosmological singularities and possible mechanisms for isotropization and long-distance correlation. While researchers in both of these areas are interested in assessing solution stability, in determining the nature of singularities, and in exploring long-time behavior, the mathematical distinction between asymptotically flat solutions and spatially compact solutions has resulted in the two research groups following somewhat different paths. Recognizing these differences, this Chapter is divided into two sub-Chapters: the first, by S. Aretakis and I. Rodnianski, focuses largely on asymptotically flat solutions; and the second, by V. Moncrief, works exclusively with spatially compact solutions.

One major question that is common to the study of both astrophysical and cosmological systems is that of *Strong Cosmic Censorship* (SCC). As noted in the Introduction to Part Three, the issue of SCC arises as a consequence of the Hawking–Penrose "Singularity Theorems", which show that large classes of solutions to Einstein's equations (both asymptotically flat and spatially compact) are characterized by geodesic incompleteness. Unbounded curvature in the solution and the extendibility of the solution outside of its maximal globally hyperbolic development can both result in geodesic incompleteness; the claim of the SCC conjecture is that generically[1] unbounded curvature is the culprit.

In this introduction, we briefly discuss one program of research into Strong Cosmic Censorship which is not covered in the two main sub-Chapters. It has roots in work from the 1960s and 1970s, became very active in the 1990s, and continues to make steady progress. This is the approach which seeks to prove SCC for a given family of solutions by first verifying that solutions in that family generically exhibit some form of "BKL behavior". The utility of this approach relies on the contention that the presence of BKL behavior in a given solution makes it simpler to estimate the curvature near the singularity of that solution. The proof of SCC in polarized Gowdy solutions [2] provides an archetypal example of how this can be done.

The thesis behind the work of Belinskii, Khalatnikov, and Lifshitz [3, 4] is that solutions of Einstein's equations generically exhibit a very special behavior near their singular regions. More specifically, this BKL behavior may be characterized by the very special evolution of the gravitational fields seen by a congruence of observers approaching the singularity. In the solution under consideration, each observer sees the metric along his/her particular path evolving closer and closer to a singular limit, modeled by a spatially homogeneous (singular) spacetime.[2] In the simplest version of BKL behavior, which has been labeled "Asymptotically Velocity-Term-Dominated" (AVTD), the limit-models seen by each observer are Kasner spacetimes (with vanishing spatial derivatives). In the more general version of the BKL conjecture, the limit-models are Bianchi type IX spacetimes. Partly as a result of studies of Bianchi type IX solutions, which found that their evolution takes the form of a series of Kasner epochs punctuated by "bounces", the general BKL behavior is sometimes referred to as "Mixmaster behavior".

For a number of years, very few researchers believed that BKL behavior will be found commonly, let alone generically, in solutions of Einstein's equations. However, during the

---

[1] The word "generic" appears both in the Strong Cosmic Censorship conjecture and in the Weak Cosmic Censorship conjecture. Of course to turn these conjectures into theorems, one needs to carefully state what "generic" means. Because of space limitations in this chapter, we do not deal with this important technicality, noting that in a general discussion, it is sufficient to define "generic" as "corresponding to an open and dense subset of the full set, with respect to some appropriate topology".

[2] These model spacetimes are solutions of the Einstein equations with spatial homogeneity. The inhomogeneous solutions of interest approach a different homogeneous solution along each observer path, varying smoothly from observer to observer.

early 1990s, Berger and Moncrief carried out a series of numerical simulations of solutions with varying amounts of spatial symmetry, and they found remarkably strong evidence for generic AVTD behavior. Their first simulations [5] examined T^3 Gowdy solutions, which are spacetimes with T^3 Cauchy surfaces, T^2 spatial isometry, and vanishing twists.[3] Although "spiky features" complicate the simulations, their work clearly suggests that T^3 Gowdy solutions generically exhibit AVTD behavior. Further numerical work [6] suggests the same for solutions with $U(1)$ isometry which are polarized, in the sense that one of the modes of the gravitational field is turned off.

This numerical evidence provided strong motivation first to prove that there are indeed T^3 Gowdy spacetimes and polarized $U(1)$-symmetric solutions with AVTD behavior, and second to show that this behavior is generic. For the first of these tasks, Fuchsian techniques have proven to be very useful. The idea is to write the metrics g of interest in the form $g = \mathcal{K} + \gamma$, where $\mathcal{K}$ represents the (singular) limit spacetime (consisting of a different model Kasner spacetime for each fixed spatial point) and where γ represents the remainder, to substitute this sum into the Einstein equations, and then to use the resulting equations to prove that γ must approach zero as the solution g approaches the singularity. Kichenassamy and Rendall [8] first carried out this analysis for real–analytic T^3 Gowdy spacetimes, and then Rendall [9] extended it to the much harder case of smooth T^3 Gowdy solutions. In both cases, their results show that for each of the possible choices of $\mathcal{K}$, there is a T^3 Gowdy spacetime which approaches that limit, and therefore has AVTD behavior. (Since there are many different Kasner spacetimes, there is an infinite-dimensional family of different possible choices of $\mathcal{K}$.) For the polarized $U(1)$-symmetric solutions, Fuchsian analyses [10] have been used to prove similar results, at least in the analytic case. Work continues to extend these results to smooth polarized $U(1)$-symmetric solutions.

Fuchsian studies show that there are many solutions with AVTD behavior in each of these families of spacetimes, but they tell us little regarding the genericity of AVTD behavior. For the simple class of polarized T^3 Gowdy solutions, proving that AVTD behavior is generic is not very difficult; it follows from the application of fairly standard long-time existence arguments (relying on an infinite sequence of energy functionals) [11]. Going beyond this class is much harder. The beautiful work of Ringström [12] shows that this can be done. It also shows that the analysis needed to do it can be fairly intricate and technical. Encouragingly, Ringström's work shows as well that once generic AVTD behavior has been verified, the proof that Strong Cosmic Censorship holds is a relatively straightforward consequence.

While numerical simulations and Fuchsian studies support the contention that polarized $U(1)$-symmetric solutions, as well as polarized T^2-symmetric solutions,[4] are generically AVTD, it is very unlikely that AVTD behavior is characteristic of more general families of

---

[3] The twists for a T^2-symmetric spacetime are defined to be $d\eta \wedge \eta \wedge v$ and $dv \wedge v \wedge \eta$, where η and v are the one-forms corresponding to a commuting pair of Killing fields generating the T^2 isometry. There are Gowdy spacetimes also with S^3 topology and $S^2 \times S^1$ spatial topology; much less is known about these, because of analytical difficulties at the "poles" [7].

[4] These have the same isometry group as T^3 Gowdy spacetimes, but have non-vanishing twists, and consequently are technically a bit more challenging than the T^3 Gowdy spacetimes.

spacetimes, such as the general $U(1)$-symmetric solutions, or the general T^2-symmetric solutions. Do these families exhibit Mixmaster behavior? Some numerical simulations indicate that this is the case [13], but others are less convincing. Indeed, based on numerical simulations some have argued that in general (unpolarized) T^2-symmetric solutions, spiky features are so prevalent that Mixmaster behavior is unlikely to prevail. As well, while the nature of AVTD behavior is fairly amenable to mathematical study, this does not appear to be the case (to date) for Mixmaster behavior.

Strong Cosmic Censorship is among the most intriguing conjectures concerning the global behavior of solutions of Einstein's equations. Ideas and techniques based on BKL behavior have facilitated the proof of SCC in a limited family of spacetimes, and may be useful in extending such proofs further. However, it is very likely that new ideas, such as those appearing elsewhere in this Chapter, will be needed to make serious progress.

References

[1] D. Christodoulou and S. Klainerman, *The global nonlinear stability of Minkowski space* (Princeton University Press, Princeton, 1994)
[2] P. Chruściel, J. Isenberg and V. Moncrief, *Class. Quant. Grav.* **7** 1671–1679 (1990).
[3] V. Belinskii, I. Khalatnikov and E. Lifshitz, *Adv. Phys.* **19** 525–573 (1970).
[4] V. Belinskii, I. Khalatnikov and E. Lifshitz, *Adv. Phys.* **31** 639–667 (1982).
[5] B. Berger and V. Moncrief, *Phys. Rev. D* **48** 4676 (1993).
[6] B. Berger and V. Moncrief, *Phys. Rev. D* **57** 7235 (1998).
[7] P. Chruściel, *Ann. Phys.* **202** 100–150 (1990).
[8] S. Kichenassamy and A. Rendall, *Class. Quant. Grav.* **15** 1339–1335 (1999).
[9] A. Rendall, *Class. Quant. Grav.* **17** 3305–3316 (2000).
[10] J. Isenberg and V. Moncrief, *Class. Quant. Grav.* **19**, 5361 (2002).
[11] J. Isenberg and V. Moncrief, *Ann. Phys.* **199** 84–122 (1990).
[12] H. Ringström, *Ann. Math.* **170** 1181–1240 (2009).
[13] B. Berger, J. Isenberg and M. Weaver, *Phys. Rev. D* **64** 084006 (2001).
[14] W. Lim, L. Andersson, D. Garfinkle and F. Pretorius, *Phys. Rev. D* **79** 123526 (2009).

9.2 The Cauchy problem in General Relativity

Stefanos Aretakis and Igor Rodnianski

9.2.1 Introduction

The Einstein equations

The *Einstein equations* are the mathematical embodiment of the theory of *general relativity* proposed by Einstein [1, 2] in 1915 as a unification of space, time and gravitation. Their definitive description consists of a $(3+1)$-dimensional Lorentzian manifold $\mathcal{M}$ and a metric g satisfying

$$R_{\mu\nu} - \frac{1}{2}Rg_{\mu\nu} = 8\pi\,T_{\mu\nu}, \tag{9.1}$$

where $R_{\mu\nu}$ and R are respectively the Ricci tensor and scalar curvature of $(\mathcal{M}, g)$ and $T_{\mu\nu}$ is the so-called *energy-momentum tensor* of matter. By virtue of (9.1), the tensor $T_{\mu\nu}$ is symmetric and divergence-free,

$$\nabla^\mu T_{\mu\nu} = 0. \tag{9.2}$$

Note that we use standard conventions of raising, lowering and summing over indices. The differential conservation law (9.2) should be implied by a system of evolution equations for the external matter fields.

Among the most popular mathematical matter models are the following.

1. The vacuum equations: $T = 0$. The Einstein-vacuum equations then simply reduce to

$$R_{\mu\nu} = 0. \tag{9.3}$$

2. The Einstein-scalar field model: $T_{\mu\nu} = \partial_\mu \phi \, \partial_\nu \phi - \frac{1}{2} g_{\mu\nu} \, \partial^\mu \phi \, \partial_\mu \phi$, where ϕ is a real-valued scalar field $\phi : \mathcal{M} \to \mathbb{R}$. The requirement (9.2) is equivalent to ϕ satisfying the wave equation

$$\Box_g \phi := \nabla^\mu \partial_\mu \phi = 0 \tag{9.4}$$

   on the curved background $(\mathcal{M}, g)$.
3. The Einstein–Maxwell equations: $T_{\mu\nu} = \frac{1}{4\pi} \left(F_\mu{}^\alpha F_{\nu\alpha} - \frac{1}{4} g_{\mu\nu} F_{\alpha\beta} F^{\alpha\beta} \right)$, where $F_{\mu\nu}$ is the electromagnetic tensor which obeys the Maxwell equations

$$\nabla^\mu F_{\mu\nu} = 0, \quad \nabla^\mu (*F)_{\mu\nu} = 0. \tag{9.5}$$

4. The perfect-fluid-matter model: $T_{\mu\nu} = (\rho + p) u_\mu u_\nu + p \, g_{\mu\nu}$, where u^μ is the four-velocity vector, p is the pressure and ρ is the proper energy density of the fluid, and all are governed by the differential conservation laws

$$\nabla^\mu T_{\mu\nu} = 0, \quad \nabla^\mu I_\mu = 0, \tag{9.6}$$

where $I^\mu = nu^\mu$ is the particle current (here n denotes the number of particles per unit volume).

The Cauchy problem in general relativity

The Einstein equations are of hyperbolic character, as shall be explained in more detail in Section 9.2.2. As a consequence, the initial value problem is the natural mathematical problem for these equations. The initial data for the vacuum equations[5] consist of the triplet (Σ_0, g_0, k_0), where Σ_0 is a 3-dimensional Riemannian manifold endowed with a Riemannian metric g_0 and k_0 is a symmetric 2-tensor field on Σ_0.

Given such an initial data set, the *Cauchy problem* consists in finding a regular $(3 + 1)$-dimensional globally hyperbolic time-oriented Lorentzian manifold $(\mathcal{M}, g)$ and an embedding $\Sigma_0 \subset \mathcal{M}$, such that $(\mathcal{M}, g)$ satisfies the Einstein-vacuum equations (9.3), Σ_0 is a

---

[5] In the general case one also needs to add initial data for the external fields.

Cauchy hypersurface of $(\mathcal{M}, g)$ such that g_0 is the restriction of g on Σ_0 and k_0 is the second fundamental form of the embedding.

As can be seen by taking the contracted Codazzi and twice-contracted Gauss equations of the embedding of Σ_0 in $\mathcal{M}$, the Einstein equations are overdetermined and the initial data set has to satisfy the *constraint equations*

$$D^j k_{0ij} - D_i \, tr \, k_0 = 0, \quad R_0 - |k_0|^2 + (tr \, k_0)^2 = 0, \tag{9.7}$$

where D is the Levi-Civita covariant derivative and R_0 the scalar curvature of (Σ_0, g_0).

In view of the general covariance of the Einstein equations, uniqueness of the above Cauchy problem can only be understood modulo diffeomorphisms.

The important class of isolated gravitating systems in general relativity is described by spacetimes which, in particular, contain asymptotically Minkowskian regions. In the context of the Cauchy problem, such spacetimes arise as developments of *asymptotically flat initial data* given by (Σ_0, g_0, k_0) with the properties that Σ_0 minus a compact set is diffeomorphic to $\mathbb{R}^3$ minus a ball and that there exists a coordinate system (x^1, x^2, x^3) such that for $r = \sqrt{(x^1)^2 + (x^2)^2 + (x^3)^2}$, the metric takes the form

$$(g_0)_{ij} = \left(1 + \frac{M}{r}\right)\delta_{ij} + O(r^{-1-\alpha}), \quad (k_0)_{ij} = O(r^{-2-\alpha}), \tag{9.8}$$

for some $M \geq 0$ and some $\alpha > 0$. The non-negativity of the *mass M* for data sets of this type which satisfy the constraint equations (9.7) was established by Schoen and Yau [3, 4] and Witten [5].

The characteristic initial value problem in general relativity

An alternative to the Cauchy problem described above is the characteristic initial value problem where initial data is prescribed on two 3-dimensional degenerate (null) hypersurfaces $C_0, \underline{C}_0$ intersecting at a 2-surface S_0; see Fig. 9.1 (or at a vertex p).

Let us assume for a moment that C_0 and $\underline{C}_0$ are embedded in a Lorentzian manifold $(\mathcal{M}, g)$. Then the normal L (resp. $\underline{L}$) of C_0 (resp. $\underline{C}_0$) is tangential to C_0 (resp. $\underline{C}_0$) and hence (by the first variational formula) the second fundamental form χ (resp. $\underline{\chi}$) relative to C_0 (resp. $\underline{C}_0$) is completely determined by the degenerate metric on C_0 (resp. $\underline{C}_0$). Furthermore, the second variation formula and the Einstein equations imply the Raychaudhuri equation

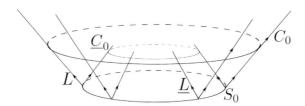

Figure 9.1 The 2-surface S_0 is the intersection of the null hypersurfaces C_0 and $\underline{C}_0$.

$$\nabla_L tr\ \chi = -\frac{1}{2}(tr\ \chi)^2 - |\hat{\chi}|^2 \tag{9.9}$$

on C_0, where $tr\ \chi, \hat{\chi}$ denote, respectively, the trace and traceless part of χ relative to the sections of a geodesic foliation of C_0; i.e.,

$$\chi = \frac{1}{2} tr\ \chi \cdot \not{g} + \hat{\chi}, \tag{9.10}$$

where $\not{g}$ denotes the induced (Riemannian) metric of the sections (conjugate definitions and relations apply for $\underline{C}_0$). Clearly, (9.9) shows that the degenerate metrics on $C_0, \underline{C}_0$ cannot be arbitrarily prescribed; that is, they have to satisfy *null constraint equations*. However, in contrast with the Cauchy problem, it turns out that the *conformal intrinsic geometry* of the metrics of $C_0, \underline{C}_0$ is not subject to any equation and hence can be freely specified. Given also the full intrinsic geometry of S_0, the null constraint equations can be solved by integrating along the null generators of $C_0, \underline{C}_0$ to determine the full geometry of $C_0, \underline{C}_0$.

The above motivates the following definition: A characteristic initial data set consists of the conformal geometry of two 3-dimensional degenerate hypersurfaces $C_0, \underline{C}_0$ intersecting at a 2-surface S_0, the initial null expansions $tr\ \chi|_{S_0}$, $tr\ \underline{\chi}\big|_{S_0}$, and the torsion 1-form ζ on S_0 (given by $\zeta(X) = g(\nabla_X L, \underline{L})$).

The absence of constraint equations for the above piece of initial data makes the characteristic initial value problem more natural than the standard Cauchy problem, at least from one point of view. A structure naturally associated with the characteristic initial value problem is the so-called *double null foliation*. This is defined as follows (see Fig. 9.2): Let u (resp. v), be functions such that $u|_{S_0} = 0$ (resp. $v|_{S_0} = 0$), and let the level sets $\underline{C}_v$ (resp. C_u), of v (resp. u), be null hypersurfaces intersecting the initial hypersurfaces C_0 (resp. $\underline{C}_0$), at embedded 2-spheres S_v (resp. $\underline{S}_u$), respectively. Let L, resp. $\underline{L}$, be normal to C_u, resp. C_v, such that $g(L, \underline{L}) = -2$ and let $e_A, A = 1, 2$, define a local basis of $S_{u,v}$. Following [6, 7], the connection coefficients with respect to the *null frame* $(L, \underline{L}, e_1, e_2)$

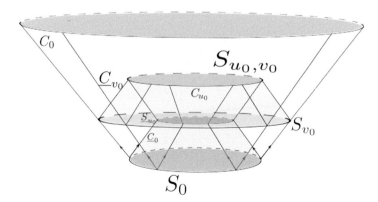

Figure 9.2 The section S_{u_0,v_0} of the double null foliation is the intersection of the null hypersurfaces $\underline{C}_{v_0}$ and C_{u_0}.

are determined by the induced Christoffel symbols on $S_{u,v}$ and the following $S_{u,v}$-tangential tensors: $\chi, \underline{\chi}, \eta, \underline{\eta}, \omega, \underline{\omega}, \zeta$. These are defined as follows: $\eta_A = g(\nabla_{\underline{L}} L, e_A)$, $\omega = -g(\nabla_L L, \underline{L})$ with conjugate definitions for $\underline{\eta}, \underline{\omega}$; the remaining components have been defined above. Similarly, we define the following curvature components:

$$
\begin{aligned}
\alpha_{AB} &= R(e_A, L, e_B, L), & \underline{\alpha}_{AB} &= R(e_A, \underline{L}, e_B, \underline{L}), \\
\beta_A &= \tfrac{1}{2} R(e_A, L, \underline{L}, L), & \underline{\beta}_A &= \tfrac{1}{2} R(e_A, \underline{L}, \underline{L}, L), \\
\rho &= \tfrac{1}{4} R(\underline{L}, L, \underline{L}, L), & \sigma &= \tfrac{1}{4} \slashed{\epsilon}^{AB} R(e_A, e_B, \underline{L}, L),
\end{aligned}
\tag{9.11}
$$

where $\slashed{\epsilon}$ denotes the induced volume form on $S_{u,v}$. If the Einstein equations $Ric(g) = 0$ are satisfied, then all the remaining curvature components can be expressed in terms of the above components.

The associated connection coefficients Γ and curvature components R satisfy transport equations along the null generators of the null hypersurfaces $C_u, \underline{C}_v$ of the schematic form

$$
\begin{aligned}
\nabla_L \Gamma &= R + \mathcal{D}\Gamma + \Gamma \cdot \Gamma, \\
\nabla_{\underline{L}} \Gamma &= R + \mathcal{D}\Gamma + \Gamma \cdot \Gamma,
\end{aligned}
\tag{9.12}
$$

where $\mathcal{D}$ denotes angular derivatives on the 2-spheres. The Raychaudhuri equation (9.9) is one of the optical structure equations. Other equations (along the C_u generators) are

$$
\begin{aligned}
\nabla_L \hat{\chi} + tr\,\chi \cdot \hat{\chi} &= -\alpha - 2\omega\hat{\chi}, \\
\nabla_L \eta &= -\beta - \chi \cdot (\eta - \underline{\eta}).
\end{aligned}
\tag{9.13}
$$

The Gauss and Codazzi equations corresponding to the embedding of the 2-spheres in the spacetime manifold give rise to elliptic equations of the schematic form

$$
\mathcal{O}\Gamma = R + \mathcal{D}\Gamma + \Gamma \cdot \Gamma,
\tag{9.14}
$$

where $\mathcal{O}$ denotes elliptic operators on the 2-spheres. For example, we have

$$
\begin{aligned}
\operatorname{div} \hat{\chi} &= -\beta + \tfrac{1}{2}\slashed{d}\, tr\,\chi - \tfrac{1}{2}(\eta - \underline{\eta}) \cdot \left(\hat{\chi} - \tfrac{1}{2} tr\,\chi \right), \\
\operatorname{curl} \eta &= \sigma + \tfrac{1}{2}\underline{\hat{\chi}} \wedge \hat{\chi}, \\
K &= -\rho + \tfrac{1}{2}\hat{\chi} \cdot \underline{\hat{\chi}} - \tfrac{1}{4} tr\,\chi\, tr\,\underline{\chi},
\end{aligned}
\tag{9.15}
$$

where K denotes the Gauss curvature of the surfaces $S_{u,v}$.

The hyperbolic radiating aspect of the vacuum equations is manifest in the following Maxwell-type system for the curvature tensor:

$$
\nabla_{[\epsilon} R_{\alpha\beta]\gamma\delta} = 0, \quad \nabla^\alpha R_{\alpha\beta\gamma\delta} = 0;
\tag{9.16}
$$

these yield the Bianchi equations for R, which schematically take the form

$$
\begin{aligned}
\nabla_L R &= \mathcal{D}R + \Gamma \cdot R, \\
\nabla_{\underline{L}} R &= \mathcal{D}R + \Gamma \cdot R.
\end{aligned}
\tag{9.17}
$$

For example,

$$\nabla_L \beta + 2 \, tr \, \chi \cdot \beta = \text{div } \alpha - 2\omega\beta + \eta\alpha.$$

9.2.2 Well-posedness of the Einstein equations

Existence and uniqueness of a maximal Cauchy development

The local well-posedness for the Cauchy problem for Einstein's equations was established by Choquet-Bruhat [8]. The proof exploited the covariance of the Einstein equations and a special gauge known as the *harmonic gauge*, in which the coordinates x^μ themselves satisfy the covariant wave equation

$$\Box_g x^\mu = 0, \quad \mu = 0, \dots, 3. \tag{9.18}$$

This, in turn, is equivalent to requiring that the components $g_{\mu\nu}$ of the metric relative to this particular coordinate system satisfy

$$\partial_\mu \left(g^{\mu\nu} \sqrt{|g|} \right) = 0. \tag{9.19}$$

The harmonic gauge provides the most direct way to view the hyperbolicity of the Einstein equations (9.3) since in that gauge, these equations take the form

$$\Box_g g_{\mu\nu} = N_{\mu\nu}(g, \partial g), \tag{9.20}$$

where ∂g denotes all space and time derivatives of g, N is quadratic in ∂g and, in terms of harmonic coordinates, $\Box_g = g^{\mu\nu} \partial^2_{\mu\nu}$. The equations (9.20) constitute a quasilinear system of wave equations for the components $g_{\mu\nu}$ on the background determined by the metric g. It is important to emphasize that a metric g satisfying (9.20) is not Ricci flat unless the associated coordinate system is harmonic. Hence, in order to solve the Einstein equations one needs to solve both systems (9.20) and (9.18) (or, equivalently, (9.19)). However, as observed by Choquet-Bruhat, the conditions (9.18), (9.19) are satisfied automatically for solutions of (9.20) provided that they are satisfied initially on Σ_0 and provided that (Σ_0, g_0, k_0) obey the constraint equations (9.7). This fact is known as propagation of the harmonic gauge. Thus to complete the initial value problem setup in harmonic gauge, we choose a local coordinate system $(t = x^0, x^1, x^2, x^3)$ on $\Sigma_0 = \{t = 0\}$ and we consider initial data $g_{\mu\nu}\big|_{t=0}$ and $\partial_t g_{\mu\nu}\big|_{t=0}$ for (9.20) to be consistent with the given data (g_0, k_0) and such that the harmonic condition is initially satisfied. The system (9.20) is solved locally by a standard iteration argument. The main difficulty is to show that the iteration converges; this is guaranteed by a priori estimates for (9.20). The original approach of Choquet-Bruhat relied on the construction of a Kirchhoff–Sobolev-type parametrix for an inhomogeneous scalar linear problem, namely

$$\Box_g \phi = 0, \quad \phi|_{t=0} = \phi^0, \quad \partial_t \phi|_{t=0} = \phi^1,$$

which, however, required a high degree of differentiability of the initial data. The metrics derived from these local problems can be patched together to form a Cauchy development of the given initial data. Subsequently, Choquet-Bruhat and Geroch [9] showed that given smooth vacuum initial data for the Einstein equations there is a unique smooth maximal Cauchy development; i.e., there is a globally hyperbolic development $(\mathcal{M}, g)$ of the data, into which all other such spacetimes embed isometrically.[6]

In 1990 Rendall [11] derived a clever method to reduce the local well-posedness of the characteristic initial value problem for the vacuum Einstein equations to that of the usual Cauchy problem. Specifically, Rendall showed that any smooth[7] characteristic initial data has a future development, bounded in the past by a neighborhood of the initial surface S in $C \cup \underline{C}$. The theorem of Choquet-Bruhat and Geroch then shows that there is a unique maximal development $(\mathcal{M}, g)$ corresponding to the given characteristic initial data. The development $(\mathcal{M}, g)$ given by Rendall's theorem does not contain a full neighborhood of the initial null hypersurfaces; only one of the 2-surface S. This problem has been overcome by Luk [12].

The maximal development of initial data is the main object of study in classical general relativity. The main problems of interest can be loosely divided into the following categories:

1. Construction of explicit solutions (e.g., Minkowski, Schwarzschild, Kerr, and Friedmann–Robertson–Walker).
2. Causality and global properties (e.g., singularity theorems, break-down, black hole uniqueness, splitting theorems).
3. Long-time existence (e.g., break-down, formation of trapped surfaces, propagation of singularities, stability, weak and strong cosmic censorships).

Local existence in H^s, $s > 2$

The differentiability requirement of Choquet-Bruhat's theorem was significantly improved in the work of Dionne [13], Fisher and Marsden [14] and Hughes, Kato and Marsden [15] via the energy method. The energy method applied to (9.20) shows that the Sobolev norm H^s of the solution $g(t)$ at time t is controlled by

$$\|g(t)\|_{H^s} + \|\partial_t g(t)\|_{H^{s-1}} \le C \exp\left(\int_0^t \|\partial g(\tau)\|_{L^\infty} \, d\tau\right)\left(\|g(0)\|_{H^s} + \|\partial_t g(0)\|_{H^{s-1}}\right).$$
(9.21)

The desired a priori estimate (and hence well-posedness) for (9.20) follows from the Sobolev embedding $H^s \subset L^\infty$ provided that $s > 5/2$.

The above analysis leads to solutions $(\mathcal{M}, g)$ of the original Einstein equations arising from arbitrary initial data $(g_0, k_0) \in H^s \times H^{s-1}$ as long as $s > 5/2$. These solutions remain in the space H^s relative to a system of coordinates (t, x) so that the metric components satisfy the conditions $g_{\mu\nu} \in C([0, T]; H_x^s)$ and $\partial_t g_{\mu\nu} \in C([0, T]; H_x^{s-1})$ on a time interval

---

[6] The original construction of the maximal development appealed to Zorn's lemma. This reliance was recently overcome by Sbierski [10].
[7] The quantitative statement behind well-posedness for the characteristic problem requires a fundamental loss in derivatives relative to the Cauchy problem.

$[0, T]$ with T dependent on the $H^s \times H^{s-1}$ norm of the initial data. The original proof in [14, 15] required one more derivative (i.e., $s > 7/2$), which accounted for the general covariance of the Einstein equations. However, it has been shown in [16] that uniqueness holds even for $s > 5/2$.

The existence result for (9.20), and consequently for the vacuum Einstein equations, can be improved when the energy estimate (9.21) is combined with a Strichartz estimate

$$\|\partial \phi\|_{L^2[0,T]L^\infty} \leq C\left(\|\phi(0)\|_{H^s} + \|\partial_t \phi(0)\|_{H^{s-1}} + \|\Box_g \phi\|_{L^1[0,T]H^{s-1}} \right). \tag{9.22}$$

If g is the Minkowski metric then the above estimate holds for all $s > 2$. This estimate was first used in the context of semilinear wave equations in [17]. In the case of general quasilinear wave equations of the form (9.20), however, one needs to establish (9.22) for rough metrics g. Specifically, since we ideally want to apply (9.22) with ϕ replaced by $g_{\mu\nu}$, we require that the constant C depends on the background metric g only through the norms $\|\partial g\|_{L^2[0,T]L^\infty}$ and, in view of the energy estimate (9.21), $\|\partial g\|_{L^\infty[0,T]H^{s-1}}$.

Such estimates were derived by Smith [18], Bahouri and Chemin [19, 20] and Tataru [21, 22]. The optimal estimate of the form (9.22) was derived in [22] for $s > 2 + 1/6$.

Smith and Tataru [23] have shown that Strichartz estimates do not hold for $s \leq 2 + 1/6$ and therefore one cannot do better than that by simply relying on Strichartz estimates on *general* backgrounds with the expected regularity. Hence, in order to show stronger well-posedness results for (9.20) one needs to exploit the fact that the metric g is itself a solution of (9.20) and take into account the structure of the non-linearity. The first result in this direction appears in [24] where the regularity requirement is improved to $s > 2 + (2 - \sqrt{3})/2$. The following more definitive result was proven in [25]:

Consider the equation (9.20) with initial data $g|_{t=0}$, $\partial_t g|_{t=0} \in H^s \times H^{s-1}$ for some $s > 2$ satisfying the constraint equations (9.7) and the harmonic gauge condition (9.18). Then there exists a unique solution g in a time interval $[0, T]$ such that $g_{\mu\nu} \in C^0([0, T]; H^s)$, with T depending only on the size of $\| g|_{t=0}\|_{H^s} + \| \partial_t g|_{t=0}\|_{H^{s-1}}$.

The above result makes use of both the vanishing of the Ricci curvature of g and the harmonic gauge condition. Other important features are the use of energy estimates along the null hypersurfaces and improvements in the conormal properties of the eikonal and null structure equations.

A parallel work by Smith and Tataru [26] obtained the H^s, $s > 2$ local well-posedness result for general quasilinear wave equations. This approach is based on the construction of a wave-packet approximation of a solution. This construction relies on the foliation by null planes and uses a gain of differentiability along each plane while avoiding reference to regularity in transversal directions. The geometry of such wave packets controls the desired Strichartz estimate.

Break-down criterion for the Einstein equations

The local existence results mentioned above provide a break-down criterion of solutions; that is to say, a solution can be extended as long as the H^s, $s > 2$, norms of the metric components in harmonic gauge remain bounded. However, these results are not geometric since they are strongly tied to a particular coordinate gauge. Moreover, they arise as an immediate corollary of *stronger* local well-posedness statements.

By virtue of (9.21), it immediately follows that break-down does not occur at $t = t_*$ unless

$$\int_{t_0}^{t_*} \|\partial g(t)\|_{L^\infty} \, dt = \infty. \tag{9.23}$$

But again this condition is non-geometric since it depends on the choice of a harmonic coordinate system.

The first geometric criterion for the break-down of solutions of the Einstein-vacuum equations appeared in the work of Anderson [27]. Assume that a (negative and monotonically increasing towards the future) time function t is defined on part of the spacetime such that its level hypersurfaces Σ_t are compact, of Yamabe type -1, and such that if k is the second fundamental form and n is the lapse function then

$$g = -n^2 \, dt^2 + \gamma_{ij} \, dx^i \, dx^j, \quad \partial_t \gamma_{ij} = -2nk_{ij}, \text{ and } tr \, k = t.$$

Then Anderson showed that break-down at $t = t_*$ of the solutions to the Einstein equations is tied to the condition that

$$\limsup_{t \to t_*^-} \|R(t)\|_{L^\infty(\Sigma_t)} = \infty, \tag{9.24}$$

where $R(t)$ denotes the Riemann curvature tensor of g. Observe that (9.24) requires one degree of differentiability more than (9.23). This was overcome in [28], where the following break-down condition was established:

$$\limsup_{t \to t_*^-} \|\mathcal{L}_T g(t)\|_{L^\infty(\Sigma_t)} = \infty; \tag{9.25}$$

here $T = n^{-1} \, \partial_t$ denotes the unit normal to Σ_t and $\mathcal{L}_T g$ is the deformation tensor of T, which satisfies the relation

$$|\mathcal{L}_T g| = |k| + |\nabla \log n| \tag{9.26}$$

and is equal to zero if T is Killing. The break-down condition (9.25) can be compared with the well-known Beale–Kato–Majda [29] criterion

$$\int_{t_0}^{t^*} \|\omega(t)\|_{L^\infty} \, dt = \infty \tag{9.27}$$

for the break-down of solutions to the incompressible Euler equations

$$\partial_t v + (v \cdot \nabla)v = -\nabla p, \quad \text{div } v = 0,$$

with smooth initial data at $t = t_0$. Here $\omega = \text{curl } v$ denotes the vorticity. The analogue of (9.23) for the Euler equations is

$$\int_{t_0}^{t^*} \|\nabla v(t)\|_{L^\infty} \, dt = \infty. \tag{9.28}$$

One can express $\nabla v = P^0(\omega)$, where P^0 is a singular integral operator which, although not a bounded map $L^\infty \to L^\infty$, is sufficient to yield the break-down criterion (9.27). On the other hand, $k, n, \nabla n$ capture many fewer degrees of freedom in determining ∂g. Indeed, as far as criterion (9.25) is concerned, the main difficulty amounts to showing that the bounds on $k, n, \nabla n$ cover all dynamic degrees of freedom of ∂g provided g satisfies the Einstein equations.

The first observation is that $k, n, \nabla n$ can used to bound both the energy and flux associated with the curvature tensor R. This can be achieved by applying a standard energy estimate based on currents constructed using the *Bel–Robinson energy-momentum tensor*

$$Q[R]_{\alpha\beta\gamma\delta} = R_{\alpha\rho\gamma\sigma} \cdot R_{\beta\ \delta}^{\ \rho\ \sigma} + (*R)_{\alpha\rho\gamma\sigma} \cdot (*R)_{\beta\ \delta}^{\ \rho\ \sigma}. \tag{9.29}$$

This quartic tensor is totally symmetric, is trace-free with respect to any pair of indices, and is divergence-free; i.e., $\nabla^\alpha Q_{\alpha\beta\gamma\delta} = 0$. The flux of these currents is an integral of the squares of certain components of R *tangential* to a null hypersurface $N^-(p)$, the boundary of the causal past of the point p, generated locally as a level set of an optical function.

Bounds on the flux of the curvature were earlier used in the large data global existence result for the $3 + 1$ Yang–Mills equation by Eardley and Moncrief [30, 31]. These bounds follow from the conservation of energy, which in turn is a consequence of the fact that $T = \partial_t$ is a Killing field on Minkowski spacetime. Eardley and Moncrief used a formula for the curvature F based on the equation $\Box F = F * F$ and the explicit representation for solutions of the inhomogeneous wave equation in Minkowski spacetime $\mathbb{R}^{3+1}$ and showed that F can be pointwise bounded by its flux and initial data.[8]

Returning now to the break-down criterion (9.25), the curvature R of the Ricci flat metric g satisfies

$$\Box_g R = R * R. \tag{9.30}$$

Based on the construction of the first-order approximate Kirchhoff–Sobolev-type covariant parametrix formula obtained in [32], one can express[9]

$$R(p) = -\int_{N^-(p;\delta)} A \cdot (R * R) + \mathcal{E} + \int_{N^-(p;\delta)} \text{Err} \cdot R, \tag{9.31}$$

where $N^-(p; \delta)$ denotes the portion of $N^-(p)$ in the time interval $[t(p) - \delta, t(p)]$. The term $\mathcal{E}$ can be bounded by $R(\tau), \nabla R(\tau)$ with $\tau \in [t(p) - \delta, t(p) - \delta/2]$, and the error term Err depends only on the extrinsic geometry of $N^-(p; \delta)$. As in the Yang–Mills setting, the

---

[8] Note that since the flux of the curvature F does not contain all the components of F, one needs the right-hand side to have some special algebraic structure. Such structure is present in the so-called Cronström gauge.

[9] A different integral equation satisfied by the curvature was also derived by Moncrief [33].

structure of $R * R$ on the right-hand side of (9.31) allows one to estimate one of the curvature terms using the curvature flux. However, in order to even make use of the curvature flux one needs to show that the null hypersurface $N^-(p;\delta)$ on which it is defined is in fact *smooth*; i.e., one needs to derive a lower bound on the radius of injectivity of $N^-(p;\delta)$. This was done in a sequence of papers [34–37] (see also [38]). Then, L^∞ bounds on R can be shown as long as one controls all the remaining geometric quantities arising in the parametrix construction, using the curvature flux and the assumed bounds on $k, n, \nabla n$. The final result then follows by appealing to Anderson's break-down criterion.

Wang [38] has derived an improved breakdown criterion which is very similar to (9.27); namely

$$\int_{t_0}^{t_*} \|\mathcal{L}_T g(t)\|_{L^\infty(\Sigma_t)} = \infty. \tag{9.32}$$

The L^2 bounded curvature theorem

By straightforward scaling considerations one might expect to make sense of the initial value problem for differentiability $s \geq s_c = 3/2$, with s_c the natural scaling exponent for L^2-based Sobolev norms. Note that for $s = s_c = 3/2$, a local-in-time existence result would imply that any smooth initial data which is small in the above critical norm would be globally smooth. However, for quasilinear hyperbolic systems, critical well-posedness results have only been established in the case of $(1+1)$-dimensional systems, or spherically symmetric solutions of higher-dimensional problems, in which case the L^2-based Sobolev norms can be replaced by bounded variation (BV)-type norms. A particularly important example of this type is the critical BV well-posedness result proved by Christodoulou [39] for spherically symmetric solutions of the Einstein equations coupled with a scalar field (see Section 9.2.4). As we shall see, this work played a crucial role in Christodoulou's celebrated work on Weak and Strong Cosmic Censorship. On the other hand, the BV-norms are inadequate in higher dimensions; the only norms which can propagate the regularity properties of the data are necessarily L^2-based.

The bounded L^2 curvature conjecture, proposed in [40], claims that the time of existence for solutions of the Einstein-vacuum equations depends only on the L^2 norms of the Riemann curvature tensor R_0 and the gradient ∇k_0 of the second fundamental, and on the lower bound on the volume radius of the initial data set.

This conjecture has been resolved in a series of papers [41–46]. A more precise version of the theorem is the following:

Let $(\mathcal{M}, g)$ be an asymptotically flat solution to the Einstein-vacuum equations together with a maximal[10] foliation by spacelike hypersurfaces Σ_t defined as level hypersurfaces of a time function t. Assume that the initial slice (Σ_0, g_0, k_0) is such that $Ric_0, \nabla k_0 \in L^2(\Sigma_0)$ and $r_{vol}(\Sigma_0, 1) > 0$.[11] Then, there is a time T and a constant C, both of which depend only

---

[10] A foliation (Σ_t, g_t) is called *maximal* if $tr_{g_t} k_t = 0$, where k_t denotes the second fundamental form of Σ_t.
[11] We define the volume radius by $r_{vol}(\Sigma, 1) = \inf_{p \in \Sigma} \inf_{r \leq 1} |B_r(p)| r^3$, with $|B_r(p)|$ the volume of the ball of center p and radius r.

on the norms $\|Ric_0\|_{L^2(\Sigma_0)}$, $\|\nabla k_0\|_{L^2(\Sigma_0)}$ *and* $r_{vol}(\Sigma, 1)$, *such that the following estimates hold on* $0 \le t \le T$:

$$\|R\|_{L^\infty_{[0,T]}L^2(\Sigma_t)} \le C, \quad \|\nabla k\|_{L^\infty_{[0,T]}L^2(\Sigma_t)} \le C \text{ and } \inf_{0 \le r \le T} r_{vol}(\Sigma_t, 1) \ge \frac{1}{C}. \tag{9.33}$$

Note that this theorem is satisfactory from the analytic point of view, since it assumes very low regularity on the metric (in fact, it is conjectured that it is sharp in the sense that one does not have such a continuation criterion for $s < 2$; see also comments below). It is satisfactory from the geometric point of view as well, since the assumptions are invariant in the sense that they do not depend on the choice of a particular coordinate system.

Equation (9.31) cannot be used to show (9.33) since the error terms on the right-hand side of (9.31) require not only control on the curvature flux but also (stronger) control on the terms (9.26). On the other hand, the special structure of the error terms of (9.31) was crucial for criterion (9.25). Hence, in order to show (9.33), one needs to express the Einstein equations in an alternative form in which, in particular, the specific (null) structure of the equations is manifest. Recall at this point that the break-down criterion (9.25) was motivated by the work of Eardley and Moncrief on the global well-posedness of the Yang–Mills equations on $\mathbb{R}^{3+1}$. It turns out that for proving (9.33) one needs to exploit the analogy between the Einstein equations and the Yang–Mills equations at a deeper level; in fact, one needs to cast the Einstein-vacuum equations in a Yang–Mills form. The idea is to give up on the search for a preferred choice of coordinates and instead express the Einstein-vacuum equations in terms of the connection 1-forms associated with moving orthonormal frames; i.e., vector fields e_α which satisfy

$$g(e_\alpha, e_\beta) = m_{\alpha\beta} = \text{diag}(-1, 1, 1, 1).$$

Specifically, we take e_0 to be the future-directed unit normal to Σ_t, and choose e_1, e_2, e_3 to be an orthonormal basis for Σ_t. The connection 1-forms (they are to be interpreted as 1-forms with respect to the external index μ, with values in the Lie algebra of $so(3, 1)$), defined by the formulas

$$(A_\mu)_{\alpha\beta} = g(\nabla_\mu e_\beta, e_\alpha), \tag{9.34}$$

satisfy the equations

$$\nabla^\mu F_{\mu\nu} + [A^\mu, F_{\mu\nu}] = 0, \tag{9.35}$$

where we denote

$$(F_{\mu\nu})_{\alpha\beta} = \left(\nabla_\mu A_\nu - \nabla_\nu A_\mu - [A_\mu, A_\nu]\right)_{\alpha\beta}. \tag{9.36}$$

Hence, we can interpret the curvature tensor R as the curvature of the $so(3, 1)$-valued connection 1-form A. We can rewrite (9.35) in the form

$$\Box_g A_\nu - \nabla_\nu\left(\nabla^\mu A_\mu\right) = J_\nu(A, \nabla A), \tag{9.37}$$

where

$$J_\nu(A, \nabla A) = \nabla^\mu \big([A_\mu, A_\nu]\big) - [A_\mu, F_{\mu\nu}].$$

The equations (9.35),(9.36) look just like the Yang–Mills equations on a fixed Lorentzian manifold $(\mathcal{M}, g)$ except, of course, that in our case A and g are not independent but rather are connected by (9.34), reflecting the quasilinear character of the Einstein equations. The analysis that follows is motivated by the work of Klainerman and Machedon [47, 48] on the well-posedness for the Yang–Mills equations in $\mathbb{R}^{3+1}$ in the energy H^1 norm (i.e., for exponent $s = 1$).

If we use a Coulomb-type gauge condition

$$\overline{div}\, \overline{A} = \overline{\nabla}^i \overline{A}_i = 0, \tag{9.38}$$

where $\overline{A}, \overline{div}$ denote the spatial component of A and the spatial divergence, respectively, then it immediately follows from (9.37) that A_0 satisfies an elliptic equation while each component $A_i, i = 1, 2, 3$ satisfies an equation of the form

$$\Box_g A_i = -\partial_i(\partial_0 A_0) + A^j\, \partial_j A_i + A^j\, \partial_i A_j + \text{l.o.t.} \tag{9.39}$$

Note the second-order term $-\partial_i(\partial_0 A_0)$ on the right-hand side. In the case of the Yang–Mills equation on flat spacetime (with corresponding wave operator $\Box$), we can remove this term by observing that, in view of the Coulomb gauge, the left-hand side is divergence-free. Indeed, by using a non-local operator $\mathcal{P}$ that projects onto divergence-free vector fields we obtain that, since $[\Box, \mathcal{P}] = 0$, one has

$$\Box A_i = \mathcal{P}(A^j\, \partial_j A_i) + \mathcal{P}(A^j\, \partial_i A_j) + \text{l.o.t.} \tag{9.40}$$

The l.o.t. can be treated by more elementary techniques (such as the use of non-sharp Strichartz estimates) and the remaining two terms can be handled by *bilinear estimates*. Such estimates go beyond Strichartz estimates and take into account the special structure of the equations signified by the presence of null forms. These forms, first introduced by Klainerman [49], require special algebraic cancellations and are designed to detect systems which decay like solutions of a linear equation (see also Section 9.2.3), and in the simplest case take the form

$$Q_0(\phi, \psi) = \partial^\alpha \psi \cdot \partial_\alpha \phi, \quad Q_{\alpha\beta}(\phi, \psi) = \partial_\alpha \phi \cdot \partial_\beta \psi - \partial_\beta \phi \cdot \partial_\alpha \psi, \quad \alpha \neq \beta, \tag{9.41}$$

for scalar functions ϕ, ψ. The simplest bilinear estimate (see [47]) for the flat D'Alembertian controls the spacetime L^2 norm of $Q(\phi, \psi)$, where ϕ, ψ are solutions to the linear homogeneous wave equation, in terms of the H^2 and H^1 norms of the initial data of ψ, ϕ, respectively. One of the main observations of [48] is that the two terms on the right-hand side of (9.40) are in fact null forms and hence can be estimated by

$$\big\|\mathcal{P}(A^j \partial_j A_i)\big\|_{L^2[0,T]L^2(\Sigma_t)} \leq C \cdot \left(\|\partial A(0)\|_{L^2(\Sigma_0)} + \int_0^T \|\Box A(t)\|_{L^2(\Sigma_t)}\, dt \right)^2. \tag{9.42}$$

The first major observation in the bounded L^2 setting is that although the commutator $[\mathcal{P}, \Box_g]$ does not vanish, it produces only null forms,[12] and therefore the problem reduces to obtaining a class of bilinear estimates, analogous to (9.42), for solutions of wave equations on background metrics which possess very limited regularity. Moreover, in view of the absence of a timelike Killing field, one is left with just applying energy estimates for $T = e_0$. However, in order to bound the spacetime terms (which are of the form $\mathcal{L}_T g * R * R$), one needs another class of *trilinear estimates*. To obtain such estimates, one follows a more geometric approach to bilinear estimates, introduced in [50], which is based on the plane-wave representation formula for solutions of scalar wave equations

$$\phi_f(t, x) = \int_{\mathbb{S}^2} \int_0^\infty e^{i\lambda\,^\omega u(t,x)} f(\lambda\omega) \lambda^2 \, d\lambda \, d\omega, \tag{9.43}$$

where f represents schematically (at the level of the Fourier transform) the initial data and where $^\omega u$ is a solution of the eikonal equation $g^{\alpha\beta}\,\partial_\alpha\,^\omega u\,\partial_\beta\,^\omega u = 0$ with appropriate initial data (obtained by solving a parabolic equation) on Σ_0. In flat space, $^\omega u$ is also a solution to the wave equation; however, this is not the case in a curved background. Hence, in order to estimate

$$Ef(t, x) = \Box_g \phi_f(t, x) = i \int_{\mathbb{S}^2} \int_0^\infty e^{i\lambda\,^\omega u(t,x)} (\Box_g\,^\omega u) f(\lambda\omega) \lambda^3 \, d\lambda \, d\omega$$

and $\phi_f(0, x)$ in terms of $\phi(0, x)$, one needs to (1) control the geometry of the level sets of $^\omega u(0, x)$ on Σ_0; (2) estimate the L^2 norm of $\phi_f(0, x)$ in terms of the L^2 norm of f; (3) control the geometry of the foliation of $\mathcal{M}$ given by the level hypersurfaces of $^\omega u$ (in particular, control $\Box_g\,^\omega u$ in L^∞); and (4) derive a spacetime L^2 estimate for Ef using the estimates of the third step.

Steps (1) and (3) yield precise control on derivatives of $^\omega u$ and $\Box_g\,^\omega u$ with respect to ω and with respect to various spacetime (tangential and transversal to $\{^\omega u = c\}$) coordinates. Recall that a first L^∞ bound for $\Box_g\,^\omega u$ (which is proportional to the trace of the second fundamental form of the null hypersurfaces $\{^\omega u = c\}$), depending only on the L^2 curvature flux, was derived in [34–36]. Steps (2) and (4) rely on dyadic decompositions of the corresponding Fourier integral operators as well as a third decomposition, which in the case of Ef is done with respect to the spacetime variable based on the geometric Littlewood–Paley theory developed in [35].

Lower bounds on the radius of injectivity of the hypersurfaces $\{^\omega u = c\}$ are crucial for the construction of parametrices and the derivation of bilinear and trilinear estimates. These bounds strictly rely on L^2 curvature bounds. Hence, although the L^2 curvature theorem is not critical with respect to the standard scaling of the Einstein equations, it is nevertheless critical with respect to the null scaling; i.e., the scaling tied with the causal boundaries (and the eikonal equation). These considerations lead one to conclude that this result is most probably sharp insofar as the minimal number of derivatives in L^2 is concerned.

---

[12] This is because the full structure of the quasilinear hyperbolic system, not just its principal part, plays a crucial role.

9.2.3 Stability problems in General Relativity

In the absence of a general "large data" result in general relativity, the problem of stability of special solutions becomes simultaneously more important and more tractable.

Stability of the Minkowski spacetime

In the category of asymptotically flat spacetimes, the first (and essentially still, the only) such problem which has been globally understood is the case where the initial data is a small perturbation of the trivial data. This is the celebrated stability of Minkowski spacetime.

The Christodoulou–Klainerman stability theorem

The monumental work of Christodoulou and Klainerman [6] in 1993 proves the following stability result:

The maximal Cauchy development of smooth, strongly asymptotically flat, sufficiently small initial data $(\Sigma_0, g_{0ij}, k_{0ij})$ *(with* $\Sigma_0 \sim \mathbb{R}^3$, $M \geq 0$ *and* $\alpha > 1/2$ *in the asymptotic expansion* (9.8), *and* g_0 *globally close to the Euclidean metric* g_{eucl} *on* $\mathbb{R}^3$) *is a causally geodesically complete vacuum spacetime asymptotically "converging" to the Minkowski spacetime in all directions.*

A semi-global version of this result was proven in [51]; this approach used the conformal method to reduce the problem of proving future stability to that of verifying the local existence of an associated symmetric hyperbolic system. Note, however, that since the smoothness properties of the conformal extensions break down when the spacelike infinity is included, this approach can obtain stability only to the future of a hyperboloidal slice terminating at null infinity.

The proof of Christodoulou and Klainerman was more flexible and manifestly invariant; in particular it did not use the conformal method or the harmonic coordinate gauge. Instead, they relied on an adaptation of the optical structure equations (see Section 9.2.1) where the double null foliation is replaced by a maximal foliation Σ_t and a null foliation C_u such that the hypersurfaces Σ_t and C_u intersect at 2-spheres $S_{t,u}$.

The optical structure equations involve the curvature components and hence have to be coupled with estimates for the curvature tensor. To derive such estimates in a geometric way, one needs to exploit the hyperbolic radiating aspect of the vacuum equations manifest in the Bianchi system (9.16).

Generalized energy estimates for R can then be derived by using the associated energy-momentum tensor. This is the Bel–Robinson tensor, defined by (9.29). Christodoulou and Klainerman used the Bel–Robinson tensor in conjunction with special geometrically constructed vector fields, designed to mimic the rotation and conformal Morawetz vector fields of the Minkowski spacetime. The error terms in the energy identities involve the deformation tensor of these vector fields and can be controlled by the estimates obtained for the connection coefficients. Hence the full system is coupled.

In view of the positive mass theorem, non-trivial data must necessarily have positive mass $M > 0$ in the asymptotic expansion (9.8). This term has the "long-range effect" of changing the asymptotic position of the cones of null geodesics relative to the maximal foliation. For this reason the optical function u needs to be suitably normalized at infinity.[13] Moreover, the positivity of M implies that the initial data do not have sufficient decay in r at spacelike infinity. This could potentially lead to a "long-range effect" problem; i.e., the decay of the solution at timelike infinity $t \to +\infty$ is affected by the slow fall-off at spacelike infinity $r \to +\infty$. This issue was resolved in [6] by taking advantage of the facts that the long-range term is spherically symmetric and that the ADM mass M is conserved along the Einstein flow. This means that appropriate derivatives[14] of the curvature tensor with respect to properly defined (non-Minkowskian) angular momentum and timelike vector fields approximately satisfy the Einstein equations but with considerably better decay properties at spacelike infinity. Generalized energy estimates were then obtained for these derivatives of the curvature. Finally, the proof of the global existence result relies on the special structure of the nonlinear terms.[15] Specifically, certain combinations of derivatives, which could lead to the formation of singularities, are absent.

Further works

Klainerman and Nicolo [52] were able to use the double null foliation to localize the analysis in a neighborhood of spacelike infinity. Zipser extended the stability result of [6] to the case of the Einstein–Maxwell equations. Bieri derived a stronger version of the global stability of Minkowski spacetime by closing all estimates using less a priori control on the regularity of the metric (one derivative of curvature as opposed to two derivatives of curvature in [6]) and assuming borderline decay in r towards infinity. The results of Bieri and Zipser are presented in [53].

   In the realm of cosmological models, stability of the de Sitter spacetime has been shown in [54] using the conformal method; these results were later generalized in [55] for odd spatial dimensions. A more general stability result covering a wider class of expanding spacetimes, without relying on the conformal method, was presented in [56]. One should also mention the proof of stability in the expanding direction of a flat-cone solution for spatially compact spacetimes [57, 58].

Applications

1. The stability of Minkowski space [6] provided the first construction (and consequently, verification of existence) of smooth, geodesically complete non-trivial solutions which become flat at infinity in any given spacelike direction. The existence of such spacetimes is far from obvious. For example, in view of the Lichnerowicz and Birkhoff theorems, one cannot look for such solutions with additional symmetries.

---

[13] The situation is further complicated by the fact that the multiplier vector fields are constructed from the Σ_t and C_u foliations.

[14] That is, suitably modified Lie derivatives with respect to the above-constructed vector fields.

[15] Recall that for both the equation of nonlinear elasticity and the equation of compressible fluids in four spacetime dimensions, there are solutions which form singularities even for arbitrarily small initial data.

2. The asymptotic structure of future null infinity $\mathcal{J}^+$ is constructed explicitly by "attaching" to each cone C_u a sphere at infinity. The set $\mathcal{J}^+$ can be viewed as a $C^{1,\alpha}$ regular conformal boundary of $\mathcal{M}$. Moreover, a complete picture of the behavior of the gravitation field at null infinity, i.e., a hierarchy of the fall-off of the curvature components towards $\mathcal{J}^+$, is obtained from Christodoulou and Klainerman's results. The fall-off rates are weaker than the rates previously derived (known as *peeling*) by Newman–Penrose under the assumption of the existence of C^2 conformal compactification. Christodoulou [59] subsequently showed that generic physically admissible data will not exhibit peeling at $\mathcal{J}^+$. On the other hand, Klainerman and Nicolo [60] have constructed a special class of initial data which are consistent with peeling.

3. Significant information regarding the passage of gravitation waves, which can be used in gravitational wave experiments, was provided. These experiments are idealized as the study of the relative displacement of two test masses m_1, m_2 with respect to a third reference mass m_0, all located and evolving on a null generator of $\mathcal{J}^+$. In a follow-up work, Christodoulou [61] has shown that the passage of gravitational waves results in a permanent displacement of m_1, m_2 relative to m_0 which are of the same magnitude in a perpendicular direction. This effect is known as the *Christodoulou memory effect*. In the linear theory, a (weaker) memory effect had previously been found by Zel'dovich and Polnarev [62]. Further applications have recently been presented by Bieri and Garfinkle [63].

Stability of Minkowski in harmonic gauge

A proof of the stability of Minkowski spacetime which technically is relatively simple has been presented in [64]. It relies on the use of harmonic coordinates, and works with asymptotically flat data with positive fall-off parameter α (see equation (9.8)). It also holds for the Einstein-scalar field equations, where the scalar field satisfies the asymptotic expansion condition

$$\phi_0 = o(r^{-1-\alpha}), \quad \phi_1 = o(r^{-2-\alpha}), \tag{9.44}$$

as well as a global smallness condition. The appeal of the harmonic gauge for the proof of the stability of Minkowski spacetime lies in the fact that the latter can be simply viewed[16] as a small-data global existence result for the quasilinear system (9.20).

To describe some of the difficulties in establishing a small-data global existence result for the system (9.20), consider a generic quasilinear system of the form[17]

$$\Box \phi_i = \sum b_i^{jk\alpha\beta} \, \partial_\alpha \phi_j \, \partial_\beta \phi_k + \sum c_i^{jk\alpha\beta} \, \phi_j \, \partial_\alpha \partial_\beta \phi_k. \tag{9.45}$$

The first sum represents the *semilinear terms* whereas the second represents the *quasilinear terms*, each with their own problems. Christodoulou [65] and Klainerman [66] showed global existence for systems of the form (9.45) if the quasilinear terms are absent and

---

[16] This statement requires additional care since a priori there is no guarantee that the obtained "global in time" solution $g_{\mu\nu}$ defines a causally geodesically complete metric. However, the latter can be established provided one has good control on the difference between $g_{\mu\nu}$ and the Minkowski metric $m_{\mu\nu}$.

[17] One can in fact also consider cubic terms, which are in fact negligible.

the semilinear terms are of the form $Q(\phi,\phi)$, where Q is given by (9.41), a condition (as noted above) known as the *null condition*. Although the Einstein equations do not satisfy the null condition (see [67]), they do satisfy a weaker notion of the null condition introduced in [68]. Specifically, a system of the form (9.45) satisfies the *weak null condition* if the corresponding *asymptotic system* (cf. [69]) has global solutions. There are two main examples which satisfy the weak null condition. The first is (see [70, 71])

$$\Box\phi = \phi\,\triangle\phi, \tag{9.46}$$

and the second

$$\Box\phi_2 = (\partial_t\phi_1)^2, \quad \Box\phi_1 = 0. \tag{9.47}$$

While every solution of (9.47) is global in time, the system fails to satisfy the classical null condition and solutions are not asymptotically free: $\phi_2 \sim \epsilon t^{-1}\ln|t|$. The semilinear terms in Einstein's equations can be shown to either satisfy the classical null condition or decouple in either (9.46) or (9.47) when expressed in a null frame.

The quasilinear terms also decouple but in a more subtle way. The influence of the quasilinear terms can be detected via the asymptotic behavior of the characteristic surfaces of the metric g. It turns out that the main features of these surfaces at infinity are determined by a particular null component of the metric. The asymptotic flatness of the initial data and the harmonic condition (9.19) give good control of this particular component, i.e. M/r, which in turn implies that the light cones associated with the metric g diverge only logarithmically $\sim M\ln t$ from the Minkowski metric. The main simplification in this approach comes from the fact that the system (9.20) coupled to (9.19) can be completely controlled by means of the generalized energy estimates, exploiting only the *exact* symmetries of Minkowski space and thus avoiding having to construct dynamically generators of the approximate symmetries of the Cauchy development $(\mathcal{M}, g)$. On the other hand, this work is less precise concerning the asymptotic behavior of the curvature components.

Extensions of this approach to other matter fields have been provided in [72, 73].

Stability of the Kerr family of black holes

The concept of a *black hole* was first encountered in explicit solutions of the Einstein vacuum equations and specifically in the celebrated one-parameter *Schwarzschild* family of solutions. These spacetimes contain a black hole region $\mathcal{B}$ with the property that observers inside $\mathcal{B}$ cannot send signals to far-away observers. More generally, following Wheeler, one defines the black hole region of an asymptotically flat spacetime $(\mathcal{M}, g)$ as the complement, if non-empty, of the causal past of future null infinity $\mathcal{I}^+$. The boundary of the black hole region is called *the event horizon* and the exterior region is known as the *domain of outer communications*. An additional distinguishing property of the future null infinity $\mathcal{I}^+$ attached to Schwarzschild is that $\mathcal{I}^+$ is *complete*. This essentially means that the null generators of $\mathcal{I}^+$ can be continued to both future and past to arbitrary values of the affine parameter. Physically, this is the statement that asymptotic observers in the radiation zone live forever.

The Schwarzschild family is a sub-family of the two-parameter *Kerr* family, which describes stationary rotating black holes. The two parameters are the mass M, which is positive, and the angular momentum a, which is subject to $|a| \leq M$. It is widely believed that if $\mathfrak{I}^+$ is complete and if there is a black hole region, then the metric in the exterior region will eventually asymptote to the exterior of a collection of Kerr black holes, each rapidly moving away from each other. A simplified version of this problem is known as the nonlinear stability of the Kerr family.

Formulation of the conjecture

In its proper rigorous formulation, the problem of nonlinear stability of the Kerr family is one of the major open problems in general relativity:

Let $|a| < M$, let $(\mathcal{M}, g_{a,M})$ denote the globally hyperbolic region of a subextremal Kerr manifold with two ends, and let $\Sigma_{a,M}$ denote a Cauchy hypersurface. Let $(\Sigma, \tilde{g}_{per}, k)$ be vacuum initial data suitably close to Kerr data on $\Sigma_{a,M}$. Let $(\mathcal{M}, g_{per})$ denote its Cauchy development. Then one can attach to $\mathcal{M}$ a complete future null infinity $\mathfrak{I}^+$, with two connected components $\mathfrak{I}_A^+, \mathfrak{I}_B^+$, such that g_{per} asymptotes in the causal pasts of $\mathfrak{I}_A^+$ and $\mathfrak{I}_B^+$ to nearby Kerr metrics g_{a_A,M_A} and g_{a_B,M_B} with a_A, a_B close to a and with M_A, M_B close to M.

One can also formulate the above conjecture so that it concerns only one of the asymptotically flat ends. To do this, one simply needs to perturb data on an initial incomplete hypersurface which contains a trapped surface.

A simplified version of the above conjecture focusses on the nonlinear stability of Schwarzschild in the context of axial symmetry. Since axially symmetric spacetimes do not radiate angular momentum, one needs to show that axially symmetric perturbations of Schwarzschild data evolve to a nearby Schwarzschild solution, in the sense of the above conjecture.

Scattering construction of black holes approaching the Kerr family

As in the case of Minkowski spacetime, the existence of vacuum black hole spacetimes approaching a Kerr solution is far from obvious. A large class of such black holes was constructed in [74] by solving a backwards scattering problem for the Einstein vacuum equations with characteristic data prescribed on the event horizon $\mathcal{H}^+$ and on null infinity $\mathfrak{I}^+$. The class admits the full "functional" degrees of freedom for the vacuum equations, and thus the solutions will in general possess no geometric or algebraic symmetries. It is essential, however, for the construction that the scattering data (and the resulting solution spacetime) converge to a stationary solution *exponentially fast* in advanced and retarded time. The exponential decay can be traced back to the celebrated *redshift effect* on the event horizon $\mathcal{H}^+$ in which the frequency of an emitter leaving the exterior region gets shifted to the red as viewed by an observer positioned to the future of the first one. In the context of the backwards evolution this is seen as blueshift. The proof uses the double null foliation (see Section 9.2.1) and relies on showing dispersive estimates for *renormalized* quantities, which are obtained by subtracting the metric components of the (final) Kerr spacetime from the metric components of the background spacetime.

In the context of the forward problem as stated in Section 9.2.3, however, the dispersion mechanism to a nearby Kerr solution has not been properly understood. In view of the supercriticality of the Einstein equations, understanding this dispersion mechanism is essential even for showing the completeness property of null infinity.

The wave equation: a poor man's linearization

The quantitative global properties (boundedness and decay) even of linear scalar waves

$$\Box_g \psi = 0 \tag{9.48}$$

on Schwarzschild and Kerr black hole backgrounds have only recently been understood. See [75–89] and references therein. For subextremal Kerr spacetimes $|a| < M$, the event horizon is responsible for the aforementioned redshift effect. Another obstruction to dispersion, known as the *trapping effect*, is the presence of null geodesics which neither cross the event horizon nor terminate at null infinity. In the case of Schwarzschild, trapping occurs only at $r = 3M$ and can be captured by pure physical-space vector fields (see [80]), to yield pointwise and energy boundedness and decay towards the future.

The evolution of linear waves on subextremal Kerr black holes is substantially harder to understand than the case of Schwarzschild. The reason for this is the much more complicated trapping, which can no longer be captured with physical-space vector fields (see, however, [76] for small a) and the existence of a third fundamental difficulty, namely that of *superradiance*. The latter corresponds to the existence of a region in the exterior of the black hole, known as the *ergoregion*, where the 'stationary' Killing vector field ∂_t becomes spacelike. This phenomenon in principle allows for the existence of exponentially growing solutions to (9.48) with finite initial energy. The nonexistence of such modes on Kerr black holes was shown by Whiting [90]. The study of individual modes is possible in view of Carter's separability of the wave equation on Kerr solutions.

One of the main new insights of [83, 87] is the use of Carter's separability of the wave equation as a geometric microlocalization, i.e., as a method to frequency-localize energy estimates in a manner particularly suited to the local and global geometry of Kerr. This allowed the authors of [83, 87] to prove definitive dispersive estimates for the whole subextremal Kerr family. This work revealed two important features of the Kerr geometry: For the high-frequency limit it is shown that superradiant frequencies are not trapped and for the bounded-frequency regime that there are no time-periodic solutions with finite initial energy.

We remark that the limit $|a| \to M$, where the redshift effect degenerates, is singular in the sense that there are translation-invariant derivatives of the solutions which blow up asymptotically in the future (see [75]).

We will not discuss the case of cosmological black hole spacetimes, but refer to [91, 92] for anti-de Sitter black holes and [93–96] for the de Sitter black hole case. For the study of black hole resonances we refer to [97, 98].

9.2.4 Large-data regimes

Weak and strong cosmic censorship

An even more ambitious problem is to understand the *global* behavior of the maximal development corresponding to *generic* initial data. The maximal Cauchy development could either terminate at singularities (such as Schwarzschild), or it might admit a regular boundary (such as Kerr), known as the *Cauchy horizon*, beyond which the spacetime can be continued, thus losing predictability from the initial data. A general theorem guaranteeing the geodesic incompleteness of the maximal Cauchy development is the following one by Penrose:

If the matter satisfies the null congruence energy condition, then the maximal development is geodesically incomplete if it admits a non-compact Cauchy hypersurface and a closed trapped surface.

The discussion below will focus on the case of the asymptotically flat spacetimes and will not address many developments that took place in the context of spatially compact or cosmological spacetimes.

A qualitative picture

In the language of partial differential equations, Penrose's theorem implies the impossibility of a large-data global existence result for all initial data in general relativity. In the absence of such a result a number of conjectures about the structure of spacetimes arising from generic data had been put forward in the 1960s by Penrose. Among them is the following:

Weak Cosmic Censorship – For generic asymptotically flat vacuum initial data, the maximal development possesses a complete future null infinity.

The completeness of null infinity can be thought of as a *global existence* result. Alternatively, the above conjecture predicts that all singularities are hidden inside black holes (and hence there are no *naked singularities* visible to null infinity).

Strong Cosmic Censorship – For generic asymptotically flat vacuum initial data, the maximal development is either complete or terminates at a totally singular boundary.

The above conjecture predicts that generically determinism does not fail and so can be thought of as a *global uniqueness* result.

Resolution for the spherically symmetric scalar field model

At this moment not much progress has been made on either of these cosmic censorship problems in the general case, with remarkable exceptions in certain cases of symmetry-reduced Einstein equations. We here focus on the definitive proof of weak and strong cosmic censorship for the Einstein-scalar field system with spherical symmetry by Christodoulou [39, 99–101]. Christodoulou considered initial data on a complete null cone C_o about a point o, extending to infinity. The initial data for this spherically symmetric problem consists of the function $\alpha_0 = \partial(r\phi)/\partial s$ on C_o, where s is the affine parameter along the null generators of C_o. Christodoulou showed the following:

The set of data leading to an incomplete null infinity has codimension at least 1 in the space of all admissible spherically symmetric data. Moreover, generically the maximal development cannot be extended as a C^0 Lorentzian manifold.

By admissible initial data we mean data of bounded variation. Well-posedness for this class of initial data was established in [39]. Moreover, a sharp extension criterion was obtained, namely that if the ratio of the mass content to the radius of the spheres tends to zero as we approach a point on the axis of symmetry from its causal past then the solution extends as a regular solution to include a full neighborhood of the point. The structure of the solutions of bounded variations was studied and it was shown that the solutions are scale-invariant at each point of the axis of symmetry.

We next provide a brief summary of the proof. It is first shown that if the axis of symmetry Γ_0 is complete, then the maximal development is also complete. Thus, one can assume that Γ_0 has a singular end point e. Let $S_{0,1}$ be the intersection of the boundary $\underline{C}_1$ of the causal past of e and C_o. Let also $\underline{C}_2$ be the incoming null hypersurface emanating from a sphere $S_{0,2}$ on C_o in the exterior, but close to $S_{0,1}$. There is a BV function f such that if we consider the data

$$\alpha_0 + cf, \quad c \in \mathbb{R}^*$$

on C_0, then there exists an outgoing null hypersurface intersecting $\underline{C}_1, \underline{C}_2$ at the spheres S_1, S_2, respectively, such that the formation of trapped surfaces criterion [100] stated below applies. It is precisely here where the singular nature of e is used and, in particular, where the infinite blue-shift along $\underline{C}_1$ is used.

Formation of trapped surfaces in spherical symmetry – Let S_1, S_2 be two spheres on a null cone C, with S_2 in the future of S_1 and let $C^{1,2}$ be the annular region of C bounded by these two spheres. Let also C_2 denote the incoming null hypersurface emanating from S_2. Then, there are positive constants c_0 and c_1 such that if the dimensionless size δ and the mass content η of $C^{1,2}$ satisfy

$$\delta \leq c_0 \quad \text{and} \quad \eta > c_1 \delta \, \log\left(\frac{1}{\delta}\right),$$

then C_2 contains a closed trapped surface.

The data on C_o are not required to be globally small (cf. Section 9.2.4). Note also that the above criterion does not impose any conditions on the incoming null hypersurface starting at $S_{1,0}$.

Returning to the proof of the main theorem, the presence of trapped surfaces implies that the spacetime admits a complete null infinity with, in fact, positive final Bondi mass. In view of an earlier result [99], the latter implies the formation of a black hole with mass equal to the final Bondi mass. Furthermore, in view of the results in [100], the region of trapped surfaces terminates at a strictly spacelike singular boundary, which completes the proof.

We note that the assumption of genericity is essential in the formulation of weak cosmic censorship, since Christodoulou [102] has constructed initial data which lead to the formation of naked singularities.

Another model with surprising features is the spherically symmetric Einstein–Maxwell-scalar field model. Dafermos [103] showed that part of the boundary of the maximal development is a Cauchy horizon, through which the metric can be continued in a C^0 manner, but at which, generically, the mass function blows up. As a consequence, the Christoffel symbols are not integrable. This analysis thus provides an appropriate version[18] of the strong cosmic censorship conjecture for this model.

The formation of trapped surfaces

It is clear from the above discussion that the formation of trapped surfaces is essential in the proof of the cosmic censorship conjectures in the framework of the spherically symmetric scalar field model. Indeed, the above analysis implies that for generic initial data, all "first singularities" in the spherically symmetric scalar field model are preceded by trapped-surface formation.

Christodoulou [7] has solved a long-standing problem of general relativity; that of the evolutionary formation of trapped surfaces in Einstein vacuum spacetimes without any symmetry assumptions. He in fact showed that trapped surfaces form in evolution from the collapse of a regular and arbitrarily dispersed initial state. Christodoulou prescribed characteristic initial data, see Section 9.2.1, for which the initial surface S_0 is a round sphere of radius r_0 and the incoming null hypersurface $\underline{C}_0$ is Minkowskian. On the outgoing null hypersurface C_0 he introduced certain large data using the so-called "*short-pulse method*". This methods seeks to introduce in a controlled way a certain large amplitude in the data (the pulse), compensated by a shortness in its characteristic length, with the latter controlled by a parameter δ. The pulse gives rise to a certain *hierarchy* of large and small quantities (i.e., connection coefficients and curvature components), which are coupled via the non-linear vacuum equations.

In sharp contrast with the spherically symmetric scalar field model, here almost all the work goes into establishing a *large-time existence theorem* for the short-pulse data. This theorem uses the double null foliation and relies on combining some of the methods introduced in the work on stability of Minkowski spacetime [6] and the above hierarchy which turns out to be propagated along the evolution of data.

The theorem describing the formation of closed trapped surfaces follows from an additional assumption on the initial data on C_0; namely the existence of a uniform lower and upper bound on the integrals

$$\int_0^\delta |\hat{\chi}|^2 \, dv \tag{9.49}$$

on the affine parameter segment $[r_0, r_0 + \delta]$ of *each* null generator of C_0. Here $\hat{\chi}$ is the shear of the sections of C_0; note that $|\hat{\chi}|^2$ is an invariant of the conformal intrinsic geometry of

---

[18] Dafermos' work illustrates that indeed much care is needed in formulating the strong cosmic censorship conjecture.

C_0 (i.e., of the initial data on C_0). The upper bound on (9.49) guarantees that there is no trapped surface in the initial data, whereas the lower bound, in combination with the information obtained as part of the existence theorem, guarantees the existence of a closed trapped surface in the maximal development. The Raychaudhuri equation implies

$$\nabla_L \, tr \, \chi \le -|\hat{\chi}|^2. \tag{9.50}$$

Therefore, if the integral of $|\hat{\chi}|^2$ along null generators of the outgoing null hypersurface $\mathcal{C}$ generated by S_{init} is large relative to $tr \, \chi|_{S_{init}}$, where S_{init} is a sphere on $\underline{C}_0$, then there is a section S_{trap} of $\mathcal{C}$ which is trapped. The mechanism that one exploits here is that, although $tr\chi|_{S_{init}} = 2/r$, we have that $|\hat{\chi}|^2 \sim 1/r^2$, and hence, one expects that given a suitable lower bound for (9.49), if the maximal development contains a null hypersurface $\mathcal{C}$ with sufficiently small values of r, then $\mathcal{C}$ contains a trapped surface.

The existence theorem, however, does not cover the whole of the maximal development, and for this reason the nature of the future boundary of the maximal development is unknown.

A subsequent insightful approach to this problem was given in [104, 105]. In this work a relaxed hierarchy is presented, which gives rise to fewer borderline error terms in the energy estimates and thus makes it easier to prove the propagation of this hierarchy. Moreover, the whole argument is closed at one lower order of differentiability. See also [106–108] for related work. In particular [107] established a dynamic formation of a trapped surface from data given on a spacelike *Cauchy* hypersurface.

An anisotropic criterion for the formation of trapped surfaces was given in [109], according to which the maximal development contains trapped surfaces even if a lower bound on (9.49) holds in a neighborhood of a *single* null generator of C_0 – and not for all null generators of C_0 as is required in [7].

Impulsive gravitational waves

Explicit solutions representing impulsive gravitational waves were first found by Penrose [110], building on earlier works [111, 112]. These are plane-symmetric spacetimes satisfying the Einstein vacuum equations such that the Riemann curvature tensor has a delta singularity across a null hypersurface. The rigorous mathematical study of the problem of impulsive gravitational waves was initiated in [113], where the following was shown:

Consider characteristic initial data such that the shear $\hat{\chi}$ has a jump discontinuity across a section S_{v_s} of the initial outgoing null hypersurface C_0 but is smooth otherwise. In particular, the data is smooth on the initial incoming null hypersurface $\underline{C}_0$. Given such data, there is a unique local solution to the Einstein vacuum equations. Moreover, the curvature has a delta singularity across the incoming null hypersurface $\underline{C}_{v_s}$ emanating from S_{v_s} and is smooth away from it.

This result can be formally compared to the works of Majda on the propagation of shocks [114, 115] for systems of conservation laws. In these works, initial data with a jump discontinuity across a surface is considered and a local existence, uniqueness and regularity result was established along with a description of the propagation of singularity. On the other hand, contrary to [114, 115], the singularity of impulsive gravitation waves is not a shock, as it propagates along the characteristics.

The work on impulsive gravitational waves uses the double null foliation. Let L be tangential to the null generators of the initial outgoing null hypersurface C_0. In view of the second variation formula we obtain (see also (9.13) with $\omega = 0$)

$$\nabla_L \hat{\chi} = -tr\, \chi \cdot \hat{\chi} - \alpha, \tag{9.51}$$

where α is the *null decomposed curvature component* given by (9.11). In view of the assumption for $\hat{\chi}$, α is defined only as a measure and hence the recently resolved L^2 bounded curvature conjecture does not suffice to prove the existence of a corresponding spacetime. The key new observation [113] is that, using the null structure of the Einstein equations, the L^2-type energy estimates for the curvature components coupled with the null transport equations for the connection coefficients can be *renormalized* and closed, avoiding the L^2-non-integrable component α of curvature. Recall that energy estimates for curvature were derived in the work on the stability of Minkowski spacetime [6] using the Bel–Robinson tensor. In view of the renormalization of the curvature components, the energy estimates in [113] were derived directly from the (null decomposed) Bianchi identities (9.16). This approach to energy estimates, which circumvents the use of the energy-momentum tensor and thus is more flexible and localized, was first introduced in the context of a new method for deriving energy decay for solutions to the wave equation on general asymptotically flat spacetimes [116] and was later adapted for the Bianchi equations by Holzegel [117].

Returning to the gravitational waves result, the main observation is that if one adds suitable expressions of the connection coefficients to the curvature components and uses the null transport equations for the latter, such as (9.51), then the renormalized Bianchi equations do not involve the singular component α. This allows one to derive energy estimates which do not couple to α. Such estimates of course couple to connection coefficients. The latter, however, always satisfy either a transport or an elliptic equation which does not contain α.

The a priori estimates derived along the above lines do not suffice to guarantee the existence and uniqueness of the solutions. Instead, one needs to approximate the data by a sequence of smooth data and show first that the corresponding solutions have a common domain of existence, which follows from the above a priori estimates, and second that they converge in this domain. For the latter, one needs to show a priori estimates for the difference of solutions. A remarkable discovery of [113] is that in the equations for the difference of the connection coefficients and curvature components, α indeed does not appear. These estimates yield uniqueness of the constructed solution among all limits of smooth solutions.

The above approach allows one to consider a much larger class of initial data, where $\hat{\chi}$ and its angular derivatives are only assumed to be in L^∞, obtaining therefore a very general existence result.

In a follow-up paper [118], spacetimes with two non-linearly interacting gravitational waves are investigated and the following is shown:

Consider characteristic initial data such that the shear $\hat{\chi}$ has a jump discontinuity across a section S_{v_s} of the initial outgoing null hypersurface C_0 but is smooth otherwise and $\underline{\hat{\chi}}$ has a jump discontinuity across a section S_{u_s} of the initial ingoing null hypersurface $\underline{C}_0$. Given such data, there is a unique local solution to the Einstein vacuum equations. Moreover, the curvature has a delta singularity across the incoming (resp. outgoing) null hypersurfaces $\underline{C}_{v_s}$ (resp. C_{u_s}) emanating from S_{v_s} (resp. S_{u_s}) and is smooth away from their union.

The construction of these spacetimes, motivated in part by the celebrated explicit solutions of Khan and Penrose [119] and Szekeres [120], relies on a suitable adaptation of the above renormalized energy estimates. In fact, a more general existence result theorem is proved, which requires only bounds of $\hat{\chi}$, $\underline{\hat{\chi}}$ in L^2 on C_0, $\underline{C}_0$, respectively, as opposed to L^∞ in [113]. A surprising application of the latter is an extension of the Christodoulou criterion for the formation of trapped surfaces, allowing non-trivial data on the initial incoming hypersurface. Indeed, although $|\hat{\chi}| \sim \delta^{-\frac{1}{2}}$ on C_0, where δ is the small parameter introduced in Section 9.2.4, the L^2 norm of $\hat{\chi}$ is of size 1 relative to δ and hence the above theorem implies the existence and uniqueness of a spacetime solution for this type of initial data, even without any smallness assumption on the initial incoming null hypersurface $\underline{C}_0$.

The importance of the above results in resolving the cosmic censorship conjectures in the general non-symmetric case remains to be seen.

References

[1] A. Einstein, *Sitzungsber. Preuss. Akad. Wiss.* **50** 778–786 (1915).
[2] A. Einstein, *Sitzungsber. Preuss. Akad. Wiss.* **50** 844–847 (1915).
[3] R. Schoen and S.T. Yau, *Commun. Math. Phys.* **65** 45–76 (1979).
[4] R. Schoen and S.T. Yau, *Commun. Math. Phys.* **79** 231–260 (1981).
[5] E. Witten, *Commun. Math. Phys.* **80**, 381–402 (1981).
[6] D. Christodoulou and S. Klainerman, *The global nonlinear stability of Minkowski space*, Princeton University Press (1994).
[7] D. Christodoulou, *The formation of black holes in general relativity*, EMS (2009).
[8] Y. Choquet-Bruhat, *Acta Math.* **88** 141–225 (1952).
[9] Y. Choquet-Bruhat and R. Geroch, *Commun. Math. Phys.* **14** 329–335 (1969).
[10] J. Sbierski, unpublished, arXiv:1309.7591.
[11] A. Rendall, *Proc. R. Soc. Lond.* A **427** 221–239 (1990).
[12] J. Luk, *Int. Math. Res. Not.* 4625–4678 (2012).
[13] P. Dionne, *J. Analyze Math.* **10** 1–90 (1962).
[14] A. Fischer and J. Marsden, *Commun. Math. Phys.* **28** 1–38 (1972).
[15] T. Hughes, R. Kato and J. Marsden, *Arch. Ration. Mech. Anal.* **63** 273–394 (1977).
[16] F. Planchon and I. Rodnianski, unpublished.
[17] G. Ponce, G., and T. Sideris, *Commun. Partial Diff. Eq.* **18** 2419–2438 (1993).
[18] H. Smith, *Ann. Inst. Fourier* **48** 797–835 (1998).
[19] H. Bahouri and J. Chemin, *Amer. J. Math.* **21** 1337–1377 (1999).
[20] H. Bahouri and J. Chemin, *Int. Math. Res. Notices* **21** 1141–1178 (1999).
[21] D. Tataru, *Amer. J. Math.* **1122** 349–376 (2000).
[22] D. Tataru, *Amer. J. Math.* **123** 385–423 (2001).
[23] H. Smith and D. Tataru, *Math. Res. Lett.* **9** 199–204 (2002).
[24] S. Klainerman and I. Rodnianski, *Duke Math. J.* **117** 1–124 (2003).

[25] S. Klainerman and I. Rodnianski, *Ann. Math.* **161** 1143–1193 (2005).
[26] H. Smith and D. Tataru, *Ann. Math.* **162** 291–366 (2005).
[27] M. Anderson, *Commun. Math. Phys.* **222** 533–567 (2001).
[28] S. Klainerman and I. Rodnianski, *J. AMS* **23** 345–382 (2010).
[29] T. Beale, T. Kato and A. Majda, *Commun. Math. Phys.* **94** 61–66 (1984).
[30] D. Eardley and V. Moncrief, *Commun. Math. Phys.* **83** 171–191 (1982).
[31] D. Eardley and V. Moncrief, *Commun. Math. Phys.* **83** 193–212 (1982).
[32] S. Klainerman and I. Rodnianski, *J. Hyper. Diff. Eq.* **4** 401–403 (2007).
[33] V. Moncrief, *J. Diff Geom: Surveys in Diff. Geom.* X (2006).
[34] S. Klainerman and I. Rodnianski, *Invent. Math.* **159** 437–529 (2005).
[35] S. Klainerman and I. Rodnianski, *Geom. Funct. Anal.* **16** 126–163 (2006).
[36] S. Klainerman and I. Rodnianski, *Geom. Funct. Anal.* **16** 164–229 (2006).
[37] S. Klainerman and I. Rodnianski, *J. AMS* **21** 775–795 (2008).
[38] Q. Wang, *Commun. Pure Appl. Math.* **65** 21–76 (2012).
[39] D. Christodoulou, *Commun. Pure Appl. Math.* **46**, 1093–1220 (1993).
[40] S. Klainerman, Proceedings of Visions in Mathematics, *Geom Funct. Anal.* Special Vol. 279–315 (2000).
[41] S. Klainerman, I. Rodnianski and J. Szeftel, unpublished, arXiv:1204.1767.
[42] J. Szeftel, unpublished, arXiv:1204.1768.
[43] J. Szeftel, unpublished, arXiv:1204.1769.
[44] J. Szeftel, unpublished, arXiv:1204.1770.
[45] J. Szeftel, unpublished, arXiv:1204.1771.
[46] J. Szeftel, unpublished, arXiv:1301.0112.
[47] S. Klainerman and M. Machedon, *Commun. Pure Appl. Math.* **46** 1221–1268 (1993).
[48] S. Klainerman and M. Machedon, *Ann. Math.* **142** 39–119 (1995).
[49] S. Klainerman, *Proceedings of the International Congress on Mathematics* (Warsaw) 1209–1215 (1983).
[50] S. Klainerman and I. Rodnianski, *J. Hyper. Diff. Eq.* **161** 279–291 (2005).
[51] H. Friedrich, *Commun. Math. Phys.* **107** 587–609 (1986).
[52] S. Klainerman and F. Nicolo, *Evolution problem in general relativity* Birkhäuser (2002).
[53] L. Bieri and N. Zipser, *Extensions of the stability theorem of the Minkowski space in general relativity* AMS/IP (2009).
[54] H. Friedrich, *J. Geom. Phys.* **3** 101–117 (1986).
[55] M. Anderson, *Ann. Inst. H. Poincaré* **6** 801–820 (2005).
[56] H. Ringström, *Invent. Math.* **173** 123–208 (2008).
[57] L. Andersson and V. Moncrief, The Einstein equations and the large scale behavior of gravitational fields, in *50 years of the Cauchy problem in general relativity* (ed. by P. T. Chruściel and H. Friedrich) Birkhäuser, 299–330 (2004).
[58] M. Reiris, Ph.D. Thesis, Stony Brook, (2005).
[59] D. Christodoulou, Ninth Marcel Grossmann Meeting (Rome 2000), ed. by V. G. Gurzadyan, World Scientific, 44–54 (2002).
[60] S. Klainerman and F. Nicolo, *Class. Quant. Grav.* **20** 3215–3257 (2003).
[61] D. Christodoulou, *Phys. Rev. Lett.* **67** 1486–1489 (1991).
[62] Y. Zel'dovich and A. Polnarev, *Astron. Zh.* **51** 30 (1974).
[63] L. Bieri and D. Garfinkle, *Phys. Rev. D* **89** 084039 (2014).
[64] H. Lindblad and I. Rodnianski, *Ann. Math.* **171** 401–1477 (2010).
[65] D. Christodoulou, *Commun. Pure Appl. Math.* **39** 267–282 (1986).
[66] S. Klainerman, *Lect. Appl. Math.* **23** 293–326 (1986).
[67] Y. Choquet-Bruhat, *C. R. Acad. Sci. Paris, Ser. A* **276** 281–284 (1973).
[68] H. Lindblad and I. Rodnianski, *C. R. Acad. Sci. Paris, Ser. A* **335** 901–906 (2003).
[69] L. Hörmander, *Pseudo-differential operators*, Lecture Notes in Mathematics, vol. **1256** Springer-Verlag, Berlin 214–280 (1987).
[70] H. Lindblad, *Commun. Pure Appl. Math.* **45** 1063–1096 (1992).
[71] S. Alinhac, *Astérisque* **284** (2003) 1–91 (2003).

[72] I. Rodnianski and J. Speck, unpublished, arXiv:0911.5501.
[73] J. Speck, to be published in Analysis and PDE.
[74] M. Dafermos, G. Holzegel and I. Rodnianski, unpublished, arXiv:1306.5364.
[75] S. Aretakis, unpublished, arXiv:1206.6598.
[76] L. Andersson and P. Blue, unpublished, arXiv:0908.2265.
[77] P. Blue and A. Soffer, *Adv. Diff. Eqs.* **8** 595–614 (2003).
[78] P. Blue and J. Sterbenz, *Commun. Math. Phys.* **268** 481–504 (2006).
[79] F. Finster, N. Kamran, J. Smoller and S. Yau, *Commun. Math. Phys.* **264** 465–503 (2006).
[80] M. Dafermos and I. Rodnianski, *Commun. Pure Appl. Math.* **62** 859–919 (2009).
[81] M. Dafermos and I. Rodnianski, *Invent. Math.* **185** 467–559 (2011).
[82] M. Dafermos and I. Rodnianski, unpublished, arXiv:1010.5132.
[83] M. Dafermos and I. Rodnianski, Proc. 12th Marcel Grossmann Meeting, ed. by T. Damour *et al.*, World Scientific 132–189 (2011).
[84] B. Kay and R. Wald, *Class. Quant. Grav.* **4** 893–898 (1987).
[85] J. Marzuola, J. Metcalfe, D. Tataru and M. Tohaneanu, *Commun. Math. Phys.* **293** 37–83 (2010).
[86] J. Metcalfe, D. Tataru and M. Tohaneanu, *Adv. Math.* **230** 995–1028 (2012).
[87] Y. Shlapentokh-Rothman, unpublished, arXiv:1302.6902.
[88] D. Tataru and M. Tohaneanu, *Int. Math. Res. Not.* **2011** 248–292 (2008).
[89] R. Wald, *J. Math. Phys.* **20** 1056–1058 (1979).
[90] B. Whiting, *J. Math. Phys.* **30** (1989), 1301 (1989).
[91] G. Holzegel, *Commun. Math. Phys.* **294** 169–197 (2010).
[92] G. Holzegel and J. Smulevici, *Commun. Pure Appl. Math.* **66** 1751–1802 (2013).
[93] J. Bony, and D. Hafner, *Commun. Math. Phys.* **282** 697–719 (2008).
[94] M. Dafermos and I. Rodnianski, unpublished, arXiv:0709.2766.
[95] R. Melrose, A. Barreto and A. Vasy, unpublished, arXiv:0811.2229.
[96] A. Vasy and S. Dyatlov, unpublished, arXiv:1012.4391.
[97] S. Dyatlov, *Commun. Math. Phys.* **306** 119–163 (2011).
[98] A. Barreto and M. Zworski, *Math. Res. Lett.* **4** 103–121 (1997).
[99] D. Christodoulou, *Commun. Math. Phys.* **109** 613–647 (1987).
[100] D. Christodoulou, *Commun. Pure Appl. Math.* **44**, 339–373 (1991).
[101] D. Christodoulou, *Ann. Math.* **149** 183–217 (1999).
[102] D. Christodoulou, *Ann. Math.* **140** 607–653 (1994).
[103] M. Dafermos, *Ann. Math.* **158** 875–928 (2003).
[104] S. Klainerman and I. Rodnianski, *Acta Math.* **208** 211–333 (2012).
[105] S. Klainerman and I. Rodnianski, unpublished, arXiv:1002.2656.
[106] M. Reiterer and E. Trubowitz, unpublished, arXiv:0906.3812.
[107] J. Wang and P. Yu, unpublished, arXiv:1207.3164.
[108] P. Yu, unpublished, arXiv:1105.5898.
[109] S. Klainerman, J. Luk and I. Rodnianski, I., unpublished, arXiv:1302.5951 (2013).
[110] R. Penrose, *General relativity (papers in honour of J. L. Synge)* 101–115 (1972).
[111] M. Brinkmann, *Proc. Natl. Acad. Sci. U.S.A.* **9** 1–3 (1923).
[112] V. Lanczos, *Z. Phys.* **21**.2 73–110 (1924).
[113] J. Luk and I. Rodnianski, unpublished, arXiv:1209.1130.
[114] A. Majda, *Mem. Amer. Math. Soc.* **43** (1983).
[115] A. Majda, *Mem. Amer. Math. Soc.* **41** (1983).
[116] M. Dafermos and I. Rodnianski, *XVIth International Congress on Mathematics and Physics*, 421–433 (2009).
[117] G. Holzegel, unpublished, arXiv:1010.3216.
[118] J. Luk and I. Rodnianski, unpublished, arXiv:1301.1072.
[119] K. Khan and R. Penrose, *Nature* **229** 185–186 (1971).
[120] P. Szekeres, *Nature* **228** 1183–1184 (1970).

9.3 Convergence and stability issues in mathematical cosmology
Vincent Moncrief

9.3.1 Introduction

Viewed on a sufficiently coarse-grained scale the portion of our universe that is accessible to observation appears to be spatially homogeneous and isotropic. If, as is usually imagined, one should be able to extrapolate these features to (a suitably coarse-grained model of) the universe as a whole then only a handful of spatial manifolds need be considered in cosmology – the familiar Friedmann–Lemaître–Robertson–Walker (FLRW) archetypes of constant positive, vanishing or negative curvature [1, 2]. These geometries consist, up to an overall, time-dependent scale factor, of the 3-sphere, $\mathbb{S}^3$, with its canonical 'round' metric, Euclidean 3-space, $\mathbb{E}^3$, hyperbolic 3-space, $\mathbb{H}^3$ and the quotient space $\mathbb{R}P(3) \approx \mathbb{S}^3/\pm I$ obtainable from $\mathbb{S}^3$ by the identification of antipodal points [3]. Of these possibilities only the sphere and its 2-fold quotient $\mathbb{R}P(3)$ are *closed* and thus compatible with a universe model of finite extent. It is not known of course whether the actual universe is spatially closed or not but, to simplify the present discussion, we shall limit our attention herein to models that are. More precisely we shall focus on spacetimes admitting Cauchy hypersurfaces that are each diffeomorphic to a smooth, connected 3-manifold that is compact, orientable and without boundary.

On the other hand if one takes literally the cosmological principle that only manifolds supporting a *globally* homogeneous and isotropic metric should be considered in models for the actual universe then, within the spatially compact setting considered here, only the 3-sphere and $\mathbb{R}P(3)$ would remain. But the astronomical observations which motivate this principle are necessarily limited to a (possibly quite small) fraction of the entire universe and are compatible with models admitting metrics that are only locally, but not necessarily globally, spatially homogeneous and isotropic. As is well-known there are spatially compact variants of all of the basic Friedmann–Lemaître–Robertson–Walker cosmological models, mathematically constructable (in the cases of vanishing or negative curvature) by taking suitable compact quotients of Euclidean 3-space $\mathbb{E}^3$ or of hyperbolic 3-space $\mathbb{H}^3$. One can also take infinitely many possible quotients of $\mathbb{S}^3$ to obtain the so-called *spherical space forms* that are locally compatible with the FLRW constant positive curvature geometry but are no longer diffeomorphic to the 3-sphere.

Still more generally, though, we shall find that there is a dynamical mechanism at work within the Einstein 'flow', suitably viewed in terms of the evolution of 3-manifolds to develop 4-dimensional, globally hyperbolic spacetimes, that strongly suggests that even manifolds that do not admit a locally homogeneous and isotropic metric *at all* will nevertheless evolve in such a way as to be *asymptotically* compatible with the observed homogeneity and isotropy. This reflects an argument which we shall sketch that, under Einsteinian evolution, the summands making up the 3-manifold *M* (in a connected sum decomposition) that do support locally homogeneous and isotropic metrics will tend to overwhelmingly dominate the spatial volume asymptotically as the universe model continues to expand and furthermore that the actual evolving (inhomogeneous, non-isotropic) metric on *M* will naturally tend to flow towards a homogeneous, isotropic one on each of these asymptotically volume-dominating summands.

We do not claim that this mechanism is yet so compelling, either mathematically or physically, as to convince one that the actual universe has a more exotic topology but only that such a possibility is not strictly excluded by current observations. However, it is intriguing to investigate the possibility that there may be a dynamical reason, provided by Einstein's equations, for the observed fact that the universe seems to be at least locally homogeneous and isotropic and that this mechanism may therefore allow an attractive logical alternative to simply extrapolating observations of necessarily limited scope to the universe as a whole.

But what are the (compact, connected, orientable) 3-manifolds available for consideration? This question has been profoundly clarified in recent years by the dramatic progress on lower-dimensional topology made possible through the advancements in Ricci flow [4]. One now knows for example that, since the Poincaré conjecture has finally been proven, any such 3-manifold M that is in fact *simply connected* must be diffeomorphic to the ordinary 3-sphere S^3. Setting aside this so-called 'trivial' manifold, the remaining possibilities consist of an infinite list of nontrivial manifolds, each of which is diffeomorphic (designated herein by $\approx$) to a finite connected sum of the following form:

$$M \approx \underbrace{S^3/\Gamma_1 \# \cdots \# S^3/\Gamma_k}_{k \text{ spherical factors}} \# \underbrace{(S^2 \times S^1)_1 \# \cdots \# (S^2 \times S^1)_\ell}_{\ell \text{ wormholes (or handles)}}$$

$$\# \underbrace{K(\pi_1,1)_1 \# \cdots \# K(\pi_1,1)_m}_{m \text{ aspherical factors}}. \tag{9.52}$$

Here k, ℓ and m are integers ≥ 0, $k + \ell + m \geq 1$ and if any of k, ℓ or m is 0 then terms of that type do not occur. The connected sum $M \# N$ of two closed connected, oriented n-manifolds is constructed by removing the interiors of an embedded closed n-ball in each of M and N and then identifying the resulting S^{n-1} boundary components by an orientation-reversing diffeomorphism. The resulting n-manifold will be smooth, connected, closed and consistently oriented with the original orientations of M and N. The above decomposition of M is only uniquely defined only provided we set aside S^3 since $M' \# S^3 \approx M'$ for any 3-manifold M'.

In the above formula if $k \geq 1$, then each Γ_i, $1 \leq i \leq k$ is a finite, nontrivial ($\Gamma_i \neq [I]$) subgroup of $SO(4)$ acting freely and orthogonally on $\mathbb{S}^3$. The individual summands S^3/Γ_i are the *spherical space forms* alluded to previously and, by construction, each is compatible with an FLRW metric of constant positive spatial curvature (i.e., k $= +1$ models in the usual notation). The individual 'handle' summands $S^2 \times S^1$ admit metrics of the Kantowski–Sachs type that are homogeneous but not isotropic and so not even locally of FLRW type.

The remaining summands in the above 'prime decomposition' theorem [5–7] are the $K(\pi,1)$ manifolds of Eilenberg–MacLane type wherein, by definition $\pi = \pi_1(M)$, the fundamental group of M and all of the higher homotopy groups are trivial; that is, $\pi_i(M) = 0$ for $i > 1$. Equivalently, the universal covering space of M is contractible and, in this case, known to be diffeomorphic to $\mathbb{R}^3$ [8]. Since the higher homotopy groups, $\pi_i(M)$ for $i > 1$, can be interpreted as the homotopy classes of continuous maps $S^i \rightarrow M$, each such map must be homotopic to a constant map. For this reason $K(\pi, 1)$ manifolds are said to be aspherical.

This general class of $K(\pi, 1)$ manifolds includes, as special cases, the 3-torus and five additional manifolds, finitely covered by the torus, that are said to be of 'flat type' since they are the only compact, connected, orientable 3-manifolds that each, individually, admits a flat metric and thus supports spatially compactified versions of the FLRW spaces of flat type (i.e., k = 0 models).

Other $K(\pi, 1)$ spaces include the vast set of compact hyperbolic manifolds $\mathbb{H}^3/\Gamma$ where here Γ is a discrete torsion-free (i.e., no nontrivial element has finite order) co-compact subgroup of the Lie group $\text{Isom}^+(\mathbb{H}^3)$ of orientation-preserving isometries of $\mathbb{H}^3$ that, in fact, is Lie-group isomorphic to the proper orthochronous Lorentz group $SO^\dagger(3, 1)$. Each of these, individually, supports spatially compactified versions of the FLRW spacetimes of constant negative (spatial) curvature (i.e., k = −1 models).

Additional $K(\pi, 1)$ manifolds include the trivial circle bundles over higher-genus surfaces Σ_p for $p \geq 2$ (where Σ_p designates a compact, connected, orientable surface of genus p) and nontrivial circle bundles over Σ_p for $p \geq 1$. Note that the trivial circle bundles $S^2 \times S^1$ and $T^2 \times S^1 \approx T^3$ are already included among the previous prime factors discussed and that nontrivial circle bundles over S^2 are included among the spherical space forms S^3/Γ for suitable choices of Γ. The circle bundles over higher-genus surfaces will reappear later as the basic spatial manifolds occurring in the so-called $U(1)$ problem. Still further examples of $K(\pi, 1)$ manifolds are compact 3-manifolds that fiber nontrivially over the circle with fiber Σ_p for $p \geq 1$. Any such manifold is obtained by identifying the boundary components of $[0, 1] \times \Sigma_p$ with a (nontrivial) orientation-reversing diffeomorphism of Σ_p.

It is known, however, that every prime $K(\pi, 1)$ manifold is decomposable into a (possibly trivial but always finite) collection of (complete, finite-volume) hyperbolic and *graph manifold* components. The possibility of such a (nontrivial) decomposition arises whenever the $K(\pi, 1)$ manifold under study admits a nonempty family $\{T_i\}$ of disjoint embedded incompressible two-tori. An embedded two-torus T^2 is said to be incompressible if every incontractible loop in the torus remains incontractible when viewed as a loop in the ambient manifold. A closed oriented 3-manifold G (possibly with boundary) is a *graph manifold* if there exists a finite collection $\{T_i'\}$ of disjoint embedded incompressible tori $T_i' \subset G$ such that each component G_j of $G \setminus \cup T_i'$ is a Seifert-fibered space.[19] Thus a graph manifold is a union of Seifert-fibered spaces glued together by toral automorphisms along toral boundary components. The collection of tori is allowed to be empty so that, in particular, a Seifert-fibered manifold itself is a graph manifold. Decomposing a 3-manifold by cutting along essential two-spheres (to yield its prime factors) and then along incompressible tori, when present, are the basic operations that reduce a manifold to its 'geometric' constituents [7].

---

[19] A Seifert-fibered space is a 3-manifold foliated by circular fibers in such a way that each fiber has a tubular neighborhood (characterized by a pair of co-prime integers) of the special type known as a standard fibered torus.

The Thurston conjecture that every such 3-manifold can be reduced in this way has now been established via arguments employing Ricci flow [4].

For comparison's sake we recall that the two-dimensional analogue of the foregoing (prime) decomposition theorem is the classical result that any compact, connected, orientable surface is S^2, T^2 or a higher-genus surface Σ_p diffeomorphic to the connected sum of p 2-tori for $p \geq 2$. These surfaces provide the spatial topologies for 'cosmological' $(2 + 1)$-dimensional Einstein gravity and, as we have mentioned, circle bundles over these provide the arenas for the $U(1)$ problem in full $(3 + 1)$-dimensional gravity.

It may seem entirely academic to consider such general, 'exotic' 3-manifolds as the composite (i.e., nontrivial connected sum) ones described above as arenas for general relativity when essentially all of the explicitly known solutions of Einstein's equations (in this spatially compact setting) involve only individual 'prime factors'. As we shall see, however, some rather general conclusions are derivable concerning the behaviors of solutions to the field equations on such exotic manifolds and astronomical observations do not logically exclude the possibility that the actual universe could have such a global topological structure. It is furthermore conceivable that the validity of central open issues in general relativity like the *cosmic censorship conjecture* could depend crucially upon the spatial topology of the spacetime under study.

9.3.2 Yamabe classification

In the following we shall focus attention on the subset of these 3-manifolds of so-called *negative Yamabe type*. By definition these admit no Riemannian metric γ having scalar curvature $R(\gamma) \geq 0$. Within the above setting a closed 3-manifold M is of negative Yamabe type if and only if it lies in one of the following three mutually exclusive subsets [9]:

1. M is hyperbolizable (that is, admits a hyperbolic metric);
2. M is a non-hyperbolizable $K(\pi, 1)$ manifold of non-flat type (the six flat $K(\pi, 1)$ manifolds are of zero Yamabe type);
3. M has a nontrivial connected sum decomposition (i.e., M is composite) in which at least one factor is a $K(\pi, 1)$ manifold. In this case the $K(\pi, 1)$ factor may be either of flat type or hyperbolizable.

The six flat manifolds comprise by themselves the subset of zero Yamabe type. These admit metrics having vanishing scalar curvature (the flat ones) but no metrics having strictly positive scalar curvature. Finally manifolds of positive Yamabe type provide the complement to the above two sets and include the stand-alone S^3, the spherical space forms S^3/Γ_i, $S^2 \times S^1$ and connected sums of the latter two types (recalling that $M' \# S^3 \approx M'$ for any 3-manifold M').

It follows immediately from the form of the Hamiltonian constraint that any solution of the Einstein field equations with Cauchy surfaces of negative Yamabe type (i.e., diffeomorphic to a manifold in one of the three subsets listed above) and strictly non-negative energy density and non-negative cosmological constant (with either or both allowed to vanish) cannot admit a maximal hypersurface. Thus such a universe model, if initially expanding, can only continue to do so (until perhaps developing a singularity) and cannot cease its expansion and 'recollapse'.

For such manifolds Yamabe's theorem [10] guarantees that each smooth Riemannian metric on M is uniquely, globally conformal to a metric γ having scalar curvature $R(\gamma) = -1$. Thus, in a suitable function-space setting [11], one can represent the conformal classes of Riemannian metrics on M by the infinite-dimensional submanifold

$$\mathcal{M}_{-1}(M) = \{\gamma \in \mathcal{M}(M) | R(\gamma) = -1\}, \tag{9.53}$$

where $\mathcal{M}(M)$ designates the corresponding space of arbitrary Riemannian metrics on M.

The quotient of $\mathcal{M}_{-1}(M)$ by the natural action of $\mathcal{D}_0(M) = \mathrm{Diff}_0(M)$, the connected component of the identity of the group $\mathcal{D}^+(M) = \mathrm{Diff}^+(M)$ of smooth, orientation-preserving diffeomorphisms of M, defines an orbit space (not necessarily a manifold) given by $\mathfrak{J}(M) = \mathcal{M}_{-1}(M)/\mathcal{D}_0(M)$. Because of its resemblance to the corresponding Riemannian construction of the actual Teichmüller space $\mathfrak{J}(\Sigma_p)$ for a higher-genus surface Σ_p [12] we refer to $\mathfrak{J}(M)$ (informally) as the 'Teichmüller space of conformal structures' of M. The actual Teichmüller space $\mathfrak{J}(\Sigma_p)$ of the higher-genus surface Σ_p is diffeomorphic to $\mathbb{R}^{6p-6}$, and hence always a smooth manifold. By contrast $\mathfrak{J}(M)$ may either be a manifold or have orbifold singularities or consist of a stratified union of manifolds representing the different isometry classes of conformal Riemannian metrics admitted by M (i.e., metrics γ with $R(\gamma) = -1$).

In certain cases, however, $\mathfrak{J}(M)$ proves to be a global, smooth and even contractible (infinite-dimensional) manifold [13] and thus to have all of the essential features (except the finite dimensionality) of an actual Teichmüller space. The infinite dimensionality of this Teichmüller-like space, $\mathfrak{J}(M)$, which will play the role of the reduced configuration space for Einstein's equations (in the vacuum case for simplicity), is of course needed to accommodate the gravitational wave degrees of freedom that are absent in $(2 + 1)$-dimensional Einstein gravity [14, 15]. One could perhaps argue that a still more natural choice for the reduced configuration space would be the analogue of Riemann moduli space, wherein one would quotient $\mathcal{M}_{-1}(M)$ by the full group, $\mathcal{D}^+(M) = \mathrm{Diff}^+(M)$ of orientation-preserving diffeomorphisms of M, instead of just its identity component. But since this construction invariably introduces orbifold singularities even in the $(2 + 1)$-dimensional problem it would also disturb the smooth character of even the favorable cases mentioned above in $3 + 1$ dimensions. For this reason we shall retain $\mathfrak{J}(M)$ as our preferred definition for the reduced configuration space, keeping in mind that the different conformal classes of M may thus not be uniquely represented.

9.3.3 Hamiltonian reduction

Hamiltonian reduction of Einstein's equations can be globally formulated most readily for manifolds of negative Yamabe type and for spacetimes that admit foliations by Cauchy hypersurfaces of constant mean curvature. A constant mean curvature (or CMC for brevity) hypersurface might be described intuitively as one with a (spatially) uniform 'Hubble constant'. This assumption allows (not only in vacuum but also with suitable matter sources) a very convenient decoupling and resolution of the Einstein constraint equations through an application of the well-known conformal method [16–18]. The inclusion of specific sources for the construction of more realistic cosmological models is hampered by the difficulty in constructing adequate models for fluid matter that evolve without singularity in the inhomogeneous situations of interest here. For simplicity of presentation we shall therefore sidestep those issues and sketch only the reduction procedure restricted to the case of vacuum spacetimes.

Any element of $\mathfrak{J}(M)$ is a conformal class that can be concretely represented by a specific metric γ having scalar curvature $R(\gamma) = -1$. To avoid over-counting, though, one should restrict the choice of such metrics to a slice for the action of $\mathcal{D}_0(M)$ on $\mathcal{M}_{-1}(M)$ or, in physicists' language, one should fix the spatial gauge. The choice of such a slice can be thought of geometrically as the choice of a (possibly only local) 'coordinate chart' for the reduced configuration space $\mathfrak{J}(M)$. If the mean curvature is designated by a (spatial) constant τ then, as is well-known, the momentum constraint equation has a general solution, for a fixed value of τ, that is parametrized by the choice of γ and by a tensor density p^{TT} that is symmetric, traceless and divergence-free (or 'transverse-traceless' in the usual terminology) with respect to γ.

A solution of the remaining, Hamiltonian constraint is then determined (for any nonzero value of τ) by solving the corresponding Lichnerowicz equation for the appropriate conformal factor $\varphi = \varphi(\tau, \gamma, p^{TT})$ that is uniquely and smoothly determined thereby as a functional of its indicated arguments [19]. The variables $\{\gamma, p^{TT}\}$, with γ restricted to a slice and p^{TT} as above, can be viewed as providing a canonical chart for the cotangent bundle, $T^*\mathfrak{J}(M)$, of the configuration space $\mathfrak{J}(M)$, which is the natural reduced phase space for the associated (reduced-) Hamiltonian system defined by Einstein's equations in this gauge.

It is not difficult to show that, relative to any time function t adapted to the CMC slicing (i.e., having CMC level surfaces), the mean curvature τ must vary monotonically. We can therefore specify a (CMC) time gauge completely by choosing t to be a particular monotonic function of τ. A natural choice to make from the present point of view is given by $t = t(\tau) = 2/(3(-\tau)^2)$ since this leads directly to the specific reduced Hamiltonian we wish to discuss. This latter functional is now given by the (dimensionless) expression

$$H_{\text{reduced}}(\tau, \gamma, p^{TT}) = (-\tau)^3 \int_M \varphi^6(\tau, \gamma, p^{TT}) d\mu_\gamma \qquad (9.54)$$

where $d\mu_\gamma$ is the Riemannian measure on M determined by the conformal metric γ and, in our sign conventions, $-\tau > 0$ for an expanding universe [20, 21]. Reexpressed in terms of the 'physical' spatial metric $g = \varphi^4 \gamma$, H_{reduced} is simply the actual spatial volume rescaled (to make it dimensionless) by the cube of $-\tau$:

$$H_{\text{reduced}}(\tau, \gamma, p^{TT}) = (-\tau)^3 \int_M d\mu_g. \qquad (9.55)$$

Through its explicit (especially via $\varphi(\tau, \gamma, p^{TT})$) dependence on the 'time' coordinate function $t = 2/(3(-\tau)^2)$, which has the natural maximal range $(0, \infty)$ – vanishing at the big bang and blowing up in the limit of infinite expansion – H_{reduced} cannot be a conserved quantity in general.[20] A straightforward calculation, which generalizes as well to other dimensions and even has a suitably defined, quasi-local analogue [20], shows however that H_{reduced} is strictly *monotonically decreasing* in the direction of cosmological expansion except for a family of continuously self-similar spacetimes [20, 21], which correspond to its critical points, for which this Hamiltonian is actually constant.

These self-similar solutions exist if and only if M admits an Einstein metric, that is a Riemannian metric $\gamma \in \mathcal{M}_{-1}(M)$ for which the Ricci tensor $\text{Ric}(\gamma)$ is a (negative) constant multiple of the metric. In three spatial dimensions the Mostow rigidity theorem excludes all but stand-alone hyperbolizable manifolds (for which the corresponding Einstein metrics are in fact hyperbolic and unique modulo isometries) while in higher dimensions (as well as in $2 + 1$ dimensions) nontrivial families of non-isometric Einstein metrics can arise to determine (finite-dimensional) submanifolds of the reduced configuration spaces $\mathfrak{J}(M) = \mathcal{M}_{-1}(M)/\mathcal{D}_0(M)$. When these self-similar solutions do arise (in $n + 1$ dimensions, $n \geq 2$) their corresponding line elements take the form

$$ds^2 = -(n/\tau^2)^2 \, d\tau^2 + [(n/(n-1))/\tau^2]\gamma_{ij} \, dx^i \, dx^j \tag{9.56}$$

where γ is a fixed (time-independent) Einstein metric with Ricci tensor $\text{Ric}(\gamma) = -\gamma/n$. We shall refer to such spacetimes as 'Lorentz cones' over their corresponding Riemannian Einstein spaces (M, γ). In a later section we shall discuss the dynamical stability of such Lorentz cones in the direction of cosmological expansion. In $2 + 1$ and $3 + 1$ dimensions, such vacuum self-similar solutions have vanishing spacetime curvature, but in higher dimensions the (moduli) spaces of such solutions can be vastly richer and the corresponding stability problems accordingly more intricate.

9.3.4 The σ constant and its cosmological significance

Since the reduced Hamiltonian is bounded from below (as the rescaled volume of CMC hypersurfaces) and is universally monotonically decaying in the direction of cosmological expansion it is natural to ask what it is decaying towards. The answer is provided by a theorem that characterizes the infimum of $H_{\text{reduced}}(\tau, \gamma, p^{TT})$ (taken, at fixed τ, over all $T^*\mathfrak{J}(M)$) in terms of a topological invariant known as the σ constant (or Yamabe invariant) of M [22, 23]:

$$\inf_{(\gamma, p^{TT}) \in T^*\mathfrak{J}(M)} H_{\text{reduced}}(\tau, \gamma, p^{TT}) = ((n/(n-1))(-\sigma(M)))^{n/2}. \tag{9.57}$$

The σ constant is, in a sense, a natural generalization of the Euler characteristic $\chi(\Sigma_p)$ of a compact surface since, when restricted to two dimensions, its definition leads to

---

[20] Mathematically one would normally formulate the reduced dynamics on a so-called contact manifold diffeomorphic to $T^*\mathfrak{J}(M) \times R^+$ but we shall instead adopt the physicists' convention of working with a time-dependent Hamiltonian on the symplectic manifold $T^*\mathfrak{J}(M)$.

$$\sigma(\Sigma_p) = 4\pi\chi(\Sigma_p) = 8\pi(1 - p). \tag{9.58}$$

More generally, for manifolds of negative Yamabe type in higher dimensions, the precise definition leads to the formula

$$\sigma(M) = -\left(\inf_{\gamma\in\mathcal{M}_{-1}(M)} \text{vol}(M,\gamma)\right)^{2/n}. \tag{9.59}$$

It has long been realized that a *graph* 3-manifold G has $\sigma(G) = 0$ since, roughly speaking, a sequence of conformal metrics seeking to achieve the indicated infimum tends to collapse its circular or Σ_p fibers. Thus no actual metric on G has a volume that realizes the σ constant; the latter can only be approached in a degenerating limit. Thanks to the recent progress in Ricci flow, however, it is now known that the σ constant of a *hyperbolizable* manifold is actually achieved by its hyperbolic metric. Using different methods some of the σ constants of positive Yamabe-type manifolds have also been computed [24].

Of most interest to us, however, is the fact that Ricci flow techniques have been used to determine the σ constant (and therefore the infimum of the reduced Hamiltonian) of the most general compact 3-manifold of *negative Yamabe type*. The result is given simply by

$$|\sigma(M)| = (\text{vol}_{-1}H)^{2/3} \tag{9.60}$$

where $\text{vol}_{-1} H$ is the volume of the hyperbolic part of M computed with respect to the hyperbolic metric normalized to have scalar curvature $= -1$ [25, 26]. In particular, any graph manifolds G, spherical space forms S^3/Γ_i or handles $S^2 \times S^1$, even if present in M, make no contribution to the sigma constant of M and hence none as well to the infimum of the reduced Hamiltonian.

Since the reduced Hamiltonian, which geometrically is nothing but the rescaled spatial volume of the expanding-universe model, is universally monotonically decaying towards its infimum and since that infimum is determined entirely by the hyperbolic component or components of M, we are naturally led to the conclusion sketched in the introduction that Einstein's equations potentially incorporate a dynamical mechanism for driving the universe model to an asymptotic state that is volume-dominated by hyperbolic components equipped with their canonical, locally homogeneous and isotropic metrics.

9.3.5 Stability results

To decide the extent to which the reduced Hamiltonian actually does decay to its infimum (or is instead perhaps obstructed from doing so) is a very demanding open problem on the global properties of solutions to the field equations. Aside from some highly symmetric examples (e.g., Bianchi models) for which one can do explicit calculations [9, 27], or in 2+1 dimensions, wherein one can verify the expected, decay-to-infimum behavior through the use of special techniques [15], available results are currently limited to stability theorems for some rather special families of 'background' solutions and to theorems which assume a priori bounds upon spacetime curvature [28–31]. An important class of solutions for which dynamical stability results can be proven directly is provided by the vacuum, self-similar 'Lorentz cone' spacetimes discussed above.

Einstein's equations, written in their conventional form, are an autonomous system of partial differential equations for the spacetime metric. When the gauge is fixed by the CMC slicing condition, however, this autonomous character is apparently broken since both the constraint and evolution equations, as well as the associated elliptic equation for the lapse function, all depend explicitly upon the mean curvature, which is now playing the role of 'time'. For the vacuum equations (or in the presence of scale-invariant matter sources), however, one can restore the autonomous character of the gauge-fixed field equations by rewriting them in terms of suitable rescaled, dimensionless variables, using appropriate powers of the mean curvature as scale factors [19–21]. The natural, dimensionless time coordinate for the rescaled equations is now given by $T = -\ln(\tau/\tau_0)$ and has maximal range $(-\infty, \infty)$ and thus serves as an effective 'Newtonian time' for this reduced, newly autonomous system.[21]

When this reformulation is carried out on a *hyperbolic* 3-manifold M (or one admitting a negative Einstein metric in higher dimensions) the resulting dynamical system has the Lorentz-cone solutions described previously as its unique fixed points [20, 21]. Since, moreover, in 3 + 1 dimensions these solutions are known to realize the infimum of the reduced Hamiltonian it is natural to ask whether these isolated fixed points (in the reduced phase space) are in fact actual attractors for the associated, reduced Einstein flow. If so then at least sufficiently nearby solutions (in a suitable function-space setting) will indeed tend asymptotically to approach the same infimum for H_{reduced} and, more significantly, the rescaled spatial metric will tend to approach the (locally homogeneous and isotropic) hyperbolic one in the limit of infinite cosmological expansion.

As a first step towards establishing this conclusion one can analyze the linearized field equations, taking an arbitrary Lorentz-cone solution as the background to perturb. While the results of such analyses confirm one's expectations [20, 21], they fall mathematically short of proving the conjectured property for the full nonlinear Einstein flow. For that purpose one needs to develop more sophisticated techniques. The vacuum field equations in 2 + 1 dimensions are so special (primarily in excluding the possibility of gravitational waves) that one can actually resolve this conjecture (affirmatively) for arbitrarily large perturbations away from the self-similar, Lorentz-cone 'backgrounds' [15, 32–34]. In 3 + 1 and higher dimensions, however, the currently available methods of stability analysis require a certain smallness condition on the nonlinear perturbations for their successful implementation. These methods proceed by defining suitable 'energy' functionals that, while positive for nontrivial perturbations, actually vanish on the backgrounds and bound the norms needed for control of the existence times of 'nearby', perturbed solutions. One aims to show that the appropriate energy functional decays asymptotically to zero, in the direction of cosmological expansion, for any solution whose 'initial values' (at some nonzero value

---

[21] Though the reduced system is autonomous, the rescaled variables are not strictly canonical, so there is no reason to expect the corresponding Hamiltonian to be conserved.

$\tau_0 < 0$ of the mean curvature) are sufficiently close to those of the background and to deduce therefrom the desired stability result.

However, even the local (i.e., short-time) existence of solutions in CMC slicing is not covered by the classical existence and uniqueness theorem for Einstein's equations [35] since this theorem assumes a *spacetime-harmonic* (or *Lorentz*-type) gauge condition to reduce the field equations to hyperbolic form. By imposing instead only a *spatial-harmonic* (or *Coulomb*-type) gauge condition to supplement the CMC time-slicing condition one arrives at an elliptic–hyperbolic system of field equations for which, however, a well-posedness theorem can also be established [36]. In $n + 1$ dimensions, for $n > 2$, this theorem requires (as does the traditional one) the metric to lie in a Sobolev space for which $s > n/2 + 1$ of its (spatial) derivatives are square integrable over M. To extend this local existence result to a global one, one needs to prove that the corresponding Sobolev norm of a solution cannot blow up in a finite time. This will be possible whenever one can make the energy arguments alluded to above work in practice.

One rather geometrically elegant implementation of this program involves defining certain, higher-order *Bel–Robinson*-type energy functionals that consist essentially of Sobolev norms of spacetime curvature. These can be employed to verify the anticipated dynamical stability for all hyperbolic 3-manifolds except the (nonempty, proper) subset admitting so-called nontrivial traceless Codazzi tensors [37]. Any member of this latter subset allows a certain finite-dimensional moduli space of nontrivial but still flat spacetime perturbations (that are not, however, of self-similar type). These are invisible to the curvature-based Bel–Robinson energies and so cannot be controlled by them. One can either fill this gap by a separate independent argument or instead develop non-curvature-based energies to handle the full range of possibilities more uniformly [21].

This latter approach can be made to work as well in higher dimensions when the background, self-similar solution is a Lorentz cone over an arbitrary (negative) Einstein metric (that need no longer be hyperbolic) provided that the spectrum of its associated elliptic, Lichnerowicz Laplacian satisfies a suitable condition [21]. In this more general setting a finite-dimensional space of Einstein metrics provides the 'center manifold' towards which the rescaled spatial metric is flowing in the limit of infinite cosmological expansion. All of the spacetimes that can be handled in this way (as sufficiently small perturbations of self-similar backgrounds that satisfy the needed spectral condition) can be shown to be causally geodesically complete in this same temporal direction. Large families of such backgrounds (and their perturbations) can be constructed by taking Riemannian products of negative Einstein spaces that satisfy the needed spectral condition and verifying that the spectral condition is automatically preserved in the process [21].

Energy arguments of the same general type as those described above had, even earlier, been shown to be applicable to $U(1)$-symmetric vacuum metrics defined on circle bundles over higher-genus surfaces [38–40]. Though limited at the outset to spacetimes having a spacelike Killing symmetry (generating the assumed $U(1)$ action) these results are especially intriguing in the challenge they provoke for an attack on the corresponding *large-data* stability problem. Large-data global existence results are currently available (in the vacuum, cosmological setting under discussion here) only for so-called Gowdy spacetimes which, by definition, have (spacelike) $U(1) \times U(1)$ isometry groups [41] or spacetimes (such as Bianchi models) that have even higher symmetry [42]. Genuine progress on the actual, large-data-$U(1)$ problem would represent a 'quantum leap' forward in one's understanding of such issues and therefore deserves a major effort.

In its basic form the vacuum $U(1)$ problem can be expressed (through a variant of Kaluza–Klein reduction) as the $(2 + 1)$-dimensional Einstein equations coupled to a wave map with (two-dimensional) hyperbolic target. The global existence problem for such wave maps on a fixed ($(2 + 1)$-dimensional) Minkowski background has recently been solved [43, 44]. In the simplest, so-called 'polarized' case, however, which requires that the bundle be trivial for its formulation, the wave map reduces to a wave equation. The global existence of such (linear) wave equations is of course already well-established even on curved (globally hyperbolic) backgrounds [45]. To handle the fully coupled $U(1)$-symmetric field equations, though, requires simultaneous control over the wave map (or wave equation) and the Teichmüller parameters of the $(2 + 1)$-dimensional Lorentz metric which now is no longer a given background. While it is not currently known how to do this, it seems encouraging that the formation of black holes in such spacetimes is obstructed by the imposed symmetry. It thus seems plausible to conjecture that every solution should exist for the maximum possible range of its geometrically defined (CMC) time and, in particular, will expand forever without developing singularities to the future.

It does not seem likely, however, that such large-data global existence questions can be settled (either for the $U(1)$ problem or, a fortiori, for the fully general non-symmetric one) by pure (higher-order) energy arguments. The reduced Hamiltonian is always at hand, and applicable to arbitrarily large data, but can only bound, in principle, an $H^1(M) \times L^2(M)$ Sobolev-type norm of the reduced phase space variables $\{\gamma, p^{TT}\}$. The best available local existence theorem (for the general, non-symmetric problem), on the other hand, requires these variables to lie in the higher order $H^2(M) \times H^1(M)$, or Bel–Robinson energy level, Sobolev space [46, 47]. But the Bel–Robinson energy, unlike the monotonically decaying reduced Hamiltonian, is *not*, a priori, under control.

There is however a rather ambitious program under development to control not only the Bel–Robinson energy but also the pointwise (or L^∞-norm) behavior of spacetime curvature through the use of what we shall informally refer to as *light-cone estimates*. We shall briefly outline one particular variant of this far-reaching program in the final section below.

9.3.6 An integral equation for spacetime curvature

It has long been realized that the Yang–Mills equations, especially when formulated in a curved background spacetime, have many similarities to the Einstein equations and thus, since methods are already at hand for bounding Yang–Mills curvature [48–51], similar techniques might well be applicable to the Einstein problem. These similarities are most pronounced when Einstein's theory is expressed in the Cartan, orthonormal frame formalism wherein the Riemann curvature tensor appears as a matrix of two-forms $\{R^{\hat{c}}{}_{\hat{c}\mu\nu}\, dx^{\mu} \wedge dx^{\nu}\}$ expressible in terms of the matrix of (Lorentz connection) one-forms $\{\omega^{\hat{a}}{}_{\hat{c}\mu}\, dx^{\mu}\}$ via

$$R^{\hat{c}}{}_{\hat{a}\mu\nu} = \theta^{\hat{c}}_{\gamma} h^{\lambda}_{\hat{a}} R^{\gamma}{}_{\lambda\mu\nu}$$

$$= \partial_{\mu}\omega^{\hat{c}}{}_{\hat{a}\nu} - \partial_{\nu}\omega^{\hat{c}}{}_{\hat{a}\mu} + \omega^{\hat{c}}{}_{\hat{d}\mu}\omega^{\hat{d}}{}_{\hat{a}\nu} - \omega^{\hat{c}}{}_{\hat{d}\nu}\omega^{\hat{d}}{}_{\hat{a}\mu}. \tag{9.61}$$

Here $h_{\hat{a}} = h^{\mu}_{\hat{a}}\frac{\partial}{\partial x^{\mu}}$ and $\theta^{\hat{a}} = \theta^{\hat{a}}_{\mu}\, dx^{\mu}$ are the orthonormal frame and co-frame fields which determine the Lorentz connection by means of the vanishing torsion condition

$$\partial_{\nu}\theta^{\hat{c}}_{\mu} - \partial_{\mu}\theta^{\hat{c}}_{\nu} + \omega^{\hat{c}}{}_{\hat{a}\nu}\theta^{\hat{a}}_{\mu} - \omega^{\hat{c}}{}_{\hat{a}\mu}\theta^{\hat{a}}_{\nu} = 0. \tag{9.62}$$

Equation (9.61) is formally identical to that for Yang–Mills curvature $\{F^{\hat{a}}{}_{\hat{c}\mu\nu}\, dx^{\mu} \wedge dx^{\nu}\}$ in terms of its connection $\{A^{\hat{a}}{}_{\hat{c}\mu}\, dx^{\mu}\}$, but Eq. (9.62) has no analogue in Yang–Mills theory, wherein the connection is the fundamental field.

The Ricci tensor also has no analogue in Yang–Mills theory, but when the contracted Bianchi identities are combined with the vanishing Ricci tensor (vacuum field equation) condition they imply the vanishing of the divergence of spacetime curvature, which is an equation of precisely Yang–Mills type. Furthermore, in each case one can compute the divergence of the associated Bianchi identity, commute covariant derivatives and impose the vanishing of the divergence of curvature to derive a natural hyperbolic equation satisfied by the corresponding curvature tensor. For the Einstein problem, expressed in the Cartan formalism, this wave equation for curvature takes the form

$$\nabla^{\alpha}\nabla_{\alpha}R^{\hat{a}}{}_{\hat{b}\mu\nu} + R_{\mu\nu}{}^{\rho\sigma}R^{\hat{a}}{}_{\hat{b}\rho\sigma} = 2R^{\hat{a}}{}_{\hat{c}\mu\sigma}R^{\hat{c}}{}_{\hat{b}\nu}{}^{\sigma} - 2R^{\hat{a}}{}_{\hat{c}\nu\sigma}R^{\hat{c}}{}_{\hat{b}\mu}{}^{\sigma}$$

$$- g^{\alpha\beta}\{\nabla_{\beta}[\omega^{\hat{a}}{}_{\hat{c}\alpha}R^{\hat{c}}{}_{\hat{b}\mu\nu} - R^{\hat{a}}{}_{\hat{c}\mu\nu}\omega^{\hat{c}}{}_{\hat{b}\alpha}]$$

$$+ \omega^{\hat{a}}{}_{\hat{c}\beta}[\nabla_{\alpha}R^{\hat{c}}{}_{\hat{b}\mu\nu} + \omega^{\hat{c}}{}_{\hat{d}\alpha}R^{\hat{d}}{}_{\hat{b}\mu\nu} - R^{\hat{c}}{}_{\hat{d}\mu\nu}\omega^{\hat{d}}{}_{\hat{b}\alpha}]$$

$$- [\nabla_{\alpha}R^{\hat{a}}{}_{\hat{c}\mu\nu} + \omega^{\hat{a}}{}_{\hat{d}\alpha}R^{\hat{d}}{}_{\hat{c}\mu\nu} - R^{\hat{a}}{}_{\hat{d}\mu\nu}\omega^{\hat{d}}{}_{\hat{c}\alpha}]\omega^{\hat{c}}{}_{\hat{b}\beta}\}, \tag{9.63}$$

where ∇_{α} designates the covariant derivative with respect to spacetime indices only, which ignores frame indices, and the 'correction' terms for the latter are reinstated explicitly through the terms involving $\omega^{\hat{a}}{}_{\hat{c}\mu}$ that have been moved over to the right.

The operator acting on curvature on the left-hand side of Eq. (9.63) has the same form as that acting on the Faraday tensor of a solution to Maxwell's equations on a vacuum background spacetime. If one pretends for the moment that the terms on the right side of Eq. (9.63) are a given 'source' for this Maxwell-like field then it is straightforward to apply the well-known Hadamard/Friedlander analysis of wave equations on curved spacetimes [45, 52] to write an integral expression for this tensor in terms of integrals over the past light cone from an arbitrary spacetime point p to an 'initial', Cauchy hypersurface and additional integrals over the intersection of this cone with the initial surface. Of course for the present problem these 'source' terms are not really given since they all involve the unknown, but, for nonlinear problems generally, wherein one could hardly expect to derive a true *representation* formula for the solution, this analysis will nevertheless yield an integral equation that can serve as the basis for making *light-cone estimates* of the unknown.

In a curved spacetime, however, where Huygens' principle fails to hold in general, the resulting Hadamard/Friedlander formulas are complicated by the appearance of integrals not only over the (3-dimensional) mantles of the light cones in question and their (2-dimensional) intersections with the initial, Cauchy surfaces but also by integrals over the (4-dimensional) interiors of those cones and their (3-dimensional) intersections with the initial hypersurfaces. It has recently been realized, however, that one can transform the conventional Hadamard/Friedlander formulas in such a way that only certain integrals over the 3-dimensional mantles of the cones involved and their 2-dimensional intersections with the initial, Cauchy surfaces actually occur [53, 54]. At first sight it might seem that one has thereby miraculously restored Huygens' principle even in a curved spacetime where one knows it shouldn't hold, but this is not the case. For purely linear wave equations for example (for which the meaning of Huygens' principle is transparent) this procedure invariably produces integrals over the cone mantles that involve the unknowns themselves, in contrast to the original Hadamard/Friedlander formulation which provides genuine, explicit representation formulas for the solutions of *linear* equations in terms of their Cauchy data (albeit ones with the aforementioned Huygens'-principle-violating complications).

The Hadamard/Friedlander formulas are most conveniently expressed in terms of normal coordinates $\{x^\nu\}$ based at the vertex of the light cone in question and defined throughout a normal neighborhood of this point [53–55]. When the Cartan formalism is employed one can most naturally fix the associated orthonormal frame (throughout such a normal neighborhood in terms of its arbitrarily chosen value at the vertex point) by a parallel propagation condition (analogous to the so-called Cronström condition often used with the Yang–Mills equations [48–51]) that takes the form

$$\langle \omega^{\hat{c}}{}_{\hat{a}}, \tilde{v} \rangle = \omega^{\hat{c}}{}_{\hat{a}\nu} x^\nu = 0. \tag{9.64}$$

Remarkably, in this gauge one can not only compute the connection explicitly in terms of curvature (as Cronström showed for the Yang–Mills problem) via

$$\omega^{\hat{c}}{}_{\hat{a}\mu}(x) = -\int_0^1 d\lambda\, \lambda x^\nu R^{\hat{c}}{}_{\hat{a}\mu\nu}(x \cdot \lambda), \tag{9.65}$$

but also express the orthonormal (co-)frame in terms of the connection (and hence the curvature) by

$$\theta^{\hat{c}}_\mu(x) = \theta^{\hat{c}}_\mu(0) + \int_0^1 d\lambda [\omega^{\hat{c}}{}_{\hat{a}\mu}(\lambda x)(\lambda x^\nu \theta^{\hat{a}}_\nu(0))]. \tag{9.66}$$

When the aforementioned reduction transformation is applied to the wave equation for spacetime curvature itself, the resulting integral equation may be expressed as

$$R^{\hat{a}}_{\ \hat{b}\alpha\beta}(x) = \theta^{\hat{e}}_{\alpha}(x)\theta^{\hat{f}}_{\beta}(x)\left\{\frac{1}{2\pi}\int_{C_p}\mu_{\Gamma}(x')\left\{\left[-\omega^{\hat{d}}_{\ \hat{e}\sigma'}(x')D^{\sigma'}\left(\kappa(x,x')R^{\hat{a}}_{\ \hat{b}\hat{d}\hat{f}}(x')\right)\right.\right.\right.$$

$$-\omega^{\hat{d}}_{\ \hat{f}\sigma'}(x')D^{\sigma'}\left(\kappa(x,x')R^{\hat{a}}_{\ \hat{b}\hat{e}\hat{d}}(x')\right)-\omega^{\hat{d}}_{\ \hat{b}\sigma'}(x')D^{\sigma'}\left(\kappa(x,x')R^{\hat{a}}_{\ \hat{d}\hat{e}\hat{f}}(x')\right)$$

$$+\omega^{\hat{a}}_{\ \hat{d}\sigma'}(x')D^{\sigma'}\left(\kappa(x,x')R^{\hat{d}}_{\ \hat{b}\hat{e}\hat{f}}(x')\right)\Big]$$

$$+\kappa(x,x')\left[-2R^{\hat{a}}_{\ \hat{c}\hat{e}\hat{d}}(x')R^{\hat{c}\ \ \hat{d}}_{\ \hat{b}\hat{f}}(x')+2R^{\hat{a}}_{\ \hat{c}\hat{f}\hat{d}}(x')R^{\hat{c}\ \ \hat{d}}_{\ \hat{b}\hat{e}}(x')\right.$$

$$\left.+R^{\hat{a}}_{\ \hat{b}\hat{c}\hat{d}}(x')R^{\ \ \hat{c}\hat{d}}_{\hat{e}\hat{f}}(x')\right]$$

$$+R^{\hat{a}}_{\ \hat{b}\hat{e}\hat{f}}(x')\left(\nabla^{\gamma'}\nabla_{\gamma'}\kappa(x,x')\right)$$

$$+\left(2\nabla^{\sigma'}\kappa(x,x')\right)\cdot\left[\omega^{\hat{d}}_{\ \hat{e}\sigma'}(x')R^{\hat{a}}_{\ \hat{b}\hat{d}\hat{f}}(x')+\omega^{\hat{d}}_{\ \hat{f}\sigma'}(x')R^{\hat{a}}_{\ \hat{b}\hat{e}\hat{d}}(x')\right.$$

$$\left.+\omega^{\hat{c}}_{\ \hat{b}\sigma'}(x')R^{\hat{a}}_{\ \hat{c}\hat{e}\hat{f}}(x')-\omega^{\hat{a}}_{\ \hat{c}\sigma'}(x')R^{\hat{c}}_{\ \hat{b}\hat{e}\hat{f}}(x')\right]\Big\}$$

$$+\frac{1}{2\pi}\int_{\sigma_p}d\sigma_p\left\{2\kappa(x,x')\left(\xi^{\sigma'}(x')D_{\sigma'}R^{\hat{a}}_{\ \hat{b}\hat{e}\hat{f}}(x')\right)\right.$$

$$+\kappa(x,x')\Theta(x')R^{\hat{a}}_{\ \hat{b}\hat{e}\hat{f}}(x')$$

$$+\kappa(x,x')\xi^{\sigma'}(x')\left[R^{\hat{a}}_{\ \hat{b}\hat{d}\hat{f}}(x')\omega^{\hat{d}}_{\ \hat{e}\sigma'}(x')\right.$$

$$+R^{\hat{a}}_{\ \hat{b}\hat{e}\hat{d}}(x')\omega^{\hat{d}}_{\ \hat{f}\sigma'}(x')+R^{\hat{a}}_{\ \hat{d}\hat{e}\hat{f}}(x')\omega^{\hat{d}}_{\ \hat{b}\sigma'}(x')$$

$$\left.\left.\left.-R^{\hat{d}}_{\ \hat{b}\hat{e}\hat{f}}(x')\omega^{\hat{a}}_{\ \hat{d}\sigma'}(x')\right]\right\}\right\}, \tag{9.67}$$

where the notation follows that of [53, 54] which, in turn, is based on that of Friedlander [45]. As promised, only integrals over the light-cone mantle C_p and over its (two-dimensional) intersection σ_p with the initial surface now occur. One can, by a further transformation, trade the derivatives of the curvature appearing in the light-cone integrals above for terms involving the divergence of the Lorentz connection which seems, superficially at least, to be an improvement [56]. But this latter formulation has always proven more problematic to estimate (even in the corresponding Yang–Mills case) than the former, so we shall here sketch what seems to be the most promising approach.

By well-known methods, which have their origins in the original studies of the Yang–Mills problem [48–51], one can bound the integrals of those terms that are purely algebraic in curvature by expressions that involve the fluxes of the Bel–Robinson energy. The latter would be controlled by the Bel–Robinson energy but, unlike in the Yang–Mills problem, this natural energy is not itself a priori under control. Of course the Bel–Robinson energy would be strictly conserved in the presence of a (conformal) Killing field, but the existence of such a field is an absurdly strong restriction to place on spacetimes of interest.

However, when the orthonormal frame fields of the Cartan formalism are subjected (without loss of generality) to the parallel-propagation gauge-fixing condition described above one can show that these fields (when parallel propagated from the vertex of a particular light-cone) satisfy the Killing equations approximately, with an error term that is explicitly expressible in terms of curvature and that tends to zero at a well-defined rate as one approaches the vertex of the chosen cone [53, 54]:

$$\theta^{\hat{a}}_{\mu;\nu} + \theta^{\hat{a}}_{\nu;\mu} = -\omega^{\hat{a}}{}_{\hat{b}\nu}\theta^{\hat{b}}_{\mu} - \omega^{\hat{a}}{}_{\hat{b}\mu}\theta^{\hat{b}}_{\nu}. \tag{9.68}$$

To handle the terms in Eq. (9.67) involving the (covariant) gradients of curvature one needs a higher-order energy for curvature, and the expression for this that seems most natural from the point of view taken herein is provided by

$$\tilde{T}^{\text{grav}}_{\mu\nu} := D_\mu R \cdot D_\nu R - \frac{1}{2}g_{\mu\nu}D_\gamma R \cdot D^\gamma R \tag{9.69}$$

where now

$$\begin{aligned} D_\mu R^{\hat{a}}{}_{\hat{b}\hat{e}\hat{f}} =&\partial_\mu R^{\hat{a}}{}_{\hat{b}\hat{e}\hat{f}} \\ &+ R^{\hat{c}}{}_{\hat{b}\hat{e}\hat{f}}\omega^{\hat{a}}{}_{\hat{c}\mu} - R^{\hat{a}}{}_{\hat{c}\hat{e}\hat{f}}\omega^{\hat{c}}{}_{\hat{b}\mu} \\ &- R^{\hat{a}}{}_{\hat{b}\hat{c}\hat{f}}\omega^{\hat{c}}{}_{\hat{e}\mu} - R^{\hat{a}}{}_{\hat{b}\hat{e}\hat{c}}\omega^{\hat{c}}{}_{\hat{f}\mu} \end{aligned} \tag{9.70}$$

with

$$R \cdot R = \sum_{\hat{a},\hat{b},\hat{e},\hat{f}} (R^{\hat{a}}{}_{\hat{b}\hat{e}\hat{f}})^2. \tag{9.71}$$

The analogous derivations can all be applied to the Yang–Mills problem and shown to yield a dramatically simplified proof of the no-blow-up of Yang–Mills curvature on a curved (globally hyperbolic background) [54], but of course the Yang–Mills problem is significantly less challenging than the gravitational one in that, for Yang–Mills fields, the orthonormal frame field and its (spacetime) curvature are part of the given background and do not require control. How best to modify the arguments in the Einstein problem to achieve the optimal results is currently under intense investigation.

It should be especially interesting to develop these techniques further and to use them to study the vital interplay between spatial topology and global evolution. Do any *spherical space form* or *handle* summands in the prime decomposition always tend to recollapse and 'pinch off' from the $K(\pi, 1)$ summands even as the model universe as a whole continues to expand? If so, would some kind of mathematical surgery be necessary (as it is in Ricci flow) to allow the evolution to continue and, if so, what implications does this have for the existence, or perhaps non-existence, of such spherical factors in the actual universe? Do the *graph manifold* components, though continuing to expand, always play a comparatively negligible role, through collapse of their *rescaled* metrics, in the asymptotic evolutions? Are the Cauchy hypersurfaces always asymptotically volume-dominated by their *hyperbolic* components with the rescaled metrics on these components asymptotically approaching homogeneity and isotropy? How is the fundamental question of *cosmic censorship* influenced by answers to these questions?

Acknowledgments

Vincent Moncrief is grateful to the Albert Einstein Institute (AEI) in Golm, Germany, the Mathematical Sciences Research Institute (MSRI) in Berkeley, California and the Clay Mathematics Institute (CMI) in Providence, Rhode Island for their warm hospitality and generous support while this article was being written. He wishes especially to acknowledge invaluable collaborations with Lars Andersson, Yvonne Choquet-Bruhat, Douglas Eardley and Arthur Fischer on much of the research described herein and to thank Michael Anderson, Robert Beig, Beverly Berger, Piotr Chruściel, Thibault Damour, Helmut Friedrich, Gregory Galloway, James Isenberg, Sergiu Klainerman, Philippe LeFloch, Geoffrey Mess, Niall Ó Murchadha, Martin Reiris, Michael Ryan, Hans Ringström, Richard Schoen, Shing-Tung Yau and James York for innumerable fruitful discussions over the years of ideas related to this material. The author's research was supported in part by NSF grants PHY-1305766 to Yale University and 0932078-000 to MSRI.

References

[1] H. Nussbaumer and L. Bieri, *Discovering the expanding universe* (Cambridge University Press, Cambridge, 2009).

[2] G. F. R. Ellis, R. Maartens and M. A. H. MacCallum, *Relativistic cosmology* (Cambridge University Press, Cambridge, 2012).

[3] J. Wolf, *Spaces of constant curvature* (American Mathematical Society, Providence, 2010). See especially the discussion of homogeneous spaces in Chapter 8.

[4] Among numerous recent references on the Ricci flow and its implications for 3-manifold topology see, for example: J. Morgan and F. Fong, *Ricci flow and geometrization of 3-manifolds*, University Lecture Series (American Mathematical Society, Providence, 2010), H. D. Cao, B. Chow, S. C. Chu and S. T. Yau (editors), *Collected papers on Ricci flow*, Series in Geometry and Topology, vol. **37** (International Press, Boston, 2003), B. Chow and D. Knopf, *The Ricci flow: an introduction*, Mathematical Surveys and Monographs (American Mathematical Society, Providence, 2004).

[5] H. Kneser, Geschlossene Flächen in dreidimensionalen Mannigfaltigkeiten, *Jahr. Deutschen Math. Vereinigung* **38** (1929) 248–260.

[6] J. Milnor, A unique decomposition theorem for 3-manifolds, *Amer. J. Math.* **84** (1962) 1–7.

[7] P. Scott, The geometries of 3-manifolds, *Bull. Lond. Math. Soc.* **15** (1983) 401–487.

[8] From recent results in Ricci flow it is now known that *no* Whitehead manifold (i.e., open 3-manifold that is contractible but not homeomorphic to $\mathbf{R}^3$) can be the universal cover of a compact 3-manifold; cf., Theorem 10 appearing in J. Porti, Geometrization of three manifolds and Perelman's proof, *Rev. R. Acad. Cien. Serie A. Mat.* **102** (2008) 101–125.

[9] A. Fischer and V. Moncrief, Hamiltonian reduction of Einstein's equations, *Encyclopedia of mathematical physics*, Vol. **2**, 607–623, J.-P. Françoise, G. Naber, Tsou S. T. (editors), (Elsevier, Amsterdam, 2006).

[10] J. Lee and T. Parker, The Yamabe problem, *Bull. Amer. Math. Soc.* **17** (1987) 37–91. See also R. Schoen, Conformal deformation of a Riemannian metric to constant scalar curvature, *J. Diff. Geom.* **20** (1984) 479–495. For the case of metrics of negative Yamabe type of most interest herein, see the original proof by N. Trudinger, Remarks concerning the conformal deformation of Riemannian structures on compact Manifolds, *Ann. Scuola Norm. Sup. Pisa (3)* **22**: (1968) 265–274.

[11] A. Fischer and J. Marsden, Deformations of the scalar curvature, *Duke Math. J.* **42** (1975) 519–547.

[12] A. Tromba, Teichmüller theory in Riemannian geometry, Lectures in Mathematics, ETH Zürich (Birkhäuser, Zürich, 1992).

[13] A. Fischer and V. Moncrief, The structure of quantum conformal superspace, *Global structure and evolution in general relativity*, Lecture Notes in Physics vol. **460**, S. Cotsakis and G. Gibbons (editors), (Springer-Verlag, New York, 1996) 111–173.

[14] V. Moncrief, Reduction of the Einstein equations in 2 + 1 dimensions to a Hamiltonian system over Teichmüller space, *J. Math. Phys.* **30** (1989) 2907–2914.

[15] V. Moncrief, Relativistic Teichmüller theory–a Hamilton–Jacobi approach to 2 + 1 dimensional Einstein gravity, *Surveys in Differential Geometry XII* (International Press, Boston, 2008) 203–249.

[16] See especially Chapter 8 of the present volume.

[17] Y. Choquet-Bruhat, *General relativity and the Einstein equations*, Oxford Mathematical Monographs (Oxford University Press, Oxford, 2009). See especially Chapter VII.

[18] R. Bartnik and J. Isenberg, The constraint equations, in *The Einstein equations and the large scale behavior of gravitational fields*, edited by P. T. Chruściel and H. Friedrich (Birkhäuser-Verlag, Basel, 2004) 1–38.

[19] A. Fischer and V. Moncrief, Conformal volume collapse of 3-manifolds and the reduced Einstein flow, in the Springer volume *Geometry dynamics and mechanics* for the 60th birthday of Jerrold Marsden edited by D. Holmes, P. K. Newton and A. Weinstein (Springer-Verlag, New York , 2002) 463–522.

[20] A. Fischer and V. Moncrief, Hamiltonian reduction and perturbations of continuously self-similar $(n + 1)$-dimensional Einstein vacuum spacetimes, *Class. Quant. Grav.* **19** (2002) 1–33.

[21] L. Andersson and V. Moncrief, Einstein spaces as attractors for the Einstein flow, *J. Diff. Geom.* **89** (2011) 1–48.

[22] R. Schoen, Variational theory for the total scalar curvature functional for Riemannian metrics and related topics, *Topics in the calculus of variations*, Lecture Notes in Mathematics vol. **1365** (Springer-Verlag, Berlin, 1989) 120–154.

[23] M. Anderson, Scalar curvature and geometrization conjectures for 3-manifolds, in *Comparison geometry* (MSRI Publications, Berkeley, 1997) 49–82.

[24] H. Bray and A. Neves, Classification of prime 3-manifolds with Yamabe invariant greater than RP^3, *Ann. Math.* **159** (2004) 407–424.

[25] M. Anderson, Geometrization of 3-manifolds via the Ricci flow, *Notices Amer. Math. Soc.* **51** (2004) 184–193.

[26] M. Anderson, Canonical metrics on 3-manifolds and 4-manifolds, *Asian J. Math.* **10** (2006) 127–164.

[27] A. Fischer and V. Moncrief, The reduced Hamiltonian of general relativity and the σ-constant of conformal geometry, *Proceedings of the 2nd Samos Meeting on Cosmology, Geometry and Relativity, Mathematical and Quantum Aspects of Relativity and Cosmology*, Lecture Notes in Physics, vol. **537** S. Cotsakis and G. Gibbons (editors) (Springer-Verlag, Berlin, 2000) 70–101.

[28] M. Anderson, On long-time evolution in general relativity and geometrization of 3-manifolds, *Commun. Math. Phys.* **222** (2001) 533–567.

[29] M. Anderson, Asymptotic behavior of future complete cosmological spacetimes, appearing in the *Spacetime Safari* issue of *Classical and Quantum Gravity* in honor of the 60th birthday of Vincent Moncrief, *Class. Quant. Grav.* **21** (2004) S11–S27.

[30] M. Reiris, The ground state and the long-time evolution in the CMC Einstein flow, *Ann. Inst. Henri Poincaré* **10** (2010) 1559–1604.

[31] M. Reiris, On the asymptotic spectrum of the reduced volume in cosmological solutions of the Einstein equations, *Gen. Rel. Grav.* **41** (2009) 1083–1106.

[32] L. Andersson, V. Moncrief and A. Tromba, On the global evolution problem in 2 + 1 gravity, *J. Geom. Phys.* **23** (1997) 191–205.

[33] R. Benedetti and E. Guadagnini, Cosmological time in (2 + 1)-gravity, *Nucl. Phys.* **B613** (2001) 330–352.

[34] L. Andersson, Constant mean curvature foliations of simplicial flat spacetimes, *Commun. Anal. Geom.* **13** (2005) 963–979. See also: T. Barbot, F. Béguin, A. Zeghib, Feuilletages des espaces temps globalement hyperboliques par des hypersurfaces à courbure moyenne constante, *C. R. Math. Acad. Sci. Paris* **336** (2003) 245–250.

[35] Y. Fourès-Bruhat, Théorème d'existence pour certains systèmes d'équations aux derivées partielles non linéaires, *Acta Math.* **88** (1) (1952) 141–225. See also: Y. Bruhat, The Cauchy problem, in *Gravitation: an introduction to current research*, L. Witten (editor) (John Wiley & Sons, San Francisco, 1962) 130–168.

[36] L. Andersson and V. Moncrief, Elliptic–hyperbolic systems and the Einstein equations, *Ann. Inst. Henri Poincaré* **4** (2003) 1–34.

[37] L. Andersson and V. Moncrief, Future complete vacuum spacetimes, appearing in the Proceedings of the Cargèse Summer School *Fifty years of the Cauchy problem in general relativity*, edited by P. Chruściel and H. Friedrich (Birkhaüser, Basel, 2004) 299–330.

[38] Y. Choquet-Bruhat and V. Moncrief, Future global-in-time Einsteinian spacetimes with U(1) isometry group, *Ann. Inst. Henri Poincaré* **2** (2001) 1007–1064.

[39] Y. Choquet-Bruhat and V. Moncrief, Future complete Einsteinian spacetimes with U(1) isometry group, *C. R. Acad. Sci. Paris*, **332**, Série I, (2001) 137–144.

[40] Y. Choquet-Bruhat, Future complete U(1) symmetric Einsteinian spacetimes, the unpolarized case, in *The Einstein equations and the large scale behavior of gravitational fields*, edited by P. T. Chruściel and H. Friedrich (Birkhäuser-Verlag, Basel, 2004). See also Chapter XVI of Ref. [17].

[41] H. Ringström, Cosmic censorship in Gowdy spacetimes, *Living Rev. Rel.* **13** (2010).

[42] For a discussion of the asymptotic behavior of the reduced Hamiltonian for spatially compactifiable, vacuum Bianchi models of negative Yamabe type see Refs. [9] and [19].

[43] W. Schlag and J. Krieger, *Concentration compactness for critical wave maps*, EMS Monographs in Mathematics (European Mathematical Society, Zürich, 2012).

[44] D. Tataru and J. Sterbenz, Regularity of wave-maps in dimension 2 + 1, *Commun. Math. Phys.* **298** (2010) 231–264.

[45] F. G. Friedlander, *The wave equation on a curved space-time* (Cambridge University Press, Cambridge, 1975). See also C. Bär, N. Ginoux, F. Pfäffle, *Wave equations on Lorentzian manifolds and quantization* (European Mathematical Society, Zürich, 2007).

[46] S. Klainerman, I. Rodnianski and J. Szeftel, The bounded L2 curvature conjecture, arXiv:1204.1767v2 [math.AP]. See also by S. Klainerman and I. Rodnianski, Rough solutions of the Einstein-vacuum equations, *Ann. Math.* **161** (2005) 1143–1193.

[47] Q. Wang, Rough solutions of Einstein vacuum equations in CMCSH gauge, arXiv:1201.0049v1 [math.AP]. This article improves the result of Ref. [36] by proving local existence for the Einstein equations in CMCSH (constant-mean-curvature-spatial-harmonic) gauge for Cauchy data g, k in the space $H^s \times H^{s-1}(M)$ for $s > 2$.

[48] D. Eardley and V. Moncrief, The global existence of Yang–Mills–Higgs fields in 4-dimensional Minkowski space I. Local existence and smoothness properties, *Commun. Math. Phys.* **83** (1982) 171–191.

[49] D. Eardley and V. Moncrief, The global existence of Yang–Mills–Higgs fields in 4-dimensional Minkowski space II. Completion of proof, *Commun. Math. Phys.* **83** (1982) 193–212.

[50] P. Chruściel and J. Shatah, Solutions of the Yang–Mills equations on globally hyperbolic four dimensional Lorentzian manifolds, *Asian J. Math.* **1** (1997) 530–548.

[51] S. Klainerman and I. Rodnianski, A Kirchhoff–Sobolev parametrix for wave equations in a Curved Spacetime, *J. Hyperbolic Diff. Eqs.* **4** (2007) 401–433.

[52] J. Hadamard, *Lectures on Cauchy's problem in linear partial differential equations* (Yale University Press, New Haven, 1923).

[53] V. Moncrief, An integral equation for spacetime curvature in general relativity, Isaac Newton Institute preprint NI05086 (2005). Published in the *Journal of Differential Geometry* volume *Surveys in differential geometry X honoring the memory of S. S. Chern* (International Press, Boston, 2006).

[54] V. Moncrief, Curvature propagation in general relativity, in preparation for *Living Reviews in Relativity* (Albert Einstein Institute, Potsdam).

[55] P. LeFloch, Injectivity radius and optimal regularity for Lorentzian manifolds with bounded curvature, *Actes Sémin. Théor. Spectr. Géom.* **26** (2007) 77–90.

[56] For the Yang–Mills equations in Minkowski spacetime the argument to bound the divergence of the connection given in [48, 49] depends upon the translational invariance of the metric and thus fails to apply in curved backgrounds.

Part Four
Beyond Einstein

Introduction

The remarkable advances summarized in the first three parts of this volume refer almost entirely to the well-established realm of classical general relativity (GR). However, Einstein [1] was quite aware of the limitations of his theory. In the context of cosmology he wrote, as early as in 1945,

"One may not assume the validity of field equations at very high density of field and matter and one may not conclude that the beginning of the expansion should be a singularity in the mathematical sense."

By now, we know that classical physics cannot always be trusted even in the astronomical world because quantum phenomena are not limited just to tiny, microscopic systems. For example, neutron stars owe their very existence to a quintessentially quantum effect: the Fermi degeneracy pressure. At the nuclear density of $\sim 10^{15}$ g/cm^3 encountered in neutron stars, this pressure becomes strong enough to counterbalance the mighty gravitational pull and halt the collapse. The Planck density is some *eighty* orders of magnitude higher! Astonishing as the reach of GR is, it cannot be stretched into the Planck regime; here one needs a grander theory that unifies the principles underlying both general relativity and quantum physics.

Early developments

Serious attempts at constructing such a theory date back to the 1930s with papers on the quantization of the linearized gravitational field by Rosenfeld [2] and Bronstein [3]. Bronstein's papers are particularly prescient in that he gave a formulation in terms of the electric and magnetic parts of the Weyl tensor and his equations have been periodically rediscovered all the way to 2002 [4]! Analysis of interactions between gravitons began only in the 1960s when Feynman extended his calculational tools from QED to general relativity [5]. Soon after, DeWitt completed this analysis by systematically formulating the Feynman rules for calculating the scattering amplitudes among gravitons and between gravitons and matter quanta. He showed that the theory is unitary order by order in perturbation theory (for a summary, see, e.g., [6]). In 1974, 't Hooft and Veltman [7] used elegant symmetry arguments to show that pure general relativity is renormalizable to 1 loop but they also found that this feature is destroyed when gravity is coupled to even a single scalar field. For pure gravity, there was a potential divergence at 2 loops because of a counter-term that is cubic in the Riemann tensor. However, there was no general argument to say that its coefficient is necessarily non-zero. A heroic calculation by Goroff and Sagnotti [8] settled this issue by showing that the coefficient is $209/(2880(4\pi)^2)$! Thus, in perturbation theory off Minkowski space, pure gravity fails to be renormalizable at 2 loops, and, when coupled to a scalar field, already at 1 loop.

The question which then arose was whether one should modify Einstein gravity at short distances and/or add astutely chosen matter which would improve its ultraviolet behavior. The first avenue led to higher derivative theories. Stelle, Tomboulis and others showed that such a theory can be not only renormalizable but also asymptotically free [9]. But it soon turned out that the theory fails to be unitary and its Hamiltonian is unbounded below. The discovery of supersymmetry, discussed in Chapter 12, suggested another avenue: with a suitable combination of fermions and bosons, perturbative infinities in the bosonic sector could be canceled by those in the fermionic sector, improving the ultraviolet behavior. This hope was shown to be realized to 2 loops by a number of authors [10]. However, by the late 1980s a consensus had emerged that all supergravity theories would diverge by 3 loops and are therefore not viable (see, e.g., [11]).

A series of parallel developments was sparked in the canonical approach by Dirac's analysis of constrained Hamiltonian systems. In the 1960s, this framework was applied to general relativity by Dirac, Bergmann, Arnowitt, Deser, Misner and others [13–17]. The basic canonical variable was the 3-metric on a spatial slice and, as discussed in Chapters 8–10, general relativity could be interpreted as a dynamical theory of 3-geometries. Wheeler therefore baptized it *geometrodynamics* [18, 19]. Wheeler also launched an ambitious program in which the internal quantum numbers of elementary particles were to arise from non-trivial, microscopic topological configurations and particle physics was to be recast as *'chemistry of geometry'*. This led to interesting discoveries at the interface of topology and general relativity but the approach did not have notable success with the particle physics phenomenology.

A distinguishing feature of the canonical approach is that in contrast to perturbative treatments it does not split the metric into a kinematic background and a dynamical fluctuation. As a result, a number of conceptual problems were brought to the forefront which revealed the deep structural differences between general relativity and more familiar field theories in Minkowski space-time. By now there is a near-universal appreciation of the importance of background independence and of the necessity of facing the ensuing complications. However, this very feature made it difficult to use the standard techniques from QED to face the mathematical difficulties associated with the infinite number of degrees of freedom of the gravitational field. Consequently, most of the work in full quantum geometrodynamics remained rather formal. Detailed calculations could be carried out in the context of quantum cosmology where one freezes all but a finite number of degrees of freedom. Initially there was hope that quantum effects would tame the cosmological singularities of general relativity. However, this hope did not materialize; even in the simplest models the big bang could not be softened without additional 'external' inputs into the theory. The program also faced a sociological limitation in that the ideas that had been so successful in QED played no role: in a non-perturbative, background independent approach, it is hard to see gravitons, calculate scattering matrices and use virtual processes to obtain radiative corrections. To use a well-known phrase [20], the emphasis on geometry in the canonical program "drove a wedge between general relativity and the theory of elementary particles." Therefore, after an initial burst of activity, the quantum geometrodynamics program became rather stagnant.

A third avenue was opened in the mid 1950s: explorations of the effects of a *classical* gravitational field on quantum matter fields. Early work by Parker explored quantum fields in the Friedmann–Lemaître–Robertson–Walker (FLRW) space-times [12]. As recent successes of inflationary scenarios discussed in Chapters 4 and 11 show, this choice was prescient. Indeed, this is the arena where we are most likely to first see the interface of gravity and quantum physics observationally. But this general area did not draw much attention until Hawking's seminal discovery in 1974 that quantum field theory (QFT) on a black hole background predicts that black holes emit quantum radiation and resemble black bodies when seen from infinity. Not only did the entire area of QFT in curved space-time experience an explosion of activity but also this discovery has served as a focal point for a great deal of research in all areas of quantum gravity over the last four decades.

Current status

Ideas developed in QFT in curved space-times have had a number of fascinating applications, ranging from the study of diverse aspects of the Casimir effect [21] to the feasibility studies of creating time machines by exploiting the violations of local energy conditions that are allowed in QFT [22]. Advances on the more fundamental side are discussed in Chapter 10. In general curved space-times, we do not have the Poincaré group to decompose fields into positive and negative frequency parts and select a canonical vacuum. Since there is no natural choice of a (Fock) representation of the canonical commutation relations, one is led to the *more general* setting of algebraic QFT. As a consequence, recent advances in QFT in curved space-times have brought to the forefront the essential conceptual ingredients of QFT that are often masked by the extraneous structure that happens to be available in Minkowski space-time. As emphasized in Chapter 10, these developments have also shown that, beyond the special context of static space-times, techniques from Euclidean QFT cannot be carried over to calculate quantities of direct physical interest in the Lorentzian theory. This is an important message also for approaches to full quantum gravity although, unfortunately, it is often overlooked. On the physical side, quantum effects on curved space-times lead to unforeseen phenomena such as evaporation of black holes and emergence of the large scale structure of the universe from pure quantum fluctuations in the very early universe. Finally, advances in describing interactions have now elevated QFT in curved space-times to the same mathematical level as that enjoyed by rigorous QFT in Minkowski space.

In quantum gravity proper, while both the perturbative and the canonical approaches reached an impasse by the early 1980s, they provided seeds for most of the subsequent developments. Although GR is perturbatively non-renormalizable, an effective field theory was developed systematically [23] and has had remarkable successes in the low energy regime, e.g., in the treatment of dynamics of compact binaries in classical GR [24] and in the computation of the leading corrections to the Newtonian potential. Therefore the problem of finding a viable quantum gravity theory can be rephrased as that of obtaining an appropriate completion of this theory in which the outstanding conceptual issues – such as the fate of the classical singularities of GR, the statistical mechanical accounting of black hole entropy, and the final stages of the black hole evaporation – can be analyzed systematically.

This quest has been undertaken in a number of directions. Each program adopts a different point of departure, treating certain aspects of the problem as more fundamental, and hoping that the remaining aspects can be handled successfully once there is a resolution of the key difficulties. Chapter 11 focuses on approaches which emphasize the dynamical nature of space-time geometry and non-perturbative methods. The point of departure is GR. However, the fundamental degrees of freedom and the short-scale dynamics in the final quantum theory are quite different from those of GR and classical space-times and gravitons emerge only in a suitable limit.

The first of these programs is Asymptotic Safety, whose goal is to provide a specific *ultraviolet completion* of the effective field theory using Weinberg's generalized notion of renormalization. On the analytical side, non-trivial fixed points of the renormalization group flow have been obtained in 4 space-time dimensions, even after allowing for 9 different gravitational couplings in addition to the coupling of gravity to matter fields. These results strongly suggest that much of the intuition derived from Goroff and Sagnotti's early results is tied to perturbation theory around Minkowski space. Consistent ultraviolet completions of the standard effective theory may well exist *without* having to invoke supersymmetry, higher dimensions or extended objects. Similarly, on the computational side it was widely believed that the renormalization group flows would inevitably lead to a 'crumpled phase' in which macroscopic space-times such as the one around us will not emerge in the infrared. Recent progress in Causal Dynamical Triangulations has shown that the 'crumpled phases' are not inevitable; theory does allow smooth, macroscopic space-times with small quantum fluctuations.

The second program discussed in Chapter 11 is Loop Quantum Gravity (LQG) which grew out of the canonical approach. Consequently, manifest background independence is at the forefront and the theory is again non-perturbative. However, emphasis is shifted from metrics to connections; the 'wedge' between general relativity and gauge theories governing other fundamental forces is removed. Indeed, the basic notions are taken directly from the Yang–Mills theory but without reference to a background space-time metric. Rather surprisingly, this strategy leads to a unique kinematic setup in which one can overcome the mathematical obstacles faced in geometrodynamics and handle the infinite number of degrees of freedom rigorously. The Hilbert space is spanned by *spin networks*.

Consequently, the fundamental excitations of geometry/gravity are not gravitons which represent quantum fluctuations on a given background geometry. Rather, they are polymer-like threads that can be woven *to create the geometry itself.* Geometry is quantum mechanical in the most direct sense: geometric operators have discrete eigenvalues. This discreteness has a deep influence on dynamics. In covariant LQG – called the Spinfoam framework – it offers relief from ultraviolet infinities by banishing degrees of freedom at scales shorter than the Planck length. A priori, there could be infrared infinities but, surprisingly, if there is a *positive* cosmological constant, they can be removed by a natural quantum deformation of the local Lorentz group. LQG has had notable success both in the cosmological and in the black hole sectors of GR. In both cases, the underlying quantum geometry plays a key role. In particular, in the cosmological sector it naturally tames all strong curvature singularities even when matter satisfies all the energy conditions. Furthermore, the framework has extended the reach of observational cosmology all the way to the Planck regime through the development of QFT on *quantum* cosmological space-times. A similar extension is now being used to study Hawking radiation on *quantum* black hole space-times.

Chapter 12 summarizes advances based on supersymmetry, extended objects, higher dimensions and holography. Chronologically, at first the emphasis in the framework was on perturbation theory in Minkowski space-time. The viewpoint was that the ultraviolet divergences of quantum gravity were signals that point particles and local quantum fields are over-idealized notions that become untenable at very high energies. Fundamental objects are strings, and interactions between them are just the simple processes of joining and splitting which naturally avoid the ultraviolet infinities of local QFT. Particles are merely excitations of strings. This immediately leads to an infinite tower of particles and fields but most of these excitations are so massive that they become relevant only at extremely high energies. Theoretical consistency implies that space-time has to be 26-dimensional without supersymmetry and 10-dimensional with supersymmetry. At first it was believed that the extra dimensions could be compact and microscopic, and that there would be severe constraints on permissible compactifications as well as on the matter content and allowed interactions to make the theory unique. Therefore it was often heralded as the 'theory of everything'.

Further research showed that these expectations were overly optimistic, but at the same time it revealed richer structures: The theory turned out *not* to be unique but various consistent theories are related by certain dualities. It is now believed that there *is* probably a single theory and the known consistent theories represent its 'corners'. However, as of now its structure remains opaque; indeed, one does not even know what the fundamental principles behind it should be. This is reflected in its very name, *'M-theory'*, where M is often said to stand for 'Mystery'. Also, most of the work to date has been carried out in the framework of first quantization and the possible role of a second quantized string field theory remains unclear. But it *is* clear that the theory contains not only strings but also higher-dimensional, extended, non-perturbative configurations, called *p* branes. This enlargement of the theory opened a gate to incorporate extremal (and some near-extremal) black holes in the string paradigm. Because of the 'non-renormalization' theorems, it is possible to calculate the number of certain string states at low energy and then correctly deduce the number of microstates of these black holes.

These considerations in turn suggested the possibility of a holographic picture in which string theory on asymptotically anti-de Sitter backgrounds could be regarded as being dual to certain conformal field theories on the boundary of these space-times. As discussed in Chapter 12, this AdS/CFT conjecture has led to a wealth of insights on the unity of the mathematical structures that underlie completely distinct physical systems such as quark–gluon plasmas and condensed matter systems. In addition, throughout its evolution, string theory has had unforeseen applications to diverse areas of mathematics. Given that these ideas extend the reach of gravitational physics to completely new territories in ways that were not even imagined before, it is fitting that the volume concludes with this Chapter.

Finally, it is important to emphasize that the content of Part Four does not do full justice to the field because space limitation did not allow us to include several promising advances. First, even in the areas covered in Chapters 10–12, in the spirit of this volume, authors focused on a few topics on which most significant advances have occurred over the last three decades or so. The second and more important omission is that several approaches to quantum gravity had to be left out entirely. In the first three parts, which are based on well-established ideas and careful observations, the choice of what to include was easier to make. In the case of quantum gravity, the subjective element is much more pronounced simply because one is now forced to leave the safety net of ideas that are firmly grounded in GR. Promising directions that were left out include: (i) Asymptotic quantization which brought out the interplay between the Bondi–Metzner–Sachs group and infrared issues in full quantum gravity [25]; (ii) The related program for calculating scattering amplitudes from past to future null infinity, using twistor methods that is now drawing a great deal of attention [26]. Twistor theory itself has provided a powerful bridge between the theory of partial differential equations and algebraic geometry which extends to (self-dual) Einstein's equations through Penrose's non-linear graviton construction [27]; (iii) The Regge calculus approach which parallels lattice QCD, but uses dynamical simplicial decompositions rather than lattices defined in a background geometry [28]; (iv) Hořava–Lifshitz gravity which sacrifices manifest local Lorentz invariance to achieve better ultraviolet behavior, hoping to recover it in the infrared limit [29]; (v) Causal sets in which one postulates that at a fundamental level one only has a discrete set of points with causal relations between them [30]; and (vi) The Vassiliev higher spin theories in which an infinite tower of *massless* higher spin fields are incorporated in a consistent manner [31]. One or more of these ideas may well lead to an ultraviolet completion of GR with desired features.

In this regard, it is instructive to look back at the events that celebrated the centennial of Einstein's birth some 35 years ago. The Princeton conference at the Institute of Advanced Study had two talks on quantum gravity; both on supergravity [32]. The Cambridge University Press volume [33] also had two Chapters, one that introduced the asymptotic safety program and the other that put all its emphasis on Euclidean quantum gravity. A year later, in his Lucasian Chair inaugural address, Hawking [34] suggested that the end of theoretical physics was in sight because $N = 8$ supergravity was likely to be the final theory. The field has evolved rather differently! Fascinating as the advances over the last three decades have been, the reader would do well to keep this historic perspective in mind.

Elephants in the room

There is no question that today we understand the interface of gravity with quantum physics much better than we did in the mid 1980s when the approaches discussed in Chapters 11 and 12 first rose to prominence. Several unworkable ideas have been weeded out and, as our summary indicates, concrete advances have provided us with novel insights. It is therefore fitting that the review articles and status reports tend to be upbeat, exuding confidence. But it is also clear that the end of quantum gravity is not yet in sight. We, the Editors, would be remiss if we do not venture to say why.

The leading approaches use diverse points of departure and have strikingly different perspectives as to what is most fundamental and what can be revisited later. The mathematical techniques they use also vary significantly. Given the difficulty of the task, this diversity is of course both healthy and essential. As our summary of early developments shows, this diversity has been a hallmark of this field for over 50 years. Today, the practitioners are even more passionate about the choices their approach makes. However, as a result, whereas other areas of gravitational science have witnessed a convergence of ideas and coming together of previously distinct communities, in quantum gravity the communities have drifted further apart. As active communications become less frequent, slowly but steadily the tendency to ignore the elephants in one's room increases. At the same time, it seems more and more natural to think of other viewpoints as untenable. To make this point more concrete, we will provide a few illustrative examples from the main programs discussed in this volume.

In the approaches based on unification, ideas that lead to mathematically rich structures play a dominant role. Consequently, as we discussed above, these approaches have led to unforeseen insights into the mathematical unity in the description of diverse physical systems. However, this success also seems to have fueled a tendency to ignore the issue of whether the central ideas behind these approaches are realized in our physical universe. In particular, there is little hesitation in building a quantum theory of gravity by demanding that it have supersymmetry, higher dimensions, infinite towers of particles and fields and a negative cosmological constant *at its very foundation*. To researchers outside quantum gravity the strategy seems surprisingly indifferent to the observations, rather akin to searching for the key under a lamppost irrespective of where it was actually lost. It is true that some of these ideas are motivated by the Kaluza–Klein theory where extra dimensions are meant to be microscopic and curled up, remaining invisible until we reach energies near the Planck scale. But in detailed explorations this primary restriction is often set aside. For example, it is rarely imposed while studying higher-dimensional solutions in these frameworks. More importantly, in the most commonly used versions of the AdS/CFT conjecture, symmetry requirements force the extra dimensions to be very large; the radius of the compactified internal spheres is *the cosmological radius*! If we lived in such a universe, the internal dimensions should be as readily observable as the 'normal' 4 space-time dimensions. Taken together, these assumptions push one to a paradigm whose relevance to the physical issues of quantum gravity in the actual universe we inhabit becomes increasingly obscure.

Approaches developed primarily by researchers from the GR community make a serious attempt to base their foundations only on the well-established principles of GR and QFT. They tend to take the gravity/geometry duality seriously and aim at first understanding the quantum nature of geometry, postponing the issue of coupling to matter to a second stage. This strategy has had notable success in the asymptotic safety program. However, in this approach exploration of the quintessentially quantum gravity issues – such as the origin of black hole entropy and the fate of the most vexing singularities of GR – is still at a preliminary stage. Furthermore, whereas at a fundamental level the program refers to renormalization group flows in the infinite dimensional space of permissible theories, in practice it seems unlikely that one would be able to go beyond finite truncations in any foreseeable future. Therefore the question of whether the program truly provides the promised ultraviolet completion is likely to remain open for a long time.

LQG has also advanced by making truncations. However, as in the concrete calculations of QED or QCD, truncations refer to physical problems of interest, such as the Planck scale physics of the very early universe or quantum properties of black holes. In these truncated sectors, the theory has had notable success. But, as discussed in Chapter 11, the issue of dynamics in full LQG is still far from being settled. Spinfoams provide a natural framework to explore it and certain Spinfoam models have led to significant advances. But a number of fundamental issues remain: Does the natural expansion used to calculate 'transition amplitudes' converge? Is a continuum limit needed and, if so, how exactly is one to take it? Do the proposals to incorporate matter in Spinfoams work in detail? An equally important open issue is to make direct contact with low energy physics. Since one begins with quantum geometry in the Planck regime and then descends to the low energy world, this problem is highly non-trivial. There are numerous preliminary, encouraging results, such as the calculation of the graviton propagator starting from a fully non-perturbative and background independent setting. But the relation to the effective field theory beyond the leading approximation remains unclear. Until there is a solid bridge linking non-perturbative dynamics to the well-developed effective theory in detail, the physical viability of the approach will remain uncertain.

Epilogue

While writing a review article on special relativity in 1907, Einstein realized that Newtonian gravity is incompatible with special relativity and set himself the task of resolving this conflict by creating a grander synthesis from which the two theories would emerge as limiting cases. Just eight years later, he arrived at the finished solution and, for the last century, physical scientists from a broad array of disciplines have been happily engaged in investigating its content! Work on unifying GR with quantum physics, on the other hand, has seen many twists and turns; periods of euphoria followed by despair at conceptual impasses, or Nature's stubborn refusal to use structures that seem compelling to the practitioners. Einstein's spectacular success is in striking contrast with the time and effort that has been devoted to this endeavor.

But it is important to note that progress of physical theories has more often mimicked the development of quantum theory rather than general relativity. More than a century has passed since Planck's discovery that launched the quantum. Yet, the theory is incomplete. We do not have a satisfactory grasp of the foundational issues, often called the 'measurement problem'. Nor do we have a single example of a mathematically complete, interacting QFT in 4-dimensional Minkowski space-time. A far cry from general relativity that Einstein offered us in 1915! Yet, no one would deny that quantum theory has been extremely successful; indeed, its scientific reach vastly exceeds that of general relativity.

Thus, while it is tempting to wait for another masterly stroke like Einstein's to deliver us a finished quantum gravity theory, it would be more fruitful to draw lessons from quantum physics. There, progress occurred by focusing not on the 'final' theory that solves all problems in one fell swoop, but on concrete physical problems where quantum effects were important. Experience to date indicates that the same will continue to be true for quantum gravity in the foreseeable future.

What could hasten progress along this path? So far individual programs have been driven by internal criteria. More significant advances could occur by critically examining common elements they share as well as tensions between the ideas that lie at their foundations.

For example, although these programs start with very different viewpoints and assumptions, they all feature a curious dimensional reduction at the Planck length [35]. In LQG, the fundamental excitations of quantum geometry have 1 spatial dimension whence Spinfoams are 2-complexes. In Asymptotic Safety one has a running spectral (space-time) dimension which (equals 4 in the infrared but) approaches 2 in the ultraviolet. In perturbative string theory, point particles are replaced by 1 (spatial)-dimensional strings, propagating on a classical space-time, say, Minkowski space. In the linearized approximation off Minkowski space, the LQG excitations have the same massless particle content as in bosonic string theory – a dilaton, an antisymmetric tensor and a spin-2 excitation – to begin with, and the graviton is extracted by imposing linearized Einstein constraints. Is there perhaps a deep reason why qualitatively similar 2-dimensional structures arise even when the starting points are so different?

The next set of issues is related to ultraviolet finiteness. Since the late 1980s there has been a strong belief in the string theory community that local interactions between point particles à la QFT would inevitably lead to ultraviolet divergences, and the theory is rendered perturbatively finite by using strings instead. There has been considerable research on this issue, especially in recent years. However, as the number of loops grows, there is an increasing number of ambiguities (because of the super-moduli measure needed) in the calculation and therefore some experts continue to believe that the issue of order by order perturbative finiteness is still open in string theory [36]. On another front, recent work on supergravity, described in Chapter 12, has shown that the $N = 8$ supergravity in 4 space-time dimensions is finite to 4 loops, contrary to the near unanimous expectations in the 1980s [11]. Some experts in supergravity have suggested that there could well be symmetries that have remained hidden so far that could make supergravity finite to all orders. Support for the general idea of hidden symmetries comes also from Vassiliev's higher spin theories. Overall, there is a small but growing community that believes that what is fundamental for finiteness is a sufficiently large and subtle symmetry group rather than extended objects. Thus, there is a healthy tension concerning the issue of what drives finiteness even at the level of perturbation theory.

The tension continues beyond perturbations. In LQG, background independence leads to a rather sophisticated quantum geometry with the property that there are no degrees of freedom below the Planck length, ensuring ultraviolet finiteness. In string theory, the sum in the perturbation expansion diverges [37], requiring a non-perturbative treatment. As we saw, a highly successful candidate is available in the sector of the theory with a negative cosmological constant: the AdS/CFT conjecture, where string theory in the bulk is equivalent to a finite field theory on the boundary.[22] Note that this AdS/CFT correspondence also provides a background independent definition of string theory in the asymptotically AdS sector. But whereas the mechanism taming the ultraviolet regime is provided by a specific quantum Riemannian geometry in LQG, in string theory it is provided by holography. Is there nonetheless a deep connection between them? If so, it may be helpful in, e.g., the analysis of the physical cosmological singularities in string theory.

The next example pertains to the gravity/gauge theory duality. As we noted above, LQG starts by reformulating GR in terms of Yang–Mills type variables and making heavy use of gauge theory notions such as holonomies, Wilson loops and quantized fluxes. The recent finiteness results in supergravity rely heavily on the relation between structures that arise in the $N = 4$ super Yang–Mills theory and $N = 8$ supergravity in 4 dimensions. In the AdS/CFT correspondence this interplay is also at the forefront; for example, string theory in a 5-dimensional asymptotically AdS background is dual to this super Yang–Mills theory (on the boundary). Is this interplay between gravity and gauge theories a beacon guiding us to a rapprochement of various approaches? Could a deeper understanding of this duality lead to a new principle which has eluded us because we have examined the issue only piecemeal, from the perspective of only one approach at a time?

The last example is provided by the analyses of the statistical mechanical origin of black hole entropy. In string theory calculations one generally considers strings with end points of branes that carry gauge fields. In LQG, the horizon 'membrane' is pierced by the polymer excitations of the bulk quantum geometry and the intrinsic geometry of quantum horizons is described by a gauge theory. Qualitatively, the two pictures appear to be similar. Yet, the detailed analyses are very different and their strengths and limitations are

---

[22] It is interesting, indeed, that the best non-perturbative definition of string theory takes us back to a local QFT, albeit on the boundary of space-time.

complementary. The string theory calculations refer to certain extremal and near-extremal black holes, while in LQG they are based on the notion of isolated horizons and therefore include all black hole and cosmological horizons. But whereas the relation to semi-classical calculations of entropy is well understood in string theory, the issue remains open in LQG because what one counts is the number of microstates of the quantum horizon geometry which are conceptually quite distinct from the quantum corrections to the Euclidean, classical action. Given the qualitative similarity of concepts underlying these calculations, can one perhaps relate them in detail? Such a bridge would enable one to export the strengths of each of these calculations to overcome the corresponding limitation of the other. Thus, a better understanding of common elements, differences and relations between different approaches could well suggest a paradigm that combines deep ideas from various approaches.

Over the past three decades, such rapprochements were hindered by periodic bouts of irrational exuberance which led individual communities to be certain that theirs was the only viable path and, by implication, nothing else was worth paying attention to. To paraphrase the biologist François Jacob, for sustained progress in any area of science, it is important that the practitioners be aware of the limits of their science and thus their knowledge. Otherwise it is easy to mix what one believes and what one knows to create a misplaced sense of certitude. Ongoing dialogues across various approaches and careful examinations of the common elements and differences between them would go a long way to avoid this trap.

References

[1] A. Einstein, *Meaning of relativity* (Princeton University Press, Princeton, 1945).

[2] L. Rosenfeld, *Annalen der Physik* **5**, 113 (1930); *Z. Phys.* **65** 587 (1930).

[3] M. P. Bronstein, *Phys. Z. Sowjetunion* **9**, 140–157 (1936); S. Deser and A. Starobinski, *Gen. Rel. Grav.* **44** 263–265 (2012).

[4] M. P. Bronstein, Zh. *Eksp. Teor. Fiz.* **6** 195 (1936); D. J. Cirlio-Lombardo, arXiv:1405.2334.

[5] R. P. Feynman, *Acta Phys. Polon.*, **24** 697–722 (1964).

[6] B. S. DeWitt, in *Magic without magic: John Archibald Wheeler*, ed. J. R. Klauder (W. H. Freeman, San Fransisco 1972).

[7] G. 't Hooft and M. J. G. Veltman, *Ann Inst. Henri Poincaré Phys. Theor.* **A20** 69 (1974).

[8] M. H. Goroff and A. Sagnotti, *Nucl. Phys.* **B266** 709 (1986).

[9] K. S. Stelle, *Phys. Rev.* **D16** 953 (1977); T. Tomboulis, *Phys. Lett.* **B97** 77 (1980).

[10] S. Deser, J. Kay and K. S. Stelle, *Phys. Rev. Lett.* **38** 527 (1977); M. T. Grisaru, P. Van Nieuwenhuizen and J. A. M. Vermaseren, *Phys. Rev. Lett.* **37** 1662 (1976).

[11] P. S. Howe and K. S. Stelle, *Int. J. Mod. Phys.* **A4** 1871 (1989).

[12] L. Parker, Ph.D. Dissertation (Harvard University, 1966); *Phys. Rev. Lett.* **21** 562(1968); *Phys. Rev.* **183** 1057(1969); *Phys. Rev.* **D3** 346 (1971).

[13] R. Arnowitt, S. Deser and C. W. Misner, in *Gravitation: an introduction to current research*, ed. L. Witten (John Wiley, New York, 1962).

[14] A. Komar, in *Relativity*, eds Carmeli M., Fickler S. I. and Witten L. (Plenum, New York, 1980).

[15] P. G. Bergmann and A. Komar, *General relativity and gravitation vol. 1, One hundred years after the birth of Albert Einstein*, ed. A. Held (Plenum, New York, 1980).

[16] A. Ashtekar and R. Geroch, Quantum theory of gravitation, *Rep. Prog. Phys.* **37** 1211–1256 (1974).

[17] K. Kuchař, in *Quantum Gravity 2, A second Oxford symposium* eds C. J. Isham, R. Penrose and D. W. Sciama (Clarendon Press, Oxford, 1981).

[18] J. A. Wheeler, *Geometrodynamics*, (Academic Press, New York, 1964).

[19] J. A. Wheeler, in *Relativity, groups and topology*, edited by C.M. DeWitt and B. S. DeWitt (Gordon and Breach, New York, 1963).

[20] S. Weinberg *Gravitation and cosmology* (John Wiley, New York, 1972).

[21] L. H. Ford and C. H. Wu, *AIP Conf. Proc.* **977** 145–159 (2008).

[22] A. Everett and T. Roman, *Time travel and warp drives: a scientific guide to shortcuts through time and space* (University of Chicago, Chicago, 2012).

[23] C. P. Burgess, *Living Rev. Rel.* **7** 5 (2004).

[24] R. A. Porto and I. Z. Rothstein, *Phys. Rev.* D**78** 044013 (2008), D**81** 029902 (2010).

[25] A.   Ashtekar, *Phys. Rev. Lett.* **46** 573 (1981); *J. Math. Phys.* **22** 2885 (1987); *Asymptotic quantization* (Bibliopolis, Naples, 1987); A. Strominger, *JHEP* **07** 152 (2014); T. He, V. Lysov, P. Mitra and A. Strominger, arXiv:1401.7026.

[26] L. Mason and D. Skinner, *Commun. Math. Phys.* **294** 827–862 (2010); F. Cachazo and Y. Geyer, arXiv:1206.6511; F. Cachazo, S. He, and Y. E. Yuan., *Phys. Rev. Lett.* **113** 171601 (2014); N. Arkani-Hamed and J. Trnka, *JHEP* **1412** 182 (2014).

[27] R. Penrose and W. Rindler, *Spinors and space-time: volume 2, spinor and twistor methods in space-time geometry* (Cambridge University Press, Cambridge, 1988).

[28] H. W. Hamber, *Quantum gravitation: the Feynman path integral approach* (Springer, Berlin, 2009).

[29] P. Hořava, *Phys. Rev.* D**79** 084008 (2009).

[30] F. Dowker, *Gen. Rel. Grav.* **45**, 1651–1667 (2013).

[31] M. Vassiliev, *Introduction to higher spin gauge theory*, Kramer's course (University of Utretcht, Utretcht, 2014).

[32] *Some Strangeness in Proportion*, H. Wolf, ed. (Addison Wesley, Reading, 1980).

[33] *General Relativity, an Einstein centennial survey*, eds. S. W. Hawking and W. Israel (Cambridge University Press, Cambridge, 1979).

[34] S. W. Hawking, *Is the end in sight for theoretical physics?: an inaugural address* (Cambridge University Press, Cambridge, 1980).

[35] S. Carlip, arXiv:1207.4503.

[36] H. Nicolai, *Physics* **2**, 70 (2009); Personal communication (2014).

[37] D. J. Gross and V. Periwal, *Phys. Rev. Lett.* **60** 1517 (1988).

10

Quantum Fields in Curved Spacetime

STEFAN HOLLANDS AND ROBERT M. WALD

10.1 The nature of quantum field theory in curved spacetime

Quantum field theory in curved spacetime (QFTCS) is the theory of quantum fields propagating in a background, classical, curved spacetime $(\mathcal{M}, g)$. On account of its classical treatment of the metric, QFTCS cannot be a fundamental theory of nature. However, QFTCS is expected to provide an accurate description of quantum phenomena in a regime where the effects of curved spacetime may be significant, but effects of quantum gravity itself may be neglected. In particular, it is expected that QFTCS should be applicable to the description of quantum phenomena occurring in the early universe and near (and inside of) black holes – provided that one does not attempt to describe phenomena occurring so near to singularities that curvatures reach Planckian scales and the quantum nature of the spacetime metric would have to be taken into account.

It should be possible to derive QFTCS by taking a suitable limit of a more fundamental theory wherein the spacetime metric is treated in accord with the principles of quantum theory. However, this has not been done – except in formal and/or heuristic ways – simply because no present quantum theory of gravity has been developed to the point where such a well defined limit can be taken in general situations. Rather, the framework of QFTCS that we shall describe in this review has been obtained by suitably merging basic principles of classical general relativity with the basic principles of quantum field theory in Minkowski spacetime. As we shall explain further below, the basic principles of classical general relativity are relatively easy to identify and adhere to, but it is far less clear what to identify as the "basic principles" of quantum field theory in Minkowski spacetime. Indeed, many of the concepts normally viewed as fundamental to quantum field theory in Minkowski spacetime, such as Poincaré invariance, do not even make sense in the context of curved spacetime, and therefore cannot be considered as "fundamental" from the viewpoint of QFTCS. By forcing one to re-think basic concepts, such as the notions of "vacuum state" and "particles," QFTCS has led to deep insights into the nature of quantum field theory – and one may hope that it will provide significant guidance towards the development of quantum gravity itself.

The fundamental ideas upon which classical general relativity is based are that (i) all aspects of spacetime structure are described by the topological and differential

Stefan Hollands and Robert M. Wald

(i.e., manifold) properties of events together with a Lorentz signature metric g, and (ii) the metric and matter fields[1] are dynamical; furthermore their evolution is locally determined. More precisely, the metric and the tensor (and/or spinor) fields describing matter satisfy partial differential equations – namely, Einstein's equation together with the equations of motion for the matter fields – that have a well-posed initial value formulation, so that these fields are uniquely determined (up to "gauge") from their initial data within a suitable domain of dependence. In particular, in classical general relativity, there is no non-dynamical, "background structure" in the laws of physics apart from the manifold structure of events. This lack of background structure in classical general relativity is usually referred to as the "covariance" or "coordinate invariance" of the theory; a "preferred set of coordinates" defined independently of the metric would provide non-dynamical, background structure.

It is much more difficult to identify the fundamental ideas upon which quantum field theory in Minkowski spacetime is based. One can attempt to formulate the quantum theory of a field in Minkowski spacetime by decomposing the field into modes and applying the rules of quantum mechanics to each mode. For a free field, each mode is an independent harmonic oscillator and one can obtain a mathematically sensible quantum field theory in this manner, although even here one encounters infinite expressions for quantities that are nonlinear in the fields. A well-known example of this general phenomenon is that one obtains an infinite expression for the total energy (and energy density) of the field, as can be seen by adding the zero-point energies and/or energy densities of the infinite number of modes. The situation is considerably worse for interacting (i.e., nonlinear) fields, wherein one immediately encounters ill defined and/or infinite expressions in the calculation of essentially all physical quantities, arising from the fact that modes of arbitrarily high energies seemingly contribute to low energy processes. Historically, it appears to have been generally assumed in its earliest days that the quantum field theory description of nature would break down at, say, the energy scale of elementary particles, and there was no reason to presume that it was a mathematically consistent theory. However, starting from the early 1950s, it was gradually understood how to give mathematically consistent rules to produce well defined expressions for physical quantities to all orders in perturbation theory for renormalizable theories such as quantum electrodynamics. This process culminated in the works of Bogoliubov, Parasiuk and of Hepp and Zimmermann, with important practical improvements (dimensional regularization) being given later by 't Hooft and Veltmann. It was also seen that the predictions of quantum field theory give truly excellent agreement with experiment – as they have continued to do through the present, LHC era. In the 1950s and 1960s, major progress was made toward putting quantum field on a mathematically sound footing via the development of the axiomatic [1], algebraic [2], and constructive [3] approaches. Nevertheless, the prevailing attitude toward quantum field theory today is not very different from what it was in its earliest days, namely, that it is not a fundamental

---

[1] One of the truly remarkable aspects of general relativity is that no new "matter field" need be introduced to describe gravitation, i.e., all physical phenomena normally attributed to "gravity" are, in fact, described by g.

theory but merely a valid description of quantum field modes up to some cutoff in energy (now assumed to be at a much higher energy scale than would have been assumed in its earliest days). At the present time, relatively little attention is generally paid to the issue of whether quantum field theory can be given a mathematically precise and consistent formulation – as compared with such issues as the "fine tuning" that would be necessary to give small values to the cosmological constant and Higgs mass if one views quantum field theory as the quantum theory of the modes of fields lying below some energy cutoff.

Our view is that it is very important to determine if quantum field theory can be given a mathematically precise and consistent formulation as a theory in its own right – and to provide such a formulation if it can be given. This is not because we believe that quantum field theory should be a "final" theory of nature; indeed, we do not believe that a quantum theory of the spacetime metric can be formulated within the existing framework of quantum field theory. However, even if quantum field theory has only a limited domain of validity, it is important to understand precisely what questions are well posed within its framework and how the answers to these questions are to be obtained. In this way, the predictions of quantum field theory can be made with clarity and precision, and hints may be provided for some of the features that might be expected to survive in a more fundamental theory that supersedes quantum field theory.

What are the "basic principles" of quantum field theory in Minkowski spacetime? The observables of the theory are the tensor fields representing the fundamental constituents of matter, together with "composite fields," such as the stress-energy tensor, derived from these matter fields. A key basic principle of quantum field theory is that each observable field, $\mathcal{O}(x)$, at each spacetime point x should be represented as an operator. These operators will satisfy nontrivial algebraic relations, such as commutation relations. However, there are two important caveats to this statement that $\mathcal{O}(x)$ is represented as an operator.

The first caveat is that, even in the case of a free Klein–Gordon scalar field ϕ – where, as already mentioned above, a quantum field theory can be formulated by ordinary quantization of the independent modes – it can be seen that one cannot make mathematical sense of $\phi(x)$ as an operator at a sharply defined point x, since modes of arbitrarily high-frequency and short-wavelength contribute to $\phi(x)$. However, $\phi(x)$ does make sense as a distribution, i.e., by "averaging" $\phi(x)$ with a smooth function of compact support, $f(x)$, one effectively eliminates the arbitrarily high-frequency and short-wavelength oscillations and thereby obtains a well defined expression for the quantum field. Thus, quantum fields are operator-valued distributions. The distributional nature of quantum fields is the source of most, if not all, of the mathematical difficulties arising in quantum field theory. Nonlinear operations involving distributions are intrinsically ill defined, and one will typically get infinite answers if one attempts to evaluate nonlinear functions of a distribution via mode expansions or other procedures.

The second caveat is that the word "operator" presumes that there is some unique underlying Hilbert space of states on which this operator will act. However, even for a free field, there are an infinite number of unitarily inequivalent representations of the fundamental commutation relations. (This contrasts sharply with the situation for a quantum mechanical system with a finite number of degrees of freedom, where the Stone–von Neuman theorem asserts that, under mild additional assumptions, all such representations are unitarily equivalent.) In Minkowski spacetime, a preferred representation normally can be chosen based upon the additional requirement that the representation contain a Poincaré invariant state ("the vacuum"). However, no criterion analogous to this can be applied in a general curved spacetime. As discussed further at the end of this section, it would therefore seem much more natural to view the algebraic relations satisfied by the field observables – rather than the choice of representation – as being fundamental. Thus, in quantum field theory, we will take as a basic principle that *the quantum fields $\mathcal{O}(x)$ are distributions valued in an algebra.*

Another basic principle of quantum field theory in Minkowski spacetime is that the fields should "transform covariantly" under Poincaré transformations. The Poincaré group is the isometry group of the metric, η, of Minkowski spacetime, but a general, curved spacetime will not admit any isometries. Nevertheless, "Poincaré invariance" may be viewed as a special relativistic version of the general relativistic requirement of "covariance," i.e., that quantum field theory in curved spacetime be constructed out of the classical spacetime metric g and the fundamental quantum fields ϕ, without any additional "background structure." Furthermore, this construction should be local in nature. We will take as a basic principle that *quantum field theory should be locally and covariantly constructed.*

An additional basic principle of quantum field theory in Minkowski spacetime is the requirement of positivity of energy. Since the notion of an energy operator for a quantum field theory in Minkowski spacetime is normally defined in terms of the transformation properties of the field under time translations, this requirement cannot be straightforwardly generalized to QFTCS. Nevertheless, one can formulate local conditions on the quantum field theory – known as microlocal spectral conditions – that correspond to positivity of energy in Minkowski spacetime and make sense in curved spacetime. We will take as a basic principle that *the quantum field theory should satisfy suitable microlocal spectral conditions.* We will give precise meaning to this statement in Appendix 10.3.3.

Finally, an additional principle of quantum field theory in Minkowski spacetime that is usually taken to be fundamental is the existence of a unique, Poincaré invariant state. However, this condition has no analog in a general curved spacetime as a condition on existence or uniqueness of states,[2] and we will not attempt to impose any condition of this nature.

Thus, we seek a formulation of quantum field theory in curved spacetime that implements the three basic principles written in italics above. In the remainder of this section, we briefly describe some of the standard approaches that have been used to formulate quantum field theory in Minkowski spacetime and explain why they do not appear suitable for the formulation of QFTCS. We will then describe the approach that we shall adopt.

Many discussions of quantum field theory are based upon a notion of "particles," and focus almost entirely on the calculation of the *S*-matrix, describing the scattering

---

[2] As argued in [4], the existence of an operator product expansion may be viewed as a generalization of this condition to curved spacetime.

of particles. For a free field in Minkowski spacetime, a notion of "vacuum state" and "particles" can be defined in a natural and precise manner. If an interacting field behaves like a free field in the asymptotic past and future, one can define asymptotic particle states. The *S*-matrix provides the relationship between the "in" and "out" particle descriptions of the states, and thereby directly yields the dynamical information about the interacting field that is most relevant to laboratory experiments in high-energy physics. However, the use of an *S*-matrix description as a fundamental ingredient in the formulation of QFTCS is unsuitable for the following reasons. Although natural notions of "vacuum state" and "particles" can be defined for a free field in stationary spacetimes, no such natural notions exist in a general curved spacetime. The difficulty is not that a notion of "particles" cannot be defined at all in a general curved spacetime but rather that many notions exist and none appears preferred. Although it may be possible (and useful) to define an *S*-matrix in spacetimes that become asymptotically stationary in a suitable manner in the past and future, many of the spacetimes of greatest interest in QFTCS are cosmological spacetimes or spacetimes describing gravitational collapse, where singularities occur in the asymptotic past and/or future. If one wishes to apply QFTCS to such spacetimes, it clearly would be preferable to formulate it in a manner that does not require one to define a notion of "particles" near singularities before one can even pose a well defined question. Furthermore, even if the spacetime of interest is suitably asymptotically stationary in the past and future, many of the most interesting physical questions are concerned with the local dynamical behavior of the fields at finite times rather than the particle-like description of states at asymptotically early and late times. For example, one may wish to know the expected stress-energy tensor of a quantum field in order to estimate the "back reaction" effects of the quantum field on the dynamics of the spacetime. An *S*-matrix would not be useful for such a calculation.

In many discussions of quantum field theory in Minkowski spacetime, Euclidean methods play an important role in both the formulation of the theory and in the calculational techniques. Minkowski spacetime can be viewed as a real 4-dimensional section of a complex 4-dimensional manifold with complex metric, which contains another 4-real-dimensional section (of "imaginary time") on which the metric is positive definite. If one can define a quantum field theory in a suitable manner on this "Euclidean section," a quantum field theory on Minkowski spacetime can then be obtained via analytic continuation. Since it is much easier to make sense of formal expressions in the Euclidean setting than in the Lorentzian setting, Euclidean methods have been employed in most of the attempts to rigorously define interacting quantum field theories and in most of the methods employed to regularize and renormalize quantities in perturbative quantum field theory [3]. Euclidean methods can be generalized so as to apply to static, curved spacetimes, where the transformation "$t \to it$" takes one from a static Lorentzian spacetime to a Riemannian space. However, a general curved spacetime will not be a real section of a complex manifold that also contains a real section on which the metric is Riemannian. Thus, although it should be possible to define "Euclidean quantum field theory" on curved Riemannian spaces [5], there is no obvious way to connect such a theory with quantum field theory on Lorentzian spacetimes. Thus, if one's goal is to define quantum field theory on general Lorentzian spacetimes, it does not appear fruitful to attempt to formulate the theory via a Euclidean approach.

Finally, by far the most prevalent approach taken towards the formulation of quantum field theory in Minkowski spacetime is to write down a formal functional integral expression for an effective action. Suitable functional derivatives of this expression are then interpreted as providing the correlation functions of the quantum field in its vacuum state. Thus, one will have defined the quantum field theory if one can make sense of this functional integral and its functional derivatives. The difficulty with using a functional integral approach to formulate QFTCS is that, in effect, it requires one to single out a preferred state in order to define the theory – namely, the state for which the correlation functions are being given. This is not a difficulty in Minkowski spacetime, where Poincaré invariance naturally selects a preferred state and, furthermore, Euclidean methods are available to make sense of the functional integral for this preferred state. However, as previously indicated above, no analogous notion of a preferred state exists in a general curved spacetime without symmetries. As in the above discussion of S-matrix approaches to the formulation of QFTCS, we do not believe that it will be fruitful to formulate QFTCS via an approach that requires one to define a preferred state in order to define the theory.[3]

For the above reasons, we shall adopt the "algebraic viewpoint" for the formulation of QFTCS. The basic idea of this approach is to take the relations satisfied by the quantum fields – such as commutation relations and field equations – as the fundamental starting point of the theory. To define the theory, one must specify the complete set of algebraic relations satisfied by the fundamental field and composite fields. As we shall see, this can be done for a free field. In addition, we can naturally define time ordered products, enabling one to give a perturbative construction of interacting quantum field theory. However, it is far less clear as to how to define appropriate algebraic relations so as to give a non-perturbative definition of interacting quantum field theory. Once the algebraic relations have been given, states are defined to be positive linear functions on the algebra of quantum fields. A construction due to Gel'fand, Nailmark and Segal (GNS) then shows that every state in this sense arises as a vector in a Hilbert space that carries a representation of the field algebra, thus connecting the algebraic notion of states with usual notions of states in quantum theory. The key point is that one can formulate QFTCS via the algebraic approach in a manner that does not require one to single out a preferred state in order to define the theory.

We begin in the next section by formulating QFTCS for a free scalar field, taking into account only the fundamental field observables. We also discuss some key applications, including the Hawking and Unruh effects and quantum effects arising from inflation. In Section 10.3, we outline how a wide class of nonlinear field observables can be defined for

---

[3] Note that this objection does not apply to the formulation of quantum field theory in (complete) Riemannian spaces, where one has unique Green functions for Laplacian-like operators, and, thus, a "preferred state" for a free field. However, as discussed above, there does not appear to be any way of relating quantum field theory in Riemannian spaces to quantum field theory on Lorentzian spacetimes.

a free scalar field. We then describe the perturbative construction of QFTCS for interacting fields. A more complete discussion of nonlinear observables and interacting QFTCS – including a discussion of open issues – is given in [6]. Throughout this chapter, we restrict our attention to scalar fields; for the treatment of Yang–Mills fields on curved spacetimes, see [7].

10.2 Free quantum fields

In this section, we provide a precise formulation of the theory of a free Klein–Gordon field in curved spacetime, insofar as the fundamental field observable, ϕ, is concerned. (Observables that are nonlinear in ϕ – i.e., "composite fields" – will be introduced in the next section.) We then discuss some key applications of the theory, namely, the Unruh effect, quantum field theory in de Sitter spacetime, the Hawking effect, and cosmological perturbations.

10.2.1 Formulation of linear QFTCS via the algebraic approach (without nonlinear observables)

We now describe how to define the quantum field theory of a real, linear Klein–Gordon field ϕ on a d-dimensional, curved, Lorentzian spacetime $(\mathcal{M}, g)$ along the lines sketched in the previous section. We begin with the classical Klein–Gordon field, which satisfies

$$(\Box_g - m^2)\phi = 0 \tag{10.1}$$

where $\Box_g = g^{\mu\nu}\nabla_\mu\nabla_\nu$ is the D'Alembertian operator associated with g. In order that (10.1) have a well-posed initial value formulation, we restrict our consideration – here and throughout this chapter – to globally hyperbolic spacetimes. By definition, a globally hyperbolic spacetime is a time-oriented spacetime that possesses a "Cauchy surface," Σ, i.e., a smoothly embedded $(d-1)$-dimensional spacelike submanifold with the property that if $\gamma : \mathbb{R} \to \mathcal{M}$ is any inextendible causal curve, then γ intersects Σ precisely once.[4] The classical Klein–Gordon equation with source j

$$(\Box_g - m^2)\phi = j \tag{10.2}$$

(where j is an arbitrary, fixed, smooth function on spacetime) has a well-posed initial value formulation on globally hyperbolic spacetimes in the following sense. Let n denote the unit normal to Σ. Then, given any pair (f_0, f_1) of smooth functions on Σ, there exists a unique solution ϕ to the Klein–Gordon equation (10.2) such that

$$\phi|_\Sigma = f_0, \quad n^\mu\nabla_\mu\phi|_\Sigma = f_1. \tag{10.3}$$

Furthermore, solutions to the initial value problem have a causal dependence upon the initial data and the source in the sense that if $x \in J^+(\Sigma)$, then the solution ϕ to the above

---

[4] It is a theorem [8] that any globally hyperbolic spacetime has the topology of a direct product, $\mathcal{M} \cong \Sigma \times \mathbb{R}$.

initial value problem will not change at x if we change the initial data (f_0, f_1) outside $J^-(x) \cap \Sigma$, or if we change the source outside of $J^-(x) \cap J^+(\Sigma)$. Here

$$J^\pm(S) \equiv \{x \in \mathcal{M} \mid \exists \text{ causal future/past-directed curve from } y \in S \text{ to } x\} \quad (10.4)$$

denotes the "causal future/past" of a set $S \subset \mathcal{M}$. Similarly, if $x \in J^-(\Sigma)$, then $\phi(x)$ will not change if we change the initial data (f_0, f_1) outside $J^+(x) \cap \Sigma$, or if we change the source outside of $J^+(x) \cap J^-(\Sigma)$. Finally, the solution ϕ depends continuously on (f_0, f_1) and j in a suitable sense.

We can define the retarded and advanced propagators $E^\pm$ of the Klein–Gordon equation as follows. For $j \in C_0^\infty(\mathcal{M})$, the advanced solution $E^- j$ is the unique solution to the Klein–Gordon equation with source j such that $E^- j(x) = 0$ for all $x \notin J^-(\operatorname{supp} j)$, with the opposite definition for the retarded solution $E^+ j$. The propagators may be viewed as maps

$$E^\pm : C_0^\infty(\mathcal{M}) \to C^\infty(\mathcal{M}), \quad (10.5)$$

or alternatively as distributional kernels on $\mathcal{M} \times \mathcal{M}$. As distributions, $E^\pm$ satisfy the differential equation

$$(\Box_g - m^2) E^\pm(x, y) = \delta(x, y), \quad (10.6)$$

where the Klein–Gordon operator acts on the first variable, x, in the sense of distributions. The support properties are

$$\operatorname{supp} E^\pm \subset \{(x, y) \in \mathcal{M} \times \mathcal{M} \mid y \in J^\pm(x)\}. \quad (10.7)$$

The advanced and retarded propagators are related by exchanging x with y. The anti-symmetric combination

$$E = E^+ - E^- \quad (10.8)$$

is called the "commutator function".[5]

As described in Section 10.1, we will formulate QFTCS for a Klein–Gordon field by defining a suitable algebra, $\mathscr{A}(\mathcal{M}, g)$ of quantum observables. In this section, we will consider only the algebra of observables generated by the fundamental field ϕ. An enlarged algebra that includes "composite fields" will be defined in the next section. The construction of $\mathscr{A}(\mathcal{M}, g)$ will take into account: (1) the distributional nature of the field $\phi(x)$, (2) the field equation, (3) the real character of ϕ and (4) the symplectic structure of the classical phase space of this theory. We construct $\mathscr{A}(\mathcal{M}, g)$ by starting with the free *-algebra generated by a unit $\mathbf{1}$ and elements $\phi(f)$, with $f \in C_0^\infty(\mathcal{M})$, and factoring by the following relations:

(1) *Linearity:* $\phi(c_1 f_1 + c_2 f_2) = c_1 \phi(f_1) + c_2 \phi(f_2)$ for all $c_1, c_2 \in \mathbb{C}$.
(2) *Field equation:* $\phi((\Box_g - m^2)f) = 0$,

---

[5] Further discussion concerning the construction and local expansion of $E^\pm$ can e.g. be found in [9].

(3) *Hermitian field:* $\phi(f)^* = \phi(\bar{f})$,

(4) *Commutator:* $[\phi(f_1), \phi(f_2)] = iE(f_1, f_2)\,\mathbf{1}$.

Item (1) incorporates the distributional character of the field; informally we write

$$\phi(f) = \int_{\mathcal{M}} \phi(x) f(x)\, dv_g \tag{10.9}$$

and we think of $\phi(x)$ as an $\mathscr{A}$-valued distribution. We refer to the algebra $\mathscr{A}$ as "abstract," because no reference has been made to any representation. In fact, we are not supposed to think, a priori, of the elements of $\mathscr{A}$ as operators on a particular Hilbert space, in the same way as an abstract Lie-algebra is defined irrespective of a particular representation. (2) incorporates the field equation in a distributional sense. (3) expresses that the field is real, and (4) implements the usual quantum mechanical relationship between classical Poisson-brackets and commutators. It also incorporates Einstein causality, because $E(x, y) = 0$ if x, y are spacelike related.

Rather than working with fields $\phi(f)$ that are smeared with test functions f on $\mathcal{M}$, we can also equivalently view ϕ as being "symplectically smeared" with solutions, F, having initial data of compact support on some Cauchy surface Σ. The correspondence is the following. If $f \in C_0^\infty(\mathcal{M})$, then $F = Ef$ is a source-free solution having initial data (10.3) of compact support. Conversely, given a solution F with initial data of compact support, there exists a test function f – unique up to addition of $(\Box_g - m^2)h$ for $h \in C_0^\infty$ – such that $F = Ef$. Defining $\phi[F] = \phi(f)$ under this correspondence between F and f, we can then informally write the field as

$$\phi[F] = \int_{\Sigma} (F\,\nabla_\mu \phi - (\nabla_\mu F)\phi) n^\mu\, dS. \tag{10.10}$$

We can think of $\pi = n^\mu\,\nabla_\mu \phi|_\Sigma, \varphi = \phi|_\Sigma$ as canonically conjugate variables and the commutation relation (4) then corresponds to $[\pi(x), \varphi(y)] = i\delta_\Sigma(x, y)\mathbf{1}$; for details, see e.g., [10] or lemma 3.2.1 of [11].

A *physical state*, ω, is simply an "expectation value functional," i.e., a linear map $\omega : \mathscr{A}(\mathcal{M}, g) \to \mathbb{C}$ satisfying the normalization condition $\omega(\mathbf{1}) = 1$, and positivity, $\omega(a^*a) \geq 0$ for all $a \in \mathscr{A}$. Any state is, by construction, specified by the collection $(W_n)_{n \geq 1}$ of its "n-point functions",

$$W_n(f_1, \ldots, f_n) \equiv \omega(\phi(f_1) \cdots \phi(f_n)). \tag{10.11}$$

The KG equation, condition (2), is translated into the fact that W_n is a distributional solution in each entry. Condition (1) implies that W_n is an n-times multi-linear functional on $C_0^\infty(\mathcal{M})$, which one normally requires to be distributional (i.e. continuous in the appropriate sense). The commutator condition (4) is translated into a linear condition which in the simplest case, $n = 2$, is

$$W_2(x, y) - W_2(y, x) = iE(x, y). \tag{10.12}$$

Positivity is translated into a rather complicated hierarchy of conditions on the multi-linear functionals W_n, the simplest of which is

$$W_2(\bar{f}, f) \geq 0 \qquad \text{for all } f \in C_0^\infty. \tag{10.13}$$

Given two states ω, ω', one can form a new state by forming any convex linear combination $\lambda \omega + (1 - \lambda)\omega'$, where $0 \leq \lambda \leq 1$. A state which cannot be written as a non-trivial convex linear combination of others is called *pure*.

The notion of algebraic state is in principle sufficient to answer all physical questions about the field observables. In particular, the specification of a state ω directly yields the expected values of all powers of $\phi(f)$ for all, say real, test functions f. It follows from the classical "Hamburger moment problem" (see e.g. [12]) that there is a unique[6] probability measure $dv(\lambda)$ such that

$$\int_{\mathbb{R}} \lambda^n \, dv(\lambda) = \omega(\underbrace{\phi(f)...\phi(f)}_{n}) \equiv m_n \qquad \text{for all } n \geq 1. \tag{10.14}$$

The probability that an observation of $\phi(f)$ will yield a value within $[a, b]$ when the field is in state ω is then given by

$$P_{a,b} = \int_{[a,b]} \lambda \, dv(\lambda) . \tag{10.15}$$

The relationship between states in the algebraic sense as defined above and the usual notion of states as vectors in a Hilbert space can be seen as follows: First, assume that we have a representation $\pi : \mathcal{A}(\mathcal{M}, g) \to \mathcal{H}$ on a Hilbert space with dense invariant domain $\mathcal{D} \subset \mathcal{H}$. If Ψ is a non-zero vector in this domain, then

$$\omega_\Psi(a) = \frac{(\Psi, \pi(a)\Psi)}{(\Psi, \Psi)} \tag{10.16}$$

defines a state in the algebraic sense. More generally, any sufficiently regular (with respect to $\mathcal{D}$) density matrix on $\mathcal{H}$ – i.e., a non-negative, trace-class operator $\rho \in \mathcal{I}_1(\mathcal{H})$ – defines an algebraic state ω_ρ. Conversely, given any algebraic state ω, there is a simple construction – known as the *GNS-construction*, or *Wightman reconstruction argument* – that yields a Hilbert space $\mathcal{H}$, a representation π of $\mathcal{A}$ on $\mathcal{H}$ with invariant domain $\mathcal{D} \subset \mathcal{H}$, and a vector $\Omega \in \mathcal{D}$ such that the algebraic state corresponding to Ω is ω. As we shall explain further below, for the case of a Gaussian state, the GNS-construction yields a Fock representation of $\mathcal{A}$, with Ω being the vacuum vector of this Fock space.

---

[6] Hamburger's theorem guarantees the existence of a measure if $\{m_n\}$ is a "positive sequence" [13], which in our case is an easy consequence of the positivity of the state ω. Uniqueness follows if the moments m_n satisfy the growth condition

$$\sum_n 1/\sqrt[2n]{m_{2n}} = \infty,$$

see e.g. [13]. For Gaussian states or finite particle states in its GNS-representation, one easily finds $m_{2n} = O(n!)$, so the measure is unique in those cases. On the other hand, if we replace $\phi(f)$ by a Wick power $\phi^k(f), k > 2$ discussed in sec. 10.3.1, we find $m_{2n} = O((kn)!)$. Thus, for general, nonlinear observables, the determination of the probability distribution from the state remains open.

The above correspondences show that the algebraic and Hilbert space formulations of QFTCS are essentially equivalent. However, there is one important difference: There are many unitarily inequivalent representations of the field algebra $\mathscr{A}$. (Two representations π and π' are said to be unitarily equivalent if there is an isometry $U : \mathscr{H} \to \mathscr{H}'$ such that $U\pi(a)U^* = \pi'(a)$ for all $a \in \mathscr{A}$.) The Hilbert space formulation requires one to choose a "preferred representation" at the outset, while the algebraic formulation does not. For this reason, we feel that the formulation of QFTCS via the algebraic approach is conceptually superior.

The algebra $\mathscr{A}$ admits states with rather pathological properties of their n-point functions W_n. However, there is a natural criterion to select physically reasonable states that is motivated from a variety of closely related considerations, including that (i) the ultra-high-frequency modes of the field should be essentially in their ground state, (ii) the short-distance singular structure of the n-point functions W_n should be similar to that of the n-point functions of the vacuum state in Minkowski spacetime, and (iii) the singular structure of the W_n's should be of "positive frequency type".

Such states are called "*Hadamard states*". As we will see in the next section, it will be necessary to restrict the discussion to Hadamard states in order to extend the action of the state to an enlarged algebra of field observables that includes all polynomials in the field and its derivatives at the same spacetime point. In particular, it will be necessary to restrict the discussion to Hadamard states when considering the perturbative expansion of interacting QFTCS. One possibility to define Hadamard states – in fact historically the first – which clearly reflects feature (ii) is as follows (see [18] for details): Concerning the n-point functions for $n \neq 2$, it is required that the connected part $\omega_n^c(\phi(x_1), ..., \phi(x_n))$ (defined in eq. (10.19) below) is smooth. For the 2-point function, it is required that there be no singularities at spacelike separations, together with the following local condition: For every convex, normal neighborhood $U \subset \mathscr{M}$, the two-point function has the form (in $d = 4$ dimensions; similar expressions hold in arbitrary d):

$$W_2 = \frac{1}{4\pi^2} \left[\frac{u}{\sigma + i0^+ t} + \left(\sum_{n=0}^{N} v_n \sigma^n \right) \log(\sigma + i0^+ t) \right]$$

$$+ \text{ some } N\text{-times continuously differentiable function } R_{N,\omega}$$

$$\equiv H_N + R_{N,\omega} . \tag{10.17}$$

Here, the spacetime arguments are understood as $(x, y) \in U \times U$, and we mean that the above formula should hold for every N, with (different) remainders $R_{N,\omega} \in C^N(U \times U)$. σ is the signed squared geodesic distance, and $t = T(x) - T(y)$ for some (in fact, any) global time function $T : \mathscr{M} \to \mathbb{R}$. The functions u, v_n are determined by certain local "transport equations" [19] and also appear in similar local forms of the advanced and retarded Green's function for the operator $\Box_g - m^2$. In particular, H_N is locally and covariantly defined in terms of the metric, and is hence the same for *any* Hadamard state. By contrast, the remainder $R_{N,\omega}$ depends on the state.

While this definition has the merit of being, in principle, completely explicit, it turned out to be of great advantage to have an alternative, equivalent characterization of Hadamard states that brings out more manifestly the "positive frequency" aspect (iii) of Hadamard states. This characterization was given first in [14, 15] and [16], and uses the language of *wavefront sets* discussed in the Appendix. The criterion is that the 2-point function have a wavefront set of the form

$$\text{WF}(W_2) \subset \{(x_1, k_1; x_2, k_2) \in T^* \mathcal{M}^2 \setminus 0 \mid k_1 \in \dot{V}^+, k_2 \in \dot{V}^-, k_1 \sim -k_2\}, \quad (10.18)$$

and, as before, that each connected n-point function $\omega_n^c(\phi(x_1), \dots, \phi(x_n))$ for $n \neq 2$ is smooth. The last requirement may actually be shown to be a consequence of the first condition [17]. Here $V^+ \subset T^* \mathcal{M}$ is the collection of all non-zero, future-directed time-like or null co-vectors (and similarly V^-), and the relation $\sim$ holds between two covectors if they are tangent to a null geodesic in $(\mathcal{M}, g)$ and are parallel transported into each other.

It should be emphasized that the Hadamard condition does not single out a particular state, but a *class* of states. Existence of a large class of Hadamard states on *any* globally hyperbolic spacetime can be established by a deformation argument [20] combined with microlocal techniques, or by methods from the theory of pseudo-differential operators [21, 22].

We now discuss two important classes of states: Gaussian states and thermal states.

Gaussian States: Gaussian states (also called "quasi-free states") are defined by the condition that the "connected n-point functions" $\omega_n^c(\phi(f_1), \dots, \phi(f_n))$ vanish for all $n > 2$, where $\omega_n^c : \mathcal{A} \times \dots \times \mathcal{A} \to \mathbb{C}$ is defined by[7]

$$\omega_n^c(a_1, \dots, a_n) \equiv \frac{\partial^n}{\partial t_1 \dots \partial t_n} \log \left\{ \omega \left(e^{t_1 a_1} \dots e^{t_n a_n} \right) \right\} \bigg|_{t_i = 0}, \quad a_i \in \mathcal{A}. \quad (10.19)$$

Thus, the n-point functions of Gaussian states can be expressed entirely in terms of their 1- and 2-point functions. For a Gaussian state, positivity will hold if and only if (10.13) is satisfied. Thus, there exists a wide class of Gaussian states.

Any Gaussian state, ω, can be expressed as the "vacuum state" in a Fock representation of $\mathcal{A}$. To see this explicitly, let W_2 be the 2-point function of ω. On the complex linear space $C_0^\infty(\mathcal{M}, \mathbb{C})$ of smooth complex-valued functions on $\mathcal{M}$ of compact support, define the inner product $\langle f | h \rangle = W_2(\bar{f}, h)$. This is hermitian and positive, $\langle f | f \rangle \geq 0$, but contains degenerate vectors, such as elements of the form $f = (\Box_g - m^2)h$. Let $\mathfrak{h}$ be the factor space of $C_0^\infty(\mathcal{M}, \mathbb{C})$, divided by the degenerate vectors. The elements in this space can be identified more concretely with a subspace of complex valued smooth solutions to the KG-equation, corresponding to "positive frequency modes", sec below. The completion of $\mathfrak{h}$, denoted by the same symbol, is a Hilbert space, usually referred to as the "1-particle space". Let $\mathcal{H}$ be the bosonic Fock space over $\mathfrak{h}$,

---

[7] Here the exponentials are to be understood in the sense of a formal series.

$$\mathcal{H} = \mathbb{C} \oplus \bigoplus_{n \geq 1} \underbrace{(\mathfrak{h} \otimes_S \cdots \otimes_S \mathfrak{h})}_{n}, \tag{10.20}$$

where $\otimes_S$ is the symmetrized tensor product. A representation, π, of $\mathscr{A}$ on $\mathscr{H}$ can then be defined by

$$\pi[\phi(f)] = a([f])^{\dagger} + a([f]) \tag{10.21}$$

where $a([f])$ is the annihilation operator associated with the equivalence class of f in $\mathfrak{h}$. The vacuum vector Ω given by the element $|1, 0, 0, \ldots\rangle$ in Fock space then corresponds to ω, and, as already mentioned above, $(\mathscr{H}, \pi, \Omega)$ is precisely the GNS triple arising from the GNS construction.

A closely related construction often used in practice to construct Gaussian pure states is the following. Suppose that we have a set of smooth, complex-valued "mode functions" $u_\xi(x)$ that are solutions to eq. (10.1), and which are labelled by $\xi \in X$ in some measure space $(X, d\mu)$. We assume that for each $f \in C_0^{\infty}(\mathscr{M}, \mathbb{R})$, the map

$$X \ni \xi \mapsto Kf(\xi) \equiv \int_{\mathscr{M}} \overline{u_\xi(x)} f(x)\, dv_g \in \mathbb{C} \tag{10.22}$$

is in $L^2(X, d\mu)$, and in fact that the (real linear) span of such vectors is dense in $L^2(X, d\mu)$. We also assume that the mode functions are such that

$$\mathrm{Im}\langle Kf_1 | Kf_2\rangle_{L^2(X, d\mu)} = \tfrac{1}{2} E(f_1, f_2) \qquad \text{for all } f_1, f_2 \in C_0^{\infty}(\mathscr{M}). \tag{10.23}$$

(These properties are equivalent to the statement that the collection of modes $(u_\xi)_{\xi \in X}$ is "complete in the KG-norm".) Then, clearly

$$W_2(f_1, f_2) = \langle K f_1 | K f_2\rangle_{L^2(X, d\mu)} \quad \Rightarrow \quad W_2(x, y) = \int_X d\mu(\xi) u_\xi(x) \overline{u_\xi(y)} \tag{10.24}$$

defines the 2-point function of a Gaussian state, which can be shown to be pure. Its GNS-representation is thus constructed as above. K is clearly well-defined on the equivalence classes $[f]$, and it can also be shown that it provides a bounded isomorphism $K : \mathfrak{h} \to L^2(X, d\mu)$. Hence, in this case, we may consider $\mathscr{H}$ as the bosonic Fock space over $L^2(X, d\mu)$, and we may informally write the representative of the field on this Fock space as

$$\pi(\phi(x)) = \int_X d\mu(\xi)[u_\xi(x) a_\xi + \overline{u_\xi(x)} a_\xi^{\dagger}] \tag{10.25}$$

This is the usual form of the field in the vacuum representation on Minkowski space, where $\xi = \mathbf{k} \in \mathbb{R}^3 = X$, $d\mu(\mathbf{k}) = d^3\mathbf{k}/(2\pi)^3$, and $u_\mathbf{k}(\mathbf{x}, t) = e^{-i\omega_\mathbf{k} t + i\mathbf{k}\mathbf{x}}/\sqrt{2\omega_\mathbf{k}}$. In this example, the mode functions have positive frequency with respect to the global time translations, and similar constructions are available also on other spacetimes with a globally defined time-like isometry.

Given two pure Gaussian states $\omega, \tilde{\omega}$ with 2-point functions $W_2, \tilde{W}_2$, respectively, one may ask when their associated GNS-representations are *unitarily equivalent*. A necessary condition is that there be a constant c such that

$$c^{-1}\mu(f,f) \le \mu'(f,f) \le c\,\mu(f,f) \tag{10.26}$$

for all $f \in C_0^\infty(\mathcal{M}, \mathbb{R})$, where μ and μ' denote the symmetric parts of W_2 and W_2'. Since it is easy to construct states violating this condition, one sees that there is in general a large class of inequivalent representations. Suppose the condition is satisfied. It immediately follows that the positive, sesquilinear[8] forms μ, μ' induce the same topology on the complex vector space $\mathcal{K}$ which is defined as $C_0^\infty(\mathcal{M}, \mathbb{C})$ modulo elements of the form $f = (\square_g - m^2)h$. $\mathcal{K}$ may be thought of as the space of complex valued, spatially compact smooth solutions to the Klein–Gordon equation, and the bilinear forms μ and μ' are well-defined and positive definite on $\mathcal{K}$.[9] The Riesz representation theorem guarantees that there is a bounded map S on $\mathcal{K}$ such that $\mu(F_1, F_2) = \mu'(F_1, SF_2)$ for all $F_1, F_2 \in \mathcal{K}$. A necessary and sufficient criterion that the two states give rise to unitarily equivalent representations is

$$\mathrm{tr}_{\mathcal{K}}|1 - S| < \infty, \tag{10.27}$$

where $|1 - S|$ is the absolute value of the operator defined using μ'. The operator $1 - S$ characterizes the difference between the 2-point functions. It is therefore plausible that if their difference is smooth – as happens e.g. if both states are Hadamard – and the spacetime has compact Cauch surfaces, then (10.27) should hold, and the representations should be unitarily equivalent. This can indeed be shown[10]. On the other hand, for the case of a non-compact Cauchy surface, the representations can be unitarily inequivalent if the 2-point functions have sufficiently different long-range behavior. Similarly, the representations can fail to be unitarily equivalent if the states are not Hadamard and have a sufficiently different short range behavior.

Thermal States: If the spacetime $(\mathcal{M}, g)$ has a complete time-like Killing vector field ξ, one can define the notion of a thermal state relative to the time evolution generated by this Killing vector field. There is an elegant version of this notion referred to as the "KMS-condition," which can be formulated directly in terms of the expectation value functional ω, without making reference to any Hilbert-space representation. The KMS-condition is formulated as follows: Let φ_t be the 1-parameter group of isometries $\varphi_t : \mathcal{M} \to \mathcal{M}$ generated by a Killing vector field ξ. We define an action $\alpha_t : \mathcal{A} \to \mathcal{A}$ of our 1-parameter family of isometries on the algebra $\mathcal{A}$ of fields on $(\mathcal{M}, g)$ by setting $\alpha_t(\phi(f_1)\cdots\phi(f_n)) = \phi(f_1^t)\cdots\phi(f_n^t)$, where $f^t(x) = f(\varphi_{-t}(x))$. Since the retarded and advanced fundamental solutions, and the field equation, are invariant under φ_t, it follows that α_t respects the algebraic relations in $\mathcal{A}$, i.e., it is an automorphism. In fact, from the composition law for the isometries φ_t it immediately follows that $\alpha_t \circ \alpha_s = \alpha_{t+s}$. In this situation, a state ω is called a *KMS-state* at inverse temperature β with respect to α_t if the following two conditions are satisfied:

---

[8] For complex valued test functions, μ and μ' are defined to be anti-linear in the first entry.

[9] It is easy to see that the closure of this space (relative to either μ or μ' can be decomposed into "positive and negative frequency solutions") as $\mathcal{K} \cong \mathfrak{h} \oplus \overline{\mathfrak{h}} \cong \mathfrak{h}' \oplus \overline{\mathfrak{h}'}$, where an overbar means the complex conjugate Hilbert space, and where $\mathfrak{h}$ respectively $\mathfrak{h}'$ are the 1-particle spaces corresponding to W_2 respectively W_2'.

[10] For Gaussian states that are not pure the generalization of unitary equivalence is "quasi-equivalence", where a corresponding criterion has been given [23]. Ref. [24] actually deals with this more general situation.

(1) For any collection of $a_i \in \mathscr{A}$, the function $\underline{t} = (t_1, \ldots, t_n) \mapsto F_{a_1, \ldots, a_n}(\underline{t})$ defined by

$$F_{a_1, \ldots, a_n}(\underline{t}) = \omega(\alpha_{t_1}(a_1) \cdots \alpha_{t_n}(a_n)) \tag{10.28}$$

has an analytic continuation to the strip

$$\mathfrak{T}_n^\beta = \{(z_1, \ldots, z_n) \in \mathbb{C}^n \mid 0 < \mathrm{Im}(z_j) - \mathrm{Im}(z_i) < \beta, \ 1 \le i < j \le n\} \tag{10.29}$$

This function is required to be bounded and continuous at the boundary.
(2) On the boundary, we have

$$F_{a_1, \ldots, a_n}(t_1, \ldots, t_{k-1}, t_k + i\beta, \ldots, t_n + i\beta)$$
$$= F_{a_k, \ldots, a_n, a_1, \ldots, a_{k-1}}(t_k, \ldots, t_n, t_1, \ldots, t_{k-1}). \tag{10.30}$$

Note that the definition of a KMS-state only assumes an algebra $\mathscr{A}$ and the existence of a 1-parameter family of automorphisms. The notion of a KMS-state is therefore not tied to the particular example $\mathscr{A} = \mathscr{A}(\mathscr{M}, g)$ and the particular 1-parameter group of automorphisms α_t considered here. It is thus a definition of a very general nature, applicable to many quantum systems, see e.g. [25] for further discussion. In the case of C^*-algebras (algebras of bounded operators), the condition for $n = 2$ implies the remaining ones, but this is not generally the case for the case of unbounded operator algebras considered here. It is however the case for the concrete algebra $\mathscr{A}$ considered here if we restrict our attention to Gaussian states. In this case, the condition $n = 2$ also implies that the state is Hadamard [26].

Let us now motivate the above technical definition by explaining its relation to the usual notion of thermal equilibrium state in statistical mechanics. Consider a self-adjoint Hamiltonian H defined on a Hilbert space with spectrum bounded from below, and suppose that $Z_\beta = \mathrm{tr}\, e^{-\beta H} < \infty$ (which cannot hold unless the spectrum of H is discrete). The standard definition of a Gibbs state is $\omega(a) = \mathrm{tr}(ae^{-\beta H})/Z_\beta$, where a is e.g. any (say) bounded operator on $\mathscr{H}$. Let $\alpha_t(a)$ be defined in this example by $\alpha_t(a) = e^{itH}ae^{-itH}$, i.e., it describes the usual time evolution of an observables a in ordinary quantum mechanics. Then, using that the spectrum of H is bounded below, we easily see that,

$$F_{a,b}(z) = Z_\beta^{-1} \mathrm{tr}(ae^{izH}be^{-izH}e^{-\beta H}) \tag{10.31}$$

is holomorphic in the strip $0 < \mathrm{Im}(z) < \beta$, because in this range $e^{-\beta H}$ provides a sufficient "damping" to make the trace finite. Furthermore, using the cyclicity of the trace, we have

$$F_{a,b}(t + is) = Z_\beta^{-1} \mathrm{tr}(e^{-\beta H}ae^{(it-s)H}be^{(-it+s)\hat{H}}) \tag{10.32}$$
$$= Z_\beta^{-1} \mathrm{tr}(e^{-\beta H}e^{(it+\beta-s))H}be^{(-it-\beta+s)H}a). \tag{10.33}$$

From the first line we see that $F_{a,b}(t + is) \to \omega(a\alpha_t(b))$ for $s \to 0^+$, while we see from the second line that $F_{a,b}(t + is) \to \omega(\alpha_t(b)a)$ for $s \to \beta^-$. Thus, (1) and (2) hold (for $n = 2$), and therefore a Gibbs state in the usual sense is a KMS-state in the sense of the above definition. The idea behind the definition of a KMS-state is to turn this statement around and define thermal equilibrium states by conditions (1) and (2).

A key technical advantage of the definition of a KMS-state is that it still makes sense when a density matrix no longer exists, as usually happens when the Cauchy surface Σ is non-compact. The standard example of this is Minkowski space (the 'thermodynamic limit'). In the standard GNS-representation $(\mathcal{H}, \pi, \Omega)$ corresponding to the vacuum state (described e.g. by the mode functions given above), the Hamiltonian does not have the property that $e^{-\beta H}$ is a trace-class operator on $\mathcal{H}$, i.e. no density matrix exists. Nevertheless, a Gaussian KMS-state can easily be defined in terms of its 2-point function, given by

$$W_2(x_1, x_2) = \omega(\phi(x_1)\phi(x_2)) = \frac{1}{(2\pi)^3} \int D_+^\beta (t_1 - t_2 - i0^+, \mathbf{p}) \, e^{i\mathbf{p}(\mathbf{x}_1 - \mathbf{x}_2)} \, d^3\mathbf{p} \,, \quad (10.34)$$

where Minkowski points are labelled by $x = (t, \mathbf{x}) \in \mathbb{R}^{1,3}$, and where D_+^β is given by

$$D_+^\beta(t, \mathbf{p}) = \frac{1}{2\omega_\mathbf{p}} \frac{\cosh(\tfrac{1}{2}\beta\omega_\mathbf{p} - i\omega_\mathbf{p} t)}{\sinh(\tfrac{1}{2}\beta\omega_\mathbf{p})} \,, \qquad \omega_\mathbf{p} = \sqrt{\mathbf{p}^2 + m^2} \,. \quad (10.35)$$

The 1-parameter family of isometries is simply given by $\varphi_T(t, \mathbf{x}) = (t + T, \mathbf{x})$, i.e. time translations. The verification of the KMS-condition boils down to the condition on the 2-point function, which in turn boils down to showing that the $z = (t + is)$-dependent distribution $F_{x,y}(t + is) = W_2(x, \varphi_{t+is}(y))$ in $x, y \in \mathbb{R}^{1,3}$ has distributional boundary values $W_2(x, \varphi_t(y))$ resp. $W_2(\varphi_t(y), x)$ for $s \to 0^+$ resp. $s \to \beta^-$. This is in turn directly seen to be a consequence of the functional relation

$$\lim_{s \to \beta^-} D_+^\beta(t - is, \mathbf{p}) = \lim_{s \to 0^+} D_+^\beta(-t + is, \mathbf{p}) \,. \quad (10.36)$$

10.2.2 Applications: Unruh effect, de Sitter space, Hawking effect, inflationary perturbations

We now discuss some concrete examples in order to illustrate the abstract ideas just given and to present some of the important applications of QFTCS.

a) Unruh effect: A relatively simple and yet very important application of QFTCS arises if we consider a "wedge" W of Minkowski spacetime and view it as a spacetime in its own right. Namely, let

$$W = \{x \in \mathbb{R}^{1,3} \mid x_1 > |x_0|\}, \quad (10.37)$$

and let W be equipped with the Minkowski metric. Of course, this is not a curved spacetime, but it is a globally hyperbolic spacetime that differs in essential ways from Minkowski spacetime, e.g., all of its timelike and null geodesics are incomplete. The spacetime (W, g) is called "Rindler spacetime".

Writing $U = -x_0 + x_1$ and $V = x_0 + x_1$, the metric of Rindler spacetime is

$$\begin{aligned} g &= dU \, dV + dx_2^2 + dx_3^2 \\ &= e^{a(u+v)} \, du \, dv + dx_2^2 + dx_3^2 \end{aligned} \quad (10.38)$$

where u and v are defined by $U = e^{au}$ and $V = e^{av}$. Further introducing (η, ξ) by

$$u = \xi - \eta, \quad v = \eta + \xi, \quad (10.39)$$

the metric takes the form

$$g = e^{2a\xi}(-d\eta^2 + d\xi^2) + dx_2^2 + dx_3^2. \tag{10.40}$$

The coordinates η and ξ are related to the original global inertial coordinates (x_0, x_1, x_2, x_3) of Minkowski spacetime by

$$x_0 = a^{-1}e^{a\xi} \sinh a\eta$$
$$x_1 = a^{-1}e^{a\xi} \cosh a\eta. \tag{10.41}$$

It is not difficult to see that the hypersurfaces, Σ_η, of constant η are Cauchy surfaces for Rindler spacetime for all $\eta \in \mathbb{R}$. Furthermore, for any $t \in \mathbb{R}$, the transformation $\varphi_t :$ $\eta \to \eta + t$ is an isometry of the Rindler spacetime, which corresponds to Lorentz boosts of Minkowski spacetime. Indeed, the key fact about Rindler spacetime is that the orbits of the Lorentz boosts are everywhere timelike and are complete in Rindler spacetime. Thus, Rindler spacetime is a static, globally hyperbolic spacetime, where the notion of "time translations" is defined by Lorentz boosts. Note that each Lorentz boost orbit in Rindler spacetime corresponds to the worldline of a uniformly accelerating observer in Minkowski spacetime. The Lorentz boost orbits become null on the boundary of Rindler spacetime and, indeed, the hypersurfaces $U = 0$ and $V = 0$ of Minkowski spacetime comprise a bifurcate Killing horizon of the Lorentz boost Killing field, with surface gravity[11] $\kappa = a$.

As on every globally hyperbolic spacetime, we quantize the field ϕ by viewing it (after smearing with a test function) as an element of the associated abstract algebra $\mathscr{A}(W, g)$ defined by the relations (1) to (4) in Section 10.2.1. Actually, in the present context, these relations are identical with those of the whole Minkowski spacetime, because the advanced and retarded propagators, and hence E, are locally the same. However, there is a difference in that the algebra $\mathscr{A}(W, g)$ contains only smeared elements of the form $\phi(f)$ for test functions f that are compactly supported in W (and in particular, away from ∂W). Thus, $\mathscr{A}(W, g)$ may be viewed as a proper subalgebra of the algebra associated with the entire Minkowski spacetime. Thus, we can obtain a state on $\mathscr{A}(W, g)$ by restricting the usual vacuum state on Minkowski spacetime, and view it as a state on Rindler space. The 2-point function of this Gaussian state is (taking $m^2 = 0$ for simplicity)

$$W_2(x, y) = \frac{1}{2\pi^2(x - y - i0^+e)^2}, \tag{10.42}$$

where e is any fixed future directed timelike vector, and the distributional boundary value prescription is understood.

Since $W_2(x, y) \neq 0$ for any $x \in W$ and $y \in W'$ (where W' denotes the "opposite wedge" $x_1 < -|x_0|$), it follows that there are correlations between field observables in W and W' and that restriction of the Minkowski vacuum to Rindler spacetime cannot yield a pure state. A key result is the following theorem, which is a special case of the Bisognano–Wichmann theorem [2] of axiomatic quantum field theory in Minkowski spacetime:

---

[11] Note that unlike in the analogous example of Schwarzschild (see below), there is in this case no canonical normalization of the Killing field.

Theorem 10.2.1 *The restriction of the Minkowski vacuum state to the Rindler algebra* $\mathscr{A}(W, g)$ *is a KMS-state with respect to the 1-parameter group of isometries given by* $\eta \rightarrow \eta + t$. *The inverse temperature of this KMS-state is given by*

$$\beta = \frac{2\pi}{a} . \tag{10.43}$$

To prove this claim, one has to verify the KMS-condition. This can be done in a completely straightforward manner. For a Gaussian state such as that given here, it suffices to verify the KMS-condition for $a_1 = \phi(f_1), a_2 = \phi(f_2)$ with f_1, f_2 having their support inside W, which in turn boils down to showing that the distribution $F_{x,y}(t + is) = W_2(x, \varphi_{t+is}(y))$ in $x, y \in W$ has distributional boundary values $W_2(x, \varphi_t(y))$ resp. $W_2(\varphi_t(y), x)$ for $s \rightarrow 0^+$ resp. $s \rightarrow \beta^-$. This is an elementary computation done by transforming the 2-point function into Rindler coordinates.

The Minkowski vacuum state is, of course, regular on the Rindler horizon and is invariant under Lorentz boosts, and it is the only Hadamard state on all of Minkowski spacetime that is Lorentz boost invariant. In fact, the uniqueness and KMS property (but not necessarily existence) of an isometry-invariant Hadamard state can be proven to hold on any globally hyperbolic spacetime with a bifurcate Killing horizon [18]. Important further examples will be provided in the next two subsections.

The above theorem has an important physical interpretation, known as the *Unruh effect*: If the field is in the Minkowski vacuum state, a uniformly accelerating observer in Minkowski spacetime – who may also be viewed as a static observer in Rindler spacetime – will "feel himself" immersed in a "thermal bath of particles" at inverse temperature (10.43). This can be explicitly seen by introducing a model "particle detector" and showing that it will be suitably excited as a result of its interaction with the quantum field [27]; for a recent treatment see e.g. [28]. This provides an excellent illustration of why notions of "vacuum state" and "particles" cannot be considered to be fundamental in the formulation of QFTCS. If the field is in the Minkowski vacuum state, an inertial observer will naturally declare that no "particles" are present, whereas the accelerating observer will naturally declare that the Rindler wedge is filled with a thermal bath of particles. However, there is no actual disagreement between these observers: They both agree that the field is in a Gaussian state with two-point function (10.42), and they will be in complete agreement on the probabilities for measuring any field observables.

b) de Sitter spacetime: Four-dimensional (global) de Sitter space dS_4 is the 4-dimensional hyperboloid defined by the equation $Y \in \mathbb{R}^5, Y \cdot Y = H^{-2}$, where $H > 0$ is the Hubble constant and the dot "·" denotes the 5-dimensional Minkowskian inner product with signature $(- + + + +)$. The metric is that induced from the ambient space. d-dimensional de Sitter space is defined in the same way.

From its definition as a hyperboloid in 5-dimensional Minkowski space, it is clear that de Sitter has the 10-dimensional group $O(4, 1)$ as its isometry group. Let us now consider a Klein–Gordon quantum field on de Sitter spacetime. Since dS_4 is globally hyperbolic, we can define the algebra of field observables $\mathscr{A}(dS_4, g)$ by the general procedure above. For $m^2 > 0$, a globally $O(4, 1)$-invariant state exists, called the Bunch–Davies (aka Hartle–Hawking, aka Euclidean) vacuum. To describe this it is convenient to introduce a function $Z : dS_4 \times dS_4 \to \mathbb{R}$ by

$$Z(x, y) = H^2 \, Y(x) \cdot Y(y) \tag{10.44}$$

in terms of the embedding $Y : dS_4 \to \mathbb{R}^5$ of de Sitter space into five-dimensional Minkowski space. This function is symmetric, de Sitter invariant, and is related to the signed geodesic distance σ by the formula

$$\cos(H\sqrt{\sigma}) = Z, \tag{10.45}$$

where the square root is taken to be imaginary for time-like separated points. The causal relationships between points can be put in correspondence with values of Z; see the conformal diagram, Fig. 10.1.

In terms of Z, the 2-point function of the Bunch–Davies state is [29–31] (in d dimensions)

$$W_2(x, y) = \frac{H^{d-2}}{(4\pi)^{d/2}} \frac{\Gamma(-c)\Gamma(c+d-1)}{\Gamma(d/2)} \, {}_1F_2\left(-c, d-1+c; d/2; \frac{1+Z-it0^+}{2}\right),$$
$$\tag{10.46}$$

where the dimensionless constant c is defined by

$$c = -\frac{d-1}{2} + \sqrt{\frac{(d-1)^2}{4} - \frac{m^2}{H^2}} \tag{10.47}$$

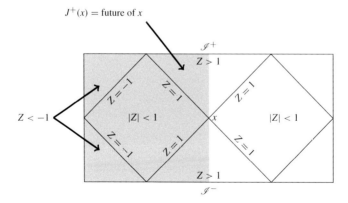

Figure 10.1 Conformal diagram and values of the point-pair invariant $Z = Z(x, y)$ as y is varied and x is kept fixed. For the sake of easier visualization, we are giving the diagram in the case of $d = 2$-dimensional de Sitter spacetime, where the left and right vertical boundaries are to be identified. For $d > 2$ dimensions, the diagram would basically consist of only the shaded "left half", with the vertical boundary lines corresponding to the north and south pole of the S^{d-1} Cauchy surface.

and the usual boundary value prescription has to be applied, with $t = Y^0(x) - Y^0(y)$. It is relatively easy to check that this 2-point function satisfies the Klein–Gordon equation in each argument, and it can be verified that its antisymmetric part satisfies (10.12). It is far less obvious that the 2-point function is positive, but this can be shown using the following rather elegant and nontrivial representation due to [30, 31] (assuming for simplicity a "principal-series scalar field" characterized by $\mu^2 := m^2 - [(d-1)^2/4]H^2 \geq 0$)

$$W_2(x,y) = \text{const.} \sum_{l=\pm} \int_{\mathbb{R}^{d-1}} (Y(x) \cdot \xi(\mathbf{k},l))^c \, (Y(y) \cdot \xi(\mathbf{k},l))^{\bar{c}} \, \frac{d^{d-1}\mathbf{k}}{\sqrt{\mathbf{k}^2 + \mu^2}}. \tag{10.48}$$

Here, $\xi(\mathbf{k},l) \in \mathbb{R}^{d+1}$ are the $(d+1)$-dimensional vectors defined by

$$\xi(\mathbf{k}, \pm) = (\sqrt{k^2 + \mu^2}, \pm\mathbf{k}, \pm\mu). \tag{10.49}$$

More precisely, W_2 is defined again as the boundary value of the analytic function obtained by adding to the time coordinate of y a small positive imaginary part. To check that it is Hadamard, one may use the relationship between the wave-front set and distributional boundary values; see Appendix 10.3.3.

The following statements hold true concerning de Sitter invariant states [29]:

- When $m^2 > 0$, then the Bunch–Davies state is the *unique* de Sitter invariant, pure, Gaussian, Hadamard state, although a 1-parameter family of states ('α-vacua') exists if the Hadamard condition is dropped.
- When $m^2 \leq 0$, no de Sitter invariant states exist, although, as emphasized, in those cases an infinite set of non-de Sitter-invariant Hadamard states still exists. In particular, the algebra $\mathscr{A}$ may *always* be defined for any value of m^2, although when $m^2 < 0$ the n-point functions of physically reasonable states will grow exponentially with time.

In de Sitter space, there is a phenomenon reminiscent of the Unruh effect which takes place in the "static chart;" see Fig. 10.2. That chart can be defined as the intersection of dS_4 with a wedge $\{|X_1| > X_0\}$ in the ambient $\mathbb{R}^5$. The static chart is again a globally hyperbolic spacetime in its own right, and can also be defined as the intersection $J^+(i^-) \cap J^-(i^+)$ of two points $i^\pm \in \mathscr{I}^\pm$ which are at the "same angle". It can be covered by the coordinate system (t, r, φ, θ) defined for $t \in \mathbb{R}$, $0 \leq r < H^{-1}$, in which the line element takes the form

$$g = -(1 - H^2 r^2)dt^2 + (1 - H^2 r^2)^{-1} dr^2 + r^2(d\theta^2 + \sin^2\theta \, d\varphi^2). \tag{10.50}$$

It can be seen from this form of the line element that, within this chart – but of course not in the full de Sitter space – the metric is static, with timelike Killing field $\xi = \frac{\partial}{\partial t}$. The corresponding flow $\varphi_s : t \mapsto t + s$ defines a 1-parameter group of isometries in the static chart, which corresponds to a boost in the X_0–X_1 plane in the ambient $\mathbb{R}^5$. The boundary $\mathcal{H} = \mathcal{H}_+ \cup \mathcal{H}_-$ is formed from two intersecting cosmological horizons, and is another example of a bifurcate Killing horizon, with surface gravity $\kappa = H$. The restriction of the Bunch–Davies state to the static chart is seen to be a *KMS-state* at inverse temperature $\beta = H/2\pi$ by the same argument as given for the Unruh effect in Rindler spacetime (see

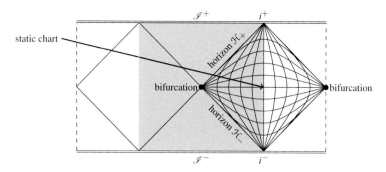

Figure 10.2 Conformal diagram for de Sitter spacetime, and the static chart. Again, we are drawing the case $d = 2$. The case $d > 2$ would correspond to the shaded square having the bifurcation surface $\cong S^{d-2}$ in the middle. The vertical boudaries of the shaded square correspond to the north and south pole of the Cauchy surface S^{d-1}.

[18]). Note that the static orbit corresponding to $r = 0$ is a geodesic, and, by de Sitter invariance, any time-like geodesic in de Sitter spacetime is an orbit of the static Killing field of some static chart. In this sense, one may say that in the Bunch–Davies state in de Sitter spacetime, every freely falling observer will "feel himself" immersed in a thermal bath of particles at inverse temperature $\beta = H/2\pi$.

Of particular interest in cosmology is the behavior of the 2-point function of the Bunch–Davies state for large time-like separation $\tau = \sqrt{-\sigma}$. Using well-known properties of the hypergeometric function (10.46), it is found that the 2-point function behaves as $e^{-(d-1)H\tau/2}$ for $\tau \gg 1$. This exponential decay reflects the exponential dispersive effects of fields on de Sitter space. It implies that the 2-point function $W_2^{\Psi}(x, y) = (\Psi, \pi(\phi(x))\pi(\phi(y))\Psi)$ of *any* Hadamard state of the form $\Psi := \pi[\phi(f_1) \cdots \phi(f_n)]\Omega \in \mathcal{H}, f_i \in C_0^{\infty}(\mathcal{M})$ in the GNS-representation $(\mathcal{H}, \pi, \Omega)$ of the Bunch–Davies state approaches that of the Bunch–Davies state [i.e. $W_2(x, y)$, see (10.46)] when we move x, y towards the distant future keeping the geodesic distance between x, y fixed. Such states are, by construction, dense in $\mathcal{H}$. The exponential decay corresponds, physically, to the "no-hair property" of de Sitter spacetime. As one can show with considerably more effort, that behavior persists for interacting quantum field theories, see [32–35].

c) Hawking effect: The algebraic formalism can be used to give a conceptually clear explanation of the Hawking effect. In fact, there are two closely related, but distinct, results that are commonly referred to as the "Hawking effect."

The first result concerns maximally extended Schwarzschild spacetime (i.e., an "eternal black hole"). As is well known, the exterior region, $r > 2M$, of Schwarzschild spacetime

$$ds^2 = -(1 - 2M/r)dt^2 + (1 - 2M/r)^{-1} dr^2 + r^2(d\theta^2 + \sin^2\theta \, d\varphi^2), \quad M > 0, \quad (10.51)$$

may be extended by introducing the Kruskal coordinates

$$U = e^{-u/4M}, \quad V = e^{v/4M},$$ (10.52)

where

$$u = t - r_*, \quad v = t + r_*,$$ (10.53)

with $r_* = r + 2M \log(r/2M - 1)$. In Kruskal coordinates, the line element takes the form

$$ds^2 = \frac{32M^3 e^{-r/2M}}{r} \, dU \, dV + r^2 (d\theta^2 + \sin^2\theta \, d\varphi^2).$$ (10.54)

By considering arbitrary U, V compatible with $r > 0$, one obtains the maximally extended Schwarzschild spacetime shown in the conformal diagram Fig. 10.3. The surfaces $U = 0$ and $V = 0$ (corresponding to $r = 2M$) comprise a bifurcate Killing horizon, $\mathcal{H}^{\pm}$, of the Killing field $\xi = \partial/\partial t$, analogous to the bifurcate Killing horizons of the boost Killing field of Minkowski spacetime and the static Killing field of de Sitter spacetime. In close analogy with those cases, there exists [36] a unique [18] Hadamard state, ω, on extended Schwarzschild spacetime that is stationary i.e., invariant under time-translation automorphisms, $\omega = \omega \circ \alpha_t$. This state is known as the "Hartle–Hawking vacuum" and is analogous to the Minkowski vacuum in Minkowski spacetime and to the Bunch–Davies vacuum in de Sitter spacetime. By the same argument as in those cases [18], when restricted to the original Schwarzschild wedge, $r > 2M$, the Hartle–Hawking vacuum is a KMS state at the Hawking temperature

$$T_H = \frac{\kappa}{2\pi} = \frac{1}{8\pi M},$$ (10.55)

where $\kappa = 1/(4M)$ is the surface gravity of the Killing horizon.

Two other states of interest on extended Schwarzschild spacetime are the "Boulware vacuum" [37], and the "Unruh vacuum" [38]. The Boulware vacuum is defined in the right wedge of extended Schwarzschild spacetime, where it is a ground state with respect to the

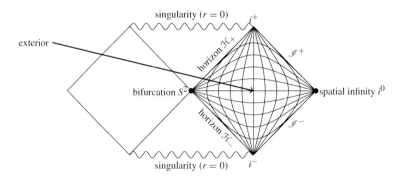

Figure 10.3 Conformal diagram of extended Schwarzschild spacetime ('eternal black hole').

timelike Killing vector field ∂_t. It is Hadamard in the right wedge, but cannot be extended as a Hadamard state beyond the right wedge, i.e. it would become singular on the past and future horizons. The Unruh vacuum is defined on the union of the right wedge and the interior of the black hole. In the right wedge, it can be thought of as a KMS state with respect to a subalgebra of $\mathscr{A}$ corresponding to the modes that are outgoing from the white hole, whereas it is the ground state with respect to a subalgebra corresponding to the modes that are incoming from past null infinity. The Unruh vacuum has been rigorously constructed in [39], and has been shown to be Hadamard on the union of the right wedge and the black hole regions. It cannot be extended as Hadamard state beyond the past horizon.

The thermal nature of the Hartle–Hawking state suggests, but does not imply, a second key result: Black holes formed by gravitational collapse will emit thermal radiation. To analyze this issue, one must consider the much more physically relevant case of a space-time in which gravitational collapse to a Schwarzschild black hole occurs, rather than the maximally extended Schwarzschild spacetime considered above. In the case of a black hole formed by collapse, one can show that if the state of the quantum field is Hadamard and if it approaches the ground state near spatial infinity, then at late times it contains quanta of *radiation appearing to emanate from the black hole*, distributed according to a Planck distribution with temperature (10.55).

We now give some details, following the argument given by Fredenhagen and Haag [40]. Let $F^T_{\nu lm}$ be a solution to the Klein–Gordon equation with smooth initial data of compact support on the gravitational collapse spacetime which has Y_{lm} angular dependence and frequency peaked sharply near $\nu > 0$ (with respect to the time-like Killing field), and which corresponds at late times to an outgoing wave reaching null infinity at retarded time centered about T. We normalize $F^T_{\nu lm}$ so that it has unit Klein–Gordon norm.[12] Then, in any state ω, the quantity $\omega(\phi[F^T_{\nu lm}]^*\phi[F^T_{\nu lm}])$ has the interpretation of being the "number of particles" in the mode $F^T_{\nu lm}$ as seen by a distant oberver at late times, as can be seen from Fock representation formulas (see Section 10.2.2 above) or by considering the behavior of model particle detectors [11, 41]. We shall show that if ω is Hadamard and approaches the ground state near spatial infinity, then

$$\lim_{T\to\infty} \omega(\phi[F^T_{\nu lm}]^*\phi[F^T_{\nu lm}]) = \frac{|D_l(\nu)|^2}{e^{2\pi\nu/\kappa}-1}, \tag{10.56}$$

where $D_l(\nu)$ is the amplitude for the absorption by the black hole of a mode of angular dependence Y_{lm} and frequency ν. This is precisely the expected number of particles that one would have for black-body radiation "emitted" by the black hole at the Hawking temperature.

To show this, we choose a partial Cauchy surface Σ_0 intersecting the future horizon $\mathcal{H}^+$ at a 2-sphere $\mathscr{S} \cong S^2$ outside of the collapsing star (see Fig. 10.4). In the future

---

[12] The Klein–Gordon inner product between two solutions F, G of compact support on a Cauchy-surface Σ is defined as $(F, G) = i\int_\Sigma (\bar{F}\nabla_\mu G - G\nabla_\mu \bar{F})n^\mu\, dS$; compare this with (10.10).

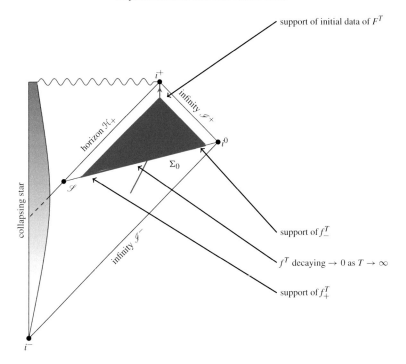

Figure 10.4   Conformal diagram of collapsing star spacetime.

domain of dependence $D^+(\Sigma_0)$ of Σ_0, the spacetime metric is precisely equal to that of the Schwarzschild metric. The conditions on the state ω imply that

(1) The Hadamard condition in the form (10.17) holds for the 2-point function $W_2(x_1,x_2)$ for x_1,x_2 in an open neighborhood of the horizon $\mathcal{H}^+$. Together with the fact that the coefficient 'u' in the Hadamard expansion satisfies $u(x_1,x_2) \equiv 1$ for all $x_1,x_2 \in \mathcal{H}^+$, one can infer that

$$W_2(x_1,x_2) = \frac{1}{2\pi^2(\sigma + it0^+)} + \text{lower order terms in } \sigma \text{ near } \mathcal{H}^+. \qquad (10.57)$$

(Here, lower order refers to the "scaling degree"; see Appendix 10.3.3.)

(2) In an open neighborhood of Σ_0, and for $r_1,r_2 \to \infty$, the 2-point function $W_2(x_1,x_2)$ approaches that of the ground state, i.e., the Boulware vacuum.

The key idea in the derivation is to replace the solution F^T_{vlm} by a test function f^T_{vlm} which is supported close to Σ_0. To do so, we let ψ be a smooth function which is equal to 1 slightly in the future of Σ_0, and equal to 0 slightly in the past of Σ_0, and we set

$$f^T_{vlm} = \Box_g[\psi F^T_{vlm}]. \qquad (10.58)$$

Then we have $\phi(f^T) = \phi[F^T]$ (see eq. (10.10)), and, because at least one derivative must hit ψ, f^T is supported near Σ_0. The key property of f^T is that it decays uniformly for large T in any region $r_1 < r < r_2$ where $r_1 > 2M$ [42]. Thus, in this limit, f^T splits approximately into two parts $f^T_+ + f^T_-$ (see Fig. 10.4), where the first part, f^T_+, is supported close to the horizon $\mathscr{S}$, and the second part, f^T_-, is supported close to spatial infinity i^0 near Σ_0. For f^T_+, one finds to the appropriate accuracy for $T \gg 1$

$$f^T_+(U, V) \sim D_l(\omega)\, \partial_V \psi(V)\, \partial_U \exp\left(\frac{i\nu}{\kappa} \log(Ue^{\kappa T})\right) \tag{10.59}$$

where, for simplicity, we have taken ψ to be a function of V only near $\mathscr{S}$. Here the logarithmic dependence on U in the exponent can be traced back to the relationship $U = \exp \kappa u$ between "affine time" U and "Killing time" u on the past horizon $\mathscr{H}^-$ of extended Schwarzschild spacetime (see (10.52)). The contribution of f^T_+ to (10.56) is given by

$$\lim_{T \to \infty} \omega(\phi[F^T_{\nu lm}]^* \phi[F^T_{\nu lm}]) = \lim_{T \to \infty} \int W_2(x_1, x_2) f^T_+(U_1, V_1, \mathbf{x}_1)^* f^T_+(U_2, V_2, \mathbf{x}_2)\, dv_1\, dv_2 \tag{10.60}$$

On the other hand assumption (2) implies that f^T_- makes no contribution in the limit $T \to \infty$, either in its direct terms or in its cross-terms with f^T_+. Since f^T_+ is supported increasingly close to $\mathscr{H}^+$, we may use assumption (1) to approximate W_2 with (10.57). Using furthermore (10.59), a short calculation then gives the thermal distribution formula (10.56).

The above argument corresponds closely to Hawking's original derivation [43], showing that the expected number of particles seen by a distant observer at late times corresponds to thermal emission by the black hole. In fact, all aspects of this radiation are thermal [44]. The precise result is that any state that is Hadamard and behaves like a ground near spatial infinity will approach the Unruh vacuum state, in the sense that the n-point functions of the state approach those of the Unruh vacuum in the exterior at late times.

We have presented the argument in the above manner to emphasize the following points: (i) The derivation of the Hawking effect does not depend on introducing a notion of "particles" near the horizon. (ii) No assumptions need to be made on the initial state other than that it is regular (Hadamard) and approaches the ground state near spatial infinity (i.e., there is no "incoming radiation" at late times). (iii) The Hawking effect follows from causal propagation of the quantum field outside of the black hole; one does not need to make any assumptions about what is happening inside of the black hole. In particular, any breakdown of known laws of physics in the high curvature regime near the singularity deep inside the black hole should not affect the validity of the derivation. (iv) The details of the collapse are not important; all that matters is that the spacetime metric asymptotically approach the Schwarzschild metric[13] by some sufficiently late "time" Σ_0.

---

[13] The results can be straightforwardly generalized to asymptotic approach to other stationary black hole geometries, with the only significant difference being that, for a rotating black hole, the horizon Killing field will now be a linear combination of a time translation and rotation at infinity; see, e.g. [43, 44] for a discussion of the Kerr case.

Nevertheless, there is one potentially disturbing aspect of this derivation. For large T, it can be seen from (10.59) that f_+^T is peaked near $U \sim \exp(-\kappa T)$, and that the locally measured frequency of f_+^T – say, as seen by observers who freely fall into the black hole from rest at infinity – diverges as $\exp(\kappa T)$ as $T \to \infty$. For any reasonable detector frequency ν, this vastly exceeds the Planck frequency for $T \gg 1/\kappa$. We cannot expect QFTCS to be a good approximation to nature on transplanckian scales, but the above derivation of the Hawking effect appears to depend upon validity of QFTCS at transplanckian frequencies. However, we claim now that this is not actually the case: For a given $F_{vlm}^T(r,t,\varphi,\theta)$ at large T, instead of doing the analysis on the partial Cauchy surface Σ_0, we can work on a partial Cauchy surface Σ_1, which is sufficiently early in time that the approximation leading to (10.59) still holds, but is sufficiently late in time that f_+^T on Σ_1 is not transplanckian. By formulating the detector response at time T as an evolution problem starting from Σ_1 rather than Σ_0, one avoids[14] any elements of the derivation that allude to phenomena at transplanckian scales.

The derivation of the emission of thermal radiation by a black hole at the Hawking temperature (10.55) brought to a culmination a remarkable relationship between certain laws of black hole physics and the ordinary laws of thermodynamics. It was already known prior to this derivation that classical black holes satisfy mathematical analogs of the zeroth, first, and second laws of thermodynamics [45], with mass, M, playing the role of energy, E; surface gravity, κ, playing the role of temperature T; and horizon area, A, playing the role of entropy, S. Even in classical physics, a hint that this relationship might be more than a mathematical analogy is provided by the fact that M and E are, in fact, the same physical quantity. However, the physical temperature of a classical black hole is absolute zero, thereby spoiling this relationship in classical physics. The fact that, when analyzed from the perspective of QFTCS, black holes have a finite temperature (10.55) proportional to surface gravity strongly suggests that these laws of black hole physics must actually be the laws of thermodynamics applied to black holes; see [11, 46] for further discussion. In particular, $A/4$ must represent the physical entropy of a black hole. As discussed in Chapters 11 and 12, ramifications of these ideas continue to be explored.

In the above derivation of thermal emission by a black hole, we considered a fixed, classical space-time corresponding to the gravitational collapse of a body to a Schwarzschild black hole. However, the quantum field has a stress-energy observable, $T_{\mu\nu}$ (see the next section), and, in semiclassical gravity, $\omega(T_{\mu\nu})$ should contribute to the right side of Einstein's equation. It is easy to see that $\omega(T_{\mu\nu})$ contributes a positive energy flux at infinity. It follows from conservation of stress-energy together with the approximate stationarity of ω at late times that $\omega(T_{\mu\nu})$ contributes a corresponding flux of negative energy[15] into the black hole. Consequently, if the black hole is isolated (so that there is no

[14] Of course, the Hadamard condition itself concerns arbitrarily short distance singularity structure and thus, in effect, involves transplanckian scales. One might therefore question the validity of QFTCS for arguing that the Hadamard condition is preserved under evolution. However, this question could equally well be raised in Minkowski spacetime.

[15] Negative energy fluxes of stress-energy or negative energy densities can occur in quantum field theory even for fields that classically satisfy the dominant energy condition, see e.g. [47, 48].

other flux of stress-energy into the black hole), the black hole will slowly lose mass as a result of the quantum field effects. An order of magnitude estimate of the mass loss of the black hole can be obtained from the Stefan–Boltzmann law

$$\frac{dM}{dt} \sim AT^4 \sim M^2 \frac{1}{M^4} = \frac{1}{M^2} \tag{10.61}$$

leading to the prediction that a black hole should "evaporate" completely[16] in a time of order M^3 (in Planck units). The prediction of black hole evaporation gives rise to an issue that has regained considerable attention recently. In the analysis of quantum field theory on the gravitational collapse spacetime shown in fig. 10.4 – where the black hole remains present forever – the field observables in an open neighborhood of any Cauchy surface comprise the entire algebra,[17] $\mathscr{A}$. Thus, a state that is pure at any initial "time" (i.e., a neighborhood of an initial Cauchy surface, say, prior to the collapse) automatically will remain pure at any final "time" (i.e., a neighborhood of a final Cauchy surface, say, after the black hole has formed). However, the field observables in a small neighborhood of a *partial* Cauchy surface – i.e., a hypersurface, such as Σ_0 in fig. 10.4, whose future domain of dependence includes the region exterior to the black hole but not the interior of the black hole – do not comprise all observables in $\mathscr{A}$ since there will be additional field observables inside the black hole. Furthermore, for any Hadamard state, there always are strong correlations between the field observables at small spacelike separations. In particular, in any Hadamard state, the observable $\phi(f^T)$ on Σ_0 in our above discussion will be highly entangled with corresponding field observables inside the black hole. Since $\phi[F^T] = \phi(f^T)$ (see (10.10)), this means that the Hawking radiation flux measured by a distant detector is highly entangled with field observables inside the black hole. If the black hole subsequently completely evaporates as discussed above, the field observables corresponding to the emitted radiation remain entangled with observables inside the (now non-existent) black hole. In particular, the subalgebra of observables that can be measured at late times – after the black hole has evaporated – do not comprise a complete set of observables, and the restriction of the state ω to this subalgebra is a mixed state. Thus, in the process of black hole formation and evaporation, an initial pure state will evolve to a final mixed state. Such an evolution does not violate any principles of quantum theory or any known laws of physics – indeed, it is derived by a straightforward application of QFTCS to a spacetime in which a black hole forms and evaporates – but it is in apparent conflict with ideas suggested by the AdS/CFT correspondence. In any case, there is a widespread belief that evolution from a pure state to a mixed state should not happen, thereby requiring a drastic modification of QFTCS in a low curvature regime, where, a priori, one otherwise would have very little reason to question its validity. In particular, in order to avoid entanglement between observables outside the black hole and inside the

---

[16] Of course, the approximate description leading to this prediction should be valid only when $M \gg M_P$, where M_P denotes the Planck mass ($\sim 10^{-5}$ g), but modifications to the evaporation process at this stage (including the possibility of Planck mass remnants) would not significantly alter the discussion below.

[17] This result, sometimes called the "time-slice property", continues to hold for the enlarged algebra $\mathscr{W}$ defined in the next section and also for the algebra of interacting fields $\mathscr{B}_I$ [49, 50].

black hole, the state must fail to be Hadamard at the event horizon of the black hole, thereby converting the event horizon to a singularity (a "firewall"). On the other hand, as we have seen above, the Hawking effect itself is crucially dependent upon the state being Hadamard arbitrarily close to (but outside of) the event horizon. Given that the event horizon is not locally determined – i.e., it requires knowledge of the future evolution of the spacetime – it would seem a daunting task for a quantum field to know, with the required precision, exactly when to stop obeying the laws of QFTCS, so that Hawking radiation is maintained but its entanglement with field observables inside the black hole is broken. Nevertheless, there is at present a widespread belief that a quantum field will somehow manage to do this – or that the currently known local laws of physics will be violated near the horizon of a black hole in some other way, so as to maintain the purity of the final state.

d) Cosmological perturbations: We would now like to investigate a massless ($m^2 = 0$) Klein–Gordon field propagating on an FLRW space-time with flat slices

$$g = -dt^2 + a(t)^2(dx_1^2 + dx_2^2 + dx_3^2) \,. \tag{10.62}$$

The isometry group of this space-time is, for general $a(t)$, the Euclidean group $E(3)$ acting on the spatial coordinates $\mathbf{x} \in \mathbb{R}^3$. We consider a scale factor of the form

$$a(t) = \begin{cases} e^{H_0 t} & \text{for } t \le t_0, \\ a_0(t/t_0)^p & \text{for } t > t_0, \end{cases} \tag{10.63}$$

describing a universe which is expanding exponentially first (inflation), followed by an era with power-law expansion factor, assumed for simplicity to last forever. To be precise, we should actually choose a scale factor that interpolates smoothly, rather than just continuously, between the epochs, but for the rough calculation this will not be needed. We could also add an earlier epoch of power law expansion prior to the exponential expansion, but this will not affect our results provided that the era of exponential expansion lasts sufficiently long.

We consider a massless quantum Klein–Gordon scalar field in the spacetime (10.63). This is a (slight) simplification of the more physically relevant problem of starting with a classical solution of the Einstein-scalar-field system with scale factor of a form approximating (10.63) – as would occur if the scalar field "slowly rolls" down an extremely flat potential – and then treating the linearized perturbations of this system as quantum fields. In that case, as explained in Chapter 3, the linearized perturbations decouple into "scalar modes" and "tensor modes," each of which behaves similarly to a scalar field in the background spacetime (10.63), see e.g. [51, 52] for more detailed reviews. Thus, consideration of a scalar field in the background spacetime (10.63) suffices to derive the general form of the power spectrum of perturbations resulting from inflation.

Consider a Gaussian, pure, $E(3)$-invariant, Hadamard state of the scalar field on the spacetime (10.63). The two-point function, W_2, of such a state may be described by a set of mode functions $u_\mathbf{k}(t, \mathbf{x}) = \chi_\mathbf{k}(t) \exp(i\mathbf{k}\mathbf{x})$ as in eq. (10.24). Let us normalize the mode functions so that the Wronskian is i times unity, i.e.,

$$i = a(t)^3 \left(\chi_{\mathbf{k}}(t) \frac{d}{dt} \overline{\chi}_{\mathbf{k}}(t) - \overline{\chi}_{\mathbf{k}}(t) \frac{d}{dt} \chi_{\mathbf{k}}(t) \right) . \tag{10.64}$$

This condition ensures that eq. (10.23) holds with $X = \mathbb{R}^3, d\mu(\mathbf{k}) = d^3\mathbf{k}/(2\pi)^3$ so that eq. (10.24) indeed defines the 2-point function of a state,

$$W_2(t_1, \mathbf{x}_1, t_2, \mathbf{x}_2) = \frac{1}{(2\pi)^3} \int \chi_{\mathbf{k}}(t_1) \overline{\chi_{\mathbf{k}}(t_2)} \, e^{i\mathbf{k}(\mathbf{x}_1 - \mathbf{x}_2)} \, d^3\mathbf{k} . \tag{10.65}$$

A quantity that (partly) characterizes W_2 in a Robertson–Walker spacetime is its "power spectrum," $P(t, \mathbf{k})$, which is defined in terms of the spatial Fourier-transform of the 2-point function at equal time t,

$$\hat{W}_2(t, \mathbf{k}, t, \mathbf{p}) = (2\pi |\mathbf{k}|)^3 \delta^3(\mathbf{k} - \mathbf{p}) \, P(t, \mathbf{k}) . \tag{10.66}$$

For an $E(3)$-invariant state, the power spectrum $P(t, k)$ depends only upon the modulus $k = |\mathbf{k}|$ of the wave number (and of course t).

We shall now show that for any $t > t_0$ (i.e., after inflation has ended), the power spectrum $P(t, k)$ for modes for which[18] $k/a(t) \ll 1/R(t)$, where R denotes the Hubble radius, $R(t) \equiv a(t)/\dot{a}(t)$, is approximately given by

$$P(t, k) \propto H_0^2 . \tag{10.67}$$

Thus, the power spectrum is "scale-free," with amplitude set by the scale of inflation. To show this, we observe that the Hadamard condition fixes the asymptotic behavior of $\chi_{\mathbf{k}}(t)$ for large $|\mathbf{k}| \to \infty$ (to all(!) asymptotic orders). In the inflationary epoch ($t < t_0$), the general solution for $\chi_{\mathbf{k}}$ giving rise to an $E(3)$-invariant state of the massless field is

$$\chi_{\mathbf{k}}(\eta) = A_k f_k(\eta) + B_k \overline{f}_k(\eta) , \tag{10.68}$$

in conformal time $\eta = \int dt/a(t)$, where

$$f_k(\eta) = \text{const.} \, \eta^{\frac{3}{2}} \, H^{(2)}_{-\frac{3}{2}}(k\eta) , \tag{10.69}$$

where $H^{(2)}_\alpha$ denotes a Hankel function, and the Wronskian condition imposes the relation $|A_k|^2 - |B_k|^2 = 1$. The Hadamard condition requires $A_k \to 1$, $B_k \to 0$ at large k. If we assume that $A_k \approx 1$, $B_k \approx 0$ for all "short-wavelength modes" (i.e., $k/a(t_1) > H_0$) at some time $t_1 < t_0$ during the inflationary cra, then we have $A_k \approx 1$, $B_k \approx 0$ for all modes[19] whose physical wavelength is smaller than $H_0^{-1} \exp[(t_0 - t_1)H_0]$ at the end of inflation. For $(t_0 - t_1)H_0 \gtrsim 60$, this encompasses all wavelengths relevant for cosmology.

---

[18] Such modes having wavelength larger than the Hubble radius in the present universe are not of observational interest. However, modes whose wavelength was larger than the Hubble radius at the end of inflation but is smaller than the Hubble radius in the present universe are highly relevant to cosmology.

[19] For $m^2 > 0$, the analogous modes are obtained by setting the index of the Hankel function to $\alpha = (\frac{9}{4} - m^2 H^{-2})^{1/2}$. The state with $A_k = 1, B_k = 0$ for all k is the de Sitter invariant Bunch–Davies state [53]. However, the choice $A_k = 1, B_k = 0$ for $m = 0$ would yield an infrared divergence in the two-point function (10.65).

The mode functions compatible with $E(3)$-invariance in the power law epoch with $a(t) \propto t^p$ have the form

$$\tilde{\chi}_{\mathbf{k}}(\eta) = \tilde{A}_k \tilde{f}_k(\eta) + \tilde{B}_k \overline{\tilde{f}}_k(\eta) , \qquad (10.70)$$

where

$$\tilde{f}_k(\eta) = \text{const. } \eta^{\frac{1-3p}{2(1-p)}} \, \mathrm{H}^{(2)}_{-\frac{1-3p}{2(1-p)}} (k\eta) , \qquad (10.71)$$

The coefficients $\tilde{A}_k, \tilde{B}_k$ are subject to the Wronskian condition $|\tilde{A}_k|^2 - |\tilde{B}_k|^2 = 1$ and are determined by matching the modes $\chi_{\mathbf{k}}$ in eq. (10.68) to $\tilde{\chi}_{\mathbf{k}}$ at time η_0. To do this, we first note that, by assumption, we are considering modes that satisfy $R(t)k/a(t) \ll 1$. During the power law epoch, we have $a(t) \propto t^p$ with $p < 1$, and $R(t) \propto t$, so the quantity $R(t)k/a(t)$ becomes even smaller as we go back in time from t to t_0. In conformal coordinates, $1 \gg R(t)k/a(t) \propto k\eta$, so this means that the modes $\tilde{\chi}_{\mathbf{k}}(\eta)$ of interest are essentially constant ("frozen") during the power law epoch for all times before t. Thus, we may assume that $\tilde{\chi}_{\mathbf{k}}(\eta) \sim \tilde{\chi}_{\mathbf{k}}(\eta_0) = \chi_{\mathbf{k}}(\eta_0)$ during that epoch, so the power spectrum at t is essentially the same as that of the state in de Sitter spacetime at time t_0 for $k\eta \ll 1$, i.e., that obtained using the modes $\chi_{\mathbf{k}}(\eta_0)$. These modes may then be approximated by $\chi_{\mathbf{k}}(\eta) \propto k^{-3/2} H_0$ for $k\eta \ll 1$, thus giving rise to the desired power spectrum (10.67). At this level of approximation, the power spectrum is independent of the power p and of the precise nature of the transition period, although the finer properties of the power spectrum would depend on such details.

As discussed in Chapter 3, for $H_0 \sim 10^{16}\,\text{GeV}$, the amplitude of the power spectrum is macroscopically large, and it provides an explanation of the observed temperature fluctuations in the cosmic microwave background as well as of "structure formation" in the universe, i.e., it produces density perturbations appropriate to act as "seeds" for the formation of clusters of galaxies and galaxies. The fact that, in the presence of exponential expansion, the short-distance quantum fluctuations of fields in the very early universe can produce macroscopically observable effects in the present universe is one of the most remarkable predictions of QFTCS.

10.3 Beyond linear quantum fields

10.3.1 Construction of nonlinear observables for a free quantum scalar field

The construction of the theory of a free quantum field in curved spacetime given in the previous section provides a mathematically consistent and satisfactory formulation of QFTCS for fields obeying linear equations of motion. However, even in this case, the theory is incomplete: The observables represented in the algebra $\mathscr{A}(\mathscr{M}, g)$ consist only of the smeared fields $\phi(f)$ and their correlation functions (10.11) in some state ω. However, $\mathscr{A}$ does not include any observables corresponding to nonlinear functions of the field ϕ, such as powers ϕ^k or the stress-energy tensor of ϕ. For this reason alone, one would like to enlarge the algebra of observables $\mathscr{A}(\mathscr{M}, g)$ to an algebra that, at the very least, includes smeared versions of all polynomial expressions in ϕ and its spacetime derivatives, as well as correlation function observables of these expressions. The definition of such an algebra, called $\mathscr{W}(\mathscr{M}, g)$, has been given first in [50], and the straightforward generalization to curved space in [54]. This algebra contains, for instance, the smeared "normal ordered products", $:\phi^k(f):_\omega$, relative to any fixed quasi-free Hadamard state, see e.g. Section 8.3 of [3] for the standard combinatorial formulae defining a normal product relative to a 2-point function such as, in this case, $W_2(x, y) = \omega(\phi(x)\phi(y))$. Their correlation functions are

$$\omega\left(\prod_{j=1}^n :\phi^{k_j}(x_j):_\omega\right) = \sum_{\text{graphs } \mathscr{G}} c_{\mathscr{G}} \prod_{(ij)\in\text{edge}(G)} W_2(x_i, x_j), \qquad (10.72)$$

where the sum is over all graphs $\mathscr{G}$ on n vertices $x_1, \ldots, x_n$ with coordination numbers $k_1, \ldots, k_n$ without "tadpoles", and where $c_{\mathscr{G}}$ is a symmetry factor. A key point, observed in [16], is that the right hand side makes sense: Even though products of distributions are in general not naturally well-defined, it can be shown, using microlocal techniques and the Hadamard condition (10.18), that the right hand side actually makes perfect sense as a distribution. (For details on the use of microlocal arguments of this sort see Appendix 10.3.3.)

Although quantities like $:\phi^k:_\omega$, or similarly e.g. the normal ordered stress tensor, $:T_{\mu\nu}(x):_\omega$, are thus perfectly well-defined as algebra-valued distributions (now in an enlarged algebra), they are not a physically acceptable definition, because their definition depends upon an arbitrary choice of Hadamard state ω. A related problem is that a state is necessarily a *globally defined* object (depending e.g. on a global choice of "positive frequency modes"), and we would like to define nonlinear observables in a *local and covariant manner*, i.e. in such a way that their definition depends only on the local geometry near a given point. This notion can be made precise by demanding, in essence, that a local and covariant quantum field, Φ, should be consistently defined on *all* globally hyperbolic spacetimes, $(\mathscr{M}, g)$. Consistency conditions arise because one can consider isometric embeddings $\psi : \mathscr{M} \to \mathscr{M}'$ (i.e., $\psi^* g' = g$) that also preserve the causal structure – so that if x_1 and x_2 cannot be connected by a causal curve in $\mathscr{M}$, then $\psi(x_1)$ and $\psi(x_2)$ cannot be connected by a causal curve in $\mathscr{M}'$. Since, as one can show, there is natural isomorphism $\alpha_\psi : \mathscr{W}(\mathscr{M}, g) \to \mathscr{W}(\mathscr{M}', g')$, the consistency condition should be [55]

$$\alpha_\psi[\Phi_{\mathscr{M}'}(f)] = \Phi_{\mathscr{M}}(f \circ \psi^{-1}). \qquad (10.73)$$

The normal ordering prescription $\Phi_{\mathscr{M}} = :\phi^k:_{\omega[\mathscr{M}]}$ is shown *not* to give a local and covariant field [54], no matter how we choose the Hadamard state $\omega[\mathscr{M}]$ for each $(\mathscr{M}, g)$. However, there is a relatively simple remedy [54]: One simply replaces ω by the locally and covariantly constructed Hadamard distribution $H_N(x_1, x_2)$ [see (10.17)]. The expansion order N must be chosen to be greater than the highest derivative in Φ (in this case $N = 1$), but is otherwise arbitrary. The resulting "locally normal ordered" field is local and covariant and can be expressed e.g. through the "reordering formula" (see e.g. Section 8.5 of [3] for such formulas)

$$: \phi^k(x) :_{H_N} = \sum_{n=0}^{\lfloor k/2 \rfloor} \frac{k!}{n!(k-2n)!} c_\omega(x)^n : \phi^{k-2n} :_\omega . \tag{10.74}$$

The function $c_\omega(x) = -\frac{1}{2}R_{N,\omega}(x,x)$ [see (10.17)] is closely related to the "Casimir effect". The definition given above is not the only way to define ϕ^k as a local and covariant field, but it is possible to show that the most general definition can only differ from this one by "lower order Wick products" and "local curvature terms" (i.e. polynomials of $\nabla_{\alpha_1} \ldots \nabla_{\alpha_N} R_{\mu\nu\sigma\rho}$ of the appropriate "dimension") [54].

10.3.2 Time ordered products

The situation with regard to defining time ordered products of polynomial expressions – as needed to define perturbative interacting quantum field theory – is similar to the above problem of defining Wick powers, although it is considerably more complicated. Let Φ_i denote a general monomial in the field ϕ, its covariant derivatives, and possibly curvature terms (considered merely as a "classical expression"). Denoting the space of such classical expressions for definiteness by $\mathbf{P}$, the time ordered product in n factors can be viewed as a map, T_n, from the n-fold tensor product $\mathbf{P}^{\otimes n} \equiv \mathbf{P} \otimes \cdots \otimes \mathbf{P}$ into distributions in n variables valued in $\mathscr{W}$. We will denote time ordered products by $T_n(\Phi_1(x_1) \otimes \cdots \otimes \Phi_n(x_n))$. For one factor, we define $T_1(\phi^k)$ (i.e. $\Phi = \phi^k$) simply by (10.74). We would then like to define $T_n, n > 1$ by "time ordering" the product of fields, so, e.g., for $n = 2$ we would like to set

$$T_2(\Phi_1(x_1) \otimes \Phi_2(x_2)) = \begin{cases} \Phi_1(x_1)\Phi_2(x_2) & \text{if } x_1 \notin J^-(x_2) \\ \Phi_2(x_2)\Phi_1(x_1) & \text{if } x_2 \notin J^-(x_1) . \end{cases} \tag{10.75}$$

(If x_1 and x_2 cannot be connected by a causal curve, then $\Phi_1(x_1)$ and $\Phi_2(x_2)$ commute, so either formula may be used.) The problem is that $\Phi_i(x_i)$ are distributions, so (10.75) provides a definition of $T_2(\Phi_1(f_1) \otimes \Phi_2(f_2))$ only when the supports of the test functions f_1 and f_2 satisfy the relations $\text{supp} f_1 \cap J^-[\text{supp} f_2] = \emptyset$ or $\text{supp} f_2 \cap J^-[\text{supp} f_1] = \emptyset$. It is not difficult to see that this enables us to straightforwardly define $T_2(\Phi_1(f_1) \otimes \Phi_2(f_2))$ whenever $\text{supp} f_1 \cap \text{supp} f_2 = \emptyset$. However, (10.75) makes no sense when $\text{supp} f_1 \cap \text{supp} f_2 \neq \emptyset$. Thus, we must extend the distribution (10.75) to the "diagonal" $x_1 = x_2$. This may seem like a relatively trivial problem, but the extension of the definition of general time ordered products $T_n(\Phi_1(x_1) \otimes \cdots \otimes \Phi_n(x_n))$ to the "total diagonal" $x_1 = \cdots = x_n$ is the main problem of renormalization theory in flat and curved spacetime [56, 57]. We refer to a definition/construction of T_n as a "renormalization scheme."

We proceed by writing down a list of properties that T_n should satisfy. We require T_n to satisfy the appropriate generalization of (10.75) (involving lower order time-ordered products) away from the total diagonal. As in the case of Wick powers (i.e. time-ordered products with one factor), we require T_n to be locally and covariantly defined, to satisfy appropriate commutation relations with Φ, to have appropriate continuous/analytic dependence on the metric, and to have appropriate scaling behavior (up to logarithmic terms) under scaling of the metric. We also require commutation of T_n with derivatives. Finally, we impose a number of additional conditions on T_n, specifically, a microlocal spectrum condition (eq. (10.89) of Appendix 10.3.3), a "unitarity" condition similar to the "optical theorem", and conditions that guarantee that the perturbatively defined interacting field – to be defined – (i) satisfies the interacting field equation and (ii) has a conserved stress-energy tensor. We refer the reader to [54, 58, 59] for a more complete and extensive discussion of all of the conditions imposed on T_n.

It was proven in [58] – key parts of which were based on [60] – that there exists a definition of T_n, that satisfies all of the above conditions. Furthermore, T_n is unique up to "appropriate local and covariant counterterms." To explain this freedom in the choice of T_n, we must introduce a considerable amount of additional notation: We denote by $\mathbf{P}(\mathscr{M}^n)$ the space of all distributional local, covariant functionals of ϕ (and its covariant derivatives), of g, and of the Riemann tensor (and its covariant derivatives), which are supported on the total diagonal (i.e. of delta-function type). Let $F = \lambda \int f\,\Phi$ be an integrated local functional $\Phi \in \mathbf{P}(\mathscr{M})$, and formally combine the time-ordered functionals into a generating functional written

$$\mathrm{T}(\exp_\otimes(F)) = \sum_{n=0}^\infty \frac{1}{n!} \mathrm{T}_n(F^{\otimes n}) \in \mathscr{W}[\![\lambda]\!]\,, \tag{10.76}$$

where $\exp_\otimes$ is the standard map from the vector space of local actions to the tensor algebra over the space of local action functionals, and $\mathscr{W}[\![\lambda]\!]$ denotes the algebra of formal power series expressions in $\mathscr{W}$. Let D denote a hierarchy D_n of linear functionals D_n : $\mathbf{P}(\mathscr{M}) \otimes \cdots \otimes \mathbf{P}(\mathscr{M}) \to \mathbf{P}(\mathscr{M}^n)$. We similarly write $\mathrm{D}(\exp_\otimes(F))$ for the corresponding generating functional obtained from D_n. We then have the following theorem:

Theorem 10.3.1 *[49, 54, 58, 60] (Uniqueness) If* T_n *and* $\hat{T}_n$ *are two different renormalization schemes, both satisfying our conditions, then they are related by*

$$\hat{\mathrm{T}}(\exp_\otimes(iF)) = \mathrm{T}\left(\exp_\otimes\left[iF + i\mathrm{D}(\exp_\otimes F)\right]\right) \tag{10.77}$$

for any $F = \lambda \int f\Phi$, $\Phi \in \mathbf{P}(\mathscr{M})$ *and* $f \in C_0^\infty(\mathscr{M})$. *The functionals* D_n *satisfy the following:*

  (i) $\mathrm{D}(e_\otimes^F) = O(\hbar)$ *if we reintroduce* $\hbar$.
 (ii) *Each* D_n *is locally and covariantly constructed from* g.
(iii) *Each* D_n *is an analytic functional of* g.
 (iv) *Each* $D_n(\Phi_1(x_1) \otimes \ldots \otimes \Phi_n(x_n))$ *is a distribution that is supported on the total diagonal (= 'contact term' = 'delta-function type').*
  (v) *The maps* D_n *are real.*
 (vi) *Each* D_n *is symmetric.*
(vii) *Each* D_n *satisfies the natural dimension constraint.*
(viii) *Derivatives can be pulled into* D_n. *This restricts the ambiguities of time ordered products of fields which are total derivatives.*

Conversely, if D_n *has these properties, then any* $\hat{T}$ *given by (10.77) defines a new renormalization scheme satisfying our conditions.*

The expressions D_n can be shown to correspond to the "counterterms" that characterize the difference between the two renormalization schemes at "n-th perturbation order".

Example: At order $n = 2$, the formula (10.77) gives (assuming for simplicity that $D_1 = 0$)

$$\hat{T}_2(\phi^2(x) \otimes \phi^2(y)) = T_2(\phi^2(x) \otimes \phi^2(y)) + T_1(D_2(\phi^2(x) \otimes \phi^2(y))) . \quad (10.78)$$

Since the scaling degree (see Appendix 10.3.3) of the delta function is 4 and the dimension of ϕ is 1, the conditions on D_2 stated in 10.3.1 imply that it must take the form

$$D_2(\phi^2(x) \otimes \phi^2(y)) = c_0 \delta(x, y) , \quad (10.79)$$

for some real constant c_0. Similarly

$$D_2(\phi^3(x) \otimes \phi^3(y)) = c_1 \delta(x, y) \phi^2(y) + (c_2 R + c_3 \Box_g + c_4 m^2) \delta(x, y) , \quad (10.80)$$

because the scaling degree of $\Box_g \delta(x, y)$ is 6, and the dimension of R is 2.

10.3.3 Interacting fields

Given the definition of local covariant Wick products (Section 10.3.1) and their time-ordered products (Section 10.3.2), one is in a position to perturbatively define the composite fields for the interacting field theory described by the Lagrangian density of the form $\mathscr{L} = \mathscr{L}_0 + \mathscr{L}_1$, where $\mathscr{L}_0$ is the free Lagrangian corresponding to the linear Klein–Gordon equation, and where, e.g., $\mathscr{L}_1 = \lambda \phi^4$ is an interaction.

The basic idea to construct the interacting fields is to initially "turn off" the interaction at some finite time in the past, so that the interacting field is equal to the local covariant Wick power in the free theory as defined in Section 10.3.1) at sufficiently early times. We then "evolve" this field forward in time into the region where the interaction is fully turned on. The resulting field is given by "Haag's series" [2, 57], eq. (10.82). Finally, we take a suitable limit where the "turn-on time" of the interaction is arbitrarily far in the past.

To implement this strategy we choose a cutoff function, θ, of compact support on $\mathscr{M}$ which is equal to 1 on an open neighborhood of some globally hyperbolic open region V with the property that $\Sigma \cap V$ is a Cauchy surface for V for some Cauchy surface Σ in $\mathscr{M}$. For $F = \lambda \int_{\mathscr{M}} \Phi(x) f(x) \, dv_g$ with $f \in C_0^\infty(\mathscr{M})$, we define the *local S-matrix* to be

$$S(F) = T(\exp_\otimes(iF)) \equiv \sum_{n \geq 0} \frac{i^n}{n!} T_n(F \otimes \cdots \otimes F) . \quad (10.81)$$

Then the interacting field, for the interacting theory with cutoff interaction $\mathscr{L}_1(\theta)$ corresponding to Φ is defined by [57]

$$\Phi(f)_{\mathscr{L}_1(\theta)} \equiv \frac{1}{i} \frac{d}{dt} S(\mathscr{L}_1(\theta))^{-1} S(t\Phi(f) + \mathscr{L}_1(\theta)) \bigg|_{t=0} . \quad (10.82)$$

Equations (10.81) and (10.82) are to be understood as formal series expressions that define the interacting field to any finite order in perturbation theory; no convergence properties are claimed. Note that the definition of $\Phi(x)_{\mathscr{L}_1(\theta)}$ has been adjusted so that it coincides with the corresponding free field $\Phi(x)$ before the interaction is "switched on," as can be seen explicitly by expressing it in terms of "totally retarded products" [49, 60, 61].

We now wish to remove the cutoff. Formula (10.82) will not, in general, make sense if we straightforwardly attempt to take the limit $\theta \to 1$, i.e. $V \to \mathscr{M}$. Indeed if θ could be set equal to 1 throughout the spacetime in eq. (10.82), then the resulting formula for $\Phi(f)_{\mathscr{L}_1(1)}$ would define an interacting field in the sense of Bogoliubov [57], with the property that the interacting field approaches the free field in the asymptotic past. However, even in Minkowski spacetime, it is far from clear that such an asymptotic limit of the interacting field will exist (particularly for massless fields), and it is much less likely that any such limit would exist in generic globally hyperbolic curved spacetimes that do not become flat in the asymptotic past. However, there exists a way of taking the limit $\theta \to 1$ such that the field remains fixed in regions of increasing size in the interior of the spacetime, see [49], which is a realization of the idea of the 'adiabatic algebraic limit' that appeared first in [60]. The key point is to sandwich $\Phi(x)_{\mathscr{L}_1(\theta)}$ between suitable unitaries (depending on θ) which in effect implement the idea of "keeping the interacting field fixed near a Cauchy surface". The collection of the so-obtained interacting fields then defines an algebra, $\mathscr{A}_{\mathrm{I}}(\mathscr{M}, g)$ of interacting fields in any globally hyperbolic spacetime generalizing the algebra $\mathscr{A}(\mathscr{M}, g)$ for the free Klein–Gordon field.

To summarize this chapter, over that last two decades, new mathematical methods and concepts have been introduced into QFTCS, which have brought the subject to a new level of clarity and rigor. In particular, it was understood how to treat, in a generally covariant way, the singularities that appear when defining composite fields within free field theories, as well as interacting fields. We believe that also flat space quantum field theory has benefited from these developments, because the framework of QFTCS serves to distinguish fundamental features of the theory from accidental ones related to Poincaré symmetry.

Acknowledgements

The research of S.H. was supported in part by ERC grant QC & C 259562. The research of R.M.W. was supported in part by NSF grant PHY 12-02718 to the University of Chicago.

Appendix: Distributions, scaling degree, and wave front sets

The objects appearing in quantum field theory such as n-point functions, time-ordered products, etc. are singular and therefore best viewed as distributions. A distribution u on a d-dimensional manifold X is a complex linear functional $u : C_0^\infty(X) \to \mathbb{C}$ for which there is a constant c_K and an $N \in \mathbb{N}_0$ for each compact $K \subset X$ such that

$$|u(f)| \leq c_K \sum_{k \leq N} \sup_{x \in K} |D^k f(x)| \qquad (10.83)$$

for any $f \in C_0^\infty$ having support within K, where D is any derivative operator on X. For instance, the delta "function" on $\mathbb{R}$ concentrated at 0, which is defined by $\delta(f) = f(0)$, evidently satisfies the above estimate with $N = 0$. The n-th derivative $\delta^{(n)}$ satisfies the criterion with $N = n$. Also, any smooth function u defines a distribution via $u(f) = \int u(x)f(x)dv$ for a given integration element dv on X. Such a distribution is called smooth, and more generally, a distribution is called smooth at $x_0 \in X$ if it can be represented in that way for f having support sufficiently close to x_0. The complement of the set of all such x_0 is called the "singular support" $\mathrm{singsupp}(u) \subset X$. The notion of singular support is not very informative since it gives no insight into the precise nature of the singularity at a given $x_0 \in \mathrm{singsupp}(u)$. This shortcoming can be dealt with by introducing more refined concepts to characterize singularities. Two such concepts of particular relevance for QFTCS are that of the *scaling degree* and that of the *wave front set*.

The *scaling degree* of a distribution u at a point $x \in X$ basically describes the "degree of divergence" at x, if any. It is defined more formally as follows. Choose an arbitrary chart (U, ψ) near x and let $u_\psi(f) = u(f \circ \psi^{-1})$ the pull-back of u to $\mathbb{R}^d$, defined for f supported in $\psi[U]$. Without loss of generality we may assume that $\psi(x) = 0$, and we define $f_\epsilon(y) = \epsilon^{-d} f(y/\epsilon)$. The scaling degree is given by

$$\mathrm{sd}_x(u) = \inf\{\delta \in \mathbb{R} \mid \lim_{\epsilon \to 0^+} \epsilon^\delta u_\psi(f_\epsilon) = 0 \text{ for all } f \text{ supported in } \psi[U]\} . \tag{10.84}$$

It is easily checked that the definition is independent of the choice of chart (U, ψ). For example, the scaling degree of the distributions $(x + i0^+)^{-n}$ on $\mathbb{R}$ at $x = 0$ is n, whereas it is 0 at any other point $x \neq 0$. The scaling degree of the n-th derivative $\delta^{(n)}$ of the delta distribution on $\mathbb{R}$ at $x = 0$ is likewise n whereas it is $-\infty$ for $x \neq 0$. The scaling degree "doesn't see logarithms": The scaling degree of $\log^n(x + i0^+)$ is 0 for any n at any point $x \in \mathbb{R}$.

The concept of *wave front set* [62] characterizes not the strength of a singularity, but rather its nature from the point of view of momentum space. To define the wave front set, assume first a distribution u of compact support contained in some chart (U, ψ) of X. Then we may define the Fourier transform in that chart by $\hat{u}_\psi(k) = u[\exp i\psi(\, . \,) \cdot k]$. If u is smooth within U, then it is easy to see that there holds

$$|\hat{u}_\psi(k)| \leq c_N (1 + |k|)^{-N} \qquad \text{for all } N \in \mathbb{N}, \tag{10.85}$$

for some constants c_N. For a general distribution supported in U, we say that $k_0 \neq 0$ is a *singular direction* if there is no open cone Γ around k_0 such that eq. (10.85) holds uniformly in Γ. If $x_0 \in X$ and if u is an arbitrary distribution, we say that $(\psi(x_0), k_0)$ is in the wave front set $\mathrm{WF}(u_\psi)$ of u_ψ if k_0 is a singular direction for χu for all cutoff functions χ supported in U such that $\chi(x_0) \neq 0$. The wave front set of u_ψ is a subset of $\psi[U] \times (\mathbb{R}^d \setminus \{0\})$. The pull-back

$$\mathrm{WF}(u) = \bigcup_{\text{charts } (U, \psi)} (\psi^{-1})^* \, \mathrm{WF}(u_\psi) \subset T^* X \setminus 0 , \tag{10.86}$$

can be shown to be invariantly defined (i.e. independent of the choice of atlas of X for the given differentiable structure), and is simply called the "wave front set".

The notion of wave front set is applied above in eq. (10.18) to characterize Hadamard 2-point functions W_2 ($X = \mathcal{M} \times \mathcal{M}$ in that example) of the free KG field. It can also be used to characterize the wave front set of an n-fold time ordered product T_n ($X = \mathcal{M} \times \ldots \times \mathcal{M}$ [16, 60] (n copies) in that case), or of the n-point functions of n interacting fields or their OPE coefficients [63].

One of the most important uses of wave-front sets in QFTCS is to characterize situations in which the product of distributions is defined. In fact, the following theorem holds: Let u, v be distributions on X. If $\mathrm{WF}(u) + \mathrm{WF}(v)$ (element-wise addition) does not contain a zero cotangent vector in T^*X, then the distributional product uv is naturally[20] defined. More generally, for a set of n distributions, if $\sum_j \mathrm{WF}(u_j)$ does not contain a zero cotangent vector, then $\prod_j u_j$ is defined.

As an example, consider the distribution $(x + i0^+)^{-1}$, whose wave front set is found to be $\{(0, k) \mid k > 0\}$. The square – and in fact any power – is therefore well defined. Next, consider $\delta(x)$, which has wave front set $\{(0, k) \mid k \neq 0\}$. Its square is therefore not well defined. One way to think about these examples is that in the first case, $(x + i0^+)^{-1}$ is, by definition, the boundary value of an analytic function. Whence its powers also are the boundary value of an analytic function, and hence automatically defined. By contrast, the distribution $\delta(x)$ is not the boundary value of an analytic function, whence its square is not automatically defined. More generally, the relationship between distributional boundary values and the wave-front set is that if $u(x + iy)$ is an analytic function in $U \times \Gamma$, where $U \subset \mathbb{R}^d$ and Γ is some open cone having finite scaling degree in y at $y = 0$ at $x \in U$ uniformly in Γ, then the wave front set of the distributional boundary value $u(x) = \lim_{y \in \Gamma, y \to 0} u(x + iy)$ is contained in

$$\mathrm{WF}(u) \subset U \times \Gamma^*, \tag{10.87}$$

where $\Gamma^* = \{k \in (\mathbb{R}^d)^* \setminus 0 \mid \langle k, y \rangle \leq 0 \; \forall y \in \Gamma\}$ is the dual cone. This criterion can also be applied in any (analytic) manifold X by localizing u in a chart (ψ, U).

This relationship between wave front set and distributional boundary values is relevant in QFTCS, because many distributions involve some "$i0^+$-prescription". The wave front set of the two-point function W_2 of a Hadamard state for instance can be determined from (10.17) and (10.87), since the $i0^+$-prescription effectively states that W_2 is given in a sufficiently small open set $U \subset \mathcal{M} \times \mathcal{M}$ by a distributional boundary value (with cone locally given by $\Gamma = \cup_{(x_1, x_2) \in U} V_{x_1}^+ \times V_{x_2}^-$). This can be used to deduce the wave front condition (10.18).

Two important applications of the above product criterion for distributions are the following. Consider first the right side of eq. (10.72) involving products of the 2-point function W_2. It is easily seen that the product criterion is satisfied, whence the expression is

---

[20] Note that there may be other exotic ways to define the product of distributions even if the wave-front criterion is not fulfilled; the key point here is that the so-defined product is continuous in some natural topology of distributions, i.e. when u, v are approximated by smooth functions in an appropriate way.

indeed well-defined. As the second example, consider the time-ordered 2-point function $W_2^T(x_1, x_2) = \omega(T_2(\phi(x_1) \otimes \phi(x_2)))$ associated with a Hadamard state ω, also called a "Feynman propagator". The wave front set of the time-ordered product is e.g. found to be

$$\mathrm{WF}(W_2^T) = \{(x_1, k_1; x_2, k_2) \in T^* \mathcal{M}^2 \setminus 0 \mid k_{1/2} \in V^\mp \text{ if } x_{1/2} \in J^+(x_{2/1}), k_1 \sim -k_2\}$$
$$\cup \{(x_1, k_1; x_2, k_2) \in T^* \mathcal{M}^2 \setminus 0 \mid k_1 = -k_2, x_1 = x_2\}. \qquad (10.88)$$

Because $\mathrm{WF}(W_2^T)$ contains e.g. the point $(x, k; x, -k)$ for any $k \neq 0$, $\mathrm{WF}(W_2^T) + \mathrm{WF}(W_2^T)$ contains the zero co-vector. The product criterion is not fulfilled, and thus the square of W_2^T cannot straightforwardly be defined as a distribution. This problem shows up precisely when one naively tries to define the product via Fourier transform e.g. for the vacuum state in Minkowski space, and is directly related to the logarithmic divergence of the "fish-graph" in Feynman diagram language. However, for $x_1 \neq x_2$ the criterion *is* fulfilled and $[W_2^T(x_1, x_2)]^2$ *can* be defined for such points. Thus, the "renormalization" required to define the time ordered product $\omega(T_2(\phi^2(x_1) \otimes \phi^2(x_2)))$ corresponds precisely to obtaining an *extension* of this distribution to the "diagonal" of $\mathcal{M} \times \mathcal{M}$, i.e. in some sense, no problems arise other than for coincident points. This is a rather non-trivial point in curved spacetime, because the behavior of null-geodesics (points where the "propagators" are singular) can be very different from that in flat spacetime.

These considerations can be generalized to the construction of higher order time ordered products. For instance, in order to define the expectation value $\omega(T_n(\otimes_j \phi^{k_j}(x_j)))$ in a Gaussian, Hadamard state for mutually distinct points $x_j \in \mathcal{M}$, one may apply the Wick product and "causal factorization" formulas. This leads to an expression in terms of a product of Feynman propagators $W_2^T(x_i, x_j)$, where $e = (ij)$ run through the edges of an abstract Feynman graph $\mathcal{G}$ with incidence number of the j-th vertex $\leq k_j$. To the product, we may again apply our criterion and conclude that it exists away from all "diagonals", i.e. for the open subset of $\mathcal{M}^n$ of points such that $x_i \neq x_j$ for all $i \neq j$. Again, the important point is that the "extension" has to be performed only on the "small" subset of diagonals in $\mathcal{M}^n$, and the potentially very complicated nature of the "null-related singularities" is taken care of by the wave front set techniques.

An important aspect of the precise analysis [58, 60] is that the wave front set of the extension can be controlled, including at the diagonals; it is characterized by the following "*microlocal spectrum property*": For Hadamard states ω, we have

$$\mathrm{WF}\left(\omega(T_n(\otimes_j \phi^{k_j}))\right) \subset \left\{(x_1, k_1; \ldots; x_n, k_n) \in T^* \mathcal{M}^n \setminus 0 \right|$$
$$k_i = \sum_{e \in \mathcal{G}: s(e)=i} p_e - \sum_{e \in \mathcal{G}: t(e)=i} p_e, \quad p_e \in V^\mp \text{ if } x_{s(e)/t(e)} \in J^+(x_{t(e)/s(e)})\right\}. \qquad (10.89)$$

Here, one is considering embeddings of the graph $\mathcal{G}$ into $\mathcal{M}$ such that its edges $e = (ij)$ are associated with null geodesics. Their cotangent null vectors are called p_e. These are future/past-oriented depending on whether the edge $e = (ij)$ (oriented so that $s(e) := i < j =: t(e)$) is future- or past-directed. An illustration is given in fig. 10.5.

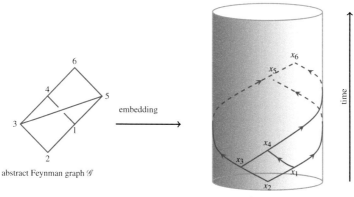

abstract Feynman graph $\mathscr{G}$

spacetime $\mathscr{M}$ with embedded graph

Figure 10.5 Shown here is the wave front set of the time-ordered products (10.89) and its relationship with embedded Feynman graphs $\mathscr{G}$ in $\mathscr{M}$. Through each line e flows a 'momentum' p_e indicated by $\rightarrow$, which is a parallel transported, cotangent null vector. At each vertex x_i the corresponding vector $k_i \in T^*_{x_i}\mathscr{M}$ in the wave front set is characterized by the 'momentum conservation rule' $k_i = \sum_{\text{in}} p_e - \sum_{\text{out}} p_e$ counting the momenta associated with the incoming vs. outgoing edges e with opposite sign.

References

[1] R. F. Streater and A. S. Wightman, *PCT, spin and statistics, and all that,* Princeton, NJ, Princeton University Press (2000).
[2] R. Haag, *Local quantum physics: fields, particles, algebras*, Berlin, Springer (1992).
[3] J. Glimm and A. Jaffe: *Quantum physics: a functional integral point of view*, Berlin, Springer (1987).
[4] S. Hollands and R. M. Wald, *Commun. Math. Phys.* **293**, 85 (2010).
[5] C. Kopper and V. F. Muller, *Commun. Math. Phys.* **275**, 331 (2007).
[6] S. Hollands and R. M. Wald, arXiv:1401.2026 [gr-qc].
[7] S. Hollands, *Rev. Math. Phys.* **20**, 1033 (2008).
[8] Antonio N. Bernal and Miguel Sánchez, *Commun. Math. Phys.* **243**, 461 (2003).
[9] C. Bär, N. Ginoux and F. Pfäffle, *Wave equations on Lorentzian manifolds and quantization*, Zürich, European, Mathematical Society (2007).
[10] J. Dimock, *Commun. Math. Phys.* **77**, 219 (1980).
[11] R. M. Wald, *Quantum field theory in curved spacetime and black hole thermodynamics*, Chicago, IL, University of Chicago Press (1994).
[12] M. Reed and B. Simon, *Fourier analysis and self-adjointness*, New York, Academic Press (1975).
[13] N. I. Akhizer, *The classical moment problem and some related questions in analysis,* Oliver and Boyd (1965).
[14] M. J. Radzikowski, *Commun. Math. Phys.* **179**, 529 (1996).
[15] M. J. Radzikowski, *Commun. Math. Phys.* **180**, 1 (1996).
[16] R. Brunetti, K. Fredenhagen and M. Kohler, *Commun. Math. Phys.* **180**, 633 (1996).
[17] K. Sanders, *Commun. Math. Phys.* **295**, 485 (2010).
[18] B. S. Kay and R. M. Wald, *Phys. Rept.* **207**, 49 (1991).
[19] B. S. DeWitt and R. W. Brehme, *Annals of Physics* **9**, 220 (1960).
[20] S. A. Fulling, F. J. Narcowich and R. M. Wald, *Annals of Physics* **136**, 243 (1981).

[21] W. Junker and E. Schrohe, *Ann. Inst. Henri Poincaré Phys. Theor.* **3**, 1113 (2002).

[22] C. Gerard and M. Wrochna, arXiv:1209.2604 [math-ph].

[23] H. Araki and S. Yamagami, *Publ. RIMS, Kyoto U.* **18** (1982).

[24] R. Verch, *Commun. Math. Phys.* **160**, 507 (1994).

[25] O. Bratteli and D. W. Robinson, *Operator algebras and quantum statistical mechanics*, Berlin, Springer (2002).

[26] H. Sahlmann and R. Verch, *Commun. Math. Phys.* **214**, 705 (2000).

[27] W. G. Unruh and R. M. Wald, *Phys. Rev. D* **29**, 1047 (1984).

[28] S. De Bievre and M. Merkli, *Class. Quant. Grav.* **23**, 6525 (2006).

[29] B. Allen, *Phys. Rev. D* **32**, 3136 (1985).

[30] J. Bros and U. Moschella, *Rev. Math. Phys.* **8**, 327 (1996).

[31] J. Bros, U. Moschella and J. P. Gazeau, *Phys. Rev. Lett.* **73**, 1746 (1994).

[32] S. Hollands, *Ann. Inst. Henri Poincaré* **13**, 1039 (2012).

[33] S. Hollands, *Commun. Math. Phys.* **319**, 1 (2013).

[34] D. Marolf and I. A. Morrison, *Phys. Rev. D* **84**, 044040 (2011).

[35] A. Higuchi, D. Marolf and I. A. Morrison, *Phys. Rev. D* **83**, 084029 (2011).

[36] K. Sanders, arXiv:1310.5537 [gr-qc].

[37] D. G. Boulware, *Phys. Rev. D* **11**, 1404 (1975).

[38] W. G. Unruh, *Phys. Rev. D* **14**, 870 (1976).

[39] C. Dappiaggi, V. Moretti and N. Pinamonti, *Adv. Theor. Math. Phys.* **15**, 355 (2011).

[40] K. Fredenhagen and R. Haag, *Commun. Math. Phys.* **127**, 273 (1990).

[41] K. Fredenhagen and R. Haag, *Commun. Math. Phys.* **108**, 91 (1987).

[42] M. Dafermos and I. Rodnianski, *Commun. Pure Appl. Math.* **62**, 859 (2009).

[43] S. W. Hawking, *Commun. Math. Phys.* **43**, 199 (1975) [Erratum *ibid.* **46**, 206 (1976)].

[44] R. M. Wald, *Commun. Math. Phys.* **45**, 9 (1975).

[45] J. M. Bardeen, B. Carter and S W. Hawking, *Commun. Math. Phys.* **31**, 161 (1973).

[46] R. M. Wald *Living Rev. Rel.* **4**, 6 (2001).

[47] C. J. Fewster, *Class. Quant. Grav.* **17**, 1897 (2000).

[48] L. H. Ford and T. A. Roman, *Phys. Rev. D* **55**, 2082 (1997).

[49] S. Hollands and R. M. Wald, *Commun. Math. Phys.* **237**, 123 (2003).

[50] M. Duetsch and K. Fredenhagen, *Commun. Math. Phys.* **219**, 5 (2001).

[51] H. Kodama and M. Sasaki, *Prog. Theor. Phys. Suppl.* **78**, 1 (1984).

[52] V. F. Mukhanov, H. A. Feldman and R. H. Brandenberger, *Phys. Rept.* **215**, 203 (1992).

[53] C. Schomblond and P. Spindel, *Ann. Inst. Poincaré Phys. Theor.* **25**, 67 (1976).

[54] S. Hollands and R. M. Wald, *Commun. Math. Phys.* **223**, 289 (2001).

[55] R. Brunetti, K. Fredenhagen and R. Verch, *Commun. Math. Phys.* **237**, 31 (2003).

[56] H. Epstein and V. Glaser, *Ann. Inst. Poincaré Phys. Theor. A* **19**, 211 (1973).

[57] N. N. Bogoliubov and D. V. Shirkov, *Introduction to the theory of quantized fields,* New York, Wiley (1980).

[58] S. Hollands and R. M. Wald, *Commun. Math. Phys.* **231**, 309 (2002).

[59] S. Hollands and R. M. Wald, *Rev. Math. Phys.* **17**, 227 (2005).

[60] R. Brunetti and K. Fredenhagen, *Commun. Math. Phys.* **208**, 623 (2000).

[61] M. Duetsch and K. Fredenhagen, *Rev. Math. Phys.* **16**, 1291 (2004).

[62] L. Hormander, *The analysis of linear partial differential operators I: distribution theory and Fourier analysis*, Berlin, Springer, 2nd edition (1990).

[63] S. Hollands, *Commun. Math. Phys.* **273**, 1 (2007).

11

From General Relativity to Quantum Gravity

ABHAY ASHTEKAR, MARTIN REUTER AND CARLO ROVELLI

11.1 Introduction

The necessity of reconciling general relativity (GR) with quantum physics was recognized by Einstein [1] already in 1916 when he wrote:

"Nevertheless, due to the inner-atomic movement of electrons, atoms would have to radiate not only electro-magnetic but also gravitational energy, if only in tiny amounts. As this is hardly true in Nature, it appears that quantum theory would have to modify not only Maxwellian electrodynamics, but also the new theory of gravitation."

Yet, almost a century later, we still do not have a satisfactory reconciliation. Why is the problem so difficult? An obvious response is that this is because there are no observations to guide us. However, this cannot be the entire story because, if there are no observational constraints, one would expect an overabundance of theories, not scarcity!

The viewpoint in approaches discussed in this Chapter is that the primary obstacle is rather that, among fundamental forces of Nature, gravity is special: it is encoded in the very geometry of space-time. This is a central feature of GR, a crystallization of the equivalence principle that lies at the heart of the theory. Therefore, one argues, it should be incorporated at a fundamental level in a viable quantum theory. The perturbative treatments which dominated the field since the 1960s ignored this aspect of gravity. They assumed that the underlying spacetime can be taken to be a continuum, endowed with a smooth background geometry, and the quantum gravitational field can be treated just like any other quantum field on this background. But the resulting quantum GR turned out to be non-renormalizable; the strategy failed by its own criteria. The new strategy is to free oneself of the background spacetime that seemed indispensable for formulating and addressing physical questions; the goal is to lift this anchor and learn to sail the open seas. This task requires novel mathematical techniques and conceptual frameworks. From the perspective of this Chapter, we do not yet have a satisfactory quantum gravity theory primarily because serious attempts to meet these challenges squarely are relatively recent. However, as our overview will illustrate, the community *has* made notable advances towards this goal in recent years.

In this Chapter, we will focus on two main programs, each of which in turn has two related but distinct parts: (i) *Loop Quantum Gravity* (LQG) whose Hamiltonian or canonical framework is well suited for cosmological issues, and whose Spinfoam or covariant framework is geared to address scattering theory [2–6]; and, (ii) the *Asymptotic Safety* paradigm which includes the Effective Average Action Framework with its functional renormalization group equation in the continuum, and the Causal Dynamical Triangulation Approach in which one uses numerical simulations à la lattice gauge theory [7–10]. (String Theory is discussed in Chapter 12 and other approaches in the Introduction to Part IV.)

A common theme in these programs is that their starting point is the physical, dynamical spacetime geometry of GR. However, as will be clear from the detailed discussion, this does not imply a conventional quantization of GR. In LQG, for example, the fundamental quanta of geometry are one-dimensional, polymer-like excitations over nothing, rather than gravitons, the wavy undulations over a continuum background. In particular, classical general relativity is recovered only in an appropriate coarse-grained limit. Another common theme is that these programs first focus on geometry and rely on non-perturbative effects – rather than specific matter couplings – to cure the ultraviolet difficulties of perturbative quantum GR. The viewpoint is that the short-distance behavior of quantum geometry is qualitatively different from that suggested by the continuum picture and it would be more efficient to first develop a detailed understanding of quantum geometry in the Planck regime and then couple matter in a second stage. In the Asymptotic Safety scenario, for example, this strategy was successfully implemented first for pure gravity, and incorporating certain matter fields afterwards did not change the basic picture [11]. Finally, as in QCD, the first priority in these programs is to uncover and explore qualitatively new, non-perturbative features of quantum gravity by focusing on just one interaction, rather than on achieving unification. Such features have already emerged. Examples are: a quantum resolution of singularities of GR [5, 12], finiteness of microstates of black hole and cosmological horizons [2, 13], and effective dimension reduction in the Planck regime [7–10].

Although these programs share several common elements, there are also some key differences in the underlying viewpoints. Let us begin with Asymptotic Safety. Recall first that, although GR is perturbatively non-renormalizable, there does exist a well-developed and powerful effective field theory [14] which, for example, has been applied with remarkable success to the long-standing problem of equations of motion of compact binaries in GR [15]. However, this theory abandons the idea of handling the Planck regime and focuses on low energy processes. Asymptotic Safety can be thought of as a specific *ultraviolet (UV) completion* of this effective field theory using Wilson's generalized notion of renormalization [16]. The idea is to avoid the notorious proliferation of undetermined couplings in the UV faced by perturbative GR by using a reliable strategy that has already been successfully tested in well-understood, *perturbatively non-renormalizable* field theories where one can, so to say, 'renormalize the non-renormalizable' [17]. This success suggests that a state of 'peaceful coexistence' of perturbative divergences with Asymptotic Safety may be possible also for gravity [7–9, 18].

In LQG the guiding principle is rather different. The viewpoint is that, just as Riemannian geometry is essential to the formulation of general relativity, an appropriate *quantum* Riemannian geometry should underlie a viable theoretical account of space, time, and gravitation that does not disregard quantum theory. To meet this goal, a specific quantum theory of geometry was constructed in detail, drawing motivation from geometric structures that underlie the phase space of GR [2–5]. In this and subsequent constructions one makes heavy use of non-perturbative techniques that have already been successful in gauge theories but with a crucial twist: now there is no reference to a background metric. This requirement of background independence is surprisingly powerful and leads to a unique kinematical framework [19, 20] on which the dynamics of the quantum theory is being built. While the Hamiltonian LQG has broad similarities with the older Wheeler–DeWitt (WDW) theory [21], the quantum nature of the underlying geometry makes a key difference leading, for example, to a natural resolution of classical singularities in cosmological models [12]. Similarly, Spinfoams provide transition amplitudes that are UV finite to any order in a natural expansion. Furthermore, a *positive* cosmological constant provides a natural mechanism for regulating their infrared (IR) behavior [5, 6].

Thus, although both LQG and Asymptotic Safety programs have similar goals, the physical concepts and mathematical techniques used in subsequent analysis are quite different. In particular, because the quantum geometry underlying LQG is fundamentally discrete, the physical degrees of freedom terminate at the Planck scale, much like in string theory. In the Asymptotic Safety program, on the other hand, there is no kinematic reason that would prevent degrees of freedom at arbitrarily small scale. A first reading of the flow equations suggests that there are physical degrees of freedom at any scale, all the way to the infinitely small. However, it is the fixed point action that determines the physical degrees of freedom in this approach. Non-perturbative renormalizability indicates that these are fewer than what one would expect classically and the mean-field considerations indicate that there are at most as many as in a theory in 2 space-time dimensions. A more thorough understanding of the fixed point is necessary to settle this important question in the Asymptotic Safety program.

This Chapter is organized as follows. Section 11.2 provides a broad-brush overview of the two programs. Since this volume is likely to draw readership from diverse quarters, we have made a special attempt to make the sub-sections self-contained. Thus a reader interested only in Asymptotic Safety can skip Sections 11.2.2 and 11.2.3 and a reader interested only in LQG can skip Section 11.2.1 without loss of continuity. Section 11.3 discusses illustrative applications to cosmology of the very early universe, black hole physics and scattering theory. While advances over the past decade are encouraging, a large number of issues remain. These are discussed in Section 11.4.

11.2 Frameworks

This section is divided into three parts. The first summarizes the main ideas and results in the Asymptotic Safety program, the second those in Hamiltonian LQG and the third those in Spinfoams.

11.2.1 Asymptotic Safety

Since GR is not renormalizable in the standard perturbation theory, it is commonly argued that a satisfactory microscopic quantum theory of the gravitational interaction cannot be set up within the realm of quantum field theory without adding further symmetries, extra dimensions or new principles such as holography. In contrast, the Asymptotic Safety program [22] retains quantum field theory without such additions as the theoretical arena and instead abandons the traditional techniques of perturbative renormalization. Moreover, as we will see, in a certain sense it even abandons the standard notion of 'quantization' because its starting point is not a given classical model to be promoted to a quantum theory.

Rather, in its modern incarnation, this program may be thought of as a systematic *search strategy among theories that are already 'quantum'*; it identifies the 'islands' of physically acceptable theories in the 'sea' of unacceptable ones plagued by short-distance pathologies. Since the approach is based on Wilson's generalized notion of renormalization [16] and the use of functional renormalization group (RG) equations, concepts from statistical field theory play an important role. They provide a unified framework for approaching the problem with both continuum and discrete methods. In this section we discuss two such complementary approaches within the Asymptotic Safety paradigm: the Effective Average Action (EAA) with its Functional Renormalization Group Equation (FRGE) [23], and Causal Dynamical Triangulations (CDT) [24].

The Functional Renormalization Group

The goal of the Asymptotic Safety program consists in giving a mathematically precise meaning to, and actually computing functional integrals over 'all' spacetime metrics of the form $\int \mathcal{D}\tilde{g}_{\mu\nu} \, \exp\left(iS[\tilde{g}_{\mu\nu}]\right)$, or

$$Z = \int \mathcal{D}\tilde{g}_{\mu\nu} \, e^{-S[\tilde{g}_{\mu\nu}]}, \qquad (11.1)$$

from which all quantities of physical interest can be deduced then. Here $S[\tilde{g}_{\mu\nu}]$ denotes the classical or, more appropriately, the bare action. It is required to be diffeomorphism invariant, but is kept completely arbitrary otherwise. In general it differs from the usual Einstein–Hilbert action. This generality is essential in the Asymptotic Safety scenario: the viewpoint is that the functional integral would exist only for a certain class of actions S and the task is to identify this class.

Following the approach proposed in [23] one attacks this problem in an indirect way: rather than dealing with the integral per se, one interprets it as the solution of a certain differential equation, a functional renormalization group equation, or 'FRGE'. The advantage is that, contrary to the functional integral, the FRGE is manifestly well defined. It can be seen as an 'evolution equation' in a mathematical sense, defining an infinite dimensional dynamical system in which the RG scale plays the role of time. Loosely speaking, this reformulation replaces the problem of defining functional integrals by the task of finding evolution histories of the dynamical system that extend to *infinitely late times*. According to the Asymptotic Safety conjecture the dynamical system possesses a fixed point which is approached at late times, yielding well defined, fully extended evolutions, which in turn tell us how to construct (or 'renormalize') the functional integral.

Let us start by explaining the passage from the functional integrals to the FRGE. Recall that in trying to put the integrals on a solid basis one is confronted with a number of obstacles:

(i) As in every field theory, difficulties arise since one tries to quantize *infinitely many degrees of freedom*. Therefore, at the intermediate steps of the construction one keeps only finitely many of them by introducing cutoffs at very small and very large distances, Λ^{-1} and k^{-1}, respectively. We shall specify their concrete implementation in a moment. The ultraviolet (UV) and infrared (IR) cutoff scales Λ and k, respectively, have the dimension of a mass, and the original system is recovered for $\Lambda \to \infty$, $k \to 0$.

(ii) Conceptually, the most severe problem one encounters when quantizing the gravitational field, one which is not shared by any conventional matter field theory, is the requirement of *background independence*: no particular spacetime (such as Minkowski space, say) should be given a privileged status. Rather, the geometry of spacetime should be determined dynamically. In the approach to Asymptotic Safety along the lines of [23] this problem is dealt with by following the spirit of DeWitt's background field method [25] and introducing a (classical, non-dynamical) background metric $\bar{g}_{\mu\nu}$ which, however, is kept absolutely arbitrary. One then decomposes the integration variable as $\tilde{g}_{\mu\nu} \equiv \bar{g}_{\mu\nu} + \tilde{h}_{\mu\nu}$, and interprets $\mathcal{D}\tilde{g}_{\mu\nu}$ as an integration over the nonlinear fluctuation, $\mathcal{D}\tilde{h}_{\mu\nu}$. In this way one arrives at a conceptually easier task, the quantization of the matter-like field $\tilde{h}_{\mu\nu}$ in a generic, but classical background $\bar{g}_{\mu\nu}$. The availability of the background metric is crucial at various stages of the construction of an FRGE. However, the final physical results do not depend on the choice of a specific background.

(iii) As in every gauge field theory, the *redundancy of gauge-equivalent field configurations* (diffeomorphic metrics) has to be carefully accounted for. Here we employ the Faddeev–Popov method and add a gauge fixing term $S_{\text{gf}} \propto \int \sqrt{\bar{g}} \bar{g}^{\mu\nu} F_\mu F_\nu$ to S, where $F_\mu \equiv F_\mu(\tilde{h}; \bar{g})$ is chosen such that the condition $F_\mu = 0$ picks a single representative from each gauge orbit. The resulting volume element on orbit space, the Faddeev–Popov determinant, we express as a functional integral over Grassmannian ghost fields $\tilde{C}^\mu$ and $\tilde{\bar{C}}_\mu$, governed by an action S_{gh}. In this way the original integral (11.1) gets replaced by $\tilde{Z}[\bar{\Phi}] = \int \mathcal{D}\tilde{\Phi} \exp\left(-S_{\text{tot}}[\tilde{\Phi}, \bar{\Phi}]\right)$. Here the total bare action $S_{\text{tot}} \equiv S + S_{\text{gf}} + S_{\text{gh}}$ depends on the dynamical fields $\tilde{\Phi} \equiv (\tilde{h}_{\mu\nu}, \tilde{C}^\mu, \tilde{\bar{C}}_\mu)$, the background fields $\bar{\Phi} \equiv (\bar{g}_{\mu\nu})$, and possibly also on (both dynamical and background) matter fields, which for simplicity are not included here.

Using the gauge fixed and regularized integral we can compute arbitrary ($\bar{\Phi}$-dependent!) expectation values $\langle \mathcal{O}(\tilde{\Phi}) \rangle \equiv \tilde{Z}^{-1} \int \mathcal{D}\tilde{\Phi}\, \mathcal{O}(\tilde{\Phi})\, e^{-S_{\text{tot}}[\tilde{\Phi}, \bar{\Phi}]}$, for instance n-point functions where $\mathcal{O}$ consists of strings $\tilde{\Phi}(x_1)\tilde{\Phi}(x_2)\cdots\tilde{\Phi}(x_n)$. For $n = 1$ we use the notation $\Phi \equiv \langle\tilde{\Phi}\rangle \equiv (h_{\mu\nu}, C^\mu, \bar{C}_\mu)$, i.e. the elementary field expectation values are $h_{\mu\nu} \equiv \langle\tilde{h}_{\mu\nu}\rangle$, $C^\mu \equiv \langle\tilde{C}^\mu\rangle$ and $\bar{C}_\mu \equiv \langle\tilde{\bar{C}}_\mu\rangle$. Thus the full dynamical metric has the expectation value $g_{\mu\nu} \equiv \langle\tilde{g}_{\mu\nu}\rangle = \bar{g}_{\mu\nu} + h_{\mu\nu}$.

The dynamical laws which govern the expectation value $\Phi(x)$ have an elegant description in terms of the *effective action* Γ. It is a functional depending on Φ similar to the classical $S[\Phi]$ to which it reduces in the classical limit. Requiring stationarity, S yields the classical field equation $(\delta S/\delta\Phi)[\Phi_{\text{class}}] = 0$, while Γ gives rise to a quantum mechanical analog satisfied by the expectation values, the *effective field equation* $(\delta\Gamma/\delta\Phi)[\langle\tilde{\Phi}\rangle] = 0$. If, as in the case at hand, $\Gamma \equiv \Gamma[\Phi, \bar{\Phi}] \equiv \Gamma[h_{\mu\nu}, C^\mu, \bar{C}_\mu; \bar{g}_{\mu\nu}]$ depends also on background fields, the solutions to this equation inherit this dependence and so $h_{\mu\nu} \equiv \langle\tilde{h}_{\mu\nu}\rangle$ functionally depends on $\bar{g}_{\mu\nu}$. Technically, Γ is obtained from a functional integral with S_{tot} replaced by $S_{\text{tot}}^J \equiv S_{\text{tot}} - \int dx\, J(x)\tilde{\Phi}(x)$. The new term couples the dynamical fields to an external, classical source, $J(x)$, and repeated functional differentiation $(\delta/\delta J)^n$ of $\ln\tilde{Z}[J, \bar{\Phi}]$ yields the n-point functions. In particular, $\Phi = \delta\ln\tilde{Z}/\delta J$. It is a standard result that $\Gamma[\Phi, \bar{\Phi}]$ equals exactly the Legendre transform of $\ln\tilde{Z}[J, \bar{\Phi}]$, at fixed background fields $\bar{\Phi}$. The importance of Γ also resides in the fact that *it is the generating functional of special n-point functions from which all others can be easily reconstructed.* Therefore, finding Γ in some quantum field theory is often considered equivalent to completely 'solving' this theory.

To calculate $\Gamma[\Phi, \bar{\Phi}]$ it is advantageous to employ a gauge breaking condition F_μ which fixes a gauge belonging to the distinguished class of the so-called *background gauges*. To see the benefit, recall that the original gauge transformations read $\delta\tilde{g}_{\mu\nu} = \mathcal{L}_v\tilde{g}_{\mu\nu}$, where $\mathcal{L}_v$ denotes the Lie derivative w.r.t. the vector field v. When we decompose $\tilde{g}_{\mu\nu} = \bar{g}_{\mu\nu} + \tilde{h}_{\mu\nu}$ we can distribute the gauge variation of $\tilde{g}_{\mu\nu}$ in different ways over $\bar{g}_{\mu\nu}$ and $\tilde{h}_{\mu\nu}$. In particular this gives rise to what is known as *quantum gauge transformations* ($\delta^{\text{Q}}\tilde{h}_{\mu\nu} = \mathcal{L}_v(\bar{g}_{\mu\nu} + \tilde{h}_{\mu\nu})$, $\delta^{\text{Q}}\bar{g}_{\mu\nu} = 0$) and *background gauge transformations* ($\delta^{\text{B}}\tilde{h}_{\mu\nu} = \mathcal{L}_v\tilde{h}_{\mu\nu}$, $\delta^{\text{B}}\bar{g}_{\mu\nu} = \mathcal{L}_v\bar{g}_{\mu\nu}$). Since the functional integral is defined by fixing an externally prescribed background metric, $\bar{g}_{\mu\nu}$, we must ensure invariance under the 'ordinary' or 'true' gauge transformations the Faddeev–Popov method deals with. Hence it is the δ^{Q}-invariance which needs to be gauge-fixed by the condition $F_\mu = 0$. Interestingly enough, there exist F_μ's, a variant of the harmonic coordinate condition, for example, which indeed fix the δ^{Q}-transformations, but at the same time are *invariant under δ^{B}-transformations*: $\delta^{\text{B}}F_\mu = 0$. They implement the background gauges, and from now on we assume that we employ one of those. Then, as a consequence, the effective action $\Gamma[\Phi, \bar{\Phi}]$ *is invariant under background gauge transformations* which include the ghosts: $\delta^{\text{B}}\Gamma[\Phi, \bar{\Phi}] = 0$ for all $\delta^{\text{B}}\Phi = \mathcal{L}_v\Phi$, $\delta^{\text{B}}\bar{\Phi} = \mathcal{L}_v\bar{\Phi}$. We emphasize that this property should not be confused with another notion of 'gauge independence' which the above $\Gamma[\Phi, \bar{\Phi}]$ actually does *not* have: It is not independent of which particular F_μ is picked from the class with $\delta^{\text{B}}F_\mu = 0$. This F_μ-dependence will disappear only at the level of observables.

Turning now to the concept of a *functional renormalization group equation* recall that the above definition of Γ is based on the functional integral regularized in the IR and UV, hence it depends on the corresponding cutoff scales: $\Gamma \equiv \Gamma_{k,\Lambda}[\Phi, \bar{\Phi}]$. It is this object for which we derive a FRGE, more precisely a closed evolution equation governing its dependence on the IR cutoff scale k. This is possible only if the IR regularization is implemented appropriately, as in the so-called *effective average action* (EAA) [26].

The EAA is related to the modified integral, $\int \mathcal{D}\tilde{\Phi} e^{-S_{\text{tot}}^J} e^{-\Delta S_k[\tilde{\Phi},\bar{\Phi}]} \equiv Z_{k,\Lambda}[J,\bar{\Phi}]$, whose second exponential factor in the integrand, containing the *cutoff action* ΔS_k, is designed to achieve the IR regularization. To see how this works, assume the integration variable $\tilde{\Phi} = (\tilde{h}, \tilde{C}, \tilde{\bar{C}})$ is expanded in terms of eigenfunctions φ_p of the covariant tensor Laplacian related to the background metric, $\bar{D}^2 \equiv \bar{g}^{\mu\nu}\bar{D}_\mu\bar{D}_\nu$. Writing $-\bar{D}^2\varphi_p = p^2\varphi_p$ we have, symbolically, $\tilde{\Phi}(x) = \sum_p \alpha_p\varphi_p(x)$. The α_p's are generalized Fourier coefficients, and so the functional integration over $\tilde{\Phi}$ amounts to integrating over all α_p:

$$Z_{k,\Lambda}[J,\bar{\Phi}] = \prod_{p^2 \in [0,\Lambda^2]} \int_{-\infty}^{\infty} d\alpha_p \exp\left(-S_{\text{tot}}^{J'}[\{\alpha\},\bar{\Phi}]\right). \tag{11.2}$$

Here $S_{\text{tot}}^{J'}$ equals $S_{\text{tot}}^J[\tilde{\Phi},\bar{\Phi}] + \Delta S_k[\tilde{\Phi},\bar{\Phi}]$ with the expansion for $\tilde{\Phi}$ inserted. In (11.2) we implemented the UV regularization by retaining only eigenfunctions (or 'modes') corresponding to $-\bar{D}^2$-eigenvalues (or squared 'momenta') smaller than Λ^2. The IR contributions, i.e. those corresponding to eigenvalues between $p^2 = 0$ and about $p^2 = k^2$, are cut off smoothly instead, namely by a p^2-dependent suppression factor arising from ΔS_k. To obtain a structurally simple FRGE, ΔS_k should be chosen quadratic in the dynamical fields. Usually one sets $\Delta S_k = \frac{1}{2}\int dx\, \tilde{\Phi}\mathcal{R}_k\tilde{\Phi}$ with an operator $\mathcal{R}_k \propto k^2 R^{(0)}(-\bar{D}^2/k^2)$ containing a dimensionless function $R^{(0)}$. In the $-\bar{D}^2$-basis we have then $\Delta S_k \propto k^2 \sum_p R^{(0)}(p^2/k^2)\alpha_p^2$ which shows that ΔS_k represents a kind of p^2-dependent mass term: A mode with eigenvalue p^2 acquires a (mass)^2 of the order $k^2 R^{(0)}(p^2/k^2)$. We require $R^{(0)}(p^2/k^2)$ to have the qualitative properties of a smeared step function which, around $p^2/k^2 \approx 1$, drops smoothly from $R^{(0)} = 1$ for $p^2/k^2 \lesssim 1$ to $R^{(0)} = 0$ for $p^2/k^2 \gtrsim 1$. This achieves precisely the desired IR regularization: In the product over p^2 in (11.2), ΔS_k equips all $\int d\alpha_p$-integrals pertaining to the *low momentum modes*, i.e. those with $p^2 \in [0, k^2]$, with a Gaussian suppression factor $e^{-k^2\alpha_p^2}$ since for such eigenvalues $R^{(0)}(p^2/k^2) \approx 1$. The *high momentum modes*, having $p^2 \in [k^2, \Lambda^2]$, yield $R^{(0)}(p^2/k^2) \approx 0$ and so they remain unaffected by ΔS_k. At least on a flat background, low (high) momentum modes $\varphi_p(x)$ have long (short) wavelengths. Therefore, when one lowers k from $k = \Lambda$ down to $k = 0$ one 'un-suppresses' modes of increasingly long wavelengths, thus proceeding from the UV to the IR. (In FRGE jargon, this is called the 'integrating out' of the high momentum modes since in older approaches the low momentum modes were completely discarded, rather than just suppressed.) This process of encoding the contribution of an increasing number of modes in a scale dependent, or 'running' functional is precisely a *renormalization in the modern sense* due to Wilson [16].

The effective average action, $\Gamma_{k,\Lambda}[\Phi,\bar{\Phi}]$, is defined to be the Legendre transform of $\ln Z_{k,\Lambda}[J,\bar{\Phi}]$ given by (11.2), with respect to J, for k, Λ, and $\bar{\Phi}$ fixed (and with $\Delta S_k[\Phi,\bar{\Phi}]$ subtracted from the result of the transformation, which is not essential here). As for the Φ, $\bar{\Phi}$-arguments, we stress that the modes classified low or high momentum are only those of the *fluctuation* field, $\tilde{\Phi}$. The externally prescribed background and source fields $\bar{\Phi}(x)$ and $J(x)$, which are also present under the integral defining $Z_{k,\Lambda}[J, \bar{\Phi}]$, have nonzero Fourier coefficients for all $p^2 \in [0, \Lambda^2]$ in general, they may contain both high and low momentum components. As a consequence, the same is true for the Φ-argument of the EAA, since J and $\Phi = \delta \ln Z_{k,\Lambda}/\delta J$ are Legendre-conjugates of one another.

The EAA, $\Gamma_{k,\Lambda}[\Phi, \bar{\Phi}]$, has a number of important features not realized in other functional RG approaches:

 (i) Since no fluctuation modes are taken into account in the $k = \Lambda \to \infty$ limit, the EAA approaches the bare (i.e., un-renormalized) action, $\Gamma_{\Lambda,\Lambda} = S_{\text{tot}}$. In the limit $k \to 0$, it yields the standard effective action (with an UV cutoff).

 (ii) It satisfies a closed FRGE, and can be computed by integrating this FRGE towards low k, with the initial condition $\Gamma_{\Lambda,\Lambda} = S_{\text{tot}}$ at $k = \Lambda$.

(iii) The functional $\Gamma_{k,\Lambda}[\Phi, \bar{\Phi}]$ is invariant under background gauge transformations $\delta^{\mathbf{B}}$ for all values of the cutoffs. This property is preserved by the FRGE: the RG evolution does not generate $\delta^{\mathbf{B}}$-noninvariant terms.

(iv) The FRGE continues to be well behaved when the UV cutoff is removed ($\Lambda \to \infty$). Denoting solutions to the UV cutoff-free FRGE by $\Gamma_k[\Phi, \bar{\Phi}]$, it reads:

$$k \, \partial_k \Gamma_k[\Phi, \bar{\Phi}] = \frac{1}{2} \text{STr}\left[\left(\Gamma_k^{(2)}[\Phi, \bar{\Phi}] + \mathcal{R}_k[\bar{\Phi}]\right)^{-1} k \, \partial_k \mathcal{R}_k[\bar{\Phi}]\right]. \qquad (11.3)$$

Here STr denotes the functional supertrace, and $\Gamma_k^{(2)}$ stands for the matrix of second functional derivatives of Γ_k with respect to Φ at fixed $\bar{\Phi}$. Since $R^{(0)}$ is essentially a step function, the derivative $\partial_k \mathcal{R}_k$ is non-zero only in a thin shell of momenta near $p^2 = k^2$, and so the supertrace on the RHS of (11.3) receives contributions only from such momenta. As a result, it is perfectly finite both in the IR and in the UV, and this is why sending $\Lambda \to \infty$ was unproblematic.

 (v) Γ_k is closely related to a generating functional for field averages over finite domains of size k^{-1}; hence the name EAA [26]. Thanks to this property, when treated as a *classical* action Γ_k can provide an effective field theory description of *quantum* physics involving typical momenta near k. This property has been exploited in numerous applications of the EAA to particle and condensed matter physics, but it plays no role in the present context. Rather, it is its interpolating property between S and $\Gamma_{k=0}$ which is instrumental in the Asymptotic Safety program.

The arena in which the RG dynamics takes place is the infinite dimensional *theory space*, $\mathcal{T}$. It consists of all well-behaved action functionals $(\Phi, \bar{\Phi}) \mapsto A[\Phi, \bar{\Phi}]$ which depend on a given set of fields and are invariant under some possible symmetry group. In metric gravity $\mathcal{T}$ comprises arbitrary $\delta^{\mathbf{B}}$ invariant functionals $A[g_{\mu\nu}, \bar{g}_{\mu\nu}, C^\mu, \bar{C}_\mu]$. The RHS of the FRGE (11.3) defines a vector field $\boldsymbol{\beta}$ on $\mathcal{T}$. Its natural orientation is such that $\boldsymbol{\beta}$ points from higher to lower momentum scales k, from the UV to the IR. (This is the direction of increasing 'coarse-graining' in which the microscopic dynamics is 'averaged' over increasingly large spacetime volumes.) The integral curves of this vector field,

$k \mapsto \Gamma_k$, are the *RG trajectories*, and the pair $(\mathcal{T}, \boldsymbol{\beta})$ is called the *RG flow*. It constitutes the dynamical system alluded to earlier.

One usually assumes that every action A in the Theory space can be expanded as $A[\Phi, \bar{\Phi}] = \sum_{\alpha=1}^{\infty} \bar{u}_\alpha P_\alpha[\Phi, \bar{\Phi}]$, where the set $\{P_\alpha\}$ forms a basis of invariant functionals. Writing the RG trajectory correspondingly, $\Gamma_k[\Phi, \bar{\Phi}] = \sum_{\alpha=1}^{\infty} \bar{u}_\alpha(k) P_\alpha[\Phi, \bar{\Phi}]$, one encounters infinitely many *running coupling constants*, $\bar{u}_\alpha(k)$, whose k-dependence is governed by an infinite coupled system of differential equations: $k \, \partial_k \bar{u}_\alpha(k) = \bar{\beta}_\alpha(\bar{u}_1, \bar{u}_2, \cdots; k)$. The dimensionful *beta functions* $\bar{\beta}_\alpha$ arise by expanding the RHS of the FRGE: $\frac{1}{2} \mathrm{STr}[\cdots] = \sum_{\alpha=1}^{\infty} \bar{\beta}_\alpha P_\alpha[\Phi, \bar{\Phi}]$. The coefficients $\bar{\beta}_\alpha$ are similar to the familiar beta functions of perturbative quantum field theory (where, however, only the finitely many beta functions of the relevant couplings are considered).

Reexpressing the RG equations in terms of dimensionless couplings $u_\alpha \equiv k^{-d_\alpha} \bar{u}_\alpha$ with d_α the canonical mass dimension of $\bar{u}_\alpha$, the resulting *FRGE in component form* is autonomous, i.e. its β-functions have no explicit k-dependence: $k \, \partial_k u_\alpha(k) = \beta_\alpha(u_1(k), u_2(k), \cdots)$. The coupling constants $(u_\alpha) \equiv u$ serve as local coordinates on $\mathcal{T}$, and the β_α's are the components of the vector field $\boldsymbol{\beta} \equiv (\beta_\alpha(u))$.

Later on *fixed points* of the RG flow will be of special interest. At a fixed point, $\boldsymbol{\beta} = 0$, so its coordinates $(u_\alpha^*) = u^*$ satisfy the infinitely many conditions $\beta_\alpha(u^*) = 0$. The fixed point's *UV critical hypersurface*, $\mathcal{S}_{\mathrm{UV}}$, or synonymously its *unstable manifold*, is defined to consist of all points in $\mathcal{T}$ which are pulled into the fixed point under the inverse RG flow, i.e. for increasing scale k. Linearizing the flow about u^* one has $k \, \partial_k u_\alpha(k) = \sum_\gamma B_{\alpha\gamma} \left(u_\gamma(k) - u_\gamma^* \right)$ with the *stability matrix* $\boldsymbol{B} = (B_{\alpha\gamma})$, $B_{\alpha\gamma} \equiv \partial_\gamma \beta_\alpha(u^*)$. If the eigenvectors of $\boldsymbol{B}$ form a basis, its solution reads $u_\alpha(k) = u_\alpha^* + \sum_\alpha C_I V_\alpha^I \left(k_0/k \right)^{\theta_I}$. Here the C_I's are constants of integration and the V^I's denote the right-eigenvectors of $\boldsymbol{B}$ with eigenvalues $-\theta_I$, i.e. $\sum_\gamma B_{\alpha\gamma} V_\gamma^I = -\theta_I V_\alpha^I$. In general $\boldsymbol{B}$ is not symmetric and the *critical exponents* θ_I are complex. Along eigendirections with $\mathrm{Re}\,\theta_I > 0$ ($\mathrm{Re}\,\theta_I < 0$) deviations from u_α^* grow (shrink) when k is lowered from the UV towards the IR; they are termed relevant (irrelevant).

A trajectory $u_\alpha(k)$ within $\mathcal{S}_{\mathrm{UV}}$, by definition, approaches $u_\alpha(k \to \infty) = u_\alpha^*$ in the UV. For the constants C_I in its linearization this implies that $C_I = 0$ for all I with $\mathrm{Re}\,\theta_I < 0$. Hence the trajectories in $\mathcal{S}_{\mathrm{UV}}$ are labeled by the remaining C_I's related to the critical exponents with $\mathrm{Re}\,\theta_I > 0$. (For simplicity we assume all $\mathrm{Re}\,\theta_I$ non-zero.) As a consequence, *the dimensionality of the critical hypersurface*, $s \equiv \dim(\mathcal{S}_{\mathrm{UV}})$, *equals the number of critical exponents with* $\mathrm{Re}\,\theta_I > 0$, i.e., the number of relevant directions.

A fixed point is called *Gaussian* if it corresponds to a free field theory. Its critical exponents agree with the canonical mass dimension of the corresponding operators. A fixed point whose critical exponents differ from the canonical ones is referred to as nontrivial or as a *non-Gaussian fixed point* (NGFP).

Asymptotic Safety

The construction of a quantum field theory involves finding an RG trajectory which is infinitely extended in the sense that it is a curve, entirely within theory space, with well defined limits $k \to 0$ and $k \to \infty$, respectively. *Asymptotic Safety is a proposal for ensuring the existence of the second limit.* Its crucial prerequisite is a nontrivial RG fixed point Γ_* on $\mathcal{T}$. Let us assume there is such a fixed point. Then it is sufficient to simply pick any of the trajectories within its hypersurface $\mathcal{S}_{UV}$ to be sure that the trajectory has a singularity-free ultraviolet behavior since it will always hit the fixed point for $k \to \infty$. There exists a $\dim(\mathcal{S}_{UV})$-parameter family of such trajectories.

Most probably an UV fixed point is not only sufficient but also necessary for an acceptable theory without divergences. Therefore, in the simplest case when there exists only one, the physically inequivalent asymptotically safe quantum theories one can construct are labeled by the $\dim(\mathcal{S}_{UV})$ parameters characterizing trajectories inside $\mathcal{S}_{UV}$. Thus the degree of predictivity of asymptotically safe theories is essentially determined by the number of *relevant* eigendirections at Γ_*. If this is a finite number $s \equiv \dim(\mathcal{S}_{UV})$, it is sufficient to measure only s of the couplings $\{u_\alpha(k)\}$ characterizing Γ_k in order to predict the infinitely many others. In particular, at $k = 0$ one obtains the standard effective action $\Gamma \equiv \Gamma_0$, which 'knows' all possible predictions.

The only input required for this construction is the theory space $\mathcal{T}$, that is the field contents and the symmetries. It fully determines the FRGE and its fixed point properties. Since $\Gamma_{k\to\infty}$ is closely related to the bare action S, the Asymptotic Safety program essentially consists in *computing* $S \sim \lim_{k\to\infty} \Gamma_k = \Gamma_*$ from the fixed point condition. In this sense the approach amounts to a selection process among quantum theories rather than the quantization of a classical system known beforehand. It has become customary to call *Quantum Einstein Gravity*, or QEG, any quantum field theory of metric-based gravity, regardless of its bare action, which is defined by a trajectory on the theory space $\mathcal{T}_{QEG}$ of diffeomorphism invariant functionals $A[g_{\mu\nu}, \bar{g}_{\mu\nu}, C^\mu, \bar{C}_\mu]$.

A priori the functional integral over 'all' metrics is only formal and plagued by mathematical problems. Knowing Γ_* and the RG flow in its vicinity, one can give a well defined meaning to it. The only extra ingredient that needs to be selected is an UV regularization for the integral. It is then possible to use the information encoded in the flow of Γ_k near Γ_* in order to determine how the ('bare') parameters, on which the integral depends, must be tuned in order to obtain a meaningful limit when the UV regulator is removed [27]. Thus the mathematical subtleties of the functional integral are overcome if the long-time behavior of the associated dynamical system on $\mathcal{T}_{QEG}$ can be controlled, e.g. by means of a fixed point. For an evolution equation as complicated as the FRGE, on an infinite dimensional theory space, it is by no means clear from the outset that this is possible, i.e. that there exist RG trajectories that extend to infinite values of the evolution parameter. An essential part of the Asymptotic Safety program consists in demonstrating that this is indeed the case, for the concrete reason that the trajectory hits a fixed point in the long-time limit.

Practical computations require a nonperturbative approximation scheme. The method of choice consists in a *truncation of theory space*. One sets all but a certain subset of couplings u_α to zero, and expands Γ_k in terms of the appropriately chosen reduced set $\{P_\alpha, \alpha = 1, \cdots, N\}$ where, as before, P_α is a basis of invariant functionals in terms of which now only the actions in the truncated theory space can be expanded. Hence the FRGE boils

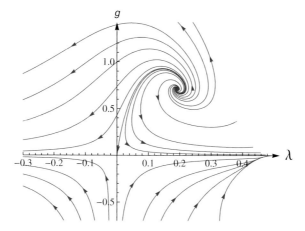

Figure 11.1 RG flow of the Einstein–Hilbert truncation on the (g,λ) plane. The arrows point towards decreasing scales k. (First obtained in [28].)

down to a system of N coupled differential equations. This amounts to a severe restriction, of course, which needs to be justified a posteriori by systematically changing and enlarging the subset chosen. This difficulty is not specific to gravity; the same strategy is followed in FRGE-based investigations of matter field theories on flat space and in statistical physics.

At the time Weinberg conjectured the possibility of Asymptotic Safety, due to the lack of nonperturbative computational techniques, a NGFP was known to exist only for a single coupling, Newton's constant, and only in $d = 2 + \epsilon$ spacetime dimensions [22]. The situation changed when the EAA-based methods became available [23]. Starting from early work on the 'Einstein–Hilbert truncation' [23, 29] and a generalization with an additional R^2-term [30], a considerable number of truncations with increasingly large subsets $\{P_\alpha\}$ were analyzed in the following decade [31]. Quite remarkably, they all agree in that *QEG indeed seems to possess a NGFP suitable for the Asymptotic Safety construction.* Although a complete proof is not within reach, by now there is highly nontrivial evidence for a NGFP on the full (un-truncated) theory space, rendering QEG nonperturbatively renormalizable [7–9]. As a representative example, Fig. 11.1 shows the phase portrait of the Einstein–Hilbert truncation [23] based upon the running action $\Gamma_k = (1/16\pi G_k) \int d^d x \sqrt{g} \, (-R(g) + 2\Lambda_k) + S_{\text{gf}} + S_{\text{gh}}$ which has $N = 2$. It involves the approximation of neglecting the k-dependence in the gauge fixing and ghost sectors which can be justified by BRST methods [23]. This ansatz contains a running Newton constant G_k and cosmological constant Λ_k, their dimensionless analogs being, in d spacetime dimensions, $g(k) \equiv k^{d-2}G_k$ and $\lambda(k) \equiv \Lambda_k/k^2$, respectively. Their beta functions $(\beta_g, \beta_\lambda) \equiv \boldsymbol{\beta}$ have been computed for any d [23]. The first steps of the calculation are reminiscent of those in perturbatively quantized general relativity [32], but this is a coincidence due to the specific ansatz for Γ_k. Moreover, β_g and β_λ are quite different from beta-functions in perturbation theory. They sum up contributions from arbitrary orders of perturbation theory and, more importantly, they contain also information about the strong power law-type renormalization effects which are not seen usually in perturbative calculations employing dimensional regularization. This is important, however, for instance in order to 'tame' the notorious quadratic (and higher) divergences due to the non-zero mass dimension of G.

Figure 11.1 shows the flow diagram obtained by solving the coupled equations $k\, \partial_k g(k) = \beta_g(g,\lambda)$ and $k\, \partial_k \lambda(k) = \beta_\lambda(g,\lambda)$ for $d = 4$ [28]. Besides a Gaussian fixed point at $g_* = 0 = \lambda_*$ there is indeed a second, *non-Gaussian* fixed point at $g_* > 0$, $\lambda_* > 0$. Both of its critical exponents have a positive real part. Hence $s \equiv \dim S_{\mathrm{UV}} = 2$. In sufficiently general truncations one finds that $s < N$, and the reduced dimensionality then allows us to predict $N - s$ couplings after s couplings have been measured. The predictions are encoded in the way the s-dimensional hypersurface S_{UV} is immersed into the (truncated or complete) theory space. There are general arguments suggesting that s should saturate at a small finite value when N is increased [22]. Concrete calculations confirmed this picture, first in $d = 2 + \epsilon$ where the '$\Lambda + R + R^2$ truncation', having $N = 3$, yields a NGFP with $s = 2$ [30]. So, given two input parameters, the third one is a prediction. One might for instance express the coefficient of the term $\int \sqrt{g} R^2$ added to the Einstein–Hilbert action in terms of g and λ. In $d = 4$, all known truncations confirmed that the projection of the flow onto the (g, λ) plane has the same structure as in Fig. 11.1, with 'perpendicular' directions added. By now, there exist very impressive analyses of $f(R)$ truncations, with f a polynomial of high degree. They do indeed display the expected stabilization of s at a small finite value when N is made large [38]. Furthermore, the first explorations of infinite dimensional truncated theory spaces were performed [33] and truly functional flows in non-polynomial $f(R)$ truncations are within reach now [34]. Trying to make the truncations more accurate it is not sufficient to generalize their $g_{\mu\nu}$-dependence only; at the same time we must also allow for a more general dependence of Γ_k on the background metric. The first results on such 'bi-metric truncations' which treat the $g_{\mu\nu}$- and $\bar{g}_{\mu\nu}$-dependence on a similar footing further support the viability of the Asymptotic Safety program [35]. The same is true for a different type of generalization, the inclusion of scale dependent surface terms into Γ_k for spacetimes with boundaries [36]. There is yet another important but technically difficult generalization that uses *non-local* terms. They are particularly important in the infrared, where they are expected to cure a problem of the Einstein–Hilbert truncation not visible in Fig. 11.1: singularities of the beta-functions at $\lambda = 1/2$ which indicate that the truncation becomes insufficient in the IR. In [37] a simple but genuinely 'functional' flow of a non-local EAA was analyzed, and it turned out to possess an infrared fixed point, i.e. a non-local 'fixed functional'.

Besides a better understanding of the RG flow in QEG, future work will also have to address the question of observables. The running couplings parameterizing the EAA have no direct physical significance in general. While under very special circumstances it might be possible to deduce observable effects directly from the k-dependence of certain couplings (by some kind of 'RG improvement'), the general strategy is to first construct the functional integral, then find interesting observables in terms of the fundamental fields, and finally compute their expectation values. In this respect the status of observables within the Asymptotic Safety program is not different from that of those in any other functional integral based approach.

Causal Dynamical Triangulations

The partition functions of standard model-like quantum field theories, analytically continued to Euclidean space and discretized, have been extensively studied by Monte Carlo techniques. It is therefore natural to apply similar ideas to gravity and to attempt a definition of the formal functional integral (11.1) as the $a \to 0$ limit of the partition function belonging to a suitably chosen statistical mechanics model, specified by a choice of dynamical variables, bare action $S[g]$ and measure $\mathcal{D}g$. Here the discretization scale a is analogous to a lattice spacing. A priori the 'lattice units' defined by a are unphysical; they can be converted to physical lengths or masses only later when it comes to computing observables.

The limit $a \to 0$ is to be taken indirectly, as follows. The statistical system has a chance of describing physics in the continuum if a can be made much smaller than any relevant physical length scale ℓ, or, more adequately, if all lengths ℓ are much larger than a. In fact, in numerical simulations where a is necessarily non-zero ($a = 1$, say) the requirement $(\ell/a) \gg 1$ is met if the free parameters of the statistical model (bare couplings) are tuned such that its correlation length diverges and ℓ, in lattice units, becomes very large. Thus the *continuum limit* $a/\ell \to 0$ amounts to $\ell \to \infty$ with a fixed (rather than $a \to 0$ and ℓ fixed). As is well known from the statistical physics of critical phenomena, for instance, the correlation length does indeed diverge at second order phase transition points. So the strategy will be to propose a plausible statistical model, compute numerically its partition function in terms of its dependence on the bare parameters, and search for points in parameter space where the correlation length diverges. If such a critical point exists one would use it to define a continuum theory and explore its properties.

The statistical systems underlying critical phenomena are conveniently analyzed in terms of their RG flow under successive 'coarse graining'. While there is considerable ambiguity in how this is done concretely, it typically boils down to a space averaging of the degrees of freedom (block spin transformation, etc.) which, in a continuum language, amounts to a step-by-step integrating out of field modes with increasing wavelengths. In this setting systems at second order phase transition points, displaying no preferred length scale, are described by fixed points of the RG flow.

This observation brings us back to Asymptotic Safety: The discrete system describes a continuum theory when its bare parameters are tuned to their fixed point values. Then the partition function Z is a sum over contributions from fluctuations whose wavelengths, in physical units, range from zero to infinity. In EAA language this amounts to specifying a *complete* trajectory Γ_k, well behaved in particular in the UV since $\lim_{k \to \infty} \Gamma_k = \Gamma_*$. One can show that Γ_* is indeed very closely related to the RG fixed point of the statistical model, and that the large-k behavior of Γ_k, with minimal additional input, can be mapped onto the RG flow of the model near the second order phase transition point [27].

The CDT approach [10, 39, 40] is a specific proposal for a statistical system representing gravity. It sums over the class of piecewise linear 4-geometries which can be assembled from 4-dimensional simplicial building blocks (with link length a) in such a way that the resulting spacetime is 'causal' in a certain technical sense. A priori the spacetimes $\mathcal{M}$ summed over have Lorentzian signature. However, to ensure the existence of a generalized Wick rotation they are restricted to be globally hyperbolic, which allows one to introduce a global proper-time foliation, $\mathcal{M} = I \times \Sigma$, where I denotes a 'time' interval and space is represented by 3-dimensional leaves Σ whose topology is not allowed to change in time. A choice extensively studied is $\Sigma = S^3$ so that at each proper-time step $t_n \in I$ the spatial geometry is represented by a triangulation of S^3. It is made up of equilateral spatial tetrahedra with positive squared side-length $\ell_s^2 \equiv a^2 > 0$. The number $N_3(t)$ of tetrahedra, and the way they are glued together to form a piecewise flat 3-dimensional manifold, will change in general when we go from $t = t_n$ to the next time slice at t_{n+1}. In order to constitute a 4-dimensional triangulation, the 3-dimensional slices must be connected in a 'causal' way, preserving the S^3-topology at all intermediate times. (This ensures that a branching of the spatial universe into several disconnected pieces (baby universes) does not occur.) For the gluing of two consecutive time slices $S^3(t_n)$ and $S^3(t_{n+1})$ it is sufficient to introduce four types of 4-simplices, namely the so-called (4,1)-simplices, each of which has 4 of its vertices on $S^3(t_n)$ and 1 on $S^3(t_{n+1})$, the (3,2)-simplices with 3 vertices on $S^3(t_n)$ and 2 on $S^3(t_{n+1})$, as well as (1,4)- and (2,3)- simplices defined the other way around. The integration over spacetimes $\mathcal{M}$ boils down to a sum over all possible ways to connect given triangulations of $S^3(t_n)$ and $S^3(t_{n+1})$ compatible with the topology $I \times S^3$, along with a summation over all 3-dimensional triangulations of $S^3(t)$, at all times t.

Denoting by ℓ_t and ℓ_s the length of the time-like and the space-like links, respectively, one has $\ell_t^2 = -\alpha\ell_s^2$, where the constant α is positive in the Lorentzian case, whence $\ell_t^2 < 0$. It was shown [39] that there exists a well defined rotation in the complex α plane ($\alpha \rightarrow -\alpha$) which, thanks to the restriction to a given foliation in the simplicial decomposition, connects the Lorentzian to the Euclidean signature, with $\ell_t^2 = |\alpha|\ell_s^2 > 0$. This turns oscillating exponentials e^{iS} into Boltzmann factors e^{-S}, so that the resulting partition function can be computed with Monte Carlo integration methods. It reads [39]:

$$Z(\kappa_0, \kappa_4, \Delta) = \sum_T \frac{1}{C_T} \exp\left(-S_{\text{Regge}}[T]\right). \tag{11.4}$$

The symmetry factor C_T equals the order of the automorphism group of the triangulation T, and S_{Regge} is the Regge-discretized Einstein–Hilbert action: $S_{\text{Regge}} = -(\kappa_0 + 6\Delta)N_0 + \kappa_4\big(N_4^{(4,1)} + N_4^{(3,2)}\big) + \Delta\big(2N_4^{(4,1)} + N_4^{(3,2)}\big)$. Here $N_4^{(4,1)}$ and $N_4^{(3,2)}$ denote the number of (4,1)- and (3,2)-simplices in T, respectively, and N_0 is the total number of vertices. The couplings κ_0 and κ_4 correspond to $1/G$ and Λ/G, respectively, and Δ parameterizes a possible asymmetry between ℓ_t and ℓ_s; it is non-zero if $|\alpha| \neq 1$.

Extensive Monte Carlo simulations of the partition function (11.4) have been performed at a number of points in the space of bare couplings $(\kappa_0, \kappa_4, \Delta)$. Three different phases were discovered, and one of them seems indeed capable of representing continuum physics.

A surface in parameter space, $\kappa_4 = \kappa_4(\kappa_0, \Delta)$, on which the 4-volume becomes large has been identified. This 'infinite' volume limit should, however, not be confused with the continuum limit. The crucial question is whether the latter can actually be realized à la Asymptotic Safety by tuning the remaining two parameters to a second order phase transition point. The answer is not known yet, but this is the topic of very active current research.

The most important result of the CDT model is that it is able to describe the emergence of a classical 4-dimensional de Sitter universe with small superimposed quantum fluctuations. The calculation is carried out in the Euclidean signature, but thanks to the above α-rotation it admits a Lorentzian interpretation. The reason this result is interesting is that it resolves a difficulty of previous attempts to address quantum gravity with dynamical triangulations: the 4-dimensional Euclidean triangulation models without the 'causality' constraint produced only states with Hausdorff dimensions $d_{\mathrm{H}} = 2$ and $d_{\mathrm{H}} = \infty$, respectively, contradicting the classical limit.

In these CDT simulations the link length a is still as large as about 2 Planck lengths so they do not yet probe the physics on sub-Planckian length scales [40]. Once simulations well beyond the Planck scale become feasible they should be able to make contact with the RG fixed point predicted by the EAA-based calculations in the continuum. Indeed, it has been shown already [41, 42] that the CDT and EAA predictions for the running spectral dimension agree quite precisely in the semiclassical regime. It is also known how, at least in principle, the information about the k-dependence of the EAA can be used to predict the expected RG running of a statistical model near the continuum limit [27]. In this respect it should also be mentioned that while most EAA studies have been performed for *Euclidean signature*, they also apply to the *Lorentzian* case almost unchanged [43]. It will be very interesting to see whether future Monte Carlo results lead to the same picture of physics near the fixed point as the FRGE studies.

Finally, CDT breaks Lorentz invariance because the spatial and temporal cut-offs are independent. There is no general argument that Lorentz invariance must be restored in the continuum limit. If it is not restored, the classical limit of the continuum theory might not be general relativity, but rather something akin to the Hořava–Lifshitz theory, which is non-Lorentz invariant [44]. Hořava–Lifshitz theory is renormalizable, and if it were physically viable, it would represent a possible solution to the quantum gravity problem.

11.2.2 Hamiltonian theory and quantum geometry

The *Asymptotic Safety* program generalizes the procedures that have been successful in Minkowskian quantum field theories (MQFTs) by going beyond traditional perturbative treatments. An avenue that is even older is canonical quantization, pioneered by Dirac, Bergmann, Arnowitt, Deser, Misner and others. Over the past 2–3 decades, these ideas have inspired a new approach, known as *Loop Quantum Gravity* (LQG).

While the point of departure is again a Hamiltonian framework, as explained in Section 11.1, there is an important conceptual shift: the idea now is to construct a *quantum* theory of geometry and then use it to formulate quantum gravity systematically. This theory was constructed in detail in the 1990s. Since then, research in LQG has progressed along two parallel avenues. In the first, discussed in this sub-section, one continues the development of the canonical quantization program, now using quantum geometry to properly handle the field theoretical issues. In the second, discussed in the next sub-section, one develops a path integral framework and defines dynamics via transition amplitudes between quantum 3-geometries. In the final picture, the fundamental degrees of freedom are *quite different* from those that would result in a 'direct' quantization of GR – they are not metrics and extrinsic curvatures but chunks, or atoms, of space with quantum attributes. Classical geometries emerge only upon coarse graining of their coherent superpositions.

This subsection is divided into three parts. The first summarizes the Hamiltonian framework that provides the point of departure, the second explains the basic structure of quantum geometry and the third sketches the status of quantum dynamics in canonical LQG.

Connection Dynamics

The key idea underlying the Hamiltonian framework used in LQG is to cast GR in the language of gauge theories that successfully describe the electroweak and strong interactions. This requires a shift from metrics to connections; Wheeler's 'geometrodynamics' [21] is replaced by a dynamical theory of spin-connections [45]. Once this is achieved, the phase space of general relativity becomes *the same* as that of gauge theories: All four fundamental forces of Nature are unified at a kinematical level. However, the dynamics of GR has two distinguishing features. First, whereas the Hamiltonian of QED or QCD uses the flat background metric, the Hamiltonian constraints that generate the dynamics of GR are built entirely from the spin connection and its conjugate momentum; the theory is manifestly *background independent*. Second, the gauge group now refers to rotations in the physical space rather than in an abstract, internal space. This is why in contrast to, say, QCD, *spacetime* geometry can now emerge from this gauge theory. As we will see, these two features have a powerful consequence: one is led to a *unique* quantum Riemannian geometry.

Fix a 3-manifold M which is to represent a Cauchy surface in space-time. The gravitational phase space Γ is coordinatized by pairs $(\underline{A}^j_a, E^a_j)$ of an SU(2) connection $\underline{A}^j_a$ and its conjugate 'electric field' E^a_j on M, where j refers to the Lie algebra su(2) of SU(2) and a to the tangent space of M. Thus, the fundamental Poisson brackets are:

$$\{\underline{A}^j_a(x), E^b_k\} = -i\kappa_N \, \delta^b_a \, \delta^j_k \, \delta^3(x, y), \tag{11.5}$$

where $\kappa_N = 8\pi \, G_N$ is the gravitational coupling constant. As remarked above, although the phase space variables have the familiar Yang–Mills form, they also admit a natural interpretation in terms of spacetime geometry. To spell it out, let us first recall from Chapter 8 that the standard Cauchy data of GR consists of a pair, (q_{ab}, K_{ab}), representing the intrinsic positive definite metric q_{ab} and the extrinsic curvature K_{ab} on Σ. If we denote by e_j^a an orthonormal triad – a 'square root' of q^{ab} – then in the Lorentzian signature we have:

$$E_j^a = \sqrt{q}\, e_j^a \qquad \text{and} \qquad \underline{A}_a^j = \Gamma_a^j - \iota K_a^j, \tag{11.6}$$

where q denotes the determinant of the metric q_{ab}, Γ_a^j is the intrinsic spin connection on M defined by e_j^a, $K_a^j = K_{ab}\, e_j^b$ and $\iota = 1$ in the Euclidean signature and $\iota = i\ (\equiv \sqrt{-1})$ in the Lorentzian signature (used in most of this Chapter). The connection $\underline{A}_a^j$ parallel transports left-handed (or unprimed) spacetime spinors. In the final solution, its curvature $\underline{F}_{ab}^j := 2\,\partial_{[a}\,\underline{A}_{b]}^j + \epsilon^{jkl}\,\underline{A}_{ak}\,\underline{A}_{bl}$ represents the (pull-back to M of the) self-dual part of the space-time Weyl curvature.[1]

As is well known, the dynamics of GR is generated by a set of constraints. While they are rather complicated and non-polynomial functionals of the geometrodynamical ADM variables, they become low order polynomials in the connection variables. In the absence of matter sources, they are [45]:

$$\mathcal{G}_j := \underline{D}_a E_j^a = 0, \ \ \mathcal{D}_a := E_j^b\,\underline{F}_{ab}^j = 0, \ \text{and} \ \mathcal{H} := \epsilon^{jkl}\left(\underline{F}_{abl} - \Lambda\eta_{abc}E_l^c\right)E_j^a E_k^b = 0. \tag{11.7}$$

The first constraint is just the familiar Gauss law of Yang–Mills theory, the second is the Diffeomorphism constraint of GR, and the third the Hamiltonian constraint. Interestingly, these are the simplest gauge invariant, local expressions one can construct from a connection and its conjugate electric field *without* reference to a background metric. Indeed, these are the *only* such expressions that are at most quartic in the canonical variables A_a^i, E_j^a. At first one might expect that it would be difficult to couple matter to gravity using these connection variables since they refer only to the self-dual part of space-time curvature. But this is not the case; one can couple spin zero, half and one fields keeping the simplicity [46, 47] and recently the framework has also been extended to include higher dimensions [48] and supersymmetry [49].

Since the constraints are polynomial in the connection variables, so are the equations of motion. Furthermore, the framework represents a small extension of GR: Since, in contrast to the ADM variables, none of the equations require us to invert the E_j^a, they remain viable even when the E_j^a become degenerate. At these phase space points one no longer has a (non-degenerate) space-time metric, but the connection dynamics continues to remain meaningful. The standard causal structures have been extended to such configurations [50]. Finally, the connection dynamics framework provides a natural setting for proofs of the positive-energy theorems à la Witten [51]; one can establish the positivity of the gravitational Hamiltonian not only on the constraint surface as in the original theorems but also in a neighborhood of the constraint surface, i.e., even 'off-shell' [52].

---

[1] For simplicity we assume that M is compact; in the asymptotically flat case, one has to specify appropriate boundary conditions at infinity and keep track of boundary terms. See, e.g., [2, 4, 45]. Since the electric field E_j^a is a density of weight 1, mathematically, it is often simpler to work with its dual $\Sigma_{ab}^j := \eta_{abc}E_j^c$, which is just a 2-form on M. Finally, as is standard in Yang–Mills theories, the internal indices $j, k, \ldots$ are raised and lowered using the Cartan–Killing metric on su(2).

As we noted after Eq. (11.6), in the Lorentzian signature the connection $\underline{A}^i_a$ is complex-valued, or, equivalently, it is a 1-form that takes values in the Lie algebra of $\mathbb{C}SU(2)$, the complexification of $SU(2)$. While this feature does not create any obstacle at the classical level, a key mathematical difficulty arises in the passage to quantum theory: Because $\mathbb{C}SU(2)$ is *non-compact*, the space of connections $\underline{A}^j_a$ is not known to carry diffeomorphism invariant measures that are necessary to construct a satisfactory Hilbert space of square integrable functions of connections. To bypass this difficulty, the mainstream strategy has been to replace the complex, left-handed connections $\underline{A}^j_a$ with *real* $SU(2)$ connections A^j_a, obtained by replacing i in (11.5) and (11.6) by a real, non-zero parameter γ [53]. Then, both the phase space variables are real and the fundamental Poisson brackets become

$$\{A^j_a(x),\, E^a_k\} = \gamma \kappa_N \delta^b_a \delta^j_k \delta^3(x, y)\,. \tag{11.8}$$

γ is known as the Barbero–Immirzi parameter [54] and taken to be positive for definiteness. As we will see in Section 11.2.2, one can introduce well-defined measures on the space $\mathcal{A}$ of these real connections A^j_a and develop rigorous functional analysis to introduce the quantum Hilbert space and operators without any reference to a background geometry. This passage from left-handed to real connections represents a systematic generalization of the Wick rotation one routinely performs to obtain well-defined measures in MQFTs. However, the rotation is now performed in the 'internal space' rather than space-time. Indeed, the *space-time* Wick rotation does not naturally extend to general curved space-times, while this internal Wick rotation does and serves the desired purpose of taming the functional integrals.

However, the strategy has two limitations. First, the form of the constraints (and evolution equations) is now considerably more complicated. But thanks to several astute techniques introduced by Thiemann [4, 55], these complications can be handled in the canonical approach, and they are not directly relevant to spin foams. The second limitation is that, while the connection A^j_a is well-defined on M and continues to have a simple relation to the ADM variables, it does not have a natural 4-dimensional geometrical interpretation in solutions to the field equations [56]. Nonetheless, one *can* arrive at the canonical pair (A^j_a, E^a_j) by performing a Legendre transform of a 4-dimensionally covariant action $S(e, \omega)$ that depends on a space-time co-tetrad e^I_μ and a Lorentz connection ω^{IJ}_μ [57].

Quantum Riemannian Geometry

The first step in the passage to quantum theory is to select a preferred class of *elementary phase space functions* which are to be directly promoted to operators in the quantum theory without factor ordering ambiguities. In geometrodynamics, these are taken to be the positive-definite 3-metric q_{ab} on M and its conjugate momentum, $P^{ab} = \sqrt{q}\,(K^{ab} - Kq^{ab})$ (integrated against suitable test fields). In connection dynamics the choice is motivated by structures that naturally arise in gauge theories. Thus, the configuration variables are now the *Wilson lines, or holonomies* h_ℓ which enable one to parallel transport left-handed spinors along 1-dimensional (curves or) *links* ℓ in M, and the conjugate momenta are the 'electric field fluxes' $E_{f,S}$ across 2-dimensional *surfaces* S (smeared with test fields f^i that take values in su(2)) [2–4, 58–60]:

$$h_\ell := \mathcal{P} \exp \int_\ell A, \qquad \text{and} \qquad E_{f,S} := \int_S d^2 S_a f^i(x) E_i^a(x). \qquad (11.9)$$

Note that the definitions do not require a background geometry; since A is an su(2)-valued 1-form, it can be naturally integrated along 1-dimensional links to yield $h_\ell \in$ SU(2), and since E is the Hodge-dual of a (su(2)-valued) 2-form the second integral is also well-defined without any background fields. However, the Poisson brackets between these variables fail to be well-defined if ℓ and S are allowed to have an infinite number of intersections. Therefore they have to satisfy certain regularity conditions. Two natural strategies are to use piecewise linear links and 2-surfaces or piecewise analytic ones (more precisely, 'semi-analytic' in the sense of [2, 19, 61]). The first choice is well-adapted to the simplicial decompositions often used in Spinfoam models while the second is commonly used in canonical LQG.

Formal sums of products of these elementary operators $\hat{h}_\ell$ and $\hat{E}_{n,S}$ generate an abstract algebra $\mathfrak{A}$. This is the analog of the familiar Heisenberg algebra in quantum mechanics and one's first task is to find its representations. The Hilbert space $\mathcal{H}_{\text{grav}}^{\text{kin}}$ underlying the chosen representation would then serve as the space of kinematical quantum states, the quantum analog of the gravitational phase space Γ of GR, the arena to formulate dynamics.

In quantum mechanics, von Neumann's theorem guarantees that the Heisenberg algebra admits a unique representation satisfying certain regularity conditions (see, e.g., [62]). However, in MQFTs, because of the infinite number of degrees of freedom, this is not the case in general: The standard result on the uniqueness of the Fock vacuum assumes *free field dynamics* [63, 64]. What is the situation with the algebra $\mathfrak{A}$ of LQG? Now, in addition to the standard regularity condition, we can and *have to* impose the strong requirement of background independence. A fundamental and surprising result due to Lewandowski, Okolow, Sahlmann, and Thiemann [19] and Fleischhack [20] is that the requirement is in fact so strong that it suffices to single out a unique representation of $\mathfrak{A}$, without having to fix the dynamics. Thus, *thanks to background independence, quantum kinematics is unique in LQG.*

This powerful result lies at the foundation of much of LQG because the unique representation it selects leads to the fundamental discreteness in quantum geometry. Therefore let us discuss the key features of this representation and compare and contrast it with representations used in MQFTs. The underlying Hilbert space $\mathcal{H}_{\text{grav}}^{\text{kin}}$ is the space $L^2(\bar{\mathcal{A}}, d\mu_o)$ of square integrable functionals of (generalized) connections with respect to a regular, Borel measure μ_o. As one would expect, the holonomy operators $\hat{h}_\ell$ act by multiplication while their 'momenta' $\hat{E}_{f,S}$ act by differentiation. There is a state Ψ_o in $\mathcal{H}_{\text{grav}}^{\text{kin}}$ which is cyclic in the sense that $\mathcal{H}_{\text{grav}}^{\text{kin}}$ is generated by repeated actions of $\hat{h}_\ell$ on Ψ_o. These properties are shared by MQFTs where the Fock space can also be represented as the space of square-integrable functionals over the space of (distribution-valued) fields on $\mathbb{R}^3$ and the vacuum plays the role of Ψ_o. In these theories, the vacuum state is Poincaré invariant and this invariance implies that the Poincaré group is unitarily implemented in the quantum theory. In LQG, the state Ψ_o is invariant under the kinematical symmetry group $SU(2)_{\text{loc}} \ltimes \text{Diff}(M)$ of connection dynamics – the semi-direct product of the local $SU(2)$ gauge transformations and diffeomorphisms of M – and this group is unitarily represented on $\mathcal{H}_{\text{grav}}^{\text{kin}}$. This fact provides a natural point of departure in the imposition of quantum constraints, discussed below.

However, the representation also has two unfamiliar features: i) $\mathcal{H}_{\text{grav}}^{\text{kin}}$ is non-separable, and, ii) while the holonomies $\hat{h}_\ell$ are well-defined operators on $\mathcal{H}_{\text{grav}}^{\text{kin}}$, the connection operators themselves do not exist (because the $\hat{h}_\ell$ fail to be continuous with respect to the links ℓ). These aspects of LQG kinematics have caused some unease among researchers outside LQG (see e.g. [65]) because it is not widely appreciated that they are *not* peculiarities of LQG but follow, in essence, just from background independence. In particular, if one seeks a representation of the properly constructed kinematical algebra of geometrodynamics in which the cyclic state is invariant under the kinematical symmetry group $\text{Diff}(M)$, *one again finds that the representation inherits these two features* [66]. Intuition derived from the $\text{Diff}(S^1)$ group used, e.g., in string theory does not carry over to higher dimensions in this respect.

Let us now discuss quantum states and operators in some detail. Recall that in MQFTs, while the characterization of the Fock space as the space of square integrable functionals of (generalized) fields is succinct, detailed calculations are most efficiently performed in a convenient basis that diagonalizes the number operators. The situation with $\mathcal{H}_{\text{grav}}^{\text{kin}}$ is analogous. More precisely, it is convenient to decompose $\mathcal{H}_{\text{grav}}^{\text{kin}}$ into orthogonal subspaces, $\mathcal{H}_{\text{grav}}^{\text{kin}} = \bigoplus \mathcal{H}_\alpha$, associated with graphs α in M with a finite number of oriented links ℓ. Next, if one labels each link ℓ of α with a non-trivial, irreducible representation $j_\ell \neq 0$ of $SU(2)$, one obtains a further decomposition [67, 68]

$$\mathcal{H}_{\text{grav}}^{\text{kin}} = \bigoplus_\alpha \mathcal{H}_\alpha = \bigoplus_{\alpha, j_\ell} \mathcal{H}_{\alpha, j_\ell} \,. \tag{11.10}$$

If α has L links, $\mathcal{H}_{\alpha, j_\ell}$ is a *finite* dimensional Hilbert space which can be identified with the space of quantum states of a system of L spins. Therefore (11.10) is called a *spin-network decomposition* of $\mathcal{H}_{\text{grav}}^{\text{kin}}$. To make this relation explicit, note first that a (generalized) connection A assigns to each link ℓ a holonomy h_ℓ and elements Ψ of $\mathcal{H}_{\text{grav}}^{\text{kin}}$ are functions of these (generalized) connections. States Ψ in $\mathcal{H}_{\alpha, j_\ell}$ are of the form

$$\Psi(A) = \psi(h_{\ell_1}, \ldots, h_{\ell_L}), \tag{11.11}$$

where ψ is a function of the L $SU(2)$ group-elements in its argument, which is square integrable with respect to the Haar measure on $[SU(2)]^L$. They know only about the action of the connection A pulled back to the L links of α. Thus, by restricting one's attention to a single graph α, one truncates the theory and focuses only on a finite number of degrees of freedom. The spirit is the same as in MQFTs. In any calculation with Feynman diagrams of a weakly coupled theory (such as low energy QED) one truncates the theory by allowing only a *finite* number of virtual particles. Similarly, in strongly coupled theories (such as low energy QCD) one truncates the theory by making a lattice approximation. In both cases, the full Hilbert space is recovered in the limit in which the degrees of freedom are allowed to go to infinity. In LQG this is achieved by taking a well-defined (projective) limit in the space of graphs [60]. Finally, the second equality in (11.10) is obtained by carrying out Fourier transforms (using the Peter–Weyl theorem) on $[SU(2)]^L$.

Such truncations are useful if the operators of interest leave the truncated Hilbert spaces invariant. This is indeed the case with geometric operators of LQG. As one would expect from the phase space description, these operators are constructed from $\hat{E}_{f,S}$ since the electric field E_i^a also serves as the orthonormal triad in the classical theory. The action of $\hat{E}_{f,S}$ on a state $\Psi \in \mathcal{H}_\alpha$ is non-trivial only if the surface S intersects one or more links of the graph α and then the action involves only group theory at the intersection [2, 4, 61]. This is just the structure one would expect from background independence! To construct geometric operators such as those corresponding to areas of 2-surfaces and volumes of 3-dimensional regions, one first expresses their classical expressions in terms of the 'elementary' phase space functions $\hat{E}_{f,S}$ and then promotes the classical expression to a quantum operator. In the intermediate stages one has to introduce auxiliary structure, but the procedure ensures that the final expressions are background independent [2–4].

Let us now consider the operator $\widehat{Ar}_{S,\alpha}$ on $\mathcal{H}_\alpha$, representing the area of a 2-surface S (without boundary) [69–71], which has played a particularly important role in LQG. Let us first suppose that the surface S intersects α only at a node n. Then, one can naturally define a *node-Laplacian operator* $\Delta_{\alpha,S,n}$ whose action on Ψ of (11.11) is an appropriate sum of the Laplacians on the copies of SU(2) associated with links ℓ_i that intersect S at n [2, 60, 71]. As one might expect, $\Delta_{\alpha,S,n}$ is a negative definite self-adjoint operator on $\mathcal{H}_\alpha$. The final expression of the area operator $\widehat{Ar}_{S,\alpha}$ is given by

$$\widehat{Ar}_{S,\alpha} = 4\pi\, \gamma\, \ell_{\mathrm{Pl}}^2 \sqrt{\Delta_{\alpha,S,n}}. \tag{11.12}$$

If there are multiple intersections n_i between α and S, $\widehat{Ar}_{S,\alpha}$ is just the sum of these operators for each n_i. The non-trivial result is that operators defined on various $\mathcal{H}_\alpha$ can be naturally *glued together* to obtain a self-adjoint operator $\widehat{Ar}_S$ on the entire $\mathcal{H}_{\mathrm{grav}}^{\mathrm{kin}}$.

The properties of $\widehat{Ar}_{S,\alpha}$ have been analyzed in detail. Its spectrum is discrete in the sense that all its eigenvectors are normalizable. In the special case when all intersections between α and S are at bi-valent nodes at which 'straight' links pierce S, the expression of eigenvalues simplifies to a form that is useful in many applications [71, 72]:

$$a_S = 8\pi\, \gamma\, \ell_{\mathrm{Pl}}^2 \sum_n \sqrt{j(j+1)}. \tag{11.13}$$

There is a smallest non-zero eigenvalue among these:

$$\Delta a_S = 4\pi\gamma\, \ell_{\mathrm{Pl}}^2 \sqrt{3}. \tag{11.14}$$

This *area gap* pays an important role in the theory. The level spacing between consecutive eigenvalues is *not* uniform but decreases *exponentially* for large eigenvalues [71]. This implies that, although the eigenvalues are fundamentally discrete, the continuum approximation becomes excellent *very rapidly*.

For the volume and length operators, the strategy is the same and the background independence of LQG again fixes the precise form of the final expressions [2, 4, 69, 73]. However, the detailed procedure is technically more complicated. The length operator has not had significant applications. The volume operator has been investigated in greater detail because it features prominently in the dynamical considerations of the canonical theory [4, 55, 74]. The problem of finding its spectrum has been cast in a form that makes it accessible to numerical studies [75]. Although the eigenvalues are discrete, there are indications that, in contrast to the area operator, the spectrum of the volume operator may not have a volume gap. This is but one indication that the quantum geometry has qualitatively different features from what one may naively expect from the classical Riemannian geometry or a naive discretization thereof.

Let us summarize. The kinematical framework of LQG is well developed, with full control on functional analysis. In particular, the infinite dimensional integrals are not formal symbols but performed with well defined measures [76, 77]. There are two key results that simplify the analysis: the uniqueness theorem [19] and the spin-network decomposition of the full Hilbert space [67, 68]. The natural truncation of the theory is achieved by restricting oneself to the Hilbert space $\mathcal{H}_\alpha$ defined by a graph α. Elements of these $\mathcal{H}_\alpha$ describe elementary quanta of geometry; to obtain classical geometries one needs to coherently superpose a large number of them.

Perhaps the simplest way to visualize the elementary quanta is to introduce a simplicial decomposition $\mathcal{S}$ of the 3-manifold M and consider a graph α which is dual $\mathcal{S}$: Each cell in $\mathcal{S}$ is a topological tetrahedron T_n, dual to a node n of α; each face F_ℓ of $\mathcal{S}$ is dual to a link ℓ. In Regge calculus, every T_n has the geometry of a tetrahedron in flat space and the curvature is encoded in the holonomies of the connection around 'bones' that lie at the intersection of any two faces of T_n. What is the situation in LQG? To bring out the similarities and contrasts, it is convenient to consider a basis Ψ_{α,v_n,a_ℓ} in $\mathcal{H}_{\alpha,j_\ell}$ that simultaneously diagonalizes the volume operator associated with the tetrahedron T_n, and the area operators associated with the faces F_ℓ, for all n, ℓ. Each of these spin-network states describes a specific *elementary* quantum geometry. One can think of the node n as a 'grain' or a 'quantum' of space captured in the (topological) tetrahedron T_n. As in Regge calculus each T_n has a well defined volume v_n and each of its faces F_ℓ has a well-defined area a_ℓ. But now the v_n, a_ℓ are *discrete*. More importantly, because the operators J_ℓ^i do not commute, T_n no longer has the sharp geometry of a geometrical tetrahedron in the Euclidean space. In particular, operators describing angles between any two distinct faces $F_\ell, F_{\ell'}$ of a T_n are *not* diagonal in the basis. Furthermore, although the area of any common face F of two adjacent tetrahedrons is unambiguous, in contrast with the Regge geometry, curvature now resides not just at the bones of tetrahedra but also along the faces; the geometry is 'twisted' in a precise sense [78]. These properties of the quantum geometry associated with the basis $\Psi_{\alpha,v_n,a_{\ell_n}}$ are closely analogous to the properties of angular momentum captured by the basis $|j, m\rangle$ in quantum mechanics: it too diagonalizes only some of the angular momentum operators, leaving values of other angular momentum observables fuzzy. Thus, each of the elementary cells in the simplicial decomposition is now a 'tetrahedron' in the same heuristic sense that a spinning particle in quantum mechanics is a 'rotating body'.

To conclude, we note that trivalent spin-networks were introduced by Roger Penrose already in 1971 in a completely different approach to quantum gravity [79]. He expressed his general view of that construction as follows: *"I certainly do not want to suggest that the universe 'is' this picture ... But it is not unlikely that essential features of the model I am describing could still have relevance in a more complete theory applicable to more realistic situations"*. In LQG one finds that the trivalent graphs α_{tri} are indeed 'too simple' because all states in the $\mathcal{H}_{\alpha_{\text{tri}}}$ have zero volume [80]. Also, we now have detailed geometric operators and find that the angles cannot be sharply specified. Nonetheless, Penrose's overall vision is realized in a specific and precise way in the LQG quantum geometry.

Quantum Einstein's equations

Recall from (11.7) that we have three sets of constraints. In the classical theory, the Gauss and the Diffeomorphism constraints generate kinematical symmetries while dynamics is encoded in the Hamiltonian constraint. In the quantum theory the physical Hilbert space $\mathcal{H}_{\text{phy}}$ is to be constructed by imposing the quantum constraints $\hat{C}\Psi_{\text{phy}} = 0$ à la Dirac. This requires one to solve two non-trivial technical problems: i) Introduce well-defined constraint operators $\hat{C}$ on $\mathcal{H}_{\text{kin}}$ starting from the classical constraint functions C; and ii) Introduce the appropriate scalar product on the solutions Ψ_{phy} to obtain $\mathcal{H}_{\text{phy}}$. The second step is non-trivial already for systems with a finite number of degrees of freedom if the constraint operator $\hat{C}$ has a continuous spectrum because then the kinematical norm of physical states Ψ_{phy} diverges. In geometrodynamics, the operators $\hat{C}$ have been defined only formally and generally the issue of the scalar product is not addressed. In LQG by contrast, the availability of a rigorous kinematical framework provides the necessary tools to address both these issues systematically.

For the kinematical constraints, both these steps have been carried out [70]. Since these constraints C_{kin} have a natural geometrical interpretation, the quantum operators $\hat{C}_{\text{kin}}$ simply implement those geometrical transformations on the kinematical (spin-network) states Ψ_{kin} in $\mathcal{H}_{\text{kin}}$. The second task, that of introducing the appropriate scalar product, is carried out using a general strategy called *group averaging* [70, 81]. The detailed implementation of these ideas is straightforward for the Gauss constraint but there are important subtleties in the case of the diffeomorphism constraint [2, 4, 61, 70]. In particular, the strategy described here allows only the exponentiated version of the diffeomorphism constraint, i.e., *finite* diffeomorphisms, and one has to specify the precise class of diffeomorphisms that are allowed.[2] The Hilbert space $\mathcal{H}^{\text{diff}}$ on which both the kinematical constraints are satisfied provides a completion of the Dirac quantization program.

---

[2] With a natural choice of this class, $\mathcal{H}^{\text{diff}}$ is *separable*, although $\mathcal{H}_{\text{kin}}$ is not. This may seem surprising at first. But the situation is completely analogous to what happens already in the quantum theory of free Maxwell theory with the Gauss constraint if one does not wish to work with an indefinite metric [66, 82].

For the Hamiltonian constraint C_H, on the other hand, the situation is still in flux. There is a non-trivial result due to Thiemann that one *can* regulate this constraint systematically on $\mathcal{H}^{\mathrm{diff}}$ [55]. By contrast, no such regularization is available for the WDW equation of geometrodynamics. But the procedure involves the introduction of additional structures in the intermediate steps, whence the final result is ambiguous. Furthermore, the *physical* meaning of the additional structures has remained unclear. Finally, recall that in GR, the Poisson bracket between Hamiltonian constraints smeared with lapse functions N and M is the diffeomorphism constraint smeared with a 'q-number' shift field $K^a = q^{ab}(ND_aM - MD_aN)$. An important question is whether this Poisson bracket structure is reflected in the quantum theory. In these regularizations, in the quantum theory the commutator of the two Hamiltonian constraints vanishes and so does the diffeomorphism constraint on the right hand side on the 'kinematical habitat' on which the calculation is carried out [83]. While this establishes consistency, one would hope for a better scheme in which neither side vanishes and the commutator structure captures the non-trivial, off-shell relation between the constraints.

Recently, a promising approach to this problem has been introduced by Laddha, Varadarajan and others [84, 85]. The first underlying idea is to take hints from earlier work on $(1 + 1)$-dimensional parameterized field theories where well-understood, unconstrained field theories are recast in an extended setting with constraints. The constraints mimic those of GR in that they are again related to space-time diffeomorphisms [86]. One finds that the techniques used in LQG provide a natural avenue to implement quantum constraints in the parameterized form, leading to the correct final quantum theory. The second and deeper observation is motivated by the fact that in connection dynamics the diffeomorphism and the Hamiltonian constraints of (11.7) can be naturally combined in the spinorial setting as $E^{aA}{}_B E^{bB}{}_C F^C_{ab\,D} = 0$ where $A \ldots D$ are spinorial indices [47]. (The trace over A and D yields the Hamiltonian constraint while the trace-free part yields the diffeomorphism constraint.) This unity suggests that, although Hamiltonian constraint generates time evolution, this action could be recast in terms of geometric operations *within the 3-manifold M*. This has been shown to be the case [85]: in the classical theory, time evolution can be re-expressed as the action of *diffeomorphism and Gauss constraints* smeared with certain 'q-number' smearing fields on M. As a consequence, an entirely new perspective emerges. These features do not carry over to geometrodynamics since there the two constraints cannot be naturally combined into a single one. In a simplified theory, where the gauge group SU(2) is replaced by the Abelian group U(1)^3, the program has been carried out and it has been shown that the algebra of constraints *closes off-shell non-trivially*. There is ongoing research to extend these results to the full theory using SU(2).

Finally there has been considerable research on coupling matter fields to gravity, particularly those that can serve as physical clocks and rods [61, 87, 88]. The idea, as in geometrodynamics, is to use the matter fields to 'deparameterize' the constraints and study the ensuing *relational* dynamics. On the conceptual side, these ideas will play a key role in the physical interpretation of canonical LQG. On the technical side, it is rather surprising that certain matter fields do make the quantum constraints manageable enabling one to extract the notion of 'evolution' from the solution to quantum constraints. Once there is a fully satisfactory implementation of the Hamiltonian constraint, these ideas will play a key role in extracting physics from canonical LQG. A qualitative understanding has already begun to emerge because the strategy of [84] to better regularize the constraints and that of [87, 88] to deparameterize the theory using matter can be seen as generalizations to the full theory of the successful strategies used in loop quantum cosmology to first obtain and then interpret the quantum theory.

A complementary approach to dynamics is provided by Spinfoams, discussed in the next subsection.

11.2.3 Covariant loop gravity: spinfoams

The covariant or *Spinfoam* formulation of LQG is built again on the quantum theory of geometry discussed in Section 11.2.2, but now the dynamics is specified by defining the transition amplitudes, order by order in a suitable expansion. This is akin to the spirit used by Feynman to build QED directly in terms of the Feynman rules, which streamlined and simplified the theory.

As with Feynman diagrams, the amplitudes defined in this manner can also be seen as given by a sum over histories. The relevant histories, however, describe spacetime as the *evolution of individual quanta of geometry*, rather than of classical configurations. Thus, a 3-geometry is still represented by a spin-network, and a 4-geometry, by a history of spin-networks. These histories are called *Spinfoams* [3, 89, 90]. The sum over Spinfoams defines transition amplitudes between quantum 3-geometries and, as discussed in Section 11.3.3, the *n*-point functions of non-perturbative quantum gravity.

The main results of covariant LQG to date are the following:

(i) The amplitudes are *finite* at every order.
(ii) At each order, the amplitudes have a well-defined classical limit, related to a truncation of classical general relativity.
(iii) The theory has been extended to include fermions and Yang–Mills fields [91].

Regarding point (i), there are two potential sources of infinities in the theory. They are the ultraviolet (UV) divergence which corresponds to the conventional infinities of perturbative Feynman diagrams, and infrared (IR) divergences that can arise from the contributions of intermediate states with large-scale geometries. The UV divergences are naturally cured by the discreteness of the underlying quantum geometry itself. The IR divergences are cured by the presence of a positive cosmological constant Λ. Therefore, interestingly, the structure of the theory is such that a cosmological constant *with a positive sign* naturally acts as a physical IR regulator. Regarding point (ii), recall that the classical limit of lattice QCD on a fixed triangulation is just the classical lattice theory. Similarly, the classical limit of covariant loop quantum gravity at a fixed order is related to Regge calculus on a finite triangulation in a precise sense. We will now discuss these issues in detail.

Transition amplitudes

For simplicity, we describe the case without fermions and Yang–Mills fields and with $\underline{\Lambda} = 0$ (thus ignoring IR problems). We later discuss the necessary modifications to incorporate $\underline{\Lambda} > 0$.

In quantum theory, one calculates the transition amplitudes between initial and final states. In LQG these are states of quantum geometry and therefore belong to the Hilbert spaces $\mathcal{H}_\alpha^{\text{diff}}$ labeled by (abstract) graphs α. In the absence of an external time parameter, there is no distinction between the initial and the final states. Therefore, it is convenient to combine the two graphs that refer to the boundary states; we denote this total graph by Γ. Then, if Γ has N nodes n and L links l, $\mathcal{H}_\Gamma^{\text{diff}}$ is spanned by states $\psi(U_l)$ which are in $L^2(\text{SU}(2))^L$ and invariant under SU(2) gauge transformations at the nodes. Thus, $\mathcal{H}_\Gamma^{\text{diff}}$ is the same as the Hilbert space of an SU(2) lattice Yang–Mills theory. The theory defines a transition amplitude for each of these states $\psi(U_l)$.

To any given order, the transition amplitudes are labeled by a *2-complex* $\mathcal{C}$, a higher-dimensional analog of a graph: it is defined as a (combinatorial) set of *faces f* meeting at *edges e*, which in turn meet at *vertices v* (see Fig. 11.2). It can be regarded as a history of a spin-network in which each link *l* of the spin-network 'evolves' to form a face *f*, each node

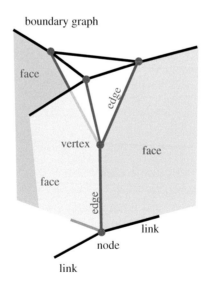

Figure 11.2 2-complex terminology. From Rovelli, "Zakopane Lectures on Loop Quantum Gravity", *Proceedings of the Quantum Gravity and Quantum Geometry School, Zakopane*, F. Hellmann *et al.* (Eds) PoS (QGQGS) (2011).

n 'evolves' to an edge *e*, and non-trivial dynamics occurs when a new vertex *v* appears. The number of vertices in $\mathcal{C}$ defines the order in the expansion of the transition amplitude.

To grasp the interplay with space-time geometry, it is useful to note that a triangulation Δ of the 4-dimensional spacetime defines a dual 2-complex: each 4-simplex of Δ corresponds to a vertex in $\mathcal{C}$, each tetrahedron of Δ to an edge in $\mathcal{C}$, and each triangle in Δ to a face of $\mathcal{C}$. For simplicity, we will only consider 2-complexes dual to triangulations. The boundary $\Gamma = \partial\mathcal{C}$ of a 2-complex is a graph that the state $\psi(U_l)$ refers to (see Fig. 11.2).

Let us fix a 2-complex with boundary graph Γ and *v* vertices. Then, the *v*th-order term in the expansion of the transition amplitude associated with a state $\psi \in \mathcal{H}_\Gamma^{\mathrm{diff}}$ is defined as the scalar product in $\mathcal{H}_\Gamma^{\mathrm{diff}}$ of the state ψ with the function $W_\mathcal{C}(U_l)$ defined as follows:

$$W_\mathcal{C}(U_l) = \int_{SU(2)} dh_{vf} \prod_f \delta(h_f) \prod_v A_v(h_{vf}). \qquad (11.15)$$

This *W* is the key object in Spinfoams because, to order *v* in our expansion, quantum dynamics is encoded in *W*. The integral in (11.15) is over one SU(2) variable h_{vf} associated with each vertex–face pair, the delta distribution is over SU(2) and its argument h_f is the (oriented) product of the variables h_{vf} around a face. In the 4d Lorentzian theory, the "vertex amplitude" A_v has the form

$$A_v(h_{vf}) = \int_{SL(2,\mathbb{C})} dg_{ve} \prod_e \sum_j (2j+1) \mathrm{Tr}_j[Y^\dagger g_e g_{e'} Y h_f]. \qquad (11.16)$$

Here the SL(2, $\mathbb{C}$) integration variables g_{ve} are associated with each edge emerging from *v* and the two edges *e* and *e′* in the trace are those bounding the face *f*. *Y* is a map from the spin *j* representation of SU(2) to the unitary representation of SL(2, $\mathbb{C}$) with continuous quantum number $\gamma(j+1)$ and discrete quantum number *j*, defined by[3]

$$Y: \; |j;m\rangle \quad \mapsto \quad |\gamma(j+1), \gamma;j,m\rangle, \qquad (11.17)$$

where, as before, γ is the Barbero–Immirzi parameter. These three equations completely define the theory. Quite remarkably, to order *v* in the vertex expansion, they encode the entire quantum dynamics.

This form of the amplitude is variously denoted the EPRL, EPRL-FK, or EPRL-FK-KKL amplitude. It was derived in [92] building on results in [93, 94] and extended to arbitrary 2-complexes in [95], and forms the basis of the 4d Lorentzian theory so far. Variants could be interesting; some of these have been considered for the Euclidean theory [96].

Let us examine the structures of this amplitude. The appearance of SL(2, $\mathbb{C}$) is not surprising: it reflects the local Lorentz invariance of GR. The appearance of the *unitary* (infinite dimensional) representations of this group should not be too surprising either, given that unitary representations of symmetry groups are ubiquitous in quantum gravity. Indeed, one may wonder why the mathematics of the infinite dimensional unitary representations of SL(2, $\mathbb{C}$) has played such a small role in the attempts to construct a quantum

---

[3] The maps extends easily from functions on SU(2) to functions of SL(2, $\mathbb{C}$).

theory of gravity so far. The map Y, on the other hand, is a new ingredient that constitutes the technical core of the Spinfoam model and deserves explanation. For this, let us first return to the classical theory discussed in Section 11.2.2. In the space-time picture, the momentum conjugate to the $SL(2, \mathbb{C})$ connection ω is [57]

$$\pi_{IJ} = \frac{1}{4\kappa_N} \left(\epsilon_{IJKL} e^K \wedge e^L + \frac{1}{2\gamma\kappa_N} e_I \wedge e_J \right). \tag{11.18}$$

On a boundary of a space-time region, the one-form normal to the boundary contracted with the tetrad gives a vector in the internal Minkowski space, which determines a preferred Lorentz frame. We can decompose π_{IJ} in this frame in the same manner in which the Maxwell field F_{IJ} decomposes in the electric and magnetic field. Simple algebra then shows that the electric and magnetic parts of π_{IJ}, denoted respectively $\boldsymbol{K}$ and $\boldsymbol{L}$, satisfy the algebraic equation

$$\boldsymbol{K} = \gamma\boldsymbol{L}. \tag{11.19}$$

This is a key equation in covariant loop quantum gravity, called the *simplicity constraint*. To ensure a correct classical limit, this constraint has to be implemented in the quantum theory in an appropriate fashion. This is precisely what the map Y does: in the quantum theory (11.19) holds on the image of this map as a weak operator equation (i.e., for all matrix elements of the operators) [97].

Classical limit

The Spinfoam dynamics presented in the last subsection was arrived at from several independent considerations: the Hamiltonian LQG [89], the fact that GR can be regarded as a constrained BF theory [98], the Ponzano–Regge and Turiev–Viro models [99–102] for quantum gravity in three dimensions and group field theory [3, 103]. Furthermore, the overall paradigm underlying Spinfoams is borne out in symmetry reduced, cosmological models, where the transition amplitudes obtained by summing over quantum geometries have been shown to be finite and in agreement with the Hamiltonian theory [104]. While these considerations provide a reasonably strong motivation, one still needs direct evidence in favor of the specific proposal (11.15). Analysis of the classical limit provides a natural avenue to test its viability.

In any quantum theory, the classical limit is obtained in a regime where quantum numbers are large. Then the relevant actions are large compared to the Planck constant and the limit can be interpreted as $\hbar \to 0$. For example, for a particle with a Hamiltonian H, in the $\hbar \to 0$ limit we have:

$$W(x, t; x', t') \sim \int [Dx] e^{\frac{i}{\hbar} S[x]} \sim A e^{\frac{i}{\hbar} S(x, t; x', t')}, \tag{11.20}$$

where the integration is over the paths from (x, t) to (x', t') and $S(x, t; x', t')$ is the Hamilton function, namely the value of the action on the solution of the classical equations of motion that starts at (x, t) and ends at (x', t'). In gravity the analogous procedure requires us to consider areas and volumes that are large compared to the Planck scale. Thus, to study the classical limit of Spinfoam dynamics, one can compare the large j limit of the transition amplitude (11.15) with the classical action. The asymptotic analysis of the vertex amplitude (11.16) is nontrivial, and has been carried out mainly by the Nottingham group [105]. For the simplest case where the 2-complex $\mathcal{C}$ has only one vertex v, the results can be summarized as follows. Recall that the amplitude is a function of the boundary quantum state ψ and quantum geometries are more general than Regge geometries. If ψ does not endow the 4-simplex Δ dual to $\mathcal{C}$ with a consistent classical geometry, the transition amplitude is suppressed exponentially. If it does, then the asymptotic form of the amplitude is given by

$$A_v \sim A \left(e^{\frac{i}{\hbar}(S_R + \frac{\pi}{4})} + e^{-\frac{i}{\hbar}(S_R + \frac{\pi}{4})} \right), \tag{11.21}$$

where S_R is the Regge action of Δ. The presence of two terms in (11.21) is a consequence of the fact that, as we saw in Section 11.2.2, the starting point of the analysis is tetrad gravity and, when the tetrad changes orientation, the first order LQG action changes sign, while the Einstein–Hilbert action does not. Consequently, for each classical metric solution we have two tetrad solutions whose action is equal in magnitude but with opposite signs. The $\pi/4$ is also well understood: it is the Maslov index that always appears in the semiclassical limit when the two classical solutions sit on different branches of the solution space [106]. Therefore the result (11.21) has the following simple interpretation: the classical limit of the transition amplitude defined by a 2-complex $\mathcal{C}$ dual to a space-time triangulation Δ is the Regge amplitude associated with that triangulation. This is precisely what one would have hoped. In this sense the proposal passes the viability criterion and it is reasonable to regard equations (11.15), (11.16) and (11.17) as providing a tentative definition of the dynamics of LQG.

This covariant formulation of LQG has some similarities with the path-integral approach based on Regge calculus [107, 108] where one sums over configurations representing a Regge discretization of general relativity. This approach was introduced already in the 1980s and has evolved considerably since then [109]. In spite of the formal structural similarity, there is an important conceptual difference between the two approaches. In Regge calculus, the lengths of the individual links can be arbitrarily small. By contrast, the geometries that are summed over in Spinfoams represent histories of *quanta* of space, whence the areas of plaquettes cannot be arbitrarily small; they are bounded below by the area-gap of LQG. This fundamental discreteness naturally removes the UV divergences and introduces the Planck scale already in the permissible histories that are summed over. Consequently, the scaling structure of the theory with respect to the Regge calculus is quite different.

We conclude this discussion by noting that the classical limit we have discussed here should not be confused with the continuum limit of the theory. The first is the standard $\hbar \to 0$ limit while the second refers to refinement, i.e., adding more and more degrees of freedom. Recall that the classical limit of lattice QCD on a fixed lattice is of course a classical lattice theory. In LQG, the lattice is replaced by a triangulation, but with the

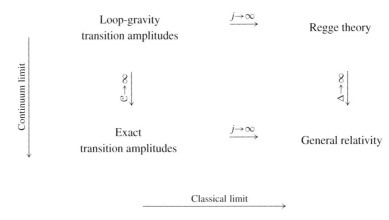

Figure 11.3 Relation between continuum limit and classical limit of the transition amplitudes.

crucial difference that its geometry is not pre-specified but constitutes the dynamical variable. Nonetheless, the situation with respect to the classical limit is similar: in covariant
loop quantum gravity, this limit is related in a precise way to Regge calculus on a finite
triangulation. As is well known, classical Regge calculus converges to full GR in the limit
in which the triangulation is refined. The structure of the theory is therefore as in the
Figure 11.3: To arrive at GR from Spinfoams, one can start from the upper left corner
of the diagram, move first to the right and then down.

Cosmological constant and IR finiteness

As we noted above, the reason behind the UV finiteness of the Spinfoam amplitude (11.15)
is intuitively simple: because of the discreteness of space at the Planck scale, there is an in-
built and natural physical cutoff preventing the standard quantum field theory divergences.
In other words, there are no degrees of freedom at arbitrary small scales. Therefore the sum
over intermediate states in a perturbation expansion does not include field configurations of
arbitrary high momentum. From this perspective, the UV divergences of standard quantum
field theory can be interpreted as pathologies introduced by the fact of neglecting the
discrete nature of space.

However, the amplitude (11.15) can have IR divergences. This can happen every time
the 2-complex has a *bubble*, i.e., a set of continuous faces with the topology of a 2-sphere.
These bubbles are the Spinfoam analog of loops in Feynman diagrams: The quantity circulating around a Feynman diagram loop is the momentum, and high momentum means
UV; while the quantity circulating around a Spinfoam bubble is the area, and high area
means IR. On a bubble, the sum over spins j in (11.15) can lead to divergent terms because
j is unbounded above. Geometrically, these divergences correspond to 'spikes, representing
large regions of space-time bounded by small hypersurfaces.'

Remarkably, these divergences disappear naturally if there is a positive cosmological
constant $\underline{\Lambda}$ in the theory. Technically, the effect of a positive $\underline{\Lambda}$ is to replace SL(2, C) with

a quantum deformation of SL(2, C). The mathematics for implementing this deformation has been developed [110, 111] and the Spinfoam amplitude with the cosmological constant has also been defined [112, 113]. The transition amplitudes of the theory with a quantum deformation of SL(2, C) are finite, and the classical limit of their vertex amplitude is still given by Eq. (11.21). But now the Regge action that appears in the classical limit has a cosmological constant $\underline{\Lambda}$, related to the deformation parameter q of the quantum group via [113, 114]

$$q = \exp(\underline{\Lambda}\hbar G). \tag{11.22}$$

Therefore the full theory now depends on *two* dimensionless parameters: q or $\underline{\Lambda}\hbar G$, and the Barbero–Immirzi parameter γ. The bare cosmological constant enters the theory as a free parameter, therefore the theory does not prescribe its value. To explore various limiting regimes, one has to calculate the behavior of physical observables, keeping appropriate combinations of these constants fixed, and let a complementary combination tend to the desired value.

QED, QCD and LQG

Similarities between the Spinfoam model defined in the last three subsections and QCD on a fixed lattice are evident: In both cases, we have a discretization of the classical theory where the connection is replaced by group elements, and a quantum theory defined by an integral over configurations of an amplitude which is a product of local quantities. The use of a triangulation in Spinfoams instead of a square lattice simply reflects the fact that a square lattice is unnatural in the absence of a flat metric. However, there is also a crucial difference. The Wilson QCD action depends on an external parameter, the lattice spacing a, while appropriate discretizations of the Einstein–Hilbert action, like the Regge action, do not. To recover the continuum theory, in QCD it is not sufficient to increase the total size of the lattice; it is also necessary to send a to zero. Equivalently, the lattice spacing a can be absorbed in the coupling constant β in front of the action, and, in order to recover the continuum limit, it is necessary to tune β to its critical value, $\beta = 0$. In gravity, instead, the Regge action (or any other admissible discretization) does *not* include a lattice spacing a (nor, therefore, a coupling constant that needs to be tuned to a critical value as $a \to 0$). The reason is simply that the lattice spacing a refers to a background geometry – the Yang–Mills theory depends on a fixed, externally given space-time metric – while in gravity the geometry is included in the dynamical variables. It can be shown in general [115] that the discretization of a reparametrization invariant theory can be defined *without* a parameter that needs to be tuned to a critical value in the continuum limit. Accordingly, in a suitable discretization of general relativity the continuum limit can be defined just by making the triangulation (or the 2-complex) increasingly finer.[4] An alternative approach to the continuum limit is discussed in [116].

---

[4] In concrete physical calculations, however, only finite triangulations suffice, as is generally the case in QCD. Similarly, in QED a finite number of Feynman graphs suffice.

Interestingly, there are also similarities between Spinfoams and perturbative QED. The nodes of the graph can be seen as quanta of space and the 2-complex can be read as a history of these quanta, showing where these quanta interact, join and split, just as real and virtual particles do in the Feynman graphs. The analogy is reinforced by the fact that the Spinfoam amplitude can actually be concretely obtained as a term in a Feynman expansion of a 'group field theory' (see for instance Chapter 9 of [3] and [96, 103]). The specific group field theory that gives the gravitational amplitude (11.15) has been derived (in the Euclidean context) in [117].

Thus, the Spinfoam paradigm shares some key features with QCD as well as QED, our two most successful, fundamental quantum theories. In addition, Spinfoams bring out a novel interplay between these theories and quantum gravity. A Feynman graph of QED is a history of quanta of a field, while the lattice used in QCD is a collection of discrete chunks of space-time. They are distinct and unrelated. But general relativity taught us that space-time itself is a field – the gravitational field – and in LQG its discrete chunks are the quanta of this field. Therefore, once we recognize that the gravitational field is both dynamical and quantum, the quantum gravity analog of the lattice used in QCD can be seen as a Feynman graph of a quantum theory, representing the history of gravitational quanta. In this sense, the Feynman graphs of QED and lattices of QCD merge in LQG via Spinfoams.

11.3 Applications

Exploration of the physical consequences of Asymptotic Safety is still at its beginning. First investigations on both cosmological [118, 119] and black-hole spacetimes [120, 121] have been performed within asymptotically safe QEG. The main idea is to employ a method often used in particle physics that goes under the name *RG improvement*. Here, it amounts to replacing G, Λ with G_k, Λ_k and identifying k with an appropriately chosen dynamical or geometrical scale. Since this identification suffers from a certain degree of ambiguity, ultimately the method will have to be to be replaced by a more precise one. Nonetheless, these investigations have already provided a first idea of the QEG effects to be expected. Because the subject is still evolving, we will discuss these ideas in the 'Outlook' section, Section 11.4.2.

On the other hand, three applications of LQG have been investigated in detail over the last 10–15 years, resulting in thousands of publications, whose results have been summarized in several detailed reviews (see, e.g., [5, 6, 12, 13, 122, 123]). In this section we will present some highlights of those developments. Even though LQG is still far from being a complete theory, advances could be made by using a *truncation strategy*: One first chooses the physical problem of interest, focuses just on that sector of the full theory which is relevant to the problem, and then uses LQG techniques to analyze it, making full use of the quantum geometry summarized in Section 11.2.2.

The section is divided into three parts. In the first we discuss the very early universe; in the second, quantum aspects of black holes, and in the third, the issue of defining n-point functions in a manifestly background independent theory.

11.3.1 The very early universe

It is evident from Chapter 3 that there has been a huge leap in our understanding of the early universe over the past two decades. However, on the conceptual front a number of issues have remained in the Planck era of the very early universe. Over the last decade these issues have been systematically addressed in Loop Quantum Cosmology (LQC).

In particular, the big bang singularity was resolved and cosmological perturbations are being analyzed following several approaches [12, 123–126]. For brevity, we will focus on one of these which provides an internally consistent paradigm starting from the Planck regime, with detailed predictions that are compatible with the WMAP and Planck data. In the first two parts of this subsection we summarize the main results and in the third we present a critical analysis of the adequacy of the truncation strategy that underlies the discussion of quantum cosmology in *any* approach.

Singularity resolution

Every expanding Friedmann–Lemaître–Robertson–Walker (FLRW) solution of GR, has a big bang singularity if matter satisfies the standard energy conditions. But scalar fields with potentials that feature in the inflationary scenarios violate these energy conditions. Therefore, initially there was a hope that the standard singularity theorems of GR [127] could be avoided in the inflationary context. However, this turned out not to be the case: Borde, Guth and Vilenkin [128] showed, *without any reference to energy conditions*, that if the expansion of a congruence of past-directed time-like or null geodesics is negative (on an average), then they are necessarily past incomplete; the finite beginning represented by the big bang in GR is not avoided. But these arguments assume a smooth, classical geometry all the way back to the big bang, which has no physical basis since quantum effects cannot be ignored in the Planck regime. Thus, although it is often heralded as reality, the big bang is a prediction of classical gravity theories in a domain in which they are *not* applicable. A key result of LQC is that the quantum geometry effects in the Planck regime lead to a natural resolution of the big bang in a wide variety of cosmological models [12, 122].

To illustrate how this comes about, consider the simple example of the $k = 0$ FLRW spacetimes with a massless scalar field ϕ as a source. It is convenient to fix a fiducial cell $\mathcal{C}$ in co-moving coordinates and plot these solutions directly in terms of physical variables of the problem, the scalar field and the volume v of $\mathcal{C}$. As the left panel of Fig. 11.4 suggests, one can regard the scalar field as a *relational* time variable, in terms of which the volume v – and hence the curvature – 'evolves'. In Bianchi models, the 'evolving' quantities would also include the anisotropies, and in, say, the Gowdy model, they would include the inhomogeneities encapsulating gravitational waves. Since the massless scalar field *does* satisfy all the energy conditions, all these solutions are singular. In the $k = 0$ FLRW case the universe either expands starting with the big bang or contracts into the big crunch singularity. Quantum cosmology was introduced in the 1970s in the hope that these classical singularities would be tamed by quantum effects [21]. However, in the WDW quantum geometrodynamics of the simple model under consideration unfortunately this hope is not

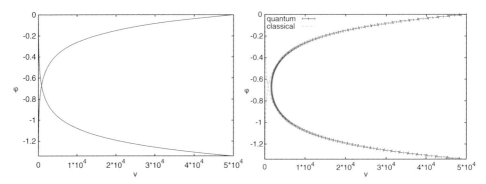

Figure 11.4 The FLRW model with a massless scalar field. *Left Panel:* In classical GR, there are two types of solutions; those which begin with a big bang and expand forever and those that start out with zero energy density and contract to a big crunch singularity. *Right Panel:* In LQG, quantum geometry effects create a novel repulsive force which starts becoming significant when the energy density and curvature are $\sim 10^{-3}$ in Planck units. The force grows with curvature and the big bang is replaced by a quantum bounce. From A. Ashtekar, *Gen. Rel. Grav.* **41**, 707–741 (2006). With kind permission from Springer Science and Business Media.

realized [129, 130]. The idea was revived some three decades later in a pioneering paper by Bojowald, who showed that the situation is quite different in LQC [131]. Subsequent conceptual completions and technical improvements of this reasoning have provided a systematic understanding of how this comes about.

First, the uniqueness theorem of LQG kinematics descends to LQC [132], making LQC inequivalent to the WDW theory already at the kinematical level. The WDW differential equation turns out not to be well-defined on the LQC Hilbert space and one has to return to the classical Hamiltonian constraint and systematically construct the corresponding quantum operator making full use of the underlying quantum geometry of LQG [133]. This construction is subtle and brings out a delicate interplay between the discreteness of *physical* areas and the mathematics underlying the definition of the Hamiltonian constraint in the connection dynamics framework [129, 134]. A detailed analytical argument [135] has established that the density operator is well defined on the physical Hilbert space with an upper bound $\rho_{max} = 3/(8\pi\gamma^2 G(\Delta a_S))$ which is *directly* controlled by the area gap Δa_S of LQG. Thus, density and curvature cannot diverge in *any* physical state. Numerical simulations showed that this upper bound is in fact reached in states which are sharply peaked at late times and, not surprisingly, it is reached precisely at the bounce. Finally, the LQC singularity resolution has also been established in the consistent histories approach [136].

The details of LQC dynamics can be summarized as follows. Let us first consider the classical solution depicted by an expanding trajectory (Fig. 11.4, left panel). Fix a point at a late time, consider a quantum state sharply peaked at that point and evolve it using the LQC Hamiltonian constraint. One then finds that the wave packet remains sharply peaked on the classical trajectory so long as the matter density or curvature is less than $\sim 10^{-3}$

of the Planck scale. Thus, in this regime there is good agreement with GR. However, if we evolve the quantum state backwards towards the singularity, instead of following the classical trajectory into the singularity – as is the case in the WDW theory – the wave packet bounces. The expectation value of the energy density now starts decreasing and once it reaches a few thousandths of ρ_{Pl}, the peak of the wave packet again follows a classical trajectory along which the universe expands as we continue to move *backward* in time (Fig. 11.4, right panel). An important feature of LQC dynamics is that while the quantum geometry effects are strong enough to resolve the big bang, agreement with GR is recovered quickly, already when the curvature has fallen by a factor only of $\sim 10^{-3}$ from the Planck scale. One can modify Einstein's equations by introducing some quantum gravity effects by hand and resolve the singularity. But such modifications generally lead to departures from GR already at the density of water! LQC naturally achieves the delicate balance: the UV pathology is tamed, leaving GR intact rather close to the Planck regime.

Although in the Planck regime the peak of the wave function deviates very substantially from the general relativistic trajectory, rather surprisingly it follows an *effective trajectory* with very small fluctuations (see Fig. 11.4). This effective trajectory was derived [137] using techniques from geometric quantum mechanics. The effective equations incorporate the leading corrections from quantum geometry. They modify the left hand side of Einstein's equations. However, to facilitate comparison with the standard form of Einstein's equations, one moves this correction to the right side through an algebraic manipulation. Then, one finds that the Friedmann equation $(\dot{a}/a)^2 = (8\pi G \rho/3)$ is replaced by

$$\left(\frac{\dot{a}}{a}\right)^2 = (8\pi G \rho/3)\left(1 - \frac{\rho}{\rho_{max}}\right). \tag{11.23}$$

At $\rho = \rho_{max}$, the right side vanishes, whence $\dot{a}$ vanishes and the universe bounces. This can occur because the LQC correction ρ/ρ_{max} *naturally* comes with a *negative* sign which gives rise to an effective 'repulsive force'. The occurrence of a negative sign is non-trivial: in the standard brane world scenario, for example, the Friedmann equation also receives a ρ/ρ_{max} correction but it comes with a positive sign (unless one makes the brane tension negative by hand) whence the singularity is not resolved. Finally, there is an excellent match between analytical results within the quantum theory, numerical simulations and effective equations. In particular, the effective equations capture the leading LQC corrections to Einstein's equations very efficiently.

This analysis has been extended to include the cosmological constant of either sign [138], the k = 1 FLRW models [134], the Bianchi I, II and IX models which include anisotropies [139, 140] and the Gowdy models which include inhomogeneities [141]. Furthermore the effective equations have been used to show that in LQC *all* curvature singularities – including, e.g., the big rip – are resolved in all FLRW models [142]. These results suggest that the quantum geometry effects of full LQG may well lead to a resolution of all space-like, strong curvature singularities of GR.

Finally, note that in all the models that have been studied in detail, singularity resolution occurs *generically* without any exotic matter or need to fine-tune initial conditions.

Furthermore, one does not have to introduce a new boundary condition such as in the Hartle–Hawking proposal. Why then does the LQC singularity resolution not contradict the standard singularity theorems of Penrose, Hawking and others? These theorems are inapplicable because *the left hand side* of the classical Einstein equations is modified by the quantum geometry corrections of LQC. What about the more recent singularity theorems that Borde, Guth and Vilenkin [128] proved in the context of inflation? They do not refer to Einstein's equations. But, motivated by the eternal inflationary scenario, they assume that the expansion is positive along any past geodesic. Because of the pre-big-bang contracting phase, this assumption is violated in the LQC effective theory.

Phenomenology: implications of the pre-inflationary dynamics

The inflationary scenario has had an impressive success in accounting for the observed 1 part in 10^5 anisotropies in the CMB. Therefore, although many of the LQC results hold in a broad class of early universe paradigms (see, e.g., [143]), for brevity and concreteness we will restrict ourselves to inflation here.

The resolution of the big bang singularity opens a natural avenue to extend this scenario to the Planck regime by systematically investigating the pre-inflationary dynamics. It is often argued that while this phase is conceptually important, it can not be relevant for observations because the near-exponential expansion during inflation would wash away all memory of prior dynamics. The reasoning is that modes seen in the CMB cannot be excited by the pre-inflationary dynamics because, when evolved back in time starting from the onset of the slow roll, their physical wavelengths λ_{phy} continue to remain within the Hubble radius $\mathfrak{R}_{\text{H}}$ all the way to the big bang. However, this argument is flawed on two accounts. First, what matters to the dynamics of these modes is the curvature radius $\mathfrak{R}_{\text{curv}} = \sqrt{6/R}$ determined by the Ricci scalar R, and not $\mathfrak{R}_{\text{H}}$, and the two scales are equal only during slow roll. Thus we should compare λ_{phy} with $\mathfrak{R}_{\text{curv}}$ in the pre-inflationary epochs. The second and more important point is that the pre-inflationary evolution should not be computed using general relativity, as is done in the argument given above. One has to use an appropriate quantum gravity theory since the two evolutions could well be very different in the Planck epoch. Therefore, modes that are seen in the CMB could have $\lambda_{\text{phy}} \gtrsim \mathfrak{R}_{\text{curv}}$ in the pre-inflationary phase. If this happens, these modes *would be* excited and the quantum state at the onset of the slow roll could be quite different from the Bunch–Davies (BD) vacuum used at the onset of the slow roll.

Now, another common assumption was that even if there are such excitations over the BD vacuum at the onset of inflation, they would have no effect because they would be diluted away during inflation. However, this is not the case: stimulated emission compensates for expansion so the excitations persist at the end of inflation [144, 145]. Indeed, the difference from the standard prediction could well be so large that the resulting power spectrum is incompatible with the amplitude and the spectral index observed by WMAP. In this case, that particular quantum gravity scenario would be ruled out. On the other hand, the differences could be more subtle: the new power spectrum for scalar modes could be the same but there may be departures from the standard predictions that involve tensor modes or higher order correlation functions of scalar modes, changing the conclusions on non-Gaussianities. In this case, the quantum gravity theory would have interesting predictions for future observational missions [145]. Thus, pre-inflationary dynamics can provide an avenue to confront quantum gravity theories with observations.

To analyze what happens during the pre-inflationary phase, in LQC one proceeds as follows. Since in the inflationary paradigm it is adequate to consider just the FLRW geometries and first order scalar (or curvature) and tensor perturbations $\mathcal{R}$, $\mathcal{T}$, one first truncates the full phase of GR to this sector, replaces the FLRW metrics with the quantum wave functions Ψ_o provided by LQC and investigates the dynamics of first order quantum perturbations $\hat{\mathcal{R}}$, $\hat{\mathcal{T}}$ on these *quantum* FLRW geometries [124, 146]. Since quantum perturbations now propagate on quantum geometries which are all regular, free of singularities, *the framework automatically encompasses the Planck regime.* What is then the status of the 'trans-Planckian issues' discussed in the context of inflation? A careful examination shows that they boil down to the following question: Is the LQC truncation scheme self-consistent? That is, is it consistent to ignore the back reaction and work just with first order quantum perturbations on quantum FLRW backgrounds? This central issue is extremely difficult to analyze in any approach to quantum gravity because it requires a careful treatment of regularization and renormalization of the stress energy tensor of quantum perturbations on FLRW *quantum* geometries.

The LQC analysis was carried out in detail using the simplest $\frac{1}{2} m^2 \phi^2$ potential that is compatible with the current observations. It has revealed three interesting features [12, 123, 124]. First, there exist quantum states of the background FLRW geometry that remain sharply peaked on solutions to effective equations all the way from the bounce till the curvature has fallen by several orders of magnitude, when general relativity is an excellent approximation. Therefore, one can focus on effective dynamics and ask if these solutions would generically encounter the phase of slow-roll inflation that is compatible with observations. It turns out that these solutions are completely determined by the value ϕ_B of the inflaton at the bounce and it is constrained to lie in a finite interval, $|\phi_B| \in [0, 7.47 \times 10^5]$. *This is the parameter space of LQC.* For definiteness, let us suppose that the inflaton and its time derivative have the same sign at the bounce. Then, the detailed analysis shows that the dynamical trajectory *will* encounter an inflationary phase compatible with observations (within the WMAP error bars) provided $\phi_B > 0.93$, i.e., in almost the entire parameter space [147].

So we can choose an effective trajectory with $\phi_B > 0.93$, select a quantum state Ψ_o which is sharply peaked on it and consider quantum fields $\hat{\mathcal{R}}$, $\hat{\mathcal{T}}$ representing scalar and tensor perturbations on the quantum geometry Ψ_o. At first the problem of studying their dynamics seems intractable. However, the detailed investigation has brought out a second non-trivial and completely unforeseen feature: assuming that the back reaction can be neglected, the dynamics of $\hat{\mathcal{R}}$, $\hat{\mathcal{T}}$ on quantum geometry Ψ_o is *identical* to that of quantum fields $\hat{\mathcal{R}}$, $\hat{\mathcal{T}}$ propagating on a smooth, classical FLRW metric $\bar{g}$ constructed from Ψ_o. This construction is quite subtle and involves rather complicated combinations of the expectation values of various operators in the state Ψ_o. Thus, although the scalar and tensor modes $\hat{\mathcal{R}}$, $\hat{\mathcal{T}}$

propagate on the quantum geometry Ψ_o, their dynamics is sensitive to only those features of Φ_o that are captured in $\bar{g}$. This $\bar{g}$ is a 'dressed' effective metric: While the metric determined by the effective equations discussed above knows only about the expectation values, $\bar{g}$ knows also about certain fluctuations, i.e., a finite number of 'higher moments' of Ψ_o. The physics behind this result can be intuitively understood in terms of a simple analogy: As light propagates in a medium, while there are many interactions between the Maxwell field and the atoms of the medium, the net effect can be neatly coded in just a few parameters such as the refractive index. In LQC, the result provides a powerful technical simplification because it enables one to 'lift' various well-developed mathematical techniques from QFT on classical FLRW spacetimes to $\hat{\mathcal{R}}$, $\hat{\mathcal{T}}$ propagating on quantum geometries Ψ_o.

However, this analysis assumes that the back reaction can be neglected. One can always start by restricting oneself to states ψ for which this assumption holds at the bounce. But there is no guarantee that the condition will continue to be satisfied under evolution, especially in the Planck regime immediately after the bounce. Does the energy density of the fields $\hat{\mathcal{R}}$, $\hat{\mathcal{T}}$ remain negligible all the way from the deep Planck regime of the bounce to the onset of slow roll, removed from the bounce by some 11 orders of magnitude in curvature? This issue can be settled only numerically. These simulations require great care because: i) the renormalization procedure subtracts two diverging terms whence even a tiny loss of precision can result in a significant error; ii) the simulation has to be carried out over a very large number of time steps; and, iii) since the background density falls rapidly, even extremely small numerical errors (of the order of one part in 10^{15}) can be comparable to the background energy density. Simulations with all the due care have been performed to establish firm *upper bounds* on the energy density in perturbations. They showed that if $\phi_{\mathrm{B}} > 1.23$, there is a natural choice of initial conditions for ψ at the bounce such that the back reaction can indeed be ignored from the bounce to the onset of inflation. Furthermore, there are analytical arguments to show that if a state ψ satisfies this condition, then all states in an open neighborhood do so. Any of these states provide a *self consistent* solution in which the initial truncation hypothesis is seen to be satisfied in the final solution. This is the third non-trivial result. Together, the three results establish that LQC does provide a self-consistent extension of the standard inflationary scenario to the Planck regime for almost all of the LQC parameter space.

What are then the phenomenological predictions of these self-consistent solutions? The power spectrum and the spectral index have been calculated and, as in the standard inflationary calculations, they agree with observations to within error bars. However, there is a small window in the LQC parameter space where certain LQC predictions differ from those of standard inflation. For example, the standard 'consistency relation' $r = -8n_t$ relating the ratio r of the tensor-to-scalar power spectra to the tensor spectral index is modified [124]. These deviations arise precisely by the mechanism we discussed above: the LQC effective dynamics of the FLRW background is qualitatively different from that of GR so that certain modes *can* have wavelengths λ larger than the curvature radius. Therefore, at the onset of inflation the LQC quantum state ψ of perturbations has excitations over the BD vacuum in these modes. This departure from the BD vacuum also has implications for the CMB and galaxy distribution [145] and observational tests for such effects have already been proposed [148]. A careful analysis of this window in the LQC parameter space is a focus of current research.

To summarize, LQC has led to a natural resolution of the initial singularity in cosmological models of direct physical interest via quantum geometry effects that replace the big bang with a big bounce [129, 131, 134]. Cosmological perturbations on these quantum geometries have been studied in detail [124–126]. There are natural choices of states at the bounce for which one obtains self consistent extensions of the inflationary scenario all the way to the Planck regime of the bounce. By combining these results with the very rich set of results on inflationary and post-inflationary dynamics, one obtains a coherent paradigm to account for large scale structure, starting right at the quantum bounce. Furthermore, in a small window of the parameter space, this analysis provides results that differ from standard inflation, thereby opening an avenue to extend the reach of observational cosmology to the Planck scale.

Is quantum cosmology justified?

As emphasized in Section 11.2.3, in our most successful theories, such as QED and QCD, the *actual* calculations of physical effects have always involved truncations. The mini and midi superspace were introduced in the 1960s in the hope that this truncation would be sufficient to capture the salient quantum effects that tame cosmological singularities. Now that this hope has been borne out, it is appropriate to reexamine the strategy and ask: Is this truncation where one ignores an infinite number of degrees of freedom not too severe?

The LQC strategy is guided by the following considerations. First, there is an analogy with Dirac's solution to the Hydrogen atom problem. From the perspective of full QED, Dirac's restriction to spherical symmetry is a drastic truncation because it removes all physical photons and ignores all but a finite number of degrees of freedom. But the results of this truncated theory are in excellent agreement with observations and we need quantum corrections from QED only when the accuracy of experiments is at the level of the Lamb shift, at which the vacuum fluctuations of the photon field cannot be ignored. The viewpoint is that the situation is similar in cosmology: An analysis of the problem in the mini-superspace approximation *that appropriately takes into account quantum geometry effects from the full theory* should provide a good approximation to the predictions of the full theory. The second source of intuition is provided by the Belinskii–Khalatnikov–Lifshitz (BKL) conjecture in GR discussed in Chapter 9. It suggests that as one approaches a generic space-like singularity in GR, the local evolution is well approximated by the Bianchi I, II and IX models. Therefore the fate of singularities in Bianchi models is of special interest. A common concern is that even if the big bang is replaced by a big bounce in the isotropic case, typically this singularity resolution would not survive in Bianchi models (primarily because the anisotropic shear terms diverge as $1/a^6$, where a is the scale factor). In LQC, by contrast, the big bang singularity is again resolved once the quantum geometry effects from full LQG have correctly been incorporated [139, 140]. Furthermore, if one traces the Hamiltonian constraint of the Bianchi I model over anisotropies, one is led precisely to the FLRW Hamiltonian constraint, bringing out the robustness of the scheme. Finally, in the CDT simulations one finds that even when one allows all fluctuations in geometry, keeping only the scale factor fixed, the behavior of the scale factor, including quantum fluctuations, is described accurately by a mini-superspace model which assumes homogeneity and isotropy from the outset [10]. Putting together these diverse results, it is not unreasonable to hope that these models adequately capture the behavior of global observables (such as the scale factor and average matter density) that would be predicted by the full theory.

What about the truncation used in treating cosmological perturbations $\hat{\mathcal{R}}$, $\hat{\mathcal{T}}$? Full LQG *will* admit states in which there are huge quantum fluctuations in the Planck regime whose physics cannot be captured by states of the type $\Psi_o \otimes \psi$ where Ψ_o is a state of the quantum FLRW geometry and ψ is the state of linear quantum perturbations $\hat{\mathcal{R}}$, $\hat{\mathcal{T}}$. It is often implicitly assumed that *all* states of the full quantum gravity will have huge fluctuations. LQC has provided concrete evidence that this need not be the case: there do exist states of the type $\Psi_o \otimes \psi$ for which truncation is *self-consistent*. These states lead to an unforeseen, tame behavior in which $\hat{\mathcal{R}}$, $\hat{\mathcal{T}}$ evolve as linear perturbations on a background quantum geometry Ψ_o carrying energy densities that are negligible compared to that in the background. The non-triviality lies in the fact that these self-consistent, truncated solutions lead to the power spectrum and spectral index that are consistent with observations. Thus, the situation is similar to that in the standard ΛCDM model where it suffices to restrict oneself to the simplest cosmological solutions. The early universe appears to be simpler than what one would have a priori imagined!

11.3.2 Black holes

As is clear from Chapter 4, black holes (BHs) serve as powerful engines that drive the most energetic astrophysical phenomena. But, as discussed in Chapter 10, they have also driven developments in fundamental physics, particularly quantum gravity, raising deep conceptual questions about the statistical mechanical origin of the Bekenstein–Hawking entropy [149, 150] and a quantum gravity description of the BH evaporation process [150]. In this subsection we will provide a brief summary of developments in LQG in this area.

Quantum horizon geometry and micro-canonical entropy

In statistical mechanics, entropy is generally associated with systems in equilibrium. BHs in equilibrium were first modeled using event horizons of stationary space-times in GR. However, in statistical mechanics equilibrium refers only to the system under consideration, and not the entire universe. Therefore, about 15 years ago, a quasi-local framework was introduced through the notion of *isolated horizons* (IHs) to better model BHs which are themselves in equilibrium, allowing for dynamical processes in the exterior [151]. Event horizons of stationary space-times as well as the cosmological horizons in de Sitter space-time are special cases of IHs. Interestingly, the first law of BH thermodynamics naturally extends to IHs, with a further advantage that mass and angular momentum in the law now refer to the BH itself, defined at the IH, rather than to the ADM quantities defined at infinity which receive contributions also from the exterior region [152]. Its form is again similar to the first law of thermodynamics, suggesting that a multiple of the area a_Δ of the IH Δ should be interpreted as entropy S_Δ. Hawking's analysis of quantum radiance provides the numerical value of the multiple, yielding the Bekenstein–Hawking formula, now for IHs: $S_\Delta = a_\Delta/(4G_N\hbar)$.

In LQG, one investigates the statistical mechanical origin of this entropy, S_Δ [13]. The shift of focus to IHs has two advantages. First, one can consider realistic, astrophysical black holes: Not only does one not have to invoke 'charges' to make BHs near-extremal, but one can even allow for distortions in the horizon geometry that may be caused by matter rings or other black holes. Second, the cosmological horizons (for which thermodynamic considerations are known to hold) are automatically incorporated. The idea is to first investigate the quantum geometry of these IHs [153, 154] and then calculate the number of quantum microstates in the specified ensemble [13, 155]. This procedure provides a statistical mechanical derivation of entropy in terms of quantum geometry. We will now summarize these developments.

As before, one carries out a truncation of the theory that is motivated by the physical problem of interest. Thus, one begins with the phase space of GR in connection dynamics, now with a spatial 3-manifold M that is asymptotically flat *and* has an internal boundary S, the intersection of M with an IH 3-manifold Δ. Detailed analysis shows that the total phase space can now be written as $\mathbf{\Gamma} = \mathbf{\Gamma}_{\mathrm{bulk}} \times \mathbf{\Gamma}_S$, where $\mathbf{\Gamma}_S$ turns out to be the phase space of an U(1) Chern–Simons theory. The IH boundary condition relates the curvature F of the U(1) Chern–Simons connection to the pull-back $\underline{\Sigma}$ of the 2-form $\eta_{abc}E_i^c r_i$, where r^i is the unit internal vector normal to S: $F = -(2\pi/a_\Delta)8\pi G_N\gamma\underline{\Sigma}$.

For this sector of GR, one has to extend the quantum geometry framework of Section 11.2.2 to allow for an inner boundary S corresponding to an IH. The bulk Hilbert space $\mathcal{H}_{\mathrm{bulk}}$ is again spanned by spin networks. However, the links of these spin-networks can now end on the boundary S, piercing it on a node (see Fig. 11.5). The surface Hilbert space $\mathcal{H}_{\mathrm{CS}}$ is now the Hilbert space of an U(1) Chern–Simons theory on the resulting punctured sphere, with the level (or, dimensionless coupling constant) $k = a_\Delta/(4\pi\gamma\ell_{\mathrm{Pl}}^2)$. The total kinematical Hilbert space is now a tensor product $\mathcal{H}_{\mathrm{kin}} = \mathcal{H}_{\mathrm{bulk}} \otimes \mathcal{H}_{\mathrm{CS}}$. States in $\mathcal{H}_{\mathrm{kin}}$ are now subject to the *quantum horizon boundary condition*, which is an operator equation:

$$(1 \otimes \hat{F})\Psi = -\frac{2\pi}{a_\Delta}8\pi G_N\gamma(\underline{\Sigma} \otimes 1)\Psi. \qquad (11.24)$$

Note that solutions to (11.24) can exist only if $\hat{F}$ on the surface Hilbert space $\mathcal{H}_{\mathrm{CS}}$ has the same eigenvalues as the triad operator $\underline{\Sigma}$ on $\mathcal{H}_{\mathrm{bulk}}$. This is a highly non-trivial condition since the two operators have been defined *completely independently* on two *distinct* Hilbert spaces. However, the framework passes this severe test because the two operators share

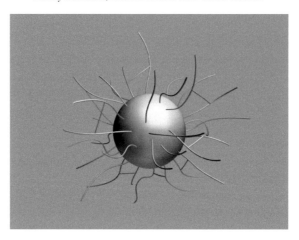

Figure 11.5 An artist's rendering of an isolated horizon punctured by spin-network links. Image credit: Alejandro Corichi.

an infinite number of eigenvalues. Finally, the physical meaning of this condition is as follows: The intrinsic curvature of the IH can fluctuate and so can the bulk geometry in its neighborhood, but they have to fluctuate in tandem, satisfying (11.24).

Next, one has to impose quantum constraints. There are some interesting subtleties which lead to mapping class groups and quantum deformation of $U(1)$ on S [153]. The final result is that what matters is only the number of punctures, not their location on S, *and* that each puncture has to be treated as 'distinguishable.' We emphasize that these are systematic implications of the constraint equations, and *not* additional inputs, as is sometimes thought. The net result is that, assuming the Hamiltonian constraint does admit a sufficient number of solutions in the bulk, LQG provides a coherent description of quantum space-times with IHs [153]. This quantum geometry is depicted in Fig. 11.5. (For more detailed summaries, see e.g. [2, 13].)

To calculate entropy, one has to fix an ensemble and count the number of states compatible with the macroscopic parameters characterizing the ensemble. This is done via the notion of multipole moments that characterize the geometry of Δ in a diffeomorphism invariant manner [154]. (In the simplest case when all moments except the mass monopole vanish, the IH is spherically symmetric with respect to *some* SO(3) action.) The ensemble is specified by requiring that all multipoles lie in a small interval around some pre-specified values. The idea is to calculate the number $\mathcal{N}$ of quantum microstates of the horizon geometry that satisfy this constraint. Its logarithm gives the microcanonical entropy of the ensemble.

To determine $\mathcal{N}$ one has to count specific types of finite sequences of half integers subject to certain constraints [13]. This problem has been investigated in detail in a series of mathematical papers by Barbero, Villaseñor and others that are of interest in their own right [155]. They combine known types of Diophantine equations with techniques involving generating functions and Laplace transforms. The final result is that the microcanonical entropy is given by

$$S_{\text{micro}} = \frac{\gamma_o}{\gamma} \frac{a_\Delta}{4\ell_{\text{Pl}}^2} + O\left(\ln\left(\frac{a_\Delta}{\ell_{\text{Pl}}^2}\right)\right), \qquad (11.25)$$

where $\gamma_o \sim 0.2$ is a root of an algebraic equation[5] [13]. Thus, in the sector of the theory where the BI parameter is set to γ_o, one recovers the Bekenstein–Hawking result to the leading order with a logarithmic correction. The Barbero–Immirzi parameter is a quantization ambiguity in LQG, rather similar to the θ-ambiguity in QCD [57]. In QCD, the value of this parameter is determined experimentally. In LQG, an experimental measurement of, say, the area gap would similarly determine γ. The LQG viewpoint is that while such measurements are completely out of reach of current technology, the Bekenstein–Hawking formula can be used as a theoretical constraint to determine γ. Note that once we set $\gamma = \gamma_o$ to get agreement with this formula for one type of IH (say spherical ones), the agreement extends to all IHs.

Semiclassical considerations and dynamical processes

The description of quantum horizons we just summarized has the advantage that it is fully background independent. But that very feature makes it difficult to relate it to the rich body of semi-classical results that have been derived in Kerr and Rindler space-times. Therefore, over the last three years, two independent avenues have been introduced to make closer contact with semi-classical results and study quantum dynamical processes. In this subsection we will briefly describe their current status.

In the first approach, developed by Ghosh, Perez and others [156], one considers the near-horizon geometry of Kerr space-times and asks the question: How would near-horizon, stationary observers describe physics *within* LQG? Denote by χ^a the Killing vector which is the null normal to the Kerr horizon Δ and consider observers $\mathcal{O}$ with 4-velocity $u^a = \chi^a/\sqrt{\chi \cdot \chi}$, at a fixed distance $d \ll R_\Delta = \sqrt{a_\Delta/(4\pi)}$ from Δ. Note that the observers $\mathcal{O}$ are approximately at rest with respect to Δ since their angular momentum is $O(d/R_\Delta)$. If one were to consider the Hamiltonian framework with a boundary at the location of the observers $\mathcal{O}$, one would find that the Hamiltonian acquires, in addition to the ADM surface integral at infinity, a 2-surface integral $H_\mathcal{O}$ at the inner boundary which, one argues, is given by $H_\mathcal{O} = a_\Delta/(8\pi Gd)$. In LQG, the corresponding operator is $\hat{H}_\mathcal{O} = \hat{\text{Ar}}_S/(8\pi Gd)$, where $\hat{\text{Ar}}_S$ is the area operator of Section 11.2.2, now associated with the intersection S of Δ with a partial Cauchy surface M used in the Hamiltonian framework. Next, since the acceleration of u^a is given to the leading order by $1/d$, one assumes that observers $\mathcal{O}$ would experience the Unruh temperature $T_U = 1/(2\pi d)$ [157]. This is supported by two

---

[5] Because the surface states on Δ are intertwined with the bulk spin-network states, a priori one can assign two meanings to the term 'pure surface terms' that are to be counted. They lead to two values ≈ 0.27 and 0.24 of γ. (See, e.g., [13].) In detailed LQG calculations this difference only changes numerical values by small amounts. But conceptually it is important to better understand and resolve this ambiguity.

independent considerations: i) if one red-shifts T_U to infinity, one obtains the Hawking temperature T_H, and, ii) detectors carried by the observers $\mathcal{O}$ coupled to $\hat{H}_{\mathcal{O}}$ would read the local temperature T_U [156].

Using these ingredients, one arrives at the following physical picture: the observers $\mathcal{O}$ would describe the punctured quantum horizon Δ as a grand canonical ensemble $\tilde{\rho}_{\mathcal{O}}[\beta, \mu; \gamma]$ of *punctures* p endowed with spin labels j_p, at an inverse temperature $\beta_U = (2\pi d)/\hbar$ and a chemical potential μ. (The dependence on the Barbero–Immirzi parameter γ comes from $\hat{H}_{\mathcal{O}}$.) As usual, this is equivalent to a canonical ensemble $\rho_{\mathcal{O}}[\beta_U; \gamma]$ in which $-T\frac{\partial S}{\partial N}|_E$ equals the chemical potential μ of the grand canonical ensemble. An explicit calculation of $-T\frac{\partial S}{\partial N}|_E$ provides μ as a function $\mu(\gamma)$ of γ. Finally, recall from Section 11.2.2 that the level spacing between eigenvalues of the area operator goes to zero exponentially for large areas. Hence the energy required to create a new puncture is arbitrarily small for large black holes. One therefore makes the final assumption that, as for photons, the physical value of the chemical potential μ should be zero. This condition determines γ uniquely and the value is precisely the γ_o arrived at by state counting in the microcanonical ensemble irrespective of the choice of d (which enters only in the local temperature that observers $\mathcal{O}$ attribute to the BH). In the resulting canonical ensemble $\rho_{\mathcal{O}}[\beta_U; \gamma_o]$ that the observers $\mathcal{O}$ would use to describe the BH, the entropy is given by $S = a_\Delta/(4\ell_{\mathrm{Pl}}^2)$ to leading order, exactly as in the microcanonical ensemble. Thus, these semi-classical considerations provide the same final result but with a novel description of the quantum horizon as a gas of punctures carrying spins. Therefore, this approach opens new avenues to describe dynamical processes, including the BH evaporation.

The second and complementary development is due to Gambini and Pullin [158] and follows a strategy that is analogous to the one used in LQC. It considers a different truncation of GR, that of spherically symmetric space-times. While this truncation was discussed in the LQG literature already in the 1990s, the *global* structure of the quantum space-times – including both the asymptotic part and the portion that is classically inside the horizon – was analyzed relatively recently.

In this symmetry reduced model, it suffices to consider spin networks with graphs along just the radial line. However, the nodes now carry additional labels that encode information about the connection and geometry in the two transverse directions. While there are close similarities with LQC, there is a major difference: Since the 3-geometry is now inhomogeneous, we have infinitely many Hamiltonian as well as diffeomorphism constraints, smeared with radial lapse and shift fields [159]. Remarkably, it is possible to express solutions to the Hamiltonian constraints in a closed form as a linear combination of the spin networks [158]. The diffeomorphism constraint can then be solved as in Section 11.2.3 by group averaging [70]. In the resulting physical Hilbert space, the ADM mass is a Dirac observable as in the classical theory. Furthermore, as in LQC, by appropriately deparameterizing the theory, one can also express the metric as a parameterized Dirac observable. As one would expect from quantum geometry, the metric is an operator-valued distribution concentrated at the nodes of the spin networks. There are semi-classical states which upon coarse graining on an appropriate scale – say, a thousand times the Planck length – yield smooth classical geometries. However, as in LQC, the quantum space-time is singularity-free and, as was anticipated by calculations within effective LQG equations for this model, the quantum space-time is 'larger' than that of classical GR. At a technical level, the fact that one can solve the infinite set of both Hamiltonian and diffeomorphism constraints is highly non-trivial.

As in LQC, it is now natural to investigate the behavior of test quantum fields on the quantum geometry of the symmetry reduced model. For this, one now truncates the theory, thereby allowing linear scalar fields on spherically symmetric space-times, again ignoring the back reaction in the first step. Then, as in LQC, the scalar field $\hat{\Phi}$ now propagates on a quantum state Ψ_o of the background geometry that, on coarse graining, yields the classical Schwarzschild geometry of a large black hole. In the interaction picture, in the approximation in which the back reaction is ignored, the field $\hat{\Phi}$ again propagates on an effective dressed quantum geometry. The main effect of the background quantum space-time on quantum field theory is to replace the partial differential equation governing $\hat{\Phi}$ with a difference equations. However, for frequencies (at infinity) which are significantly smaller than the Planck frequency, there is negligible difference from the thermal spectrum at infinity. This is not surprising because the Hawking radiation is robust with respect to the near-horizon microstructure of space-time [160]. But conceptually the underlying discreteness of quantum geometry does have one important effect: it removes the UV divergences encountered in the Boulware and Unruh vacua at the horizon [158].

This recent development has provided a coherent framework to describe Hawking radiation from first principles using the strategy of truncating LQG to the physical problem of interest. At a technical level, as we indicated, there is a close similarity with the framework used in LQC. On the physical side, on the other hand, there is a difference. Since the issue of the back reaction of *quantum* perturbations is significant only in the very early universe, in LQC one could analyze this issue systematically and show that the truncation used is physically self-consistent. For black holes, on the other hand, it is the back reaction that drives the evaporation process. Therefore, the truncation used so far for black holes is not adequate to systematically analyze the issue of information loss.

11.3.3 *n-point functions in a diffeomorphism invariant theory*

As Wightman emphasized already in the 1950s, in MQFTs the *n*-point functions

$$W(x_1, \ldots, x_n) = \langle 0|\phi(x_n) \ldots \phi(x_1)|0\rangle \tag{11.26}$$

completely determine the theory [161]. In particular, one can calculate the scattering amplitudes from these distributions. However, since they make an explicit reference to the Minkowski metric, it is far from being a priori clear that these ideas can be extended in a meaningful manner to non-perturbative quantum gravity. Indeed, at first it may appear that, because manifolds do not admit non-trivial, diffeomorphism invariant *n*-point distributions, a background independent framework cannot lead to non-trivial *n*-point functions either.

However, as we will see, this argument is too naive. The n-point functions refer to a state and in gravity that state can naturally encode information about a specific geometry which can then appear in the expressions of these distributions. In particular, LQG does lead to non-trivial n-point functions. Furthermore, to the leading order they have been shown [162–165] to agree in the appropriate sense with the n-point functions calculated in the effective low energy quantum general relativity [14] referred to in Section 11.2.1. These calculations have created a bridge from the rather abstract and unfamiliar background independent framework of LQG to notions and techniques used in concrete calculations in familiar MQFTs.

To spell out the construction, let us first return to MQFTs and recall that an n-point function can be written as a path integral:

$$W(x_1,\ldots,x_n) = \langle 0|\phi(x_n), \ldots, \phi(x_1)|0\rangle = \int D\phi \; \phi(x_n)\ldots\phi(x_1)e^{iS[\phi]}. \qquad (11.27)$$

For simplicity, consider the two-point function. We can organize the integration in (11.27) as follows. Select an arbitrary *compact* region $\mathcal{R}$, as in Fig. 11.6, such that the points x, x' of interest lie on its boundary b. Denote by $W(\varphi)$ the integral over fields ϕ defined only on $\mathcal{R}$ with the boundary value φ on b, and by $\Psi_b[\varphi]$ the integral over fields defined only on the exterior region $M - \mathcal{R}$, again with the boundary value φ on b (and appropriate fall-off at infinity). Then, we have:

$$W(x,x') = \int D\varphi \; W[\varphi]\varphi(x)\varphi(x')\Psi_b[\varphi_b]. \qquad (11.28)$$

From the perspective of the region $\mathcal{R}$, this expression can be interpreted as providing the 2-point function for the boundary state $\Psi_b[\varphi]$, the transition amplitude ($\sim e^{iS_{\mathcal{R}}}$) being given by $W[\varphi]$.

In the form (11.28), the functional integral can be taken over to quantum gravity using the Spinfoam transition amplitudes $W_\mathcal{C}$ introduced in Section 11.2.2. But there are crucial differences in the underlying structures that are needed in MQFTs versus LQG. In MQFTs, in addition to the value φ of the field on the boundary b, we must *also* use the background metric to fix the shape and geometry of the boundary b, and the space-time distance

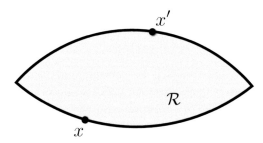

Figure 11.6 Geometrical interpretation of the transition amplitude.

between x' and x. Only then can we calculate the transition amplitude $W[\varphi]$ associated with the region $\mathcal{R}$, and the boundary state $\Psi_b[\varphi]$:

$$W[\varphi] = \int D\phi \; e^{i \int_{\mathcal{R}} L[\phi]} \quad \text{and} \quad \Psi_b[\varphi] = \int D\phi \; e^{i \int_{M-\mathcal{R}} L[\phi]}. \tag{11.29}$$

But in LQG, these expressions cannot and do not make reference to a background metric. As we saw in Section 11.2.2, a signature of the non-triviality of the LQG construction is that the transition amplitude $W[\varphi]$ is a function only of the field φ. It does not depend on any background structures; it refers only to the dynamical fields and the physical process of interest.

How can one then make contact with the effective field theory calculations which do refer to a background geometry? This is achieved through the state $\Psi_b[\varphi]$. Because of the functional integration involved in its definition, $\Psi_b[\varphi]$ depends on the dynamics of the theory as well as the boundary conditions at infinity. But as in MQFTs, these ingredients are invisible to the calculation of the 2-point function (11.28); what matters directly is the state $\Psi_b[\phi]$ itself. But the key difference from MQFTs is that, because we are considering gravitational fields, *this state now encodes information about the geometry.* For appropriate boundary conditions at infinity, $\Psi_b[\varphi]$ would be peaked at a certain classical (intrinsic and extrinsic) 3-geometry of b. Therefore, it assigns to b a certain shape and size and also selects a classical 4-geometry in $\mathcal{R}$ that extremizes the classical action for the given boundary geometry φ at b. The relative position of the n arguments of the n-point function is well defined with respect to that 4-geometry. In particular, for comparison with the effective theory, one picks a region $\mathcal{R}$ of Minkowski space-time and approximates it with a triangulation that is sufficiently fine to capture the relevant dynamical scale of the phenomenon of interest. One then determines the intrinsic and extrinsic geometry of the boundary of this region, and picks a quantum state $\Psi \in \mathcal{H}_\Gamma$ of gravity peaked on these data. Then the two points x and x' sit on nodes of the boundary graph. The operators E_ℓ associated with these nodes and links provide the geometric interpretation of the quanta in the boundary state.

This construction resolves a longstanding confusion in quantum gravity, referred to at the very beginning of our discussion. Formally, if $S[g]$ is a generally covariant action and $D[g]$ a generally covariant measure, the distribution

$$W(x_1, \ldots, x_n) = \int D[g] \, g(x_1) \ldots g(x_n) e^{iS[g]} \tag{11.30}$$

would be trivial, i.e., independent of the position of the points $(x_1, \ldots, x_n)$ (as long as they do not overlap). Therefore (11.30) are not the physically interesting n-point functions in a generally covariant theory. In particular, they are completely unrelated to the n-point functions used in the effective field theory: while the (11.30) refer only to the n points $x_1 \ldots x_n$ in the space-time manifold, those in the effective theory depend on *physical distances* between these points computed in the background metric. In the LQG construction, the n-point functions *do know* about the physical distances between x_n in the background geometry determined by the boundary state $\Psi_b[\varphi]$. However, under a space-time diffeomorphism that sends $\mathcal{R} \to \mathcal{R}'$, $b \to b'$ and $(x_1, \ldots x_n) \to (x'_1, \ldots, x'_n)$ and $\varphi \to \varphi'$ the boundary state also transforms covariantly, $\Psi_b(\phi) \to \Psi_{b'}(\phi')$, whence the geodesic distances between any two points x_i, x_j on the boundary and their images x'_i, x'_j are preserved. Therefore, the final results are diffeomorphism invariant.

Calculation of n-point functions have been performed in this framework, using states and operators of the canonical theory and the transition amplitude (11.15) provided by Spinfoams [162–164]. The result is that, in a suitable semiclassical limit (to terms $\mathcal{O}(\hbar)$) the two-point function *exactly matches* [163] with the one obtained from Lorentzian Regge calculus [100]. In turn, this limit is consistent with the effective field theory [14]. Thus, although the basic notions and techniques appear to be very different from those used in perturbative treatments, the final results show that, as in the Asymptotic Safety scenario, a peaceful co-existence with low energy results is possible. A program to make a systematic connection with effective field theory has been initiated recently [165]. Some radiative corrections have also been computed using more refined 2-complexes, containing bubbles [166]. Ideally, one would hope that the low energy behavior of the n-point functions computed in the non-perturbative theory would agree with the effective theory, while at high energies the LQG calculations would provide an UV completion of the non-renormalizable perturbation theory. Whether this is the case is still an open question.

The conceptual non-triviality of results to date lies in the fact that they provide a streamlined approach to compare the non-perturbative theory with background dependent effective theories, without sacrificing the underlying diffeomorphism invariance. Reconciling the two had been a long-standing open issue in quantum gravity.

11.4 Discussion

In this section we will first summarize the main ideas and results, putting them in a broader context, and then discuss open issues that remain.

11.4.1 Summary

The approaches discussed in this Chapter are rooted in well-established physics: the principles of GR and QFT. The viewpoint is that ideas that have no observational support should not constitute an integral part of the foundation of quantum gravity, even when they can lead to rich mathematical structures. In particular, these approaches do not rely on a negative cosmological constant, or extended objects or specific matter content involving towers of fields and particles.[6] The primary goal is not unification with other fundamental forces. Rather, the emphasis is on qualitatively new insights into *quantum* spacetimes that can emerge from non-perturbative techniques. In classical GR, the dynamical nature of geometry led to new phenomena – such as gravitational waves, black hole horizons and the big bang – that could not even be imagined before. As earlier Chapters in this volume

---

[6] In the same spirit, they do not *demand* supersymmetry or higher dimensions, but the methods used in these approaches have been extended to incorporate these possibilities [11, 48, 49].

vividly bring out, these notions have had deep impact on the subsequent developments in astrophysics, cosmology, computational physics, and geometric analysis. In these developments one can often use perturbative techniques, but they have to be built around the novel nonlinear configurations and one must use qualitatively new boundary conditions, dictated by full general relativity. The expectation is that the situation will be similar in the quantum domain with the new, unforeseen features of quantum geometry. Results to date, e.g. on the UV finiteness and the physics of the very early universe, which are discussed in Sections 11.2 and 11.3, provide concrete evidence in favor of this expectation.

We saw in Section 11.2.1 that the new notion of the Effective Average Action (EAA) in the continuum can be used to give a meaning to the basic functional integral by reformulating the problem as a question about solutions of a functional flow equation [23]. These renormalization group trajectories possess a well defined ultraviolet limit, allowing one to reconstruct the functional integral from them if they hit a non-Gaussian fixed point in this limit. We summarized the evidence for the existence of such a non-trivial fixed point [23, 28–30]. We then discussed CDT, a lattice approach based on statistical mechanics ideas [10, 39, 40, 167].

Over the last two decades, concrete progress has occurred by carrying out suitable, finite dimensional truncations of the infinite dimensional theory space $\mathcal{T}$. The initial truncation had only two coupling constants, G_N and Λ, corresponding to the Einstein–Hilbert and the cosmological constant terms. By now, the truncations have reached a mature level, allowing for a considerable number of different coupling constants in the gravitational sector. Not only has the non-trivial fixed point persisted but there is also consistency with the two-dimensional truncation in a precise sense. The analysis has also been extended beyond pure gravity and several matter couplings have been investigated in detail [11]. These results have provided highly non-trivial evidence in support of Asymptotic Safety.

The CDT approach has led to an unforeseen result that had eluded earlier lattice simulations: In the Euclidean signature, de Sitter space-time with small fluctuations was shown to emerge from Monte Carlo simulations using the discretized Einstein–Hilbert action. The primary importance of this demonstration is not so much that it is the de Sitter space-time that resulted but rather that the result has an interpretation as a 4-dimensional classical geometry in the first place. Indeed, previous dynamical triangulation simulations had led only to 'crumpled' or 'polymer-like' phases rather than the one corresponding to a smooth macroscopic geometry.

In LQG, the emphasis is again on non-perturbative methods. But while in the EAA framework one introduces a background metric $\bar{g}_{ab}$ in the intermediate stages, splits the physical metric $\tilde{g}_{\mu\nu}$ as $\tilde{g}_{\mu\nu} = \bar{g}_{\mu\nu} + h_{\mu\nu}$ and interprets $\mathcal{D}\tilde{g}_{\mu\nu}$ as an integration over the nonlinear fluctuations, $\mathcal{D}\tilde{h}_{\mu\nu}$, the LQG framework is *manifestly* background independent. Quantum geometry, developed in the canonical framework [2–5], provides well-defined techniques to carry out path integrals directly in terms of Spinfoams [3, 5, 6] which represent physical, quantum spacetimes, without any split.

In Section 11.2.3 we discussed LQG dynamics via a Spinfoam model [92–95]. This model has attracted a great deal of attention because it represents a notable confluence of ideas from apparently distinct directions: Canonical LQG [2, 4], Regge calculus [100–102], topological field theories and group field theory [3, 96, 103]. The number of simplices that feature in the underlying 4-geometries provides a mathematically natural expansion parameter to calculate transition amplitudes. These amplitudes are UV finite to any order in this expansion [92–95] and, in the presence of a positive cosmological constant, there is an elegant procedure involving a quantum deformation of $SL(2, \mathbb{C})$ (the double cover of the local Lorentz group) that provides a natural infrared regulator [112, 113]. Finally, in Section 11.3.3 we summarized the construction of n-point functions in the semi-classical limit of this model. The leading term in the 2-point function reproduces the low energy graviton propagator in a precise sense [162–164]. These developments have begun to create a bridge [165] from the background independent, non-perturbative framework of LQG to effective field theories that encompass low energy scattering processes in quantum gravity. Thus, LQG offers a well-defined set of fundamental equations describing quantum space-time, free of UV and IR divergences in a natural expansion, with substantial evidence that GR will emerge in a suitable limit.

Over the past two decades, LQG has also been used to analyze long-standing issues which originally constituted the main motivation for quantum gravity. As in Asymptotic Safety, progress has occurred by truncating the full theory appropriately and analyzing the truncated sectors in detail. But now truncations are motivated directly by each physical problem under consideration. In the cosmological truncation, discussed in Section 11.3.1, not only are the strong curvature singularities naturally resolved by the quantum geometry effects [12, 129, 131, 134, 139, 141, 142] but also the standard paradigms have been extended all the way to the bounce by facing the Planck regime squarely, using quantum field theory on *quantum* cosmological space-times [124–126, 146]. Furthermore there is a small window in the parameter space where the theory can be confronted with future observations. In Section 11.3.2, we summarized the current status of quantum black holes in LQG. Using the notion of isolated horizons and techniques from quantum geometry one can treat all black hole and cosmological horizons in one go, without having to restrict oneself to extremality [153, 154]. More recently, semi-classical considerations have brought the LQG description closer to the more familiar treatments in terms of energy and temperatures measured by suitable families of near-horizon observers [156]. Finally, there is now a novel approach to investigate the quantum evaporation process, using quantum field theory on *quantum* space-times describing black holes in LQG [158].

11.4.2 Outlook

Every quantum gravity program faces two types of issues: i) those which are *internal* to any given program which must be resolved before one has a conceptually complete, coherent theory with the correct low energy limit in four space-time dimensions; and, ii) those which are *common* to all programs, addressing the long-standing physical questions. As our summary illustrates, concrete advances have occurred on both these fronts. However, a number of important challenges remain. In particular, so far *none* of the approaches to quantum gravity satisfies the 'internal' criterion of completeness.

We will now illustrate these challenges and ensuing opportunities through examples. Strategies summarized here are necessarily provisional; our primary intent is only to provide a general idea of the directions that are being currently pursued.

• **Infrared Issues:** In the EAA approach of the Asymptotic Safety program there exist trajectories admitting the non-trivial UV fixed points which are known to reduce to GR in the low energy limit. In the CDT approach, because of the usual limitations on the size of lattices that can be handled in simulations, the smallest physical length a of the link is still about $2\ell_{Pl}$, and the infrared regime corresponds to $\sim 20\ell_{Pl}$ [10]. The IR behavior is already illuminating in that not only does a classical de Sitter geometry with small quantum fluctuations arise in the Euclidean signature, but this occurs even for universes of radius $\sim 20\ell_{Pl}$![7] An important open issue is whether other physically interesting spacetimes can be recovered from CDT.

In full LQG, as we saw in Sections 11.2.3 and 11.3.3, the low energy limit is recovered in a certain well-defined sense. However, there is considerable room for improvement. In particular, although the boundary states currently used are well-motivated, being peaked on the metric as well as the extrinsic curvature of the boundary induced by the Minkowski metric, there is still considerable ambiguity in their choice, which descends to the transition amplitude and n-point functions. Conceptually, this is not a problem because both these quantities are, by definition, functions of the boundary state. But different boundary states would give rise to different sub-leading terms, making comparison with the effective theory ambiguous. A principle to select *canonical* states corresponding to Minkowski and de Sitter space-times is still lacking. A second important limitation is that most of the results on classical and semi-classical limits we summarized have been obtained using only one simplex. There is substantial ongoing work that considers refinements, allowing a large number of simplexes in the interior, keeping the boundary state peaked on the classical geometry of interest. These results will either firmly establish the infrared viability of the specific Spinfoam model that is currently being used or suggest better alternatives.

• **Matter couplings:** In the cosmological truncation of LQG, matter fields have been incorporated and their effect has been analyzed in detail [12, 122, 123]. In full connection dynamics, matter couplings have been discussed exhaustively at the classical level [4, 46, 47] and the framework has also been extended to incorporate supersymmetry [49]. However, at the quantum level, so far only formal schemes have been laid out in the full theory [4, 91]. Interestingly, one can arrive at a unification which is 'dual' to the Kaluza–Klein scheme in the following sense: One can continue with four space-time dimensions but enlarge the *internal* group to a product of the Lorentz group (associated with gravity) with groups associated with Yang–Mills fields governing other interactions [168]. Whether these

---

[7] Interestingly, the same scale arose completely independently in LQC: in the $k = 1$ Lorentzian FLRW cosmology, for example, the dynamics of the quantum wave functions is accurately described by GR once the radius of the universe exceeds $8\ell_{Pl}$ even in the case when the universe grows to a radius only of $23\ell_{Pl}$ before undergoing a recollapse à la GR [134].

ideas are fully compatible with particle physics phenomenology is, however, still unclear. More generally, constructing a detailed quantum theory with matter coupling represents a challenging and fertile area for LQG research in coming years.

In the Asymptotic Safety program, by contrast, there is already very substantial work on incorporating matter. It has provided interesting constraints on the number of fermions and gauge fields that can be accommodated within this scenario, constraints that are satisfied by the standard model of particle physics [11]. Furthermore, insights on the quantum nature of geometry provided by results to date are likely to have implications for the ultraviolet issues in field theories of other interactions as well. Indeed, there are already indications that the coupling to asymptotically safe gravity might cure certain notorious problems in the matter sector [169, 170], and it is conceivable that the coupled system is more predictive than the standard model of particle physics without gravity. There are for example scenarios in which the Higgs mass [171] and the fine-structure constant [169] are computable quantities. These promising ideas are likely to be more fully developed in the coming decade.

• **New Physics:** In both approaches, there is a large number of avenues that will be pursued to explore new physics. We will present just a few illustrative examples.

In the Asymptotic Safety program one is naturally led to an EAA-based 'quantum geometry' of space-time which goes beyond Riemannian geometry in a specific sense: in general, the metric is scale dependent. So a single (smooth) manifold is furnished not with just one metric, but rather a family, $\{\langle g_{\mu\nu}\rangle_k, \ 0 \le k < \infty\}$, where $\langle g_{\mu\nu}\rangle_k$ is a solution of the effective field equation following from Γ_k. This general framework [172] was used, for instance, to demonstrate that under certain conditions the EAA, while defined in the continuum, can give rise to a dynamically generated minimum length scale. It was also used to analyze the fractal-like properties of the 'quantum spacetimes' which follow from the EAA [29, 30]. In particular a running spectral dimension has been defined and computed [173, 174]. One finds that there is a dimensional reduction from four macroscopic to two microscopic dimensions.[8] As a consequence, the graviton propagator is modified near the UV fixed point [29, 118]. These novel features have interesting implications for the early universe and black holes [118–121] which provide interesting avenues for future research.

The applications of LQG discussed in Section 11.3 also provide a number of interesting directions to explore new physics. First, as we discussed in Section 11.3.1, there is a small window in the parameter space where LQC leads to new predictions [123, 124, 126, 148]. It needs to be analyzed in much greater detail, keeping in mind the planned astronomical surveys. While the a priori probability that this window is realized in Nature is small, if the initial observations were to favor it, it would be possible to use novel avenues to confront the theory with observations in detail, precisely because the window is small. On the conceptual front, there are a number of issues concerning the specification of initial conditions at the bounce. So far the focus has been on establishing the *existence* of initial conditions that lead to a self-consistent extension of standard inflation to the Planck regime. But the

---

[8] It is interesting that this general phenomenon also occurs in LQG, where the 4-dimensional space-time continuum arises from coarse graining of a 2-complex representing the evolution of the fundamental quanta of geometry.

issue of uniqueness is quite open except for some preliminary ideas involving a quantum extension of Penrose's Weyl curvature hypothesis [124]. These will be explored in detail in the coming years. If one uses inflation, observations inform us that the entire observable universe should originate from a ball of radius less than $10\ell_{Pl}$ at the bounce. But standard inflation does not explain why there was an extraordinary homogeneity at this scale. The repulsive force of LQC that dominates near the bounce provides a novel avenue to explore this issue. LQC models that have been analyzed in detail indicate that in the Planck regime the net effect of this repulsion is to dilute the wrinkles in the curvature, thereby enforcing homogeneity and isotropy at this scale. It is important to translate these physical ideas into detailed calculations also because they imply that the repulsive force would wash away the memory of the pre-bounce phase as far as observations are concerned, making it natural to specify initial conditions at the bounce. Finally, in the self-consistent solutions, it has been possible to argue that while the LQG effects are critical for the background FLRW quantum geometry, they can be ignored for perturbations since the energy density in perturbations is so small. It is important to carry out detailed calculations to investigate new physics that may emerge in more general situations from a full LQG treatment of perturbations.

There are similar challenges and opportunities in the investigation of quantum properties of black hole and cosmological horizons. While there is a detailed understanding of the microscopic quantum geometry of horizons in equilibrium [153], the relation between the number of these microstates and the more familiar semi-classical calculations of entropy [175] via path integrals has begun to receive attention only recently. This is a key open issue. More generally, the intriguing relation between the microscopic geometry of quantum horizons and the semi-classical ideas [156] discussed in Section 11.3.2 remains to be explored in detail. Finally, as we saw in Section 11.3.2, recently, a new window has been opened to investigate the Hawking effect within LQG [158]. This important development offers many opportunities for detailed calculations that will lead us to a deeper understanding of the evaporation process.

• **Beyond Truncations:** Recall that in both approaches discussed in this Chapter, concrete progress could be made by studying the appropriate truncations of the full theory. As emphasized towards the end of Sections 11.2.3 and 11.3.1, this is the common situation in fundamental physics: all the concrete calculations in QED, QCD and scenarios of the early universe, for example, involve truncations. Nonetheless, from the conceptual viewpoint, a central question remains: Is there an underlying coherent theory without reference to truncations that is being approximated in these calculations?

In the Asymptotic Safety program, a conceptual framework to address this question is provided by the infinite dimensional theory space $\mathcal{T}$. At a fundamental level, one should find the renormalization group flows in $\mathcal{T}$ and then investigate whether, in concrete physical problems, finite dimensional truncations carried out to date provide a trustable approximation. However, this lofty goal is far too ambitious for now. Progress is likely to occur by further enlarging the reach of truncations. In particular, a simplified version of the (particle physics) standard model in which the gauge fields are assumed to be Abelian has already been incorporated in the EAA program [11]. An important goal which may be within reach in the foreseeable future would be to extend these calculations to include the full standard model.

What is the situation with Spinfoams? Results to date have focused on finite simplicial decomposition of the space-time manifold. The key open question is whether one should take a suitable limit by successively refining the decomposition in a *well-controlled fashion*, or whether one should sum these contributions, *appropriately avoiding the obvious redundancy*. Does the final transition amplitude remain finite in either case? In three space-time dimensions, the refinement limit does converge, and yields the correct result. Similarly, one can recast LQC in the Spinfoam framework and show that the sum converges and yields a result that agrees with the Hamiltonian theory [104]. These calculations are helpful but do not provide deep insight because these theories do not have local degrees of freedom. Therefore currently there is a great deal of activity in the full 4-dimensional theory. In particular, generalized renormalization group flows are being studied by Dittrich and others to constrain the refinement procedures and investigate the phase diagrams that result, and group field theory is being used by Oriti and others to carry out the sum systematically. Interestingly, the two procedures have quite different conceptual underpinnings. In the first, the viewpoint is more akin to that in the study of condensed matter systems using statistical mechanics, where the atomic structure is fundamental and phonon fields are convenient tools to encode collective behavior of atoms. In LQG, the quanta of geometry play the role of atoms while continuum quantum fields are the rough analogs of phonons. In the second approach, quantum fields are more fundamental as in particle physics, and one uses well-established methods with the goal of summing a perturbative expansion. But quantum gravity introduces a key difference: the quantum fields are now defined on a group manifold rather than space-time. It is fortunate that the central issue of whether there is a coherent theory underlying Spinfoam truncations is being analyzed from very different, if not opposing, perspectives. Since this central issue is deep and difficult, it is essential to have variety.

Acknowledgments

We thank Alejandro Corichi for permission to use Fig. 11.5. This work was supported in part by the NSF grant PHY-1205388 and the Eberly research funds of Penn state.

References

[1] A. Einstein, *Sitzungsber. Preuss. Akad. Wiss.* 688 (1916); G. E. Gorelick, in *Studies in the history of general relativity*, edited by J. Eisenstaedt and A. J. Kox (Birkhäuser, Boston, 1992).

[2] A. Ashtekar and J. Lewandowski, *Class. Quant. Grav.* **21** R53–R152 (2004).

[3] C. Rovelli, *Quantum gravity* (Cambridge University Press, Cambridge, 2004).

[4] T. Thiemann, *Introduction to modern canonical quantum general relativity* (Cambridge University Press, Cambridge, 2007).

[5] F. Hellmann *et al.* eds. *Proceedings of the Quantum Gravity and Quantum Geometry School, Zakopane*, PoS (QGQGS) (2011), http://pos.sissa.it/cgi-bin/reader/conf.cgi?confid=140.

[6] A. Perez, *Living Rev. Rel.* **16** 3 (2013).
[7] M. Niedermaier and M. Reuter, *Living Rev. Rel.* **9** 5 (2006).
[8] M. Reuter and F. Saueressig, *New. J. Phys.* **14** 055022 (2012).
[9] R. Percacci, in *Approaches to quantum gravity*, D. Oriti ed (Cambridge University Press, Cambridge, 2009), arXiv:0709.3851.
[10] J. Ambjørn, A. Görlich, J. Jurkiewicz and R. Loll, *Phys. Rep.* **519** 127 (2012).
[11] R. Percacci, arXiv:1110.6389; P. Dona, A. Eichhorn, R. Percacci, arXiv:1311.2898.
[12] A. Ashtekar and P. Singh, *Class. Quant. Grav.* **28** 213008 (2011).
[13] J. F. Barbero, J. Lewandowski and E. J. S. Villaseñor, in *Proceedings of the 3rd Quantum Gravity and Quantum Geometry School*, PoS (QGQGS) (2011), http://pos.sissa.it/cgi-bin/reader/conf.cgi?confid=140.
[14] J. F. Donoghue, *Phys. Rev. Lett.* **72** 2996 (1994); C. P. Burgess, *Living Rev. Rel.* **7** 5 (2004).
[15] R. A. Porto and I. Z. Rothstein, *Phys. Rev.* **D78** 044013 (2008), **D81** 029902 (2010).
[16] K. Wilson and J. Kogut, *Phys. Rep.* **12** 75 (1974).
[17] K. Gawedzki and A. Kupiainen, *Phys. Rev. Lett.* **55** 363 (1985).
[18] M. Niedermaier, *Phys. Rev. Lett.* **103** 101303 (2009); *Nucl. Phys.* **B833** 226 (2010).
[19] J. Lewandowski, A. Okolow, H. Sahlmann and T. Thiemann, *Commun. Math. Phys.* **267** 703–733 (2006).
[20] C. Fleischhack, *Commun. Math. Phys.* **285** 67–140 (2009).
[21] J. A. Wheeler, in *Battelle rencontres,* edited by J. A. Wheeler and C. M. DeWitt, (W. A. Benjamin, New York, 1972).
[22] S. Weinberg, in *General relativity, an Einstein centenary survey*, Edited by S. W. Hawking and W. Israel (Cambridge University Press, Cambridge, 1979).
[23] M. Reuter, *Phys. Rev.* **D57** 971–985 (1998).
[24] J. Ambjørn and R. Loll, *Nucl. Phys.* **B536** 407–434 (1998).
[25] B. S. DeWitt, *The global approach to quantum field theory* (Oxford University Press, Oxford, 2003).
[26] C. Wetterich, *Phys. Lett.* **B301** (1993) 90; M. Reuter and C. Wetterich, *Nucl. Phys.* **B417** 181 (1994).
[27] E. Manrique and M. Reuter, *Phys. Rev.* **D70** 025008 (2009).
[28] M. Reuter and F. Saueressig, *Phys. Rev.* **D65** 065016 (2002).
[29] O. Lauscher and M. Reuter, *Phys. Rev.* **D65** 025013 (2002).
[30] O. Lauscher and M. Reuter, *Phys. Rev.* **D66** 025026 (2002); *Class. Quant. Grav.* **19** 482 (2002).
[31] A. Codello, R. Percacci and C. Rahmede, *Ann. Phys.* **324** 414 (2009).
[32] G. 't Hooft and M. J. G. Veltman, *Ann. Inst. Henri Poincaré Phys. Theor.* A**20** 69 (1974); M. H. Goroff and A. Sagnotti, *Nucl. Phys.* **B266** 709 (1986).
[33] M. Reuter and H. Weyer, *Phys. Rev.* **D80** 025001 (2009).
[34] M. Demmel, F. Saueressig and O. Zanusso, *JHEP* **1406** 026 (2014) and references therein.
[35] E. Manrique and M. Reuter, *Annals of Physics* **325** 785 (2010) E. Manrique, M. Reuter and F. Saueressig, *Annals of Physics* **326** 440 (2011); *Annals of Physics* **326** 463 (2011).
[36] D. Becker and M. Reuter, *JHEP* **07** 172 (2012).
[37] M. Reuter and F. Saueressig, *Phys. Rev.* **D66** 125001 (2002).
[38] A. Codello, R. Percacci and C. Rahmede, *Int. J. Mod. Phys.* A**23** 143 (2007); P. Machado and F. Saueressig, *Phys. Rev.* **D77** 124045 (2007).
[39] J. Ambjørn, J. Jurkiewicz and R. Loll, *Nucl. Phys.* **B610** 347 (2001); *Phys. Rev.* **D72** 064014 (2005).
[40] J. Ambjørn, A. Görlich, J. Jurkiewicz and R. Loll, *Phys. Rev. Lett.* **100** 091304 (2008); *Phys. Rev.* **D78** 063544 (2008).
[41] O. Lauscher and M. Reuter, *JHEP* **10** 050 (2005).
[42] M. Reuter and F. Saueressig, *JHEP* **1112** 012 (2011).
[43] S. Rechenberger and F. Saueressig, *JHEP* **1303** 010 (2013).
[44] P. Hořava, *Phys. Rev.* **D79** 084008 (2009).

[45] A. Ashtekar, *Phys. Rev. Lett.* **57** 2244–2247 (1986); *Phys. Rev.* D**36** 1587–1603 (1987).
[46] A. Ashtekar, J. Romano and R. S. Tate, *Phys. Rev.* D**40** 2572–2587 (1989).
[47] A. Ashtekar, *Lectures on non-perturbative canonical gravity*, Notes prepared in collaboration with R. S. Tate (World Scientific, Singapore, 1991).
[48] N. Bodendorfer, T. Thiemann and A. Thurn, *Class. Quant. Grav.* **30** 045001 (2013), **30** 045002 (2013), **30** 045003 (2013), **30** 045004 (2013).
[49] N. Bodendorfer, T. Thiemann and A. Thurn, *Phys. Lett.* B **711** 205–211 (2012); *Class. Quant. Grav.* **30** 045006 (2013), **30** 045007 (2013).
[50] H. J. Matschull, *Class. Quant. Grav.* **13** 765–782 (1996).
[51] E. Witten, *Commun. Math. Phys.* **80** 381–402 (1981).
[52] A. Ashtekar and G. T. Horowitz, *J. Math. Phys.* **25** 1473–1480 (1984).
[53] J. F. Barbero, *Phys. Rev.* D**51** 5507–5510 (1996).
[54] G. Immirzi *Nucl. Phys. Proc. Suppl.* **57** 65–72 (1997).
[55] T. Thiemann, *Phys. Lett.* B**380** 257–264 (1996); *Class. Quant. Grav.* **15** 839–873 (1998), **15** 875–905 (1998), **15** 1207–1247 (1998), **15** 1281–1314 (1998).
[56] J. Samuel, *Class. Quant. Grav.* **17** L141–L148 (2000).
[57] S. Holst, *Phys. Rev.* D**53** 5966–5969 (1996); S. Mercuri, *Phys. Rev.* D**77** 024036 (2008); G. Date, R. Kaul and S. Sengupta, *Phys. Rev.* D**79** 044008 (2009).
[58] R. Rovelli and L. Smolin, *Nucl. Phys.* B**331** 80–152 (1990).
[59] A. Ashtekar and C. J. Isham, *Class. Quant. Grav.* **9** 1433–1467 (1990).
[60] A. Ashtekar and J. Lewandowski, *J. Geo. Phys.* **17** 191–230 (1995).
[61] K. Giesel and H. Sahlmann, in *Proceedings of the 3rd Quantum Gravity and Quantum Geometry School*, PoS (QGQGS) (2011), http://pos.sissa.it/cgi-bin/reader/conf.cgi?confid=140.
[62] G. M. Emch, *Algebraic methods in statistical mechanics and quantum field theory* (Wiley-Interscience, New York, 1972).
[63] L. Garding and A. S. Wightman, *Proc. Nat. Acad. Sci. U.S.A.* **40** 622–626 (1956); R. Haag, *Danske Vid. Selsk. Mat.-fys. Medd.* **29** No.12 (1955).
[64] I. E. Segal, *Illinois J. Math.* **6** 500–523 (1962).
[65] H. Nicolai, K. Peeters and M. Zamaklar, *Class. Quant. Grav.* **22** R193 (2005).
[66] A. Ashtekar, *Gen. Rel. Grav.* **41** 1927–1943 (2009); International loop quantum gravity seminar, http://relativity.phys.lsu.edu/ilqgs/ashtekar022707.pdf.
[67] C. Rovelli and L. Smolin, *Phys. Rev.* D**52** 5743–5759 (1995).
[68] J. C. Baez, In *The interface of knots and physics* edited by L. Kauffman (American Mathematical Society, Providence, RI, 1996).
[69] C. Rovelli and L. Smolin, *Nucl. Phys.* B**442** 593–622 (1995); Erratum: *Nucl. Phys.* B**456** 753 (1995).
[70] A. Ashtekar, J. Lewandowski, D. Marolf, J. Mourão and T. Thiemann, *J. Math. Phys.* **36** 6456–6493 (1995).
[71] A. Ashtekar and J. Lewandowski, *Class. Quant. Grav.* **14** A55–A81 (1997).
[72] S. Frittelli, L. Lehner and C. Rovelli *Class. Quant. Grav.* **13** 2921–2932 (1996).
[73] A. Ashtekar and J. Lewandowski *Adv. Theor. Math. Phys.* **1** 388–429 (1997).
[74] K. Giesel and T. Thiemann, *Class. Quant. Grav.* **23** 5667–5692 (2006), **23** 5693–5772 (2006).
[75] J. Brunnemann and D. Rideout, *Class. Quant. Grav.* **25** 065002 (2008), **27** 205800 (2010); J. Brunnemann, in *Proceedings of the 3rd Quantum Gravity and Quantum Geometry School*, PoS (QGQGS) (2011), http://pos.sissa.it/cgi-bin/reader/conf.cgi?confid=140.
[76] A. Ashtekar and J. Lewandowski, in *Knots and quantum gravity* edited by J. C. Baez (Oxford University Press, Oxford, 1994).
[77] J. C. Baez *Lett. Math. Phys.* **31** 213–223 (1994).
[78] L. Freidel and S. Speziale, *Phys. Rev.* D**82** 084040 (2010).
[79] R. Penrose, in *Quantum Theory and beyond* ed T. Bastin (Cambridge University Press, Cambridge, 1971).
[80] R. Loll, *Phys. Rev. Lett.* **75** 3048 (1995).
[81] D. Giulini and D. Marolf, *Class. Quant. Grav.* **16** 2479–2488 (1999), **16** 2489–2505 (1999); A. Ashtekar, L. Bombelli and A. Corichi, *Phys. Rev.* D**72** 025008 (2005).

[82] W. Thirring and H. Narnhofer, *Rev. Math. Phys.* Special Issue **1** 197–211 (1992).
[83] R. Gambini, J. Lewandowski, D. Marolf and J. Pullin, *Int. J. Mod. Phys.* D**7** 97–109 (1998).
[84] C. Tomlin and M. Varadarajan, *Phys. Rev.* D**87** 044039 (2013); M. Varadarajan, *Phys. Rev.* D**87** 044040 (2013).
[85] A. Henderson, A. Laddha and C. Tomlin, *Phys. Rev.* D**88** 044028 (2013), **88** 044029 (2013); A. Laddha, arXiv:1401.0931.
[86] A. Laddha and M. Varadarajan, *Phys. Rev.* D**78** 044008 (2008), **83** 025019 (2011); A. Laddha and V. Bonzom, *SIGMA* **8** 50 (2012).
[87] K. Giesel, T. Thiemann, *Class. Quant. Grav.* **27** 175009 (2010); M. Domagała, K. Giesel, W. Kamiński and J. Lewandowski, *Phys. Rev.* D**82** 104038 (2010).
[88] T. Pawłowski and V. Husain, *Phys. Rev. Lett.* **108** 141301 (2012).
[89] M. P. Reisenberger and C. Rovelli, *Phys. Rev.* D**56** 3490–3508 (1997).
[90] J. C. Baez, *Class. Quant. Grav.* **15** 1827–1858 (1998).
[91] E. Bianchi, M. Han, E. Magliaro, C. Perini, C. Rovelli and W. Wieland, *Class. Quant. Grav.* **30** 235023 (2013).
[92] J. Engle, E. Livine, R. Pereira and C. Rovelli, *Nucl. Phys.* B**799** 136–149 (2008).
[93] L. Freidel, L. and K. Krasnov, *Class. Quant. Grav.* **25** 125018 (2008).
[94] E. Livine, and S. Speziale, *Phys. Rev.* D**76** 84028 (2007).
[95] W. Kamiński, M. Kisielowski and L. Lewandowski, *Class. Quant. Grav.* **27** 95006 (2010).
[96] A. Baratin and D. Oriti, *Phys. Rev.* D**85** 044003 (2012).
[97] Y. Ding, and C. Rovelli, *Class. Quant. Grav.* **27** 205003 (2010).
[98] J. Baez, An introduction to Spinfoam models of BF theory and quantum gravity, in *Geometry and quantum physics, proceedings of the 38th Internationale Universitaätswochen für Kern- und Teilchenphysick*, H. Gausterer, L. Pittner and H. Grosse, eds. Lecture Notes in Physics, vol. 543, pp. 25–94.
[99] G. Ponzano and T. Regge, Semi-classical limit of Racah coefficients, in *Spectroscopic and group theoretic methods in physics*, F. Bloch *et al.* eds. (North Holland, Amsterdam, 1968).
[100] T. Regge, *Nuovo Cimento* **19** 558–571 (1961).
[101] R. M. Williams, *Class. Quant. Grav.* **3** 853–869 (1986).
[102] J. W. Barrett, *Class. Quant. Grav.* **4** 1565–1576 (1987).
[103] D. Oriti, arXiv:gr-qc/0607032.
[104] A. Ashtekar, M. Campiglia and A. Henderson, *Class. Quant. Grav.* **27** 135020 (2010).
[105] J. Barrett, R. Dowdall, W. Fairbairn, F. Hellmann and R. Pereira, *Class. Quant. Grav.* **27** 165009 (2010).
[106] H. Haggard, PhD Thesis, http://bohr.physics.berkeley.edu/hal/pubs/Thesis/ (2011).
[107] H. W. Hamber and R. M. Williams, *Phys. Lett.* B**157** 368 (1985).
[108] H. W. Hamber, *Nucl. Phys.* B**400** 347–389 (1993).
[109] H. W. Hamber, *Quantum gravitation: the Feynman path integral approach* (Springer, Berlin, 2009).
[110] E. Buffenoir, and P. Roche, *Commun. Math. Phys.* **207** 499–555 (1999).
[111] K. Noui and P. Roche, *Class. Quant. Grav.* **20** 3175–3214 (2003).
[112] W. Fairbairn and C. Meusburger, *J. Math. Phys.* **53** 22501 (2012).
[113] M. Han, *Phys. Rev.* D**84** 64010 (2011).
[114] Y. Ding, and M. Han, arXives:1103.1597.
[115] C. Rovelli, in *Quantum gravity and quantum cosmology*, G. Calcagni, L. Papantonopoulos, G. Siopsis and N. Tsamis. Lecture Notes in Physics, Vol. 863, pp. 57–66 (Springer, Berlin, 2013).
[116] B. Bahr, B. Dittrich and S. Steinhaus, *Phys. Rev.* D**83** 105026 (2011).
[117] T. Krajewski, J. Magnen, V. Rivasseau, A. Tanasa, Adrian and P. Vitale, *Phys. Rev.* D**82** 124069 (2010).
[118] A. Bonanno and M. Reuter, *Phys. Rev.* D**65** 043508 (2002).
[119] A. Bonanno and M. Reuter, *JCAP* **0708** 024 (2007).
[120] A. Bonanno and M. Reuter, *Phys. Rev.* D**62** 043008 (2000).
[121] A. Bonanno and M. Reuter, *Phys. Rev.* D**73** 083005 (2006).

[122] M. Bojowald, Loop quantum cosmology. *Living Rev. Rel.* **8** 11 (2005).

[123] A. Barrau, T. Cailleteau, J. Grain and J. Mielczarek, *Class. Quant. Grav.* **31** 053001 (2014).

[124] I. Agullo, A. Ashtekar and W. Nelson, *Phys. Rev. Lett.* **109** 251301 (2012); *Phys. Rev.* D**87** 043507 (2013), *Class. Quant. Grav.* **30** 085014 (2013).

[125] M. Fernández-Méndez, G. A. Mena Marugán, and J. Olmedo, *Phys. Rev.* D **86** 024003 (2012).

[126] L. Linsefors, T. Cailleteau, A. Barrau and J. Grain, *Phys. Rev.* D **87** 123509 (2013); L. Linsefors and A. Barrau, *Quant. Grav.* **31** 015018 (2014).

[127] S. W. Hawking and G. F. R. Ellis, *Large scale structure of space-time* (Cambridge University Press, Cambridge, 1973).

[128] A. Borde, A. Guth and A. Vilenkin, *Phys. Rev. Lett.* **90** 151301 (2003).

[129] A. Ashtekar, T. T. Pawłowski and P. Singh, *Phys. Rev. Lett.* **96** 141301 (2006); *Phys. Rev.* D**74** 084003 (2006).

[130] D. Craig and P. Singh, *Phys. Rev.* D**82** 123526 (2010).

[131] M. Bojowald, *Phys. Rev. Lett.* **86** 5227–5230 (2001).

[132] A. Ashtekar and M. Campiglia, *Class. Quant. Grav.* **29** 242001 (2012).

[133] A. Ashtekar, *Gen. Rel. Grav.* **41** 707–741 (2009).

[134] A. Ashtekar, T. T. Pawłowski, P. Singh, and K. Vandersloot, *Phys. Rev.* D**75** 024035 (2007); L. Szulc, W. Kamiński and J. Lewandowski, *Class. Quant. Grav.* **24** 2621 (2007).

[135] A. Ashtekar, A. Corichi and P. Singh, *Phys. Rev.* D**77** 024046 (2008).

[136] D. Craig and P. Singh, *Class. Quant. Grav.* **30** 205008 (2013).

[137] V. Taveras, *Phys. Rev.* D**78** 064072 (2008).

[138] E. Bentivegna and T. T. Pawłowski, *Phys. Rev.* D**77** 124025 (2008); T. Pawłowski and A. Ashtekar, *Phys. Rev.* D**85** 064001 (2012).

[139] A. Ashtekar and E. Wilson-Ewing, *Phys. Rev.* D**79** 083535 (2009); D**80** 123532 (2009).

[140] E. Wilson-Ewing, *Phys. Rev.* D**82** 043508 (2010).

[141] M. Martín-Benito, L. J. Garay and G. A. Mena Marugán, *Phys. Rev.* D**78** 083516 (2008); D. Brizuela, G. A. Mena Marugán and T. Pawłowski, *Class. Quant. Grav.* **27** 052001 (2010); M. Martín-Benito, G. A. Mena Marugán, E. Wilson-Ewing, *Phys. Rev.* D**82** 084012 (2010).

[142] P. Singh, *Class. Quant. Grav.* **26** 125005 (2009).

[143] T. Cailleteau, P. Singh and K. Vandersloot, *Phys. Rev.* D**80** 124013 (2009); E. Wilson-Ewing, *JCAP* **1303** 026 (2013); P. Singh and B. Gupt, *Phys. Rev.* D**89** 063520 (2014). arXiv:1309.2732.

[144] L. Parker, *Phys. Rev. Lett.* **21** 562 (1968); *Phys. Rev.* D**183** 1057 (1969).

[145] R. Holman and A. Tolley, *JCAP* **0805** 001 (2008); I. Agullo and L. Parker, *Phys. Rev.* D **83** 063526 (2011); *Gen. Rel. Grav.* **43** 2541–2545 (2011). J. Ganc, *Phys. Rev.* D **84** 063514 (2011).

[146] A. Ashtekar, W. Kamiński and J. Lewandowski, *Phys. Rev.* D**79** 064030 (2009).

[147] A. Ashtekar and D. Sloan, *Gen. Rel. Grav.* **43** 3619–3656 (2011).

[148] J. Ganc and E. Komatsu, *Phys. Rev.* D **86** 023518 (2012); I. Agullo and S. Shandera, *JCAP* **1209** 007 (2012).

[149] J. D. Bekenstein, *Phys. Rev.* D**7** 2333–2346 (1973); D**9** 3292–3200 (1974).

[150] S. W. Hawking, *Nature* **248** 30–31 (1974); *Commun. Math. Phys.* **43** 199–220 (1975).

[151] A. Ashtekar and B. Krishnan, *Living Rev. Rel.* **7** 10 (2004).

[152] A. Ashtekar, C. Beetle and J. Lewandowski, *Phys. Rev.* D**64** 044016 (2001).

[153] A. Ashtekar, J. Baez, A. Corichi and K. Krasnov, *Phys. Rev. Lett.* **80** 904–907 (1998); A. Ashtekar, J. Baez, A. Corichi and K. Krasnov, *Adv. Theor. Math. Phys.* **4** 1–95 (2000).

[154] A. Ashtekar, J. Engle, T. Pawłowski and C. Van Den Broeck, *Class. Quant. Grav.* **21** 2549–2570 (2005); A. Ashtekar, J. Engle and C. Van Den Broeck, *Class. Quant. Grav.* **22** L27–L34 (2005).

[155] M. Domagała and J. Lewandowski, *Class. Quant. Grav.* **21** 5233 (2004); I. Agullo, J. F. Barbero G., E. F. Borja, J. Diaz-Polo, and E. J. S. Villaseñor, *Phys. Rev. Lett.* **100** 211301 (2008); *Phys. Rev.* D**82** 084029 (2010).

[156] A. Ghosh, and A. Perez, *Phys. Rev. Lett.* **107** 241301 (2011); E. Frodden, A. Ghosh and A. Perez, *Phys. Rev.* D**87** 121503 (2013); A. Ghosh, K. Noui and A. Perez arXiv:1309.4563; E. Bianchi and A. Satz, *Phys. Rev.* D**12** 12403 (2013).

[157] W. Unruh, *Phys. Rev.* D**14** 870 (1976).

[158] R. Gambini and J. Pullin, *Phys. Rev. Lett.* **110** 211301 (2013); *Class. Quant. Grav.* **31** 115003 (2014); arXiv:1312.5512; R. Gambini, J. Olmedo and J. Pullin, arXiv:1310.5996.

[159] M. Bojowald and R. Swiderski, *Class. Quant. Grav.* **23** 2129–154 (2006).

[160] W. G. Unruh and R. Schützhold, *Phys. Rev.* D**71** 024028 (2005); D**78** 041504 (2008); T. Jacobson and D. Mattingly, D**61** 024017 (2000).

[161] A. Wightman, *Phys. Rev.* **101** 860 (1959).

[162] E. Bianchi and Y. Ding, *Phys. Rev.* D**86** 104040 (2012).

[163] E. Bianchi, E. Magliaro and C. Perini, *Nucl. Phys.* B**822** 245–269 (2009).

[164] C. Rovelli and M. Zhang, *Class. Quant. Grav.* **28** 175010 (2011).

[165] M. Han, *Phys. Rev.* D**89**, 124001 (2014).

[166] A. Riello, *Phys. Rev.* D**88** 24011 (2013).

[167] I. Khavkine, R. Loll and P. Reska, *Class. Quant. Grav.* **27** 185025 (2010).

[168] S. Chakraborty and P. Peldan, *Phys. Rev. Lett.* **73** 1195–1198 (1994).

[169] U. Harst and M. Reuter, *JHEP* **1105** 119 (2011).

[170] R. Percacci and D. Perini, *Phys. Rev* D**68** 044018 (2003).

[171] M. Shaposhnikov and C. Wetterich, *Phys. Lett.* B**683** 196 (2010).

[172] M. Reuter and J. Schwindt, *JHEP* **01** 070 (2006); *JHEP* **01** 049 (2007).

[173] O. Lauscher and M. Reuter, *JHEP* **10** 050 (2005).

[174] M. Reuter and F. Saueressig, *JHEP* **12** 012 (2011).

[175] A. Sen, *JHEP* **04** 156 (2013); arXiv:1402.0109.

12

Quantum Gravity via Supersymmetry and Holography

HENRIETTE ELVANG AND GARY T. HOROWITZ

This chapter offers a survey of ideas and results in the approach to quantum gravity based on supersymmetry, strings, and holography.

Extra spatial dimensions appear naturally in this approach, so to set the stage we begin in Section 12.1 with a discussion of general relativity in more than four spacetime dimensions. In higher dimensions, one encounters a richness of structure with no parallel in 4D. Even in vacuum gravity, this includes black hole solutions with non-spherical horizon topologies, black hole non-uniqueness, and regular multi-horizon black holes. We give an overview of such solutions and their properties, both in the context of Kaluza–Klein theory and for asymptotically flat boundary conditions.

A very interesting extension of general relativity is to include matter in such a way that the action becomes invariant under supersymmetry transformations. Supersymmetry is a remarkable symmetry that relates bosons and fermions. It is the only possible extension of the Poincaré group for a unitary theory with non-trivial scattering processes. Supersymmetry is considered a natural extension of the standard model of particle physics; the study of how supersymmetry is broken at low energies, and its possible experimental consequences, is an important active research area in particle physics. Furthermore, independently of its potential phenomenology, supersymmetry offers strong calculational control and that makes it a tremendously powerful tool for analyzing fundamental properties of quantum field theories.

When supersymmetry and general relativity are combined, the result is supergravity. The metric field is accompanied by a spin-3/2 spinor field and this gives a beautiful and enticing playground for advancing our understanding of quantum gravity. Supergravity theories exist in spacetime dimensions $D \leq 11$ and they provide a natural setting for studies of charged black holes. Certain extremal limits of charged black holes in supergravity are invariant under supersymmetry; such 'supersymmetric black holes' are key for understanding the statistical mechanical nature of black hole thermodynamics, specifically the microstates responsible for the Hawking–Bekenstein entropy. An example of a supersymmetric black hole is the extremal Reissner–Nordström solution.

Section 12.2 begins with a brief introduction to supersymmetry and supergravity, followed by a survey of supersymmetric black holes and their properties. We then discuss perturbative quantization of gravity as a quantum field theory, an approach in which the metric field is quantized in a flat-space background and the resulting gravitons are point-like spin-2 particles. It is well-known that this approach leads to ultraviolet divergences, starting at 2-loop order in pure gravity, that – unlike the corresponding infinities in gauge theories – cannot be cured since general relativity is non-renormalizable in perturbation theory about Minkowski space. However, in supergravity, the symmetry between bosons and fermions results in crucial cancellations in graviton scattering amplitudes and this can delay the occurrence of the ultraviolet divergences to higher-loop order. It has even been suggested that with maximal supersymmetry, perturbative supergravity in 4D may be free of such ultraviolet divergences. We will offer a short description of these ideas and related results.

A profound solution to the problem of these perturbative ultraviolet divergences in quantum gravity is to treat gravitons as extended one-dimensional objects: strings. Then the short-distance behavior is regulated by the finite extent of the string and scattering processes are free of ultraviolet divergences. Thus string theory is a very promising framework for a quantum theory of gravity. String theory naturally incorporates supersymmetry, and general relativity – and its extension to supergravity – emerges as a low-energy effective theory. String theory predicts extra spatial dimensions, so compatibility with observations requires that either these extra dimensions are compact and small (incorporating Kaluza–Klein theory) or that we live on a $(3 + 1)$-dimensional subspace of this higher-dimensional spacetime.

Section 12.3 is dedicated to an introduction to string theory. We begin with an overview of perturbative string theory and then turn to nonperturbative aspects, specifically quantum black holes. One of the remarkable features of string theory is that it provides a precise microscopic description of black hole entropy and Hawking radiation for certain black holes, specifically the supersymmetric or near-supersymmetric black holes. We describe in detail the precision-counting of black hole microstates in string theory and its match to the Bekenstein–Hawking entropy.

At the nonperturbative level, there are arguments that quantum gravity might be holographic: this is the notion that the physics in a region of space is completely described by degrees of freedom living on the boundary of this region. This idea was originally proposed based on considerations of black hole entropy. The entropy of a black hole scales with its area, in striking contrast to most systems which have an entropy proportional to their volume. This suggests that everything that happens inside the black hole might be encoded in degrees of freedom at the horizon. A precise formulation of holography emerges from string theory and is called "gauge/gravity duality". In Section 12.4, we discuss holography and present a detailed account of the motivation for gauge/gravity duality and the evidence in its favor. Recent years have seen applications of gauge/gravity duality to a wide variety of problems in physics, including black holes, quark confinement, hydrodynamics, and condensed matter physics. We give a brief survey of these results.

Section 12.5 contains some concluding remarks. It is our hope that this chapter will convey the depth and richness of the subjects mentioned above and motivate the reader to pursue further information in the references provided throughout the text.

12.1 Gravity in D dimensions

At first sight, general relativity in a D-dimensional spacetime looks much like 4D general relativity. The Einstein equation takes the same form,

$$G_{MN} = 8\pi G_D T_{MN}, \qquad (12.1)$$

in which G_D is the D-dimensional Newton's constant and the Einstein tensor is given in terms of the Ricci tensor as $G_{MN} = R_{MN} - \frac{1}{2}g_{MN}R$. The spacetime indices M, N run over $0, 1, 2, \ldots, D-1$. The field equations can be derived using the variational principle from the D-dimensional Einstein–Hilbert action

$$S_{\text{EH}} = \frac{1}{16\pi G_D} \int d^D x \sqrt{-g}\, R \; + \; S_{\text{matter}}, \qquad (12.2)$$

where the stress-energy tensor is $T_{MN} = -(2/\sqrt{-g})\delta S_{\text{matter}}/\delta g^{MN}$.

Given the similarities, one expects that solutions to the 4D Einstein equation have straightforward generalizations to higher dimensions. This is indeed the case, for example the Schwarzschild metric generalizes to $D > 4$ dimensions as a solution to the vacuum Einstein equations. The metric of the D-dimensional *Schwarzschild–Tangherlini solution* [1] found in 1963 is

$$ds^2 = -\left[1 - \left(\frac{r_0}{r}\right)^{D-3}\right] dt^2 + \left[1 - \left(\frac{r_0}{r}\right)^{D-3}\right]^{-1} dr^2 + r^2 \, d\Omega_{D-2}^2, \qquad (12.3)$$

where $d\Omega_d^2$ is the line element for a d-dimensional round unit sphere, S^d. The horizon is located at $r = r_0$ and has topology S^{D-2}. Black hole thermodynamics works the same in higher dimensions as in 4D. This includes the first law and the area theorem. For the D-dimensional Schwarzschild–Tangherlini black holes, the ADM mass, temperature (calculated from the surface gravity κ), and horizon 'area' are[1]

$$M = \frac{(D-2)\Omega_{D-2}r_0^{D-3}}{16\pi G_D}, \qquad T = \frac{\kappa}{2\pi} = \frac{D-3}{4\pi r_0}, \qquad A = \Omega_{D-2}r_0^{D-2}, \qquad (12.4)$$

where Ω_d is the volume of the unit d-sphere. The entropy is $S = A/(4G_D)$. These quantities satisfy $dM = T\, dS$, and the Smarr relation $(D-3)M = (D-2)TS$.

There are of course also important differences as D varies. For example, it is well known that there is no 3D vacuum black hole,[2] so the 4D Schwarzschild solution does not generalize as we go down in dimension. Might it be that there are black hole solutions in $D > 4$ that do not exist in 4D? The answer turns out to be yes.

That gravity has richer structure in higher dimensions is apparent already from solutions to the linearized Einstein equation. For example, a gravitational wave in D dimensions has $D(D-3)/2$ degrees of freedom. This formula counts the well-known 2 polarizations of a gravitational wave in a $D = 4$ dimensional spacetime. But it also tells us that in

---

[1] We will refer to the horizon volume as 'area' although of course it is the volume of a $(D-2)$-dimensional manifold.
[2] With a negative cosmological constant, there is a 3D black hole [2].

3 dimensions there are no propagating modes of gravity. Going up in dimensions, we learn that a gravitational wave in 5D carries 5 degrees of freedom, in 6D it is 9 degrees of freedom, and so on. This hints that gravitational dynamics has important dimensional dependence and that going up in spacetime dimension gives 'more freedom' and new phenomena may be found. In this section, we give several examples of the rewards of studying gravity in spacetime dimensions $D > 4$.

12.1.1 Kaluza–Klein theory

One motivation for studying higher-dimensional gravity is that it offers a method for unifying gravity with other forces, as first explored by Kaluza and Klein in the early 1920s [3, 4]. The idea is that pure gravity in 5 dimensions can be viewed as a Maxwell-scalar-gravity system in 4 dimensions. To see how it works, we write an ansatz for the 5D metric,

$$ds_{5D}^2 = g_{MN}\, dX^M\, dX^N = e^{\phi/\sqrt{3}}\, g_{\mu\nu}\, dx^\mu\, dx^\nu + e^{-2\phi/\sqrt{3}}\big(dy + A_\mu\, dx^\mu\big)^2, \tag{12.5}$$

where $M, N = 0, 1, 2, 3, 4$ and $\mu = 0, 1, 2, 3$. Let us assume that $g_{\mu\nu}$, ϕ and A_μ are independent of $X^4 = y$ and that the y direction is a circle S^1 of radius R. Evaluating the 5D Ricci tensor with this ansatz, one finds (after partial integration) that the 5D Einstein–Hilbert action can be written

$$\begin{aligned} S &= \frac{1}{16\pi G_5} \int d^5X \sqrt{-g_5}\, R_{5D} \\ &= \frac{1}{16\pi G_4} \int d^4x \sqrt{-g}\Big(R_{4D} - \tfrac{1}{2}\,\partial_\mu \phi\, \partial^\mu \phi - \tfrac{1}{4}e^{-\sqrt{3}\phi} F_{\mu\nu} F^{\mu\nu}\Big), \end{aligned} \tag{12.6}$$

where $F_{\mu\nu} = \partial_\mu A_\nu - \partial_\nu A_\mu$ is a Maxwell field strength and the 4D indices μ, ν are lowered/raised with the 4D metric $g_{\mu\nu}$ and its inverse. The 4D Newton's constant is related to the 5D one by

$$G_4 = G_5/(2\pi R). \tag{12.7}$$

Let us now consider the equations of motion derived from (12.6). The 4D description gives the Einstein equation coupled to a Maxwell field and the scalar ϕ called the 'dilaton'. In addition, we have the matter equations of motion: the Maxwell equation, with its non-minimal coupling to ϕ, and the scalar equation of motion sourced by the Maxwell field. This appears to be a rather complex gravity-electromagnetic system. However, from the 5D point of view, this is nothing but vacuum gravity: the 5D equation of motion is just the vacuum Einstein equation. Thus 'lifting' the 4D system to 5D unifies electromagnetism with gravity! This is a very clean and beautiful example of unification of forces.

In the *Kaluza–Klein ansatz* (12.5), we assumed that the metric components were independent of the S^1 direction y. Generally, we could express the dependence of y in terms of the Fourier modes of the 4D fields, e.g.

$$\phi(x, y) = \sum_n \phi_n(x)\, e^{iny/R}. \tag{12.8}$$

It follows from the ϕ equation of motion that the modes with non-zero n have mass-terms of the order $|n|/R$. If the radius of the Kaluza–Klein circle R is very small compared to energies we are interested in (or that otherwise appear in the system), then in a low-energy long-wavelength approximation these modes do not contribute. Thus we can truncate the $n \neq 0$ modes and focus only on the massless modes. This means that we are effectively taking the fields to be independent of y, and that is exactly the Kaluza–Klein ansatz.

The S^1 *Kaluza–Klein reduction* (also known as dimensional reduction) from 5D to 4D described above – or, equivalently, the unifying lift from 4D to 5D – can be generalized to D dimensions with minor changes in the numerical coefficients for the dilaton dependence. It also has generalizations to reduction on other manifolds than a circle, for example on a torus $S^1 \times S^1$ or a sphere S^d. The required key property is that the lower-dimensional equations of motion are consistent truncations of the higher-dimensional ones.

The Kaluza–Klein reduction is the prototype for *compactifications* of a higher-dimensional system to a lower-dimensional one. Compactifications play a central role in many areas of high-energy theoretical physics, both in field theory and in string theory. For now, we focus on classical aspects of higher-dimensional gravity so we will consider various solutions to the D-dimensional Einstein equations, in the context of the Kaluza–Klein ansatz as well as more generally.

12.1.2 Black strings

As a simple, but nonetheless quite interesting, example of a Kaluza–Klein spacetime, we consider black strings. Choose the Schwarzschild solution as the 4D solution in the Kaluza–Klein ansatz. The ansatz (12.5) then has $\phi = 0$ and $A_\mu = 0$ and it tells us that the metric

$$ds^2_{5D} = -\left(1 - \frac{2G_4M}{r}\right)dt^2 + \left(1 - \frac{2G_4M}{r}\right)^{-1}dr^2 + r^2\,d\Omega_2^2 + dy^2 \qquad (12.9)$$

solves the 5D vacuum Einstein equation. At any constant y-slice, the geometry described by (12.9) looks like a 4D Schwarzschild black hole with mass M, so it describes a continuous uniform string of Schwarzschild black holes: it is called a *homogeneous black string*. When y is a circle, the topology of the black string horizon is $S^2 \times S^1$.

Now, let us think of the black string as a uniform distribution of mass along the circle parameterized by y. Suppose this distribution of mass is perturbed a little: then there will be regions of higher mass-density and regions of lower mass-density. The denser regions will tend to attract more matter and grow while the lower-density regions are depleted. This indicates that the black string has a classical instability. Indeed, when the radius of the circle R is (roughly) larger than the Schwarzschild radius $2GM$, the homogeneous black string solution (12.9) is unstable to spherical linear perturbations, as demonstrated first by Gregory and Laflamme [5]. The evolution of the *Gregory–Laflamme instability* is exactly as our intuition indicates: the black string horizon becomes non-uniform along y as some parts of the string bulge while others shrink when the mass concentrates/depletes the corresponding regions. The S^2 of the constant y-slices is not perturbed, it remains round.

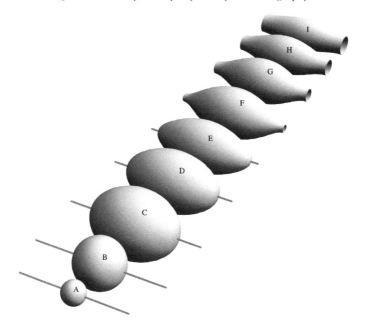

Figure 12.1 Embedding plots of $D = 5$ Kaluza–Klein black hole horizons: A–E show the localized black holes and F–I are non-uniform black strings. (Plots from [6].)

What is the endstate of the Gregory–Laflamme instability? The intuitive description of the instability as mass concentrations in certain regions of the y direction indicates that a localized black hole forms. Such a localized black hole would be like the $D = 5$ Schwarzschild–Tangherlini black hole (12.3), but placed in a spacetime with a compact S^1 direction. We can easily imagine this when the black hole is much smaller than the S^1; the black hole does not 'know' that the y-direction is compact, so it is essentially unaffected. However, if the Schwarzschild radius of the black hole is comparable to the size of the S^1, there will be a significant backreaction that deforms the shape of the horizon (while keeping its S^3 topology). Such *localized black hole* solutions have been constructed numerically; Fig. 12.1 shows the change of horizon shape as the mass of the black hole is increased for fixed size of the Kaluza–Klein circle at infinity.

For the localized black hole to be an endstate of the Gregory–Laflamme instability requires the horizon topology to transition from the $S^2 \times S^1$ of the string to the S^3 of the localized black hole. Classically, the horizon cannot bifurcate without forming a naked singularity at the pinch-off [7]. In classical gravity the pinch-off is not reached in finite affine time along the null generators of the horizon [8], but a numerical analysis [9] indicates that a naked singularity forms in finite asymptotic time as the horizon pinches. In fact, the numerical work [9] reveals that the horizon develops in an approximately self-similar fashion at late times: the black string becomes a string of 5D black holes of various sizes connected by thin strings. These thin strings are themselves subject to the Gregory–Laflamme instability and this results in further clumping, thus giving a self-similar evolution. (This is similar to the behavior in a low-viscosity fluid stream: the Rayleigh–Plateau instability causes a cascade of spherical beads to develop in a self-similar manner along the stream [10].)

Since a naked singularity forms without fine-tuning of the initial data, this constitutes a *violation of cosmic censorship.* Classical gravity can no longer be trusted near the singularity and it is expected that quantum gravity effects must be included to understand the evolution. However, the most natural outcome is simply that the horizon bifurcates and the endstate of the Gregory–Laflamme instability is a localized black hole.

The homogeneous black string and the localized black holes are not the only static black hole solutions to the 5D vacuum Einstein equations with Kaluza–Klein boundary conditions. As one increases the mass of a localized black hole on a circle of fixed asymptotic size $L = 2\pi R$, there is a critical mass $G_5 M/L^2 \sim 0.12$ where the horizon merges across the Kaluza–Klein circle and for larger masses one has a new *inhomogeneous black string solution* whose horizon topology is $S^2 \times S^1$. It has been constructed numerically [11–13]; embeddings of the horizon are illustrated in Fig. 12.1.

The 5D vacuum solutions discussed here – the homogeneous and inhomogeneous black strings and the localized black hole – all have Kaluza–Klein asymptotics: at large r, these 5D vacuum solutions approach 4D Minkowski spacetime times the Kaluza–Klein circle S^1. Figure 12.2 indicates the different solution branches in a "phase diagram" where solutions are compared for fixed size of the Kaluza–Klein circle at infinity. Note that there can be more than one solution with the same mass; so we have *black hole non-uniqueness* in 5D Kaluza–Klein spacetimes! As shown in the phase diagram in Fig. 12.2, the inhomogeneous black string joins the homogeneous black string at the onset of the Gregory–Laflamme instability. This is expected due to the existence of a static inhomogeneous perturbation at this point. It is also clear from Fig. 12.2 that the entropy of the inhomogeneous black string is smaller than that of the homogeneous black string of the same mass, so the area theorem implies that it could not have been a viable endstate of the Gregory–Laflamme instability,

It interesting to note that there is no positive energy theorem for Kaluza–Klein spacetimes; in fact there exist solutions with arbitrarily low energy[3] [15, 16]. Moreover, the simplest Kaluza–Klein spacetime, 4D Minkowski space times a circle, is actually unstable. It can undergo decay by nucleation of *Kaluza–Klein bubbles*, which are obtained by a double analytic continuation of the 5D Schwarzschild–Tangherlini black hole [17]. This instability can be removed, and a positive energy theorem proven, by including fermions with periodic boundary conditions on the S^1 [18]. These fermions are naturally included in the supersymmetric theories we discuss later. The existence of Kaluza–Klein bubbles actually allow for even more classical solutions to the 5D Einstein equation with Kaluza–Klein boundary conditions: these are static, analytically known solutions that describe combinations of black strings, black holes, and Kaluza–Klein bubbles [19, 20].

---

[3] For a definition of energy in Kaluza–Klein theory, see [14].

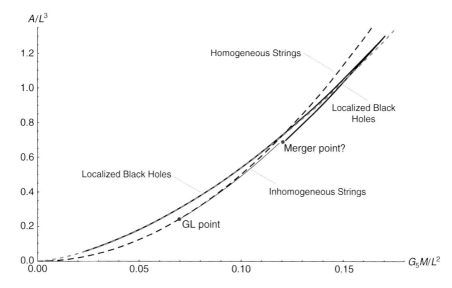

Figure 12.2 Phase diagram for $D = 5$ Kaluza–Klein black holes. The horizon 'area' is plotted versus the black hole mass for fixed length $L = 2\pi R$ of the Kaluza–Klein circle at asymptotic infinity. The localized black hole curve (solid) starts off as $A \propto M^{2/3}$ at small mass, since for small $G_5 M/L^2$ it behaves as the asymptotically flat 5D Schwarzschild–Tangherlini black hole. The dashed curve for the homogeneous black string is $A \propto M^2$, reflecting the area dependence of the 4D Schwarzschild black hole of the string. The inhomogeneous black string branch (also solid) begins at the point $G_5 M \approx 0.7 L^2$ where the Gregory–Laflamme instability first sets in. The numerics make it plausible that the branch of inhomogeneous black strings merge near $G_5 M \approx 1.2 L^2$ with the localized black hole branch, as indicated in the plot. (Plot from [6].)

In our presentation of 5D Kaluza–Klein gravity, we have encountered a richness of structure: linearly unstable black strings, black hole non-uniqueness, and violation of cosmic censorship. It turns out that this also carries over to black holes in asymptotically flat 5D spacetimes. This is the subject of the next section.

12.1.3 *Asymptotically flat black holes in* $D = 5$ *vacuum gravity*

We have already described a black hole solution in D-dimensional asymptotically flat space, namely the Schwarzschild–Tangherlini solution (12.3). Black holes in 4D vacuum gravity are characterized by their mass M and angular momentum J. The 4D rotating black hole described by the Kerr solution can be generalized to $D > 4$ dimensions: the rotating black hole solutions of the $D > 4$ dimensional vacuum Einstein equation were found analytically in 1986 [21] and are called *Myers–Perry black holes*.

In 4D spacetime, angular momentum is often associated with an axis of rotation, but this is an artifact of having three spatial directions. It is more general to associate angular momentum with *planes of rotation*; in three spatial dimensions a plane is characterized by its normal vector, but this is not true in higher dimensions. The independent planes of rotation in D dimensions can be characterized by the $\lfloor(D-1)/2\rfloor$ independent generators of the Cartan subalgebra of the D-dimensional rotation group $SO(D-1)$. Thus, in addition to its mass M, the 5D Myers–Perry black hole is characterized by two angular momenta J_1 and J_2 associated with rotations in two independent planes, say (x^1x^2) and (x^3x^4). The 5D Myers–Perry metric is

$$ds^2 = -dt^2 + \frac{\mu r^2}{\Delta}\Big(dt + a_1\cos^2\theta\, d\phi_1 + a_2\sin^2\theta\, d\phi_2\Big)^2$$
$$+ \frac{\Delta}{(r^2+a_1^2)(r^2+a_2^2)-\mu r^2}\, dr^2$$
$$+ (r^2+a_1^2)(\sin^2\theta\, d\theta^2 + \cos^2\theta\, d\phi_1^2) + (r^2+a_2^2)(\cos^2\theta\, d\theta^2 + \sin^2\theta\, d\phi_2^2)\,,$$
(12.10)

where

$$\Delta = (r^2+a_1^2)(r^2+a_2^2)\left(1 - \frac{a_1^2\cos^2\theta}{r^2+a_1^2} - \frac{a_2^2\sin^2\theta}{r^2+a_2^2}\right).$$
(12.11)

The topology of the horizon is S^3 but, with rotation turned on, its shape is no longer round, but pancaked in the planes of rotation. The mass and angular momentum are

$$M = \frac{3\pi\,\mu}{8G_5}\,, \quad J_i = \frac{\pi\,\mu}{4G_5}a_i\,.$$
(12.12)

The first law of thermodynamics is now $dM = T\,dS + \Omega_1\,dJ_1 + \Omega_2\,dJ_2$ with $\Omega_{1,2}$ the angular velocities of the horizon. This is a straightforward generalization of the first law for Kerr black holes. Just as for Kerr black holes, there is an upper bound on the magnitude of the angular momentum for given mass: $M^3 \geq \big(\tilde{J}_1^2 + \tilde{J}_2^2 + 2|\tilde{J}_1\tilde{J}_2|\big)$, where $\tilde{J}_i = \sqrt{27\pi/(32G_5)}J_i$. When both angular momenta are non-vanishing, the 5D Myers–Perry solution approaches a smooth solution describing an extremal black hole with $T=0$, just like Kerr black holes. However, when one of the angular momenta vanishes, say $J_2 = 0$, the maximally rotating 5D Myers–Perry black hole becomes singular. It is smooth for $\tilde{J}_1^2 < M^3$, but as the angular momentum is increased from $\tilde{J}_1 = 0$ to the maximum value, the horizon 'area' decreases monotonically to zero as the horizon flattens out in the plane of rotation.

Now it turns out that the Myers–Perry black holes are not the only regular black hole solutions to the 5D vacuum Einstein equation: there is another class of solutions called *black rings*. The black rings have horizon topology $S^2 \times S^1$ and the metrics are known analytically [22, 23]. To get some intuition for what a black ring is, recall the black string: suppose you take a 4D Schwarzschild black hole times a line, but instead of wrapping the string on a Kaluza–Klein circle, close it into a round ring in an asymptotically flat spacetime. Then we have a black hole with $S^2 \times S^1$ topology, i.e. a black ring. The difference between a black ring and a black string is that the S^1 of the ring is contractible, whereas the S^1 of the string is not.

A ring-like distribution of mass in space is going to collapse upon itself, so clearly no black ring solution can exist without something balancing its gravitational self-attraction.

In vacuum, a static ring can be constructed in 5D asymptotically flat space, but it suffers from a conical excess angle inside the plane of the ring; the excess is needed to support the ring-shaped horizon topology. However, the black ring can be balanced against self-collapse by giving it angular momentum in the plane of the ring. For a given ADM mass M, the minimum angular momentum needed is $\tilde{J}^2 > \frac{27}{32}M^3$ [22, 24] and when this bound is satisfied there are black ring solutions that are smooth everywhere outside and on the horizon.

As solutions to the vacuum Einstein equations, black hole thermodynamics is valid for black rings too, with entropy proportional to the three-dimensional 'area' of the horizon. Figure 12.3 is a plot of entropy versus angular momentum J for smooth 5D Myers–Perry black holes and the black rings. The black holes and rings in this plot have rotation only in one plane, and for the ring, this is the plane of its S^1, as needed to balance it. Let us highlight some remarkable features:

- There are *two branches of black rings*, the 'lower' one consists of 'fat' flattened-out black rings, while the higher entropy branch are thin black rings. The two branches meet at the cusp where the angular momentum takes its minimal possible value, $\tilde{J}^2/M^3 = 27/32$.
- In the range $27/32 \leq \tilde{J}^2/M^3 < 1$, three distinct black hole solutions exists: one Myers–Perry black hole and two black rings, thin and fat. This is an exciting (and historically the first) example of *black hole non uniqueness* in an asymptotically flat spacetime. Of course, this is completely different from 4D in which the Kerr black hole famously is the only smooth asymptotically flat stationary black hole vacuum solution.

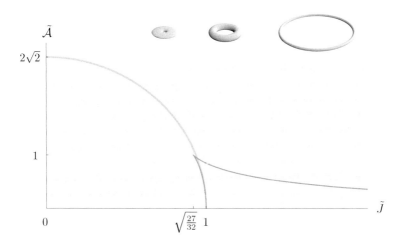

Figure 12.3 Phase diagram for asymptotically flat $D = 5$ black holes with angular momentum in one plane: 5D Myers–Perry (light line), black rings (dark line), and black saturns (gray-shaded). The plot shows for fixed mass scale $M = 1$ horizon area (entropy) $\mathcal{A}$ versus angular momentum J in rescaled units $\tilde{\mathcal{A}} = \sqrt{27/(256\pi G_5^3)}\mathcal{A}$ and $\tilde{J} = \sqrt{27\pi/(32G_5)}J$. (Plots from [25, 26].)

- For the thin black rings, there is *no upper bound on the magnitude of the angular momentum*. As J^2/M^3 increases, the ring's S^1 radius grows and the ring becomes very thin. A small part of the ring will look like a piece of a boosted black string [25] – given that it is thin, one would expect it to undergo Gregory–Laflamme instability. As the instability develops, there will be gravitational radiation from the time-varying quadruple moment of the rotating bumps on the ring, but the time scale of radiating away these bumps cannot compete [25] with the time scale of the horizon pinch, so the likely endstate of the instability of an ultra-spinning black ring is a pair of black holes flying apart in such a way that the angular momentum is preserved. The pinch of the horizon would go through a naked singularity, so this would also constitute a violation of cosmic censorship.

Now, just as the Myers–Perry black holes can carry angular momentum in the two independent planes of $\mathbb{R}^4$, so can black rings. As an intuitive picture, consider starting with a black string made from a Kerr black hole (instead of a Schwarzschild black hole) times a line. Then bend this Kerr string into a ring and set it into rotation in the plane of the S^1 of the ring. That gives a rotating black ring with angular momentum also on the S^2 of the ring cross-section. Saying the words is easy, but the construction of the *doubly-rotating black ring* as a solution to the 5D Einstein equation is less trivial. An exact solution does exist [27]; it was constructed using the *'inverse scattering method'* [28–30]. An analysis of the physical properties of the doubly-rotating black ring can be found in [31].

The inverse scattering method is an integrability technique that uses Lax pairs to generate new solutions to non-linear partial differential equations with input of a known solution. The method was originally used to study solitonic waves in shallow water. In the late 1970s it was realized [28] that such techniques can also be applied to the 4D Einstein equations for co-dimension-2 systems. For example, one can use inverse scattering to generate the full Kerr solution from flat Minkowski space [30]. Much more recently, the inverse scattering method was applied to generating solutions in 5D gravity and several new solutions were found. Let us survey them.

Consider the balance of the singly spinning black ring: for a given mass M and J above the lower bound, the ring adjusts its radius to achieve the needed balance. One can simulate this with a Newtonian model of a rotating rubber band that wants to contract to zero size: balance between the band tension and the centrifugal force is obtained for only one special radius. Now suppose the ring (or rubber band) is placed in an external central attractive potential. It then has to rotate a bit faster to be balanced at the same radius. This situation can happen too in 5D general relativity: we can imagine a rotating black ring balanced around a central 5D Schwarzschild–Tangherlini black hole. Remarkably, there exists an exact analytic solution to the 5D Einstein vacuum equations that realizes this *Black Saturn* configuration [32]. It is completely smooth everywhere outside and on the two horizons [32, 33]. The Black Saturn solution was constructed using the inverse scattering technique.

Black Saturn displays a number of novel properties [32]:

- It offers *2-fold continuous black hole non-uniqueness*: for given ADM mass M (the total mass of the hole–ring system) and total ADM angular momentum J, there are continuously many ways of distributing the mass and angular momentum among the two black objects of the saturn system. The ring and hole can be co-rotating or counter-rotating.

- It shows that the 5D Schwarzschild–Tangherlini solution (12.3) is *not* the unique solution with zero ADM angular momentum. One can arrange the ring and hole of Black Saturn to be counter-rotating in such a way that the total system has zero angular momentum at infinity. This leaves the freedom of the mass distribution between the hole and the ring, thus leaving a continuous 1-parameter family of $J = 0$ solutions which are degenerate (as far as asymptotic data goes) with the 5D Schwarzschild–Tangherlini black hole (but have smaller entropy).
- The Black Saturn system illustrates frame-dragging effects as can be seen by studying the effect of, say, the rotating ring on the central black hole.
- No Black Saturn configuration has higher entropy than that of the 5D static Schwarzschild–Tangherlini black hole, but it can come arbitrarily close. In fact the whole gray-shaded region of phase space with $S < S_{\text{Schw}}$ and $\tilde{J} \geq 0$ in Fig. 12.3 is filled out by a continuum of Black Saturn configurations [26].

In our solar system, the planet Saturn has more than one ring and this is also possible for Black Saturn. One can use the inverse scattering method to construct exact Black Saturn solutions with any number of rings rotating in the same plane. And why keep the 'planet'? Drop the black hole at the center and simply just have a *multi-ring system*; the simplest case with two rings in the same plane is called a *di-ring* solution [34, 35]. And we can take this even further: why should the rings be in the same plane? After all, we have two independent planes of rotation, so how about arranging two rotating rings in the two orthogonal planes? Such a system is known as a *bi-ring* (or bicycling rings) and the exact solution has been constructed with the inverse scattering method [31, 36].

The solutions we have discussed above are "special" in the sense that they have more symmetry than is strictly needed: they have three commuting Killing vectors ∂_t, ∂_{ϕ_1}, and ∂_{ϕ_2}. The black hole rigidity theorem [38, 39] requires only one rotational isometry, and even before these theorems were established it was conjectured that such less-symmetric non-static stationary black hole solutions exist [37, 39]. This has been demonstrated by the construction [40] of so-called *helical black rings* using asymptotic matching methods in the limit of large angular momenta. Helical black rings have the same horizon topology as the ring, $S^2 \times S^1$, but are shaped as a "slinky" bent into a ring.

The richness of black holes in 5D Einstein vacuum gravity is clearly remarkable and unparalleled in 4D. Let us now briefly discuss what is known (and not known) about black holes in asymptotically flat spacetimes with $D > 5$.

12.1.4 *Asymptotically flat black holes in* $D > 5$ *vacuum gravity*

Examples of black holes in higher-dimensional Kaluza–Klein theory are easy to obtain as direct products of lower-dimensional vacuum solutions times circles. For instance, we get a 6D black string from 5D Schwarzschild–Tangherlini times S^1. Or, taking the product of 4D

Schwarzschild and a torus $S^1 \times S^1$, we get a 6D *black membrane*. Of course this generalizes: the product of n-dimensional Schwarzschild–Tangherlini times a p-torus gives a *black p-brane* in a spacetime with $D = (n + p)$ dimensions. One can also construct stationary solutions such as a 6D rotating 'black cylinder' as the product of a black ring and a circle.

The asymptotically flat black holes we met in 5D generalize to higher dimensions. This includes the Myers–Perry rotating black holes. Consider a D-dimensional Myers–Perry black hole with rotation just in one plane. The metric is [41]

$$ds^2 = -dt^2 + \frac{\mu}{r^{D-5}\rho^2}\Big(dt + a\sin^2\theta\, d\phi\Big)^2 + (r^2 + a^2)\sin^2\theta\, d\phi^2$$

$$+ \rho^2\, d\theta^2 + r^2\cos^2\theta\, d\Omega_{D-4}^2 + \frac{\rho^2}{\Delta}\, dr^2, \qquad (12.13)$$

where

$$\rho^2 = r^2 + a^2\cos^2\theta \quad \text{and} \quad \Delta = r^2 + a^2 - \frac{\mu}{r^{D-5}}. \qquad (12.14)$$

The mass is $M = (D-2)\Omega_{D-2}\mu/(16\pi G_D)$ and the angular momentum is $J = 2Ma/(D-2)$. While the angular momentum of the $D = 5$ Myers–Perry black hole is bounded from above, it turns out that there is no such bound on the angular momentum for $D > 5$: for given mass of the black hole, the angular momentum can be arbitrarily large. In this ultra-spinning limit, the black hole flattens out in the plane of rotation. At some stage it becomes very similar to a thin black membrane and is then expected [41] to undergo an instability much like the Gregory–Laflamme instability. In several cases, this instability has been seen numerically. If the instability mode preserves the isometry of the generators of the rotation, then one can imagine that it causes a pinch that splits the rotating spherical black hole into a black ring! Or a black saturn. Or a multi-ring saturn-like system. (Generally, the fewer the disconnected horizons, the higher the entropy of the system.) One can also imagine that there exist 'lumpy' black holes whose horizons are topologically spherical analogs of the inhomogeneous black strings discussed in Section 12.1.2. The possibilities and discussion of the phase diagram can be found in [42].

In $D \geq 5$, there are also black rings and di-rings and bi-rings and saturns; but beyond 5D, no exact solutions are currently known for these. Black rings (and helical black rings) in $D > 5$ have been constructed using the *blackfold method* [43–46]. (There have also been numerical constructions [47] of $D > 5$ black rings.) The blackfold method exploits the fact that black holes in higher dimensions can have more than one characteristic scale. For example, for a black ring, there is one scale associated with the size R_{S^1} of the S^1 of the ring and another R_{S^2} with the S^2. In the ultra-rotating regime where the ring is thin, the scales are separated, we have $R_{S^2} \ll R_{S^1}$. The blackfold methods exploit such separation of scales to solve the Einstein equations in a matched asymptotic expansion. The method has also been used to construct black holes with more exotic horizon topologies, for example products of odd-spheres. For an overview of possibilities for the horizon topologies in $4 \leq D \leq 11$, see table 1 in [40].

So far, we have discussed only black holes in higher-dimensional Einstein vacuum gravity. It is natural to introduce matter fields and also consider charged black holes. This can be done in a general context, but with the aim of constructing the quantum theory, in the following we will focus on supergravity theory.

12.2 Supergravity

Supersymmetry is a symmetry that mixes bosons and fermions. It is the only possible extension of Poincaré spacetime symmetry for a unitary theory with non-trivial scattering processes [48, 49]. Not only does this make supersymmetry a natural candidate for physics beyond the standard model of particle physics and a beautiful path to the unification of forces, it also provides an extremely powerful tool for understanding gauge theories – and black holes! The combination of general relativity with supersymmetry is *supergravity*. We will briefly introduce the ideas of supersymmetry and supergravity and then discuss their impact on our understanding of charged black holes and how it improves the perturbative quantum theory.

12.2.1 Supersymmetry

Let us begin with a simple, but concrete, example of supersymmetry. Consider in 4D flat space the Lagrangian for a 2-component Weyl fermion χ and a complex scalar field ϕ interacting via Yukawa terms and a quartic scalar interaction:

$$\mathcal{L} = i\chi^\dagger \bar{\sigma}^\mu \, \partial_\mu \chi - \partial_\mu \bar{\phi} \, \partial^\mu \phi + \tfrac{1}{2} g \, \phi \, \chi \chi + \tfrac{1}{2} g^* \, \bar{\phi} \, \chi^\dagger \chi^\dagger - \tfrac{1}{4} |g|^2 \, |\phi|^4 \,. \tag{12.15}$$

The bar on ϕ denotes the complex conjugate and we have the 2×2 matrices $\sigma^\mu = (1, \sigma^i)$ and $\bar{\sigma}^\mu = (1, -\sigma^i)$, where σ^i are the Pauli matrices. In addition to the usual Poincaré symmetry, $\mathcal{L}$ also has a symmetry that mixes the fermions and bosons:

$$\begin{aligned} \delta_\epsilon \phi &= \epsilon^\alpha \chi_\alpha \,, & \delta_\epsilon \bar{\phi} &= \epsilon^\dagger_{\dot\alpha} \chi^{\dagger\dot\alpha} \,, \\ \delta_\epsilon \chi_\alpha &= -i\sigma^\mu_{\alpha\dot\beta} \, \epsilon^{\dagger\dot\beta} \, \partial_\mu \phi + \tfrac{1}{2} g^* \bar\phi^2 \epsilon_\alpha \,, & \delta_\epsilon \chi^\dagger_{\dot\alpha} &= i \, \partial_\mu \bar\phi \, \epsilon^\beta \sigma^\mu_{\beta\dot\alpha} + \tfrac{1}{2} g\phi^2 \, \epsilon^\dagger_{\dot\alpha} \,. \end{aligned} \tag{12.16}$$

This is an example of a *supersymmetry transformation*. The anti-commuting constant spinor ϵ is an infinitesimal supersymmetry parameter (a fermionic analogue of the infinitesimal angle θ of a rotation transformation).

The anti-commuting conserved Noether supercharges Q and $Q^\dagger$ resulting from supersymmetry give symmetry generators that extend the Poincaré algebra to a graded Lie algebra. The commutator of two supersymmetry transformations is a translation, $[\delta_{\epsilon_1}, \delta_{\epsilon_2}] \sim (\epsilon^\dagger_1 \sigma^\mu \epsilon_2) \, \partial_\mu$, so this induces the algebra

$$\{Q^\dagger, Q\} \sim P^\mu \,, \quad \{Q, Q\} = 0 \,, \quad \{Q^\dagger, Q^\dagger\} = 0 \,. \tag{12.17}$$

In addition, one has $[Q^{(\dagger)}, P^\mu] = 0$ and $[Q, M^{\mu\nu}] \sim Q$.

In the quantum theory, the fields ϕ and χ in (12.15) create a spin-0 or spin-1/2 particle from the vacuum, respectively. Since the fields are related by supersymmetry, so are the corresponding particles. The algebra outlined above implies that $P^2 = P_\mu P^\mu$ commutes with the supersymmetry generators, so this means that particles related by supersymmetry – i.e. in the same *supermultiplet* – must have the same mass. In our example (12.15), the boson and fermion are both massless. The supercharges act on a particle with spin s by relating it to a particle with spin $s \pm \frac{1}{2}$.

Supersymmetry implies that the number of on-shell bosonic and fermonic degrees of freedom are equal.[4] In our example (12.15), the complex scalar encodes two real degrees of freedom and the on-shell Weyl fermion similarly gives two real degrees of freedom (the positive and negative helicity states of the massless spin-1/2 fermion). Similarly, the two helicity states ± 1 of a massless vector boson, such as a photon or gluon, are matched by the two $\pm 1/2$ helicity states of its supersymmetric partner fermion, a photino or gluino.

Extended supersymmetry means that one has $\mathcal{N}$ pairs of supersymmetry charges $Q^A, Q_A^\dagger$ with $A = 1, 2, \ldots, \mathcal{N}$. In that case, the supersymmetry algebra allows for the possibility of a central charge extension $\{Q_\alpha^A, Q_\beta^B\} \sim \epsilon_{\alpha\beta} Z^{AB}$, where Z^{AB} is antisymmetric. The model we described above in (12.15) has $\mathcal{N} = 1$ supersymmetry. In a supersymmetric theory, we distinguish the internal 'flavor' symmetries that commute with the supersymmetry generators from the *R-symmetries* that do not. An $\mathcal{N} = 1$ supersymmetric theory may have a $U(1)$ R-symmetry, while theories with extended supersymmetry can have non-abelian R-symmetry, typically $SU(\mathcal{N})$, that rotates the supercharges among each other.

In a 4D theory with spin no greater than 1, the maximal admissible amount of supersymmetry is $\mathcal{N} = 4$. The 4D $\mathcal{N} = 4$ supersymmetric theory turns out to be unique: it is the maximally supersymmetric extension of Yang–Mills theory and it is known as "$\mathcal{N} = 4$ *super Yang–Mills theory*" (SYM). Its spectrum of particles consists of the gluon, $(\mathcal{N} =)$ 4 spin-1/2 gluinos, and 6 scalars. With the two helicity states of the gluon and the 6 real scalars, this amounts to 8 bosonic degrees of freedom and $4 \times 2 = 8$ fermonic degrees of freedom. $\mathcal{N} = 4$ SYM has truly remarkable properties, for example the beta-function vanishes at all orders in perturbation theory, so there is no running of the gauge coupling. The theory is conformal, meaning that its super-Poincaré symmetry is enhanced to the superconformal group $SU(2, 2|4)$. The bosonic part of this group is the 4D conformal group $SU(2, 2) \sim SO(4, 2)$ and $SU(4) \sim SO(6)$ R-symmetry. The fermonic part is generated by 16 supersymmetry generators Q^A and $Q_A^\dagger$ and 16 superconformal generators S^A and $S_A^\dagger$. The $\mathcal{N} = 4$ SYM theory plays a key role in many modern developments in high-energy physics. The theory can also be obtained by keeping only the massless modes of 10-dimensional $\mathcal{N} = 1$ super Yang–Mills theory after Kaluza–Klein reduction on a 6-torus.

Above we introduced supersymmetry in the context of flat Minkowski space and used a constant spinor, $\partial_\mu \epsilon = 0$, as the parameter in the supersymmetry transformations.

---

[4] Off-shell counting of bosonic and fermonic degrees of freedom also matches with the inclusion of auxiliary fields. See for example textbooks such as [50] and [51].

This is called *global* or *rigid supersymmetry*. The next possibility to consider is 'gauged' supersymmetry, i.e. *local supersymmetry*, where the supersymmetry parameter depends generally on the local spacetime coordinates, $\epsilon = \epsilon(x)$: the result is *supergravity*.

12.2.2 Supergravity

Supergravity is the wonderful combination of supersymmetry and general relativity. A general feature of supergravity is that the gravitational field $g_{\mu\nu}$ is partnered with a Rarita–Schwinger field ψ_μ; thus the spin-2 graviton is paired with a spin-3/2 gravitino. In spacetime dimensions $D = 2, 3, 4$ (mod 8), the gravitino field ψ_μ can be Majorana (real),[5] and in these cases the most fundamental structure of supergravity can be described in terms of the following action [51][6]

$$S = \frac{1}{16\pi G_D} \int d^D x \sqrt{-g} \left[R - \overline{\psi}_\mu \gamma^{\mu\nu\rho} D_\nu \psi_\rho \right].$$ (12.18)

Here, $\gamma^{\mu\nu\rho} = \gamma^{[\mu}\gamma^\nu\gamma^{\rho]}$ is the fully antisymmetric product of three γ-matrices of the D-dimensional Clifford algebra. The gravitino covariant derivative,

$$D_\nu \psi_\rho = \partial_\nu \psi_\rho + \frac{1}{4}\omega_{\nu ab}\gamma^{ab}\psi_\rho,$$ (12.19)

is given in terms of the torsion-free spin-connection $\omega_{\nu ab}$. (The Christoffel connection is not needed because of the contraction with the antisymmetric gamma-matrix.) Using the vielbein e_a^ν, we have

$$\omega_\mu^{ab} = 2e^{\nu[a}\partial_{[\mu}e_{\nu]}^{b]} - e^{\nu[a}e^{b]\rho}e_{\mu c}\,\partial_\nu e_\rho{}^c.$$ (12.20)

Consider the local supersymmetry transformation

$$\delta_\epsilon e_\mu^a = \frac{1}{2}\bar{\epsilon}\gamma^a\psi_\mu, \qquad \delta_\epsilon \psi_\mu = D_\mu\epsilon.$$ (12.21)

It is instructive to outline how (12.21) acts on the action (12.18); for full detail, see [51]. First, recalling the derivation of Einstein's equation from the action principle, we are familiar with the result of varying the Einstein–Hilbert part of the action:

$$\delta_\epsilon\left(\sqrt{-g}\,R\right) \;\rightarrow\; \sqrt{-g}\,\left(R_{\mu\nu} - \frac{1}{2}g_{\mu\nu}R\right)\left(-\bar{\epsilon}\,\gamma^\mu\psi^\nu\right).$$ (12.22)

Next, the variation of the spinors in the spin-3/2 kinetic term gives – after partial integration and careful tracking of the order of the fermion fields – a term proportional to $\gamma^{\mu\nu\rho}D_\mu D_\nu\psi_\rho$. Antisymmetrization of the Lorentz-indices on the covariant derivatives

---

[5] In other dimensions, one uses a Dirac or symplectic Majorana spinor; the supersymmetry transformations (12.21) are then modified accordingly to ensure that $\delta_\epsilon e_\mu^a$ is real. For an overview of spinor representations in D-dimensions, see Table 3.2 in [51].

[6] Here and henceforth, we use Greek letters $\mu, \nu \ldots$ for the D-dimensional coordinate frame indices.

allows us to replace $[D_\mu, D_\nu]$ with the Riemann curvature tensor; explicitly we have, at linear order in the gravitino field,

$$\delta_\epsilon \left(-\sqrt{-g}\, \overline{\psi}_\mu \gamma^{\mu\nu\rho} D_\nu \psi_\rho \right) \Big|_{\text{lin. } \psi} \;\rightarrow\; \frac{1}{4} \sqrt{-g}\, \overline{\epsilon} \gamma^{\mu\nu\rho} \gamma^{ab} R_{\mu\nu ab} \psi_\rho \,. \qquad (12.23)$$

Now the product of gamma-matrices can be expanded on a basis of rank $r = 1, 3, 5$ antisymmetric products of gamma-matrices $\gamma^{\rho_1 \cdots \rho_r} = \gamma^{[\rho_1} \cdots \gamma^{\rho_r]}$. Upon contraction with the Riemann tensor, the rank-5 term $\gamma^{\rho\mu\nu ab} R_{\mu\nu ab}$ vanishes thanks to the Bianchi identity. The rank-3 terms vanish as a result of application of the Bianchi identity and the symmetry properties of the Ricci tensor. Finally, one is left with two rank-1 contributions:

$$\delta_\epsilon \left(-\sqrt{-g}\, \overline{\psi}_\mu \gamma^{\mu\nu\rho} D_\nu \psi_\rho \right) \Big|_{\text{lin. } \psi} \;\rightarrow\; \sqrt{-g} \left(R_{\mu\nu} - \frac{1}{2} g_{\mu\nu} R \right) \left(\overline{\epsilon}\, \gamma^\mu \psi^\nu \right) . \qquad (12.24)$$

This cancels the variation of the Einstein–Hilbert term (12.22) and we therefore see that the action (12.18) is invariant under the supersymmetry transformation (12.21) to *linear order in the fermions*. The cancellation of the variations (12.22) and (12.24) was first demonstrated in [52]. It is remarkable how the proof of linearized supersymmetry relies on a delicate interplay between fundamental identities: the commutator of covariant derivatives in Riemannian geometry, spin and the Clifford algebra, and Fermi-statistics and its connection to the anti-commutation of the fields without which the Majorana gravitino kinetic term would be a total derivative. Invariance at non-linear order is dimension-dependent and can require additional fields and other terms in the action. For minimal $\mathcal{N}=1$ supergravity in 4D, no other fields are needed, but the action (12.18) must be supplemented by terms quartic in the gravitino field [52]. Local supersymmetry was also demonstrated in [53].

As with global supersymmetry, there are equal numbers of fermionic and bosonic on-shell degrees of freedom in supergravity theories. Massless particles in D-dimensional spacetime are characterized by the irreducible representations of the 'little group' $SO(D-2)$ (the part of the Lorentz group that leaves the null momentum vector invariant). The graviton is symmetric and traceless, so that amounts to $D(D-3)/2$ bosonic degrees of freedom; this is the same counting as the number of independent polarizations of a gravitational wave in D dimensions, as discussed early in Section 12.1. A Majorana gravitino in the vector–spinor representation of $SO(D-2)$ has $(D-3)2^{\lfloor (D-2)/2 \rfloor}$ degrees of freedom. So, for $D = 4$, the graviton and the Majorana gravitino each have 2 degrees of freedom and hence the $\mathcal{N}=1$ supergravity multiplet in 4D consists precisely of the graviton and the gravitino.

One can couple other fields to supergravity as 'matter' supermultiplets. For example in 4D we can add to the $\mathcal{N} = 1$ supergravity action N_v copies of $\mathcal{N} = 1$ vector multiplets (consisting of a gauge boson and its gaugino partner) or N_χ copies of $\mathcal{N}=1$ chiral multiplets (1 spin-1/2 fermion and 1 complex scalar) while preserving $\mathcal{N} = 1$ supersymmetry. If a model has only the supergravity multiplet and no matter multiplets, we call it *pure supergravity*.

Next, consider extended supergravity, i.e. supergravity theories with more than one gravitino field. The $D = 4$, $\mathcal{N} = 2$ pure supergravity theory [54] has 2 bosonic degrees of freedom for the graviton, 2×2 fermionic degrees of freedom from the two gravitinos, and finally 2 more bosonic degrees of freedom from a spin-1 graviphoton. One can couple to it extra $\mathcal{N}=2$ vector supermultiplets (1 gauge boson, 2 gauginos, 1 complex scalar) and still preserve $\mathcal{N}=2$ supersymmetry of the full action.

We will be describing black holes in higher dimensions, so let us next consider supergravity in five dimensions. The on-shell 5D graviton has 5 degrees of freedom. There is no spinor representation that can match this, so in 5D we cannot have a simple $\mathcal{N} = 1$ supergravity multiplet consisting of just a graviton and a gravitino. Instead, we can take the gravitino to be a symplectic Majorana spinor with 2×4 degrees of freedom and include a graviphoton with 3 degrees of freedom in the on-shell supermultiplet. This is the field content of minimal supergravity theory in 5D and it has $\mathcal{N} = 2$ supersymmetry [55]. The bosonic action for minimal supergravity in 5D is not just Einstein–Maxwell theory, but also has a Chern–Simons term $A \wedge F \wedge F$.

For a theory whose highest spin particle is the spin-2 graviton,[7] the maximal amount of supersymmetry allowed in 4D is $\mathcal{N} = 8$. This is easy to see by working down from the highest helicity state of $+2$ and reducing the helicity by 1/2 at each application of the supersymmetry charge. After application of $\mathcal{N} = 8$ supercharges, one reaches the helicity -2 state of the graviton. Thus having more than 8-fold supersymmetry would give states with spin higher than 2 in 4D.

$\mathcal{N} = 8$ supersymmetry in 4D gives a uniquely determined supergravity theory, simply called '$\mathcal{N} = 8$ *supergravity*'. Its spectrum of $2^8 = 256$ massless states is organized into fully antisymmetric rank r representations of the global $SU(8)$ R-symmetry: the two states of the graviton, 8 pairs of gravitinos, 28 pairs of graviphotons, 56 pairs of spin-1/2 graviphotinos, and 70 scalars.

Supergravity with a spin-2 graviton as the highest spin state exists in dimensions $D \leq 11$. To see how the bound on the spacetime dimension arises, start in $D = 11$ where the minimal spinor is a 32-component Majorana spinor. Upon dimensional reduction on a 7-torus, an 11D Majorana gravitino gives eight Majorana gravitinos in 4D. Indeed, the dimensional reduction of 11D supergravity on a 7-torus is $\mathcal{N} = 8$ supergravity theory in 4D. The minimal spinor representation in $D > 11$ has more than 32 components (e.g. for $D = 12$ it has 64), so starting with a gravitino in $D > 11$ and reducing toroidally to 4D gives $\mathcal{N} > 8$ gravitinos in 4D. If this were a 4D supergravity theory, it would have states with spin greater than 2. Thus we conclude that we cannot have supergravity in $D > 11$.

In $D = 11$, the gravitational field $g_{\mu\nu}$ encodes 44 on-shell degrees of freedom. The 11D gravitino is a Majorana spinor in the vector–spinor representation, so it has 128 degrees of freedom. Matching the fermionic and bosonic degrees of freedom requires an antisymmetric 3-form field $A^{(3)}_{\mu\nu\rho}$: it contains precisely the needed 84 bosonic degrees of freedom. The 11D supergravity theory is unique and the bosonic part of the action is [61]

---

[7] Theories with states of spin higher than 2 have been constructed in anti-de Sitter space (AdS), see [56, 57] and the newer review [58]. This is particularly interesting in connection with the gauge–gravity duality, see for example [59, 60].

$$S = \frac{1}{16\pi G_{11}} \int d^{11}x \left[\sqrt{-g} \left(R - \frac{1}{4!} F^{(4)}_{\mu\nu\rho\sigma} F^{(4)\,\mu\nu\rho\sigma} \right) - \frac{\sqrt{2}}{3} A^{(3)} \wedge F^{(4)} \wedge F^{(4)} \right],$$
(12.25)

where $F^{(4)}_{\mu\nu\rho\sigma}$ are the components of the 4-form field strength $F^{(4)} = dA^{(3)}$. The Chern–Simons term is needed for supersymmetry. The fermionic terms include the standard kinetic gravitino term of (12.18), but also terms coupling the gravitinos to $F^{(4)}$. The 3-form potential $A^{(3)}$ naturally encodes the electric charge of a membrane in 11 dimensions; its electric charge is captured by Gauss' law $Q_{\mathrm{E}} \propto \int_{S^7} \star F^{(4)}$, where $\star$ indicates the 11D Hodge dual and the S^7 is transverse to the membrane. Similarly, $Q_{\mathrm{M}} \propto \int_{S^4} F^{(4)}$ calculates the magnetic charge of an object extended in 5 spatial directions in 11D. Thus, the fundamental objects carrying electric and magnetic charges in 11D supergravity are 2-branes and 5-branes: they are called M2- and M5-branes and we will meet them again later in our discussion of string theory and M-theory in Section 12.3.

There are two distinct $\mathcal{N} = 2$ supergravity theories in 10D: they are called Type IIA and Type IIB and differ by whether the supersymmetry generators have different chirality (Type IIA) or the same chirality (Type IIB). Type IIA can be obtained as the Kaluza–Klein reduction of 11D supergravity on a circle. Both Type IIA and Type IIB supergravity contain an antisymmetric 2-form potential $B_{\mu\nu}$. The objects that are electrically charged, $Q_{\mathrm{E}} \propto \int_{S^7} \star H$, under the corresponding 3-form flux $H = dB$ are 1-dimensional: they are strings! Indeed, it turns out that the Type II supergravity theories are low-energy limits of superstring theories with $\mathcal{N}=2$ supersymmetry.

Upon Kaluza–Klein compactification of the 11D supergravity theory to lower dimensions, one obtains many other interesting D-dimensional supergravity theories. For example, the 5D minimal $\mathcal{N} = 2$ supergravity theory described above is a certain truncation of 11D supergravity on a 6-torus. And, as noted earlier, $\mathcal{N}=8$ supergravity in 4D arises from 11D supergravity by reduction on a 7-torus.

In contemporary applications, compactifications of 11D supergravity, or 10D Type IIA/IIB supergravity, on curved manifolds are very important. When a Kaluza–Klein reduction of supergravity is performed on a manifold with positive curvature, such as a p-sphere S^p, the resulting lower-dimensional theory is 'gauged' supergravity. One can think of the 'gauging' as having the gravitinos charged under the gauge fields. Gauged supergravity typically comes with a non-trivial scalar potential – or in the simplest cases a negative cosmological constant. Whereas Minkowski space is the simplest 'vacuum' solution for ungauged supergravity, anti-de Sitter space (AdS) is the simplest solution in gauged supergravity. As an example, Type IIB supergravity on an S^5 gives a 5D gauged supergravity theory [62] that plays a central role in studies of the gauge–gravity duality. We discuss this further in Section 12.4.

12.2.3 Charged black holes, BPS bounds, and Killing spinors

In Section 12.1 we discussed higher-dimensional black holes as solutions to the vacuum Einstein equations. It is very interesting to study classical solutions in supergravity, in particularly those with special supersymmetric properties, as we now describe.

Denoting generic bosonic and fermonic fields by B and F, supersymmetry transformations generically take the schematic form [51]

$$\delta_\epsilon B = \bar\epsilon f(B) F + O(F^3) \quad \text{and} \quad \delta_\epsilon F = g(B) \epsilon + O(F^2), \tag{12.26}$$

where f and g are functions of the bosonic fields and their derivatives. We are interested in classical solutions (of the supergravity equations of motion) that the supersymmetry transformations (12.26) leave invariant. Typically, we consider solutions that have only non-trivial bosonic fields, i.e. all the fermion fields are set to zero, $F = 0$. Since the supersymmetry variations of bosons (12.26) are proportional to the fermion fields, they automatically vanish, $\delta_\epsilon B = 0$, on a purely bosonic solution. On the other hand, we get non-trivial constraints from the condition that fermion variations vanish, $\delta_\epsilon F = 0$. In the simplest form (12.21), the constraint is $0 = \delta_\epsilon \psi_\mu = D_\mu \epsilon$, so it requires the existence of a covariantly constant spinor ϵ. More generally, there will be other fields involved in the condition $\delta_\epsilon F = 0$; we will see examples shortly. The spinors that solve the constraints arising from setting the supersymmetry variations of the fermion fields to zero are called *Killing spinors*. If a classical solution has n parameters characterizing its Killing spinors, it is said to preserve n supersymmetries.

The existence of Killing spinors has important implications. For example, the bispinor products of Killing spinors $\bar\epsilon_1 \gamma^\mu \epsilon_2$ are Killing vectors associated with the ordinary (bosonic) symmetries of the spacetime solution. And very importantly, the Killing spinor equations imply a set of first order equations consistent with the equations of a motion, making it easier to find exact solutions.

Another important implication is that the existence of Killing spinors implies that certain energy bounds are saturated. These are called BPS bounds after Bogomol'nyi, Prasad, and Sommerfield [63, 64]. Thus solutions with Killing spinors are often called *BPS solutions*.

As a simple example, Witten's proof of the positive energy theorem [65] shows that the ADM mass is positive $M \geq 0$ with equality precisely when there exists a covariantly constant spinor, $D_\mu \epsilon = 0$. In particular, Minkowski space has a covariantly constant Killing spinor and obviously it has $M = 0$. It is a BPS solution in pure $\mathcal{N} = 1$ supergravity.

Pure $\mathcal{N} = 1$ *gauged* supergravity in 4D has a negative cosmological constant $\Lambda = -3/L^2$ and the Killing spinor equation is

$$0 = \delta_\epsilon \psi_\mu = D_\mu \epsilon - \frac{1}{2L} \gamma_\mu \epsilon. \tag{12.27}$$

Four-dimensional anti-de Sitter space (AdS$_4$) admits such a Killing spinor, so AdS with radius L is a BPS solution in gauged supergravity.

Recall from Section 12.2.2 that the bosonic sector in pure $\mathcal{N} = 2$ supergravity in 4D consists of the gravitational field $g_{\mu\nu}$ and the graviphoton field A_μ. The purely bosonic part of the action turns out to be Einstein–Maxwell theory. The vanishing of the supersymmetry transformation of the gravitino fields in this theory gives a Killing spinor equation of the form[8]

---

[8] Here we are setting $G_4 = 1$ for simplicity.

$$\hat{D}_\mu \epsilon \equiv D_\mu \epsilon - \tfrac{1}{4} F_{\nu\rho} \gamma^\nu \gamma^\rho \gamma_\mu \epsilon = 0. \qquad (12.28)$$

An argument similar to Witten's [65] shows that the mass M and electric and magnetic charges, Q and P, of regular solutions to the equations of motion of $\mathcal{N} = 2$ supergravity in 4D satisfy the bound [66]

$$M \geq \left(Q^2 + P^2\right)^{1/2}. \qquad (12.29)$$

Equality holds precisely when the solution admits a Killing spinor $\hat{D}_\mu \epsilon = 0$.

The bound (12.29) looks very familiar: it is precisely the bound the Reissner–Nordström black hole must satisfy in order to have a smooth horizon! Thus, the *extremal Reissner–Nordström black hole is a BPS solution of $\mathcal{N}{=}2$ supergravity in 4D*.

The temperature of the extremal Reissner–Nordström black hole is zero. However, extremality in the sense of zero temperature – or coinciding inner and outer horizons – does not necessarily mean that the solution is BPS. For example, the extremal Kerr black hole is not BPS: without an electromagnetic charge, a solution with $M > 0$ cannot saturate the BPS bound (12.29) and hence it does not admit a Killing spinor. Similarly, no Kerr–Newman black hole with $J \neq 0$ saturates the bound (12.29). Hence, there are no asymptotically flat rotating BPS black holes in 4D ungauged supergravity.

Given our discussion in Section 12.1.3 of vacuum solutions describing black holes and black rings in 5D, it is natural to ask if they have charged cousins. For simplicity, we first answer this question in the context of minimal 5D supergravity (described briefly in Section 12.2.2) and then generalize. In 5D, the equivalent of the Reissner–Nordström black hole is a static, charged black hole with a round S^3 horizon. In its simplest form, it is an electrically charged solution of the equations of motion in minimal 5D supergravity. It has an extremal limit in which the inner and outer horizon coincide; in this limit, the solution is supersymmetric and saturates the appropriate 5D BPS bound $M \geq \frac{\sqrt{3}}{2} Q$ [67].

The 5D version of the Kerr–Newman solution is a charged version of the Myers–Perry black hole described in Section 12.1.3. It can have angular momenta J_1 and J_2 in both the two independent planes of 5D spacetime. The solution was first constructed in [68] and, in its simplest version, it is a solution to minimal 5D supergravity. The BPS limit of this charged rotating black hole is called the BMPV black hole [69]. The BMPV black hole has $M = \frac{\sqrt{3}}{2} Q$ and – unlike in 4D – it can still carry angular momentum provided that the magnitudes are the same in the two planes of rotation, $|J_1| = |J_2|$.

Black rings can also carry charges [24] and they have limits in which they are BPS solutions. BPS black rings were first constructed as exact solutions in minimal 5D supergravity [70]. These BPS rings have $M = \frac{\sqrt{3}}{2} Q$, but – contrary to the BMPV black holes – the angular momentum J_1 in the plane of the ring must be strictly greater than the angular momentum of the S^2 (the orthogonal plane): $|J_1| > |J_2|$. Charged black rings have a new feature: they carry a non-conserved 'dipole' charge [24, 70] associated with application of Gauss' law with an S^2 surrounding a piece of the ring. This measures a string-like charge density along the S^1 of the black ring; since this ring is a contractible circle, the 'dipole' charge is not conserved, but it impacts the solution non-trivially and is required for smoothness of the horizon.

The BMPV black hole and the charged black rings described above have a natural generalization [69, 71–73] in which they carry conserved charges of three distinct gauge fields of a 5D supergravity theory obtained from reduction of Type IIB supergravity in 10D on a 5-torus. The 'minimal' solutions are recovered in the limit where the three charges are equal. The BPS black holes with three different charges play a key role in Section 12.3 when we discuss how string theory offers a precise microscopic account of black hole entropy.

12.2.4 Perturbative quantum gravity

The focus of this section is on the application of standard quantum field theory in flat spacetime to scattering of gravitons, the spin-2 particles associated with the quantization of the gravitational field $g_{\mu\nu}$. More precisely, we expand the gravitational field around a flat space background: $g_{\mu\nu} = \eta_{\mu\nu} + \kappa h_{\mu\nu}$, where $\kappa^2 = 8\pi G_D$. The fluctuating field $h_{\mu\nu}$ is the *graviton field*. Consider pure gravity without matter and expand the Einstein–Hilbert action in powers of $\kappa h_{\mu\nu}$:

$$
\begin{aligned}
S_{\text{EH}} &= \frac{1}{2\kappa^2} \int d^D x \sqrt{-g}\, R \\
&= \int d^D x \left[h\, \partial^2 h + \kappa\, h^2\, \partial^2 h + \kappa^2\, h^3\, \partial^2 h + \kappa^3\, h^4\, \partial^2 h + \cdots \right].
\end{aligned}
\tag{12.30}
$$

Since the Ricci-scalar R involves two derivatives, every term in the expansion has two derivatives. There are infinitely many terms, with increasingly delightful assortments of index-structures; in (12.30) we have written them schematically as $h^{n-1}\, \partial^2 h$. There are no mass terms in (12.30), so the particles associated with quantization of the gravitational field $h_{\mu\nu}$ are massless: they have spin-2 and are the *gravitons*.

It is interesting to study graviton scattering processes, but we have to gauge fix the action (12.30) before extracting the Feynman rules. A standard choice is the *de Donder gauge*, $\partial^\mu h_{\mu\nu} = \frac{1}{2} \partial_\nu h_\mu{}^\mu$, which brings the quadratic terms in the action to the form

$$
h\, \partial^2 h \;\rightarrow\; -\frac{1}{2} h_{\mu\nu} \Box h^{\mu\nu} + \frac{1}{4} h_\mu{}^\mu \Box h_\nu{}^\nu .
\tag{12.31}
$$

The propagator resulting from these quadratic terms is

$$
P_{\mu_1\nu_1,\mu_2\nu_2} = -\frac{i}{2} \left(\eta_{\mu_1\mu_2} \eta_{\nu_1\nu_2} + \eta_{\mu_1\nu_2} \eta_{\nu_1\mu_2} - \frac{2}{D-2} \eta_{\mu_1\nu_1} \eta_{\mu_2\nu_2} \right) \frac{1}{k^2} .
\tag{12.32}
$$

The external lines in graviton Feynman diagrams have two Lorentz-indices that must be contracted with graviton polarization vectors. In 4D, the polarizations encode the two helicity $h = \pm 2$ physical graviton states. They can be constructed as products of spin-1 photon polarization vectors $\epsilon_{\pm}^{\mu}(p_i)$. Picking a basis where $\epsilon_{\pm}^{\mu}(p_i)^2 = 0$, the graviton polarizations

$$e_-^{\mu\nu}(p_i) = \epsilon_-^{\mu}(p_i)\epsilon_-^{\nu}(p_i), \qquad e_+^{\mu\nu}(p_i) = \epsilon_+^{\mu}(p_i)\epsilon_+^{\nu}(p_i). \qquad (12.33)$$

are automatically symmetric and traceless.

The infinite set of 2-derivative interaction terms $h^{n-1}\,\partial^2 h$ yields complicated Feynman rules for n-graviton vertices for *any* $n = 3, 4, 5, \ldots$. Together with the 3-term de Donder propagator (12.32), this is a clear indication that calculation of tree-level graviton scattering amplitudes from Feynman diagrams is highly non-trivial. Nonetheless, it turns out that the final result for the on-shell amplitudes can be written in a relatively simple form (for an overview, see [74]). In fact, there is an interesting relationship between tree-level graviton amplitudes and gluon amplitudes in Yang–Mills theory. For the case of 4-particle amplitudes this relationship is

$$M_4^{\text{tree}}(1234) = -s A_4^{\text{tree}}[1234] A_4^{\text{tree}}[1243], \qquad (12.34)$$

where M_n^{tree} denotes a tree n-graviton amplitude and A_n^{tree} a (color-ordered) tree n-gluon amplitude. The prefactor is the kinematic invariant Mandelstam variable $s = -(p_1 + p_2)_{\mu}(p_1 + p_2)^{\mu}$. In four dimensions, the relation between the scattering states in (12.34) is

$$\text{graviton}^{\pm 2}(p_i) - \text{gluon}^{\pm 1}(p_i) \otimes \text{gluon}^{\pm 1}(p_i), \qquad (12.35)$$

where $\pm$ indicates the helicity state.

There are similar (although somewhat more involved, see Appendix A of [77]) expressions for M_n^{tree} in terms of sums of products of two A_n^{tree} for all n. These are called the *KLT relations* after Kawai, Lewellen and Tye [76] who first derived such relations between closed and open string amplitudes; the field theory relations, such as (12.34), are obtained in the limit where the string tension goes to infinity and the string behaves as a point particle. We discuss string theory in Section 12.3.

From the point of view of the Lagrangian (12.30), the KLT relations are very surprising. Field redefinitions and clever gauge choices can bring the Feynman rules into a KLT-like form; see [78–80] and the review [81]. More recently, another form of the relation between gravity and gauge theory amplitudes has been found: it is known as *BCJ duality relations*, named after Bern, Carrasco, and Johansson [82]. In contrast to the KLT relations, the BCJ relations can also be applied at loop level; not only do they offer a powerful alternative to the gravitational Feynman rules, they also hint at a possible deeper structure in perturbative gravity. The study of the surprisingly rich and enticing mathematical structure of scattering amplitudes in both Yang–Mills theory and in gravity is currently an exciting area of research (see for example [74, 75]).

Let us now discuss the behavior of graviton loop amplitudes in the high-energy (ultraviolet, UV) limit. Consider for example a 1-loop diagram with m external gravitons and only cubic interactions. The numerator of the loop-integrand can have up to $2m$ powers of momenta since each graviton interaction vertex has two derivatives and, with m

propagators, this naively gives

$$\text{gravity 1-loop diagram} \sim \int^{\Lambda} d^4k \, \frac{(k^2)^m}{(k^2)^m} \sim \Lambda^4. \tag{12.36}$$

This is power-divergent as the UV cutoff Λ is taken to ∞ for all m. On the other hand, for Yang–Mills theory, the interactions are at most 1-derivative, so the integral (12.36) (now with m external gluons) has at most k^m in the numerator, and hence it is manifestly UV finite for $m > 4$.

However, the power-counting is too naive. There can be cancellations within each diagram. Moreover, individual Feynman diagrams should not necessarily be taken too seriously since they are not gauge invariant. So cancellations of *UV divergences* can take place in the sum of diagrams, rendering the on-shell amplitude better behaved than naive power-counting indicates.

Actually, pure gravity in 4D is finite at 1-loop order [83]: all the 1-loop UV divergences cancel! This can be seen from the fact that the only viable 1-loop counterterm in pure gravity in 4D must be quadratic in the Riemann tensor, but by a field redefinition such a term can be completed to the Gauss–Bonnet term which is a total derivative. At 2-loop order, pure gravity indeed has a divergence [84, 85]. In Yang–Mills theory, divergences are treated with the procedure of renormalization. However, in gravity, it would take an infinite set of local counterterms to absorb the divergences and hence the result is unpredictable: pure gravity is a non-renormalizable theory.

From the point of view of renormalization, the theory described by the Einstein–Hilbert action is naturally regarded as an *effective field theory* that cannot be extrapolated to arbitrarily high energy. To see this, note that the 4D gravitational coupling $\kappa \sim G_4^{1/2}$ has dimension of $(\text{mass})^{-1}$. Perturbative calculations rely on an expansion in the small dimensionless coupling $E\kappa$, where E is the energy scale of the scattering process. Thus, perturbation theory is valid only at energies much smaller than $G_4^{-1/2} \sim M_{\text{Planck}} \sim 10^{19}$ GeV. In other words, perturbation breaks down at high energies and from this point of view Einstein gravity is an effective field theory. As a classical effective field theory, general relativity is hugely successful and captures classical gravitational phenomena stunningly, as shown by experimental tests.

Viewing gravity as an effective theory, we can study the perturbative amplitudes. The tree-level amplitudes capture the classical physics and there are no UV divergences to worry about. Could we imagine adding matter fields to cure the 2-loop divergence in pure gravity? Gravity with generic matter is 1-loop divergent [83, 86], but it turns out that any 4D theory of pure ungauged supergravity is finite at 1- and 2-loop order [87–90]. *Supersymmetry helps to tame the UV divergences.* This is in part due to cancellations between the boson and fermion loops. To date, only one explicit example of a UV divergence has been calculated in an amplitude in a pure ungauged supergravity theory in 4D, namely at 4-loop order in $\mathcal{N} = 4$ supergravity [91].

It has been proposed [92] that maximal supergravity, $\mathcal{N} = 8$, in 4D could perhaps be ultraviolet finite. Explicit calculations [92–96] have demonstrated finiteness of 4-graviton amplitudes up to and including 4-loop order, while symmetry arguments have established that no divergences can occur until 7-loop order [97–101]. This can also be analyzed using superspace methods, see [102] and references therein. The known symmetries do not constrain the divergences past 7-loop order, so it seems that UV finiteness would require the theory to have some hitherto hidden structure. Perhaps the relationships with Yang–Mills theory can clarify this.

From a field theory perspective, gravity needs a suitable UV completion, i.e. another theory that reduces to general relativity in the low-energy limit. This is true for several reasons. First, as we have just discussed, perturbation theory breaks down at high energies. Second, even if each order in a supergravity perturbation theory were UV finite, the perturbation series is not likely convergent; hence nonperturbative information is needed for a complete theory. Finally, the UV completion is also needed in order to make sense of microscopic quantum properties of nonperturbative objects such as black holes. A very successful candidate for such a UV complete theory of quantum gravity is string theory, which is the subject of our next section.

12.3 String theory

String theory combines the ideas of the previous two sections (higher dimensions and supersymmetry) and reduces to supergravity in a low-energy limit. In this section we give an overview of string theory, focussing on gravitational aspects of this large subject.[9] We will see that it has many remarkable properties, including providing a theory of black hole microstates which reproduce both the Hawking–Bekenstein entropy and Hawking radiation.

12.3.1 Perturbative string theory

String theory starts with the idea that particles are not really point-like, but excitations of a one-dimensional string. The string can be closed (topologically a circle) or open (a line segment). For now, we will focus on closed strings. As a string travels through spacetime, it traces out a two-dimensional worldsheet. To describe the dynamics of this worldsheet, we introduce local coordinates (σ, τ) on the worldsheet and X^μ on spacetime, so the position of the worldsheet is given by $X^\mu(\sigma, \tau)$. If we introduce a nondynamical worldsheet metric γ_{ab}, then the string action can be written as follows:

$$S[X^\mu, \gamma_{ab}] = \frac{1}{4\pi \ell_s^2} \int d\tau \, d\sigma \, \sqrt{-\gamma} \gamma^{ab} \, \partial_a X^\mu \, \partial_b X^\nu \, g_{\mu\nu} . \qquad (12.37)$$

Here, ℓ_s is a new length scale in string theory (called the "string length") which determines the string tension $T = 1/(2\pi \ell_s^2)$. The worldsheet metric γ_{ab} is essentially "pure gauge" since S is invariant under both worldsheet diffeomorphisms and Weyl rescalings. One can

---

[9] For an introduction to string theory, see [103]. More complete references include [104, 105].

remove it by solving its equation of motion $\delta S/\delta\gamma_{ab} = 0$ to find that γ_{ab} is proportional to the induced metric on the worldsheet. Substituting this back into S reduces the action to just the area of the induced metric. In other words, the worldsheet must be an extremal surface. This is a useful way to picture the motion of classical strings but, to quantize the string in flat spacetime, it is much more convenient to work with (12.37), since it is quadratic in the dynamical fields X^μ.

There are several different approaches to quantize this string (e.g. light cone gauge or covariant quantization) which differ in how one treats the gauge freedom, but they all agree on the physical results. The first thing one discovers is that the theory is consistent only in 26 spacetime dimensions. This surprising result arises in different ways depending on the approach one uses to quantize the string. If one completely fixes the gauge by going to a light cone gauge, one breaks manifest Lorentz invariance. Lorentz invariance is recovered only in $D = 26$. In a covariant quantization, there are negative norm states created by the operators associated with $X^0(\sigma, \tau)$. These states are removed by constraints (associated with the gauge invariance) in $D = 26$.

The physical spectrum of this bosonic string includes a scalar tachyon with $m^2 < 0$. The existence of this tachyon shows that 26-dimensional Minkowski space is unstable in this theory. Although it is possible that this theory has a stable ground state, it has never been found. Instead, one proceeds by adding fermions, ψ^μ, to the worldsheet and makes the two-dimensional theory supersymmetric. Quantizing this superstring in flat spacetime, one now finds that it is consistent in ten spacetime dimensions.

The $D = 10$ spectrum is now tachyon-free and consists of the following massless bosonic modes: a symmetric traceless "graviton" $h_{\mu\nu}$, a scalar "dilaton" ϕ, an antisymmetric "Kalb–Ramond" field $B_{\mu\nu}$, and some "Ramond–Ramond" fields $F_{\mu\cdots\nu}$ which are higher rank generalizations of a Maxwell field. In addition, there is an infinite tower of higher mass and higher spin states. Finally, there are fermionic partners for each of these bosonic states so that the complete spectrum is invariant under spacetime supersymmetry. Note that we impose only worldsheet supersymmetry, but nevertheless, the spacetime spectrum turns out to be supersymmetric.[10] At large mass, the spectrum is highly degenerate, with an entropy proportional to the mass. This can be understood by thinking of a highly excited string as a random walk on a discrete grid, made up of segments of length ℓ_s. Due to the tension, a string with n segments has mass $M \sim n/\ell_s$. If the string can move in p directions at each step, then the total number of configurations is[11] p^n, so the entropy is $S = n \log p \sim M\ell_s$.

The existence of the graviton is the first indication that a theory of strings has something to do with gravity. Much stronger evidence comes from quantizing the string in a static curved spacetime. As we have said, the action (12.37) is classically invariant under rescaling γ_{ab}, but quantum mechanically there is a conformal anomaly. To calculate this anomaly, one analytically continues the spacetime and worldsheet metrics to Euclidean

---

[10] This involves imposing a consistency condition known as the GSO projection [106].
[11] It is actually slightly less than this if we require the string to return to its starting point and form a closed loop, but this correction is subleading at large mass.

signature,[12] and expands $g_{\mu\nu}(X)$ in Riemann normal coordinates about a point X_0. The anomaly can then be computed perturbatively in powers of ℓ_s/L, where L is a typical length scale of the curvature. To leading order, the conformal anomaly vanishes if the spacetime satisfies Einstein's equation: $R_{\mu\nu} = 0$. If one couples the string to other background fields corresponding to other massless bosonic modes of the string, one recovers the equations of supergravity [107]. This is an important point: *in string theory, the full classical equations of motion for the spacetime fields come from demanding conformal invariance of the quantized worldsheet theory.* It is a remarkable and deep fact about string theory that Einstein's equation arises as a condition on the background fields (which are like coupling constants) in a two-dimensional quantum field theory. In general, these equations receive higher order corrections involving higher derivative terms, but they are usually negligible unless the curvature is of order the string scale. Thus general relativity (or supergravity) arises as the leading order classical equations of motion in string theory.

One of the early successes of string theory was that it provided a perturbatively finite quantum theory of gravity. To describe this we have to introduce string interactions. The basic assumption of string theory is that strings interact via a simple splitting and joining interaction. By quantizing a single string above, we have described a first quantized string. One might have thought that the next step would be to construct a string field theory in which the first quantized states $\Phi[X^{\mu}(\sigma), \psi^{\mu}(\sigma)]$ are promoted into field operators and one introduces interactions by adding cubic terms to the action. Progress has been made in this direction (see, e.g., [108–110]), but perturbative scattering amplitudes in string theory are usually computed in a first-quantized framework using analogs of Feynman diagrams, see Fig. 12.4.

This is most well developed starting with Minkowski spacetime, but it can be extended to other static backgrounds. One again analytically continues both the spacetime and world-sheet metrics to Euclidean signature. The scattering amplitudes are obtained by summing over worldsheet topology, and computing the path integral

$$\int DX^{\mu} \, D\psi^{\mu} \, D\gamma_{ab} \, e^{-S[X^{\mu},\psi^{\mu},\gamma_{ab}]} \tag{12.38}$$

for each topology. One usually takes the external strings to infinity, so the amplitude corresponds to an S-matrix element. By a conformal transformation, the external strings can then be mapped to points on a compact Riemann surface, with the state of the string represented by an operator inserted at that point. After fixing the γ_{ab} gauge freedom of diffeomorphisms and Weyl rescalings, there remains a finite-dimensional "moduli space" of metrics to integrate over for each worldsheet topology.

The analog of the loop expansion in ordinary quantum field theory is the genus expansion of the string worldsheet. The reason for this is that the dilaton couples to the string via the scalar curvature $\mathcal{R}$ of the string worldsheet. In other words, one adds to the string

---

[12] To evaluate a path integral, one often analytically continues to Euclidean space to convert the oscillating integrand into a convergent integral.

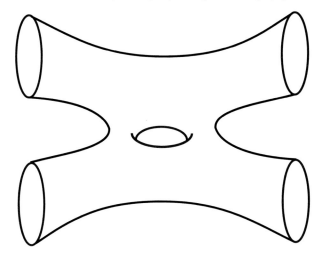

Figure 12.4 A one-loop string diagram.

action (12.37) the term $(1/(4\pi)) \int \phi \mathcal{R}[\gamma]$. By the Gauss–Bonnet theorem, if the dilaton is a constant ϕ_0, this contributes $\phi_0 \chi$ to the action, where χ is the Euler number of the string worldsheet. This is related to the genus g by $\chi = 2(1 - g)$. In the path integral (12.38) the net effect is to weigh different worldsheet topologies by $g_s^{-\chi}$, where the string coupling g_s is determined by the dilaton: $g_s \equiv e^{\phi_0}$.

As an example, when the external states are all gravitons, the tree-level amplitude for two strings to scatter into two strings reproduces the $2 \rightarrow 2$ graviton scattering of general relativity in a limit where strings become point-like. Newton's constant is not an independent parameter in string theory, but determined in terms of g_s and ℓ_s. The ten-dimensional Newton's constant is $G_{10} = 8\pi^6 g_s^2 \ell_s^8$.

Unlike quantum field theory, in which the number of Feynman diagrams grows rapidly with the order of perturbation theory, string theory has only one diagram at each loop order (for a fixed number of external legs). The statement that string theory is UV finite is the observation that for each loop order, the integral (12.38) has no UV divergences. Intuitively, this is because the string is an extended object and there is no special point where the interaction takes place. The extent of the string regulates the short-distance divergence in graviton scattering amplitudes that was discussed in Section 12.2.4. When these amplitudes were first computed in the 1980s, there were some remaining subtleties associated with the fermions and the integration over "supermoduli" space. However, these issues have all been resolved [111, 112].

While it is remarkable that string theory picks out a unique spacetime dimension, it is clearly too large compared to our everyday observations of a 4D spacetime. As discussed earlier, a standard way to make contact with the real world is to compactify six of the dimensions. To preserve spacetime supersymmetry in the remaining four noncompact dimensions, the internal space is highly constrained. It turns out that it must be a complex manifold with Ricci flat Kahler metric [113]. Such spaces are called *Calabi–Yau spaces*. Simple examples include the 6-torus, T^6, $T^2 \times K3$, and a quintic hypersurface in $\mathbb{CP}_4$. The realization of the importance of Calabi–Yau spaces in string theory started a very fruitful collaboration between algebraic geometers and string theorists which has continued for over twenty-five years.

A common problem in earlier studies of Kaluza–Klein compactification was to obtain chiral fermions (as observed in the standard model of particle physics) in the lower-dimensional spacetime. Even if one starts with chiral fermions in the higher-dimensional spacetime, the reduced spacetime always had an equal number of left-handed and right-handed fermions. String theory has several ways to cure this problem. The simplest is to use the fact that the higher-dimensional theory is not just pure gravity. If one starts with non-zero gauge fields in the higher-dimensional spacetime, the lower-dimensional theory can have chiral fermions. In Calabi–Yau compactifications, various properties of fermions are simply related to the topology of the internal manifold (e.g., the number of generations is related to the Euler number).

Strings sense spacetime very differently from point particles. In particular, two geometrically different spacetimes can be indistinguishable in string theory. A simple example of this is flat spacetime with one direction compactified to a circle of radius R. The part of the string spectrum that depends on R includes the usual momentum modes with energy $O(1/R)$, but there are also winding modes with energy $O(R/\ell_s^2)$. If we change the radius to ℓ_s^2/R, this spectrum is completely invariant. The winding and momentum modes simply trade places. In fact, one can show that all interactions are also invariant, and strings cannot tell the difference between these two spacetimes. This is the simplest example of *T-duality* [114]: whenever the solution is independent of a periodic spacelike direction, one can change variables in the path integral (12.38) and rewrite the action in terms of a different set of background fields (which include inverting the radius of the periodic spacelike direction). Since a change of variables does not change the physics, the two backgrounds are equivalent in string theory.

Another example of different geometries being equivalent in string theory is *mirror symmetry* [115, 116]: for solutions of the form $M_4 \times K$, where M_4 is four-dimensional Minkowski space and K is a Calabi–Yau manifold, one can change a sign of a certain charge in the worldsheet theory and the interpretation changes from strings moving on $M_4 \times K$ to strings moving on $M_4 \times \tilde{K}$, where $\tilde{K}$ is a geometrically and topologically different Calabi–Yau space. Since the sign of the charge is arbitrary from the worldsheet standpoint, these two compactifications are equivalent in string theory. As one striking application, mirror symmetry was used to count the number of holomorphic curves of various degrees in a given Calabi–Yau manifold [117], reproducing and greatly generalizing results that had been obtained by mathematicians.

Mirror symmetry has been used to show that spacetime topology change is possible in string theory. A given Calabi–Yau space usually admits a whole family of Ricci flat metrics, so one can construct a solution in which the four large dimensions stay approximately flat and the geometry of the Calabi–Yau space changes slowly from one Ricci flat metric to another. In this process the Calabi–Yau space can develop a curvature singularity resulting from a topologically non-trivial S^2 being shrunk down to zero area. In the mirror geometry, there is no singularity and the evolution can be continued. In the original description, the evolution corresponds to continuing through the geometrical singularity to a nonsingular Calabi–Yau space on the other side with different topology [118]. It should perhaps be emphasized that examples like this show that area is not quantized in string theory. In many supersymmetric examples, the area of certain surfaces in the internal space can vary continuously. They give rise to massless scalar fields in the noncompact directions that are known as "moduli".

In the mid 1980s, *five* different perturbative string theories were constructed that were all consistent in ten spacetime dimensions. There was a theory of open strings called Type I with $\mathcal{N} = 1$ supersymmetry, and two theories of closed strings with $\mathcal{N} = 2$ supersymmetry [119]. The latter two differed in whether the two supersymmetry generators had different chirality (Type IIA) or the same chirality (Type IIB). As the names suggest, the low-energy limits of these two string theories are the $\mathcal{N} = 2$ supergravity theories in 10D mentioned in Section 12.2.2. In addition, there were two theories of closed strings with $\mathcal{N} = 1$ supersymmetry which required either $E_8 \times E_8$ or $SO(32)$ gauge groups in ten dimensions (called *heterotic strings*) [120]. These gauge groups were required by anomaly cancellation. In fact, it was this discovery [121] that sparked an explosion of interest in string theory in 1984. It seemed remarkable that string theory picked not only a unique spacetime dimension, but also an essentially unique gauge group. A decade later it was realized that the five perturbative string theories are all related by a series of "dualities" (which include T-duality), so there was really only one theory with different weak coupling limits. This new insight was possible due to an improved understanding of some nonperturbative aspects of string theory which we now discuss.

12.3.2 Nonperturbative aspects of string theory and quantum black holes

In the mid 1990s, it was discovered that string theory is not just a theory of strings. There are other extended objects called branes. The name comes from membranes which are two-dimensional, but branes exist in any dimension: p-branes are $(p + 1)$-dimensional extended objects. Branes are nonperturbative objects with a tension that is inversely related to a power of the coupling g_s. The most common type of brane is called a *D-brane* and it has a tension $T \propto 1/g_s$. So one could never see these objects in perturbation theory in g_s. Even though they are very heavy, the gravitational field they produce is governed by $G_{10}T \sim g_s$ so, as $g_s \to 0$, there should be a flat space description of these objects, and it was found by Polchinski [122]. At weak coupling, a D-brane is a surface in spacetime on which open strings can end. The D stands for "Dirichlet" and refers to the boundary conditions on the ends of the open strings. In fact, D-branes were discovered by applying T-duality to a theory of open strings. Open string worldsheets have boundaries, and by looking at how the Euler number changes when open strings interact, one finds that the open string coupling constant, g_o, satisfies $g_o^2 = g_s$. The tension of a D-brane can be understood by viewing it as a soliton of the open string theory: $T \propto 1/g_o^2 \propto 1/g_s$. The endpoints of the open strings move freely along the brane but cannot leave the brane unless they join and form a closed string. The massless states of an open string include a spin-1 excitation, so every D-brane comes with a $U(1)$ gauge field. When N D-branes coincide in spacetime, the open strings stretching from one to another also become massless. This enhances the resulting gauge group from $U(1)^N$ to $U(N)$. These D-branes are also sources for the p-form "Ramond–Ramond" fields F_p.

D-branes have found applications in string phenomenology, i.e., the attempt to connect string theory with standard four-dimensional particle physics. String theory clearly unifies the graviton with many matter degrees of freedom. One difficulty is that one often has too many light degrees of freedom. In particular, the moduli associated with the size of certain surfaces in the internal directions correspond to 4D massless scalars which we do not see. One way to make the models more realistic is, roughly speaking, to wrap a brane around the surface in the internal space (which will try to contract it) and also add flux associated with F_p (which will try to expand it). Under certain circumstances these forces balance at one size of the surface [123]. In terms of the lower-dimensional theory, the scalar field now has a large mass and has no low-energy dynamical effects.

Soon after D-branes were discovered, evidence was found for a strong–weak coupling duality called *S-duality*. The evidence included comparing the masses of certain BPS states whose masses are fixed by the charges. Using these dualities, it was argued that all the perturbative string theories are related. In addition, it was proposed that there is an eleven-dimensional theory called *M-theory*, whose dimensional reduction on a circle of radius $R = g_s \ell_s$ yields Type IIA string theory [124]. In particular, D0-branes have just the right properties to represent the momentum modes around this circle. M-theory can thus be viewed as a strong coupling limit of Type IIA string theory. Its low-energy limit is eleven-dimensional supergravity. As discussed in Section 12.2.2, this theory has only a metric, 4-form field strength and spin-3/2 field. The bosonic action is given in (12.25). The 4-form can carry electric charges of 2-branes and magnetic charges of 5-branes, so we deduce that M-theory is not a theory of strings, but a theory including 2-branes and 5-branes, which are called M2- and M5-branes.

Putting all the dualities together, M-theory can be viewed as having six different weak coupling limits corresponding to the five different 10D perturbative string theories, and 11D supergravity (see Fig. 12.5). We do not yet understand the fundamental principles underlying M-theory. The best description we have of this theory is in terms of a matrix model, i.e., a quantum theory describing a collection of matrices depending only on time [125]. In this description, space emerges from the properties of the matrices.

One of the main successes of string theory is its ability to reproduce the Hawking–Bekenstein entropy of certain black holes as well as Hawking radiation, from a microscopic quantum theory. For many years it was thought that strings could not explain the entropy of Schwarzschild black holes since the string entropy is proportional to the mass, whereas the entropy of Schwarzschild black holes (in $D = 4$) goes like the mass squared. However,

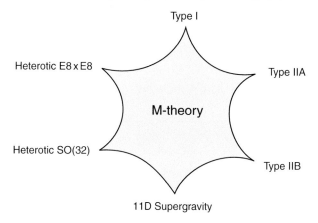

Figure 12.5 A schematic view of different weak coupling limits of M-theory.

this is misleading. The string entropy is $S_{\text{string}} \sim M\ell_s$ whereas the black hole entropy is $S_{\text{BH}} \sim G_4 M^2$ and the four-dimensional Newton's constant is $G_4 \sim g_s^2 \ell_s^2$. So before one can ask if strings can explain black hole entropy, one must first specify the string coupling. The natural point to compare them is when the Schwarzschild radius is of order the string scale, since this is when the Schwarzschild solution starts to receive stringy corrections. At this point, the black hole entropy is [126]

$$S_{\text{BH}} \sim M r_0 \sim M\ell_s \sim S_{\text{string}}, \tag{12.39}$$

so the entropies agree. Since $r_0 = 2G_4 M \sim \ell_s$ implies $g_s \sim (M\ell_s)^{-1/2}$, the coupling remains small for a large-mass black hole. This argument continues to hold in $D > 4$, and can be generalized to include charges and angular momentum. It leads to a simple *correspondence principle* between black holes and strings [127]: when the curvature at the horizon becomes of order the string scale, the typical black hole state becomes an excited string state with the same charges and angular momentum. This provides a simple picture for the endpoint of Hawking evaporation: when black holes evaporate down to the string scale, they turn into highly excited strings which then continue to decay down to an unexcited string which is just another elementary particle.

For certain black holes, the entropy can be reproduced exactly in string theory. This was first shown for nonrotating, extremal 5D black holes [128] and soon generalized to near-extremal [129, 130] and rotating [69] 5D black holes. Similar results hold for extremal [131, 132] and near-extremal [133] 4D black holes. Due to the importance of this result, we now describe one example in some detail. (The following discussion is based on [134].) The Type IIB supergravity action includes the following terms:

$$S = \frac{1}{16\pi G_{10}} \int d^{10}x \sqrt{-g} \left[e^{-2\phi}[R + 4(\nabla\phi)^2] - \frac{1}{12}F_3^2 \right], \tag{12.40}$$

where F_3 is a Ramond–Ramond 3-form field strength and ϕ is the dilaton. This action is written in terms of the so-called string metric, which is the metric that appears in (12.37), i.e., the metric that the strings directly couple to. We will compactify four directions x_i ($i = 6, 7, 8, 9$) on circles of length[13] 2π. In the resulting six-dimensional spacetime, F_3 has both electric and magnetic type charges, each carried by one-dimensional extended objects:

$$Q_1 = \frac{1}{4\pi^2 g_s} \int e^{2\phi} * F_3 , \qquad Q_5 = \frac{1}{4\pi^2 g_s} \int F_3 . \tag{12.41}$$

The integrals are over an S^3 surrounding the object, and $*$ denotes the six-dimensional Hodge dual. The factor of $e^{2\phi}$ is needed in the first integral since $*F_3$ by itself is not a closed form after dimensional reduction to six dimensions. The labels come from the fact that Q_1 is the charge carried by D1-branes and Q_5 is the charge carried by D5-branes (which wrap the T^4). The charges are normalized so that Q_i are integers which simply count the number of branes of each type.

One can show that the following is a black brane solution to the equations of motion coming from the action (12.40):

$$ds^2 = f(r)^{-1} \left[-dt^2 + dx_5^2 + \frac{r_0^2}{r^2} \left(\cosh \sigma \, dt + \sinh \sigma \, dx_5 \right)^2 + \left(1 + \frac{g_s Q_1}{r^2} \right) dx_i \, dx^i \right]$$

$$+ f(r) \left[\left(1 - \frac{r_0^2}{r^2} \right)^{-1} dr^2 + r^2 \, d\Omega_3^2 \right] , \tag{12.42}$$

where

$$f(r) = \left(1 + \frac{g_s Q_1}{r^2} \right)^{1/2} \left(1 + \frac{g_s Q_5}{r^2} \right)^{1/2} . \tag{12.43}$$

The matter fields take the form:

$$e^{-2\phi} = \left(1 + \frac{g_s Q_5}{r^2} \right) \left(1 + \frac{g_s Q_1}{r^2} \right)^{-1} , \tag{12.44}$$

$$F_3 = 2 g_s Q_5 \epsilon_3 + 2 g_s Q_1 e^{-2\phi} * \epsilon_3 , \tag{12.45}$$

where ϵ_3 is the volume form on a unit S^3 and $*\epsilon_3$ is its six-dimensional dual.

If x_5 is periodically identified with period $2\pi R$, then momentum in this direction should be quantized: $P = n/R$. From (12.42) one finds

$$n = \frac{r_0^2 R^2 \sinh 2\sigma}{2 g_s^2} . \tag{12.46}$$

If one Kaluza–Klein reduces to five dimensions, one has a spherical black hole with three charges Q_1, Q_5, n.

---

[13] To simplify the presentation, for the remainder of this Section we will set $\ell_s = 1$.

The total energy of this solution is[14]

$$E = \frac{RQ_1}{g_s} + \frac{RQ_5}{g_s} + \frac{n}{R} + \frac{Rr_0^2 e^{-2\sigma}}{2g_s^2}. \qquad (12.47)$$

The extremal limit corresponds to $r_0 \to 0, \sigma \to \infty$ with n fixed. In this limit the last term vanishes. The remaining three terms are exactly what one would expect from Q_1 D1-branes wrapping the S^1 with radius R, Q_5 D5-branes wrapping the T^4 with unit radii and the S^1 with radius R, and momentum $P = n/R$. This is a consequence of the fact that the extremal solution is BPS, so the energy is uniquely determined by the charges.

The Bekenstein–Hawking entropy is

$$S_{\mathrm{BH}} = \frac{A}{4G_{10}} = 2\pi \sqrt{Q_1 Q_5} \, \frac{r_0 R \cosh \sigma}{g_s}. \qquad (12.48)$$

String theory can reproduce this entropy in both the extremal and near-extremal limits by counting states at weak coupling, i.e., in flat spacetime. In the extremal case, one expects the number of states to be independent of coupling since the solution is supersymmetric and the states are BPS. (It was a surprise to find that this agreement continued to hold for slightly near-extremal solutions also.) So we want to count the number of bound states of D1-branes and D5-branes with given momentum in a flat spacetime compactified on a T^4 with unit radii and an S^1 with radius R. Since the D5-branes wrap the small T^4, low-energy excitations move only along the S^1, and can be described by an effective (1+1)-dimensional theory.

In the extremal limit, the entropy reduces to

$$S_{\mathrm{BH}} = 2\pi \sqrt{Q_1 Q_5 n}. \qquad (12.49)$$

Note that the dependence on all continuous parameters such as R and g_s has dropped out. An examination of the low-energy excitations of Q_1 D1-branes and Q_5 D5-branes shows that there are $4Q_1 Q_5$ massless bosons. These can be viewed as open strings connecting one of the D1-branes with one D5-brane. The factor of 4 arises since there are two possible orientations for the strings and for each orientation, the ground state is two-fold degenerate. When these D1–D5 strings are excited, the D1–D1 and D5–D5 open strings become massive and do not contribute to the entropy. This is a reflection of the fact that branes are bound together.

Free fields in 1+1 dimensions have independent right and left moving modes. BPS states with non-zero momentum correspond to exciting only the right moving modes of these massless fields. So one can simply count the number of states of $4Q_1 Q_5$ bosonic fields (plus an equal number of fermionic fields required by supersymmetry) on a circle of radius R with total right moving momentum $P = n/R$. In the limit of large n, the answer is e^S, where

---

[14] To compute the ADM energy, one should first conformally rescale the metric by a function of ϕ so the action (12.40) contains the standard Einstein–Hilbert term. The result is called the Einstein metric.

$$S = 2\pi \sqrt{Q_1 Q_5 n},\tag{12.50}$$

in perfect agreement with the Bekenstein–Hawking entropy (12.49).

Now suppose we add a small amount of energy to the system keeping the charges fixed, so it is no longer extremal. To maximize the entropy, the energy will excite the lightest modes. If R is large, the lightest modes are just the momentum modes. So the energy excites some additional left and right moving modes. At weak coupling there is very little interaction between the modes, but it is not zero. However, if $r_0^2 \ll g_s Q_1, g_s Q_5$ and $g_s^2 n/R^2 \ll g_s Q_1, g_s Q_5$, we are in a "dilute gas" regime where interactions are negligible. This corresponds to a near-extremal black hole. In this case, the entropy computed at weak coupling is just the sum of the entropies of the left and right moving modes:

$$S = 2\pi \sqrt{Q_1 Q_5} \left(\sqrt{n_R} + \sqrt{n_L} \right),\tag{12.51}$$

where n_R and n_L are defined by setting $P = (n_R - n_L)/R$ and requiring that $(n_R + n_L)/R$ be the contribution to the energy from the momentum modes. From (12.46) and (12.47) we get

$$n_R = \frac{r_0^2 R^2 e^{2\sigma}}{4g_s^2}, \qquad n_L = \frac{r_0^2 R^2 e^{-2\sigma}}{4g_s^2}.\tag{12.52}$$

Substituting into (12.51) we see that the counting at weak coupling precisely reproduces the entropy of the near-extremal black hole (12.48).

We now include some interactions between the left and right moving modes. Occasionally, a left-moving mode can combine with a right-moving mode to form a closed string which can leave the brane. This corresponds to the decay of an excited configuration of D-branes and is the weak-coupling analog of Hawking radiation. Given the remarkable agreement between the entropy of the black hole and the counting of states of the D-branes, the next step is to ask how the radiation emitted by the D-brane compares to Hawking radiation. In both cases, the radiation is approximately thermal with the same temperature. This is expected since the entropy as a function of energy agrees in the two cases. What is surprising is that the overall rate of radiation agrees [135]. What is even more remarkable is that the deviations from the blackbody spectrum also agree [134]. On the black hole side, these deviations arise since the radiation has to propagate through the curved spacetime outside the black hole. This contains potential barriers which give rise to frequency-dependent greybody factors. On the D-brane side there are deviations since the modes come from separate left and right moving sectors on the D-branes, with different effective temperatures. The calculations of these deviations could not look more different. On the black hole side, one solves a wave equation in a black hole background. The solutions involve hypergeometric functions. On the D-brane side, one does a calculation in D-brane perturbation theory. Remarkably, the answers are identical. It is worth emphasizing that it is not just the dependences on a few parameters which agree. One is comparing the decay rate as a function of frequency and the entire functional form agrees on both sides. It is as if the black hole knows that its states are described by an effective $(1+1)$-dimensional field theory with left and right moving modes.

As mentioned earlier, this precise agreement between D-branes and black holes holds for black holes in 4D as well as 5D,[15] but always in the near-extremal limit. Far from extremality, one still expects string theory to provide a microscopic description of black holes, but one can no longer rely on the weak-coupling D-brane picture. The extrapolation to strong coupling now produces significant changes in the properties of the quantum states. One approach to this problem is discussed in the next section.

12.4 Holography and gauge/gravity duality

We begin in Section 12.4.1 by briefly reviewing some general arguments suggesting that quantum gravity might be holographic. In Section 12.4.2, we formulate gauge/gravity duality, a precise form of holography that emerges from string theory. We also discuss the evidence for this duality and its consequences. Section 12.4.3 offers an overview of some applications of gauge/gravity duality.

12.4.1 Holography

The suggestion that quantum gravity might be holographic was originally motivated by black hole entropy [139, 140]. The idea was simply that the fact that a black hole has an entropy that scales with its area is in striking contrast to most systems which have an entropy that is proportional to their volume. This suggests that everything that happens inside the black hole could be somehow encoded in degrees of freedom at the horizon.

In trying to find a more precise formulation of holography, it is natural to consider space-times which asymptotically approach anti-de Sitter (AdS) spacetime.[16] There are several reasons for this. First, a static slice in AdS is a constant negative curvature hyperboloid, so the area of a sphere at proper radius R grows exponentially with R. Thus, compared to flat space, there is "more room at infinity" for the holographic description to live. Second, the conformal boundary at infinity is timelike, so a holographic description at infinity could live on an ordinary spacetime. Finally, black hole thermodynamics is better behaved in AdS since the negative curvature acts like a confining box. In particular, large black holes in AdS have positive specific heat and can be in thermal equilibrium with their Hawking radiation [142].

A general argument for holography in quantum gravity was given recently by Marolf [143]. His argument can be made for either asymptotically flat or asymptotically AdS boundary conditions, but we focus on the AdS case here. Consider first perturbative quantum gravity about AdS. At linear order, any field at a point in the interior can be evolved back and expressed as an integral over operators along a surface at large radius. Thus, the set of boundary operators defined at large radius form a complete set of operators at linear order. The same argument can be made at each order in perturbation theory. This is not yet a

---

[15] For extensions to black rings, see for example [136–138].
[16] See [141] for an early study of quantum field theory in AdS that set the stage for many later developments.

statement about holography, but more analogous to expressing the value of a field in terms of initial data at an earlier time. However, the Hamiltonian itself is a boundary operator. This is a unique feature of diffeomorphism invariant theories coming from the fact that the Hamiltonian can be expressed as a surface integral at infinity. Since the Hamiltonian generates time evolution, we can express any boundary operator $\mathcal{O}$ at one time in terms of a boundary operator at a different time by solving

$$\frac{d}{dt}\mathcal{O}(t) = i[H, \mathcal{O}(t)]. \tag{12.53}$$

It follows that, at least in perturbation theory, any observable in the bulk can be expressed in terms of a boundary observable at one fixed time. This is a statement of holography.

At the full nonperturbative level, the argument for the completeness of boundary observables is less clear, not least because of the difficulty in defining observables in full quantum gravity. However, the Hamiltonian is still presumably a boundary observable so, given a set of boundary observables at one time, one can always evolve them using (12.53) and relate them to observables at a later time. Thus any information available at the boundary at one time is also available at any later time. In particular, this is true for an evaporating black hole. This result can be called "boundary unitarity". A similar argument can be made even for asymptotically flat spacetimes using observables defined on spacelike and null infinity.

12.4.2 Gauge/gravity duality

A much more precise formulation of holography was found by Maldacena by studying extremal and near-extremal black holes in string theory [144]. The holographic theory turns out to be an ordinary (supersymmetric) gauge theory. The equivalence between the gravitational and gauge theories is often called gauge/gravity duality. The motivation for this duality is the following. Consider a stack of N D3-branes in Type IIB string theory. At weak coupling, $g_sN \ll 1$, the excitations are described by open strings on the brane and closed strings off the brane. The massless states consist of $\mathcal{N} = 4$ supersymmetric $U(N)$ gauge theory (described in Section 12.2.1) on the brane and Type IIB supergravity off the brane. In a low-energy limit, one has only long wavelength supergravity modes, but keeps all modes of the gauge theory since this theory is conformally invariant. These two sectors decouple at low energy since the dimensionless coupling to gravity is $G_{10}E^8$.

When $g_sN \gg 1$, the backreaction of the branes becomes important and produces the following metric:

$$ds^2 = H^{-1/2}\left[-dt^2 + dx_1^2 + dx_2^2 + dx_3^2 \right] + H^{1/2}\left[dr^2 + r^2\, d\Omega_5^2 \right], \tag{12.54}$$

where

$$H(r) = 1 + \frac{L^4}{r^4}, \qquad L^4 = 4\pi g_sN\ell_s^4. \tag{12.55}$$

The branes no longer appear explicitly, but are represented by a non-zero flux of a 5-form field strength F_5. This metric describes an extremal black brane. There is a smooth degenerate horizon at $r = 0$. The excitations are closed strings in the above spacetime. Low-energy excitations consist of either arbitrary closed strings which are very close to the horizon (and hence have a large redshift) or massless closed strings far from the horizon with very low frequency. The latter sector is again long-wavelength modes of IIB supergravity. In a low-energy limit, these two sectors decouple, since the absorption cross section of the black hole goes to zero as $\omega \to 0$. The near-horizon limit of (12.54) is obtained by dropping the "1" in H, i.e., setting $H = L^4/r^4$. The resulting spacetime is the product of S^5 with radius L and five-dimensional anti-de Sitter spacetime (AdS_5) with radius of curvature L:

$$ds^2 = \frac{r^2}{L^2}\Big[-dt^2 + dx_1^2 + dx_2^2 + dx_3^2 \Big] + \frac{L^2\, dr^2}{r^2}. \tag{12.56}$$

We see that at both weak and strong coupling, the low-energy description of a stack of D3-branes has two decoupled sectors. In each case, one sector is low-energy supergravity modes. It is thus natural to identify the other two sectors. This means that we have a system which at weak coupling looks like supersymmetric Yang–Mills (SYM) and at strong coupling looks like strings in $AdS_5 \times S^5$. But SYM also exists at strong coupling, and string theory exists at weak coupling, so these two descriptions must be equivalent. Thus we are led to the following remarkable conjecture:

Four-dimensional $\mathbb{N} = 4$ supersymmetric U(N) gauge theory is equivalent to IIB string theory with $AdS_5 \times S^5$ boundary conditions.

This is the simplest (and most well-studied) example of *gauge/gravity duality*, the equivalence between a theory of (quantum) gravity and a nongravitational gauge theory. We will discuss some generalizations later. String theory with boundary conditions $AdS_5 \times S^5$ has two dimensionless parameters $(g_s, L/\ell_s)$, and the gauge theory has two dimensionless parameters, (g_{YM}, N). They are related by $4\pi g_s = g_{YM}^2$ and $(L/\ell_s)^4 = g_{YM}^2 N$.

The coordinates (12.56) do not cover all of AdS_5 but only the so-called "Poincaré patch". As it stands, the duality is between string theory on spacetimes asymptotic to (12.56) and SYM on Minkowski spacetime, where the Minkowski space can be viewed as the conformal boundary of (12.56). However, it is easy to extend the duality to all of AdS_5, where the conformal boundary is now the Einstein static universe, $S^3 \times R$, and the SYM lives on this spatially compact space. The interior of AdS_5 is often called the "bulk".

Recall that Newton's constant in ten dimensions is given in string theory by $G_{10} \sim g_s^2 \ell_s^8$. It follows from (12.55) that $L^4 \sim N\ell_p^4$, where ℓ_p is the Planck length. Thus if N is $O(1)$, the curvature in the bulk is of order the Planck scale everywhere. This is an interesting regime from the standpoint of quantum gravity. The gauge theory is relatively simple, but it is hard to give a physical interpretation of any SYM observables in terms of the dual gravitational theory.

In the opposite limit when N is large, we have $L \gg \ell_p$, so typical curvatures are much smaller than the Planck scale and quantum gravity effects are suppressed. In the gauge theory it is convenient to consider the 't Hooft limit: $N \to \infty$, $g_{YM} \to 0$ with $\lambda \equiv g_{YM}^2 N$ held fixed. The 't Hooft coupling λ then acts as the natural coupling constant in this limit and only Feynman diagrams that can be drawn on a plane contribute. On the string theory side, when $\lambda \gg 1$ we have $L \gg \ell_s$, so stringy excitations in the bulk are suppressed and one can work with just supergravity modes. Note that in this limit where the gravity side is simple, the gauge theory is strongly coupled and poorly understood. Conversely, when $\lambda \ll 1$, the gauge theory is weakly coupled and well understood, but the gravity side is very stringy. It is this strong/weak coupling aspect of the duality which allows two very different sounding theories to be equivalent.

Since $G_{10} \sim g_s^2 \ell_s^8 \sim L^8/N^2$, and L is held fixed in the 't Hooft limit, it follows that the gravitational backreaction of all states becomes negligible unless the energy of the state grows at least as fast as N^2. So all states with $E < O(N^2)$ can be described by fields propagating on AdS. Since there are $O(N^2)$ degrees of freedom in the gauge theory, it is natural to consider states with energy $E \sim O(N^2)$ since they correspond to exciting each degree of freedom by an amount independent of N.

Note that a four-dimensional gauge theory is describing a ten-dimensional theory of gravity. So it is an extreme type of hologram which encodes six extra spatial dimensions. For perturbations of $AdS_5 \times S^5$ one can show (by comparing representations of $SO(6)$) that information about position on S^5 is encoded in products of the six scalar operators ϕ^i in $\mathcal{N} = 4$ SYM. The radial dimension in AdS_5 is related to an energy scale in the dual gauge theory. This is suggested by the fact that in Poincaré coordinates (12.56) the metric is invariant under $r \to ar$, $(t, x_i) \to (t, x_i)/a$. So small radius corresponds to large distance or low energy in the gauge theory.

The claim that a four-dimensional gauge theory could describe all of ten-dimensional string theory sounds so crazy that one might think that one could disprove it quite easily. Let us try. Having extra spatial dimensions usually leads to more quantum states (since, e.g., one can have momentum modes in more directions). So we will compare the entropy in the two theories at a high temperature T. In the gauge theory, the entropy is given by the usual formula for a thermal gas with N^2 degrees of freedom

$$S_{SYM} \sim N^2 T^3 V_3, \tag{12.57}$$

where V_3 is the volume of the three-dimensional space. On the gravity side, it would seem that one could exceed this including only the massless modes of the string. A thermal gas of massless particles in ten dimensions has energy density proportional to T^{10} so, from the first law, its entropy density is proportional to T^9. It would appear that at sufficiently high temperature, the entropy in the bulk vastly exceeds that on the boundary.

Of course this estimate is incorrect since it ignores the fact that the gas will collapse to form a black hole. In addition to the usual spherical black holes, AdS has black holes in which the horizon geometry is flat. Since we have estimated the entropy of SYM on flat space, these "planar black holes" are the appropriate comparison. They take the form

$$ds^2 = \frac{r^2}{L^2}\left[\left(1 - \frac{r_0^4}{r^4}\right)dt^2 + dx_i\,dx^i\right] + \left(1 - \frac{r_0^4}{r^4}\right)^{-1}\frac{L^2\,dr^2}{r^2} + L^2\,d\Omega_5^2, \qquad (12.58)$$

and have a temperature $T = 3r_0/(4\pi L^2)$. Their entropy is

$$S_{\text{BH}} = \frac{A}{4G_{10}} \sim \frac{L^8 T^3 V_3}{G_{10}} \sim N^2 T^3 V_3, \qquad (12.59)$$

which agrees with the estimate from the gauge theory. In fact, this shows that the gauge theory has enough microstates to reproduce the black hole entropy. Gauge/gravity duality relates the black hole to a thermal state in the dual field theory.

It is difficult to compare the numerical coefficient in the above comparison of the entropy. This would require an exact calculation of the number of states in the gauge theory at strong coupling. If one instead does the calculation at weak coupling, one gets an answer which differs from the black hole entropy by a factor of 3/4: $S_{\text{BH}} = (3/4)S_{\text{SYM}}$ [145]. It should be emphasized that this difference is not a problem for gauge/gravity duality: $e^{S_{\text{SYM}}}$ is the number of SYM states at weak coupling, and the number of states of a given energy is expected to decrease with coupling. This is because the form of the Yang–Mills Hamiltonian implies that increasing the coupling will increase the potential energy of each SYM state, and hence lower the total number of states at fixed energy. The duality predicts that at strong coupling, the entropy will agree with the black hole. It is surprising that what appears to be a complicated QFT calculation only changes the answer by a factor of 3/4. This prediction is waiting to be verified by a direct calculation.

This situation should be contrasted with that at the end of the previous subsection. In that case, we again compared a black hole entropy with a calculation at weak coupling and found precise agreement. The key difference is that that black hole was supersymmetric (or nearly supersymmetric), so the mass of each state is fixed by the charge. The entropy is then independent of the coupling constant, and a precise comparison is possible.

Only gauge invariant observables can be compared on both sides of the duality. A large class of such observables can be compared as follows [146, 147]. For every field Φ in the bulk there is a corresponding operator $\mathcal{O}$ in the dual gauge theory. Supergravity fields correspond to simple gauge invariant operators constructed as a single trace of a local product of the super Yang–Mills fields. (An early check of the duality was that there was indeed a one-to-one correspondence between supergravity fields and suitable SYM operators.) Asymptotic values of the bulk fields act as sources for the dual operator in the following sense: the string theory partition function with boundary condition[17] $\Phi \to \Phi_0$ should equal the field theory partition function with action

$$S[\Phi_0] = S_{\text{SYM}} + \int \Phi_0\,\mathcal{O}. \qquad (12.60)$$

---

[17] More precisely, Φ typically vanishes asymptotically, and $\Phi \to \Phi_0/r^\Delta$, where Δ is related to the mass of the bulk field.

In other words

$$Z_{\text{string theory}}(\Phi \rightarrow \Phi_0) = \int DA \, D\phi \, e^{iS[\Phi_0]} \equiv Z_{\text{SYM}}[\Phi_0]. \qquad (12.61)$$

In the 't Hooft limit with large λ, the left hand side can be approximated by just the super-gravity fields, and further approximated by the exponential of the action of the classical solution with the boundary condition $\Phi \rightarrow \Phi_0$. By taking a derivative with respect to the source Φ_0, one can show that the expectation value of $\mathcal{O}$ is related to a subleading term in the asymptotic behavior of the classical solution.

Gauge/gravity duality is a conjecture. It has not yet been proven. Since $\mathcal{N} = 4$ SYM is a complete nonperturbative quantum theory, one might be tempted to *define* nonperturbative string theory in terms of the dual gauge theory and claim the duality is true by definition. But this is much too quick. Non-trivial consistency conditions must be met, showing that everything we know about string theory, including the space of classical solutions (with AdS boundary conditions) and perturbation theory about them, is reproduced in the SYM theory.

Although there is no proof, by now there is overwhelming evidence that the conjectured gauge/gravity duality is correct. The early evidence included the fact that the symmetries on the two sides agree: $\mathcal{N} = 4$ supersymmetric gauge theory in $D = 4$ is conformally invariant, so it is invariant under $SO(4,2)$. As described in Section 12.2.1, the theory includes 6 scalars that transform in the fundamental representation of the R-symmetry $SO(6)$. $AdS_5 \times S^5$ has an isometry group which is precisely $SO(4,2) \times SO(6)$. The supergroups also agree. More non-trivial checks came later and include a vast number of calculations in which a physical quantity is computed on the two sides of the duality. Although the two calculations often look very different, the final answers agree. We mention a few examples below:

- Wilson loops in the gauge theory are natural (nonlocal) gauge invariant operators. Given a curve $\mathcal{C}$ one considers[18] $W = \text{Tr} \, P \exp[\int_{\mathcal{C}} A]$, where P denotes path ordering and Tr denotes trace in the adjoint representation of the gauge group. The expectation value $\langle W \rangle$ of these Wilson loops can be calculated on the gravity side by considering string world-sheets in spacetime that end on the loop $\mathcal{C}$ at infinity. The area of the string worldsheet is then related to $\langle W \rangle$. In certain cases, one can compute $\langle W \rangle$ exactly in the gauge theory and find complete agreement with the gravity calculation [148].
- Renormalization group (RG) flow is the quantum field theory process of integrating out the high-energy modes to obtain a new effective theory at lower energy. $\mathcal{N} = 4$ SYM is conformally invariant, so there is no scale, and the RG flow is trivial. However, one can add relevant operators, for example mass-terms, to this theory and find that RG flow leads to a different conformal field theory at low energy with fewer degrees of freedom. On the gravity side, this corresponds to modifying the boundary conditions at infinity and finding a new static solution to Einstein's equation. One finds that at small radius, this

---

[18] The objects that can be computed holographically are actually slight generalizations of the usual Wilson loop in that they include the six scalars in the gauge theory.

new solution approaches (12.56) but now with L replaced by a new AdS radius $\tilde{L}$. There is detailed agreement between the CFT one gets at low energy and the new AdS [149]. In particular the new AdS length scale, $\tilde{L}$, is related to the number of degrees of freedom in the new low-energy dual theory in just the same way as in the asymptotic region.

- All the states of supergravity in the bulk have precise descriptions in the dual gauge theory. What about the excited string states? In general, it is hard to identify the dual of these states, but this has been done in a certain limit [150]. If one starts with a null geodesic wrapping the S^5, one can take a Penrose limit and obtain a 10D plane wave. In the gauge theory, this corresponds to considering states of the form $\text{Tr}\,[Z^J]|0\rangle$, where J is a large angular momentum and $Z = \phi_5 + i\phi_6$ (ϕ_i are the six scalars of $\mathcal{N} = 4$ SYM). The complete spectrum of the string in the bulk plane wave background is exactly reproduced in the gauge theory by replacing some of the Z terms with ϕ_i, $i = 1, 2, 3, 4$, or $D_\mu Z = \partial_\mu Z + [A_\mu, Z]$. It is as if the Z terms in the gauge theory create a string with transverse oscillations generated by ϕ_i and $D_\mu Z$.

- If one restricts the discussion to a class of states preserving 1/2 of the supersymmetry, one can make the correspondence between the gravity and gauge theory much more explicit [151]. Let us work with global AdS so the field theory lives on $S^3 \times R$. We will actually restrict the discussion to fields that are independent of S^3 and hence reduce to $N \times N$ matrices. In fact, we consider only states created by a single complex matrix, so they can be described by a one-matrix model. (In terms of the six scalars in the gauge theory, this is again $Z = \phi_5 + i\phi_6$.) This theory can be quantized exactly in terms of free fermions, and the states can be labeled by arbitrary closed curves on a plane. (The plane represents phase space and the closed curves denote the boundaries of regions that are occupied.) The states are all invariant under $SO(4) \times SO(4)$, where the first factor corresponds to rotations on the S^3 and the second factor corresponds to rotations of the remaining four scalars $\phi_1, \ldots, \phi_4$ in the gauge theory.

On the gravity side, one considers solutions to ten-dimensional supergravity involving just the metric and self-dual 5-form F_5. The field equations are simply $dF_5 = 0$ and

$$R_{\mu\nu} = F_{\mu\alpha\beta\gamma\delta}F_\nu{}^{\alpha\beta\gamma\delta}. \tag{12.62}$$

There exists a large class of stationary solutions to (12.62), which have an $SO(4) \times SO(4)$ symmetry and can be obtained by solving a linear equation [151]. These solutions are nonsingular and have no event horizons, but can have complicated topology. They are also labeled by arbitrary closed curves on a plane. This provides a precise way to map states in the field theory into bulk geometries. Only for some "semi-classical" states is the curvature below the Planck scale everywhere, but the matrix/free-fermion description readily describes all the states in this class, of all topologies, within a single Hilbert space.

- The above examples check gauge/gravity duality in the large N limit where the bulk is described by supergravity. There is also evidence that the duality remains true at finite values of N. A striking example is the 'string exclusion principle' [152]. Graviton states on S^5 arise in the gauge theory from acting on the vacuum with an operator involving traces of products of the ϕ^i. However, these fields are $N \times N$ matrices, so the traces cease to be independent for products of more than N fields. This leads to an upper bound on the angular momentum J on S^5:

$$J/N \leq 1. \tag{12.63}$$

From the point of view of supergravity this is mysterious, because the graviton states exist for arbitrary J. However, there is an elegant resolution in string theory [153] using something called the "Myers effect" [154]. This is the fact that a stack of D-branes in an external field can become polarized and take the shape of a sphere. In a certain limit, the graviton can be viewed as a stack of D0-branes, and one can show that when it moves sufficiently rapidly on the sphere, it will blow up into a spherical D3-brane, and $J = N$ turns out to be the largest D3-brane that will fit in the space-time. Thus the same bound is found on both sides of the duality, and this is a nonperturbative statement in N: it would be trivial in a power series expansion in $1/N$.

So far we have discussed gauge/gravity duality with asymptotic $AdS_5 \times S^5$ boundary conditions. This is the most well-studied example of gauge/gravity duality, but many other examples exist. Applying a similar argument to a stack of M2-branes shows that M-theory on asymptotically $AdS_4 \times S^7$ is equivalent to a $(2 + 1)$-dimensional field theory of gauge fields and matter with a Chern–Simons action for the gauge fields, called ABJM theory [155]. Applying the argument to a stack of M5-branes shows that M-theory on asymptotically $AdS_7 \times S^4$ space-times is equivalent to a still-mysterious $(5 + 1)$-dimensional field theory describing low-energy excitations of the M5-branes. These dualities can also be extended in other ways, such as replacing the S^n with other Einstein spaces. The corresponding change in the gauge theory is known in many cases [156]. One can also put the gauge theory on any space-time, not just Minkowski space and the Einstein static universe. In these cases, the boundary condition in the bulk is a space-time which is only locally asymptotically AdS.

Let us mention an example of a non-trivial check of the AdS_4 duality. For certain $(2+1)$-supersymmetric gauge theories, one can calculate exactly the Euclidean partition function of the theory on a squashed S^3 [157]. This is possible using a technique known as localization. One can then find the dual Euclidean gravitational solution which asymptotically approaches the squashed S^3. If one computes the gravitational action and compares it with the gauge-theory partition function one finds $Z_{\text{gauge}} = e^{-S_{\text{grav}}}$. Each side is a function of the squashing parameter, and the two functions agree exactly. Once again, the calculations on each side look completely different, but the final answers agree. Similarly, the maximization of the free energy ('f-maximization') of the 3D ABJM field theory [155] on S^3 has a matching gravitational dual description [158].

12.4.3 Applications

Given the overwhelming evidence for gauge/gravity duality, we now assume its validity and ask what it can teach us. The duality can be used in both directions to learn about quantum gravity and also about aspects of nongravitational strongly coupled physics. The following applications all represent fields of active research.

Quantum black holes: An immediate consequence of gauge/gravity duality is that the process of forming and evaporating a small black hole must be unitary. This is because it can be mapped to a process in the dual gauge theory in which states evolve by a standard Hamiltonian. However, the details of this map are not yet clear. In particular, it is still unknown how the information gets out of the black hole, and what is wrong with Hawking's original semiclassical argument [159] that black hole evaporation would lead pure states to evolve to mixed states.

For over a decade after the discovery of gauge/gravity duality, many people believed that the information could be restored by keeping track of subtle correlations in the Hawking radiation that were missed in a semi-classical treatment. However, it was shown in [160] that this can never work: small corrections to Hawking's calculations are not sufficient to get the information out. One alternative that has been suggested is "fuzzballs" [161]. This is the idea that the standard black hole solutions describe ensemble averages, and individual pure states do not have horizons. They are instead described by classical solutions (or more generally quantum states) that extend out to the would-be horizon. In support of this idea, many stationary, nonsingular supergravity solutions that have the same mass and charge as an extremal black hole have been constructed. It remains to be seen whether all black hole microstates can be realized in this way.

Another alternative has been proposed recently. A key assumption in Hawking's argument that black hole evaporation would not be unitary was that the horizon locally looks like flat space to an in-falling observer. Since we now believe that the evolution is unitary, people have started to question this assumption. It has been suggested [162] that someone falling into an evaporating black hole would hit a "firewall" at the horizon and burn up. It was argued that this would be true even for a large black hole as long as it has evaporated to at least half its original mass. This has caused considerable controversy. The only thing that is clear is that the following three apparently plausible statements are inconsistent: (1) Information is not lost in black hole evaporation, i.e., pure states evolve to pure states. (2) Observers falling into a large black hole pass through the horizon unaffected. (3) Quantum field theory in curved space-time is a good approximation outside a large black hole. At the moment there is no resolution in sight. The fundamental questions raised by Hawking 40 years ago are still unanswered.

Quark confinement: String theory began in the late 1960s as a model of hadrons. As a result of quark confinement, quarks often act like they live at the ends of a string. (It was only in the 1970s that string theory was reinterpreted as a theory of quantum gravity.) In the light of this, it is interesting that gauge/gravity duality provides a simple geometric picture of quark confinement. The idea is that the potential between two quarks on the boundary can be computed in terms of a string in the bulk which ends on the quarks. Since strings have a tension, they want to minimize their length. Given the geometry of *AdS*, such strings do not stay in the asymptotic region, but extend into the bulk. Suppose the bulk geometry smoothly caps off at some radius, e.g., because a circle pinches off there.[19] Then the string

---

[19] It is easy to construct bulk solutions with this property, see e.g. [164].

connecting two quarks separated by a large distance $\mathcal{L}$ on the boundary drops quickly to this minimum radius, moves a distance $\mathcal{L}$ at that radius, and then returns to the boundary. This means that the length of the string increases linearly with $\mathcal{L}$, resulting in a linearly growing potential between the quarks, i.e., the quarks are confined. It is remarkable that a complicated strongly coupled quantum field theory effect such as quark confinement can be given such a simple geometric description. Since we do not currently have a gravitational dual of pure QCD, one cannot yet use holography to argue for quark confinement in the standard model.

Hydrodynamics: The long wavelength limit of any strongly coupled field theory is expected to be described by hydrodynamics. It has been shown that general relativity indeed reproduces standard (relativistic) hydrodynamics in the boundary theory. To see this, one uses the boundary stress tensor, which can be defined for any asymptotically AdS space-time [163, 165]. Under gauge/gravity duality, this is equal to the expectation value of the stress tensor in the dual field theory. One then starts with the planar black hole representing a system in equilibrium at temperature T, and adds long wavelength perturbations. The boundary stress tensor is conserved and takes the form of a perfect fluid plus corrections involving derivatives of the 4-velocity. The first derivative term represents viscosity. Dissipation in the dual theory simply corresponds to energy flowing into the black hole.

One can show that the ratio of the shear viscosity η to the entropy density s is a universal constant for any theory with a gravity dual [166]:

$$\frac{\eta}{s} = \frac{1}{4\pi}. \tag{12.64}$$

This number is very low compared to values for ordinary fluids. Remarkably, when experiments at the RHIC and at the LHC collide heavy ions together, they produce a quark–gluon plasma which has very low viscosity. The measured viscosity is in fact close to the value predicted from the gravity dual. This is difficult to explain using traditional methods.

The connection between gravity and fluid dynamics raises an interesting question. Turbulence is a common feature of fluids, but perturbations of black holes are expected to decay and not show turbulent behavior. In recent work [167] it has been shown that under certain conditions, black holes in AdS do show turbulent behavior.

Condensed matter: Given the success with heavy ion collisions, people became more ambitious and started to apply gauge/gravity duality to study properties of finite density quantum matter, i.e., the subject of condensed matter. Despite the fact that there is no obvious analog of the large N limit in this case, classical gravity analogs of several condensed matter phenomena have been found. The advantage of this duality is that it allows one to calculate transport properties of strongly correlated systems at finite temperature. This is difficult to do using standard condensed matter techniques, but is easy to do holographically. One starts with a black hole in *AdS* which represents the equilibrium system at temperature T. To compute transport using a linear response, one simply perturbs the black hole.

We focus on one example: superconductivity. In standard superconductors, pairs of electrons with opposite spin can combine to form a charged boson called a Cooper pair. Below a critical temperature, these bosons condense and the DC conductivity becomes infinite. To construct a "holographic superconductor", i.e., the gravitational dual to a superconductor, we need just gravity coupled to a Maxwell field and a charged scalar field. A charged black hole corresponds to a system at temperature equal to the Hawking temperature, T, and non-zero charge density (or chemical potential). To represent a non-zero condensate, one needs a static charged scalar field outside the black hole. This is like black hole "hair". So, to describe a superconductor, we need to find a black hole that has hair at low temperatures, but no hair at high temperatures. More precisely, we need the usual Reissner–Nordström AdS black hole (which exists for all temperatures) to be unstable with respect to forming hair at low temperature. At first sight, this is not an easy task, since it contradicts our usual intuition that black holes have no hair.

A surprisingly simple solution to this problem was found by Gubser [168]. He argued that a charged scalar field around a charged black hole in AdS would have the desired property. Consider

$$ S = \int d^4x \, \sqrt{-g} \left(R + \frac{6}{L^2} - \frac{1}{4} F_{\mu\nu} F^{\mu\nu} - |\nabla \Psi - iqA\Psi|^2 - m^2 |\Psi|^2 \right). \qquad (12.65) $$

This is just general relativity with a negative cosmological constant $\Lambda = -3/L^2$, coupled to a Maxwell field and charged scalar with mass m and charge q. It is easy to see why black holes in this theory might be unstable with respect to forming scalar hair: for an electrically charged black hole, the effective mass of Ψ is $m_{\text{eff}}^2 = m^2 + q^2 g^{tt} A_t^2$. But the last term is negative, so there is a chance that m_{eff}^2 becomes sufficiently negative near the horizon to destabilize the scalar field. Detailed calculations confirm that scalar hair does indeed form at low temperature [169]. Why wasn't such a simple type of hair noticed earlier? One reason is that this does not work for asymptotically flat black holes. In that case, the scalar field simply radiates away some of the mass and charge of the black hole in a form of superradiance.

The conductivity can be calculated by perturbing this black hole with boundary conditions on the Maxwell field at infinity that correspond to adding a uniform electric field. The induced current is read off from a subleading term in the perturbation. One finds that at low temperature when the scalar hair is present, the DC conductivity is infinite, showing that one really does have a superconductor [169]. A similar calculation can be done in five dimensions, corresponding to a $(3 + 1)$-dimensional superconductor. However the four-dimensional bulk calculation is appropriate for some "high temperature" superconductors such as the cuprates, in which the superconductivity is associated with two-dimensional CuO planes.

To make the model more realistic, we can add the effects of a lattice by requiring that the chemical potential be a periodic function. This corresponds to a periodic asymptotic boundary condition on A_t. One then numerically finds the rippled charged black holes with this boundary condition. One can then perturb this solution and compute the conductivity as a function of frequency. At high temperature (when the scalar field is zero), the result shows a finite DC conductivity followed by a power law fall-off $|\sigma(\omega)| = B/\omega^{2/3}+C$ [170]. Exactly this same type of power law fall-off (but without the constant offset C) is seen in certain cuprates in their normal phase before they become superconducting. This behavior is not understood from standard condensed matter arguments. It is believed to be a result of strong correlations. Experiments show that the coefficient B and the exponent 2/3 are temperature independent and do not change even when T drops below the superconducting transition temperature. Similarly, one finds no change in the power law fall-off on the gravity side, when one lowers the temperature of the black hole into the superconducting regime [171].

Entanglement entropy: An important quantity in condensed matter is the entanglement entropy. Given a quantum state of a system and a subregion A, one can construct the density matrix ρ_A by tracing over all degrees of freedom outside A. The entanglement entropy is then defined to be

$$S_{EE} = -Tr\, \rho_A \ln \rho_A. \tag{12.66}$$

This is a measure of long-range correlations and has proven useful in a variety of applications including identification of exotic ground states. Unfortunately, it is difficult to calculate in general interacting theories. A simple formula has been given for S_{EE} using gauge/gravity duality [172]. For a static space-time, consider the minimal surface Σ which ends on the boundary of A at infinity. The conjecture is that $S_{EE} = A_\Sigma/(4G)$, where A_Σ is the area of Σ. In other words, the entanglement entropy is given by a formula which is very similar to the Hawking–Bekenstein entropy of a black hole. Since Σ extends out to infinity, its area is infinite. But S_{EE} is also infinite if one includes modes of arbitrarily short wavelength which cross the boundary of A. There is a prescription to regulate this divergence both in the gravity side and in dual field theory. Using this prescription, this conjecture has passed a large number of non-trivial tests [173]. A derivation using Euclidean gravity has recently been provided [174]. It has even been suggested that quantum entanglement might be a key to reconstructing the bulk space-time [175].

12.5 Conclusion

We have seen that combining supersymmetry with general relativity to form supergravity improves the behavior of perturbative graviton scattering amplitudes, but is not expected to provide a UV-complete quantum theory. The situation is much improved in string theory, which not only has perturbatively finite scattering amplitudes, but also provides a complete description of the microstates of certain black holes, and gives a precise form of holography for certain boundary conditions.

In a review of such a large subject, several topics are inevitably left out. One is string phenomenology – the attempt to find a choice of compactification and fluxes so that the low energy theory looks like some extension of the standard model of particle physics (coupled to gravity). Another area that we have not had space to describe consists of attempts to model de Sitter space-time in string theory. This is useful for stringy models of both inflation and the current acceleration of the universe due to dark energy.

There remain many directions for future research. One concerns space-time singularities in string theory. It has been shown that string theory can resolve certain types of singularities, but these tend to be static timelike singularities. It is not known in detail how string theory resolves the naked singularities arising from the instability of black strings and black branes. More importantly, very little is known about the most significant singularities of general relativity, such as the big bang and the singularity inside black holes. It is not clear that string theory "resolves" such singularities in the sense that there is another semiclassical space-time on the other side. For example, in the case of the big bang, it is possible that time emerges from a more fundamental description in much the same way that space emerges in our current theories of holography. In addition, we need to better understand the dictionary relating quantum gravity with anti-de Sitter boundary conditions to the dual gauge theory. Some elements of this dictionary are known, but many more remain to be understood. A key test will be to understand how the information comes out of an evaporating black hole. Another area where progress is needed is to extend holography to asymptotically flat or de Sitter spacetimes.

String theory has had an amazing history. It began as a theory of hadrons and was reinvented as a theory of quantum gravity based on 10D strings. It can be understood as a weak-coupling limit of an 11D theory, M-theory, which includes membranes and 5-branes. And with anti-de Sitter boundary conditions, superstring theory is believed to be equivalent to a supersymmetric gauge theory. Since string theory has all the ingredients both of the standard model of particle physics and of quantum gravity, it unifies the fundamental forces and is therefore a candidate for a 'theory of everything'. Thus, it was long hoped that string theory would produce a single unique 'vacuum' that would unify our understanding of particle physics, gravity, and cosmology, i.e. that it would explain all parameters in particle physics and beyond in a unified framework. It was later realized that string theory does not produce such unique predictions, and we will have to make choices among the various vacua.

However, string theory has recently surprised us again. Rather than "just" being a unified framework for particle physics and quantum gravity, investigations of gauge/gravity duality have revealed totally unexpected connections with other areas of physics. String theory is now expanding into the realm of nuclear physics, hydrodynamics and condensed matter. We are clearly only beginning to explore the depth and potential applications of this remarkable subject.

Acknowledgements

It is a pleasure to thank N. Engelhardt, G. Hartnett, and T. Olson for comments on the draft of this chapter. G.H. is supported in part by the National Science Foundation under Grant No. PHY12-05500. H.E. is supported by NSF CAREER Grant PHY-0953232.

References

[1] F. R. Tangherlini, "Schwarzschild field in *n* dimensions and the dimensionality of space problem," *Nuovo Cim.* **27** (1963) 636.

[2] M. Banados, C. Teitelboim and J. Zanelli, "The Black hole in three-dimensional space-time," *Phys. Rev. Lett.* **69** (1992) 1849 [hep-th/9204099].

[3] T. Kaluza, "On the Problem of Unity in Physics," *Sitzungsber. Preuss. Akad. Wiss.* (1921) 966.

[4] O. Klein, "Quantum Theory and Five-Dimensional Theory of Relativity (in German and English)," *Z. Phys.* **37** (1926) 895 [*Surveys High Energy Phys.* **5** (1986) 241].

[5] R. Gregory and R. Laflamme, "Black strings and *p*-branes are unstable," *Phys. Rev. Lett.* **70** (1993) 2837 [hep-th/9301052].

[6] G. T. Horowitz and T. Wiseman, "General black holes in Kaluza–Klein theory," in *Black holes in higher dimensions* (G. Horowitz ed.), Cambridge: Cambridge University Press (2012); arXiv:1107.5563 [gr-qc].

[7] S. W. Hawking and G. F. R. Ellis, *The large scale structure of space-time*, Cambridge: Cambridge University Press, (1973).

[8] G. T. Horowitz and K. Maeda, "Fate of the black string instability," *Phys. Rev. Lett.* **87** (2001) 131301 [hep-th/0105111].

[9] L. Lehner and F. Pretorius, "Final State of Gregory–Laflamme Instability," arXiv:1106.5184 [gr-qc].

[10] V. Cardoso and O. J. C. Dias, "Rayleigh–Plateau and Gregory–Laflamme instabilities of black strings," *Phys. Rev. Lett.* **96** (2006) 181601 [hep-th/0602017].

[11] T. Wiseman, "Static axisymmetric vacuum solutions and nonuniform black strings," *Class. Quant. Grav.* **20** (2003) 1137 [hep-th/0209051].

[12] H. Kudoh and T. Wiseman, "Properties of Kaluza Klein black holes," *Prog. Theor. Phys.* **111** (2004) 475 [hep-th/0310104].

[13] B. Kleihaus, J. Kunz and E. Radu, "New nonuniform black string solutions," *JHEP* **0606**, 016 (2006) [hep-th/0603110].

[14] S. Deser and M. Soldate, "Gravitational Energy in Spaces With Compactified Dimensions," *Nucl. Phys. B* **311** (1989) 739.

[15] D. Brill and H. Pfister, "States of Negative Total Energy in Kaluza–Klein Theory," *Phys. Lett. B* **228** (1989) 359.

[16] D. Brill and G. T. Horowitz, "Negative energy in string theory," *Phys. Lett. B* **262** (1991) 437.

[17] E. Witten, "Instability of the Kaluza–Klein Vacuum," *Nucl. Phys. B* **195** (1982) 481.

[18] X. Dai, "A positive energy theorem for spaces with asymptotic SUSY compactification", *Commun. Math. Phys.* **244** (2004) 335.

[19] H. Elvang and G. T. Horowitz, "When black holes meet Kaluza–Klein bubbles," *Phys. Rev. D* **67** (2003) 044015 [hep-th/0210303].

[20] H. Elvang, T. Harmark and N. A. Obers, "Sequences of bubbles and holes: New phases of Kaluza–Klein black holes," *JHEP* **0501** (2005) 003 [hep-th/0407050].

[21] R. C. Myers and M. J. Perry, "Black Holes in Higher Dimensional Space-Times," *Annals of Physics* **172** (1986) 304.

[22] R. Emparan and H. S. Reall, "A Rotating black ring solution in five-dimensions," *Phys. Rev. Lett.* **88** (2002) 101101 [hep-th/0110260].

[23] R. Emparan and H. S. Reall, "Black Rings," *Class. Quant. Grav.* **23** (2006) R169 [hep-th/0608012].

[24] H. Elvang, "A Charged rotating black ring," *Phys. Rev. D* **68** (2003) 124016 [hep-th/0305247].

[25] H. Elvang, R. Emparan and A. Virmani, "Dynamics and stability of black rings," *JHEP* **0612** (2006) 074 [hep-th/0608076].

[26] H. Elvang, R. Emparan and P. Figueras, "Phases of five-dimensional black holes," *JHEP* **0705** (2007) 056 [hep-th/0702111].

[27] A. A. Pomeransky and R. A. Sen'kov, "Black ring with two angular momenta," hep-th/0612005.

[28] V. A. Belinsky and V. E. Zakharov, "Integration of the Einstein Equations by the Inverse Scattering Problem Technique and the Calculation of the Exact Soliton Solutions," *Sov. Phys. JETP* **48** (1978) 985 [*Zh. Eksp. Teor. Fiz.* **75** (1978) 1953].

[29] V. A. Belinsky and V. E. Sakharov, "Stationary Gravitational Solitons with Axial Symmetry," *Sov. Phys. JETP* **50** (1979) 1 [*Zh. Eksp. Teor. Fiz.* **77** (1979) 3].

[30] V. Belinski and E. Verdaguer, *Gravitational solitons*, Cambridge: Cambridge University Press (2001).

[31] H. Elvang and M. J. Rodriguez, "Bicycling Black Rings," *JHEP* **0804** (2008) 045 [arXiv:0712.2425 [hep-th]].

[32] H. Elvang and P. Figueras, "Black Saturn," *JHEP* **0705** (2007) 050 [hep-th/0701035].

[33] P. T. Chruściel, M. Eckstein and S. J. Szybka, "On smoothness of Black Saturns," *JHEP* **1011** (2010) 048 [arXiv:1007.3668 [hep-th]].

[34] H. Iguchi and T. Mishima, "Black di-ring and infinite nonuniqueness," *Phys. Rev. D* **75**, 064018 (2007) [Erratum *ibid. D* **78**, 069903 (2008)] [hep-th/0701043].

[35] J. Evslin and C. Krishnan, "The Black Di-Ring: An Inverse Scattering Construction," *Class. Quant. Grav.* **26**, 125018 (2009) [arXiv:0706.1231 [hep-th]].

[36] K. Izumi, "Orthogonal black di-ring solution," *Prog. Theor. Phys.* **119** (2008) 757 [arXiv:0712.0902 [hep-th]].

[37] H. S. Reall, "Higher dimensional black holes and supersymmetry," *Phys. Rev. D* **68** (2003) 024024 [Erratum *ibid. D* **70** (2004) 089902] [hep-th/0211290].

[38] S. Hollands, A. Ishibashi and R. M. Wald, "A Higher dimensional stationary rotating black hole must be axisymmetric," *Commun. Math. Phys.* **271** (2007) 699 [gr-qc/0605106].

[39] V. Moncrief and J. Isenberg, "Symmetries of Higher Dimensional Black Holes," *Class. Quant. Grav.* **25** (2008) 195015 [arXiv:0805.1451 [gr-qc]].

[40] R. Emparan, T. Harmark, V. Niarchos and N. A. Obers, "New Horizons for Black Holes and Branes," *JHEP* **1004** (2010) 046 [arXiv:0912.2352 [hep-th]].

[41] R. Emparan and R. C. Myers, "Instability of ultra-spinning black holes," *JHEP* **0309** (2003) 025 [hep-th/0308056].

[42] R. Emparan and H. S. Reall, "Black Holes in Higher Dimensions," *Living Rev. Rel.* **11** (2008) 6 [arXiv:0801.3471 [hep-th]].

[43] R. Emparan, T. Harmark, V. Niarchos, N. A. Obers and M. J. Rodriguez, "The Phase Structure of Higher-Dimensional Black Rings and Black Holes," *JHEP* **0710**, 110 (2007) [arXiv:0708.2181 [hep-th]].

[44] R. Emparan, T. Harmark, V. Niarchos and N. A. Obers, "Essentials of Blackfold Dynamics," *JHEP* **1003** (2010) 063 [arXiv:0910.1601 [hep-th]].

[45] R. Emparan, T. Harmark, V. Niarchos and N. A. Obers, "World-Volume Effective Theory for Higher-Dimensional Black Holes," *Phys. Rev. Lett.* **102** (2009) 191301 [arXiv:0902.0427 [hep-th]].

[46] J. Camps and R. Emparan, "Derivation of the blackfold effective theory," *JHEP* **1203** (2012) 038 [Erratum *ibid.* **1206** (2012) 155] [arXiv:1201.3506 [hep-th]].

[47] B. Kleihaus, J. Kunz and E. Radu, "Black rings in six dimensions," *Phys. Lett. B* **718**, 1073 (2013) [arXiv:1205.5437 [hep-th]].

[48] S. R. Coleman and J. Mandula, "All Possible Symmetries of the S Matrix," *Phys. Rev.* **159** (1967) 1251.

[49] R. Haag, J. T. Lopuszanski and M. Sohnius, "All Possible Generators of Supersymmetries of the S Matrix," *Nucl. Phys. B* **88** (1975) 257.

[50] J. Wess and J. Bagger, *Supersymmetry and supergravity*, Princeton, NJ: Princeton University Press (1992).

[51] D. Z. Freedman and A. Van Proeyen, *Supergravity*, Cambridge: Cambridge University Press (2012).

[52] D. Z. Freedman, P. van Nieuwenhuizen and S. Ferrara, "Progress Toward a Theory of Supergravity," *Phys. Rev. D* **13** (1976) 3214.

[53] S. Deser and B. Zumino, "Consistent Supergravity," *Phys. Lett. B* **62** (1976) 335.

[54] S. Ferrara and P. van Nieuwenhuizen, "Consistent Supergravity with Complex Spin 3/2 Gauge Fields," *Phys. Rev. Lett.* **37** (1976) 1669.

[55] E. Cremmer, "Supergravities in Five Dimensions," in *Superspace and supergravity*. Proceedings, Nuffield Workshop, Cambridge, UK, June 16–July 12, 1980," Eds. S. W. Hawking and M. Rocek, Cambridge: Cambridge University Press (1981).

[56] C. Fronsdal, "Massless Fields with Integer Spin," *Phys. Rev. D* **18**, 3624 (1978).

[57] E. S. Fradkin and M. A. Vasiliev, "Cubic Interaction in Extended Theories of Massless higher spin fields," *Nucl. Phys. B* **291**, 141 (1987).

[58] M. A. Vasiliev, "Higher spin gauge theories: star product and AdS space," in Shifman, M.A. (ed.): *The many faces of the superworld*, pp. 533–610 [hep-th/9910096].

[59] I. R. Klebanov and A. M. Polyakov, "AdS dual of the critical O(N) vector model," *Phys. Lett. B* **550**, 213 (2002) [hep-th/0210114].

[60] M. R. Gaberdiel and R. Gopakumar, "An AdS$_3$ Dual for Minimal Model CFTs," *Phys. Rev. D* **83**, 066007 (2011) [arXiv:1011.2986 [hep-th]].

[61] E. Cremmer, B. Julia and J. Scherk, "Supergravity Theory in Eleven-Dimensions," *Phys. Lett. B* **76** (1978) 409.

[62] M. Gunaydin, L. J. Romans and N. P. Warner, "Gauged $N = 8$ Supergravity in Five-Dimensions," *Phys. Lett. B* **154** (1985) 268.

[63] E. B. Bogomol'nyi, "Stability of Classical Solutions," *Sov. J. Nucl. Phys.* **24**, 449 (1976) [*Yad. Fiz.* **24**, 861 (1976)].

[64] M. K. Prasad and C. M. Sommerfield, "An Exact Classical Solution for the 't Hooft Monopole and the Julia-Zee Dyon," *Phys. Rev. Lett.* **35** (1975) 760.

[65] E. Witten, "A Simple Proof of the Positive Energy Theorem," *Commun. Math. Phys.* **80** (1981) 381.

[66] G. W. Gibbons and C. M. Hull, "A Bogomol'nyi Bound for General Relativity and Solitons in $N = 2$ Supergravity," *Phys. Lett. B* **109** (1982) 190.

[67] G. W. Gibbons, D. Kastor, L. A. J. London, P. K. Townsend and J. H. Traschen, "Supersymmetric selfgravitating solitons," *Nucl. Phys. B* **416** (1994) 850 [hep-th/9310118].

[68] M. Cvetic and D. Youm, "General rotating five-dimensional black holes of toroidally compactified heterotic string," *Nucl. Phys. B* **476**, 118 (1996) [hep-th/9603100].

[69] J. C. Breckenridge, R. C. Myers, A. W. Peet and C. Vafa, "D-branes and spinning black holes," *Phys. Lett. B* **391** (1997) 93 [hep-th/9602065].

[70] H. Elvang, R. Emparan, D. Mateos and H. S. Reall, "A Supersymmetric black ring," *Phys. Rev. Lett.* **93** (2004) 211302 [hep-th/0407065].

[71] I. Bena and N. P. Warner, "One ring to rule them all . . . and in the darkness bind them?," *Adv. Theor. Math. Phys.* **9** (2005) 667 [hep-th/0408106].

[72] H. Elvang, R. Emparan, D. Mateos and H. S. Reall, "Supersymmetric black rings and three-charge supertubes," *Phys. Rev. D* **71** (2005) 024033 [hep-th/0408120].

[73] J. P. Gauntlett and J. B. Gutowski, "General concentric black rings," *Phys. Rev. D* **71** (2005) 045002 [hep-th/0408122].

[74] H. Elvang and Y.-t. Huang, "Scattering Amplitudes," arXiv:1308.1697 [hep-th].

[75] N. Arkani-Hamed, J. L. Bourjaily, F. Cachazo, A. B. Goncharov, A. Postnikov and J. Trnka, "Scattering Amplitudes and the Positive Grassmannian," arXiv:1212.5605 [hep-th].

[76] H. Kawai, D. C. Lewellen and S. H. H. Tye, "A Relation Between Tree Amplitudes of Closed and Open Strings," *Nucl. Phys. B* **269**, 1 (1986).

[77] Z. Bern, L. J. Dixon, M. Perelstein and J. S. Rozowsky, "Multileg one loop gravity amplitudes from gauge theory," *Nucl. Phys. B* **546**, 423 (1999) [hep-th/9811140].

[78] Z. Bern and A. K. Grant, "Perturbative gravity from QCD amplitudes," *Phys. Lett. B* **457**, 23 (1999) [hep-th/9904026].

[79] Z. Bern, L. J. Dixon, D. C. Dunbar, A. K. Grant, M. Perelstein and J. S. Rozowsky, "On perturbative gravity and gauge theory," *Nucl. Phys. Proc. Suppl.* **88**, 194 (2000) [hep-th/0002078].

[80] W. Siegel, "Two vierbein formalism for string inspired axionic gravity," *Phys. Rev. D* **47**, 5453 (1993) [hep-th/9302036].

[81] Z. Bern, "Perturbative quantum gravity and its relation to gauge theory," *Living Rev. Rel.* **5**, 5 (2002) [gr-qc/0206071].

[82] Z. Bern, J. J. M. Carrasco and H. Johansson, "New Relations for Gauge-Theory Amplitudes," *Phys. Rev. D* **78**, 085011 (2008) [arXiv:0805.3993 [hep-ph]].

[83] G. 't Hooft and M. J. G. Veltman, "One loop divergencies in the theory of gravitation," *Ann. Inst. Henri Poincaré Phys. Theor. A* **20**, 69 (1974).

[84] M. H. Goroff and A. Sagnotti, "Quantum Gravity At Two Loops," *Phys. Lett. B* **160**, 81 (1985).

[85] A. E. M. van de Ven, "Two loop quantum gravity," *Nucl. Phys. B* **378**, 309 (1992).

[86] S. Deser and P. van Nieuwenhuizen, "One Loop Divergences of Quantized Einstein–Maxwell Fields," *Phys. Rev. D* **10**, 401 (1974).

[87] M. T. Grisaru, P. van Nieuwenhuizen and J. A. M. Vermaseren, "One Loop Renormalizability of Pure Supergravity and of Maxwell–Einstein Theory in Extended Supergravity," *Phys. Rev. Lett.* **37**, 1662 (1976).

[88] M. T. Grisaru, "Two Loop Renormalizability of Supergravity," *Phys. Lett. B* **66**, 75 (1977).

[89] E. Tomboulis, "On the Two Loop Divergences of Supersymmetric Gravitation," *Phys. Lett. B* **67**, 417 (1977).

[90] S. Deser, J. H. Kay and K. S. Stelle, "Renormalizability Properties of Supergravity," *Phys. Rev. Lett.* **38**, 527 (1977).

[91] Z. Bern, S. Davies, T. Dennen, A. V. Smirnov and V. A. Smirnov, "The Ultraviolet Properties of $N = 4$ Supergravity at Four Loops," arXiv:1309.2498 [hep-th].

[92] Z. Bern, L. J. Dixon and R. Roiban, "Is N = 8 supergravity ultraviolet finite?," *Phys. Lett. B* **644**, 265 (2007) [hep-th/0611086].

[93] Z. Bern, J. J. Carrasco, L. J. Dixon, H. Johansson, D. A. Kosower and R. Roiban, "Three-Loop Superfiniteness of $N = 8$ Supergravity," *Phys. Rev. Lett.* **98**, 161303 (2007) [hep-th/0702112].

[94] Z. Bern, J. J. M. Carrasco, L. J. Dixon, H. Johansson and R. Roiban, "Manifest Ultraviolet Behavior for the Three-Loop Four-Point Amplitude of $N = 8$ Supergravity," *Phys. Rev. D* **78**, 105019 (2008) [arXiv:0808.4112 [hep-th]].

[95] Z. Bern, J. J. M. Carrasco, L. J. Dixon, H. Johansson and R. Roiban, "The Ultraviolet Behavior of $N = 8$ Supergravity at Four Loops," *Phys. Rev. Lett.* **103**, 081301 (2009) [arXiv:0905.2326 [hep-th]].

[96] Z. Bern, J. J. M. Carrasco, L. J. Dixon, H. Johansson and R. Roiban, "The Complete Four-Loop Four-Point Amplitude in $N = 4$ Super-Yang–Mills Theory," *Phys. Rev. D* **82**, 125040 (2010) [arXiv:1008.3327 [hep-th]].

[97] H. Elvang, D. Z. Freedman and M. Kiermaier, "A simple approach to counterterms in $N = 8$ supergravity," *JHEP* **1011**, 016 (2010) [arXiv:1003.5018 [hep-th]].

[98] J. M. Drummond, P. J. Heslop and P. S. Howe, "A Note on $N = 8$ counterterms," arXiv:1008.4939 [hep-th].

[99] H. Elvang and M. Kiermaier, "Stringy KLT relations, global symmetries, and $E_{7(7)}$ violation," *JHEP* **1010**, 108 (2010) [arXiv:1007.4813 [hep-th]].

[100] N. Beisert, H. Elvang, D. Z. Freedman, M. Kiermaier, A. Morales and S. Stieberger, "E7(7) constraints on counterterms in $N = 8$ supergravity," *Phys. Lett. B* **694**, 265 (2010) [arXiv:1009.1643 [hep-th]].

[101] M. B. Green, J. G. Russo and P. Vanhove, "Modular properties of two-loop maximal supergravity and connections with string theory," *JHEP* **0807**, 126 (2008) [arXiv:0807.0389 [hep-th]].

[102] G. Bossard, P. S. Howe, K. S. Stelle and P. Vanhove, "The vanishing volume of $D = 4$ superspace," *Class. Quant. Grav.* **28**, 215005 (2011) [arXiv:1105.6087 [hep-th]].

[103] B. Zwiebach, *A first course in string theory*, Cambridge: Cambridge University Press (2009).

[104] J. Polchinski, *String theory. Vol. 1. An introduction to the bosonic string*, Cambridge: Cambridge University Press (1998); *String theory. Vol. 2. Superstring theory and beyond*, Cambridge: Cambridge University Press (1998).

[105] K. Becker, M. Becker and J. H. Schwarz, *String theory and M-theory: a modern introduction*, Cambridge: Cambridge University Press (2007).

[106] F. Gliozzi, J. Sherk, and D. Olive, "Supersymmetry, supergravity, and the dual spinor model," *Nucl. Phys. B* **122** (1977) 253.

[107] C. G. Callan, Jr., E. J. Martinec, M. J. Perry and D. Friedan, "Strings in Background Fields," *Nucl. Phys. B* **262** (1985) 593.

[108] E. Witten, "Noncommutative Geometry and String Field Theory," *Nucl. Phys. B* **268** (1986) 253.

[109] A. Sen, "Universality of the tachyon potential," *JHEP* **9912** (1999) 027 [hep-th/9911116].

[110] N. Moeller and W. Taylor, "Level truncation and the tachyon in open bosonic string field theory," *Nucl. Phys. B* **583** (2000) 105. [hep-th/0002237].

[111] E. Witten, "Superstring Perturbation Theory Revisited," arXiv:1209.5461 [hep-th].

[112] E. Witten, "More On Superstring Perturbation Theory," arXiv:1304.2832 [hep-th].

[113] P. Candelas, G. T. Horowitz, A. Strominger and E. Witten, "Vacuum Configurations for Superstrings," *Nucl. Phys. B* **258** (1985) 46.

[114] A. Giveon, M. Porrati and E. Rabinovici, "Target space duality in string theory," *Phys. Rept.* **244** (1994) 77 [hep-th/9401139].

[115] W. Lerche, C. Vafa and N. P. Warner, "Chiral Rings in $N = 2$ Superconformal Theories," *Nucl. Phys. B* **324** (1989) 427.

[116] B. R. Greene and M. R. Plesser, "Duality in Calabi–Yau Moduli Space," *Nucl. Phys. B* **338** (1990) 15.

[117] P. Candelas, X. C. De La Ossa, P. S. Green and L. Parkes, "A Pair of Calabi–Yau manifolds as an exactly soluble superconformal theory," *Nucl. Phys. B* **359** (1991) 21.

[118] P. S. Aspinwall, B. R. Greene and D. R. Morrison, "Calabi–Yau moduli space, mirror manifolds and space-time topology change in string theory," *Nucl. Phys. B* **416** (1994) 414 [hep-th/9309097].

[119] J. H. Schwarz, "Superstring Theory," *Phys. Rept.* **89** (1982) 223.

[120] D. J. Gross, J. A. Harvey, E. J. Martinec and R. Rohm, "The Heterotic String," *Phys. Rev. Lett.* **54** (1985) 502.

[121] M. B. Green and J. H. Schwarz, "Anomaly Cancellation in Supersymmetric $D = 10$ Gauge Theory and Superstring Theory," *Phys. Lett. B* **149** (1984) 117.

[122] J. Polchinski, "Dirichlet Branes and Ramond–Ramond charges," *Phys. Rev. Lett.* **75** (1995) 4724 [hep-th/9510017].

[123] S. Kachru, R. Kallosh, A. D. Linde and S. P. Trivedi, "De Sitter vacua in string theory," *Phys. Rev. D* **68** (2003) 046005 [hep-th/0301240].

[124] E. Witten, "String theory dynamics in various dimensions," *Nucl. Phys. B* **443** (1995) 85 [hep-th/9503124].

[125] T. Banks, W. Fischler, S. H. Shenker and L. Susskind, "M theory as a matrix model: a conjecture," *Phys. Rev. D* **55** (1997) 5112 [hep-th/9610043].

[126] L. Susskind, "Some speculations about black hole entropy in string theory," in Teitelboim, C. (ed.): *The black hole*, pp. 118–131 [hep-th/9309145].

[127] G. T. Horowitz and J. Polchinski, "A correspondence principle for black holes and strings," *Phys. Rev. D* **55** (1997) 6189 [hep-th/9612146].

[128] A. Strominger and C. Vafa, "Microscopic origin of the Bekenstein–Hawking entropy," *Phys. Lett. B* **379** (1996) 99 [hep-th/9601029].

[129] C. G. Callan and J. M. Maldacena, "D-brane approach to black hole quantum mechanics," *Nucl. Phys. B* **472** (1996) 591 [hep-th/9602043].

[130] G. T. Horowitz and A. Strominger, "Counting states of near extremal black holes," *Phys. Rev. Lett.* **77** (1996) 2368 [hep-th/9602051].

[131] J. M. Maldacena and A. Strominger, "Statistical entropy of four-dimensional extremal black holes," *Phys. Rev. Lett.* **77** (1996) 428 [hep-th/9603060].

[132] C. V. Johnson, R. R. Khuri and R. C. Myers, "Entropy of 4-D extremal black holes," *Phys. Lett. B* **378** (1996) 78 [hep-th/9603061].

[133] G. T. Horowitz, D. A. Lowe and J. M. Maldacena, "Statistical entropy of nonextremal four-dimensional black holes and U duality," *Phys. Rev. Lett.* **77** (1996) 430 [hep-th/9603195].

[134] J. M. Maldacena and A. Strominger, "Black hole grey body factors and d-brane spectroscopy," *Phys. Rev. D* **55** (1997) 861 [hep-th/9609026].

[135] S. R. Das and S. D. Mathur, "Comparing decay rates for black holes and D-branes," *Nucl. Phys. B* **478** (1996) 561 [hep-th/9606185].

[136] M. Cyrier, M. Guica, D. Mateos and A. Strominger, "Microscopic entropy of the black ring," *Phys. Rev. Lett.* **94**, 191601 (2005) [hep-th/0411187].

[137] I. Bena and P. Kraus, "Microscopic description of black rings in AdS/CFT," *JHEP* **0412**, 070 (2004) [hep-th/0408186].

[138] I. Bena and P. Kraus, "Microstates of the D1–D5–KK system," *Phys. Rev. D* **72**, 025007 (2005) [hep-th/0503053].

[139] G. 't Hooft, "Dimensional reduction in quantum gravity," gr-qc/9310026.

[140] L. Susskind, "The World as a hologram," *J. Math. Phys.* **36** (1995) 6377 [hep-th/9409089].

[141] S. J. Avis, C. J. Isham and D. Storey, "Quantum Field Theory in anti-De Sitter space-time," *Phys. Rev. D* **18** (1978) 3565.

[142] S. W. Hawking and D. N. Page, "Thermodynamics of Black Holes in anti-De Sitter space," *Commun. Math. Phys.* **87** (1983) 577.

[143] D. Marolf, "Unitarity and Holography in Gravitational Physics," *Phys. Rev. D* **79** (2009) 044010 [arXiv:0808.2842 [gr-qc]].

[144] J. M. Maldacena, "The large N limit of superconformal field theories and supergravity," *Adv. Theor. Math. Phys.* **2** (1998) 231 [hep-th/9711200].

[145] S. S. Gubser, I. R. Klebanov and A. W. Peet, "Entropy and temperature of black 3-branes," *Phys. Rev. D* **54** (1996) 3915 [hep-th/9602135].

[146] S. S. Gubser, I. R. Klebanov and A. M. Polyakov, "Gauge theory correlators from noncritical string theory," *Phys. Lett. B* **428** (1998) 105 [hep-th/9802109].

[147] E. Witten, "Anti-de Sitter space and holography," *Adv. Theor. Math. Phys.* **2** (1998) 253 [hep-th/9802150].

[148] N. Drukker and D. J. Gross, "An Exact prediction of $N = 4$ SUSYM theory for string theory," *J. Math. Phys.* **42** (2001) 2896 [hep-th/0010274].

[149] D. Z. Freedman, S. S. Gubser, K. Pilch and N. P. Warner, "Renormalization group flows from holography supersymmetry and a c theorem," *Adv. Theor. Math. Phys.* **3** (1999) 363 [hep-th/9904017].

[150] D. E. Berenstein, J. M. Maldacena and H. S. Nastase, "Strings in flat space and pp waves from $N = 4$ superYang–Mills," *JHEP* **0204** (2002) 013 [hep-th/0202021].

[151] H. Lin, O. Lunin and J. M. Maldacena, "Bubbling AdS space and 1/2 BPS geometries," *JHEP* **0410** (2004) 025 [hep-th/0409174].

[152] J. M. Maldacena and A. Strominger, "AdS(3) black holes and a stringy exclusion principle," *JHEP* **9812** (1998) 005 [hep-th/9804085].

[153] J. McGreevy, L. Susskind and N. Toumbas, "Invasion of the giant gravitons from anti-de Sitter space," *JHEP* **0006** (2000) 008 [hep-th/0003075].

[154] R. C. Myers, "Dielectric branes," *JHEP* **9912** (1999) 022 [hep-th/9910053].

[155] O. Aharony, O. Bergman, D. L. Jafferis and J. Maldacena, "$N = 6$ superconformal Chern–Simons-matter theories, M2-branes and their gravity duals," *JHEP* **0810** (2008) 091 [arXiv:0806.1218 [hep-th]].

[156] I. R. Klebanov and E. Witten, "Superconformal field theory on three-branes at a Calabi–Yau singularity," *Nucl. Phys. B* **536** (1998) 199 [hep-th/9807080].

[157] D. Martelli and J. Sparks, "The gravity dual of supersymmetric gauge theories on a biaxially squashed three-sphere," *Nucl. Phys. B* **866** (2013) 72 [arXiv:1111.6930 [hep-th]].

[158] D. Z. Freedman and S. S. Pufu, "The Holography of F-maximization," arXiv:1302.7310 [hep-th].

[159] S. W. Hawking, "Breakdown of Predictability in Gravitational Collapse," *Phys. Rev. D* **14** (1976) 2460.

[160] S. D. Mathur, "The Information paradox: A Pedagogical introduction," *Class. Quant. Grav.* **26** (2009) 224001 [arXiv:0909.1038 [hep-th]].

[161] S. D. Mathur, "Fuzzballs and the information paradox: A Summary and conjectures," arXiv:0810.4525 [hep-th].

[162] A. Almheiri, D. Marolf, J. Polchinski and J. Sully, "Black Holes: Complementarity or Firewalls?," *JHEP* **1302** (2013) 062 [arXiv:1207.3123 [hep-th]]; S. L. Braunstein, "Black hole entropy as entropy of entanglement, or it's curtains for the equivalence principle," [arXiv:0907.1190v1 [quant-ph]] published as S. L. Braunstein, S. Pirandola and K. Zyczkowski, "Better late than never: information retrieval from black holes," *Phys. Rev. Lett.* **110**, 101301 (2013) for a similar prediction from different assumptions.

[163] V. Balasubramanian and P. Kraus, "A Stress tensor for Anti-de Sitter gravity," *Commun. Math. Phys.* **208** (1999) 413 [hep-th/9902121].

[164] G. T. Horowitz and R. C. Myers, "The AdS/CFT correspondence and a new positive energy conjecture for general relativity," *Phys. Rev. D* **59** (1998) 026005 [hep-th/9808079].

[165] S. de Haro, S. N. Solodukhin and K. Skenderis, "Holographic reconstruction of space-time and renormalization in the AdS/CFT correspondence," *Commun. Math. Phys.* **217** (2001) 595 [hep-th/0002230].

[166] P. Kovtun, D. T. Son and A. O. Starinets, "Viscosity in strongly interacting quantum field theories from black hole physics," *Phys. Rev. Lett.* **94** (2005) 111601 [hep-th/0405231].

[167] A. Adams, P. M. Chesler and H. Liu, "Holographic turbulence," *Phys. Rev. Lett.* **112** (2014) 15, 151602 [arXiv:1307.7267 [hcp-th]].

[168] S. S. Gubser, "Breaking an Abelian gauge symmetry near a black hole horizon," *Phys. Rev. D* **78** (2008) 065034 [arXiv:0801.2977 [hep-th]].

[169] S. A. Hartnoll, C. P. Herzog and G. T. Horowitz, "Building a Holographic Superconductor," *Phys. Rev. Lett.* **101** (2008) 031601 [arXiv:0803.3295 [hep-th]].

[170] G. T. Horowitz, J. E. Santos and D. Tong, "Optical Conductivity with Holographic Lattices," *JHEP* **1207** (2012) 168 [arXiv:1204.0519 [hep-th]].

[171] G. T. Horowitz and J. E. Santos, "General Relativity and the Cuprates," *JHEP* **1306** (2013) 087 [arXiv:1302.6586 [hep-th]].

[172] S. Ryu and T. Takayanagi, "Holographic derivation of entanglement entropy from AdS/CFT," *Phys. Rev. Lett.* **96** (2006) 181602 [hep-th/0603001].

[173] T. Nishioka, S. Ryu and T. Takayanagi, "Holographic Entanglement Entropy: An overview," *J. Phys. A* **42** (2009) 504008 [arXiv:0905.0932 [hep-th]].

[174] A. Lewkowycz and J. Maldacena, "Generalized gravitational entropy," *JHEP* **1308** (2013) 090 [arXiv:1304.4926 [hep-th]].

[175] M. Van Raamsdonk, "Building up space-time with quantum entanglement," *Gen. Rel. Grav.* **42** (2010) 2323 [*Int. J. Mod. Phys. D* **19** (2010) 2429] [arXiv:1005.3035 [hep-th]].

Index

Printed in the United States
By Bookmasters